Springer Geography

Advisory Editors

Mitja Brilly, Faculty of Civil and Geodetic Engineering, University of Ljubljana, Ljubljana, Slovenia

Nancy Hoalst-Pullen, Department of Geography and Anthropology, Kennesaw State University, Kennesaw, USA

Michael Leitner, Department of Geography and Anthropology, Louisiana State University, Baton Rouge, USA

Mark W. Patterson, Department of Geography and Anthropology, Kennesaw State University, Kennesaw, USA

Márton Veress, Department of Physical Geography, University of West Hungary, Szombathely, Hungary

The Springer Geography series seeks to publish a broad portfolio of scientific books, aiming at researchers, students, and everyone interested in geographical research.

The series includes peer-reviewed monographs, edited volumes, textbooks, and conference proceedings. It covers the major topics in geography and geographical sciences including, but not limited to; Economic Geography, Landscape and Urban Planning, Urban Geography, Physical Geography and Environmental Geography.

Springer Geography—now indexed in Scopus

Assefa Melesse · Berhan Gessesse · Worku Zewdie
Editors

Abbay River Basin

Biophysical Setting, Environmental Degradation, Hydropolitics and Development Potential

Editors
Assefa Melesse
Department of Earth and Environment
Institute of Environment
Florida International University
Miami, FL, USA

Berhan Gessesse
Remote Sensing Department
Ethiopian Space Science and Geospatial Institute
Addis Ababa, Ethiopia

Worku Zewdie
Remote Sensing Department
Ethiopian Space Science and Geospatial Institute
Addis Ababa, Ethiopia

ISSN 2194-315X ISSN 2194-3168 (electronic)
Springer Geography
ISBN 978-3-031-65240-0 ISBN 978-3-031-65241-7 (eBook)
https://doi.org/10.1007/978-3-031-65241-7

© The Editor(s) (if applicable) and The Author(s), under exclusive license to Springer Nature Switzerland AG 2025

This work is subject to copyright. All rights are solely and exclusively licensed by the Publisher, whether the whole or part of the material is concerned, specifically the rights of translation, reprinting, reuse of illustrations, recitation, broadcasting, reproduction on microfilms or in any other physical way, and transmission or information storage and retrieval, electronic adaptation, computer software, or by similar or dissimilar methodology now known or hereafter developed.
The use of general descriptive names, registered names, trademarks, service marks, etc. in this publication does not imply, even in the absence of a specific statement, that such names are exempt from the relevant protective laws and regulations and therefore free for general use.
The publisher, the authors and the editors are safe to assume that the advice and information in this book are believed to be true and accurate at the date of publication. Neither the publisher nor the authors or the editors give a warranty, expressed or implied, with respect to the material contained herein or for any errors or omissions that may have been made. The publisher remains neutral with regard to jurisdictional claims in published maps and institutional affiliations.

This Springer imprint is published by the registered company Springer Nature Switzerland AG
The registered company address is: Gewerbestrasse 11, 6330 Cham, Switzerland

If disposing of this product, please recycle the paper.

Preface

The Blue Nile River basin is the main source of the Nile River with a drainage area of 324,530 km^2. Eighty-six percent of the annual flow of the Nile comes from the Blue Nile River basin (59%), from the Barro-Akobo-Sobat sub-system (14%) and from the Tekeze/Atbara/Gash sub-system (13%). The remaining 14% comes from the equatorial lakes after losses of evaporation in the Sudd region and Machar marshes. Annual rainfall varies from over 2200 mm in the humid southwest highlands to 500–800 mm in the drier northeast lowlands. Around 80% of annual flows occur during the June-September wet season from heavy seasonal rains. The basin is also characterized as highly degraded, and the level of soil erosion is estimated to be one of the highest in the country. Soil erosion and land degradation are major environmental issues due to topography, deforestation, overgrazing and unsustainable agricultural practices. Subsistence agriculture is the dominant land use and livelihood for the basin's population which is heavily rural and agrarian. The basin is also highly vulnerable to the impacts of climate change including rainfall variability and frequent droughts.

The book outlines the various aspects of the basin from physical geography to future developments in the 25 chapters. Geological formations, rock types and structures present in the Abbay Basin and the geological history of the region, the distribution of different rock units and their impact on the landscape and hydrology of the area are shown. The landforms and processes of the basin, landscape and river systems and the underlying geological and environmental factors that influence their formation are shown.

Agricultural practices, land use patterns, cropping systems, soils and soils fertility, forest resources and mapping, surface and ground water resources, climate change impact and flow variability, drought, land management, Gran Ethiopian Renaissance Dam (GERD) potential, soil erosion and land degradation, tourism development potentials, economic contributions and future development potential of the basin are addressed.

The book will be a useful resource and reference for researchers, planners, managers and scientists.

Miami, USA	Assefa Melesse
Addis Ababa, Ethiopia	Berhan Gessesse
Addis Ababa, Ethiopia	Worku Zewdie

Contents

Part I Geographical Snapshot of Abbay Basin

1 **Abbay Basin: A Geographical Overview** 3
   Berhan Gessesse, Assefa M. Melesse, and Worku Zewdie

2 **Geology of the Abbay Basin** 17
   Balemwal Atnafu Alemu

3 **Geomorphology of the Abbay Basin from Geographical Perspective** 33
   Mehretie Belay Ferede

Part II Economic Development Potential

4 **Agriculture in the Abbay Basin** 67
   Mezegebu Getnet and Shimelis Asseffa

5 **Tourism Development Potentials and Economic Contributions of the Grand Ethiopian Renaissance Dam (GERD)** 89
   Endalkachew Teshome and Amare Nega Wondirad

Part III Soils, Forest and Water Resources of the Basin

6 **Soils of the Abbay Basin** 109
   Ashenafi Ali, Wondwosen Tena, Assefa Abegaz, Teklu Erkossa,
   Kiflu Gudeta, Degefe Tibebe, Berhan Gessese, Wuletawu Abera,
   Terefe Mekete, Amsalu Tilaye, and Lulseged Tamene

7 **Forest Resources in the Abbay Basin** 163
   Aramde Fetene

8 **Hydrology of the Abbay River Basin, Ethiopia** 193
   Getachew Tegegne and Assefa M. Melesse

9	**Abbay Basin's Regional Groundwater Flow System** .... Mebruk Mohammed Nurhusein	209

Part IV Natural Resources Degradation in the Basin

10	**Land Use/Land Cover Changes, Drivers, and Implications** .... Gizachew Kabite Wedajo	235
11	**Soil Erosion and Sediment Yield Status of the Abbay Basin** .... Gizachew Kabite Wedajo, Berhan Gessesse, Worku Zewdie, Wubetu Anley, and Seyoum Eshetie	261
12	**Remote Sensing-Based Agricultural Drought Monitoring in the Eastern Abbay Basin** .... Gebremariam Adane Getu and Worku Zewdie	291
13	**Quantifying Spatiotemporal Drought Dynamics Under Climate Change in the Abbay River Basin, Ethiopia** .... Getachew Tegegne and Assefa M. Melesse	321

Part V Water Use and Watershed Management

14	**Irrigation Development in the *Abbay* River Basin, Ethiopia** .... Sisay Demeku Derib	343
15	**The Hydropolitics and Legal Dimension of Ethiopia's Right to Utilize the Abbay River** .... Firehiwot Sintayehu, Yusuf Ali Mohammed, and Melak Melkamu	375
16	**Land Management and Productivity in the Abbay Basin** .... Mengistie Mersha	397
17	**Lesson Learned from Over 50 years of Watershed Management in the Abbay Basin, Ethiopia** .... Ermias Teferi, Tibebu Kassawmar, Woldeamlak Bewket, and Gete Zeleke	417

**Part VI Recent Advances in Remote sensing for Basin
          Management**

18	**Enhancing Earth Observation Application to Derive Hydrological Modelling Parameters in the Abbay Basin** .... Berhan Gessesse, Gebeyehu Abebe, Wubetu Anley, and Worku Zewdie	443
19	**Linking Earth Observation to Crop Area Mapping and Yield Estimation in the Abay Basin** .... Gebeyehu Abebe and Berhan Gessesse	471

20	Recent Development and Innovative Tools for Climate and Hydrological Data Collection and Analysis in the Abbay Basin 497
	Yonas Getaneh, Wuletawu Abera, Getachew Tesfaye Ayehu, Degefie Tibebe, and Lulseged Tamene

21	Root-Zone Soil Moisture Prediction in Rainfed Systems Using Satellite-Derived Product: The Case of Abbay River Basin in Ethiopia 529
	Getachew Tesfaye Ayehu, Tsegaye Tadesse, Berhan Gessesse, Wuletawu Abera, Degefie Tibebe, Yonas Getaneh, and Lulseged Tamene

22	Downscaling ESA CCI Soil Moisture Using Sentinel-1 SAR Data: A Case Study in the Abbay River Basin in Ethiopia 553
	Getachew Tesfaye Ayehu, Wuletawu Abera, Degefie Tibebe, Yonas Getaneh, and Lulseged Tamene

Part VII Future Research and Development Direction in the Basin

23	Interactive Web Tool for Mapping Soil Organic Carbon Dynamics 577
	Worku Zewdie, Degefie Tibebe, Abraraw Assefa, Assefa Abegaz, Berhan Gessesse, Ashenafi Ali, Wuletawu Abera, Lulseged Tamene, and Amsalu Tilaye

24	A Synthesis of Literature on the Grand Ethiopian Renaissance Dam (GERD) Using Text Mining Approaches 599
	Wuletawu Abera, Melkamu Beyene, Aminu Mohammed, Teshome Alemu, Temtim Assefa, Miftah Hassen, Mekdelawit Messay, and Lulseged Tamane

25	Future Research, Planning and Management Directions 621
	Worku Zewdie and Berhan Gessesse

Index 631

Part I
Geographical Snapshot of Abbay Basin

Chapter 1
Abbay Basin: A Geographical Overview

Berhan Gessesse, Assefa M. Melesse, and Worku Zewdie

Abstract The Nile River, the "Longest River in Africa," spans 6,695 km and is home to around 257 million people. It is shared by eleven African countries, including Burundi, Rwanda, Uganda, Democratic Republic of the Congo, Kenya, the United Republic of Tanzania, South Sudan, Sudan, Ethiopia, Eritrea, and Egypt. The two major tributaries, White Nile and Blue Nile (Abbay River), make up the basin, and the Abbay River accounting for over 75% of the Nile flow. The Abbay Basin covers 173,000 km^2 in Ethiopia flows approximately 1,400 km and the main topics of investigation in this book. The basin supports substantial populations, irrigation and hydropower projects, ecosystems conservation and agricultural activities. Major hydrological uses include large-scale irrigation projects and hydropower dams like the Grand Ethiopian Renaissance Dam near the Sudan border. However, land degradation, soil erosion, and sediment loads are high in the basin, leading to loss of storage capacity in reservoirs. Climate change could also alter the hydrology of the Abbay Basin, potentially leading to increased floods and droughts. Management of the Abbay River faces challenges due to its highly variable flows, large land degradation processes and sediment loads, and its transboundary nature. This book provides an overview of the geographical overview, biophysical setting, natural resource degradation, hydro politics, and development potential of the basin. It also highlights recent advances in Earth Observation datasets, geospatial solutions, models, tools, and services utilization in the basin. Accordingly, the purpose of this book is to compile information about the biophysical setting, geo-environmental resource conditions, major economic activities, sociopolitical characteristics, legal rights to

---

B. Gessesse (✉) · W. Zewdie
Remote Sensing Department, Ethiopian Space Science and Geospatial Institute, Addis Ababa, Ethiopia
e-mail: berhan.gessesse@ssgi.gov.et

W. Zewdie
e-mail: worku.zewdie@ssgi.gov.et

A. M. Melesse
Department of Earth and Environment, Institute of Environment, Florida International University, Miami, FL, USA
e-mail: melessea@fiu.edu

© The Author(s), under exclusive license to Springer Nature Switzerland AG 2025
A. Melesse et al. (eds.), *Abbay River Basin*, Springer Geography,
https://doi.org/10.1007/978-3-031-65241-7_1

utilize the Abbay River as well as the contribution of remote sensing and advanced tools applications for Abbay Basin characterization, management and development.

Keywords Nile River · White Nile · Abbay River Basin · Earth Observation · Ethiopia

Background

With a length of 6695 km, the Nile River is known as the "Longest River in Africa" and is the home of ~257 million people (FAO and IHE Delft 2020). The basin is shared by eleven African countries namely Burundi, Rwanda, Uganda, Democratic Republic of the Congo, Kenya, the United Republic of Tanzania, South Sudan, Sudan, Ethiopia, Eritrea, and Egypt (FAO 2020). White Nile and Blue Nile are the two major tributaries of the Nile River. The White Nile is considered as the upstream part of the basin. It begins from south of the Equatorial Lakes region. On the other hand, Ethiopia has significant surface and ground water resources including large areas of wetlands, 22 lakes, and 12 major river basins. As a result of this, one of the major tributaries of the Nile, Blue Nile River, is originated from Ethiopia and it is the source of most of the water of the Nile and silt to the downstream countries (Molden et al. 2010). After joining these two rivers at the confluence (i.e., Khartoum City), they make Nile River and the Nile flows north through northern Africa to enter the Mediterranean Sea.

In general, Ethiopia supplies 90% of the water and 96% of the total sediment carried by the Nile, with many tributaries including the Gilgel Abbay, Megech, Ribb, Gumera, Beshlo, Woleka, Jemma, Muger, Guder, Chemoga, Fincha, Dedessa, Dabus, Angar, Dura and Beles rivers, the Abbay River supplying 59% of water resources (Melesse et al. 2011) and the remaining water here comes from the Tekezé, Atbarah, Sobat, and minor tributaries, all of which originate from the Ethiopian and Eritrean highlands.

Although Blue Nile Basin includes the Abbay, Baro Akobo, and Tekeze basins, the focus of this book is only the Abbay River Basin.

It originates from the Ethiopian Highlands mainly from the Lake Tana, which is the largest lake in Ethiopia. According to Melesse et al. (2011), the Abbay Basin is one of the two major tributaries of the Nile River accounting over 75% of the Nile flow. This basin covers an area of about 173,000 km^2 in Ethiopia and the river flows approximately 1400 km (Conway 2000). Flowing south and southward from Lake Tana, the Abbay river crosses diverse and undulating topography and landscapes.

Annual rainfall across the basin is highly variable, ranging from over 2200 mm in the humid southwestern highlands to 500–800 mm in the lowlands near Sudan Border. Most rainfall occurs during the June–September wet season. Due to the heavy seasonal rainfall and steep landscape, the Abbay has a rapid hydrologic response with overland flow and common flooding characteristics. Around 80% of annual discharge occurs during the wet rainy season. The annual discharge of the

Abbay is estimated at 52 cubic km on average. Flows vary from 300 cubic meters per second (m^3/s) in the dry season to over 5000 m^3/s during the peak of the rains.

The Abbay River Basin in Ethiopia generally supports substantial populations, significant irrigation and hydropower projects, various ecosystems and agricultural activities all of which are dependent on the basin's resources. Besides, the major hydrological uses are large-scale irrigation projects in the upper and middle basins and hydropower dams like the Grand Ethiopian Renaissance Dam near the Sudan border. However, land degradation in the form of soil erosion and sediment loads are high, estimated at 137 million tons annually (Melesse et al. 2011). This causes loss of storage capacity in reservoirs due to sedimentation. Similarly, climate change could alter Abbay hydrology through impacts on precipitation, runoff, and evapotranspiration. This may lead to increased floods and droughts. Thus, the management of this important transboundary river basin must take into account concerns including soil erosion, land degradation, climate change, and conflicting water needs. The enormous amount of sediment production only occurs during the Ethiopian rainy season when rainfall is especially high in the Ethiopian highlands (Berihun et al. 2022; Dile et al. 2018; Haregeweyn et al. 2017; Hurni et al. 2005).

Overall, the Abbay River is characterized by highly variable flows, large sediment loads, and transboundary nature that create hydrological management challenges for Ethiopia and downstream nations. In this regard, scientific investigations on the Abbay Basin were offered to the users that centered on various aspects of the basin (Yesuph and Dagnew 2019; Haregeweyn et al. 2017; Gelagay and Minale 2016; Bewket and Teferi 2009). In general, the Abbay Basin supplies water for home consumption, industrial processing, irrigated agricultural and livestock production, hydropower generation, extensive eco-tourism, fisheries development potential, and a significant biosphere reserve center in Ethiopia to name just a few.

Hydrology of the Nile River Basin has been studied by various researchers, These studies encompass various areas including stream flow modeling, sediment dynamics, teleconnections and river flow, land use dynamics, climate change impact, groundwater flow modeling, hydrodynamics of Lake Tana, water allocation, and demand analysis (Melesse et al. 2011, 2014; Chebud and Melesse 2013; 2009a, b; Setegn et al. 2009a, b).

The book elaborates on the geographical overview, biophysical setting, natural resource degradation, hydro-politics, and development potential of the Abbay Basin. In this regard, Melesse et al (2014) reported the hydrology of the basin, the degradation of land and water, the influences of climate change, the potentials of the basin for watershed services, and transboundary water management. Another study also focused on the hydrology and water budget, the contribution of geospatial technologies on rainfall estimation and watershed modeling, the effect of climate variability on the hydrology of the basin, and the management of the water resources of the basin.

Awulachew et al. (2007; 2012) also documented several aspects of the Nile Basin including water and agriculture governance as well as livelihoods of the society residing in the Abbay Basin. In their study, they emphasized the dominant challenges and opportunities regarding the Nile River Basin management and development.

Moreover, they assessed the population of the basin in line with their agricultural activities and access to water. From agricultural activity point of view, they highlighted the crop-water-livestock nexus. In this respect, they evaluated policy, institution, and technological intervention for enhancing production and productivity and the use of the basin water. Very recently, Abtew (2021) also reported the contribution of Grand Ethiopian Renaissance Dam, the different activities in the basin, and the sociopolitical disputes with the utilization of the natural resources of the basin. He evaluated different treaties at different times and their implications on the equitable utilization of the Nile water. Furthermore, it assessed the transboundary flood variabilities and sustainable water resource management. Abtew and Dessu (2019), on the other hand, discussed the different aspects of the Grand Ethiopian Renaissance Dam (GERD) and the Abbay Basin hydrology. They stressed on GERD site as well as the dam design, filling and operation with the utilization of the water resources.

However, some of the manuscripts focused on the hydrology of the basin and some characteristics of climate change that impact the water and the GERD construction. Accordingly, it has not given sufficient attention regarding the contribution of Earth Observation and geospatial science and technologies, and data for efficient monitoring and utilization of the resources of the Abbay Basin. In particular, the satellite imageries that have vital contributions in measuring the biophysical setting as well as the condition of resource degradation, hydrology, hydropoltics, water budget, evapotranspiration, the natural resources, agricultural activities, and the status of water and natural resource degradation of the basin are not well-addressed. Besides, the availability of consolidated data and scientific studies on several aspects of the basin are not presented in an organized manner. Thus, there is a need to compile the available information in the form of reference book for wider audiences.

In view of that, the Abbay Basin's biophysical setting, hydrology, degradation of its natural resources, hydro-politics, and agricultural as well as tourism development potential are the main topics discussed in this book. Besides, the status of recent advances of Earth Observation datasets, geospatial solutions, models, tools and services utilization in the basin is the other concern of the book. Leading scientists, researchers, and academicians in the fields of natural resource management, hydrology, hydro-politics, rain-fed and irrigation agriculture engineering fields, development fields, geospatial sciences, and sociopolitical disciplines of the basin presented a variety of research findings in this book. Besides, its main emphasis is on leveraging the advantages of open-source satellite data, algorithms, systems and tools application for biophysical setting, hydrology, natural resource, hydro-politics, and development potential characterization and monitoring, as well as visualizing the results for evidenced-based decision-making for the benefit of the society in the Abbay Basin.

Purpose

The purpose of this book is to explore information about the biophysical setting, the status of geo-environmental resources, the major economic activities setting, sociopolitical characteristics, and legal rights to utilize the Abbay River, as well as the contribution of remote sensing and recent advanced tools for Abbay Basin characterization, management and development. Accordingly, compiling such a book is essential to provide an authoritative supplementary reference text for readers. The specific objectives of the book were to:

- synthesize previous studies to extract profound knowledge and develop consolidated reference book that can be used by different users;
- generate empirical data and knowledge on what has been done so far using geospatial data and technology for sustainable utilization of resources of the Abbay Basin;
- serve as a worthy reference book for researchers, scientists, engineers and policymakers who wish to keep up with new developments in the Abbay Basin;
- highlight the contribution of satellite remotely sensed imageries, techniques, and tools to create and disseminate high-quality information about the Abbay Basin and
- avail direction to help students, researchers, and other interested groups in developing unique perspectives on the function and value of satellite and geospatial data applications, as well as developing cutting-edge tools and techniques for basin resources management and development.

Scope of the Book

The book covered topics such as physical geography, surface, and groundwater water resources, agricultural development, the state of geo-environmental resources, land resource degradation, land management practices, the current national and regional hydro-politics, as well as potential development agenda in the Abbay Basin. Furthermore, the book is the first of its kind in Ethiopia, providing comprehensive information about remote sensing applications as well as the latest state-of-the-art tools applicable for biophysical resources, natural resource status, land degradation magnitudes and land management option characterization, as well as important development and legal dimensions for utilizing Abbay Basin for Ethiopians' benefit. Besides, the book is addressing policy and use-right decision about how to properly manage and develop the Abbay Basin's resources regionally shared natural resources and environment. Furthermore, the book will advise policymakers on which exogenous conditions to be aware of when managing basin-level natural resources shared with other up-and-down-stream partner countries, as well as institutional design and governance mechanisms to establish in order to improve basin-level resource management and development effectiveness.

Target Audience

In general, physical geographers, hydrologists, geological and agricultural irrigation engineers, climate scientists, remote sensing scientists, social and economic geographers, demographers, natural resource conservationists and managers, environmental disaster experts, political geographers and scientists and decision-makers at regional and federal levels to name just a few are major users of the book. Accordingly, the intended national- and international-level institutions beneficiaries of the book are summarized as follows.

a. **Ministries, Bureaus, and Regulatory Government Organs**: The proposed book will be useful to federal ministries such as the Ministry of Water and Energy, the Ministry of Irrigation and Lowlands, the Ministry of Agriculture, the Ministry of Mines, and the concerned water and agriculture-related Regional, Zonal, and Woreda Bureaus. The book also benefits regulatory organs at various level such as the environment, forest, and climate commission at the federal and environmental protection and land administration authorities at the regional level.

b. **Education Sectors, Universities and Research Institutes**: High school, undergraduate and graduate students; educators; researchers; technical experts and scientists interested basin-level resource management studies, and conducting research using remote sensing applications as part of their educational training and research activities would find the book useful.

c. **Donors and Non-governmental Organizations**: International and local funders as well as non-governmental organizations (NGOs) are heavily involved in environmental restoration and rehabilitation programs. These organizations are mandated for implementing novel methodologies and approaches to monitor natural resource degradation and management at basin scale. As a result, this book will give a wealth of knowledge for various stakeholders in order to implement integrated basin-level natural resource degradation assessment and management methods in a long-term way. In general, this book will fulfill the demand of different groups that have an interest in the water dynamics, ecological status, legal, political aspects of the Abbay Basin and the contribution of geospatial technologies for basin-level resource assessment and monitoring. It is also expected to satisfy the requirement of higher educations for teaching graduate and postgraduate students and practitioners in hydrology, remote sensing, and natural resource conservation.

Structure of the Book

This book is organized into seven sections. The geographical snapshot of the Abbay Basin is covered in Section I. This section contains three chapters. In particular, Section II highlights the basin's potential for economic growth and it comprises two chapters. Section III focuses on the soils, forests, and hydrological resources status and distribution and this section covers four chapters. Section IV addresses natural resource degradation in the basin and this section encloses four chapters. Water utilization and watershed management is covered in Section V, while Section IV discusses current developments in tools, models, and remote sensing applications for the monitoring and management of the Abbay Basin. Sectionss V and VI contain four chapters and five chapters, respectively. Finally, Section VII examines the future research and development direction in the Abbay Basin and it comprehends three chapters.

In Chap. 1, overview of the issues related to the geographical snapshot of Abbay Basin, the justification of producing such a very comprehensive book, the purpose, scope, target audiences, as well as structure of the book are shown. Chapter 2 provides a thorough summary of the Abbay Basin's geological formation processes. The author of the book outlined that the geologic evolution of the basin includes three unique temporal fingerprints. The ancient Precambrian basement served as the starting point for the geological formation, and the second point deals with the development of the Phanerozoic sedimentary succession in the Horn of Africa. The third phase, which is now underway, is focused on the genesis of a massive Cenozoic uplift and the associated volcanism. Precambrian basement, Palaeozoic clastics, Adigrat Sandstone Formation, Gohatsion Formation, Mugher Mudstone Formation, Debre Libanos Sandstone Formation, and Trap volcanics have all been identified as units associated with this period.

In Chap. 3, it is discussed how the structural characteristics of landforms and geomorphic processes have a big influence on the environment of Abbay Basin. This chapter provided further evidence that the East African Orogen and the collision and splitting of East and West Gondwana Land were responsible for the structural geomorphology of the Abbay Basin. Additionally, it showed that headward erosion over the Afro-Arabian Lava Dom caused the Gondwana continent to rift and fault along a northwest trending Mesozoic fault, resulting in the Abbay fault. The five main physical regions such as the Lake Tana basin, Amarasaint Massifs, Mid-central Plateau, Southwest Upland, and the Abay-Dinder Lowlands are presented in this chapter.

Chapter 4 addressed the summary of key scientific and technological concerns in the evolution of agricultural methods in the Abbay Basin. A significant source of agricultural exports and imports comes from the basin. In the basin, in comparison with potential, the current/actual production level is quite low. Even though, close to 77.6% of the basin's agricultural area is moderately to highly suited for farming, agriculture is negatively impacted by severe soil degradation, inadequate systems for supplying inputs, and basin-wide climate change and variability. Additionally, this chapter illustrated the state of rain-fed and irrigated agriculture conditions in the Abbay Basin.

The prospects for tourism development in the Grand Ethiopian Renaissance Dam (GERD) and its economic impact at national scale is documented in Chap. 5. Besides, the impact of the Grand Ethiopian Renaissance Dam (GERD) on the growth and development of tourism in the Abbay Basin is equally significant. In light of this, this essay investigates the possibilities for tourist growth and the economic benefits of

GERD through tourism. The findings of this chapter confirmed that the GERD would offer important insights into the potential for tourism development in the basin.

The soils of the Abbay Basin are described in Chap. 6. This chapter provides a summary of the most recent soil spatial information of the Abbay Basin. This chapter also evaluated and summarized previous top soil characteristics and nutrient mapping investigations. As a result, the spatial soil information has been updated and presented. In order to target soil and land management actions and increase agricultural output in the basin, this enhanced soil spatial information will thus have significant potential.

The state and condition of the Abbay Basin forest resource are analyzed and reported in Chap. 7. The author argued that the forests in the Abbay Basin play a crucial role in promoting sustainable development goals, delivering essential ecological benefits, protecting soil and water resources from degradation, and providing priceless resources for building needs and biomass energy production. According to the chapter, the Abbay Basin is home to a variety of plant communities, including Combretum-Terminalia woodlands and grassy wooded areas, a variety of dry evergreen Afromontane forests and grasslands, Afro-Alpine plant species, the Ericaceous Belt, moist evergreen Afromontane forests that thrive in humid climates, as well as wetland habitats like freshwater marshes, floodplains, and lakeshore vegetation. In some areas of the basin, riverine and bamboo forests are also highly prevalent.

Information regarding the hydrology of the Abbay Basin is summarized in Chap. 8. According to the authors of this chapter, Abbay Basin produces unreliable hydrological modeling results, and these sources of uncertainty for hydrologic modeling in the basin are caused by water diversion for small-scale irrigation, the building of infrastructure upstream of the watershed, the presence of artificial open reservoirs, the abstraction of upstream water supplies, poor watershed management, and observational errors in the hydro meteorological variables. Furthermore, the chapter outlined that Abbay Basin has considerable hydrologic forecast uncertainty due to inadequate spatial coverage and uneven distribution of hydrometric gauging stations. Thus, lowering the forecast uncertainty of hydrologic modeling is one of the key topics for future research in the Abbay River Basin. Emerging technology for the observation of hydro meteorological data can be used to achieve this.

On the other hand, the Abbay Basin's groundwater flow system is described in Chap. 9. This chapter's main goal is to characterize the Abbay Basin aquifers' regional groundwater flow components. Regional groundwater flow systems are capable of supplying a relatively consistent source of water and are less impacted by seasonal fluctuations in groundwater recharge. To assess the notion of inter- and intra-boundary groundwater flow for the basin, the chapter analyzes two conceptual models. The author also discussed the direction of regional groundwater flow, access depths, and potentiometric surface maps that were estimated using the model's output.

In Chap. 10, the magnitude, trends, drivers, and implications of land use/land cover dynamics of the Abbay Basin are documented. The Abbay Basin has undergone significant changes in land use and land cover during the last two decades. The findings of this chapter demonstrated significant spatiotemporal changes in land use and cover over the past few decades in the basin. Population increase, resettlement initiatives, careless use of land resources, and development activities including extensive agricultural investment and reservoir building are some of the main contributors. On the other hand, the human-induced changes to land use and land cover that were seen in the Abbay Basin exacerbated soil erosion, increased the production of sediment, changed hydrologic processes, reduced ecosystem services, and had an impact on the socioeconomic situation of the region as well as the entire nation.

The status of soil erosion and sediment production in the Abbay Basin is covered in Chap. 11. It also discusses the scope and causes of soil erosion, the on-and off-site effects of soil erosion, sustainable land management practices, and challenges and opportunities associated with reducing soil erosion and sediment yield. The chapter confirmed that the basin is extremely sensitive to surface runoff and soil erosion that adversely influence agricultural productivity and the availability of water resources due to the basin's mountainous topography, high erosive rainfall, and management techniques, and agriculturally dominant land use system. In order to reduce the consequences of soil erosion and sediment generation, several land management measures have been developed and put into practice in the upstream portion of the Abbay Basin during the past few decades.

Chapter 12 provides remote sensing-based agricultural drought monitoring in the eastern Abbay Basin. This chapter revealed that moderate to extreme drought conditions in 2009 and 2015 over most parts of the study area were observed. Besides, the chapter also confirmed that remote sensing data's contribution to monitoring and assessing agricultural drought risk areas and vulnerable societies is paramount. The authors suggested that the combined analysis of VHI is recommended rather than using single indices to monitor agricultural drought in NWZ and in other places with similar agro ecology. On the other hand, Chap. 13 provides information on climate change impacts on the spatiotemporal dynamics of drought in the Abbay Basin. This chapter investigated the impact of climate change on meteorological, agricultural, and hydrological droughts in the Abbay Basin. The analysis of the SPI indices-based drought results showed that the risk of drought in the Abbay River Basin will increase in the middle future and potentially decline in the far future under both SSP2-4.5 and SSP5-85. Besides, the intensity–duration–frequency (IDF) curve under climate change revealed that the frequency of drought occurrence would be more than 20% and 15% in 2031–2060 under SSP2-4.5 and SSP5-8.5, respectively.

Chapter 14 examines the irrigation potential of Abbay Basin. The chapter outlined that the basin has ample potential to enhance agricultural production and productivity in in the basin using surface water as well as groundwater resources for irrigation practices.

Chapter 15 provides insight into the hydro-politics and legal dimension of Ethiopia's right to utilize the Abbay River. The chapter clearly addressed that Ethiopia contributes 86% of the Nile waters; however, it has not been able to access its equitable and reasonable share from the river basin so far. More specifically, the chapter intended to address the legal and political basis by which Ethiopia can ensure its equitable and reasonable share from the Abbay River Basin by outlining various basin-wide treaties, international watercourse laws, and political discourse. The authors also suggested that Ethiopia continues to push for its interest in ensuring equitable and reasonable shares of the Abbay River by employing counter hegemony strategies of deconstructing Egypt's securitization through discourses such as "the right to development", cooperation, and pan-Africanism.

Chapter 16 discussed the land management and land productivity characteristics of the Abbay Basin. This chapter aimed at for presenting the current condition of land resources; the type, nature and magnitude of land degradation; impacts of land degradation; land management practices; determinants of applying land management practices; and productivity of the land in terms of certain crop yields focusing on the "*Bechet Watershed*" of the Abbay Basin at a certain depth. Besides, the authors confirmed that the implementation of land management practices such as terracing and tree planting are believed to have significant impacts on the future fate of the Great Ethiopian Renaissance Dam. Finally, the chapter acknowledged the efforts of land management practices so far implemented in the basin and recommends how such practices will be strengthened. Parallel to this, lesson learned from over 50 years of watershed management in the Abbay Basin is presented in Chap. 17. Specifically, the chapter drew the efforts made by the Government of Ethiopia and development partners to promote watershed management interventions that incorporated various technical, institutional, and policy improvements in Ethiopia in general and Abbay Basin in particular. The authors argued that over the past five decades, Ethiopia has shifted toward more participatory and livelihood-focused approaches, as well as adopting an integrated watershed management paradigm. These changes demonstrate a willingness to learn from previous experiences and the advantages of adapting approaches based on lessons learned along the way.

Chapter 18 explores a brief overview of the contribution of Earth Observation application to derive hydrological modeling parameters in the Abbay Basin. To make informed decision on the management of water resources, hydrological model simulation needs adequate and almost real-time datasets. Earth observation (EO) data could help with this problem by providing data and information for the finest decision support services and systems that are spatially oriented. Unfortunately, the authors claimed that the use of EO for managing and improving water resources continues to be a significant barrier for many poor nations. Ethiopia has not taken full advantage of this opportunity, and numerous obstacles still pose a serious challenge for the application of EO data for the management and development of water resources. Nevertheless, EO application has occasionally increased in the Abbay Basin, and EO products could significantly contribute to the effective and efficient use of hydrological parameters and information retrieval as well as to the improved decision-making process in the water sector.

Chapter 19 outlined linking Earth Observation to crop area mapping and yield estimation in the Abbay Basin. In this way, repeated and synoptic views of important agricultural metrics, such as crop nutrition and growth status, may be obtained using multisource images from Earth observation satellites, frequently in conjunction with auxiliary data from ground observation. For this, several techniques have been developed for employing a variety of Earth observation data. The approaches range from completely empirical, such as optical/microwave-based and the synergistic methods combining optical and microwave data, to more complicated ones, such as assimilation of Earth Observation data with crop growth model by continuous infusion of EO data into the model system. The chapter provides a thorough and systematic review of the various approaches to crop monitoring and yield estimation methods and their potential applications from the perspective of the Abbay Basin.

Recent development and innovative tools used for climate and hydrological data collection and analysis for the Abbay Basin are synthesized and presented in Chap. 20. With a water volume flow of 54 BCM per year, the Abbay Basin is Ethiopia's biggest river basin. However, there is still a dearth of quality- and quantity-relevant hydroclimate data as well as research facilities, which has hampered the advancement of scientific understanding in the area. This chapter focused on demonstrating cutting-edge cloud-based computer platforms (like Google Earth Engine) and script-based data analysis tools (like R-based packages) methods for climatological and hydrological data sources and analysis tools that might enhance the rigor and precision of research in the Abbay Basin. In addition, the authors noted newly developed satellite-based hydroclimate data sources that can be used as viable alternatives to deal with the issue of inadequate data.

Chapter 21 conducts a summarizing review of root-zone soil moisture prediction in rain-fed systems using satellite-derived product for Abbay Basin. To advance the application of remote sensing technology and its contributions to hydrology and agriculture, it is essential to be able to get satellite-derived root zone soil moisture (RZSM) data from close-up observations. Thanks to the development of microwave remote sensing, it is now possible to deduce the quantity of soil moisture in large basins from satellite data. However, estimations of soil moisture from satellite data are typically given for the first few centimeters (5 cm) of the soil column. The method is developed by fusing the polynomial regression model and the cumulative density function (CDF) matching method. The authors claimed that remote sensing technology might yet provide chances and alternatives for estimating RZSM at various scales and with better spatial and temporal resolutions.

Chapter 22 provides an overview of downscaling European Space Agency (ESA) Climate Change Initiative (CCI) using sentinel-1 SAR Data for the Abbay Basin. Using high-resolution Sentinel-1 SAR data and the SMAP baseline downscaling method, the coarse-scale soil moisture product in this work was downscaled to a high-resolution soil moisture dataset. The observed and downscaled soil moisture data from the automated weather station (AWS) were compared. The downscaled soil moisture estimations and the in situ measured and AWS observed soil moisture are in good agreement. The conclusion of this study, which shows the ability of Sentinel-1 SAR data and the downscaling method to break down the coarse-resolution soil moisture data into smaller scales in Ethiopia's Abbay River Basin, is therefore highly encouraging.

On the other hand, suggested tools and techniques for soil organic carbon-interactive web mapping of the Abbay Basin is presented in Chap. 23. The authors argued that continuous monitoring of the soil condition helps to know the status of soils for remedial actions. In this regard, remote sensing data and tools contribute a lot to filling the gaps observed in conventional soil monitoring approaches. This chapter addresses soil maps and produced interactive soil maps for observing carbon stocks and predicting attainable soil organic carbon over the Abbay Basin for the coming fifty years. The chapter also outlined that the users can register on the web to obtain both datasets based on their area of interest from the Kebele to micro watershed levels over the years. This interactive web map plays a significant role in easily availing carbon content information for researchers and policymakers for making better decisions.

Chapter 24 documented a synthesis of literature on Grand Ethiopian Renaissance Dam (GERD) using text mining approaches. In the past 11 years, there have been a number of multidisciplinary studies covering a wide range of topics, including hydrology, geology, ecology, socioeconomics, politics, and legal issues related to the dam and the basin, because of this project's interest from both the scientific and political communities. The literature on GERD was reviewed and synthesized using a novel method that was more automated, efficient, and objective. The publications from riparian states and other nations are noted in this chapter for their modest differences in focus. These results can enhance GERD scholarly debate and direct policymakers and government organizations in prioritizing their GERD project-related initiatives.

Chapter 25 summarizes the case studies, conclusions from the whole collection of chapters, and possible advancements in spatial approaches for integrated modeling of Abbay Basin characterization, monitoring, and management. This will help to strengthen the relationship between basin management scientific researches, regional, national, and global policy, and management decision-making. Additionally, this chapter offers closing thoughts and suggestions for additional research directions in the future.

References

Abtew W (2021) The Grand Ethiopian Renaissance Dam: evaluation of filling and operation plans and negotiations. In: Melesse AM, Abetew W, Moges SA (eds) Nile and Grand Ethiopian Renaissance Dam: past, present and future. Springer Geography, Cham, Switzerland

Abtew W, Dessu SB (2019) The Grand Ethiopian Renaissance Dam on the Blue Nile. Springer International Publishing. https://doi.org/10.1007/978-3-319-97094-3

Awulachew SB, Yilma AD, Loulseged M, Loiskandl W, Ayana M, Alamirew T (2007) Water resources and irrigation development in Ethiopia. International Water Management Institute (IWMI), Colombo, Sri Lanka, p 66 (IWMI Working Paper 123). https://doi.org/10.3910/2009.305

Awulachew SB, Smahktin V, Molden D, Peden D (eds) (2012) The Nile River Basin: water, agriculture, governance and livelihoods, 1st edn. Routledge. https://doi.org/10.4324/9780203128497

Berihun ML, Tsunekawa A, Haregeweyn N, Tsubo M, Fenta AA, Ebabu K, Sultan D, Dile YT (2022) Reduced runoff and sediment loss under alternative land capability-based land use and management options in a sub-humid watershed of Ethiopia. J Hydrol Reg Stud 40:100998

Bewket W, Teferi E (2009) Assessment of soil erosion hazard and prioritization for treatment at the watershed level: case study in the Chemoga watershed, Blue Nile basin, Ethiopia. Land Degrad Dev 20(6):609–622

Chebud YA, Melesse AM (2009a) Numerical modeling of the groundwater flow system of the Gumera sub-basin in Lake Tana basin, Ethiopia. Hydrol Process Spec Issue Nile Hydrol 23(26):3694–3704

Chebud YA, Melesse AM (2009b) Modeling lake stage and water balance of Lake Tana, Ethiopia. Hydrol Process 23(25):3534–3544

Chebud Y, Melesse AM (2013) Stage level, volume, and time-frequency change information content of Lake Tana using stochastic approaches. Hydrol Process 27(10):1475–1483. https://doi.org/10.1002/hyp.9291

Conway D (2000) The climate and hydrology of the Upper Blue Nile River. Geogr J 166:49–62

Dile YT, Tekleab S, Ayana EK, Gebrehiwot SG, Worqlul AW, Bayabil HK, Yimam YT, Tilahun SA, Daggupati P, Karlberg L, Srinivasan R (2018) Advances in water resources research in the Upper Blue Nile basin and the way forward: a review. J Hydrol 560:407–423

FAO (2020) WaPOR database. Food and Agriculture Organization of the United Nations. Database accessed on 29 May 2020. https://wapor.apps.fao.org/home/WAPOR_2/1

Food and Agriculture Organization (FAO), IHE Delft Institute for Water Education (2020) Water accounting the Nile Riveir basin: remote sensing for water productivity. Rome. https://doi.org/10.4060/ca9895en

Gelagay HS, Minale AM (2016) Soil loss estimation using GIS and Remote sensing techniques: a case of Koga watershed, Northwestern Ethiopia. Int Soil Water Conserv Res 4(2):126–136

Haregeweyn N, Tsunekawa A, Poesen J, Tsubo M, Meshesha DT, Fenta AA, Nyssen J, Adgo E (2017) Comprehensive assessment of soil erosion risk for better land use planning in river basins: case study of the Upper Blue Nile River. Sci Total Environ 574:95–108

Hurni H, Tato K, Zeleke G (2005) The implications of changes in population, land use, and land management for surface runoff in the upper Nile basin area of Ethiopia. Mt Res Dev 25(2):147–54

Melesse AM, Bekele S, McCornick P (2011) Introduction: hydrology of the Niles in the face of climate and land-use dynamics. In: Melesse AM (ed) Nile River basin: hydrology, climate and water use. Springer Science+Business Media B.V. Springer Dordrecht Heidelberg London New York. https://doi.org/10.1007/978-3-319-02720-3_1

Melesse A, Abtew W, Setegn S (2014) Introduction. In: Melesse A, Abtew W, Setegn S (eds) Nile River basin. Springer, Cham. https://doi.org/10.1007/978-3-319-02720-3_1

Molden D, Oweis T, Steduto P, Bindraban P, Hanjra M, Kijne J (2010) Improving agricultural waterproductivity: between optimism and caution. Agric Water Manage 97:528–535. https://doi.org/10.1016/j.agwat.2009.03.023

Setegn SG, Srinivasan R, Dargahi B, Melesse AM (2009a) Spatial delineation of soil erosion prone areas: application of SWAT and MCE approaches in the Lake Tana basin, Ethiopia. Hydrol Process Spec Issue Nile Hydrol 23(26):3738–3750

Setegn SG, Srinivasan R, Melesse AM, Dargahi B, (2009b) SWAT model application and prediction uncertainty analysis in the Lake Tana basin, Ethiopia. Hydrol Process 24(3):357–367

Setegn SG, Bijan Dargahi B, Srinivasan R, Melesse AM (2010) Modelling of sediment yield from Anjeni gauged watershed Ethiopia using SWAT. JAWRA 46(3):514–526

Yesuph AY, Dagnew AB (2019) Soil erosion mapping and severity analysis based on RUSLE model and local perception in the Beshillo Catchment of the Blue Nile Basin, Ethiopia. Environ Syst Res 8:17

Chapter 2
Geology of the Abbay Basin

Balemwal Atnafu Alemu

Abstract The geologic evolution found in the Abbay River Basin has three temporally distinctive signatures. One points towards the formation of the old Precambrian basement, second towards the evolution of the Phanerozoic sedimentary succession in the Horn of Africa and finally towards the formation of an enormous Cenozoic uplift and volcanism related to it. The units related to this history, the Precambrian basement, Palaeozoic clastics, Adigrat Sandstone formation, Gohatsion Formation, Carbonate Unit, Mugher Mudstone formation, Debre Libanos Sandstone formation and Trap volcanics, have been described. The description and distribution show that the Abbay River Basin geology is quite different from the geology of the Nile River Basin as a whole. A brief summary of the geologic resources associated with them is presented.

Keywords Precambrian basement · Phanerozoic · Cenozoic · Adigrat Sandstone · Gohatsion Formation · Mugher Mudstone · Debre Libanos Sandstone · Trap volcanics

Introduction

Geomorphologically, Ethiopia is presently subdivided into the Northwestern Plateau and the Southeastern Plateau separated by the main Ethiopian rift (MER). A northern extension of the MER is an extensive depression called the Afar. The Abbay drainage basin dominates the Northwestern Ethiopian Plateau.

Springs of Sakala join Gilgel Abbay river to flow north into Lake Tana before the river continues its journey. The Abbay River originates in such a way. Tectonically, Chorowicz et al. (1998) suggested that the formation of Lake Tana in its present location was controlled by converging north, northeast, and east–west trending grabens. Additionally, Lake Tana has an escarpment to its west. The Abbay River crosses different geological formations and units along its Ethiopian course from the starting

---

B. Atnafu Alemu (✉)
School of Earth Science, Addis Ababa University, Addis Ababa, Ethiopia
e-mail: balemwal.atnafu@aau.edu.et

© The Author(s), under exclusive license to Springer Nature Switzerland AG 2025
A. Melesse et al. (eds.), *Abbay River Basin*, Springer Geography,
https://doi.org/10.1007/978-3-031-65241-7_2

point to intersection of the Blue Nile River in Sudan. The first 150 km of its course lies on lava flows of Tertiary age, after which river flows through Mesozoic and Palaeozoic sediments until the Precambrian is reached approximately 330 km from Lake Tana (Jepsen and Athearn 1964). Quaternary sediments have attained a notable thickness around the Sudanese border. The geology of the Abbay River Basin is a regional composite product of the geological framework in the Horn of Africa and has resulted from a relationship to the vast geological time and the diversity of the geological environments. The Abbay River is also referred to as the part of the Blue Nile in relationship to the Nile Drainage Basin as a whole. More recently, Abdelsalam (2018) included it as a part of a drainage system coined the Ethiopia–East Sudan Nile. Although there are overlaps in the geological history of the Nile countries, a unique combination of geology is different and fascinating in Ethiopia from the neighbouring countries. While the Blue Nile River from its outlet at Lake Tana passing the Ethiopian Plateau and crossing the plains of Sudan to meet the White Nile, it passes over 800 km length and carves a mighty gorge (sometimes termed the Grand Canyon of Africa).

From the geologic setup of Ethiopia, this chapter is an attempt to provide a synthesis of the particular geology of the Abbay River system as a more peculiar part of the more famous and extensive larger Nile River Basin. The information in this synthesis is an extensive work in the Abbay River Basin by colleagues, myself and students I have advised and supervised for their MSc and PhD degrees. In addition, more than 60 years of publications on the different aspects of the geology of the basin have a great input.

A reflection of the roles of tectonics at the local, regional and global scales provides a unique opportunity to understand the geology of the Abbay Sedimentary Basin (Ayalew et al 2021, Gani et al 2009, Schandelmeier and Reynolds 1997, Scotese et al 1999).

The age range of the rocks in the Abbay River Basin is between 600 million years ago and today. The rock units in the basin are represented by a gross composite stratigraphic log (Fig. 2.1).

The Precambrian

The term Precambrian refers to the time or rocks that are older than 540 Ma. Works have synthesized the data from the Precambrian history in the Nile Basin (Whiteman 1971; Williams and Faure 1980; Said 1993).

The Precambrian in the Horn of Africa refers to rocks that have an age of up to one billion years or so. These rocks have been highly eroded igneous and metamorphic rocks. The data geologists have as an evidence for these Precambrian rocks has been reviewed (Asrat et al. 2001). The review asserts that the rocks were modestly well-dated. These ancient rocks are the ones which have formed the foundation or basement. Over this basement, horizontal sedimentary formations were formed and these sedimentary formations have been later overlain by volcanic rocks.

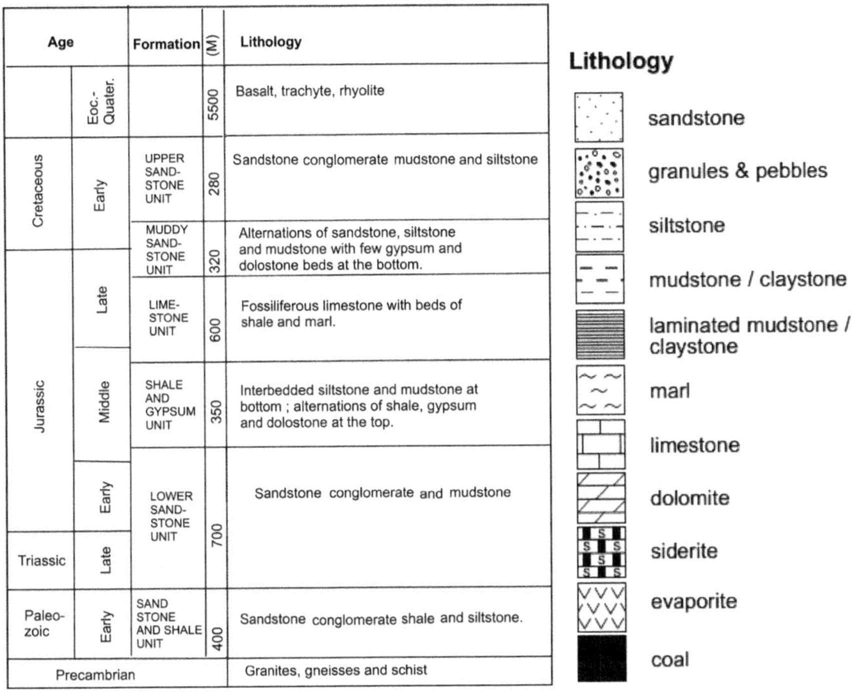

2 Geology of the Abbay Basin

Age		Formation	(M)	Lithology
	Eoc.-Quater.		5500	Basalt, trachyte, rhyolite
Cretaceous	Early	UPPER SANDSTONE UNIT	280	Sandstone conglomerate mudstone and siltstone
Jurassic	Late	MUDDY SANDSTONE UNIT	320	Alternations of sandstone, siltstone and mudstone with few gypsum and dolostone beds at the bottom.
Jurassic	Late	LIMESTONE UNIT	600	Fossiliferous limestone with beds of shale and marl.
Jurassic	Middle	SHALE AND GYPSUM UNIT	350	Interbedded siltstone and mudstone at bottom ; alternations of shale, gypsum and dolostone at the top.
Jurassic	Early	LOWER SANDSTONE UNIT	700	Sandstone conglomerate and mudstone
Triassic	Late			
Paleo-zoic	Early	SANDSTONE AND SHALE UNIT	400	Sandstone conglomerate shale and siltstone.
Precambrian				Granites, gneisses and schist

Lithology

- sandstone
- granules & pebbles
- siltstone
- mudstone / claystone
- laminated mudstone / claystone
- marl
- limestone
- dolomite
- siderite
- evaporite
- coal

Fig. 2.1 Gross composite stratigraphy of the Abbay River basin

As the Precambrian Complex, we refer to the East African Orogen (EAO) for the zone of folded, faulted and metamorphosed major phases of mountain building that extended from Arabia and Egypt in the north to Mozambique and Madagascar in the south. The intense ductile and brittle deformation mentioned above was not homogenous throughout the zone. In fact the EAO diversity included the accretion of continental crust but also the accretion of oceanic crust (Fritz et al. 2013).

The Arabian–Nubian Shield in the north and the Mozambique Belt (MB) in the south are two loose terms in subdividing the zone.

The East African Orogeny lasted from approximately 850 Ma to approximately 550 Ma. From approximately 850 to 550 Ma (Fritz et al. 2013) the East African Orogeny consisted of a number of orogenic episodes. Beyond orogeny other geological processes were important during this time interval in this area. These geological processes were associated metamorphism, a number of phases of erosion and several intrusive igneous rocks like granites. In turn, these were subject to intense heat and pressure and thus were metamorphosed to gneisses and schists.

The Precambrian basement rocks formed are hosts to many precious and semi-precious metals and minerals that have been sought after from prehistoric times onwards.

Focusing on the area of the Abbay River Basin, we should confine our attention here to only rocks back to the last billion years. The Precambrian rocks in the Abbay

River Basin have been reasonably studied by early geologists in Ethiopia and the Horn of Africa (Kazmin 1973, 1975; Senbeto and de Wit 1981; Asrat et al. 2001; Fritz et al. 2013).

The Phanerozoic

The previous part gave an overview of what happened in the first half of the last billion years in the Horn of Africa. Following the previous upheavals, geological indications gave a fairly calm time. This calm time was only relative to the past hundreds of thousands of years. The Abbay River Basin, Ethiopia and the East African Orogen rotated and later gradually drifted southwards as part of Gondwanaland. Thus, they did not remain still.

The Precambrian mountains in the Abbay River Basin were slowly being worn down by erosion by various agents—water, wind and probably ice. The considerable amount of sediments washed down from the basin was carried away by rivers towards other regions and the oceans beyond. The worn down remnants of the ancient mountains became the Precambrian basement.

Younger than 540 Ma, there is a formal stratigraphic term called the Phanerozoic Eon, which refers to a time when complex living organisms became diverse and widespread.

For the coming discussion parts, a basic introduction of the subdivision of geologic time is helpful. Related terms that would crop up in the discussion are also added. The Phanerozoic eon is subdivided into three formal eras—Palaeozoic, Mesozoic and Cenozoic. The first is referring to the ancient life, the second time interval is referring to the middle life and the third refers to the new or recent life. This subdivision was based on the evolution of life within the Phanerozoic. The eras were further subdivided to periods, epochs and ages.

In the Abbay River Basin, the Palaeozoic has left restricted sedimentation records. The Mesozoic was dominated by clastic sedimentation and marine deposition in the Blue Nile Sedimentary Basin In an extensive part of Ethiopia the Cenozoic era represents an enormous amount of volcanic extrusion. Towards the end of the era Quaternary clastic sedimentation is recorded in the basin.

The pioneering work of Blanford (1870), who accompanied General Napier to Maqdala and to overthrow of Emperor Tewodros in the northern part of Abyssinia, determined a number of geologic units in northern Ethiopia that include Basement, Adigrat Sandstone, Antalo Limestone, Upper Sandstone and flood basalts (Ashenge and Termaber groups). The next huge synthesis of Dainelli (1943) has presented a quite simple general Phanerozoic sedimentary stratigraphy of the Northwestern Plateau. It consists of lower sandstone, an intermediate carbonate—marly unit and an upper sandstone unit. A detailed geological work and its resources of the Blue Nile Basin was made by Jepsen and Athearn (1964). They prepared a geological map and described the units under the Precambrian rocks, the Paleozoic sediments, the Mesozoic sediments, the volcanic series (Traps), the younger (Aden) volcanic series and younger continental sediments. Later works (Assefa 1981, 1991; Russo et al. 1994; Atnafu 2003; Dawit 2010, 2016; Wolela 2014; Chernet et al. 2020) addressed the details of this gross stratigraphy.

The Abbay River Canyon of central Ethiopia contains excellent exposures of up to a couple of thousands of metres thick siliciclastic, evaporite and carbonate sedimentary successions, a large part of which accumulate in a rift basin—the Blue Nile Basin The basin is one of a number of transtensional pull-apart basin that were initially formed during Karoo rifting in the Permian to Triassic times (Schandelmeier and Reynolds 1997). During the Mesozoic break-up of Gondwana, the Karoo rifts were reactivated, and this basin evolved to NW–SE trending, nearly parallel half grabens separated by transfer zones (Bosellini 1989, 1992).

In relation to the sedimentary evolution of the Abbay River Basin, Russo et al. (1994) suggested five geodynamic stages and events must be considered: (1) A peneplain stage representing the situation before the Gondwanaland break-up, (2) Intracontinental rift stage as a consequence of the Gondwana break-up, (3) Post-rift stage related to a deposition over the entire Horn of Africa, (4) The start of the marine flooding of the craton and (5) Drowning of the craton. The final stage is in association with a major regional transgression as a result of the Gondwana break-up.

The major geologic events that have occurred in the Abbay Sedimentary Basin during the Phanerozoic have been summarized in Table 2.1.

The Palaeozoic

The geological record of the Palaeozoic in Ethiopia is limited. Primarily, the beginning represents the situation before any tectonic effect. It was a time of extensive denudation and some restricted siliciclastic sedimentation.

Of these exposed limited Palaeozoic sediments, those in the southeastern part of the Abbay River Basin have been presented as 3 units (Dawit and Bussert 2009; Dawit 2010, 2016)- Pre-Adigrat I, II, and III formations. These authors assert that these units represent sedimentation at the initial stages of intracratonic basin formation during the middle–late Palaeozoic (Catuneanu et al. 2005, Dawit and Bussert 2009). The uppermost formation is equivalent to the Karoo sediments (Dawit 2010, Worku and Astin 1992).

These three units represent sedimentation under various depositional environments. Their palaeoenvironmental reconstruction includes deposition of the sediments in aeolian environments and alluvial flood plain systems (Dawit and Bussert 2009). Attempts are being made by these authors (personal communication with one author) to correlate some of the pre-Adigrat units in the Abbay basin with siliciclastic products in north Ethiopia which are asserted of being glaciogenic successions (Bussert 2010). No published work has been reported of Palaeozoic glaciogenic sediments in the Abbay basin yet.

Table 2.1 Phanerozoic geologic time and the major events in the Abbay River basin

The Phanerozoic time scale

Eon	Era	Period	Epoch	Some distinctive events in the Abbay River basin	Years before present
Phanerozoic Eon	Cenozoic	Quaternary	Holocene	• Sparse volcanism and sedimentation	< 2.58 million years
			Pleistocene		
		Neogene	Pliocene		3 - 66 million years
			Miocene	• Phases of enormous volcanism • Inter-trappean sedimentation	
		Paleogene	Oligocene	• Phases of enormous volcanism • Inter-trappean sedimentation	
			Eocene		
			Paleocene		
	Mesozoic	Cretaceous		• Regression of the marine water from the Abbay basin • (Mugher Mudstone and Debre Libanos Sandstone formations.)	66 - 540 million years
		Jurassic		• Marine transgression from the southeast • (Gohatsion Formation and the Carbonate Unit)	
		Triassic		• Break-up of Gondwana • Basin formation (NW–SE grabens) • Adigrat Sandstone	
	Palaeozoic	Permian		• Karoo rifting • Pre-Adigrat sandstones • Glaciation	- 252 - 540 million years
		Carboniferous		• Karoo rifting	

(continued)

Table 2.1 (continued)

The Phanerozoic time scale					
Eon	Era	Period	Epoch	Some distinctive events in the Abbay River basin	Years before present
		Devonian		• Erosion and peneplanation	
		Silurian		• Erosion and peneplanation	
		Ordovician		• Glaciation	
		Cambrian		• Erosion and peneplanation of the Precambrian orogeny	
Precambrian				• Tectonism, mountain building and metamorphism	540 million years–1 billion years

The Mesozoic

The Mesozoic record in Ethiopia is quite extensive. The Abbay River Basin, in the central part of the northwestern Ethiopian Plateau, has an extensive and exclusively sedimentary record. Sedimentary rocks range from clastic rocks to biochemical and chemical rocks. Mesozoic sediments have been categorized into five formations and have been depicted by various geologists in a geological map. Jepsen and Athearn (1964) had the first attempt focusing on the Abbay units but details have followed successively.

The Lower Sandstone (The Adigrat Sandstone)

The Adigrat Sandstone Formation forms an unconformable contact with the underlying formation. What are underlying this Triassic to Middle Jurassic sandstone formation are either the pre-Adigrat units or the Precambrian basement rocks (Russo et al. 1994; Dawit and Bussert 2009; Wolela 2008). Starting from earlier works on this siliciclastic unit, a dominantly fluvial origin has been interpreted. More recent works have proposed a storm-dominated shoreface to a barrier/inlet spit. The discussion is in its controversial stage (Wolela 2008; Dawit 2010, 2016), and more detailed facies analysis is needed.

Russo et al. (1994) have suggested that Adigrat Sandstone corresponds to the post-rift stage of the sedimentary evolution of the basin. The facies, thickness and age of the formation vary considerably (Bosellini 1989) as it is considered over the entire craton. Thus, depositional environments could have produced fanglomerates,

Photo 2.1 Adigrat sandstone cliff at the Abbay River Gorge. The Millennium Abbay Bridge can be seen at the *right* (2010)

lacustrine, deltaic and even shallow marine environments. The discussion on the depositional environment of Adigrat Formation in the Abbay basin should be viewed from such a context.

The Adigrat Sandstone Formation and its correlable units are extensive in the Horn of Africa, East Africa and South Arabia. In the Abbay Basin the formation has a dominantly reddish colour and forms a prominent cliff by the river (Photo 2.1).

The Gohatsion Formation

The Middle Jurassic evaporite-dominated succession (Gohatsion Formation), locally attaining 450 m in thickness, is presumed to have been deposited in the early rift stage and marks the initial marine transgression over the Ethiopian Craton (Bosellini 1989; Russo et al. 1994).

A formal subdivision of the Gohatsion Formation has been subdivided into four separate members (Assefa 1981), from bottom to top a mudstone member, a lower claystone member, a gypsum member and an upper claystone member. The formation was interpreted (Assefa 1981; Russo et al. 1994) to have formed in a peritidal depositional environment (Photo 2.2).

A more recent work (Chernet et al. 2020) on a number of stratigraphic sections of the formation in the Abbay River Basin provided a detailed facies analysis and detailed distribution of the rocks. The palaeoenvironmental reconstructions showed

Photo 2.2 Layers of gypsum and dolomite of the Gohatsion formation in the Abbay basin

that the environments changed with time in the formation. A barrier tidal environment was responsible for the fine clastic unit at the base; a marine sabkha for the gypsum-dominated member and finally a coastal shelf with seasonal flooding for the uppermost member. Sediment delivery was from the northwest and a marine influence was from southeast.

The Carbonate Unit

The Gohatsion Formation in the Abbay River Basin is conformably overlain by a thick limestone-marl unit. The upper part of this carbonate unit in the Abbay River Basin has been named the Lagajima Limestone without being formally defined, and it is also considered to be a correlable unit with the Antalo Formation. This formation was defined in the Mekele Basin which is part of the Northwestern Plateau.

The unit is fossiliferous and is biostratigraphically dated between the middle and late Mesozoic (Russo et al. 1994). The basal part of the unit represents the transgressive phase of the ocean. The overlying beds of limestone and marls are an indication of the flooding of the ocean of the Abbay Basin, as a part of the former and aborted rift basins. This drowning of the East African Craton is a major transgression caused by the Gondwana break-up. This also resulted in the formation of the African continental margin (Bosellini 1989; Russo et al. 1994).

The carbonate unit is made up of various types of limestones ranging from fine micritic limestones to oolitic limestones and small carbonate build-ups. These limestones are interbedded with thick marly intervals. Most of them are fossiliferous as micro and macrofossils (Photo 2.3). Detailed sedimentologic work is being done on the Abbay River Basin carbonate sediments to understand the distribution of the

Photo 2.3 (**a**) Coquina from the carbonate unit in the Abbay basin (2010) (**b**) An exposure of the carbonate unit in the Abbay River Basin

carbonates in relationship to the carbonates in the Horn of Africa. Clear indications of succession are closely related to changes of sea level (Russo et al. 1994).

The Upper Sandstone

The retreat of the Neo-Tethyan Ocean is represented by two units dominated by siliciclastic sediments. They were originally referred to as the Upper Sandstone Unit (to differentiate the unit from the Lower Sandstone Unit in the Abbay Basin, the Adigrat Sandstone). Assefa (1991) later differentiated the unit into two formal formations—the underlying Mugher Mudstone and the overlying Debre Libanos Sandstone Formation. Several depositional sub-environments for these sediments range from transitional supratidal to continental fluvial environments.

The Mugher Mudstone

Overlying the carbonate unit, the base of the Mugher Mudstone consists of alternating gypsum, dolomite and shale in the type area of the Abbay River Basin (Assefa 1991). The upper part of the formation is interbedded with mudstone, siltstone and fine- and medium-grained sandstone. The Mugher Mudstone has been reported and described in the Abbay River as far north as Mertuleariam and Mekaneselam in a recent work (Getaneh and Atnafu 2020). Schmidt and Werner (1997) indicated that these early Cretaceous sediments are coastal plain sediments with vertebrates (Goodwin et al. 2019)

One focal research point in northeast Africa is the withdrawal of the Tethyan Sea. The nature and timing of this sea is still attractive and the deposition of clastic/ evaporitic rocks are a source of information. Deposition of clastic rocks during the Jurassic–Cretaceous periods in Ethiopia is a source of vital information on the nature and timing of the withdrawal of the Tethyan Sea from Northeast Africa. The Mugher Mudstone, controversially dated in the Jurassic-Cretaceous periods, is one of these clastic rocks in the Abbay River Basin, Ethiopia. The lowest beds of the formation show characteristics of supratidal flat and shallow lagoonal deposits (Assefa 1991).

The Debre Libanos Sandstone

The Debre Libanos Sandstone, overlying the Mugher Mudstone, is composed of two subfacies (Assefa 1991): fining-upwards sandstones and thick massive sandstones. The fining-upwards part is generally thick and persistent. Overlying the Mugher Mudstones, according to Assefa (1991) these fine- to medium-grained sandstones accumulated in sandy-braided rivers.

More recently (Wolela 2009; Mohammedyasin and Wudie 2019) this formation has been informally grouped into the following facies. From the base we have greenish to greyish black shales. Overlying these fine-grained white sandstones and minor thin layers of siltstones exist. Following these fine- and medium-grained white sandstones with sparsely interbedded reddish siltstones are layered. Finally, medium-grained red sandstones and pebble-supported conglomerates are exposed at the top. This variety may give you the sense of the nature of the Debre Libanos Sandstone.

The Cenozoic

The Palaeozoic and Mesozoic sedimentation culminated with the Cenozoic volcanism and faulting associated with the initiation of rifting in eastern Africa. The greater portion of the northwestern Ethiopian Plateau is covered by volcanic rocks (Fig. 2.2). These volcanic rocks are clearly stratified and are collectively named as Traps originating from a Swedish word meaning stairs, representing the volcanic activity on older sedimentary or metamorphic rock formations.

Mohr (1962) divided the Cenozoic volcanic rocks of Ethiopia into the Trap and Aden Series The term Trap Series refers to plateau flood basalts with intercalations of rhyolite. The term Aden Series refers to the post-rift (late Miocene to Quaternary) volcanic rocks of the main Ethiopian rift (MER), Afar Depression, and sparse volcanic rocks in localized rift zones on the plateau.

A new stratigraphic scheme has been developed for the northeastern part of the volcanic succession (Zanettin and Justin-Visentin 1973). This scheme divided the extensive volcanics into fissure-fed series and centre-fed shield volcanoes. The volcanic plateau is covered by large numbers of low-angle shield volcanoes (Tarmaber Formation, Mohr 1963), a conspicuous feature of the Ethiopian volcanic province.

The Northwestern Plateau may have formed of the following supportable ideas proposed. It could have been a result of a regional tectonic uplift. This uplift is possibly due to an uprising mantle plume. This resulted in the formation of a topography related to the extrusion of a large igneous province. Associated with this is the uplift of the present Ethiopian plateau from an elevation approximately at sea-level to

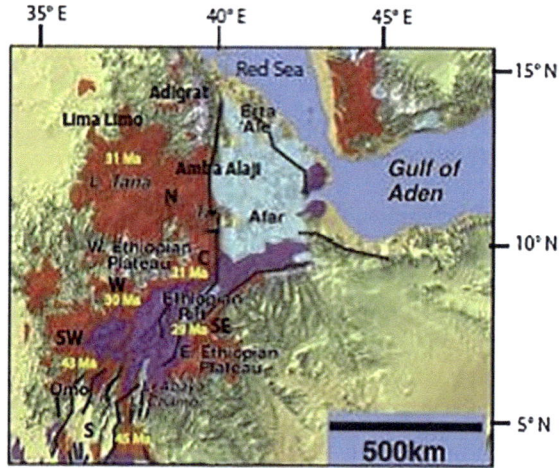

Fig. 2.2 Topographic relief map showing the areal extent of the flood basalts (brownish red) in Ethiopia (after Ayalew et al. 2021)

an elevation as high as 3000 m. It also caused the uplift of Jurassic marine carbonate units from sea-level to above 2000 m above sea-level now.

A composite stratigraphic section of the Abbay River Basin has a 2000 m thick sedimentary succession. Abbate et al. (2015) point out that overlying this sedimentary succession we have a maximum thickness of up to 1000 m of Oligocene flood basalts. Such enormous volcanism has occurred from vast regional tectonic events. Such events include the break-up of Arabia from Africa and the related formation of the Red Sea (Hofmann et al. 1997; Pik et al. 1998; Kieffer et al. 2004; Abbate et al. 2015).

Interbedded in the volcanics, the eastern part of the Abbay River Basin is known for peculiar Opal gems hosted in ignimbirite and forming sedimentary-like layers (Rondeau et al. 2010). Intertrappean intervals, in the northern and eastern part of the Abbay River Basin have carbonaceous rich sediments.

Geologic Resources in the Abbay River Basin

The whole geologic succession in the Abbay River Basin is a resource. Detailed economic analyses are needed to make the particular usage of different units or layers. To give a general indication on the range of the potential resources, the following brief paragraphs may give indications. The Precambrian rocks that are exposed in the southeastern part of the basin have and are being used for their precious metallic minerals. The same terrain also consists of attractive construction rocks from this area that are economically useful resources.

Mesozoic sediments and sedimentary rocks are currently being used. Extensive sands are the raw materials of the glass industry. The gypsum unit in the Gohatsion Formation is extensively used by small quarrying activities. A number of large and small cement factories are functional. The raw materials for these industries are almost all available in the basin. The limestone in the basin can be beautiful building stones and additionally, the limestone is useful in agricultural soils to reduce soil acidity. Farmlands in the Abbay basin are under soil acidity problems, and this problem is challenging agricultural productivity in Ethiopia.

The Cenozoic volcanics are quite extensive and are being used in construction activities. Coal intercalations in this enormous succession are showing indications. Further studies may come with more extensive reserves. Gemstones are also of great quality intercalated in the sediments associated with the ignimbrites. Volcanic rocks are also sources of rich and fertile soils.

Conclusions

This synthesis of the geology in the Abbay River Basin shows how different the geologic diversity is compared to the geology of the Nile Basin as a whole. Quite numerous geologic works focusing on the Nile Basin have detailed the geologic history of the basin. Three large time intervals in terms of metamorphic, sedimentary and volcanic rocks are summarized here. I assume that a different view of the geology of the Abbay Basin is visualized from this perspective.

References

Abbate E, Bruni P, Sagri M (2015) Geology of Ethiopia: a review and geomorphological perspectives. In: Billi P (ed) Landscapes and landforms of Ethiopia, world geomorphological landscapes. Springer Science, Business Media Dordrecht, pp 33–64

Abdelsalam MG (2018) The Nile's journey through space and time: a geological perspective. Earth Sci Rev 177:742–773

Asrat A, Barbey P, Gleizes G (2001) The Precambrian in Ethiopia: a review. Afr Geosci Rev 8(3):271–288

Assefa G (1981) Gohatsion Formation. A new lias-malm lithostratigraphic unit from the Abay River basin, Ethiopia. Geosci J 63–88

Assefa G (1991) Lithostratigraphy and environment of deposition of the late Jurassic—early cretaceous sequence of the central part of northwestern plateau, Ethiopia. N Jb Geol Palaont Abh 182(3):255–284

Atnafu B (2003) Facies and diagenetic development of Jurassic carbonates in the Blue Nile basin, Ethiopia. Ph.D. Thesis, Friedrich Alexander Universität Erlangen—Nürnberg

Ayalew D, Getaneh W, Pik R, Atnafu B, Zemelak A, Belay E (2021) Stratigraphic framework of the northeastern part of the Ethiopian flood basalt province. Bull Volcanol 83(57):1–13. https://doi.org/10.1007/s00445-021-01482-z

Blanford WT (1870) Observations on the geology and zoology of Abyssinia. MacMillan, London, p 487

Bosellini A (1989) The continental margins of Somalia: their structural evolution and sequence stratigraphy. Memoirie di Scienze Geologiche XII:373–458

Bosellini A (1992) The continental margins of Somalia: structural evolution and sequence stratigraphy. In: Watkins JS, Zhiqiang F, McMillen KJ (eds) Geology and geophysics of continental margins, vol 53, pp 185–205

Bussert, R (2010) Exhumed erosional landforms of the Late Palaeozoic glaciation in northern Ethiopia: Indicators of ice-flow direction, palaeolandscape and regional ice dynamics Gondwana Research Vol 18, Issues 2–3, pp 356–369. https://doi.org/10.1016/j.gr.2009.10.009

Catuneanu O, Wopfner H, Eriksson PG, Cairncross B, Rubidge BS, Smith RMH, Hancox PJ (2005) The Karoo basins of south-central Africa. J Afr Earth Sci 43:211–253

Chernet SG, Atnafu B, Asrat A (2020) Stratigraphic and lithofacies analysis of the Gohatsion Formation in the Blue Nile basin, central Ethiopia: implications for depositional setting. J Afr Earth Sci 162. https://doi.org/10.1016/j.jafrearsci.2019.103693

Chorowicz J, Collet B, Bonavia FF, Mohr P, Parrot JF, Korme T (1998) The Tana basin, Ethiopia: intra-plateau uplift, rifting and subsidence. Tectonophy 295(3–4):351–367

Dainelli G (1943) Geologia dell'Africa Orientale. (3 vols. text, 1 vol. maps). Reale Accademia Italia, Roma, 1916 pp

Dawit L.E (2010) Adigrat sandstone in Northern and central Ethiopia: stratigraphy, facies, depositional environments and palynology. Unpublished PhD thesis. Technische Universität Berlin, p 166

Dawit L.E (2016) Palaeoclimatic records of late Triassic palaeosols form central Ethiopia. Palaeogeogr Palaeoclimatol Palaeoecol 449:127–140

Dawit L. E, Bussert R (2009) Stratigraphy and facies architecture of Adigrat sandstone, Blue Nile basin, central Ethiopia. Zbl Geol Palaont Teil I 3(4):217–232

Fritz H, Abdelselam MG, Ali K, Bingen B, Collins AS, Fowler AR, Gebreab W, Hauzenberger CA, Johnson PR, Kusky TM, Macey P, Muhongp S, Stern RJ, Viola G (2013) Orogen styles in the East African Orogen: a review of the Neoproterozoic to Cambrian tectonic evolution. J Afr Earth Sci 86:65–106

Gani ND, Abdelsalam MG, Gera S, Gani MR (2009) Stratigraphic and structural evolution of the Blue Nile basin, northwestern Ethiopian plateau. Geol J 44:30–56. https://doi.org/10.1002/gj.1127

Getaneh W, Atnafu B (2020) Geochemistry and lithostratigraphy of the Mugher Mudstone: insights into the late Jurassic-early Cretaceous clastic sedimentation in Ethiopia and its surroundings. J Afr Earth Sci 164. https://doi.org/10.1016/j.jafrearsci.2020.103770

Goodwin MB, Irmis RB, Wilson GP, DeMar Jr DG, Melstrom K, Rasmussen C, Atnafu B, Alemu T, Alemayehu M, Chernet SG (2019) The first confirmed sauropod dinosaur from Ethiopia discovered in the Upper Jurassic Mugher Mudstone. Jo Afr Earth Sci 159:103571. https://doi.org/10.1016/j.jafrearsci.2019.103571

Hofmann C, Courtillot V, Feraud G, Rochette P, Yirgu G, Ketefo E, Pik R (1997) Timing of the Ethiopian flood basalt event and implications for plume birth and global change. Nature 389:838–841

Hunegnaw A, Sage L, Gonnard R (1998) Hydrocarbon potential of the intracratonic Ogaden basin, SE Ethiopia. J Pet Geol 21(4):401–425

Jepsen DH, Athearn MJ (1964) General geology map of the Blue Nile River basin Ethiopia. Unpublished geological map, U.S. Department of Interior/Department of water resources Ethiopia, Addis Ababa, Ethiopia

Kazmin V (1973) Geological map of Ethiopia, scale 1:2,000,000. Ethiopian Institute of Geological Survey, Addis Ababa, Ethiopia

Kazmin V (1975) Explanation note to the geological map of Ethiopia. Unpublished technical report. Geological Survey of Ethiopia, Addis Ababa, Ethiopia, p 18

Kieffer B, Arndt N, Lapierre H, Bastien F, Bosch D, Pecher A, Yirgu G, Ayalew D, Weis D, Jerram AD, Keller F, Meugniot C (2004) Flood and shield basalts from Ethiopia: magmas from the African super swell. J Petrol 45:793–884

Mohammedyasin MS, Wudie G (2019) Provenance of the Cretaceous Debre Libanos sandstone in the Blue Nile basin, Ethiopia: evidence from petrography and geochemistry. Sed Geol 379:46–59

Mohr P (1962) The geology of Ethiopia. University College of Addis Ababa Press, Ethiopia, 268 pp

Mohr P (1963) Geological map of Horn of Africa, scale 1:2,000,000. Philip and Tacey, London

Pik R, Daniel C, Coulon C, Yirgu G, Hofmann C, Ayalew D (1998) The northwestern Ethiopian plateau flood basalts: classification and spatial distribution of magma types. J Volcanol Geotherm Res 81:91–111

Rondeau B, Fritsch E, Mazzero F, Gauthier JP, Tok BC, Eyassu B, Gaillou E (2010) Play-of-color opal from Wegel Tena, Wollo province, Ethiopia. Gems Gemology 46(2):90–105

Russo A, Assefa G, Atnafu B (1994) Sedimentary evolution of the Abbay river (Blue Nile) basin, Ethiopia. N Jb Geol Paläont Mh 5:291–308

Said R (1993) The River Nile—geology, hydrology and utilization. Pergamon Press, London

Schandelmeier H, Reynolds PO (eds) (1997) Palaeogeographic-palaotectonic atlas of Northeastern Africa and adjacent areas. A.A Balkema, Rotterdam, Netherlands, p 160

Schmidt D, Werner C (1997) Early cretaceous coastal plain sediments of the Mugher mudstone formation, Abay River basin, Ethiopia. Zbl Geol Palaeont Teil I H.1/2:293–309

Scotese CR, Boucot AJ, McKerrow WS (1999) Gondwanan palaeogeography and palaeoclimatology. J Afr Earth Sci 28(1):99–114

Senbeto C, de Wit MJ (1981) Plate tectonics and metallogenesis: some guidelines to Ethiopian mineral deposits. Bulletin No. 2, Ethiopian Institute of Geological Surveys, Addis Ababa

Whiteman AJ (1971) The geology of the Sudan Republic. Clarendon Press, Oxford

Williams MAJ, Faure H (1980) The Sahara and the Nile—quaternary environments and prehistoric occupation in northern Africa. A.A. Balkema, Rotterdam

Wolela A (2008) Sedimentation of the triassic-jurassic Adigrat sandstone formation, Blue Nile (Abbay) basin, Ethiopia. J Afr Earth Sci. https://doi.org/10.1016/jafrsci.2008.04.001

Wolela A (2009) Sedimentation and depositional environments of the barremian-cenomanian Debre Libanos sandstone, Blue Nile (Abbay) basin, Ethiopia. Cretac Res 30(1133):1145

Wolela A (2014) Diagenetic contrast of sandstones in hydrocarbon prospective Mesozoic rift basins (Ethiopia, UK, USA). J Afr Earth Sci 99:529–553

Worku T, Astin TR (1992) The karroo sediments (late Palaeozoic to early Jurassic) of Ogaden basin, Ethiopia. Sediment Geol 76:7–21

Zanettin B, Justin-Visentin E (1973) Serie di vulcaniti etiopiche. I: la serie dell'altipiano etiopico centro-orientale. Boll Soc Geol It 92:313–327

Chapter 3
Geomorphology of the Abbay Basin from Geographical Perspective

Mehretie Belay Ferede

Abstract Geomorphic processes and landforms with their structural attributes impose significant impact on environmental systems. A literature review study was conducted to concisely document the geomorphic processes and structural landforms of the Abbay Basin in Ethiopia. A total of 75 papers available online were reviewed and statistical data were extracted from 41% of the reviewed papers. The data collected through this process were then systematically organized into three themes and analyzed using explanatory transcriptions to convert the raw data into useful information. The analyzed information indicated that the structural geomorphology of the Abbay Basin resulted from continental collision and fragmentation of East and West Gondwana and the formation of the East African Orogen. The review results indicate that the Abbay fault evolved by rifting and faulting of the Gondwana continent along a northwest trending Mesozoic fault through headward erosion over the *Afro-Arabian Lava Dom*. River Abbay crosses varied geomorphic landscapes ranging from < 500 to > 4000 m above mean sea level. Five major geographic regions (Lake Tana basin, Amarasaint Massifs, Mid-central plateau, southwestern upland and the Abbay-Dinder lowlands) makeup the entire geomorphic region of the Abbay Basin. The geomorphic structure of the area is rugged favouring the development of fast flowing streams and rivers. Most of the streams and rivers rising over the rugged mountains erode large amounts of soil and deposits along the lower courses and plain areas. The sediments in turn affect the reservoirs, irrigation and hydropower dams in downstream areas. The implementation of watershed-based soil and water conservation measures that reflect the varied geomorphic structures in watersheds is suggested to minimize both on-site and off-site impacts.

Keywords Geomorphology · Geomorphic processes · Geomorphic regions · Abbay basin · Ethiopia

---

M. Belay Ferede (✉)
Department of Geography and Environmental Studies, Bahir Dar University, Bahir Dar, Ethiopia
e-mail: belaymehrete@yahoo.com

Introduction

The Blue-Nile River (hereafter named the Abbay River) is the largest river in Ethiopia in terms of its irrigation and hydroelectric power potential. Its irrigation potential reaches 815,581 hectares (ha), while the hydropower potential is estimated at 78,820 Giga Watt Hours (GWH) per year (Ahmed and Ismail 2008; Yilma and Awulachew 2009; Shobary et al. 2021). Abbay forms the second largest basin (with approximately 202,994 km^2 catchment area) in Ethiopia (ENTRO 2006). It roughly covers > 20% of the Ethiopian land surface (Abdel-Aziz 2009; FDRE and ABA 2016). Its annual runoff is estimated at 54.5 billion cubic metres (BCM) after crossing the Ethio-Sudan border (with joining Gelego, Rihad and Dinder rivers). The total annual discharge of Abbay at the Sudan border is 49 BCM. With this annual volume of water, Abbay contributes approximately 50% of the total water flow in the country (ENTRO 2006) and approximately 60% of the total flow to the Aswan High Dam (Sutcliffe and Parks 1999).

The Abbay Basin hosts more than 50% of the country's population and volcanic agricultural soils. Nevertheless, the geomorphology of the Abbay Basin is not well-studied and documented and little is known about the diverse geomorphic and geographic features by the local community and even by academicians. The time as well as development process of the Abbay fault (gorge) is still not vividly understood (Blackburn 2016) even by scholars and academicians.

Many papers have been published on the Abbay Basin. However, most focus on issues other than geomorphology. Examples of such papers focus on geology and stratigraphy (Gani et al. 2009; Abbate et al. 2015; Abdelsalam 2016; Williams 2016; Alemu et al. 2019; Woodward et al. 2022); soil erosion and sedimentation (Betrie et al. 2011; Dessie et al. 2014; Gessesse 2014; Steenhuis et al. 2014; Tesfaye 2022); water and sediment management (Ahmed and Ismail 2008; Abdel-Aziz 2009); Land and water management (Abtew 2014; Setegn et al. 2014; Adem et al. 2020; Duguma and Duguma 2022); and on climatology and hydrology (Sutcliffe and Parks 1999; Wale 2008; Abdo et al. 2009; Dile et al. 2013; Adem et al. 2014; Enyew et al. 2014; Fenta et al. 2014; Kigobe et al. 2014; Moges and Gebremichael 2014; Nigussie et al. 2014; Gebre 2015; Taye et al. 2015; Roth et al. 2018; Lemann et al. 2019; Meresa and Getachew 2019; Yenehun et al. 2021; Wubneh et al. 2023). Other papers focus on Nile River challenges (Abtew and Melesse 2014a, b; Melesse et al. 2014; Dile et al. 2018), water governance and hydropolitics (El-Fadel et al. 2003; Abtew and Melesse 2014a, b; Deneke 2014; Kitaw and Yitayew 2014; Morbach et al. 2014; Mulat et al. 2014; Stoa 2014; Paisley and Henshaw, in press), and biomass and hydro-epidemiology (Ayana et al. 2014a; Wimberly and Midekisa 2014).

Studies on geomorpholgy and geomorphic processes are few and mostly emphasize specific issues or microbasins (Chorowicz et al. 1998; Ayalew and Yamagishi 2004; Poppe et al. 2013; Ayana et al. 2014b; Blackburn 2016; Meshesha 2016; Yibeltal et al. 2019), cover more areas of Ethiopia (Pik et al. 1998; Coltorti et al. 2007, 2015; Billi 2015b; Billi et al. 2015) or focus on areas outside of the Abbay Basin (Billi 2015a; Mège et al. 2015). Most of these geomorphic studies focus on western highlands in general or on other specific locations, such as northern or eastern Ethiopia. The study by Ayalew and Yamagishi (2004) focused on the Abbay gorge but was limited to the assessment of slope failure only.

Generally, most of the studies cited above focus on the geology, stratigraphy, climatology and hydrology of the Abbay Basin. Papers on the geomorphology of the Abbay Basin are rare or focus on specific geomorphological issues, as indicated in the case of Ayalew and Yamagishi (2004) above. The geomorphological landscapes of the Abbay Basin are therefore not well-studied and documented.

This paper thus tries to describe the geomorphology of the Abbay Basin based on three major themes through a review of past studies. Under the first theme, geomorphic processes involved in the development and evolution of the different landscapes and geomorphic features (including development and incision of the Abbay gorge) are presented. The second theme presents the typical and dominant geomorphic and geographic features of the Abbay Basin. The third theme presents the hydrology and surface planation processes and their impacts in the Abbay Basin.

Geographical Setting and Methods

Geographical Setting of the Abbay Basin

The Abbay River rises from the western highlands of Ethiopia and flows towards Sudan and Egypt. It transpires at the southeastern end of Lake Tana (Fig. 3.1) and divides the city of Bahir Dar into eastern and western parts. From Bahir Dar, Abbay descends into a deep canyon named Abbay Gorge [a colossal gorge similar to the Grand Canyon of the Colorado River of the USA (Williams 2016)] after flowing some 30 km to the south. Abbay then flows southwards and turns west towards Sudan (Gani et al. 2009; Billi et al. 2015; Williams 2016). After winding for over 900 km (Ahmed and Ismail 2008; Alemu et al. 2019; Woodward et al. 2022) in a crescent-shaped journey, Abbay crosses the Ethio-Sudan border and retains a northwards course. It traverses approximately 735 km over the clay plains of eastern Sudan to reach Khartoum by flowing 1635 km from Lake Tana (Ahmed and Ismail 2008). Abbay joins the Victoria Nile at Khartoum and then River Tekeze (Atbara) in the city of Atrbara in Sudan (Ahmed and Ismail 2008; Williams 2016).

The Abbay Basin roughly extends for 09° 35′ north–south (from 07° 40′ N to 16° 05′ N latitudes) and for 07° 23′ east–west (from 32° 27′ E to 39° 50′ E longitudes) (ENIRO 2006; Fig. 3.1). The total estimated area of the Abbay Basin is 313,657 km^2 (see Table 3.1). The largest portion ($\approx$64%) of this total area mainly occurs in Ethiopia (Table 3.1).

Nearly 93,565 km^2 ($\approx$47%) of the Abbay Basin in Ethiopia falls within the Amhara National Regional State (ANRS). The other 31% (62,478 km^2) is held in the Oromia region while 44,676 km^2 ($\geq$ 22%) of the Abbay Basin falls within the Banishangul region (Table 3.1). According to ENTRO (2006), approximately 60%

Fig. 3.1 Map showing the location of the Abbay Basin. *Source* ENTRO (2006)

of the ANRS, 40% of Oromia, and 95% of the Banishangul region fall within the Abbay Basin.

As shown in Table 3.1, the Sudan part of the Blue-Nile basin accounts for 36% of the total Abbay Basin. Areas sharing the Abbay Basin in Sudan include Sennar (30,106 km^2), Blue Nile (19,162 km^2), El-Gezira (15,686 km^2), Gadrife (38,829 km^2) and Khartoum (6880 km^2). Two important rivers rising from Ethiopia join Abbay in

3 Geomorphology of the Abbay Basin from Geographical Perspective

Table 3.1 Area share of the Blue-Nile basin in Ethiopia and Sudan by region

№	Regional state	Area (km^2)	Area (in %)*	Area (in %)**	Country
1	Amhara National Regional State (ANRS)	93,565	46.61	30	Ethiopia
2	Oromiya	62,478	31.13	20	
3	Banishangul-Gumuz	44,676	22.26	14	
	Total	202,994	100	64	
4	Sennar	30,106	27.21	10	Sudan
5	Blue-Nile	19,162	17.32	6	
6	El Geziera	15,686	14.17	5	
7	Gedarif	38,829	35.09	13	
8	Khartoum	6,880	6.22	2	
	Total	110,663	100	36	

Sudan. The first of these is Dinder, with a catchment area of 15,339.7 km^2 and the second is Rahad covering an area of 8290 km^2 (see Table 3.2).

On its way to Sudan, Abbay is joined by several tributaries. For instance, Beshilo and Welaka join Abbay in its eastern part. These rivers rise from the Wollo Massifs and have 12,926.7 and 5793.3 km^2 mean basin areas, respectively (Table 3.2). Rivers such as Fettam, Temcha, Chemoga and Birr rising from the Choke Mountain catchments join Abbay in the northern part. The catchment areas of these and other rivers rising from Gojjam and joining the Abbay are reported as north and south Gojjam and Wombera catchments (Table 3.2), perhaps due to their numerous numbers to account for each individually. Jamma (15,687.7 km^2), Guder (5313 km^2), Beles (11,265 km^2), Muger (7804.3 km^2), Fincha (3850.3 km^2), Dabus (16,564 km^2) and Didessa (13,253.7 km^2) also join the Abbay River in the southern and eastern parts. These rivers rise from the highlands of Shoa and Wellega. River Beles rises from the Alefa-Taqusa region of west Gondar and passes across the Jawi and Pawi lowlands then joins Abbay at its northwestern corner (Ahmed and Ismail 2008). This river is among the largest tributaries of Abbay. Didessa joins Abbay from the south (in Wellega). Dabus is another river joining Abbay from Wellega.

Methods

This paper develops through a review of previous studies focusing on geomorphology, geology, climatology and hydrology. As papers focusing on the geomorphology of the Abbay Basin are scarce, the review mainly relays on geology, climatology, hydrology and related publications available online. A total of 75 papers were reviewed in detail and used as sources of information. The expressive and meaningful information was combined and organized into themes. Three themes were identified

Table 3.2 Tributaries of Abbay with their basin areas

Basin	Basin area (in km^2) by source				
	Meresa and Getachew (2019)	Shobary et al. (2021)	World Bank (2008)	ENTRO (2006); Yilma and Awulachew (2009)	Mean
Beles	4106	12,327	14,200	14,426	11,265
Dabus	10,139	13,716	21,032	21,367	16,564
Guder	524	6594	7011	7123	5313
Anger	3742	7475	7901	8027	6786
Beshelo		12,085	13,242	13,453	12,926.7
Jamma		15,248	15,782	16,033	15,687.7
Welaka		4448	6415	6517	5793.3
Mugger		6907	8188	8318	7804.3
Didessa		17,855	1963	19,943	13,253.7
Fincha		3308	4089	4154	3850.3
Dinder		16,000*	14,891	15,128	15,339.7
Rahad		8200**	8269	8401	8290
South Godjam		27,635	16,762	17,029	20,475.3
Lake Tana		15,935	15,054	15,294	15,427.7
North Gojam		12,045	14,389	14,618	13,684
Wombera		16,859	12,957	13,163	14,326.3

Source Adapted from ENTRO (2006)
*From the Ethiopian Abbay; **From the Sudan Abbay

to address the intended objectives. Accordingly, data related to geomorphic processes were assembled into one category. Information linked to the structural landforms was also compiled into another theme. Runoff, surface movement of materials and sedimentation related data were also categorized as third themes.

The systematized data were then transformed into useful information by an explanatory piece of writing through descriptive recitations. However, problems were faced during the data categorization process. Statistical differences were encountered among the reviewed papers. In particular, data related to the sizes of drainage basins, flow volumes, erosion and sedimentation rates were contradictory, perhaps due to measurement differences among the different studies (see Table 3.3). Reporting contradictory data is perceived as inconvenient and unfair. To avoid this problem, data gathered from the different sources were tabulated, added and converted into mean values to provide readers with pertinent information about the geomorphic features of the Abbay Basin. Finally, the calculated data were reported through explanatory analysis. Trials are also made to integrate the testified data with the results of past studies to validate the reported information. Over 30 papers are used as sources of

Table 3.3 Example of differing information about the area coverage and surface runoff of the Abbay Basin

Area coverage in (km²)		Surface runoff (BCM)	Contribution to Nile flow (%)	Source
Ethiopia	Total			
199,812		50	62	Meresa and Getachew (2019)
250,000	324,000	54.8		Shobary et al. (2021)
200,000				World bank (2008)
41,900	324,530	54.8		Ahmed and Ismail (2008)
202,994				Abdel Aziz (2009)
202,884	313,547	75	65	ENTRO (2006)
	311,437			FDRE and ABA (2016)
			60	Sutcliffe and Parks (1999)

statistical information. The authors' main themes and years of publication of these papers are given in Table 3.4.

Results and Discussion

Geomorphic Processes that Created Structural Landforms

The formation of the Abbay Basin is part of the formation of the entire Ethiopian landscape. The Ethiopian geomorphology was also the outcome of the continental collision and fragmentation of the earliest continents (Abbate et al. 2015; Williams 2016). Many writers (e.g. Abbate et al. 2015) indicate that the development of many of the geomorphic and topographic features of the Abbay Basin was the result of the collision and fragmentation of East and West Gondwana that created the super Gondwana continent and the East African Orogen (Fig. 3.2). This event was followed by planation of the East African Orogen through the forces of running water, wind and glaciers. The long period of the Carboniferous-Permian glaciation when Ethiopia was placed at the edge of the southern hemisphere (*glaciation of Pangaea*); faulting and rifting of Gondwana and the subsequent marine transgression from the Palaeotethys Ocean[1] during the Jurassic-cretaceous periods; the great uplift of the Arabo-Ethiopian landmass and the subsequent outpouring of the thick basaltic lava

---

[1] The Palaeotethys Ocean later created the Indian Ocean and other adjacent water bodies, such as the Mediterranean Sea.

Table 3.4 Summary of studies used as sources of data and information

Author(s)	Main theme	Journal/publisher
ENTRO (2006)	Water Atlas of Blue Nile sub-basin	
Abbate et al. (2015)	Geology of Ethiopia	Book Chapter
Billi et al. (2015)	Ethiopian Rivers	Book Chapter
Billi (2015a, b)	Geomorphology and geology	Book Chapter
World Bank (2008)	Project appraisal report	Project report
Meresa and Getachew (2019)	Climate change and river flow	Journal of Water and Climate Change
Poppe et al. (2013)	Geomorphology	Journal of Maps
Abdel-Aziz (2009)	Water and sediment management for the Blue Nile Basin	Conference paper
Gani et al. (2009)	Stratigraphic and structural evolution	Geological Journal
Shobary et al. (2021)	Land use land cover	IOP Conference paper
Yilma and Awulachew (2009)	Atlas of Blue-Nile Basin	
Williams (2016)	Geology	Book chapter
McCartney et al. (2010)	Water resource development	Research report
Meshesha (2016)	Human impact on geomorphology	Thesis
Yenehun et al. (2021)	Hydrology	Water
Enyew et al. (2014)	Climate change and hydrology	Journal of Geology and Geosciences
Gebre (2015)	Runoff	Hydrology current Research
Ayana et al. (2014a)	Bathymetry	Book chapter
Dile et al. (2013)	Water resources	Journal of Hydrology
Abtew and Melesse (2014a, b)	Nile River (hydrology)	Book chapter
Ahmed and Ismail (2008)	Sedimentation	Research report
Abdo (2008)	Climate change and hydrology	Thesis
Betrie et al. (2011)	Sedimentation	Hydrology and Earth System Sciences
Wale (2008)	Hydrological balance	Thesis
Yitbarek et al. (2019)	Water Hyacinth	International Journal of Science and Technology
Woodward et al. (2022)	River Nile evolution and environment	Book chapter
Adem et al. (2020)	Hydrology and land management	Water

(continued)

Table 3.4 (continued)

Author(s)	Main theme	Journal/publisher
Zimale et al. (2018)	Hydromechanics	Journal of Hydrology and Hydromechanics
Dagnew et al. (2017)	Sedimentation and soil erosion	Land Degradation and Development
Guzman et al. (2013)	Sedimentation	Hydrolysis and Earth System Sciences
Guzman et al. (2017)	Sedimentation	Hydrological Processes

over the Mesozoic sedimentary layers during Oligocene–Miocene times; and the global climate change and the recurrent dry and wet events of the Neogene ages (e.g. see Abbate et al. 2015) are associated processes that caused the formation of the geomorphic features of the Abbay Basin and related landforms in Ethiopia (e.g. see the Karoo Rift, Fig. 3.3).

The initial processes were the earliest volcanism and mountain building events, i.e. solidification of the Earth and formation of the oldest rocks prior to 530 Ma (Williams 2016). The rocks created during that time make up the base (lower) geologic surface of the Abbay Basin as in any other part of Ethiopia (see Fig. 3.4). Typical Proterozoic metamorphic rocks (containing valuable minerals such as copper, gold, nickel and platinum) mainly occur as outcrops in only a few locations of the Abbay Basin (e.g. parts of western lowlands along the Ethio-Sudan border). These formations are commonly called basement rocks because they form the base geology of the basin. These Proterozoic processes were more affected and modified by later geologic

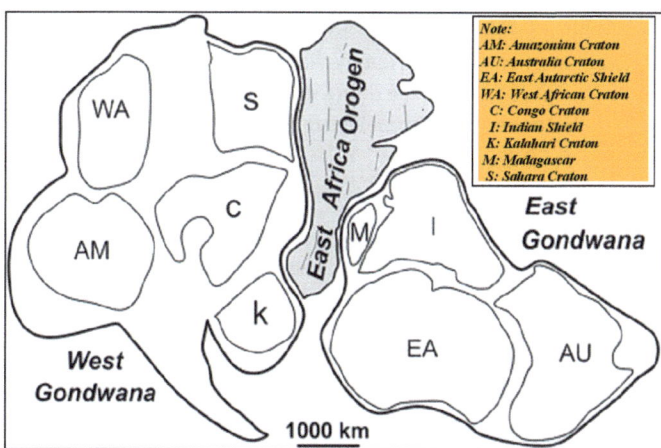

Fig. 3.2 Sketch showing East and West Gondwana continents and the East African Orogen. *Source* Adapted from the modified sketch of Meert and Lieberman (2008; in Abbate et al. 2015)

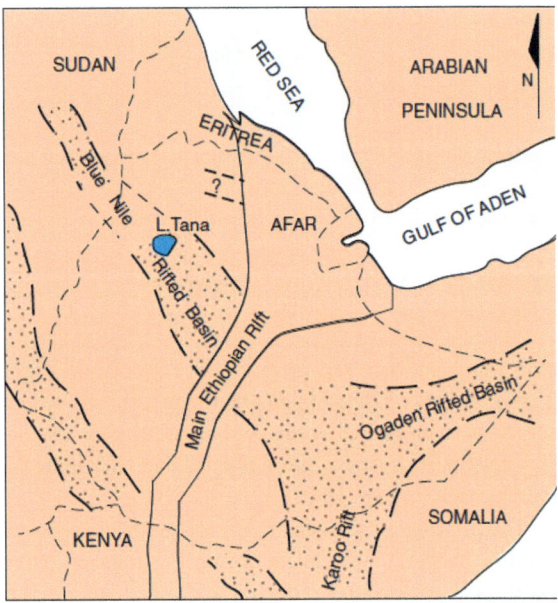

Fig. 3.3 Example of the Abbay Rifted Basin (Karoo Rift). *Source* Williams (2016)

events of the Palaeozoic denudation cycles (planation of the Precambrian surface) ≈286–505 Ma (Williams 2016).

The erosional process of the Palaeozoic age was then followed by continental collision and fragmentation (at ≈250 Ma) that caused continental faults, rifts and broad-based valleys and marked climatic changes and basaltic lava flows. The formation of a north–south approaching crack (the *Karoo rift*) in the southern part of Ethiopia and the north–west extending fault including the Abbay fault (see Fig. 3.3) initiated the invasion of Horn of Africa by water from the Palaeotethys Ocean (Gani et al. 2009). This event resulted in the deposition of three layers of sedimentary rocks over the smooth Proterozoic surface. The three sedimentary rocks are the Adigrat

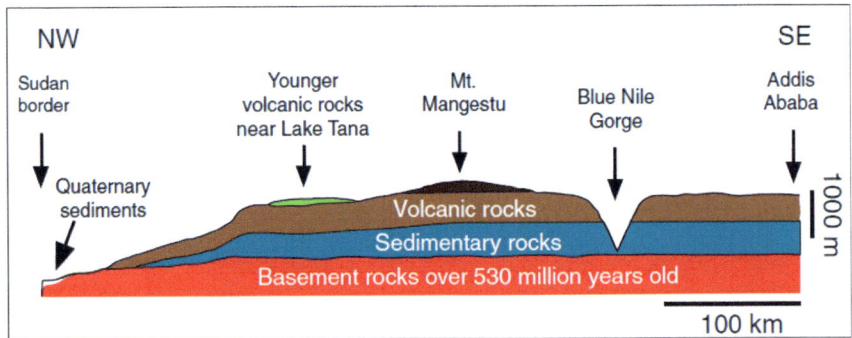

Fig. 3.4 Schematic geological cross-section through western highlands. *Source* Williams (2016)

sandstone (deposited 250–200 Ma), Antalo limestone (deposited 150 Ma) and upper sandstone deposited when the sea regressed. The old sedimentary rocks were then overlain by continental flood basalts (CFBs) (Pike et al. 1998; Coltorti et al. 2007, 2015); starting from approximately 45 Ma and recent fluvial/lacustrine deposits in the latest times (see the rock sequence from Fig. 3.4).

For instance, the figure taken from Williams (2016) shows that older rocks are forming the base geology at the Millennium Bridge in the Abbay gorge (Figs. 3.4 and 3.5). These older rocks were then overlain by the layers of the three Mesozoic sedimentary rocks in the middle part while passing from the Abbay River course to the top of the volcanic plateau. The upper surface in the area; as is observed in the sketch (Fig. 3.5) is capped by recent volcanic rocks, perhaps ≈30 Ma. Nevertheless, when directly passing from the Millennium Bridge to Goha Tsion, the tertiary lava is directly resting over the Antalo limestone (*mud*, *gypsum* and *dolomite*) and not over the upper sandstone. This is because the upper sandstone is missing or was not deposited in that location. This is what is known as 'unconformity' in geology, as shown by the snaky line (Fig. 3.5) at the joining site of the sedimentary formations and the stair-stepped lava rocks. As indicated in Williams (2016), there was a time-break of > 100 years between the deposition of the Antalo limestone and the out-powering of volcanic lava. The upper sandstone in another location appears to occur between the Antalo limestone and the tertiary lava at the Jamma River (as upper sandstone) and as Mugher mudstone in the Mugher River valley.

The structural geomorphology of most parts of the Abbay Basin is generally believed to have evolved and shaped during the four geological eras as indicated in the foregoing discussions. The impression is further validated by other studies including (Coltorti et al. 2007, 2015; Gani and Gani 2007; Abbate et al. 2015).

According to Coltorti et al. (2007), the lower surface of the Ethiopian landscape is believed to be made-up of the oldest basement complex rocks of the Proterozoic ages. These basement complex rocks are mainly composed of exceedingly changed ultramafic plutonic formations (e.g. schists and sandstones). According to these authors, the oldest rocks were created at approximately 20 km deep in the crust but they were later uplifted or reduced into an extensive flat planation surface during the Palaeozoic era. During the Mesozoic era (≈250 million years ago), the older Precambrian rocks were overlain by the three layers of sedimentary rocks mentioned above. Finally, they were also capped by the CFBs during the Oligocene and Miocene epochs. The depth of the CFB lava is estimated to be 1500–2000 m thick in most of the highland areas of the Abbay Basin (Abbate et al. 2015; Williams 2016); but reported to be 300 m thick in the Abbay gorge around the Goha Tsion.

Gani et al. (2009) assessed the stratigraphic and structural evolution of the Blue-Nile basin through the use of field and remote sensing data. According to these authors, the formation of the Abbay gorge is associated with the Mesozoic faulted basin of central and east Africa. They classify the development of the Abbay Basin into three successive stages. The first stage according to Gani et al. (2009) was the Palaeozoic era (the time before the denudation of the old basement rocks). The second stage was the time when the Gondwana continent experienced rifting and faulting and the succeeding transgression and regression of the sea in the Horn of Africa. This

Fig. 3.5 Sketch showing the rock sequence from the Millennium Bridge to Goha-Tsion. *Source* Williams (2016)

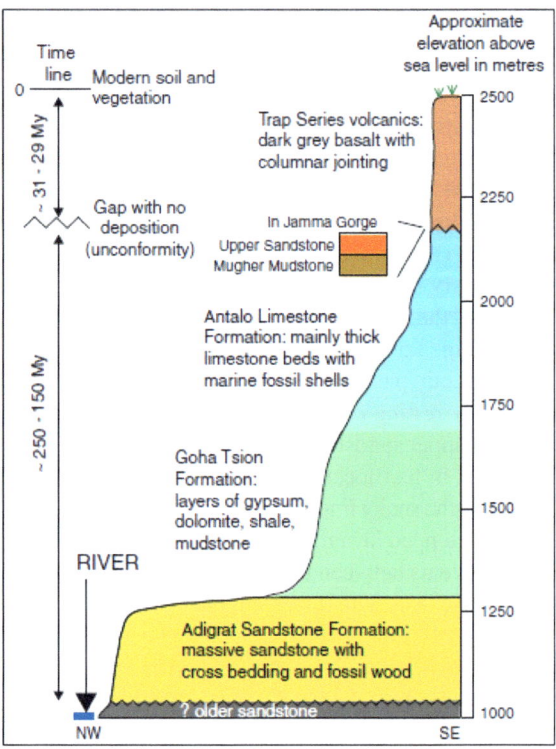

was the Mesozoic era when the three layers of sedimentary rocks were deposited over the old basement surface. The third stage was the eruption of volcanic lava in the tertiary and quaternary periods after Mesozoic sedimentation. According to Gani et al. (2009), Abbay was created during the rifting and faulting of the Gondwana continent along a northwest (NW)-extending Mesozoic fault over which the bulk of the Mesozoic sediments settled down. Gani and Gani (2007) further argued that the enduring opening of the Abbay Basin followed a three-stage dome building process: a gentle level of uplift from 29 to 10 Ma; a swift level of swelling (uplifting) at 10 Ma; and a dramatic plateau rising at approximately 6 Ma. These are linked to the trap series and Aden volcanic lava formations connecting to the Afar 'Mantle Plume' (Gani et al. 2009).

Abbate et al. (2015) suggested that the formation of the tertiary lava and the swelling of the Ethiopian Plateau to over 1000 m resulted in the development of raised highlands at approximately 30 Ma. According to these authors, the event reversed the geomorphic configuration and the drainage pattern of East Africa, which eventually shifted the regional water-divide. Rivers that were formerly flowing towards the southeast (towards the Indian Ocean) steadily turned to the Mediterranean Palaeo-Nile system. In support of this premise, Abbate et al. (2015) remarked that shield volcanoes that overlaid the northern Ethiopian Plateau started to control the circular courses of the Abbay and Tekeze Rivers starting from the first Miocene times. According these authors, the courses of the Abbay and Tekeze Rivers changed to be 'annular and stable' by the continued erosion due to the increased highland rising. The rivers retained their recent deeper canyons at their upstream reaches through the progression of abrupt gradients and carved valleys.

A study on Palaeobotanical data ($\approx$130 m thick fossil remnants) *interbedded* with lavas of the upper and middle traps gathered from the Lake Tana areas (*Chilga intertrappean beds*) (Abbate et al. 2015; and Williams 2016) signifies that the geomorphic structure of the Abbay Basin comprises a complex set of rock sequences (ranging from Precambrian basement to recent quaternary volcanism and sediments with sedimentary layers in between). The Oligocene to Miocene trap volcanic rocks dominate the highlands of the Abbay Basin, while Palaeozoic to Quaternary sedimentary rocks and Proterozoic basement formations characterize the lower courses of the river Abbay and the plains along the Ethio-Sudan border. Plio-Quaternary volcanism was also evolved in some pockets of the Lake Tana basin (Abbate et al. 2015).

The original direction of flow and source of the Abbay River is not accurately known; rather, it poses puzzling questions among geographers. Historical travellers from Europe such as Priest Pedro Paez (*the Portuguese Jesuit*) who travelled to the locality of Gish Abbay in 1613 with the then Ethiopian king Susenyous (Meshesha 2016); and in 770 James Bruce claimed that they discovered the source of the Abbay (Meshesha 2016). People living around Sekella also claim that Gish-Abbay is the source of the Abbay River. Nonetheless, this claim is legendary, a fairy tale and doubting for it has no a convincing proof. Of course, Gilgel-Abbay emerges from the village named Gish-Abbay but enters Lake Tana (see Fig. 3.6) and does not directly join the Abbay River. It is a subsequent tributary of the Abbay River. Williams (2016) claims that as numerous rivers and streams ($\geq$ 40, McCartney et al. 2010; Ayana et al. 2014a; Enyew et al. 2014; Gebre 2015; Yenehun et al. 2021) enter Lake Tana, the waters of the Abbay emerge from different sources and locations. Williams (2016) further argues that the sources of the other numerous streams and rivers would have the right to enjoy 'an equal claim' to be source of the Abbay. Lake Tana contributes only 8% of the water of the Nile at Khartoum and a very slight portion of that water emerges from Gilgel-Abbay. The immense volume of the Abbay water emerges from other subsequent rivers, such as the Beles and Didessa. Such giant rivers can also have 'their own' claim to be sources of the Abbay (Williams 2016). The issue, therefore, demands detailed geomorphological investigation.

Regarding the incision and historical development of the Abbay River, Williams (2016) argues that it occurred through 'headward erosion', rifting and volcanic arch formation. To this author, Abbay starts as a fairly small stream by flowing along the western scarp of a growing 'Afro-Arabian Dome' by incising a rift along the arc during the mid-Eocene (~40 Ma). Then, the stream slowly continues curving its course upstream through vertical and headward erosion as uplifting of the 'Afro-Arabian' landscape commences through earth movements, cracking of the land and lava flows. Through such a gradual incision, the Abbay gorge reached the Bure-Nekemte pass and the site of the Millennium Bridge by 25 Ma and within 15 and 20 Ma, respectively. Currently, the Abbay gorge has almost approached but has not

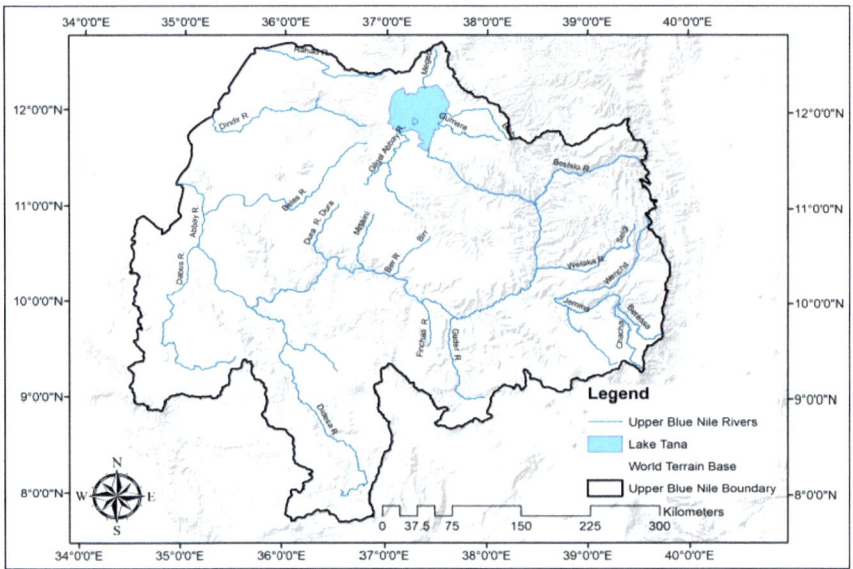

Fig. 3.6 Abbay River basin in Ethiopia. *Source* Ayele et al. (2019)

fairly reached Lake Tana. Its headward growth is slowed by the quaternary hard basaltic rock at the site of the Tis Isat Falls. The magnificent scenery of the Tis Isat Falls has recently been reduced because part of the water from Lake Tana is diverted for hydroelectric power production. Abbay and its feeding streams are estimated to remove ≈90,000 km^3 of the rock and 72% of the mineral materials from the highlands of Ethiopia.

Major Geomorphic/Topographic Features in the Abbay Basin

Referring to Table 3.5, 24% of the total area of the Abbay Basin falls within 342–500 m amsl. Based on the Ethiopian traditional altitude classification, this area is considered a desert and semi-desert landscape. This type of landscape is rare in the Ethiopian part of the Abbay Basin. The typical landscape characterizes the clay plains of Sudan to the northwest of the Roseires reservoir in the Sennar, El Jeizera and Khartoum regions (Fig. 3.7). This is probably an arid and semiarid area largely dependent on the Roseires and Sennar reservoirs.

The other 35% of the area falls within 500–1500 m amsl (Table 3.5) altitudes. This is a lowland area dominated by pediment plains of the Gedarif area and large part of the Ethio-Sudan border of the Abbay Basin. In Ethiopia, these lowlands form the bottom-hills of the western highlands. Some of the hills are characterized by sloping scarps mainly in the Abbay gorge of Ethiopia closer to the Sudan border. The upper

Table 3.5 Landscape regions of the Abbay basin

№	All Abbay basin*		Ethiopian Abbay basin**		
	Altitudinal range (m amsl)	Area coverage (in %)	Altitudinal range (m amsl)	Area (km²)	% of basin
1	342–500	24.00	500–1000	34,442	16.97
2	500–1000	22.00	1000–1500	42,454	20.91
3	1000–1500	13.00	1500–2000	54,583	26.89
4	1500–2000	18.00	2000–2500	40,432	19.92
5	2000–2500	13.00	2500–3000	24,259	11.95
6	2500–3000	7.00	3000–3500	6065	2.99
7	3000–3500	2.63	3500–4000	734	0.36
8	> 3500	0.37	> 4000	25	0.01
Total		100		202,994	100.00

*Adapted from ENTRO (2006)
**Adapted from Abdel-Aziz (2008)

surface of the relief ≤ 1500 m amsl is mainly composed of sedimentary formations of Mesozoic and Cenozoic ages.

The landscape between 1500 and 2500 m amsl accounts for 31% of the whole Abbay Basin and approximately 47% of the total landscape in the Ethiopian Abbay Basin. This mid-land topography dominates the Ethiopian Abbay region. The area partly forms the volcanic highlands of western Ethiopia that are mainly overlain by tertiary lava rocks. Some exceptions in the area occur over the Lake Tana lowlands where Aden series lava and recent sediments jammed part of the landscape. Much of the area in this relief region forms part of the highlands where the headwaters of the Abbay River are rising. The vast plateau of Godjam, the tablelands of north and west Shoa, the highlands of southwest Wollo and the Debre-Tabor highlands are included under this region. The upper surface geology of these plateau regions is mainly composed of the Alaji rhyolites, Tarmaber, Ashangi, or Aiba basalts. The highlands of the Wellega and Banishangul-Gumuz regions are included in these plateau groups, but they are geologically different from the aforementioned volcanic landscapes. The landscapes in the Wellega and Banishangul-Gumuz regions are mainly characterized by old basement rocks of Proterozoic age with a small portion capped with tertiary lava (e.g. see Williams 2016).

The geomorphic region mentioned above hosts the most important agricultural land and volcanic soils that are intensively cultivated in the basin. Over 60% of the population in the Abbay Basin is also settled in these geomorphic regions. Similar geomorphic regions are rare in the Abbay Basin of Sudan. The typical landscape in the Abbay Basin of Sudan is the plateau area to the south of the Roseires reservoir in the Blue-Nile Administrative Region (Fig. 3.7).

The area above 250 m amsl comprises approximately 10% of the entire Abbay Basin and accounts for some 15% of the Abbay Basin in Ethiopia (Table 3.5). The area generally forms the upper plateau landscape of the Abbay Basin of Ethiopia.

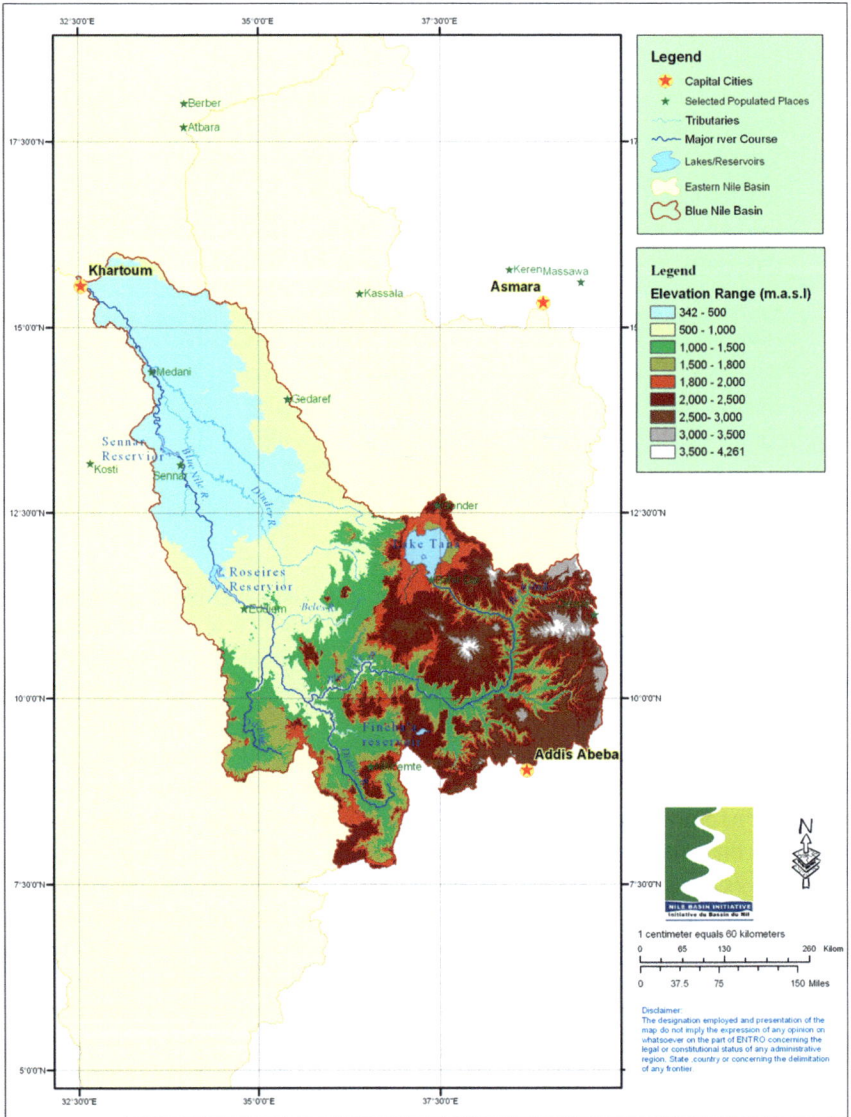

Fig. 3.7 Topographic regions of the Abbay basin. *Source* ENTRO (2006)

The most important are the volcanic mountains of north Shoa, the Choke Mountain in central Godjam and Amharasaint Massifs in Wollo and the Guna massifs to the east of Lake Tana. These volcanic highlands are characterized by the Alaji rhyolites, Tarmaber shield volcanoes (11–23 Ma), Ashangi and Aiba basaltic lava formations. Most of them are results of the Hawaian type shield volcano of tertiary ages that are as thick as 100 km and form 1000 to 2000 m elevations (Abbate et al. 2015; Williams

2016). They are the major sources of the Abbay head streams and tributaries. As a result, they are highly dissected and divided into irregular massifs that greatly hinder communication. They are characterized by steeper scarp slopes adjacent to the river gorges. Based on the foregoing information, it is possible to classify the Abbay Basin of Ethiopia into five major geographic regions (Lake Tana basin, Amarasaint Massifs, Mid-Central plateau, southwestern upland and Abbay-Dinder lowlands).

The Lake Tana Basin

The Lake Tana basin comprises the northernmost part of the Abbay Basin covering an average drainage area of 15,469 km^2 (Table 3.6). The dominant geomorphic features within the Lake Tana basin include the lake and adjacent swampy lowlands, quaternary basalts and sediments, shield volcanoes, trap series volcanism, Mesozoic sandstones and limestones, Precambrian basement rocks, escarpments, and faulted and rifted areas (Fig. 3.8).

The Lake and Adjacent Swampy Lowlands

The Lake Tana occurs at the northernmost end of the Abbay Basin at an average altitude of 1807 m amsl (Fig. 3.8; Table 3.6). The mean surface area of the lake measures 3308 km^2. The typical length and width of Lake Tana are 82 and 66 km, respectively. The average depth of the lake is 8.3 m while its maximum depth is 14.3 m. Lake Tana was estimated two-times larger than the present size and its water level was 75 m higher approximately 15,000 years ago (Williams 2016). The total annual water volume is estimated to be 28 BCM (World Bank 2008). The mean annual flow volume from the lake is estimated to be 3.7 BCM by the World Bank (2008); 3576 million cubic metres (MCM) by Yilma and Awulachew (2009); 3753 MCM by McCartney et al. (2010); and 3732 MCM by Abtew and Melesse (2014a, b). Figure 3.9 shows a fluctuating water flow for months from August to November (1980–2000). The minimum water flow from Lake Tana was from 1984 to 1986, which may have been due to droughts. The maximum flows were in 1988, 1994 and 1998 rising over 1500 MCM.

Lake Tana is fed by several rivers joining the lake from its southern, northern and eastern parts (Ahmed and Ismail 2008; Poppe et al. 2013). The little Abbay (Gilgel Abay) is the largest of all in the catchment area ($\approx$4589 km^2). The mean annual water flow from Gilgel-Abbay is calculated to be 2492 MCM (Table 3.7). This river emerges from the Gish-Abbay hills in west Godjam and joins Lake Tana in its southern part (see Fig. 3.8) after flowing for an average of 84 km. In its eastern part, Lake Tana is fed by the Gumara and Rib Rivers which both rise from the Guna Massifs. These rivers have catchment areas of approximately 1620 km^2 and 2014 km^2, respectively. The corresponding mean annual flow volume from these rivers is calculated to be 1203.3 MCM and 510.4 BCM. Megech joins Lake Tana in its northern edge and contributes $\geq$ 195 MCM annual water flow to the lake. It stretches for approximately 43 km from its source to Lake Tana, covering an approximately 990 km^2 catchment area. The Gummara and Rib Rivers flow for approximately 110 and 98 km, respectively to reach Lake Tana (see again Table 3.7). These and other rivers contribute roughly 93% of the water flow into Lake Tana (McCartney et al. 2010; Gebre 2015). Nevertheless,

Table 3.6 Lake Tana: basic information

Area (km²)	Length (km)	Width (km)	Depth		Height (m amsl)	Basin area (km²)	Source
			Max	Av.			
3156	84	66	14	9			Williams (2016)
3600	78	67	15	8	1800	16,500	Ahmed and Ismail (2008)
3000	78	67	14		1788	15,294	Yilma and Awulachew (2009)
3156			14	7.2	1786	16,000	Abtew and Melesse (2014a, b)
3060	84	66	14	7.2	1786		Wale (2008)
3000	74	65	15		1800	16,000	Ayana et al. (2014b)
3100	84	66	15		1800	15,300	Enyew et al. (2014)
3156					1786	15,321	McCartney et al. (2010)
3077				9		15,077	Yenehun et al. (2021)
3150			14	9	1875	15,123	World Bank (2008)
5041					1785		Poppe et al. (2013)
					1829		Woodward et al. (2022)
3200	90	65	14	9	1840		Yitbarek et al. (2019)
						15,077	Meshesha (2016)
						15,935	Shobary et al. (2021)
						15,000	Gebre (2015)
						15,000	Dile et al. (2013)
3308	81.7	66	14.3	8.3	1807	15468.9	Mean

Max: maximum; Av: average

3 Geomorphology of the Abbay Basin from Geographical Perspective

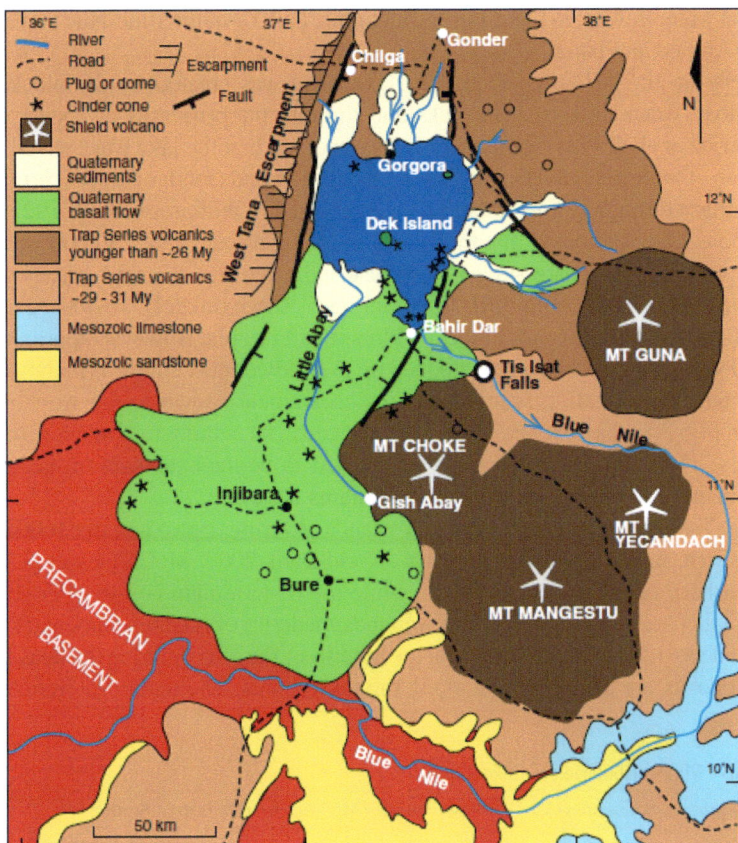

Fig. 3.8 Map showing the geology and geomorphology of the Lake Tana basin. *Source* Geological Survey of Ethiopia (1996; in Williams 2016)

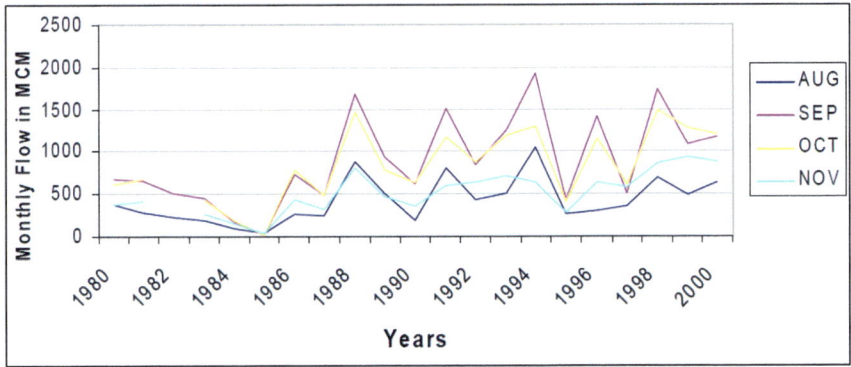

Fig. 3.9 Water flow at Abbay station, Bahir Dar (1980–2000). *Source* Yilma and Awulachew (2009)

the total water flow from Lake Tana is negligible (< 10% of the Blue-Nile total annual flow) (Ahmed and Ismail 2008).

Poppe et al. (2013) reported that Lake Tana was created by the blockage of the Abbay water flow by lava rocks and through the processes of epeirogenesis. Chorowicz et al. (1998) on the other hand remarked that the lake was created by the collision of three grabens (Debre-Tabor, Dengel-Ber and Gondar grabens) prior to the mid-tertiary through *deep-slip* faulting. As reported in Williams (2016), a 92 m sediment core study indicated that the lake could be approximately 250,000 years old. Additional sediment core studies have also indicated that the lake was swarming vast areas from 95,000 to 130,000 years ago. Approximately 16,400 years ago, Lake Tana dried out because of the dry climate in Ethiopia. However, approximately 15,000 years ago, the lake again flooded up to its lowermost edges. Williams (2016) added that the retreating scarp slope (the escarpment shown to the west of Lake Tana in Fig. 3.8) may expand to the lake and cause its culmination in a few millions of years. The indicated scarp slope is believed to be the extension of the western escarpment of the Semien Mountains (Williams 2016).

Lake Tana is enclosed by swampy lowlands covering approximately 1600 km^2 in its northern, western and eastern parts (World Bank 2008; McCartney et al. 2010). The northern plain is named Dembia while the eastern plain is named Fogera. All these plains are seasonally flooded wetlands made-up of recent sedimentary rocks. Fluvial deposition is very common at the mouths of the rivers joining the lake. Lacustrine deposits are also common along the lake shores. The natural morphology of

Table 3.7 Major tributaries of Lake Tana

Tributary	Basin area (km^2)	Flow volume (MCM)	Length (km)	Source
Gilgel Abbay	4100			Abdo (2008)
		4000		Enyew et al. (2014)
	5004			Dile et al. (2013)
	4557.8	1753.5	83.8	Wale (2008)
	4694	1721.3		Yilma and Awulachew (2009)
			100	Shobary et al. (2021)
	4589	2492	84	Mean
Ribb	2013.8	510.4 (BCM)	98.3	Wale (2008)
Megech	990.4	195.2	42.6	
Koga	280	176		Yilma and Awulachew (2009)
Gummara	1768.3	1229.5	86.7	Wale (2008)
	1595	1177	132.5	Yilma and Awulachew (2009)
	1496			Meshesha (2016)
	1619.8	1203.3	109.6	Mean

wetlands has recently been affected by the invasion of exotic weeds (water *hyepaths/ Eichhornia crassipes*) (Yitbarek et al. 2019). Finally, Lake Tana is characterized by smaller islands and peninsulas reaching approximately 30 (Abitew and Melesse 2014). Some of the islands are currently hosting ancient monasteries (sacred religious places/orthodox churches) including Daga Estifanos, Kibran Gebreal, Nagra Sellasie, Tana Cherkos and Ura Kidane-Mihret (see Williams 2016). The lake is renowned among the 250 topmost biodiverse lake ecosystems of global relevance (McCartney et al. 2010; Enyew et al. 2014; Yenehun et al. 2021).

The Volcanic Highlands

The volcanic highlands of the Lake Tana basin encompass the landscape resulting from recent volcanism (quaternary volcanism) and the tertiary lava domes. The quaternary lava formation comprises the landscape created by the Aden series volcanic lava flow at approximately 2 Ma. This area covers the monotonous landscape to the south of Lake Tana and west of the trap series lava and shield volcano formations (Fig. 3.8).

The trap series volcanism covers the vast area to the north and east of Lake Tana. Shield volcanoes cover the highest parts of the lake catchment in the eastern and southern parts of the basin. These include the Guna-Debre-tabor and Choke Mountains. The Guna Massifs are located to the north of the Choke plateau, east of Lake Tana (see Fig. 3.8). They are dissected by tributaries of the Abbay and Tekeze Rivers. They form a closer neighbourhood with the Lake Tana basin in the west and are located northwest of the Amarasaint Massifs. Mt. Guna (4086 m masl, Poppe et al. 2013) is the highest peak in the Massif region developed at approximately 10.7 Ma (Williams 2016). These volcanic massifs divide the rivers that flow towards the Tekeze River from those flowing towards Abbay and Lake Tana.

The Choke Plateau forms an extended tableland to the south of the Lake Tana lowland in central Godjam (Fig. 3.8). The relief is comparatively less dissected compared to the Guna Massifs. The plateau makes its focus at the heart of the Choke Mountains. The area is engraved by the Abbay River, which effectively separates it from the Shoan Plateau in the south and southeast, the Amarasaint Massif in the northeast and the Guna Massifs in the north. The well-known Mt. peaks in this plateau region include the following: Mt. Mangestu (evolved at ≈22.4 Ma) (Williams 2016), and Mt. Choke (3504 m amsl) (Poppe et al. 2013).

Mesozoic sedimentary layers and Precambrian basement rocks are also found along the middle and lower courses of the Abbay. Mesozoic sedimentary rocks occur to the south of Choke Mts. while the Proterozoic basement rocks occur to the southwest of the Aden series lava formations. The Aden volcanic lava formations are indented by numerous cinder cones and volcanic plugs. Volcanic plugs also occur in the trap series lava areas to the east of Lake Tana. The area is also characterized by escarpments and rifted lines at different locations (see Fig. 3.8).

Amharasaint Massifs

These are volcanic highlands found in southwest Wollo between the Beshillo and Wonchit Rivers. The known highest point in the region is Amba-Farit (3975 m masl). These mountain groups make a water divide between the Abbay and Awash Rivers.

The Mid-Central Plateau

This is a '**Dome-shaped**' plateau (Williams 2016) occurring in northern and western Shoa and serves as a water divide between the Awash and Abbay River basins. It is named the Mid-Central Plateau because it is located in the *very centre* of the country. It extends westwards into western Wellega through Horo-Guduru and forms a falcate shape which causes the Abbay to turn and drain northwestwards. The Mid-Central Plateau is separated from the plateau of Godjam by the Abbay gorge in the north; from the southeastern highlands by the Awash River and the Great Ethiopian Rift Valley; and from the highlands of Keffa by the Ghibe River. It forms a water divide between the Abbay and Awash, Abbay and Ghibe, and Awash and Ghibe drainage systems (Zewdu 2008).

Southwestern Upland

The southwestern upland is located to the south of the Abbay River in southwest Ethiopia. The area mainly covers the largest areas of Wellega, parts of the Banishangul-Gumuz Region and some areas of Gambela. The dominant rock cover in the area includes the Trap series volcanism and the Proterozoic basement rocks representing the Mzambique belt formations and the Arabian Nubian shield. The Trap series volcanic formations cover vast areas in the region and are indented with volcanic plugs. The best example of these tertiary volcanoes is the Tulu Welel created at approximately 7–10 Ma (Fig. 3.9; Williams 2016).

The Mozambique Belt formations mainly contain silicic rocks (Silica-rich rocks). Rocks of the Arabian-Nubian shield contain metamorphic rocks developed from the remnants of marine sediments and 'volcanic island arcs'. These rocks are composed of *mafic* minerals. The old Precambrian formations are rich in metamorphic rocks (serpentinite, migmatite and gneiss) as well as precious minerals such as gold and platinum (see Williams 2016). The old Precambrian formation in the region is characterized by granitic and ultramafic intrusions (see Figure). WNW–ESE and N–S trimming shear formations are also observed in the area. The Tulu Dimtu Ridge joins the WNW–ESE and N–S trending shear zones, as shown in the map (Fig. 3.10).

The Abbay-Dinder Lowlands

The Abbay-Dinder lowlands run north–south along the Ethio-Sudan border and are bounded by Tekezze-Angereb lowlands in the north and by Baro-Akobo lowlands in the south. They cover Quara (Alefa-Taqusa plains), Jawi and pawi lowlands and the Assosa plains. These lowlands are characterized by residual hills composed of metamorphic rocks (granitic and gneiss). The remnant ridges viewed prominently from the adjacent setting because they resisted a long period of erosion. They are thus observed standing as 'inselbergs' over the vast rocky basement plain landscapes

Fig. 3.10 Geological and geomorphological map showing the southwestern upland. *Source* Modified from Alemu and Hailu (2013; in Williams 2016)

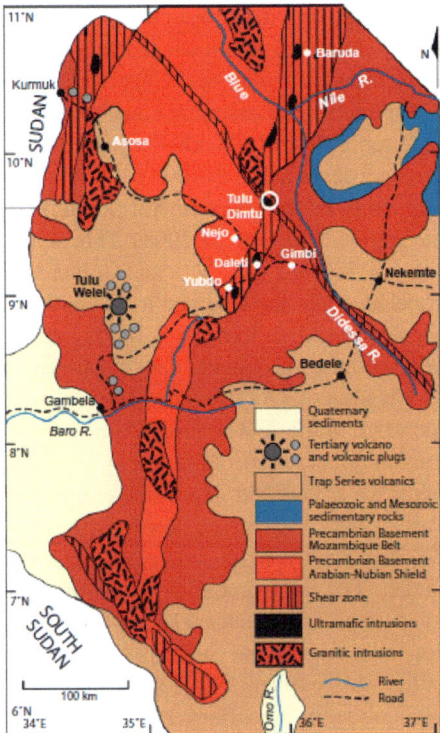

(Abbate et al. 2015). The general elevation of some of these residual hills ranges from 500 to 1000 mamsl.

Surface Runoff, Planation and Sedimentation

The diverse geomorphic landscape of the Abbay Basin is characterized by elevated mountains and broad-based plateaus broken by numerous streams and rivers. Most of the rivers cut deeper gorges and have steep-sided valleys. This makes most rivers erosive and sources of degradation. Adem et al. (2020) remarked that 30% of the rainfall in the Gomit watershed, northwestern Ethiopia flows out through widely rifted tertiary lava faults to other basins. Soil erosion on the rugged mountains and high rainfall areas is thus very intense. Sedimentation in the lower reaches of the flood plains and water resource development areas (in dam building areas) is also a growing problem in both Ethiopia and the riparian countries of Sudan and Egypt (Betrie et al. 2011). Siltation and sedimentation at reservoirs and dams is very serious (ENTRO 2006). Ahmed and Ismail (2008) complain that the sediment of the Nile principally originates from the Ethiopian highlands due to eroded soils from extensively cultivated less managed fields. These authors argue that 85% of the annual water flow to Sudan and Egypt originates from the highlands of Ethiopia and is transported to downstream countries by the Abbay, Tekeze and Baro Rivers.

A report from the World Bank indicated that the Lake Tana area of the Abbay Basin is facing degradation (> 100 t h^{-1} y^{-1} soil erosion; sedimentation; and eutrophication) that led to harm by ephemeral flooding, water pollution, loss of fish and effects on navigation (World Bank 2008). This report remarks that the burden of sedimentation reaches 5–250 t ha^{-1} y^{-1} and is increasing with worrying rates. The report then concludes that the lake and its resources are ecologically fragile. Another study by Meshesha (2016) likewise testifies that the landscape of the upper Abbay Basin is affected by anthropogenic encroachment and climate change.

Abdel-Aziz (2009) reported that approximately 95% of the sediment and 71% of the water at the Aswan High Dam during the July to October months emerges from the Ethiopian Rivers of the Abbay and Tekeze. Abbay alone contributes roughly 60% of the water flow and 72% of the sediment load at the Aswan High Dam (Sutcliffe and Parks 1999). The surface flow from the highland groups where the Abbay head waters originate are the main sources of the sediments deposited downstream in Egypt and Sudan. The large amounts of soil moving by the Abbay waters have been proven to cause pressing water resource management problems in neighbouring Nile Basin countries (Egypt and Sudan) (Ahmed and Ismail 2008). The water discharge graph of the Abbay River (1920–2006; Fig. 3.11) similarly shows that the volume of water draining from the Abbay River is enormous (ranging from ≤ 30,000 to ≥ 70,000 MCM). The highest discharge was in 1929 while the lowest was in 1972. The graph shows the occurrence of great fluctuation of flow during wetter and drought years. The high amount of discharge particularly during the wetter years may carry huge sediment loads that could pose both onsite and offsite impacts which may require great concern among user countries.

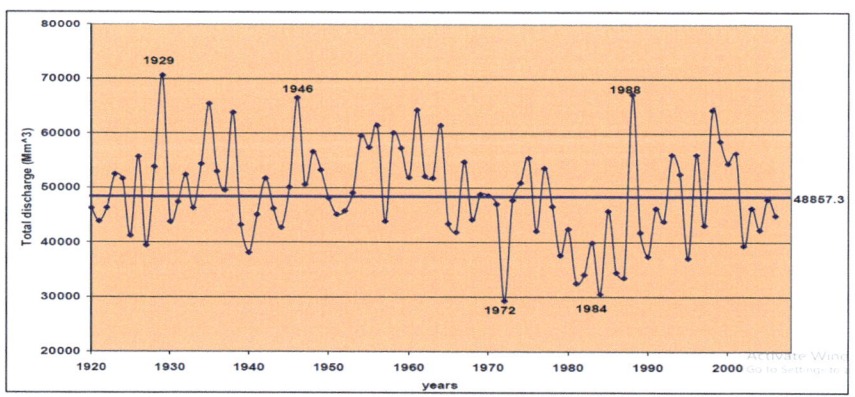

Fig. 3.11 Hydrograph of the Abbay in MCM (1920–2006). *Source* Ahmed and Ismail (2008)

River runoff sizes ranging from 681 to 9 mm y^{-1} are for instance reported for catchments representing the central highlands of Ethiopia and the lowlands of the Sudan border. River runoff over the Ethiopian highlands was also calculated to range 92–156 mm y^{-1} (Table 3.8) for three stations by Abdel-Azi (2009). Tesfaye (2022) remarked that the levels of soil removal and sediment deposition over the Abbay Basin are very high reaching 16–67.4 t ha^{-1} y^{-1} and 4.2–18 t ha^{-1} y^{-1}, respectively. The soil erosion and sedimentation rates over the watersheds similarly reach 8.3–100 t ha^{-1} y^{-1} and 1.1–43.4 t ha^{-1} y^{-1} according to the same author. Approximately 15,050–150 tonnes per km^2 of soil erosion rates were also reported perhaps for other larger catchments by other authors (Table 3.9). Abdel-Azi (2009) calculated the average annual soil removal from the Abbay Basin to be 375.5 million tonnes from contour maps and 320 million tonnes through application of the Revised Universal Soil Loss Equation (see again Table 3.9). Mean annual sediment rates ranging between 4.2 Mg h^{-1} y^{-1} and 49.4 Mg h^{-1} y^{-1} are reported for different microwatersheds by previous studies (e.g. see Table 3.9). Abdel-Aziz determined 140 million tonnes of sediment deposition at El-Diem. Approximately 44% of this sediment load is also estimated by the same author to come from the Abbay Basin.

The foregoing evidence indicates that the rivers originating from the rugged mountains move enormous amounts of soil, water and sediment from the highlands of the Abbay Basin to the downstream areas and cause siltation of dams and reservoirs. They remarked that they cause on-site and off-site impacts. Betrie et al. (2011) proposed reforestation and installation of stonebunds and filtering strips at the outlet of river basins and subwatersheds. Gessesse (2014) recommended rill and ditch control practices. Tilahun et al. (2014) suggested the implementation of soil erosion control measures on upland areas with thinner soil covers and on the downstream 'bottomlands'. Steenhuis et al. (2014) advised the plantation of trees on eroded and barren lands. Ahmed et al. (2022) recommend site-specific soil and water conservation (SWC) measures considering geological formations. Tesfaye (2022) suggests that basin-wide and efficient SWC practices could minimize both the on-site and off-site impacts of soil erosion. Implementation of watershed-based SWC measures over the adjacent uplands is thus required to gradually minimize the soil erosion and sedimentation problems in the Abbay Basin highland areas.

Table 3.8 Rainfall and river runoff at three different locations of the Abbay Basin in Ethiopia

Altitude	Latitude	Longitude	Rainfall runoff coefficients	Rainfall mm y^{-1}	River runoff mm y^{-1}
1800	8.04	39.02	0.20	689	137.80
1870	13.32	39.01	0.12	767	92.04
530	7.48	34.24	0.10	1568	156.80

Source Abdel-Aziz (2009)

Table 3.9 Surface runoff, soil loss and sedimentation at different sites in the Abbay Basin

Station/site		Soil erosion (million tonnes)	Sediment deposition million tonnes /year	Source
Climate data from 142 stations		375.5		Abdel-Aziz (2009)
Calculated using RUSLE		320		
EL-Diem			140	
Rosaries reservoir			30	
Nile tributaries from Ethiopia			160–180	Ahmed and Ismail (2008)
Ethio-Sudan border			140	
At the Beles-Abbay confluence			138.8	
At Khartoum			99.04	
Watershed	River runoff (mm/year)	Average soil loss (t/km^2)	Mean annual sediment load (Mg ha^{-1} y^{-1})	Source
Anjeni	680.50	15,050	6.3	Guzman et al. (2017)
Bahir Dar	19.50	1390		Abdel-Aziz (2009)
Jiga	18.30	2129		
Nekemte	13.40	150		
Sudan border	9.00	700		
Microwatershed	Area (ha)	Mean annual sediment load (Mg ha^{-1} y^{-1})		Source
Gomit	369	4.2		Adem et al. (2020)
Gilgel Abay	166,500	35.4		Zimale et al. (2018)
Gummara	128,100	49.4		
Rib	128,900	24.6		
Megech	50,000	12.2		
Debre Mawi	95	11.6		Dagnew et al. (2017)

Conclusions

A literature survey study was conducted to precisely document the geomorphic processes and structural landforms of the Abbay Basin in Ethiopia. Statistical data extracted from 31 papers were systematically organized into three themes and analysed using explanatory notes. The reviewed information indicated that the River Abbay flows across varied geomorphic regions ranging from rugged, high and humid highlands in Ethiopia to relatively dry and arid plains in Sudan. The entire landscape of the Abbay Basin traverses over 4000 m elevations ranging from 344 m amsl in the downstream areas of Sudan to > 4000 m amsl in the upstream areas of Ethiopia. The lowlands in Sudan drop to < 700 m amsl, while the lowlands of Ethiopia fall between 500 and 1500 m amsl. Reviewed sources indicated that the structural geomorphology of the Abbay Basin was the result of the collision and fragmentation of ancient continents (east and west Gondwana and the east African Orogen). The Abbay fault also evolved by rifting and faulting of the Gondwana continent along a northwest trending Mesozoic fault through headward cutting over the *Afro-Arabian Basaltic Dom.*

The structural geomorphology of the Abbay Basin is characterized by high and rugged mountains capped with shield volcanoes; vast mid-land plateaus capped with tertiary and quaternary lava rocks; deep river gorges; swampy lowlands covered with recent lacustrine and fluvial deposits; and lowlands of Mesozoic sedimentary and Precambrian formations. The Abbay Basin of Ethiopia is generally classified into five major geomorphic/geographic regions: the Lake Tana basin, Amarasaint Massifs, Mid-central plateau, southwestern upland and Abbay-Dinder lowlands. The lake Tana basin comprises the: (i) Aden volcanic series formations to the south of Lake Tana, north of the Abbay gorge and northwest of the Choke mountains; (ii) swampy lowlands of the Lake Tana (Dembia, Fogera and Kunzula plains); (iii) trap series lava capped plateau in the northern and eastern parts of the basin, (iv) Guna and Choke shield volcanoes; and (v) Mesozoic sedimentary and Precambrian formations in the Abbay gorge and at the western margin of the basin. The Amharasaint Massifs occupy the area east of the Tana basin; to the north of the Shoan (mid-central) plateau. The mid-central plateau covers the eastern and southern parts of the Abbay Basin. The southwestern upland occupies the southwestern part of the Abbay Basin in Wellega. The Abbay-Dinder lowlands occupy the area west of the Lake Tana basin comprising the Alfea-Taqusa, Jawi-Pawi and Assossa lowlands.

River Abbay is fed by several tributaries rising from its volcanic uplands, as mentioned above. However, most streams and rivers are agents of erosion. Enormous amounts of water (> 54 BCM); and large quantities of soils and sediments are carried away by the rivers. The sediments deposited by rivers impact water reservoirs, irrigation and hydropower dams in downstream areas. Managing the watersheds in the Guna, Choke, Amharasaint, mid-central and southwestern uplands recommended minimizing the risks of soil erosion, siltation and sedimentation problems.

Acknowledgements The preparation of this paper was initiated by the Space Science and Geospatial Institute (SSGI). The SSGI is therefore accredited for its encouragement and coordination in the development of this baseline material. Professor Assefa Melesse and Dr. Berhan Gessesse are acknowledged for their assistance in reviewing and providing critical improvements during the book chapter development.

References

Abbate E, Bruni P, Sagri M (2015) Geology of Ethiopia: a review and geomorphological perspectives. In: Billi P (ed) World geomorphological landscapes. Springer Science+Business Media Dordrecht 2015, Heidelberg, New York, London, pp 33–64

Abdel-Aziz TM (2009) Water and sediment management for the Blue Nile basin. Thirteenth International Water Technology Conference (IWTC), Hurghada, Egypt, p 14

Abdo KS (2008) Assessment of climate change impacts on the hydrology of Gilgel-Abbay catchment in lake Tana Basin, Ethiopia. Msc Thesis, International Institute for Geo-Information Science and Earth Observation (ITC). Enschede, The Netherlands, p 86

Abdo KS, Fiseha BM, Rientjes THM, Gieske ASM, Haile A.T (2009) Assessment of climate change impacts on the hydrology of Gilgel-Abay catchment in Lake Tana basin, Ethiopia. Hydrol Process 23:3661–3669

Abdelsalam MG (2016) The Nile's journey through space and time: a geological perspective. Paper submitted to *earth science reviews.* Oklahoma State University, p 96

Abtew W (2014) Land and water in the Nile basin. In: Melesse AM, Abtew W, Setegn SG (eds) Nile River basin. Springer International Publishing Switzerland, pp 119–129

Abtew W, Melesse AM (2014a) The Nile River basin. In: Melesse AM, Abtew W, Setegn SG (eds) Nile River basin. International Publishing Switzerland, pp 7–21

Abtew W, Melesse AM (2014b) Transboundary Rivers and the Nile. In: Melesse AM, Abtew W, Setegn SG (eds) Nile River basin. Springer International Publishing Switzerland, pp 565–579

Adem AA, Melesse AM, Tilahun SA, Setegn SG, Ayana EK, AbeyouWale A, Assefa TT (2014) Climate change projections in the Upper Gilgel Abay River catchment, Blue Nile basin Ethiopia. In: Melesse AM, Abtew W, Setegn SG (eds) Nile River basin. Springer International Publishing Switzerland, pp 363–388

Adem AA, Addis GG, Aynalem DW, Tilahun SA, Mekuria W, Azeze M, Steenhuis TS (2020) Hydrogeology of volcanic highlands affects prioritization of land management practices. Water (12):1–32 (2702)

Ahmed AI, Ismail UHAE (2008) Sediment in the Nile River system. UNESCO, Khartoum, Sudan, p 104

Alemu TB, Page JD, Abdelsalam MG, Atnafu B (2019) Stratigraphic controls on the morphotectonic evolution of the Nile Gorge, Ethiopia. Arab J Geosci 12:705

Ayalew L, Yamagishi H (2004) Slope failures in the Blue Nile basin, as seen from landscape evolution perspective. Geomorphology 57:95–116

Ayana EK, Zimale FA, Collick AS, Tilahun SA, Elkamil M, Philpot WD, Steenhuis TS (2014a) Monitoring state of biomass recovery in the Blue Nile basin using image-based disturbance index. In: Melesse AM, Abtew W, Setegn SG (eds) Nile River basin. Springer International Publishing Switzerland, pp 237–252

Ayana EK, Philpot WD, Melesse AM, Steenhuis TS (2014b) Bathymetry, lake area and volume mapping: a remote-sensing perspective. In: Melesse AM, Abtew W, Setegn SG (eds) Nile River basin. Springer International Publishing Switzerland, pp 253–267

Ayele T, G/Yohannes A, Girma Y, Sahlu D, Alamrew T (2019) Baseline database on hydrology and water use and related report in the Blue Nile upstream of GERD. Water and cooperation within the Nile River basin (WACONI) contract document, p 17

Betrie GD, Mohamed YA, van Griensven A, Srinivasan R (2011) Sediment management modelling in the Blue Nile basin using SWAT model. Hydrol Earth Syst Sci 15:807–818

Billi P (2015a) Geomorphology of ephemeral streams in the Kobo basin. In: Billi P (ed) Landscapes and landforms of Ethiopia, world geomorphological landscapes. Springer Science+Business Media Dordrecht Heidelberg, New York, London, pp 213–225

Billi P (2015b) Geomorphological landscapes of Ethiopia. In: Billi P (ed) Landscapes and landforms of Ethiopia, world geomorphological landscapes. Springer Science+Business Media Dordrecht Heidelberg, New York, London, pp 1–32

Billi P, Golla S, Tefferra D (2015) Ethiopian rivers. In: Billi P (ed) Landscapes and landforms of Ethiopia, world geomorphological landscapes. Springer Science+Business Media Dordrecht Heidelberg, New York, London, pp 89–116

Blackburn NC (2016) Apatite helium thermochronology of the Blue Nile Canyon, Ethiopian Plateau. Master's Thesis, Western Kentucky University, p 65

Chorowicz J, Collet B, Bonavia F, Mohr P, Parrot JF, Korme T (1998) The Tana basin, Ethiopia: intraplateau uplift, rifting and subsidence. Tectono Phys 295:351–367

Coltorti M, Dramis F, Ollier C (2007) Planation surfaces in northern Ethiopia. Geomorphology 89(3–4):287–296

Coltorti M, Firuzabadi D, Borri A, Pierlorenzo Fantozzi P Pieruccini P (2015) Planation surfaces and the long-term geomorphological evolution of Ethiopia. In: Billi P (ed) Landscapes and landforms of Ethiopia, world geomorphological landscapes. Springer Science+Business Media Dordrecht Heidelberg, New York, London, pp 117–136

Dagnew DC, Guzman CD, Zegeye AD, Akal AT, Moges MA, Tebebu TY, Mekuria W, Ayana EK, Tilahun SA, Steenhuis TS (2017) Sediment loss patterns in the sub-humid Ethiopian highlands. Land Degrad Dev 28:1795–1805

Deneke TT (2014) Processes of institutional change and factors influencing collective action in local water resources governance in the Blue Nile basin of Ethiopia. In: Melesse AM, Abtew W, Setegn SG (eds) Nile River basin. Springer International Publishing Switzerland, pp 477–497

Dessie M, Verhoest NEC, Pauwels VRN, Admasu T, Poesen J, Adgo E, Deckers J, Nyssen J (2014) Analysing runoff processes through conceptual hydrological modelling in the Upper Blue Nile basin, Ethiopia. Hydrol Earth Syst Sci Discuss 11:5287–5325

Dile YT, Berndtsson R, Setegn SG (2013) Hydrological response to climate change for Gilgel Abay River, in the Lake Tana basin—Upper Blue Nile basin of Ethiopia. PLoS ONE 8(10):e79296. https://doi.org/10.1371/journal.pone.0079296

Dile YT, Tekleab S, Ayana EK, Gebrehiwot SG, Worqlul AW, Bayabil HK, Yimam YE, Tilahun SA, Daggupati P, Louise Karlberg L, Srinivasan R (2018) Advances in water resources research in the Upper Blue Nile basin and the way forwards: a review. J Hydrol 560:407–423

Duguma TA, Duguma GA (2022) Assessment of groundwater potential zones of Upper Blue Nile River basin using multi-influencing factors under GIS and RS environment: a case study on Guder watersheds, Abay basin, Oromia region, Ethiopia. Geofluids 2022:1–26

Eastern Nile Technical Regional Office/ENTRO (2006) Water Atlas of the Blue Nile Subbasin, p 49

El-Fadel M, El-Sayegh Y, El-Fadl K, Khorbotly D (2003) The Nile River basin: a case study in surface water conflict resolution. J Nat Resour Life Sci Educ 32:107–117

Enyew BD, Van Lanen HAJ, Van Loon AF (2014) Assessment of the impact of climate change on hydrological drought in Lake Tana catchment, Blue Nile basin, Ethiopia. J Geol Geosci 3(6):1–17

Federal Democratic Republic of Ethiopia (FDRE), Abbay Basin Authority (ABA) (2016)

Fenta AA, Rientjes T, Alemseged Tamiru Haile AT, Reggiani P (2014) Satellite rainfall products and their reliability in the Blue Nile basin. In: Melesse AM, Abtew W, Setegn SG (eds) Nile River basin. Springer International Publishing Switzerland, pp 51–67

Gani ADS, Abdelsalam MG, Gera S, Gani MR (2009) Stratigraphic and structural evolution of the Blue Nile basin, Northwestern Ethiopian Plateau. Geol J 44:30–56

Gani NDS, Gani MR (2007) Blue Nile incision on the Ethiopian Plateau: pulsed plateau growth, Pliocene uplift, and hominin evolution. GSA Today 17(9):10.1130. https://doi.org/10.1130/GSAT01709A.1

Gebre SL (2015) Application of the HEC-HMS model for runoff simulation of Upper Blue Nile River basin. Hydrol Curr Res 6(2):1–8

Gessesse GD (2014) Assessment of soil erosion in the Blue Nile basin. In: Melesse AM, Abtew W, Setegn SG (eds) Nile River basin. Springer International Publishing Switzerland, pp 193–218

Guzman C, Tilahun S, Zegeye A, Steenhuis T (2013) Suspended sediment concentration discharge relationships in the (sub) humid Ethiopian Highlands. Hydrol Earth Syst Sci 17:1067–1077

Guzman CD, Zimale FA, Tebebu TY, Bayabil HK, Tilahun SA, Yitaferu B, Rientjes THM, Steenhuis TS (2017) Modelling discharge and sediment concentrations after landscape interventions in a humid monsoon climate: the Anjeni watershed in the highlands of Ethiopia. Hydrol Process 31:1239–1257

Kigobe M, Wheater H, McIntyre N (2014) Statistical downscaling of precipitation in the Upper Nile: use of generalized linear models (GLMs) for the Kyoga basin. In: Melesse AM, Abtew W, Setegn SG (eds) Nile River basin. Springer International Publishing Switzerland, pp 421–449

Kitaw M, Yitayew M (2014) Water governance in the Nile basin for hydropower development. In: Melesse AM, Abtew W, Setegn SG (eds) Nile River basin. Springer International Publishing Switzerland, pp 499–515

Lemann T, Roth V, Zeleke G, Subhatu A, Kassawmar T, Hurni H (2019) Spatial and temporal variability in hydrological responses of the Upper Blue Nile basin, Ethiopia. Water 11(21):1–35

McCartney M, Alemayehu T, Shiferaw A, Awulachew SB (2010) Evaluation of current and future water resources development in the Lake Tana basin, Ethiopia. International Water Management Institute (IWMI), Colombo, Sri Lanka, p 39. Research Report 134. https://doi.org/10.3910/2010.204

Mège D, Purcell P, Pochat S, Guidat T (2015) The landscape and landforms of the Ogaden, Southeast Ethiopia. In: Billi P (ed) Landscapes and landforms of Ethiopia, world geomorphological landscapes. ©Springer Science+Business Media Dordrecht 2015, pp 323–348. https://doi.org/10.1007/978-94-017-8026-1_20

Melesse AM, Abtew W, Setegn SG (2014) A scarce and shared resource: hydrologic threats, trends, and challenges in the Nile River basin. In: Melesse AM, Abtew W, Setegn SG (eds) Nile River basin. Springer International Publishing Switzerland, pp 1–4

Meresa HK, Gatachew MT (2019) Climate change impact on river flow extremes in the Upper Blue Nile River basin. J Water Clim Change

Meshesha MA (2016) Human impact on hydrogeomorphology of Gumara River upper Blue Nile basin, Ethiopia. PhD Thesis, Bahir Dar University, Ethiopia, p 130

Moges SA, Gebremichael M (2014) Climate change impacts and development-based adaptation pathway to the Nile River basin. In: Melesse AM, Abtew W, Setegn SG (eds) Nile River basin. Springer International Publishing Switzerland, pp 339–361

Morbach M, Ribbe L, Pedroso R (2014) Supporting the development of efficient and effective river basin organizations in Africa: what steps can be taken to improve transboundarywater cooperation between the Riparian States of the Nile? In: Melesse AM, Abtew W, Setegn SG (eds) Nile River basin. Springer International Publishing Switzerland, pp 597–636

Mulat AG, Moges SA, Ibrahim Y (2014) Impact and benefit study of Grand Ethiopian Renaissance Dam (GERD) during impounding and operation phases on downstream structures in the Eastern Nile. In: Melesse AM, Abtew W, Setegn SG (eds) Nile River basin. Springer International Publishing Switzerland, pp 543–564

Nigussie TA, Fanta A, Melesse AM, Quraishi S (2014) Modelling rainfall erosivity from daily rainfall events, Upper Blue Nile basin, Ethiopia. In: Melesse AM, Abtew W, Setegn SG (eds) Nile River basin. Springer International Publishing Switzerland, pp 307–335

Paisley RK, Henshaw TW. Transboundary governance of the Nile River basin: past, present and future. Environ Dev (in press)

Pik R, Deniel C, Coulon C, Yirgu G, Hofmann C, Ayalew D (1998) The northwestern Ethiopian Plateau flood basalts: classification and spatial distribution of magma types. J Volcanol Geoth Res 81:91–111

Poppe L, Frankl A, Poesen J, Teshager A, Mekete D, Enyew A, Deckers J, Nyssen J (2013) Geomorphology of the Lake Tana basin, Ethiopia. J Maps. https://doi.org/10.1080/17445647.2013.801000

Roth V, Tatenda Lemann T, Zeleke Z, Subhatu AT, Nigussie TK, Hurni H (2018) Effects of climate change on water resources in the upper Blue Nile basin of Ethiopia. Heliyon 4:1–28

Setegn SG, Melesse AM, Rayner D, Dargahi B (2014) Climate change impact on water resources and adaptation strategies in the Blue Nile River basin. In: Melesse AM, Abtew W, Setegn SG (eds) Nile River basin. Springer International Publishing Switzerland, pp 389–404

Shobary AS, Elsharkawy A, El-Hanafy HEM, Moussa OM (2021) Updating land use and land cover classes of Blue Nile basin using landsat-8 images. IOP Conf Ser Mater Sci Eng 1172:012014

Steenhuis TS, Tilahun SA, Tesemma ZK, Tebebu TY, Moges M, Zimale FA, Worqlul AW, Alemu ML, Ayana EK, Mohamed YA (2014) Soil erosion and discharge in the Blue Nile basin: trends and challenges. In: Melesse AM, Abtew W, Setegn SG (eds) Nile River basin. Springer International Publishing Switzerland, pp 130–147

Stoa R (2014) International water law principles and frameworks: perspectives from the Nile River basin. In: Melesse AM, Abtew W, Setegn SG (eds) Nile River basin. Springer International Publishing Switzerland, pp 581–595

Sutcliffe JV, Parks YP (1999) The hydrology of the Nile. International Association of Hydrological Sciences, p 192

Taye MT, Willems P, Block P (2015) Implications of climate change on hydrological extremes in the Blue Nile basin: a review. J Hydrol Reg Stud 4:280–293

Tesfaye G (2022) Soil erosion and sedimentation rate in the Blue Nile River basin of Ethiopia. Am J Environ Prot 11(1):1–5

Tilahun SA, Guzman CD, Zegeye AD, Ayana EK, Collick AS, Yitaferu B, Steenhuis TS (2014) Spatial and temporal patterns of soil erosion in the semihumid Ethiopian highlands: a case study of Debre Mawi watershed. In: Melesse AM, Abtew W, Setegn SG (eds) Nile River basin. Springer International Publishing Switzerland, pp 149–163

Yenehun A, Dessie M, Azeze M, Nigate F, Belay AS, Jan Nyssen J, Adgo E, Van Griensven A, Van Camp M, Walraevens K (2021) Water resources studies in headwaters of the Blue Nile basin: a review with emphasis on Lake water balance and hydrogeological characterization. Water 13:1–28

Yibeltal M, Tsunekawa A, Haregeweyn N, Adgo E, Meshesha DT, Masunaga T, Tsubo M, Paolo Billi P, Ebabu K, Fenta AA, Mulatu Liyew Berihun ML (2019) Morphological characteristics and topographic thresholds of gullies in different agro-ecological environments. Geomorphology 341:15–27

Yilma AD, Awulachew SB (2009) Characterization and Atlas of the Blue Nile basin and its subbasins. International Water Management Institute, p 253

Yitbarek M, Belay M, Bazezew A (2019) Determinants of manual control of water hyacinth expansion over the Lake Tana, Ethiopia. Int J Sci Technol (STECH) Ethiopia 8(1):1–14

Wale A (2008) Hydrological balance for Lake Tana upper Blue Nile, Ethiopia. Master Thesis. International Institute for Geo-Information Science (ITC) and observation, Enschede, Netherlands

Williams FM (2016) GeoGuide: understanding Ethiopia, geology and scenery. Springer International Publishing Switzerland, p 356

Wimberly MC, Midekisa AA (2014) Hydro-epidemiology of the Nile basin: understanding the complex linkages between water and infectious diseases. In: Melesse AM, Abtew W, Setegn SG (eds) Nile River basin. Springer International Publishing Switzerland, pp 219–233

Woodward JC, Macklin MG, Krom MD, Martin AJ, Williams MAJ (2022) The River Nile: evolution and environment. In: Gupta A (ed) Large rivers: geomorphology and management, 2nd edn. Wiley

World Bank (2008) Tana & Beles integrated water resources development project. Project Appraisal Document, Report No: 43400-ET, p 131

Wubneh MA, Worku TA, Chekol BZ (2023) Climate change impact on water resources availability in the kiltie watershed, Lake Tana subbasin, Ethiopia. Heliyon 9:1–15

Zewdu F (2008) Basics of geomorphology: a teaching material for geomorpholog. Bahir Dar University, Ethiopia, p 214

Zimale FA, Moges MA, Alemu ML, Ayana EK, Demissie SS, Tilahun SA, Steenhuis TS (2018) Hydromechanics. Budgeting suspended sediment fluxes in tropical monsoonal watersheds with limited data: The Lake Tana basin. J Hydrol Hydromechanics 66:65–78

Part II
Economic Development Potential

Chapter 4
Agriculture in the Abbay Basin

Mezegebu Getnet and Shimelis Asseffa

Abstract Ethiopia is the second most populous country in Africa, with more than 114 million people. The capacity to feed its rapidly growing population largely depends on agricultural production in diverse agro-climatic regions. Agriculture is undermined by severe land degradation, poor input supply systems, and climate change and variability. This chapter tries to show the status and challenges in rainfed and irrigated agriculture in the Abbay basin. The basin is a major contributor of agricultural products for consumption and export. There are distinct orientations of the production system in the basin. The commercially oriented production around North Gondar mainly of sesame and soybean targets the export market and local agro-processing industries. The cereal-based high-potential production system produces surplus and supplies agricultural food and raw materials for urban and local markets. Food-insecure areas are characterized by a low level of productivity associated with land degradation, climate, and poor socioeconomic status of the community with food gap months. The current/actual production level is very low compared with the potential. There is a huge average yield gap of approximately 11.5 t ha^{-1} for maize, 6.6 t ha^{-1} for wheat, 4.9 t ha^{-1} for sorghum, and 3.3 t ha^{-1} for millet. Furthermore, approximately 77.6% of the basin's agricultural land is moderately to highly suitable for agriculture, showing the potential to improve productivity and production in the commercially oriented production, cereal-based high potential production system. Furthermore, there has been a remarkable increase in irrigated area over the past decades; however, irrigation is dominantly practiced in a traditional way, challenging the sustainable utilization of water resources, particularly under severe stress due to increasing demand and climate change, which is anticipated to result in long-term changes in productivity and production. However, the effort to close the prevailing yield gap is challenged by poor soil fertility status, low level of input use, poor linkage of stakeholders, and limited access to finance. Therefore, improving the soil health conditions and the seed and fertilizer supply system, increasing financial access,

M. Getnet (✉)
Stichting Wageningen Research Ethiopia, Addis Ababa, Ethiopia
e-mail: mezegebu.debas@gmail.com

S. Asseffa
Institution of Technology, Hawassa University, Hawassa, Ethiopia

improving the efficiency of irrigation and extension service delivery, and creating market linkages are key to transforming the agricultural sector.

Keywords Agriculture · Agroecology zone · Major crops · Abbay basin · Ethiopia

Introduction

Ethiopia, one of the populated countries in Africa with a population of 114 million, is affected by land degradation (Gashaw et al. 2014; Abera et al. 2019). A rapidly growing population, inappropriate land management, rigid land tenure, industrialization, and urbanization have significantly impacted land use patterns. Ethiopia has diverse agroecologies endowed with abundant natural resources. About 74 million hectares of arable land available in 18 major agroecological zones ranging in altitudes between 148 and 4620 m above sea level that makes the country suitable for growing diversified crops (ATA 2019).

Different farming systems have evolved in the basin as the result of the diverse landscapes, climatic condition, and the dynamics of farmers' decisions in response to changing livelihood opportunities (Malcolm et al. 2001). Approximately 80% of the population of Ethiopia lives in rural areas. Agriculture is the principal economic sector, contributing 35% to the gross domestic product (GDP), 65% of employment, and over 80% of the national export value (World Bank 2019; Central Statistics Agency 2018).

Agriculture in the Abbay Basin feeds the rapidly increasing population and generating a large amount of foreign earnings through the export of commercial commodities. This is achieved mainly from the rain-fed system; however, there is a substantial increase in the contribution of the irrigated system in the basin. The Abbay Basin, i.e., the Ethiopian part of the Blue Nile Basin, is a major contributor of the hydrology of the Nile region. It generates more than 60% of the total Nile River streamflow that reaches Sudan and Egypt.

Based on MoWR data, the irrigated agriculture in the Abbay Basin has the potential to implement 211 irrigation projects (90 small, 69 medium and 52 large scale). The total estimated potential irrigable land is approximately 815,580 ha, comprising of 45,860 ha of small-scale, 130,395 ha of medium-scale, and 639,330 ha of large-scale development (Awulachew et al. 2007). Feeding the increasing population under various climate, biophysical and socioeconomic-related constraints while simultaneously minimizing the negative effects downstream remains to be a challenge for the future of agriculture in the Abbay basin. This chapter discusses the agricultural production system, the prevailing yield gaps, and the challenges for sustainable production in the Abbay Basin.

Resource Base

Administrative Zones and Agroecologies in the Abbay Basin

The Abbay Basin covers a catchment area of 199,812 km^2 in which 21 administrative zones within Amhara, Oromia and Benishangul-Gumuz regional states have a stake (Fig. 4.1). The basin is divided into twelve subbasins, namely, the Anger, Beles, Dabus, Debre Markos, Mota, Lake Tana, Didesa, Dindir/Rahid, Fincha, Guder, Muger, and Jemma, subbasins. The main river in the basin is the Blue Nile (Abbay), which rises in Lake Tana, and its elevation varies between 500 and 4261 m. The total mean annual flow from the river basin is estimated to be 54.8 Bm^3. Abbay is the most important river basin in Ethiopia hosting about 25% of the population, covering about 20% of Ethiopia's land area, contributing approximately 50% of the total average annual runoff, and more than 40% of the country's agricultural production (Awulachew et al. 2007).

The Abbay Basin is endowed with diverse agro-ecologies and production systems that have created different livelihoods and food systems with unique characteristics. This diversity has huge implications for the diversity and productivity of agricultural commodities. The northwestern part is characterized by warm moist and warm submoist lowland agro-ecologies, whereas the southwestern part is dominated by warm subhumid lowlands, both accounting for approximately 43.1% of the basin area (Fig. 4.2). The central and eastern parts are characterized by Tepid moist mid-highlands and Tepid submoist mid-highlands, respectively, accounting for approximately 34.2% of the basin.

Soils

The western part of the Abbay River basin is dominated by Nitisols, covering approximately 37% of the area. The majority of the eastern region is dominated by Leptisols, accounting for 24% of the area. The central part is dominantly covered with Vertisols (13%) and Luvisols (12%), and the central part is dominated by Leptisols, comprising 24% of the area, respectively (Table 4.1, Fig. 4.3). Nitosols, Liptosols, and vertisols, which account for 73%, are said to have limiting factors for agriculture in which the former is affected by acidity, whereas the latter two are affected by depth- and drainage-related problems (Erkossa et al. 2014).

Land Use

The mixed farming in the highlands and pastoralism and agro-pastoralism the lowlands were the main types of land use farming systems (Erkossa et al. 2009a,

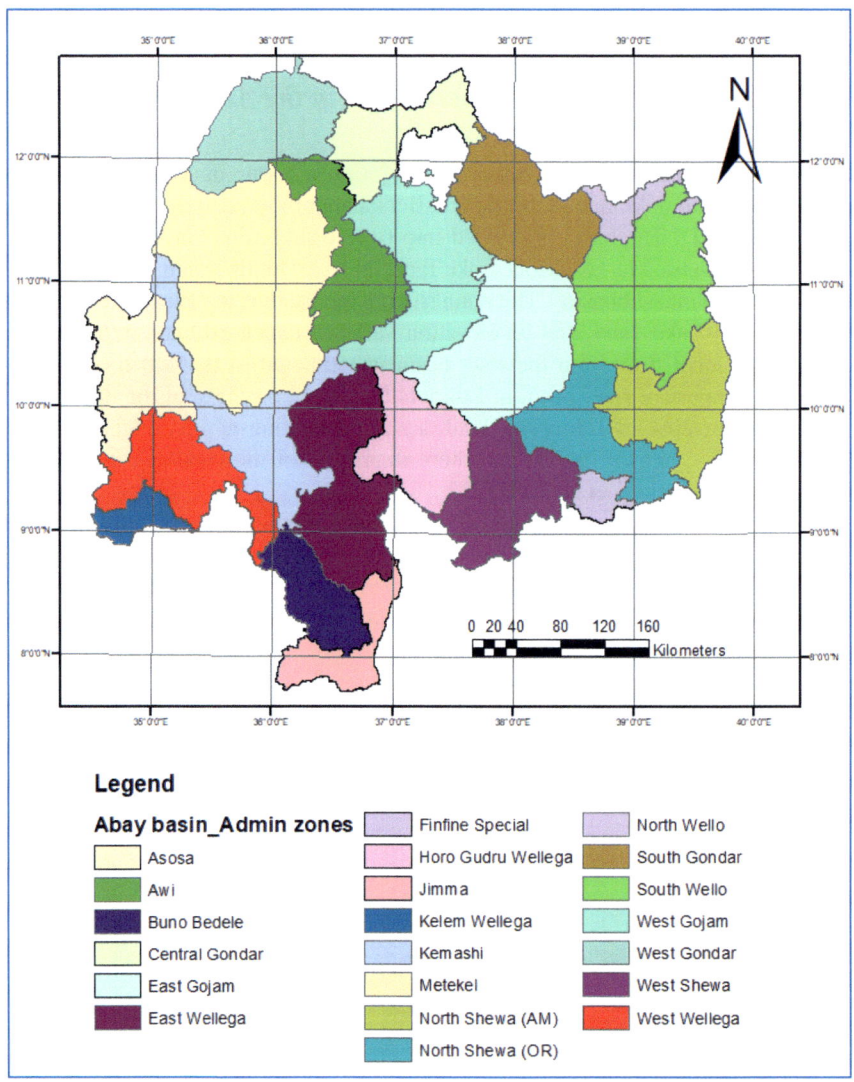

Fig. 4.1 Administrative zones in the Abbay River basin

b). The majority of the Abbay Basin is covered with cropland (48%) and trees and shrubs (35%), followed by grassland (15%). Cropland is the dominant land cover covering major part of the central and eastern parts, whereas trees and shrubs are predominant in the western and southwestern part (Fig. 4.4).

Unplanned land-use and mining culture of cultivation is practiced in the basin, as in any other parts of the county associated with poor land-use policy instruments. This resulted in continuous deforestation, loss of soil fertility, increased soil erosion,

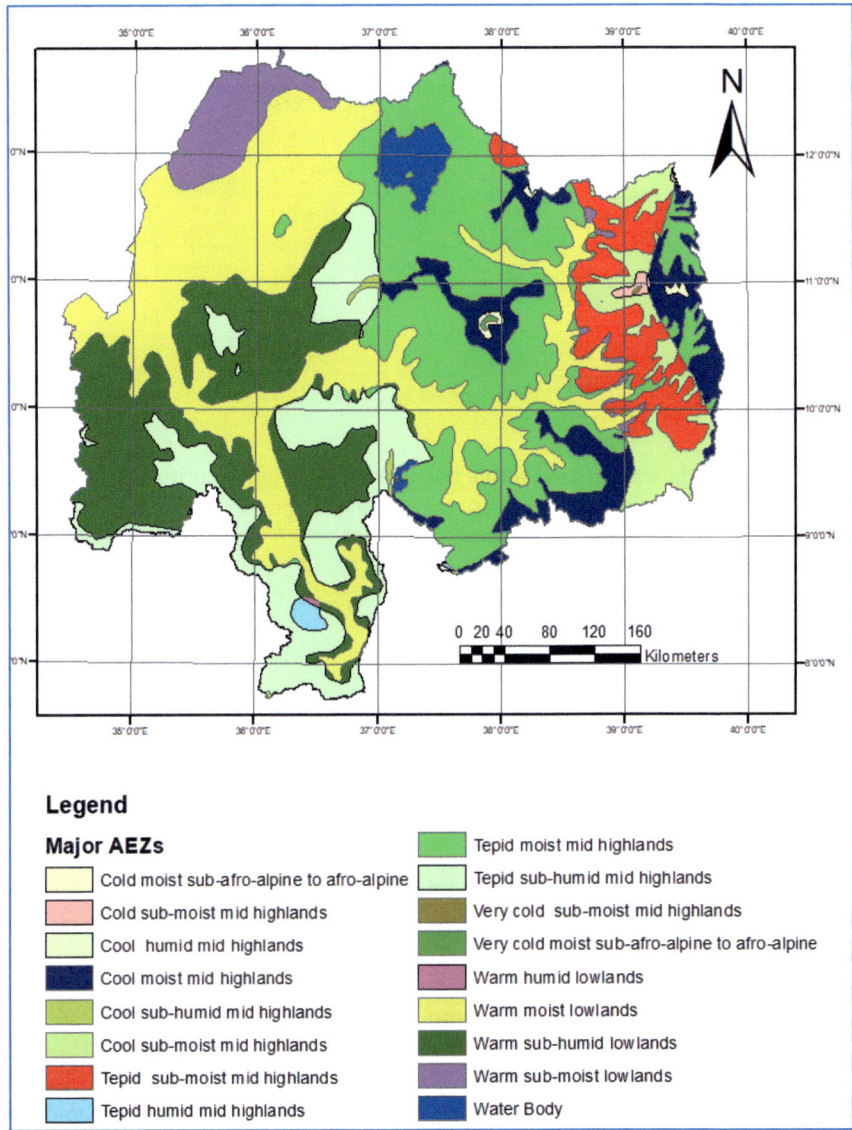

Fig. 4.2 Major agroecological zones in the Abbay Basin of Ethiopia. Figures in the text are based on authors calculation from this map

and land degradation, especially in the highlands (Zeleke and Hurni 2001; Bewket 2002; Awulachew et al. 2010) where crop production is the main livelihood of the community. Some initiatives, such as land titling, have been in place with the aim of mitigating the problem of land degradation and promoting adaptation strategies.

Table 4.1 Total area of major sols in the Abbay River basin

Soil type	Area	Percent
Alfisols	837,105	4
Cambisols	1,102,669	6
Fluvisols	119,275	1
Leptosols	4,785,746	24
Luvisols	2,364,141	12
Nitosols	7,417,151	37
Phaeozems	270,690	1
Vertisols	2,647,625	13
Water Body	354,728	2

Source Erkossa et al. (2014)

Some evidence shows that land titling has improved tenure security enhanced land management participation and reduced land-related disputes (Behaylu 2015).

Agricultural Production System

Small-scale farming is the main livelihood system in the Abbay River basin, and it can be described as crop-livestock-based mixed agriculture in the highlands, and agro-pastoralism in the lowland areas (Tibebe et al. 2022). The crop categories cultivated in the basin include cereals, legumes, root crops, oil crops, vegetables, fruits targeting to meet food security as well as market. There are diverse features of the production systems in the basin based on how agricultural products are targeted. The first is the commercial-oriented production around the northwestern part of the basin, mainly of sesame and soybean targeting the export market as well as local agro-processing industries. Sesame is an important export commodity as a source of currency for the country. Kelali et al. (2014) reported that the average yield of sesame in 2013 was 450 kg/ha with a postharvest loss of approximately 12%, whereas the application of improved production technologies can lead to yields of 800 kg per hectare. The seed system in sesame is poor because it is neither supported in the formal seed system of the country nor attracts private seed producers as it is required in very small amounts to cover a hectare of land.

The second is a cereal-based high-potential production system in which most farmers produce surplus and supply agricultural products for urban and local markets. This includes the Agricultural Growth Program (AGP) Weredas in the basin. The major driver for this production system is the local and national market for production more than household consumption. However, there is a huge yield gap in the cereal-based high potential production system (Section "Land Suitability For Agriculture") mainly because of the low level of fertilizer use compared with the recommended rate (Tamene et al. 2017). The high-potential areas are usually characterized by limited

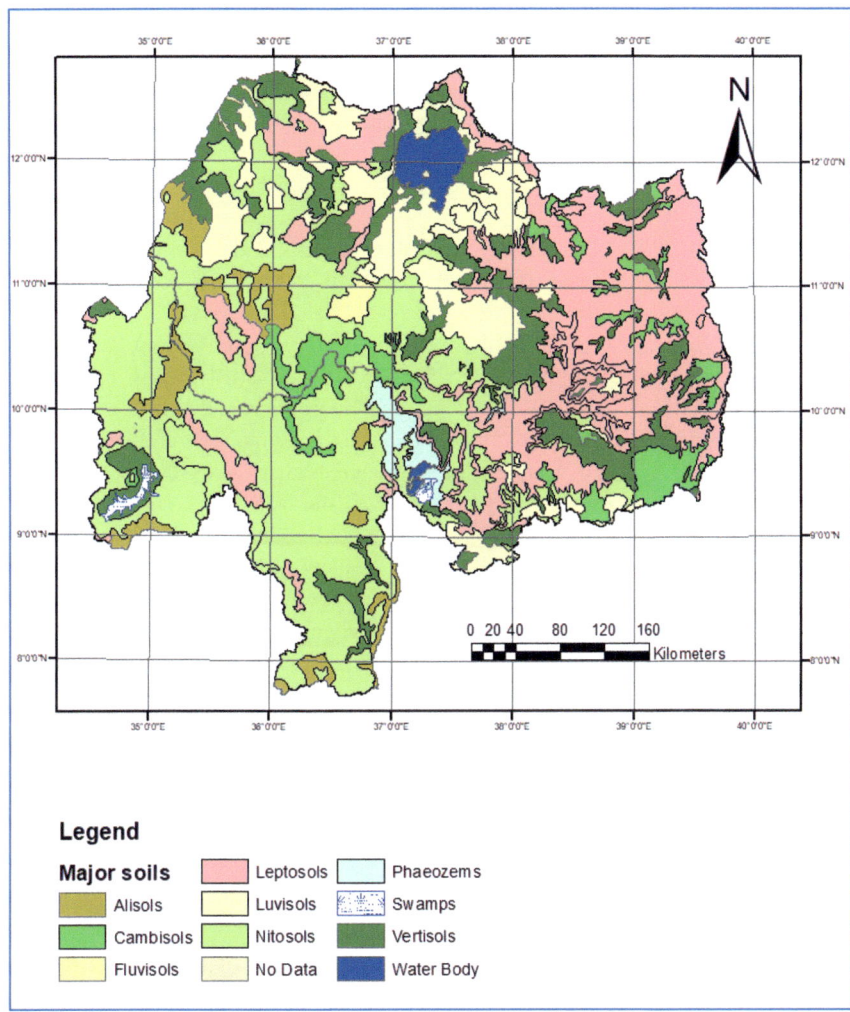

Fig. 4.3 Major soils in the Abbay River basin

diversity of crops resulting in poor dietary diversity, often showing a high rate of malnutrition as opposed to the relatively high level of food security achieved with this farming system.

The third represents the low potential areas that are usually known as Productive Safety Net Program (PSNP) Weredas. This farming system is characterized by higher yield gaps and longer food gap months because of the low level of productivity associated with land degradation, variable climate with longer dry spells, and socioeconomic risks associated with limited access to finance and low-income level of the community. The main objective of production in this type of farming system

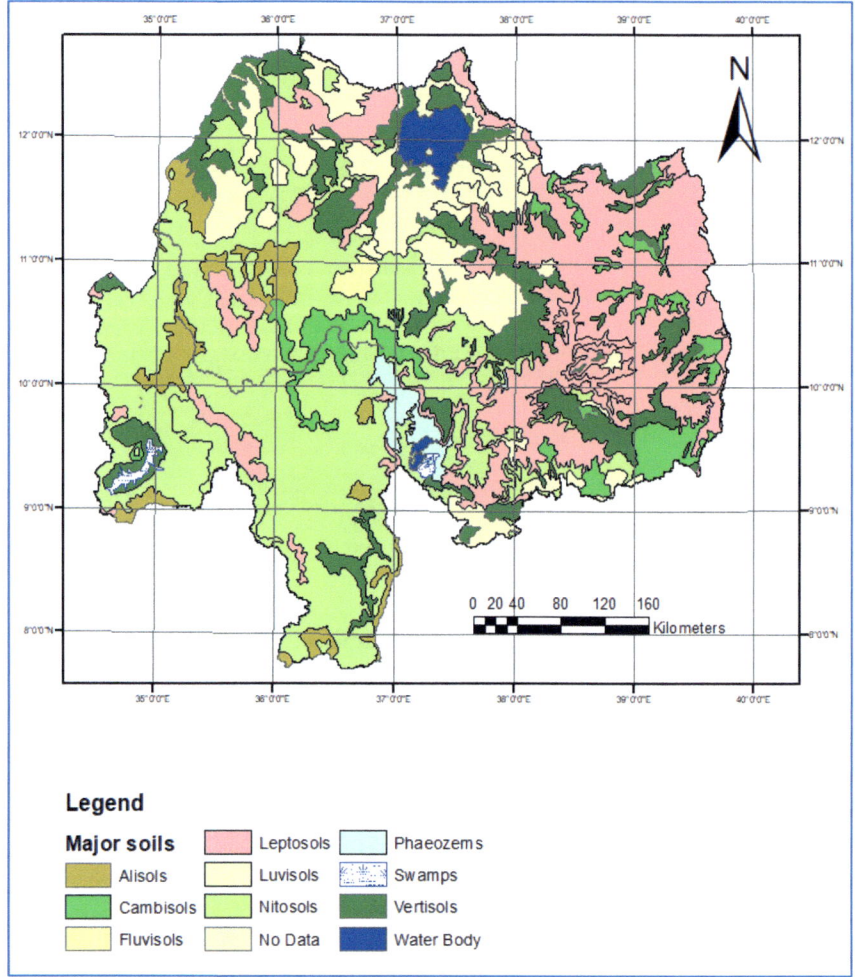

Fig. 4.4 Land cover map of the Abbay River basin adapted by authors from the Water and Land Resource Center, 2019

characterized by food insecurity is to achieve food security. However, the diversity of production in this farming system is better than that in the other two systems.

Livelihood Zones and Major Crops in the Abbay Basin

Over 80% of the populations in the basin mainly depend on small-scale rain-fed agriculture for their livelihood (Erkossa et al. 2009a, b). More than 80% of the cultivated land is under a rain-fed system, and the remaining land is cultivated under irrigation and residual moisture (Abera 2017). The sorghum-based livelihood is dominant in the northwest part of Amhara and northern part of Benishangul Gumuz (Fig. 4.5). This part of the basin is also characterized by vegetation such as shrubs and soils of the Nitosol. It covers some lowland areas in the South Wollo and North Wollo zones of the Amhara region. Maize, millet, sesame, and pulses are also produced in this farming system. Maize-and tef-based livelihoods are dominant in the central and western parts of the basin. Wheat, maize, flax, barley, and nug are common crops in this farming system. The wheat-based system dominates the eastern and central parts of the basin and some part of which is also covered by barley, tef, maize, millet, and pulses. Rice-based livelihoods are rapidly expanding surrounding Lake Tana, whereas-based commercial-oriented production of sesame and soybean characterize the areas west of Lake Tana.

Sustainable Land Management and Rural Land Certification

Studies assessed the impacts of sustainable land management (SLM) programs in the Amhara region, which makes up the majority of the Abbay basin. Kato et al., (2019), Schmidt and Tadesse, (2019) revealed that the program has (i) helped to adopt various best management practices at the plot level; (ii) significantly increased plot-level adoption of SLM practices, particularly of soil bunds and stone terraces; (iii) contributed to improved water security for both crop and livestock production; and (iv) provided households in SLM-supported learning watersheds with more access to groundwater for irrigation; and (v) increased income from livestock products compared to households in control watersheds. The study attributed the positive impacts of SLM and complementary interventions on livestock income to three key factors. These are improved water security conditions in the learning watersheds and access to better animal forage planted along SLM structures.

Rural land administration and certification have been implemented to enhance smallholder farmer tenure security. Evidence from first-stage land registration shows a positive improved investment land productivity and land rental market activities, resulting in the initiation of another phase of land registration and certification involving technically advanced land survey methods and computer registration (Bezu and Holden 2014). An important incentive for increased farmers' and landowners' motivation to adopt sustainable land and water management practices in individual fields was to increase tenure security. This increases farmer confidence in investing in long-term solutions.

Land Suitability for Agriculture

The start of the growing season ranges between early April and mid-June. The start of sowing varies depending on the location and the crop/cultivar choice (Table 4.2).

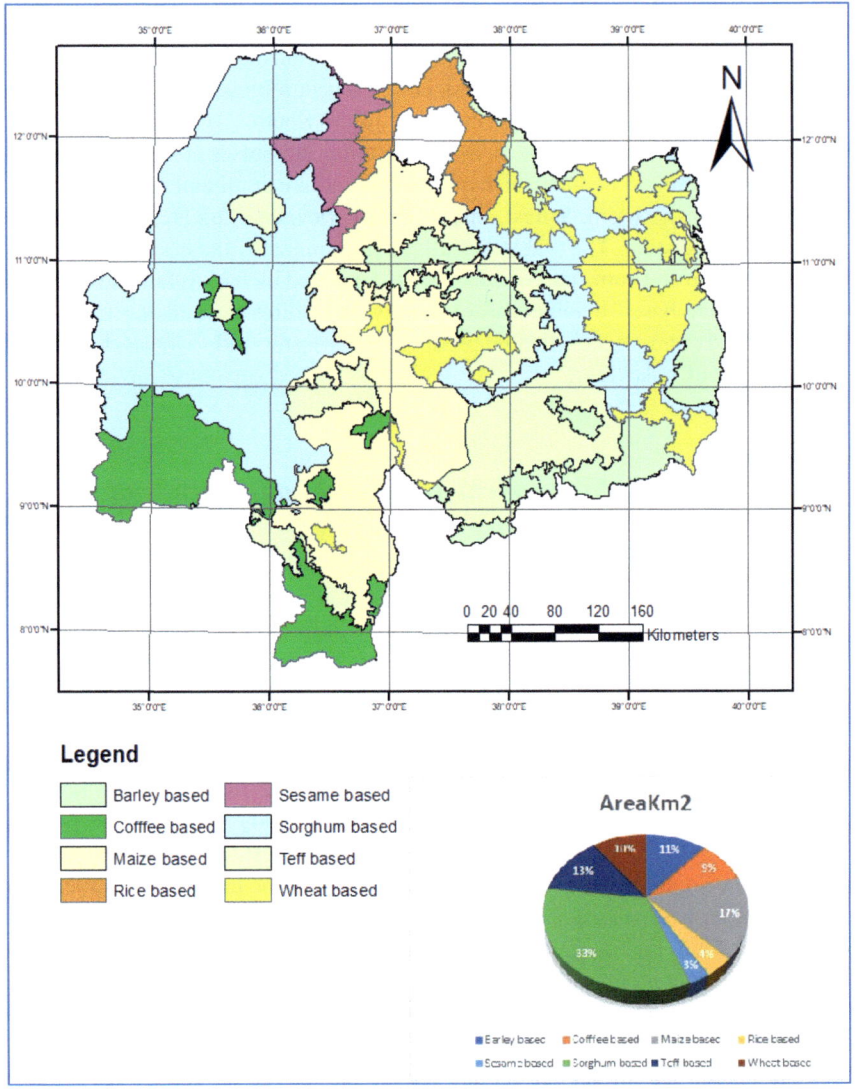

Fig. 4.5 Major livelihood zones in the Abbay basin. Adapted by authors based on FEWSNet.2018

The end of the main growing season ranges from early October to early November. The variation in the length of growing period (LGP) across the basin provides an opportunity to grow diverse crops and varieties as a guide for the selection of short-medium-or late-maturing cultivars. Furthermore, this spatial difference in the LGP along with spatial differences in the inherent fertility of the soil and potential differences in management are responsible for the spatial variation in suitability and actual productivity.

4 Agriculture in the Abbay Basin

Table 4.2 Sowing windows and thermal time requirements of major crops at some reference stations in the Abbay basin

Location name	Latitude	Longitude	Elevation (m)	Maize			Wheat			Sorghum			Millet		
				Start	End	GDD	Start	End	GDD	Start	End	GDD	Start	End	GDD
Adet	11.27	37.48	2240	1-May	30-May	1470	15-Jun	30-Jul	2470				15-May	30-Jun	1270
Ambo	8.96	37.835	2100	5-May	3-Jun	1600	15-Jun	15-Jul	2290						
Assosa	10.07	34.52	1419	16-Apr	15-May	1750	15-Jun	15-Jul	2450	1-Jun	30-Jun	1420	1-Jun	30-Jun	1580
Ayira	9.06	35.33	1700	1-Apr	30-Apr	1750				15-May	15-Jun	1630	15-May	30-Jun	1420
Bahir Dar	11.58	37.38	1790	15-May	14-Jun	1200				10-May	15-Jun	1090	10-May	15-Jun	1050
Bako	9.07	37.03	1650	1-May	14-Jun	1620									
Debremarkos	10.33	37.736	2470	1-Apr	30-Apr	1090									
Gore	8.02	35.53	1880	1-Apr	30-Apr	1400									
Jimma	7.84	36.43	2574	1-Apr	30-Apr	1580									
Nekemte	9.09	36.54	2110	15-May	13-Jun	1260				15-May	13-Jun	1190	15-May	15-Jun	1080
Pawe	11.31	36.403	1100	16-Apr	15-May	1470							1-Jun	10-Jul	1970
Shambu	9.57	37.12	2367	16-Apr	15-May	1000							1-May	31-May	900
Sheno	39.3	9.344	2848				1-Jul	31-Jul	1680						
Gondar	37.47	12.59	2052							15-May	14-Jun	1100	15-May	25-Jun	1030
Kobo	39.63	12.15	1500							16-Jun	15-Jul	1480	16-Jun	15-Jul	1460
Kombolacha	39.72	11.1	1840							1-Apr	30-Apr	1320			
Pawe	36.40	11.31	1100							16-Apr	15-May	1900			
Shambu	37.12	9.57117	2367							16-Apr	15-May	1160			

GDD = Growth degree days; *Source* Authors compilation based on Global Yield Gap Atlas (2022)

Table 4.3 Land suitability for agriculture in the Abbay River basin

Suitability	Area (km^2)	Percentage
Highly suitable	57,050	28.6
Moderately suitable	97,812	49.0
Marginally suitable	12,378	6.2
Unsuitable	11,978	6.0
Unavailable*	20,594	10.3
Total area	199,812	100.0

*Protected areas, forest cover and water bodies. *Source* Yalew et al. (2016)

Yalew et al. (2016) identified nine criteria and weights to determine agricultural land suitability including soil type, soil depth, soil water content, soil stoniness, slope, elevation and proximity to towns, roads, and water sources. Based on these criteria, the majority of the agricultural land in the basin falls under moderately to highly suitable for agriculture (Table 4.3).

Actual Production and Yield Gaps of Major Crops in the Abbay Basin

Although an increase in cereal productivity has been observed over the past years, the current productivity of most cereals represented as actual yield (Ya) in the Abbay basin is still less than 3 t ha^{-1} (Table 4.4). Crop productivity is very low in the Abbay basin, as in most areas of Ethiopia. The current/actual production level is very low compared with the potential. There is a huge average yield gap of approximately 11.5 t ha^{-1} for maize, 6.6 t ha^{-1} for wheat, 4.9 t ha^{-1} for sorghum, and 3.3 t ha^{-1} for millet. Furthermore, approximately 77.6% of the basin's agricultural land is moderately to highly suitable for agriculture, showing the potential to improve productivity and production.

Irrigation Potential and Practices in Abbay Basin

Irrigation has multiple advantages such as improving food security and rural employment, by increasing agricultural productivity (Bryan et al. 2013) by closing the prevailing yield gap required to meet the growing demands for food and nutrition in the face of the growing population in Ethiopia (Awulachew et al. 2005; Lipton et al. 2003). It also plays a vital role in the improvement of land values (Mequanent and Mingist 2019). The monthly maximum crop water requirement and cropping pattern vary widely within the Abbay River basin (Fig. 4.6). For example, the maximum

4 Agriculture in the Abbay Basin

Table 4.4 Yield gaps of major crops in selected locations within the Abbay basin

Yield gap	Maize			Wheat			Sorghum			Millet		
	Ya	Yp	Yg	Ya	Yp	Yg	Ya	Yp	Yg	Ya	Yp	Yg
Adet	3.2	17.0	13.8	2.1	9.5	7.4				1.5	5.5	4
Ambo	3.0	16.2	13.2	2.2	8.6	6.4						
Assosa	2.8	18.1	15.3							1.2	6.3	5.2
Ayira	2.9	18.8	15.9	1.6	8.0	6.4	2.4	9.7	7.3	1.7	6.0	4.3
Bahir Dar	2.8	11.5	8.7				1.7	6.3	4.6	1.6	3.2	1.6
Bako	3.3	15.8	12.5									
Debre markos	2.8	12.6	9.8									
Gonder							2.0	7.6	5.7	1.9	3.7	1.8
Kobo							1.5	2.2	0.6	1.2	1.2	0.0
Kombolcha							2.0	2.8	0.8			
Nekemte	3.3	12.9	9.5				2.3	9.8	7.5	1.8	7.2	5.4
Pawe	3.0	12.8	9.8				1.8	5.8	4.1	1.7	5.0	3.4
Shambu	3.5	10.2	6.7				2.6	11.1	8.5	1.9	6.1	4.2
Sheno				1.9	8.1	6.2						

Ya = actual yield; Yp = potential yield; Yg = yield gap. *Source* Authors' compilation based on the GYGA atlas

monthly crop water requirement is approximately 1100 mm at the Dindir subbasin in the lower Dindir and Rahid irrigation projects, whereas the CWR is 439 mm at the Debre Markos subbasin (Leh irrigation project) and JEMMA, Debre Guracha irrigation project (Shenkut 2006).

Irrigated area has increased remarkably over the past decades. For example, area of irrigated croplands has increased from 55 to 65% between 1986 and 2016 (Abera et al. 2021), whereas irrigation consumption in the basin increased from 0.380 km^3 year^{-1} to 0.798 km^3 year^{-1} in the same period. However, despite the huge potential (Table 4.5), the irrigation practice in the Abbay River basin is very low compared with other basins located in the lower Nile basin areas in Sudan and Egypt. For example, Ethiopia has a large potential to develop surface irrigation; however, only 5% of the potentially available land has been developed (Worqlul et al. 2015).

However, irrigation development in Ethiopia is generally characterized by poor practices and technologies. The irrigation practice is generally surface with poor irrigation efficiency and low productivity. This in turn resulted in poor sustainability of most irrigation schemes. The use of modern technologies is generally limited and almost absent in small-scale irrigation schemes.

Increasing productivity in irrigation systems should also attain increased resource use efficiency for sustainability. Promoting the efficient use of fertilizers, use of improved varieties and quality seeds, promoting market-oriented production pattern

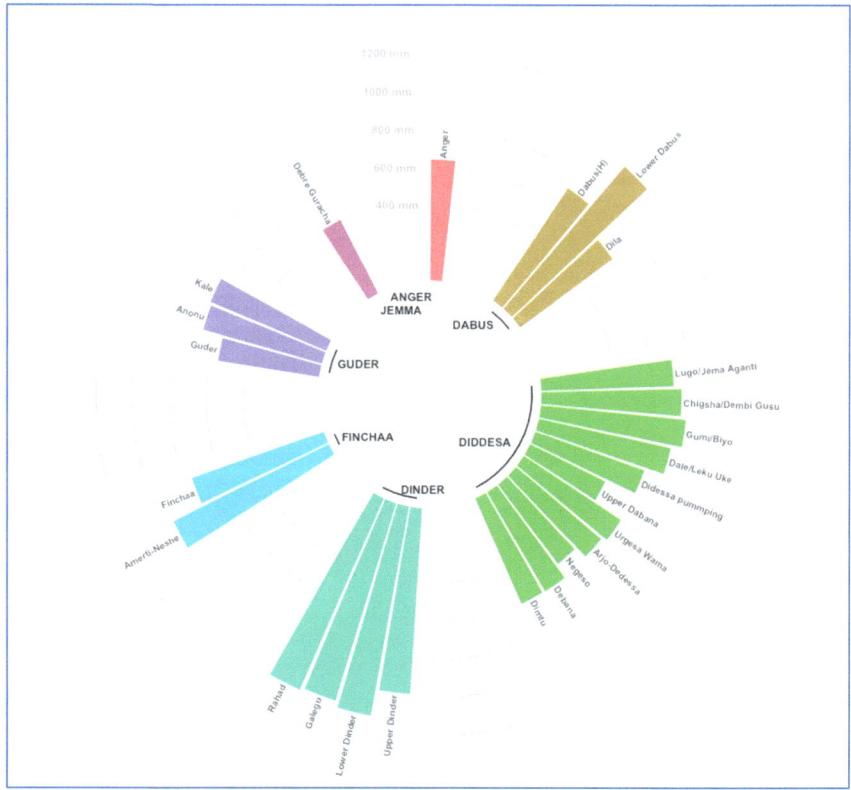

Fig. 4.6 Monthly crop water requirement at selected weather stations in Abbay basin

Table 4.5 Irrigation potential of the Abbay River basin

River basin	Area (km^2)	Runoff (Bm^3)	Potential irrigable area (ha)	Gross hydroelectric potential (GWH/year)	Estimated ground water potential (BM3)
Abbay	199,812	54.8	815,581	78,820	1.8

Source Integrated River Basin Master Plan Studies (MoWR 1996, 1997, 1998); Irrigable land from the IWMI irrigation database (based on MoWR data)

and cropping intensity, and improving irrigation water management and use efficiency are key for the resilience and sustainability of the system. Furthermore, strengthening the input supply system, improving financial access for input and irrigation facilities, improving market linkage, promoting the scaling of irrigation innovations by strengthening partnerships between stakeholders, managing irrigation facilities, and ensuring the security of land tenure are critical.

Challenges for Agriculture in the Abbay Basin

Land Degradation

Land degradation and declining soil health have been recognized as major constraints on agricultural development in the Abbay River Basin. Rapid population growth coupled with inappropriate land, water, and production practices have led to serious soil erosion and land degradation in many areas of the basin. This resulted in poor water retention of soils often causing intermittent drought during dry spells and dry seasons and increasing the vulnerability of agriculture to weather uncertainties and drought. Runoff resulted in nutrient losses of 10–120 kg ha^{-1} year^{-1}, siltation of downstream reservoirs and productivity losses in the uplands (Adimassu et al. 2017; Gebrehiwot et al. 2013). Heavy soil erosion (380 million tons annually) from upland areas of the Upper Blue Nile basin would have serious siltation problems at the Great Ethiopian Renaissance Dam reservoir, reducing its live storage capacity (Hurni et al. 2015).

Low Soil Fertility

The soils in many parts of the basin are characterized by deficiency of nutrients that are critical for sustained high crop yields. Long years of mining type of cultivation associated with a low level of input use resulted in serious fertility degradation. The level of use of external inputs is low mainly because of limited access, high costs, and risk aversion. Nutrient depletion has led to soil exhaustion and poor soil fertility. Integrated soil fertility management (ISFM) has been promoted in most areas of the basin. The site-specific nature, the huge biomass requirement to prepare organic fertilizers, and the bulkiness of the rate required to cover a given size of land limited the scalability of ISFM technologies and constrained the ability to boost agricultural production particularly for field crops despite the huge potential for yield increase in most agricultural lands.

Small Land Holdings and Insecure Land Tenure

The per capita land holding of households is increasingly becoming small and fragmented associated with fast-growing population. These forces farmers to cultivate steep slopes that are not suitable for cultivation. Despite the implementation of practices that aim to promote the security of land tenure for enhancing sustainable land management and agricultural productivity, land scarcity has continued to result in poor land-use practices, such as more intensive use of land, shortening or absence of fallow periods, and abandonment of shifting cultivation. Furthermore, land fragmentation makes it difficult to adopt mechanized systems for yield-increasing measures and to implement market-oriented production.

Poor Infrastructure and Institutional Linkage

The poor transportation infrastructure and market access in rural areas reduce motivation for market-oriented production and investment for inputs. The existing commercially oriented production of Sesame dominating the northwestern part of Tana is facing with an increasing trend of replacement by Soybean mainly because Sesame production is labor intensive, and it has a very narrow time window for harvesting, whereas any delay could result in substantial amount of loss due to shattering. Consequently, area covered by Soybean in this commercially oriented has increased dramatically over the past years at the expense of Sesame production. However, the market opportunity for the Soybean dropped following the increase in production.

The low level of storage and postharvest management practices not only resulted in a substantial amount of loss from pests but also forced producers to sell their products at low prices at the time of harvest. Strengthening the institutional and market linkage is critical for sustainability of increased production. The farmers' training centers established in almost all Kebeles in the basin lack the necessary facilities to support the provision of timely and site-specific advisory services. Most kebeles have three development agents, each supporting the crop, natural resource, and livestock sectors at the kebele level. However, the capacity of DAs to implement coordinated actions at the Kebele level is very limited. The linkage between stakeholders involved in the input supply chain, production, processing, and marketing of agricultural products is poor, resulting in a lengthy process of input procurement and distribution, a high cost of production due to inefficiencies, and a limited opportunity for value addition.

Climate Change and Variability

The temporal and spatial patterns of precipitation in the Abbay Basin is predicted to experience significant changes (Conway 2005; Kim et al. 2008; Elshamy et al. 2009; Kim and Kaluarachchi 2009; Beyene et al. 2010; Girma 2012; Koch and Cherie 2013). Projections showed increase in mean annual temperature and decrease in precipitation in most parts of the basin (Mengistu et al. 2021). Climate projections from two contrasting GCMs (wet scenario from MIROC5 and dry scenario from CSIRO) for the Upper Blue Nile Basin (UBNB) show annual precipitation change in the early (2011–2040), middle (2041–2070), and late century (2071–2100) varies between − 18.3% and +13.6% of the baseline of 1350 mm (Lazin et al. 2022). The eastern part of the basin experiences normal to moderate variability in the annual and Kiremt season rainfalls but high variability and declining trend for 73% of the Belg season rainfall (Gonfa et al. 2022).

Problematic Input Supply System

The lack of availability of high-quality seeds, low usage of agricultural technology, and high postharvest losses resulted in a low level of agricultural production. The input supply system mainly of seed and fertilizer is problematic because the majority of the crops are not included in the formal seed system. Substantial efforts have been made to promote community-based production of early generation seed (EGS) to solve the problem of a shortage of quality seed. However, the poor regulatory system made the system vulnerable to adulteration and smuggling, lack of sufficient basic seed from research.

The fertilizer supply chain also has multifaceted problems at various stages of the supply chain, including annual demand assessment, bid processes, foreign currency, inefficient logistics and transportation, distribution, and financial access to farmers. The fertilizer made available every year is much less than the demand, and farmers often apply suboptimal amounts, resulting in low productivity. The supply system needs to be made efficient to deliver inputs in a timely manner to close the prevailing yield gap.

Conclusions and Recommendations

Agriculture continues to be the major contributor for livelihood of millions of people in the Abbay River basin, and a major contributor to the national market and economy. The basin has huge potential for agriculture; however, the current productivity is very low, resulting in high yield gaps. Yield gaps are associated with a low level of input use, increasing degradation of soil fertility and land quality, and land fragmentation that hinders the adoption of improved technologies, including mechanization, poor postharvest handling, market linkage, and poor strategic planning to adapt to climate- and socioeconomic-related vulnerabilities. There is an urgent need to continue to provide the rapidly growing population with a sufficient and healthy diet. This is, however, challenged by the limited availability of agricultural land for further expansion to ensure increased production. The most plausible option is, therefore, narrowing the prevailing yield gap through increased productivity of the existing agricultural land. This is achieved through the following:

- Wide use of context-specific recommendations of agricultural technologies, including fertilizer recommendations, integrated soil fertility management, and integrated pest management that suit the biophysical and socioeconomic situations of localities.

- Transform the agricultural research system from dominantly plot/field scale studies toward generating system level innovations that would provide solutions for food system transformation, including improved input supply systems, resilient and sustainable production practices, access to finance, postharvest management, and consumption of safe and nutritious food, among others.
- Improve the input supply system including:
  - improve the efficiency of production/purchase and timely supply of agricultural inputs and technologies to boost production and productivity,
  - improve the production of sufficient and quality/healthy early generation seeds for major crops, improve producers' access to finance and inputs,
  - improve the governance across sectors involved in the seed system and fertilizer supply chain,
  - planning, purchase, distribution, and financial access for timely availability of fertilizer in all seasons
- Improve the use of advanced weather models and climate information system supporting fine spatiotemporal scale to adapt with climate related vulnerabilities affecting agriculture activities.
- Improve planning and decision-making of farmers and stakeholders supporting the extension system by blending bottom up (involving baseline and rapid appraisals) and top down (use of earth observations and spatial mapping) through multistake holder engagement and use of ICT-based decision-making tools.
- Improve the extension service by involving the private sector along with the public system also by enabling digital information delivery and feedback system.
- Improve the value chain and market performance so that agriculture in the basin can operate as a sustainable business.
- Support diversified production of crops to increase the availability of nutrient-dense foods as well as increase the resilience of production to risks associated with climate change and variability.

References

Abera M (2017) Agriculture in the Lake Tana subbasin of Ethiopia. In: Stave K, Goshu G, Aynalem S (eds) Social and ecological system dynamics. AESS interdisciplinary environmental studies and sciences series. Springer, Cham. https://doi.org/10.1007/978-3-319-45755-0_23

Abera W, Tamene L, Tibebe D, et al (2019) Characterizing and evaluating the impacts of national land restoration initiatives on ecosystem services in Ethiopia. Land Degrad Dev 31:37–52

Abera A, Verhoest NEC, Tilahun S, Inyang H, Nyssen J (2021) Assessment of irrigation expansion and implications for water resources by using RS and GIS techniques in the Lake Tana basin of Ethiopia. Environ Monit Assess 193:13. https://doi.org/10.1007/s10661-020-08778-1

Adimassu Z, Langan S, Johnston R, Mekuria W, Amede T (2017) Impacts of soil and water conservation practices on crop yield, run-off, soil loss, and nutrient loss in Ethiopia: review and synthesis. Environ Manage 59:87–101. https://doi.org/10.1007/s00267-016-0776-1

ATA (Agricultural Transformation Agency) (2019) Annual Report 2018–19. http://www.ata.gov.et/wp-content/uploads/2019/12/ANNUALREPORT-2011.pdf

Awulachew SB, Merrey DJ, Kamara AB, Van Koopen B, Penning De Vries F, Boelle E (2005) Experiences and opportunities for promoting small-scale/micro irrigation and rainwater harvesting for food security in Ethiopia. IWMI Working Paper 98

Awulachew SB, Yilma AD, Loulseged M, Loiskandl W, Ayana M, Alamirew T (2007) Water resources and irrigation development in Ethiopia. International Water Management Institute, Colombo, Sri Lanka, p 78 (Working Paper 123)

Awulachew SB, Ahmed A, Haileselassie A, Yilma A, Bashar K, Mccartney M, Steenhuis T (2010) Improved water and land management in the Ethiopian highlands and its impact on downstream stakeholders dependent on the Blue Nile

Behaylu A (2015) Debates on land titling. Am J Environ Prot 4(4):182–187

Beyene T, Lettenmaier DP, Kabat P (2010) Hydrologic impacts of climate change on the Nile River Basin: implications of the 2007 IPCC scenarios. Clim Change 100:433–461. https://doi.org/10.1007/s10584-009-9693-0

Bewket W (2002) Land cover dynamics since the 1950s in Chemoga watershed, Blue Nile basin, Ethiopia. Mt Res Dev 22:263–269

Bezu S, Holden S (2014) Are rural youth in Ethiopia abandoning agriculture? World Dev 64:259–272

Bryan E, Ringler C, Okoba B, Roncoli C, Silvestri S, Herrero M (2013) Adapting agriculture to climate change in Kenya: household strategies and determinants. J Environ Manage 114:26–35

Central Statistics Agency (2018) Area and production of major crops, Agricultural sample survey for 2017/18, statistical bulletin 589, Addis Ababa Ethiopia

Conway D (2005) From headwater tributaries to international river: observing and adapting to climate variability and change in the Nile Basin. Glob Environ Change 15:99–114. https://doi.org/10.1016/j.gloenvcha.2005.01.003

Elshamy ME, Seierstad IA, Sorteberg A (2009) Impacts of climate change on Blue Nile flows using bias-corrected GCM scenarios. Hydrol Earth Syst Sci 13:551–565. https://doi.org/10.5194/hess-13-551-2009

Erkossa T, Awulachew SB, Hagos F (2009a) Characterization and productivity assessment of the farming systems in the upper part of the Nile basin. Ethiop J Nat Resour 11(2):149–167

Erkossa T, Awulachew SB, Haileslassie A, Denekew A (2009b) Impacts of improving water management of smallholder agriculture in the Upper Blue Nile basin. International Water Management, CP 19 Project Workshop Proceedings

Erkossa T, Haileslassie A, Macalister C (2014) Enhancing farming system water productivity through alternative land use and water management in vertisol areas of Ethiopian Blue Nile basin (Abay). Agric Water Manage 132:120–128

Famine Early Warning Systems Network (FEWSNet) (2018) Ethiopia livelihoods zones map, January 2018 Ethiopia—livelihood zone map: Mon, 2018–01–01 | Famine Early Warning Systems Network (fews.net)

Gashaw T, Bantider A, G/Silassie H (2014) Land degradation in Ethiopia: causes, impacts and rehabilitation techniques. J Environ Earth Sci.4:9

Girma MM (2012) Potential impact of climate and land use changes on the water resources of the upper Blue Nile Basin, p 119. PhD Thesis, Freie Universität Berlin

Gebrehiwot SG, Seibert J, Gärdenäs AI, Mellander P, Bishop K (2013) Hydrological change detection using modelling: half a century of runoff from four rivers in the Blue Nile basin. Water Resour Res 49(6):1–10

Gonfa KH, Alamirew T, Melesse AM (2022) Hydro-climate variability and trend analysis in the Jemma sub-basin, upper Blue Nile River, Ethiopia. Hydrology 9:209. https://doi.org/10.3390/hydrology9120209

GYGA (2022) Global yield gap atlas. https://www.yieldgap.org/web/guest/contact-us

Hurni K, Zeleke G, Kassie M, Tegegne B, Kassawmar T, Teferi E, Moges A, Tadesse D, Ahmed M, Degu Y, Kebebew Z (2015) Economics of land degradation (ELD) Ethiopia case study. Soil degradation and sustainable land management in the rainfed agricultural areas of Ethiopia:

An assessment of the economic implications. Report for the Economics of Land Degradation Initiative, 94

Kato E, Mekonnen D, Tiruneh S, Ringler C (2019) Sustainable land management and its effects on water security and poverty: evidence from a watershed intervention program in Ethiopia. IFPRI Discussion Paper 1811. International Food Policy Research Institute (IFPRI), Washington, DC. https://doi.org/10.2499/p15738coll2.133144

Kelali K, Misker M, Kormelinck AG (2014) Sesame yields and (post-)harvest losses in Ethiopia: evidence from the field. Sesame Business Network. www.sbnethiopia.org

Kim U, Kaluarachchi JJ (2009) Climate change impacts on water resources in the upper blue Nile River Basin, Ethiopia. J Am Water Resour Assoc 45:1361–1378. https://doi.org/10.1111/j.1752-1688.2009.00369.x

Kim U, Kaluarachchi JJ, Smakhtin VU (2008) Climate change impacts on hydrology and water resources of the Upper Blue Nile River Basin, Ethiopia. Res Report 126. International Water Management Institute (IWMI), Colombo, Sri Lanka

Koch M, Cherie N (2013) June SWAT-modeling of the impact of future climate change on the hydrology and the water resources in the upper blue Nile river basin, Ethiopia. In: Proceedings of the 6th International Conference on Water Resources and Environment Research. ICWRER, vol 6, no 6, pp 488–523

Lazin R, Shen X, Moges S, Anagnostou E (2022) The Role of Renaissance dam in reducing hydrological extremes in the Upper Blue Nile basin: current and future climate scenarios. J Hydrol 616. https://doi.org/10.1016/j.jhydrol.2022.128753

Lipton M, Litchfield J, Faures JM (2003) The effects of irrigation on poverty: a framework for analysis. Water Pol 5(5–6):413–427

Malcolm H, Dixon J, Gulliver A, Gibbon D (2001) Farming systems and poverty: improving farmers' livelihoods in a changing world. FAO and World Bank, Rome, and Washington D.C.

Mengistu D, Woldeamlak B, Alessandro D, Hans-Juergen P (2021) Climate change impacts on water resources in the Upper Blue Nile (Abay) River Basin, Ethiopia. J Hydrol 592:125614

Mequanent D, Mingist M (2019) Potential impact and mitigationmeasures of pump irrigation projects on Lake Tana and itsenvirons, Ethiopia. Heliyon 5:e03052. https://doi.org/10.1016/j.heliyon.2019.e03052

MoWR (1996, 1997, 1998) Integrated River Basin Master Plan Studies, carried out during 1997–2007. GoE

Schmidt E, Tadesse F (2019) The impact of sustainable land management on household crop production in the Blue Nile Basin, Ethiopia. Land Degrad Dev 30:777–787. https://doi.org/10.1002/ldr.3266

Shenkut M (2006) Multipurpose development of the Eastern Nile, one–system inventory report on water resource related data and information Ethiopia. ENTRO: Eastern Nile Technical Regional Office, Ms Consultancy, Addis Ababa, Ethiopia

Tamene L, Amede T, Kihara J, Tibebe D, Schulz S (eds) (2017) A review of soil fertility management and crop response to fertilizer application in Ethiopia: towards development of site- and context-specific fertilizer recommendation. CIAT Publication No. 443. International Center for Tropical Agriculture (CIAT), Addis Ababa, Ethiopia, p 86. http://hdl.handle.net/10568/82996

Tibebe D, Teferi E, Bewket W and Zeleke G (2022) Climate induced water security risks on agriculture in the Abbay river basin: a review. Front. Water 4:961948. https://doi.org/10.3389/frwa.2022.961948

Water and Land Resource Center (2019) National Geospatial Database System EthioGIS-3. Database. http://www.wlrc-eth.org/

World Bank (2019) Sustainable land management project, implementation completion and results report. World Bank Report [Cited 15 July 2020]. http://documents1.worldbank.org/curated/en/470921571491240529/pdf/Ethiopia-Sustainable-Land-Management-Project.pdf

Worqlul AW, Collick AS, Rossiter DG, Langan S, Steenhuis TS (2015) Assessment of surface water irrigation potential in the Ethiopian highlands: the Lake Tana basin. Catena 129:76–85. https://doi.org/10.1016/j.catena.2015.02.020

Yalew SG, van Griensven A, Mul ML, van der Zaag P (2016) Land suitability analysis for agriculture in the Abbay basin using remote sensing, GIS and AHP techniques, model. Earth Syst Environ 2:101. https://doi.org/10.1007/s40808-016-0167-x

Zeleke G, Hurni H (2001) Implications of land use and land cover dynamics for mountain resource degradation in the Northwestern Ethiopian highlands. Mt Res Dev 21:184–191

Chapter 5
Tourism Development Potentials and Economic Contributions of the Grand Ethiopian Renaissance Dam (GERD)

Endalkachew Teshome and Amare Nega Wondirad

Abstract Ethiopia is an ancient country that possesses the highest number of UNESCO-registered objects on the continent. As a country that defended its independence, Ethiopia has maintained its uniqueness in many respects. Tourism was one of the key economic pillars until the outbreak of the pandemic. Given its enormous contribution to economic development, the government increasingly pays more attention to the sector. Ethiopia is the second-most populous country in Africa after Nigeria, with an approximate population of 120 million. The growing industry and the expansion of cities and towns raise energy needs and make power shortages a chronic problem for much-needed economic progress. To alleviate this challenge, the country launched the construction of the self-financed Grand Ethiopian Renaissance Dam (GERD) in 2011. While generating power for domestic consumption and export purposes is the fundamental objective of this mega-dam, its implications for tourism development are also profound. This paper looks at the potential for tourism development and the economic benefits of the Grand Ethiopian Reminiscence Dam (GERD). The research findings provide valuable insights into the available tourism development potential, the work that needs to be done in advance to optimize the area's resource potential, and the appropriate development model that needs to be adopted to properly develop tourism in the area and unleash its full potential.

Keywords GERD · Tourism development potential · Ecotourism · Tourism infrastructure · Water-based tourism · New tourism destination development · Poverty reduction · Ethiopian tourism

---

E. Teshome
Department of Tourism Management, College of Business and Economics, The University of Gondar, Gondar, Ethiopia
e-mail: endalkachew.teshome@uog.edu.et

A. N. Wondirad (✉)
Tourism and Heritage Department, College of Humanities and Social Sciences, United Arab Emirates University, Al Ain, United Arab Emirates
e-mail: amarewondirad@gmail.com

© The Author(s), under exclusive license to Springer Nature Switzerland AG 2025
A. Melesse et al. (eds.), *Abbay River Basin*, Springer Geography,
https://doi.org/10.1007/978-3-031-65241-7_5

Introduction

Travel and tourism have become some of the fastest-growing sectors that make profound socioeconomic and environmental contributions worldwide (UNWTO 2022a, b). In numerous developing countries, tourism has been recognized as an avenue for economic growth, job creation, and poverty alleviation (Honey and Gilpin 2009; Teshome et al. 2022a, b). The environmental protection role of tourism is also widely discussed in the literature (Cobbinah et al. 2017; Jamaliah and Powell 2018; Masud et al. 2017; Teshome et al. 2015). The unique attributes of tourism in creating a wide range of value chains and multiplier effects in developing countries attract attention from policymakers and researchers alike. Furthermore, developing countries have comparative advantages in terms of tourism resources and can benefit from nature-based and cultural attractions that are unique to specific destinations (Wondirad 2020). Internationally, the tourism sector has been a major foreign exchange earner and employer (UNWTO 2022b). For example, tourism is a leading export earner for one-third of the world's poorest countries (Honey and Gilpin 2009). A decade ago, the UNWTO projected that tourism would grow twice as fast in the developing world compared to developed nations and recommended that developing countries make substantial investments and turn their tourism sector into a key driver of economic development and ecological conservation (UNWTO 2011). It is in this vein that nearly 80% of poverty alleviation and economic development programs in Africa underscore the essential part of tourism in tackling poverty and enhancing communities' livelihoods (Rogerson 2012).

The travel and tourism sector has also become a vital economic pillar in many African countries. Particularly in some African countries, such as South Africa, Botswana, Mauritius, Kenya, Morocco, Tunisia, and Egypt, tourism constitutes a significant portion of their annual GDP and has become an important driver for economic growth and development (World Economic Forum 2021). According to the UNWTO (2019) report, Africa hosted 67 million international tourists (5% of the global share), generating approximately US$38 billion in revenue, while 72.4 million international tourists visited the continent in 2019 (UNWTO 2020). Nevertheless, despite its wide-ranging tourism resources, the overall tourism performance of the continent remains poor. The sector has been forecasted to create employment opportunities for more than 16 million people directly and indirectly in Sub-Saharan Africa in 2021 alone, demonstrating the key role of tourism in tackling poverty and thereby improving people's livelihoods in developing regions (Wondirad et al. 2021). However, in addition to being one of the hardest-hit sectors due to the unprecedented COVID-19 outbreak, the future recovery and growth of travel and tourism is going to be suppressed in emerging destinations (Connell 2021) because of the global economic downturn triggered by the pandemic and now due to the Russia–Ukraine war. To mitigate the far-reaching consequences of the pandemic on the travel and tourism sector, destination management organizations and the hospitality industry in developing countries need to formulate new and innovative mechanisms and destination development and management approaches, along with proactive stakeholder engagement (UNWTO 2022a). In particular, developing countries should carefully reform their tourism sector development and management and design short-, medium-, and long-term plans to rescue their fragile tourism sector in the face of the current global crisis (Connell 2021).

Ethiopia is an antique country located in the Horn of Africa. Ethiopia is the tenth largest country in Africa, rich and unique in culture, and diverse in nature (Asress et al. 2013). As an ancient country, Ethiopia has a long and uninterrupted history that spans thousands of years (Feseha 2012) and has retained its unique social and cultural identities, values, belief systems, and historical legacies from the destruction of colonization (Frost and Shanka 2002). Ethiopia has an extensive range of tourist destinations with immense potential for eco- and nature-based tourism development (Teshome et al. 2022a, b; Young 2012). In this respect, scholars described Ethiopia as a country that has historical and cultural accounts as well as wildlife resources in East Africa (Cotgreave 1998); a climate and scenery as pleasant as fairly nation in Southern Africa (Fazzini et al. 2015); and ethnic diversity and distinctiveness as stimulating as nations in West Africa (Sukkar 2004).

The above description affirms that Ethiopia can be a one-stop destination if it properly develops, manages, and markets its unparalleled tourism resources, and it has the potential to stand out from other African countries as it combines almost everything. However, in contrast to its massive tourism endowments, the benefits that the country is currently receiving from the tourism sector would be considered low for numerous reasons. Until the recent political upheavals in the country and the unprecedented outbreak of the coronavirus in late 2019, Ethiopian tourism had been experiencing consistent and steady growth. Prior to the outbreak of the novel coronavirus (COVID-19), tourism contributed 9.4% to Ethiopia's GDP and created 2.2 million jobs (8.3% of total employment) in the country, demonstrating the substantial impact of the sector (WTTC 2019). Based on data compiled by the World Bank, Ethiopia received 933,000 and 849,000 international tourists in 2017 and 2018, respectively (World Bank 2022). As WTTC's report also shows, in 2018, travel and tourism contributed 3.5 billion USD to the country's economy, and nearly 80% of this revenue was generated by leisure tourists (WTTC 2019).

The popularization of tourism portfolio from a tourism commission to a full ministerial level in 2005, the formulation and implementation of a new tourism policy in 2009, and progressive investment in the countrywide carrier and airport expansion are some of the factors that contribute to the growth of the sector (Wondirad 2018). Among the five principal pillars of the national tourism policy, developing existing and new tourist attractions and products is listed as a priority to improve the nation's tourism sector. Moreover, the country's 2015–2025 sustainable tourism master plan underlines the significance of product development as one of the core elements and emphasizes the key role of tourism product development in invigorating destination competitiveness and extending visitors' length of stay (UN Economic Commission for Africa 2015). Furthermore, developing countries need more tourism investment that takes the Sustainable Development Goals (SDGs) into consideration to increase the benefits of tourism and successfully satisfy the constantly changing consumer demand. In this regard, the importance of the Grand Ethiopian Renaissance Dam

(GERD) is profound, especially in terms of diversifying the tourism product base, enhancing competitiveness, redistributing the current tourist flows to this part of the country, and extending visitors' stays in the country. GERD is particularly crucial in promoting domestic and diaspora tourism given that every citizen, in the country and abroad, makes substantial moral backing and financial contributions to support the construction of this megaproject, which is destined to rejuvenate the country's overall development. In fact, beyond its wide-ranging socioeconomic and political implications, the successful completion of the GERD is a symbol of national pride for the current generation and a reassertion of the Ethiopian people's resilience.

Literature Review

Tourism Potential of the Grand Ethiopian Renaissance Dam

The literature documents that several dam reservoirs have been created in Anatolia, Babylon, Egypt, Greece, Mesopotamia, Rome, Persia, and Syria since ancient times (Angelakıs et al. 2020). The eldest reservoir in the world was constructed in 2900 BCE on the Nile River (Duda-Gromada 2012). These and other artificial reservoirs constructed in different countries were used for tourism and recreation. For example, the Hoover Dam and Grand Coulee Dam in Washington (Duchemin 2009), Alqueva Dam in Portugal (Dias-Sardinha and Ross 2015), southern Alberta (McNaughto 1994), dam reservoir in France (Germaine and Lespez 2017), and High Aswan Dam in Egypt Tol et al. 2008) can be mentioned. Moreover, dams in northeastern Thailand (Sinthusiri et al. 2014), Brno Location in the Czech Republic (Raus 2017), Poland (Duda-Gromada 2012), the Tehri Dam in India (Naithani and Saha 2019), the reservoir of Bagré Dam in Burkina Faso (Armah et al. 2010), and the Kariba Dam in Zambia (Hughes 2006) are among the human-made reservoirs with profound tourism and recreational significance. Dam reservoirs are most often anthropogenic and could be used as a natural resort, a space for leisure, relaxation, sports activities, and events, among others (Havlíˇcek et al. 2022).

They are also popular destinations for both local and international visitors, providing a wide variety of recreational opportunities, including swimming, fishing, water sports, boating, camping, and wildlife observation (Bonnet et al. 2015), and offer tax revenue and employment. For instance, the Grand Coulee Dam in Washington provides theaters, hotels, restaurants, and museums to organize activities such as projecting shows on the wall of the dam (Grcheva 2016); the reservoir of Bagré Dam in Burkina Faso is a famous natural resort promoting ecotourism, and in Zambia, around the Kariba Dam, there are approximately twenty hotels that provide service to their guests and create employment opportunities for the surrounding communities (Hughes 2006). In France, dam reservoirs raise the tourist destinations of the region and improve the visitor experience (Germaine and Lespez 2017).

Although Ethiopia is known as the *water tower* of East Africa due to the high topographic characteristics with a comparatively enormous portion of water resources such as highland fresh lakes, many rivers and high precipitation in Africa, only a fraction of this potential has been used thus far − 1% at the end of the twentieth century (Grcheva 2016). Given such a huge potential, river dam reservoirs were built at Koka in 1960, Legedadi in 1967, Tendaho in 2014, and Genale Dawa III in 2017 in the Awash basin, and Fincha in 1973, Gilgel Gibe I in 2004, Gilgel Gibe III in 2014, Koysha in 2016, Fincha in 1973, Tekeze in 2009, Amerti and Neshe in 2012 (Kruger et al. 2019), and the Grand Ethiopian Renaissance Dam (GERD) in 2016 in the Nile basin were built in the twenty-first century, with the main goals of power supply, irrigation, and water supply.

Of the above mentioned twelve larger dam reservoirs in the country, the Grand Ethiopian Renaissance Dam (GERD) is the largest in the country, with a maximum capacity of over 74 billion cubic meters. On its completion, the dam will create an artificial lake that covers an area of 172,250 km^2 for the use of hydroelectric power, commercialized fishing, and irrigation and can also provide outstanding landscape futures that revitalize wildlife composition and diversity in the area (Grcheva 2016). The scenery values created by the dam, especially in the island mountains, would be breathtaking and even unique from a tourism attraction viewpoint (Teshome et al. 2022a, b).

In addition to the provision of hydroelectric power, irrigation, and commercialized fishing services, the Grand Ethiopian Renaissance Dam creates a large human-made lake, beaches, seventy small islands, lakeshores, and wetlands (Abbay-Basin-Authority 2016) that offer endless opportunities for popular recreational and sport activities, such as bathing, walking, hiking, biking, picnicking, hunting, sport fishing, kayaking, paddling, speedboat riding, rowing, bird watching, and photography experiences. The various water-based and inland activities attract different age categories, education levels, and personal preferences of international and domestic tourists (Ajayi and Amole 2022). Among the various activities, the most universally appreciated events could be walking, bathing, picnicking, fishing, photography, and the artistic practice (Hudson 2006). These practices are consistent with many other large reservoirs found around the globe (Duchemin 2009; Duda-Gromada 2012; Hughes 2006; Tol et al. 2008).

Due to the increase of water sports enthusiasts who explore both soft and hard water-based activities, the demand for water adventure tourism is increasing globally and adventures such as canoeing, sailing, motor boating, motorized sports, surfing, water rafting, kayaking, boating and sailing adventures, scuba and free diving, and snorkeling (Verghese 2022). Therefore, the Grand Ethiopian Renaissance Dam can provide adventurous water-based activities and experiences. Indeed, the lake and the islands created as a result of the dam will become a paradise and a preferred living space for those who live in or around the area (Grcheva 2016). These activities mostly attract the youthful and adventurous crowd, on top of providing hotels and smooth cruises over the islands for those who prefer to relax (Raus 2017). However, nature trails should be inviting with a series of panoramic viewpoints and at the same time clearly show where the tourist may walk and explore different landscapes.

As more people look for water sports and travel to participate in fascinating water-based activities, the demand for aquatic adventure is rising (Verghese 2022). For these popular activities, the Grand Ethiopian Renaissance Dam can be the best destination that can absorb a large number of visitors per day since it is located in a subtropical region that provides favorable sunny and warm climatic conditions (Melese and Belda 2021). Based on carrying capacity standards developed by the United Nations World Tourism Organization for water, rural, and recreational activities, expected visitors per day per hectare-based activities were fishing/sailing 5–30, speed boating 5–10, and water skiing 100–200 (UNWTO 1983). The Grand Ethiopian Renaissance Dam water body can provide 86,125,000–516,750,000 fishing/sailing visitors, 86,125,000–17,2250,000 speed boating visitors, and 1,722,500,000–3,445,000,000 water skiing visitors per day without any carrying capacity problem. This enormous capacity meets the growing needs of water adventure tourism, a trend that is becoming more and more popular around the world. Visitors may participate in adventure sports and other activities that give an extra element of excitement to their holiday (Verghese 2022), and the revenues collected from adventure visitors can improve the lifestyle and livelihood of the indigenous communities in the area and the overall infrastructure of the region.

In addition to water-based recreational activities, the landscape created by the dam provides a variety of unique tourism resource potentials, including natural and cultural tourist attractions. The spectacular mountain ranges in and around the artificial lake can offer panoramic views and other recreational and touristic activities, such as horseback riding, hiking, trekking, climbing, bird watching, wildlife viewing, cultural experiences, and volunteering (Teshome et al. 2022a, b).

Around the Grand Ethiopian Renaissance Dam, there is a large coverage of plants, varieties of large mammals, and bird species since it is affiliated with the Dedessa natural forest and Dinder-Alatash-Bejimiz Parks complex. The dam area is also a center of endemism for Chlorophytum species of evergreen perennial flowering plants, with a description of Chlorophytum species pseudocaulesp, and three species of Chlorophytum are endemic to the Benishangul Gumuz Region (Awas and Nordal 2007). The best conditions for such endemism are likely to have been in the most topographically and geologically complex areas in the lower reaches and at the mouth of the largest river system in western Ethiopia (Demissew et al. 2005). The availability of endemic species can attract ecotourists and can spark scientific or research visits (Teshome et al. 2022a, b). People who live around the dam also have unique settlement patterns and sociocultural features, such as religious practices, ways of life, dietary habits, agricultural activities, processes of food acquisition, and marriage practices (Erkihun 2015). Archeological sites around the dam show the ancient indigenous Gumuz people's unique cultural practices and lifestyle (Fernández et al. 2007). This unique cultural diversity also attracts many culture enthusiasts across the globe (Teshome et al. 2022a, b).

The combination of water leisure sports with wildlife-based tourism and cultural attractions makes the GERD and its environs a center of gravity and an important tourism hub in the region. However, this was most often not taken into full consideration during the planning and construction phases of the dam (Abbay-Basin-Authority

2016). Indeed, experience shows that in many counties, tourism development in the dam/reservoir is frequently associated with a secondary benefit (Duda-Gromada et al. 2010; Grcheva 2016).

Economic Contribution of Water-Based Tourism Development

Water and tourism are both crucial components of a green economy and have a significant impact on economic growth (Table 5.1). Tourism is a major component of economic development in many countries that have freshwater lakes (Mangan et al. 2013). The global water adventure tourism market is projected to reach US$ 845.8 billion by 2032, up from US $156.9 billion in 2022, and income shows an increase of 16.9% during the forecast period (Verghese 2022). Thirty percent of the world's tourists head for the Mediterranean coast of warm and sunny beaches and other water-based activities (Honey and Krantz 2007).

Water-based recreations in the Mediterranean attract hundreds of millions of visitors annually, accounting for between seventy percent and eight percent of the gross national product in France, Italy, and Spain (Ksinner and Zalewski 1995). The authors added that tourism represents ten percent of the gross national product in Italy and Spain, between 12 and 15% in Malta, and in Cyprus and Israel ten percent of the workforce (Ksinner and Zalewski 1995). Such income from tourism can trigger local economic development, and this economic development leads to poverty reduction (María José Zapata et al. 2011). In 2019, the water adventure tourism industry generated approximately 0.2% of global gross domestic product (Future Market Insight 2022).

Tourism can provide significant economic benefits in terms of revenue earned, taxation raised, and jobs created, even for unskilled workers (Teshome et al. 2015). For example, an estimated 50 million people are expected to spend nearly $10 billion on the observation and photography of wetland-dependent birds each year (Yua et al. 2018). This amount of income is particularly important in rural areas (Teshome et al. 2020), where this economic sector is often an alternative for agricultural activities such as fishing, animal rearing, farming, and forestry. People who live in tourist destinations engage in various income-generating activities, such as resort and lodge investments in the seventy small islands and the provision of services such as accommodation, tourist guides, equipment rental, transporters, scouting/guard service, and vendors (Teshome et al. 2015). Apart from creating new jobs and new sources of income, tourism encourages entrepreneurship, lifestyle modernization, and the creation of new infrastructure to improve the livelihood of the community in the destination (Manzoor et al. 2019; Teshome et al. 2020). Potential economic, social, and environmental benefits of using water as a tourism resource were also discussed (Teshome et al. 2020).

Table 5.1 Review of large dam reservoir economic contribution

Author	Year	Title	Results
Bonnet et al.	2015	The economic benefits of multipurpose reservoirs in the United States-Federal Hydropower Fleet	Common spending amounts range from $10 per visitor per day for local visitors participating in day-use activities like bird watching or hiking, to $40 per visitor per day or more for nonlocal visitors participating in leisure or multiday recreation activities
John Loomis	2002	Quantifying recreation use values from removing dams and restoring free-flowing rivers: a contingent behavior, travel, cost, and demand model for the Lower Snake River	The river recreation uses value estimates of $192–310 million are 6–10 times larger than the current reservoir recreation benefits ($31.6 million)
Adamowicz, Louviere, and Williams	1992	Little bow river project/ Highwood River Diversion plan: impacts on recreation	Consumer surplus ranges from $0.46 to $3.99 per trip, depending on distance traveled
Kulshreshtha	1991	Estimated water value for water-related recreation in Saskatchewan	Mean daily consumers' surplus of $3.90 with a range from $0.71–$14.67. Short run value of water of $0.83–$770/dam3. Long run value of $0–$151.69/dam3
UMA engineering	1989	Demand for water based recreation in the South Saskatchewan River basin	Daily consumers' surplus is in the range of $1.46–$8.18. Total consumer surplus of $20,000–$458,000
Thompson, Sen and Scace	1987	Bow River Recreation Study	Mean expenditures of $22.14 per day, and a mean consumer surplus of $7.61 per day. Total annual expenditures of $2.2 million and total consumers' surplus of $0.85 million
O'Grady, Kulshreshtha and Brockman	1978	The value of water based recreation Saskatchewan	Daily willingness to pay value in the range of $1.82–$10.19

Source Loomis (2002), McNaughto (1994)

Methodology

To successfully accomplish its objective, this study uses a qualitative research methodology. Qualitative research has become a well-established research approach to investigate phenomena in the social sciences, including tourism, due to its capability to provide rich and in-depth information from multiple viewpoints and its appropriateness to understand the "how" and "why" questions that cannot be dealt with a quantitative approach (Jennings 2012; Lune and Berg 2017). As Jennings (2012, p. 309) further elaborated, in contrast to quantitative research, qualitative research subscribes to an emic approach that "acknowledges the subjectivity of

Table 5.2 Categories of key informants

No	Participant category	Number of participants
1	Tourism industry players	6
2	Government entities	3
3	Tourism professionals	3
4	Local communities	3
	Total	15

Source Researchers (2022)

the researcher within research processes as well as shapes research to include the researcher as a subjectively embedded being in the research." In contrast, quantitative research adopts etic approaches to research and tends to frame the entire research process from a natural stance with an outsider perspective (Wondirad et al. 2020).

Purposive sampling was used in this study, and sample sizes are frequently chosen in accordance with the concept of theoretical saturation, which refers to the point at which further data cannot be collected without contributing new information to the research topics. The parameters we select allow us to concentrate on those we believe would be most likely to have experience with, knowledge about, or insights into the research issue. Overall, we recruited 15 key participants who represented the government, the tourism sector, tourism professionals, and local communities (Table 5.2).

The authors used a combination of in-depth key informant interviews and literature reviews (secondary data) to carry out this pioneering study successfully. Additionally, the authors had the chance to travel to the GERD and its surroundings and gain first-hand information through personal observations. The observations of the researchers are essential for confirming the information gathered through key informant interviews and secondary data sources.

We contend that using a qualitative research approach is highly justified because this study is the first of its kind as a scientific research effort regarding the relationship between GERD and its role in enhancing tourism development (Easterby-Smith et al. 2015). Furthermore, given the condition of the study objectives outlined in the introduction, a qualitative research approach appears to be more appropriate, and we decided to employ this approach to examine the multifaceted roles of GERD in revitalizing the country's tourism sector. As a result, a qualitative research approach enhances the scientific rigor and credibility of research findings (Maxwell 2012) by enabling the in-depth extraction of information using a variety of data collection techniques, such as in-depth interviews and field observations (Wondirad 2020). Thematic analysis was used to present qualitative data.

Results and Discussion

This section of the paper presents the results and discussion in reconciliation with the literature. As the only non-colonized nation and the melting pot of diverse ethnic groups and tribes, Ethiopia currently possesses the highest number of UNESCO World Heritage properties on the entire continent (Wondirad 2018). Furthermore, as a country that hosts different religious practices, Ethiopia is featured as a symbol of religious tolerance and peaceful coexistence (Østebø et al. 2021). In terms of wildlife tourism resources, due to its varied ecological zones, Ethiopia has a wide range of wildlife species, several of which are endemic to the country (Amare 2015; Berhanu and Teshome 2018). The potential for tourism within Ethiopia has increased significantly over the previous two years as a vital component of the government's national reform strategy, with various natural assets created to supplement the already existing historical and cultural legacy (Ahmed 2022). The Grand Ethiopian Renaissance Dam's favorable climatic conditions, freshwater, and topographic and geographic features create diverse and outstanding tourist attractions that appeal to many interests. To this end, a participant who represents the tourism sector elaborates that:

> In my opinion, GERD has huge potential for the development of sustainable tourism given its multifaceted economic, environmental, sociocultural, political, and historical importance. This mega-dam can also be a magnet for both domestic recreational purposes and international tourist attractions, given the international nature of the Nile River. Furthermore, the dam could be an ideal spot for educational trips and researchers interested in hydroelectric dams, engineering, and marine biodiversity. (Tourism industry representative, August 2022)

The attractions include the astonishing nature of the waterscape and landscape, pleasing weather conditions, floral and aquatic and terrestrial faunal resources, and cultural, anthropological, and archeological resources. It's rare to find such a distinctive assortment of attractions in one location, both globally and on the African continent (Tilahun 2021). The majority of what makes the dam significant internationally and potentially attractive to tourists from across the world is the spectacular beauty and geography, as well as the well-known name of the Blue Nile River, the longest river on Earth. However, visitor arrivals are determined by not only natural attractions and cultural attractions but by safety, security, infrastructure, facilities, and the overall perception of the destination. The national capacity of tourism-specific security personnel, including community policing, rangers, scouts, and other officers, needs to be upgraded in terms of logistical facilities and a variety of skill development programs, despite the fact that tourist safety and security is part of the overall national home security system. Therefore, to attract international and domestic visitors' standard infrastructure, safety, security, facilities, and amenities shall be well developed ahead of time to properly cater to the needs of visitors, as highlighted in the following excerpt by an academician echoing the sentiments of other participants.

> The first and foremost task to be done is infrastructural and facility development, such as roads, accommodation facilities, health centers, and communication networks with high security levels for both tourists and the dam. To ensure this, a coordinated effort from local,

regional, and national institutions is instrumental. Moreover, before tourism is advanced in the area, environmental sustainability and ecosystem service indicators should be identified to guarantee ecologically friendly tourism in this tranquil and wilderness area. Furthermore, to optimize the recreational value of the dam and its surroundings, well thought out and meticulous destination planning that integrates tourism with other economic activities in the corridor is paramount. (Tourism professionals' representative, August 2022)

The huge artificial lake, vast wetlands, and seventy stunning islands created by the Grand Ethiopian Renaissance Dam can provide a fascinating landscape for sports activities such as bathing, walking, hiking, biking, picnicking, hunting, sport fishing, kayaking, paddling, speedboat riding, rowing, bird watching, and photography. Along with the great potential for these multiple visitor activities, the Grand Ethiopian Renaissance Dam can be one of the resort towns planned by the Federal Government to attract seven million tourists to the country by 2030.

The surroundings of GERD are endowed with numerous bird species that can attract ornithologists. The geographical distribution of these bird species is restricted only around the dam and the western border of Ethiopia. Among the various bird species, the following can be mentioned: black-headed bush shrike (*Laniarius erythrogaster*), red-eyed puff back (*Dryoscopus senegalisis*), white-rumped serin (*Serinus leucopygius*), yellow-crowned canary (*Serinus canicollis*), thick-billed serin (*Seritus burtoni*), red-winged pytilia (*Pytilia phoenicoptera*), and black-faced fire finch (*Logonosticta vinacea*). Furthermore, splendid glossy starling (*Lamprotomis splendidus*), yellow-billed ox pecker (*Buphagus africanus*), olive sunbird (*Nectarinia olivacea*), brown babbler (*Turdoides plebejus*), moustached warbler (*Melocichla mentalis*), red-chested swallow (*Hirundo lucida*), bar-tailed trogon (*Apoloderma narina*), long-tailed nightjar (*Coprimulgus climacurus*), African green pigeon (Treno colua), black-crowned crane (balearica pavonina), Denham's bustard (*Neotis denhami*), and stone partridge (*Pilopachus petrosus*) have strong potential to attract bird watchers and create research opportunities for ornithologists. Cormorant (*Phalacrocoraxm carbo*), egret (*Mesophoyx intermedia*), fulvous whistling-duck (*Dendrocygna bicolor*), open-billed stork (*Anastomus lamelligerus*), sacred ibis (*Threskiornis aethiopicus*), common crane (*Grus grus*), and wattled crane (*Grus carunculatus*) are among the frequent marsh bird species seen during bird-watching activities. Although the dam is located not far from different transponder protected areas, such as the Dinder-Alatash-Bejimiz National Park complex, approximately forty species of mammals, including the Patas monkey (*Erythrocebus patas*), Anubis baboon (*Papio Anubis*), elephant (*Loxodonta Africana*), giraffe (*Giraffa camelopardalis*), hippopotamus (*Hippopotamidae*), buffalo (*Syncerus caffer*), roan antelope (*Hippotragus equinus*), and hartebeest (*Alcelaphus buselaphus*), have been recorded (Friis 1992). This demonstrates that the dam is blessed with a variety of regionally limited bird species to draw ornithologists and bird watchers, as well as diverse fauna for ecotourism operations involving wildlife viewing.

In accordance with the literature Manzoor et al. (2019), Nicholas and Odhiambo (2021), Teshome et al. (2020), UN Economic Commission for Africa (2015), if Ethiopia's tourism development is managed responsibly and sustainably, it can significantly contribute to the development of the nation. In fact, at the global level, the tourism sector is strongly linked to various important pillars of sustainable development and is regarded as a key sector for achieving the United Nations Sustainable Development Goals (SDGs). In particular, tourism is directly connected to goals 8, 12, and 14 of the SDGs, with the potential to contribute to several others (Scheyvens and Cheer 2022).

The role of GERD in this regard is crucial since the tourist industry is one that largely depends on a number of facilities, infrastructure, and support services for its growth and development, including transportation infrastructure, ICT, and energy, among others. Accommodations (tourist courts, tourist cottages, tourist houses, trailer parks, trailer courts, motels, motor hotels, hotels, resorts, and lodges), dining establishments, and travel and tour operating services are all considered part of the tourism infrastructure amenities such retail centers, museums, parks, theaters, information centers, health services or care, emergency, postal, financial, and personal services for tourists.. To use the immense resource potential, the country needs to develop its tourism infrastructure in parallel with promoting its untapped tourism capacity to the world using all available means of communication and media outlets, including the internet.

As the literature underlines (Tamene and Wondirad 2019; Teshome et al. 2020), the travel and tourism sector, especially in developing countries, is dominated by micro, small, and medium enterprises (MSMEs). Therefore, as research participants commonly underscore, the realization of the Grand Ethiopian Renaissance Dam will be a stimulating factor for the proliferation of various MSMEs along the route. This adds another tourism route to the existing five tourism routes in the country and the development of multiple spots along the way (Tafesse 2016). The economic implications of these spots and MSMEs are profound in multiple facets, such as employment creation for local residents, elongating tourist stays in the region, and distributing tourists to this part of the country. In this respect, a participant who represents local communities expresses his views as follows:

> From economic aspects, the GERD will make massive contributions through various touristic services, such as hotels and resorts, transportation provision, tour operating and guiding companies, attracting tourism-related investments, and many others. That, in turn, improves the livelihood of local communities, increases the revenue for the government through taxes and entrance fees, creates permanent and seasonal employment opportunities, especially for women and youths, and helps to diversify the economic base in the area. (Local community representative, September 2022).

Such income from tourism services can trigger local economic development, and this economic development leads to poverty reduction in local communities that live in and around the GERD and the country. This finding is in line with previous research output which indicates that tourism has been forecasted to create employment opportunities for more than 16 million people directly and indirectly in Sub-Saharan Africa in 2021, demonstrating the key role of tourism in tackling poverty and thereby improving people's livelihoods in developing regions (Wondirad et al. 2021).

Conclusions

This chapter addresses an immediate research gap by establishing empirical evidence regarding the multiple implications of the Grand Ethiopian Renaissance Dam (GERD) for tourism development. To that end, this piece of work will be a foundation for future research endeavors in numerous frontiers pertaining to GERD and its multifarious developmental implications at both the regional and national levels. As unanimously pointed out by key informants, the GERD has enormous natural tourism resources with massive potential to develop, especially eco- and nature-based tourism, which has wide-ranging implications both regionally and nationally. These potentials include unique and abundant wildlife, beaches, beautiful waterscapes, landscapes, pleasing climatic conditions, and cultural, anthropological, and archeological significance in addition to providing a fascinating place for sports activities, such as bathing, walking, hiking, biking, picnicking, hunting, sport fishing, kayaking, paddling, speedboat riding, rowing, bird watching, filming, and photography experiences.

As far as the potential tourist attractions are concerned, the GRED, coupled with the unique landscape and the creation of 70 islands, could make this place unique and one of the top tourist destinations in the country. The possible economic contribution of the dam includes participation in small- and medium-scale tourism enterprises such as bed and breakfast, cafeterias, coffee shops, bars, accommodations, traditional markets, and petrol stations; employment in government offices; handicrafts production and distribution; and providing various services such as tour guides, equipment rentals, and transportation.

In conclusion, although the main purpose of the GERD is to generate much-needed electric power to satisfy the exponentially growing energy consumption in the country, this iconic project has a multitude of implications for Ethiopia, including tourism development. The dam itself and the water body created, along with the islands within the dam reservoir, will be the center of gravity for a broad range of tourism typologies, such as ecotourism, scientific/research tourism, diaspora tourism, ecotourism, film tourism, and marine tourism, inter alia. Therefore, the tourism development endeavor in this remarkable and multidimensional destination should be guided by the principles of sustainable tourism development. Sustainable tourism development stresses the significance of maintaining biodiversity, recognizing the aesthetic values of the natural and built environment, celebrating local cultures, livelihoods, and customs, participating in local communities in the entire development process, ensuring equitable distribution of the costs and benefits of tourism, and maintaining the carrying capacity of destinations, among other things.

Recommendations

Several recommendations were provided by research participants regarding the tourism development approach given the unique nature of the GERD as well as what must be done ahead to fully utilize the potential that might arise from the completion of this megaproject. In this respect, participants highlighted the vitality of the adoption of a sustainable tourism development model that pays adequate attention to ecological preservation, cultural conservation, economic improvement, and social advancement. All stakeholders in tourism and other relevant fields should prepare well in advance and try to put all the required facilities and infrastructure on the ground to fully tap the immense potential that this national flagship project brings to the area and the country alike. Infrastructures such as roads, hiking trails, biking routes, viewpoints, and walking routes should be carefully developed in a manner that can support future tourist or consumer demand both in terms of quality and in terms of capacity.

Regarding economic benefits, numerous opportunities can be created, and to capitalize on these opportunities, the government, in collaboration with nongovernmental organizations and private sector financial institutions, should create a conducive environment for local communities so that they can smoothly enter the tourism market and be active economic players. Entrepreneurial skill-based training and capacity-building schemes are also equally important to bring the surrounding communities, who have zero experience in tourism, up to the required level of participation. Moreover, successive awareness-creation campaigns regarding the available economic and livelihood opportunities to local communities in and around the GERD are extremely important. Furthermore, to make GERD and its environs more visible on tourism maps of the country and unleash its full potential, a coordinated marketing effort using various media outlets, including digital technology, is crucial. From a future research standpoint, the researchers strongly underline the significance of developing tourism in this region with the consistent support of relevant academic research that informs policies and various decisions pertinent to visitor management, natural resource management, cultural preservation, and community participation.

References

Abbay-Basin-Authority (2016) The great Ethiopian Rennaisance Dam. Abbay Basin Authority, High Council Secretariat and International Relations Directorate, BahirDar

Ahmed A (2022) Welcoming speech by Ethiopian Prime Minister Abiy Ahmed at the opening of the 35th ordinary session of the AU assembly [Press release]

Ajayi AO, Amole OO (2022) Open spaces and wellbeing: the impact of outdoor environments in promoting health. Cities Health. https://doi.org/10.1080/23748834.2021.2011537

Amare A (2015) Wildlife resources of Ethiopia: opportunities, challenges and future directions: from ecotourism perspective: a review paper. Nat Resour 6(6):405

Angelakıs AN, Zaccaria D, Krasilnikoff J, Salgot M, Bazza M, Roccaro P, Fereres E (2020) Irrigation of world agricultural lands: evolution through the millennia. Water 12(85):1–50. https://doi.org/10.3390/w12051285

Armah FA, Yawson DO, Yengoh GT, Odoi JO, Afrifa EKA (2010) Impact of floods on livelihoods and vulnerability of natural resource dependent communities in northern Ghana. Water 2:120–139

Asress MB, Simonovic A, Komarov D, Stupar S (2013) Wind energy resource development in Ethiopia as an alternative energy future beyond the dominant hydropower. Renew Sustain Energy Rev 23:366–378

Awas T, Nordal I (2007) Benishangul Gumuz region in Ethiopia: a centre of endemism for chlorophytum—with a description of C. pseudocaulesp. nov. (Anthericaceae). Kew Bull 62

Berhanu K, Teshome E (2018) Opportunities and challenges for wildlife conservation: the case of Alatish National Park, Northwest Ethiopia. Afr J Hospitality Tourism Leisure 7(1):1–13

Bonnet M, Witt A, Stewart K, Hadjerioua B, Mobley M (2015) The economic benefits of multipurpose reservoirs in The United States-federal hydropower fleet *environmental sciences division*. Oak Ridge National Laboratory, U.S. Department of Energy Wind and Water Program

Cobbinah PB, Amenuvor D, Black R, Peprah C (2017) Ecotourism in the Kakum Conservation Area, Ghana: local politics, practice and outcome. J Outdoor Recreat Tour 20:34–44

Connell J (2021) COVID-19 and tourism in Pacific SIDS: lessons from Fiji, Vanuatu and Samoa? Round Table 110(1):149–158

Cotgreave P (1998) Helping Ethiopia's Wildlife. J Zool 245(1):119–120. https://doi.org/10.1111/j.1469-7998.1998.tb00079

Demissew S, Nordal I, Herrmann C, Friis I, Awas T, Stabbetorp O (2005) Diversity and endemism of the western Ethiopian escarpment a preliminary comparison with other areas of the Horn of Africa. Biol Skr 55:315–330

Dias-Sardinha I, Ross D (2015) Perceived impact of the Alqueva dam on regional tourism development. Tourism Plann Dev 12(3):362–375. https://doi.org/10.1080/21568316.2014.988880

Duchemin M (2009) Water, power, and tourism: Hoover Dam and the making of the New West. Univ Calif Press Assoc Calif Hist Soc 86(4):60–89

Duda-Gromada K (2012) Use of river reservoirs for tourism and recreation case study: Solińskie lake in Poland. Pol J Nat Sci 27(4):367–376

Duda-Gromada K, Bujdosó Z, David L (2010) Lakes, reservoirs and regional development through some examples in Poland and Hungary. GeoJournal Tourism Geosites 5(1).

Easterby-Smith M, Thorpe R, Jackson PR (2015) Management and business research. Sage Publications Ltd., London

Erkihun A (2015) Mapping the socio-cultural landscape of the Gumuz community of Metekel, Northwestern Ethiopia. Afr J Hist Cult 7(12):209–218. https://doi.org/10.5897/AJHC2015.0251

Fazzini M, Bisci C, Billi P (2015) *The climate of Ethiopia*: world geomorphological landscapes. Springer, Dordrecht

Fernández VM, Torre I, Luque L, González-Ruibal A, López-Sáez JA (2007) A late stone age sequence from West Ethiopia: the sites of K'aaba and Bel K'urk'umu (Assosa, Benishangul-Gumuz Regional State). J Afr Archaeol 5(1):91–126. https://doi.org/10.3213/1612-1651-10087

Feseha M (2012) The fundamentals of community based ecotourism development in Ethiopia: Mulugeta Feseha. Addis Ababa University, Addis Ababa

Friis I (1992) Forests and forest trees of Northeast Tropical Africa, vol XV (KEW BULLETIN). HMSO, UK, pp 29–41

Frost FA, Shanka T (2002) Regionalism in tourism-the case for Kenya and Ethiopia. J Travel Tour Mark 11(1):35–58

Future Market Insight (2022) Water adventure tourism market. Water adventure tourism market outlook (2022–2032). Retrieved on 28 Sept 2022

Germaine MA, Lespez L (2017) The failure of the largest project to dismantle hydroelectric dams in Europe? (Sélune River, France, 2009–2017). Water Altern 10(3):655–676

Grcheva I (2016) Water urbanism in transboundary regions: the Nile basin and the Grand Ethiopian Renaissance Dam (Degree of Master of Science in Human Settlements), KU Leuven, Heverlee (Belgium)

Havlíˇcek M, Dostál I, Pavelková R (2022) Water reservoirs as a driver of anthropogenic changes in landscape and transport networks: the Czech Republic experience. Water 14(1870):1–18. https://doi.org/10.3390/w14121870

Honey M, Krantz D (2007) Global trends in coastal tourism. Center on Ecotourism and Sustainable Development

Honey M, Gilpin R (2009) Tourism in the developing world: promoting peace and reducing poverty, vol 23. United States Institute of Peace

Hudson BJ (2006) Waterfalls, tourism and landscape. Geography 91(1):3–12. https://doi.org/10.1080/00167487.2006.12094145

Hughes DM (2006) Whites and water: how Euro-Africans made nature at Kariba Dam, heritage in Southern Africa. J South Afr Stud 32(4):823–838

Jamaliah MM, Powell RB (2018) Ecotourism resilience to climate change in Dana Biosphere Reserve, Jordan. J Sustain Tourism 26(4):519–536

Jennings GR (2012) Qualitative research methods. Edward Elgar Publishing Limited, Cheltenham

Kruger W, Stuurman F, Alao O (2019) Ethiopia country report *energy and economic growth research programme (W01 and W05) PO number*, vol 5. Power Futures Lab., Energy and Economic Growth, pp 1–41

Ksinner J, Zalewski J (1995) Functions and values of Mediterranean wetlands. In: Skinner J, Crivelli AJ (eds) Conservation of Mededitrranean Wtlands MedWet. France, pp 1–81

Loomis J (2002) Quantifying recreation use values from removing dams and restoring free-flowing rivers: a contingent behavior travel cost demand model for the Lower Snake River. Water Resour Res 38(6):21–28. https://doi.org/10.1029/2000WR000136

Lune H, Berg BL (2017) Qualitative research methods for the social sciences. Pearson

Mangan T, Brouwer R, Das Lohano H, Nangraj GH (2013) Estimating the recreational value of Pakistan's largest freshwater lake to support sustainable tourism management using a travel cost model. J Sustain Tour 21(3):473–486. https://doi.org/10.1080/09669582.2012.708040

Manzoor F, Wei L, Asif M, Zia ul Haq M, ur Rehman H (2019) The contribution of sustainable tourism to economic growth and employment in Pakistan. Int J Environ Res Public Health 16(19):3785

María José Zapata C, Hall M, Lindo P, Vanderschaeghe M (2011) Can community-based tourism contribute to development and poverty alleviation? Lessons from Nicaragua. Curr Issues Tourism 14(8):725–749

Masud MM, Aldakhil AM, Nassani AA, Azam MN (2017) Community-based ecotourism management for sustainable development of marine protected areas in Malaysia. Ocean Coast Manage 136:104–112

Maxwell JA (2012) Qualitative research design. Sage publications

McNaughto RB (1994) Economic benefits of recreation sites on irrigation reservoirs in Southern Alberta. Can Water Resour J 19(1):3–16

Melese KB, Belda TH (2021) Determinants of tourism product development in Southeast Ethiopia: marketing perspectives. Sustainability 13(23):13263

Naithani S, Saha AK (2019) Changing landscape and ecotourism development in a large dam site: a case study of Tehri dam, India. Asia Pac J Tourism Res 24(3):193–205

Nicholas M, Odhiambo (2021) Tourism development and poverty alleviation in sub-Saharan African countries: an empirical investigation. Dev Stud Res 18(1):396–406

Østebø T, Tronvoll K, Østebø MT (2021) Religion and the 'Secular shadow': responses to covid-19 in Ethiopia. Religion 51(3):339–358

Raus L (2017) Tourism development in the area of the Brno dam. Brno University (MA)

Rogerson CM (2012) The tourism-development nexus in sub-Saharan Africa-progress and prospects. Afr Insight 42(2):28–45

Scheyvens R, Cheer JM (2022) Tourism, the SDGs and partnerships. J Sustain Tour 30(10):2271–2281

Sinthusiri O, Yodmalee B, Laoakka S (2014) Culture and ecotourism management of dams in North-eastern Thailand. Asian Cult Hist 6(2):205–214. https://doi.org/10.5539/ach.v6n2p205

Sukkar YA (2004) Some reflections on tourism in the less developed countries-the case of Ethiopia. In: A presentation at forum of Barcelona

Tafesse A (2016) The historic route in Ethiopian Tourism development. Afr J Hospitality Tourism Leisure 5(2):1–13

Tamene K, Wondirad A (2019) Economic impacts of tourism on small-scale tourism enterprises (SSTEs) in Hawassa City, Southern Ethiopia. Int J Tourism Sci 19(1):38–55

Teshome E, Worku A, Astary M (2015) Community based ecotourism as a tool for biodiversity conservation in Wunania-Kosoye natural attraction site, Ethiopia. Ethiop Renaissance J Soc Sci Humanity (ERJSSH) 2(1):43–55

Teshome E, Ayalew G, Workie A (2020) Community based ecotourism development in Meqdela Amba, Ethiopia: current development barriers. J Hospitality Manag Tourism 11(2):12–20

Teshome E, Aberaw G, Tesgera D, Abebe F (2022a) The untold tourism potentials of Bela Mountain, for community-based-ecotourism development, ecosystem conservation and livelihood improvement, Waghimera Zone, Ethiopia. Environ Dev Sustain 1–22

Teshome E, Dereje M, Asfaw Y (2022b) Potentials, challenges and economic contributions of tourism resources in the South Achefer district, Ethiopia. Cogent Soc Sci 8(1):2041290

Tilahun M (2021) Review impending of ecotourism for sustainable development in Ethiopia. J Earth Sci Clim Change 12:590

Tol RSJ, Strzepek K, Yohe G, Rosegrant MW (2008) The value of the high Aswan Dam to the Egyptian economy. Ecol Econ 166:117–126. https://doi.org/10.1016/j.ecolecon.2007.08.019

UN Economic Commission for Africa (2015) Sustainable tourism master plan 2015–2025 the Federal Democratic Republic of Ethiopia. Addis Ababa. Retrieved on 05 July 2022 from https://hdl.handle.net/10855/23486

UNWTO (1983) Risks of saturation or tourist carrying capacity overload in holiday destinations. World Tourism Ornanization, Madrid

UNWTO (2019) International tourism highlights. Retrieved from https://www.eunwto.org/doi/pdf/10.18111/9789284421152

UNWTO (2020) International tourism highlights. Retrieved from https://www.eunwto.org/doi/epdf/10.18111/9789284422456/

UNWTO (2022a) International tourism consolidates strong recovery amidst growing challenges. Retrieved on 08 Dec 2022 from https://www.unwto.org/news/international-tourism-consolidates-strong-recovery-admidst-growing-challenges

UNWTO (2022b) World tourism barometer, vol 20

UNWTO (2011) Tourism towards 2030. UNWTO, Madrid

Verghese S (2022) Future Market Insights is registered in the state of Delaware as future market insights. Water adventure tourism market outlook (2022–2032), 8 Oct 2022

Wondirad A (2018) Stakeholder collaboration for sustainable ecotourism development in Southern Ethiopia. Ph.D. thesis, the Hong Kong Polytechnic University

Wondirad A (2020) Saving tourism is more than saving lives; when tourism dies, billions go hungry, and nature will be under threat! Retrieved on 12 July 2022 from https://www.press.et/english/?p=24108&fbclid=IwAR3BKNFz_t0iWhtdEwz25zZWm9EUsj4HI8eI08nuicpUddtHqSdvqeBBuc [Press release]

Wondirad A, Tolkach D, King B (2020) Stakeholder collaboration as a major factor for sustainable ecotourism development in developing countries. Tourism Manag 78:104024

Wondirad A, Kebete Y, Li Y (2021) Culinary tourism as a driver of regional economic development and socio-cultural revitalization: evidence from Amhara National Regional State, Ethiopia. J Destin Mark Manage 19:100482

World Bank (2022) International tourism, number of arrivals—Ethiopia. Retrieved on 02 July 2022 from https://data.worldbank.org/indicator/ST.INT.ARVL?locations=ET

World Economic Forum (2021) Travel & Tourism Development Index 2021. Rebuilding for a sustainable and resilient future. Retrieved on 11 July 2022 from https://www3.weforum.org/docs/WEF_Travel_Tourism_Development_2021.pdf

WTTC (2019) Ethiopia records the biggest growth in world travel and tourism. https://www.ethioembassy.org.uk/ethiopia-records-biggest-growth-in-world-travel-and-tourism/. Accessed 13 July 2022

Young J (2012) Ethiopian protected areas: a "snapshot". A reference guide for future strategic planning and project funding, vol 46

Yua X, Mingju E, Suna M, Xue Z, Lu X, Jiang M, Zou Y (2018) Wetland recreational agriculture: balancing wetland conservation and agro-development. Environ Sci Policy 87:11–17

Part III
Soils, Forest and Water Resources of the Basin

Chapter 6
Soils of the Abbay Basin

Ashenafi Ali, Wondwosen Tena, Assefa Abegaz, Teklu Erkossa, Kiflu Gudeta, Degefe Tibebe, Berhan Gessese, Wuletawu Abera, Terefe Mekete, Amsalu Tilaye, and Lulseged Tamene

Abstract Soil resource spatial information including physical and chemical characteristics is required for various applications. However, there have been limitations on the availability of up-to-date, complete and reliable basin-wide soil spatial information for the Abbay basin. Soil data for the basin have come from obsolete sources. Even the recent information generated by soil profile-based studies is fragmented and hardly accessible or require harmonization, processing, and synthesis to meet coherent and up-to-date basin-wide soil data demands. This chapter presents a reviewed, synthesized, and updated wall-to-wall soil spatial information and knowledge of the basin. In addition, existing soil spatial information and legacy topsoil properties and nutrient mapping studies were reviewed and synthesized. This is believed to have tremendous potential for targeting soil and land management options and boosting agricultural productivity. The synthesized soil spatial information in this chapter will lay a foundation for developing content complete 3D gridded soil spatial information compatible with modeling applications in the Abbay basin.

Keywords Spatial soil information · Subbasins · Agroecological zones · Geolandforms · Reference soil group · Soil properties

---

A. Ali (✉) · A. Abegaz · D. Tibebe
Department of Geography and Environmental Studies, Addis Ababa University, Addis Ababa, Ethiopia
e-mail: ashenafi.ali@aau.edu.et

W. Tena
Department of Plant Biology and Biodiversity Management, Addis Ababa University, Addis Ababa, Ethiopia

T. Erkossa
Duetsche Gesselschaft Fuer Internationale Zusammenarbeit, GIZ, Addis Ababa, Ethiopia

K. Gudeta · T. Mekete
Soil Resource Development, Ministry of Agriculture, Addis Ababa, Ethiopia

D. Tibebe · W. Abera · A. Tilaye · L. Tamene
Alliance of Bioversity International and CIAT, Addis Ababa, Ethiopia

B. Gessese
Departement of Remote Sensing, Space Science and Geospatial Institute, Addis Ababa, Ethiopia

© The Author(s), under exclusive license to Springer Nature Switzerland AG 2025
A. Melesse et al. (eds.), *Abbay River Basin*, Springer Geography,
https://doi.org/10.1007/978-3-031-65241-7_6

Introduction

Information on soil's physical, chemical and biological characteristics is fundamental for strategic planning of land uses and managements and sustained agricultural production (Gebreselassie et al. 2014; Dinssa and Elias 2021). However, geospatially explicit information on soil-landscape resources is either lacking or only partially available in many areas of Ethiopia (Leenaars et al. 2020a; Ali et al. 2022). Since there have been restrictions on the availability of reliable basin-wide soil spatial information, only some of the data that is accessible has come from a variety of sources in the Abbay basin (BCEOM 1998; Aster and Seleshi 2009; Erkossa et al. 2009; WLRC 2016a).

A 3D gridded quantitative spatial soil data is crucial for various modeling applications. In addition, various earth systems (soil carbon dynamics, greenhouse gas emission, climate change) and crop models require more soil properties including soil hydraulic properties, than those commonly made from legacy soil datasets (Teferi 2015; Dile et al. 2018; Han et al. 2019; WLRC 2019). However, most of the legacy soil data from the basin fail to meet such requirements. Simple pedotransfer functions (PTFs) can be used to derive soil hydraulic properties from easy-to-measure parameter such as soil texture and organic matter content. However, soils of tropical regions, including the Abbay basin, are largely underrepresented in the development of the widely used PTFs (Wösten et al. 2013; Teshome and Kibret 2015; Teferi 2015).

In the basin and national studies layering the basin, many different approaches were used in digital soil spatial predictions (Abegaz et al. 2016, 2020; Leenaars et al. 2016; ATA 2016, 2018, 2019; Leenaars et al. 2020a, b; Takele and Iticha 2020; Abera et al. 2021; Abegaz et al. 2022; Ali et al. 2022; Yeneneh et al. 2022). In the landscapes of the basin, simultaneous use of proximal and remote sensing products has been employed to estimate several soil parameters, including soil texture (Taye et al. 2018; Tiruneh et al. 2022) soil functional characteristics employing Landsat and SRTM-based methods (Vagen et al. 2013), soil moisture using MODIS products (Gebretsadik 2014), and radar-based soil moisture monitoring and residual soil moisture retrieval (Getachew et al. 2019, 2020).

In parallel, there have been fragmented soil spatial information sources from subbasin and area-based soil resource studies, topsoil properties, and fertility assessment missions (ORLEPB 2014a, 2014b; Alem et al. 2015; Adugna and Abegaz 2016; WLRC 2016b; Getahun and Selassie 2017; Deressa et al. 2018; OIDA 2018; ORLEPB 2011; Assefa et al. 2020; Deressa et al. 2020; Desta et al. 2021; Fekadu 2020; Dinssa and Elias 2021). These studies generated a large soil profile dataset along with soil maps mainly using a conventional approach. With this method, modal soil profiles are used to generate polygon-based qualitative soil spatial information. This soil information requires harmonization processing and synthesis to generate coherent, up-to-date, reliable, and spatially complete basin-wide soil spatial information.

This chapter presented a review and synthesis of recent studies conducted across the basin. Accordingly, the existing knowledge on wall-to-wall soil spatial information of the basin is updated. In addition, fragmented soil's physicochemical characteristics and soil nutrient mapping studies on the Abbay basin were compiled, interpreted, and synthesized based on the newly developed agroecological zones and landforms of the basin.

Description of Soil Forming Factors of the Basin

Relief

Topographic features such as slope gradients, length, positions, and altitude are among the major differentiating factors for soil types and their distributions (Ali et al. 2022; Regassa et al. 2023a, b, c). The Abbay Basin is characterized by rugged topography and a great elevation difference of approximately 3743 m, between the highest peak (4243 m) and lowest point (about 500 m a.s.l.) (Aster and Seleshi 2009; Ludwin et al. 2013; WLRC 2019; Tibebe et al. 2022). Since Abbay basin lacks extensive wall-to-wall topographic feature characterization in the form of geolandforms, we reformulated the geolandforms of the Abbay basin after Leenaars (2019), who made a 50-m spatial resolution geolandform characterization of Ethiopia using a combination of 30 m SRTM-DEM derived slope classes, topographic wetness index, and relief intensity.

Our analysis and aggregation identified five physiographic units with their associated geolandforms (Fig. 6.1; Table 6.1). The major units were midlands (49%) and highlands (29%) with flat, slope, and summit geolandforms. Lowland plains with flat and slope geolandform, and complex units with bottom geolandforms cover about 10 and 6% of the basin, respectively. Approximately 5% of the basin land mass was identified as a mountain unit with slope and summit geolandforms. Midlands with straight-flat geolandforms cover 10% of the total area while midlands with slope geolandforms characterized by 15–30% and 30–60% slope classes cover about 9 and 8% of the basin, respectively. Highlands with straight-flat, straight summit, and lowland plains with straight-flat geolandforms cover about 7, 6, and 5% of the basin land mass, respectively.

Agro-climatic and Agroecological Zones

The variation in climate over space and time and its impacts on soil formation show significant spatial variation in the soils of Abbay basin (Dile et al. 2018; Ali et al. 2022; Tibebe et al. 2022). Tibebe et al. (2022) carried out a detailed climatic element

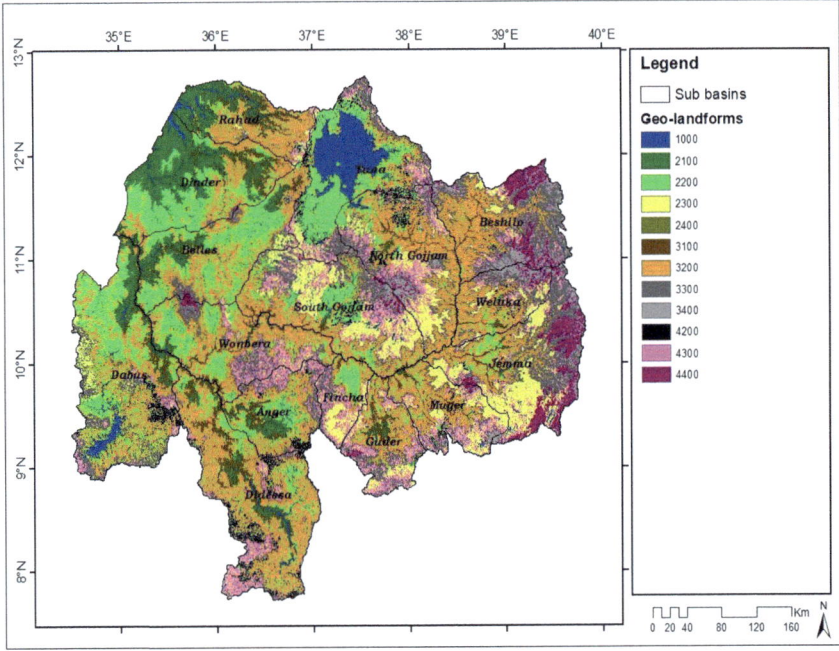

Fig. 6.1 Geolandforms of the Abbay basin (Adapted from Leenaars 2019), aggregate analysis and interpretation by the authors. *Note* 1000 = mixed bottom (6.3%); 2100 = lowlands plains flat (6.3%); 2200 = midlands flat (14.6%); 2300 = highlands flat (8.2%); 2400 = mountains flat (0.001%); 3100 = lowlands plains slope (3.8%); 3200 = midlands slope (32.6%); 3300 = highlands slope (12.4%); 3400 = mountains slope (2.1%); 4200 = midlands summit (2.1%); 4300 = highlands summit (8.7%); 4400 = mountains summit geolandforms (2.8%)

characterization of the basin including their temporal and spatial patterns disaggregated by a mosaic of subbasins. In addition, the authors synthesized and updated the agro-climatic zones of Abbay basin by superimposing elevation, length of growing season, and thermal zones. Based on gridded mean monthly temperature, rainfall, and evapotranspiration data at 4 km resolution from 1983 to 2017, data from about 300 meteorological stations and MODIS land surface satellite data, the length of the growing season and thermal zones were determined.

About 18 agro-climatic zones were defined for the basin based on six thermal zones (hot, warm, tepid, cool, cold, and very cold) and seven length of growing period (LGP) classes (arid, semi-arid, submoist, moist, subhumid, humid, and perhumid) (Abegaz et al. 2022; Tibebe et al. 2022), and were classified them into 11 zones (Fig. 6.2; Table 6.2). In terms of area coverage, the top four large agro-climatic zones were warm moist (33%), tepid moist (25%), tepid subhumid (13%), and warm subhumid (12%). Agro-climatic zones with smaller area coverage, in increasing order, were cold to very cold submoist-humid (0.3%), warm humid (0.6%), and warm submoist (0.7%). The agro-climatic zones showed spatial patterns across the basin including within the different subbasins. Warm moist zones dominantly occur

Table 6.1 Geolandforms and physiographic units of the Abbay basin

Major geolandforms	Physiographic regions	Geomorphic features	Area (%)	Major geolandforms	Physiographic units	Geomorphic features	Area (%)
Slope (51%)	Lowland plains	0.5–5 (% slope)	0.04	Bottom (6.3%)	Complex	Lowest/moist/wet/undefined	6.40
		5–10	0.72	Flat (29.17%)	Lowland plains	Straight	4.52
		10–15	0.65			Convergent	0.97
		15–30	0.90			Divergent	0.78
		30–60	0.60		Midlands	Straight	10.27
		>60	0.11			Convergent	2.35
		Convergent	0.45			Divergent	2.02
		Divergent	0.33		Highlands	Straight	6.13
	Midlands	0.2–5	0.14			Convergent	1.08
		5–10	3.16			Divergent	1.04
		10–15	4.26	Summit (13.58%)	Midlands	Straight	1.57
		15–30	8.53			Convergent	0.26
		30–60	7.75			Divergent	0.31
		>60	2.02		Highlands	Straight	6.88
		Convergent	3.43			Convergent	0.77
		Divergent	3.34			Divergent	1.02
	Highlands	1–5	0.015		Mountains	Straight	1.99
		5–10	0.44			Convergent	0.33
		10–15	0.72			Divergent	0.45
		15–30	3.71				

(continued)

Table 6.1 (continued)

Major geolandforms	Physiographic regions	Geomorphic features	Area (%)	Major geolandforms	Physiographic units	Geomorphic features	Area (%)
		30–60	4.12				
		>60	1.74				
		Slope (51%)	0.92				
		Divergent	0.78				
	Mountains	5–10	0.001				
		10–15	0.01				
		15–30	0.66				
		30–60	0.96				
		>60	0.23				
		Convergent	0.13				
		Divergent	0.13				

Source Leenaars (2019), authors' analysis

in the basin's western region, particularly in Dinder, Belles, and Dabus subbasins. The majority of tepid moist zones were identified in north and south Gojjam, Weleka, Muger, and Jemma subbasins while tepid subhumid units dominantly occurred in Guder and Fincha subbasins.

Following agro-climatic zone reclassification, agroecological zone delineation (AEZ delineation) was carried out by combining the 11 agro-climatic zone with four aggregated major geolandforms prepared at 50-m spatial resolution data in a GIS environment. The major geomorphic units have four major classes, viz., bottom, flat, slope, and summit (Fig. 6.3).

The four geomorphologic units combined with 11 agro-climatic zones are ideally supposed to give 44 AEZs. However, in this process, only 42 actual zones were identified in the basin (Table 6.2). In terms of area coverage, the top three large agroecological zones were warm moist slope (19%), warm moist flat (12%), and tepid moist slope (12%).

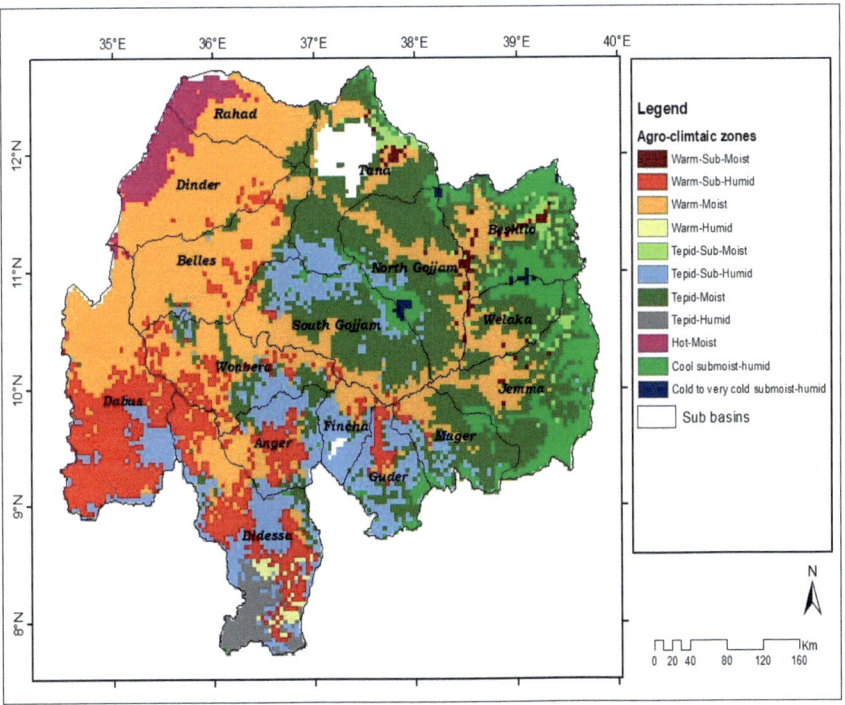

Fig. 6.2 Agro-climatic zones of the Abbay basin (Adapted from Tibebe et al. 2022, 2023)

Table 6.2 Matrix of eleven agro-climatic zones and four major geolandforms

Agro-climatic zones	Golandforms spatial coverage (%)				
	Bottom	Flat	Slope	Summit	Total (%)
Warm moist	1.72	12.32	19.00	0.21	33.24
Tepid moist	1.00	7.20	11.76	5.00	24.96
Tepid subhumid	0.34	2.59	6.28	3.78	12.99
Warm subhumid	1.02	3.55	7.84	0.18	12.60
Cool submoist-humid	0.13	1.10	4.01	3.54	8.78
Hot moist	0.41	2.15	0.26	NA	2.82
Tepid humid	0.004	0.21	0.76	0.80	1.81
Tepid submoist	0.006	0.16	0.92	0.13	1.27
Warm submoist	0.09	0.10	0.50	0.01	0.70
Warm-humid	0.05	0.09	0.41	0.003	0.56
Cold to very cold submoist-humid	0.0004	NA	0.18	0.08	0.26
Total area (%)	4.86	29.48	51.93	13.73	100

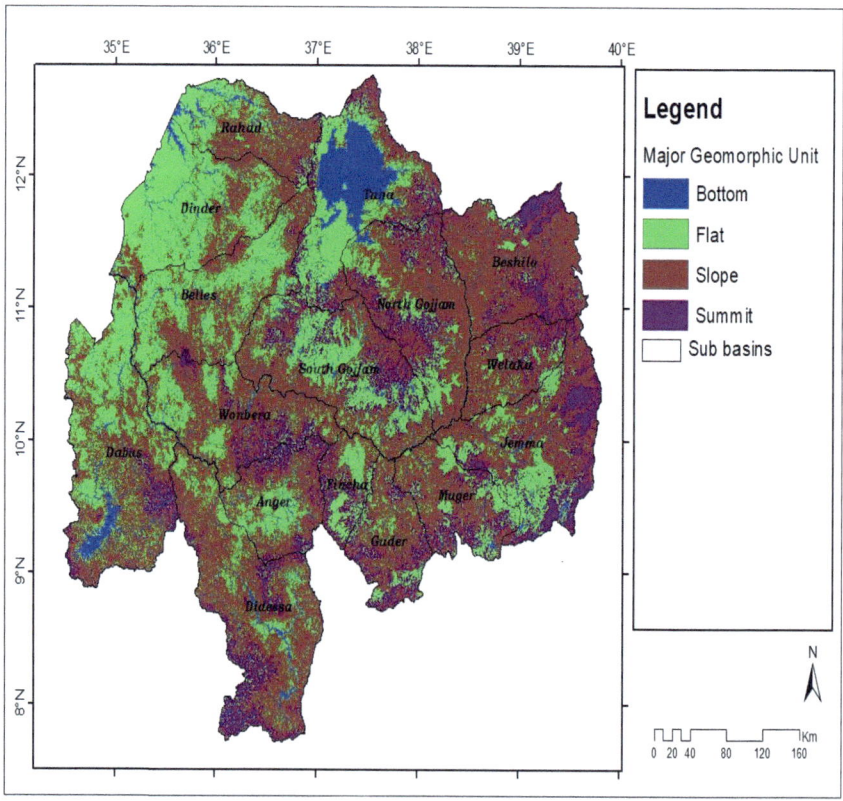

Fig. 6.3 Aggregated major geolandforms (Adapted from Leenaars 2019)

Geology/Parent Materials

The basin contains groups of igneous, sedimentary, and metamorphic rocks (Aster and Seleshi 2009). The igneous rocks occur as cappings and comprise younger and older volcanics. The younger volcanics consist of basaltic lava flows interbedded with ash and tuff which are often weathered. Older volcanics usually consist of massive basalt. Sedimentary rocks belonging to the Mesozoic era and are found overlying the metamorphics, which is mainly exposed at mid slopes of the Abbay River gorge. Metamorphic rocks were belonging to the Precambrian and volcanic rocks belonging to the Tertiary and Quaternary origins and are primarily present in the basin's eastern and northern regions forming the highland plateau.

Vegetation and Land Use/Land Cover

The main land use and land cover categories found in the Abbay basin include cultivated land, grassland, forestland, bush and shrub land, settlement, and waterbodies (Aster et al. 2009; Teklu 2009, Gebrehiwot et al. 2013; Tibebe et al. 2022). According to WLRC (2016a), croplands (agricultural lands) were the largest proportion in the Abbay Basin, covering an area of approximately 37.9% of the basin. Woodlands, shrub/bushlands, grassland, and forestlands were the second, third, fourth, and fifth largest land cover types covering approximately 18, 14.9, 12.6, and 10.8% of the basin, respectively. Other land cover types of the basin include barrenlands (4.6%), waterbody (0.5%), wetland (0.3%), settlements (0.3%), and afro alpines (0.2%).

Soil Resource Spatial Information

Overview of Soil Spatial Information

Different sources have provided spatial data on the basin's soil resources, including information on the physical and chemical characteristics of the soil (Dile et al. 2018). Basin-wide spatial soil data have been sourced from national (BCEOM 1998; Erkossa et al. 2009; Berhanu et al. 2013; Elias 2016; ATA 2016, 2018, 2019; Ali et al. 2022; Regassa et al. 2023a), continental/regional soil atlas of Africa (Jones et al. 2013), global soil grids (Poggio et al. 2022) and Harmonized World Soil Database (FAO and IIASA 2023) information providers. These data sources vary in approach and method, and provide mosaic of soil information, which requires careful selection of soil data source based on for the intended purpose. For instance, the Abbay basin master plan study at 1:250,000 scale by BCEOM (1998) identified eleven soil types as per the revised 1988 FAO-UNESCO-ISRIC legend of the soil map of the world.

Accordingly, 96% of the entire terrestrial land mass is covered by Leptosols (24%), Alisols (21%), Nitisols (16%), Cambisols (10%), Luvisols (9%), and Acrisols (5%). Furthermore, Erkossa et al (2009) reported eight soil types, namely Nitosols (39.08%), Leptosols (23.68%), Luvisols (14.2%), Vertisols (12.05%), Cambisols (5.07%), Alisols (4.17%), Phaeozems (1.22%), and Fluvisols (0.54%). A similar soil map updating mission by WLRC (2016a) identified 17 RSGs and 59 soil units in the Grand Ethiopian Renaissance Dam (GERD) basin, excluding two subbasins (Rahad and Dinder) of the Abbay basin. Accordingly, the dominant IUSS WRB 2014 RSGs of GERD basin were Leptosols (25.6%), Nitisols (21%), Vertisols (11.85%), Cambisols (11.07%), Luvisols (8.62%), Alisols(6.56%), Acrisols(6.13%), Lixisols (2.21%), Fluvisols (1.45%), and Regosols (1.32%).

Recently, a reference soil group (RSG) map of Ethiopia (250 m × 250 m spatial resolution) developed following the IUSS WRB (2015) soil classification/correlation system identified fifteen RSG, in the basin (Ali et al. 2022; Ali et al. 2024). In this countrywide soil spatial information, layering the basin, five RSGs, viz. Vertisols (40%), Nitisols (20%), Leptosols (19%), Luvisols (11%), and Cambisols (8%) constitute 98% of the basin terrestrial land. Planosols (0.001%), Andosols (0.01%), Gleysols (0.08%), and Lixisols (1.92%) which had not been recognized by previous studies were identified and mapped, although they have a small spatial coverage.

Synthesis of Soil Resource Spatial Information

Since the 2010s significant area-based, subbasin level, semi-detail, and detail soil characterization, classification, and mapping studies have been conducted in the basin (ORLEPB 2014a, b; Elias 2016; WLRC 2016a, b; Getahun and Selassie 2017; Taye et al. 2018; Deressa et al. 2020; Takele and Iticha 2020; Dinssa and Elias 2021; Sori et al. 2021; Tiruneh et al. 2022). Soil data these studies has the potential to enhance the basin's soil spatial data and give fresh, enhanced insights into the scope and spatial distribution of the soil resources for more effective soil and land management decisions. Accordingly, the chapter presents polygon-based semi-detail, detail and very detailed soil resource spatial information and aggregated to 1:250,000 scale.

Most of the soil classification systems employed by these recent IUSS WRB (2006) and IUSS WRB (2015). In this chapter, the classification systems were harmonized into IUSS WRB (2015), Jones et al. (2013), based on 745 representatives sampled soil profiles out of a compilation of 2950 profile points.

The updated soil resource spatial information for the basin is presented in Fig. 6.4. In the updated map, 17 dominant RSGs and 64 s-level soil units were identified in the basin. Approximately 96% of the Abbay basin landmass covered by 10 RSGs, viz. Nitisols (21.79%), Leptosols (21.65%) Cambisols (15.82%), Vertisols (11.69%), Acrisols (9.51%), Luvisols (8.83%), Fluvisols (2.50%), Lixisols (1.91%), Alisols (1.32%), and Regosols (1.29%).

Approximately, 1% of the terrestrial landmass is covered by Arenosols (0.2%), Ferralsols (0.19%), Calcisols (0.17%), Phaeozems (0.13%), Gleysols (0.08%),

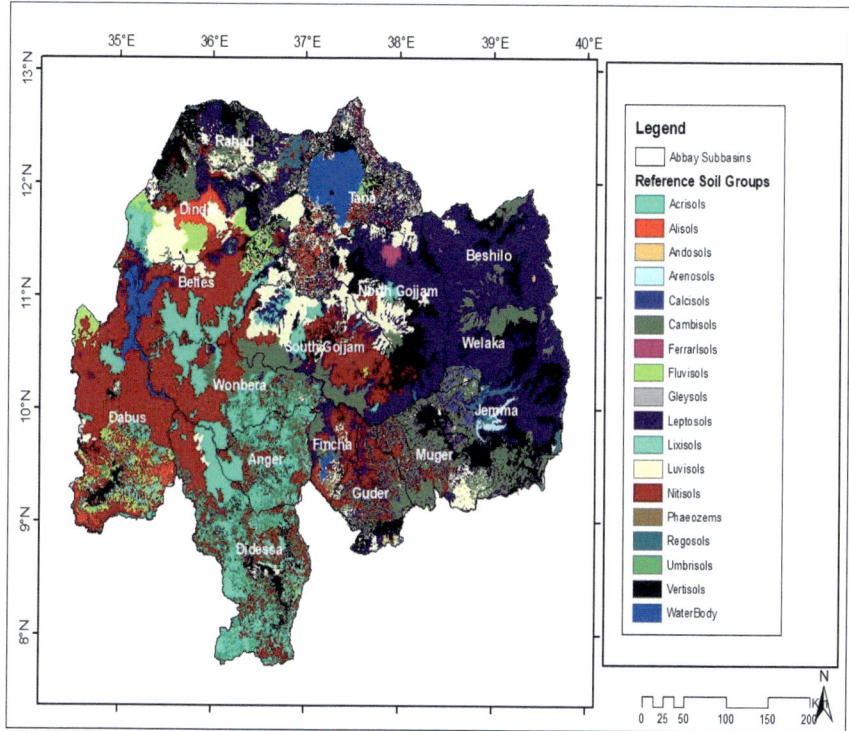

Fig. 6.4 Reference soil groups (RSGs) of Abbay basin

Andosols (0.06%), and Umbrisols (0.00003%). Compared with previous basin-wide soil studies, the extent and spatial distribution of the top ten ranked reference soil groups uncovered new insights into the soil resource diversity of the basin. Our analysis also revealed that Lixisols, Ferralsols, Calcisols, Gleysols, Andosols, and Umbrisols with diverse spatial coverage are new RSGs confirmed to occur in the basin unlike in previous basin-wide same scale soil mapping efforts. Twenty-three second-level soil units aggregated into ten RSGs covered approximately 90% of the basin terrestrial land area. The top ten ranked second-level soil units include Rhodic Nitisols (14.93%), Eutric Leptosols (14.41%), Haplic Acrisols (8.05%), Haplic Vertisols (6.18%), Eutric Cambisols (5.79%), Leptic Cambisols (4.62%), Lithic Leptosols (4.44%), Haplic Luvisols (3.88%), Vertic Cambisols (3.21%), and Pellic Vertisols (2.98%).

Description of Synthesized Dominant Soil Types

A description of the top ten ranked dominant soil types and associated soil units of the respective RSGs area coverage is presented. The soil description is synthesized after overlay analysis of the reference soil groups over major physiographic units and geolandforms disaggregated by dominant geomorphic features (Table 6.3).

Acrisols

They are characterized by their argic B-horizon, and dominance of stable, low-activity clays covering approximately 9.51% of the Abbay basin landmass. With the exception of the Beshilo and Rahad subbasins, these soils are widely spread throughout the subbasins. The two main second-level soil units are Haplic Acriosls (8.05%) and Chromic Acrisols (1.47%).

Alisols

They are moderately well-drained to well-drained soils with a reddish brown to dark brown color covering approximately 1.32% of the total land area of the basin. The dominant soil units were Haplic Alisols (1.24%), Chromic Alisols (0.07%), and Rhodic Alisols (0.01%).

The soils are claye in the surface occurring across 0–1% slope facets. These soils have very deep effective soil depth of > 150 cm, weak fine to medium subangular blocky structure, dark reddish brown (5YR3/2) surface color, very strongly acidity, low total and available phosphorous, medium cation-exchange capacity, and low base saturation. These soils require soil acidity management interventions for optimum crop production.

Cambisols

Cambisols are formed under different environments as confirmed by their occurrence in all the subbasins of the basin. These soils have a wide range of variation among themselves. The majority of Cambisols are medium-textured soils with high porosity, high-water-holding capacity, good structural stability, and a high proportion of weatherable minerals in the silt and sand fractions. They occupy about 15.81% of the basin's total land area. The dominant soil units were Eutric Cambisols (5.79%), Leptic Cambisols (4.62%), Vertic Cambisols (3.21%), Dystric Cambisols (1.28%), Chromic Cambisols (0.35%), Rhodic Cambisosls (0.27%), Lithic Cambisols (0.2%), Gleyic Cambisols (0.07%), and Skeletic Cambisols (0.0%).

Fluvisols

Fluvisols dominantly develop on flood plains, riverbanks, and lake deposits in the Dabus, Belles, Dinder, Tana, South Gojjam, and Didessa subbasins. They regularly receive fresh sediments during rainy seasons and form a stratified material that forms stratified layers with clear or diffuse boundaries between horizons depending on the nature and particle distribution of the transported materials. The majority of the transported materials lack a sizable proportion of organic matter. The characteristics

6 Soils of the Abbay Basin

Table 6.3 Reference soil group occurrence across geolandforms of the Abbay basin

RSGs	Geolandforms												Total
	1000	2100	2200	2300	2400	3100	3200	3300	3400	4200	4300	4400	
Acrisols	0.32	0.49	1.58	0.58	–	0.40	3.53	1.05	0.06	0.34	1.11	0.05	9.51
Alisols	0.07	0.25	0.46	0.03	–	0.03	0.26	0.13	–	0.05	0.03	–	1.32
Cambisols	0.48	0.65	1.13	0.93	–	0.90	7.01	1.78	0.38	0.32	1.55	0.69	15.81
Fluvisols	0.32	0.30	0.91	0.07	–	0.03	0.65	0.12	–	0.07	0.04	–	2.50
Leptosols	0.55	0.73	0.97	0.48	–	1.01	9.49	5.17	1.11	0.30	1.25	0.60	21.65
Lixisols	0.10	0.03	0.22	0.06	–	0.04	0.83	0.28	0.00	0.13	0.22	–	1.91
Luvisols	0.47	0.48	2.08	1.08	–	0.14	1.80	1.04	0.32	0.19	1.06	0.19	8.84
Nitisols	0.79	1.34	5.65	1.59	–	0.85	7.94	1.40	0.02	0.54	1.60	0.08	21.80
Regosols	0.05	0.03	0.15	0.03	–	0.04	0.63	0.10	0.04	0.07	0.10	0.04	1.29
Vertisols	1.32	1.37	1.42	3.21	0.00	0.31	0.62	0.62	0.17	0.08	1.47	1.08	11.67
Total	4.46	5.67	14.57	8.07	0.00	3.75	32.75	11.69	2.09	2.08	8.44	2.73	96.30

Note 1 1000 = mixed bottom; 2100 = lowland plains flat; 2200 = midlands flat; 2300 = highlands flat; 2400 = mountains flat; 3100 = lowlands plains slope; 3200 = midlands slope; 3300 = highlands slope; 3400 = mountains slope; 4200 = midlands summit; 4300 = highlands summit; 4400 = mountains summit geolandforms

Note 2 The total area coverage of soil units under a given geolandforms is lower than the total area of that geolandforms, because in soil unit mapping, some areas are excluded (e.g., water body, settlements, bare land)

of Fluvisols are influenced by the type of transferred soil. They are mostly found as surface or subsurface depositions or a mixture of both soil horizons, and hence, are characterized by a weak horizon variation (A-C). Their structure varies from single grain to crumby or massive depending on the particle size distribution and land management, while their texture is dependent on the nature and origin of the deposited sediments. They cover approximately 2.5% of the landmass of the Abbay basin. The dominant soil units in the basin are Eutric Fluvisols (1.68%), Dystric Fluvisols (0.81%), and Satgnic Fluvisols (0.01%).

Leptosols

Leptosols are genetically young soils with weak profile development over a variety of continuous rock types, or unconsolidated materials, fewer than 20% (by volume) of which are fine earth. They occupy about 21.65% of the basin's total landmass area. The dominant soil units identified in the basin include Eutric Leptosols (14.42%), Lithic Leptosols (4.44%), Hyperskeletic Leptosols (1.53%), Rendzic Leptosols (0.81%), Dystric Leptosols (0.2%), Mollic Leptosols (0.13%), Umbric Leptosols (0.09%), Calcaric Leptosols (0.02%), and Skeletic Leptosols (0.0003%).

Luvisols

Luvisols are derived mostly from the trap series of volcanic rocks and basalts in the basin. The primary pedogenetic process of Luvisols is the formation of their argic horizon by translocation of clay from the surface soil to the subsoil without marked leaching of base cations, or advanced weathering of high-activity clays. Luvisols typically have a brown to dark brown surface horizon over a (grayish) brown to strongly brown or red argic subsurface. They have a moderately deep to deep profile, with granular or crumb surface soil structures that are porous and well aerated, and their argic horizon has a stable blocky structure, contributing to a good water-holding capacity. These soils cover 8.85% of the total land area of the basin. Luvisols are good agricultural soils due to their deep rooting depth and large nutrient reserve. They are widely distributed across the subbasins and intensively used for crop production and grazing. The dominant soil units include Haplic Luvisols (3.89%), Chromic Luvisols (2.33%), Rhodic Luvisols (1.21%), Vertic Luvisols (0.59%), Gleyic Luvisols (0.56%), Cambic Luvisols (0.12%), Ferric Luvisols (009%), and Leptic Luvisols (0.06%).

Lixisols

Lixisols are derived mainly from parent materials with a fine texture that is unconsolidated, severely weathered and leached. They are distinguished by a buildup of clay in the argic horizon, the presence of low-activity clays, and a moderate to high base saturation. Most of them are deep soils with a medium texture. Low to medium water-holding capacity and weak aggregate stability are other characteristics. They are located throughout the Didessa, Dabus, South Gojjam, Wonbera, Anger, and North Gojjam subbasins and occupy about 1.92% of its total area of the basin. The dominant soil units include Haplic Lixisols (1.85%) and Chromic Lixisols (0.01%) distributed in different physiographic regions and geolandforms.

Nitisols

Nitisols are formed from basic iron-rich rocks such as basalt. Most Nitisols exhibit an A-B(t)-C horizon sequence. These soils are characterized by having dusky red or dark red color and by having clayey to fine clayey texture. They are deep, well-drained soils with a typical nutty or polyhedral blocky structure and shiny ped faces with high total porosity. They cover approximately 21.79% of the basin's land area. The dominant soil units were Rhodic Nitisols (14.93%), Umbric Nitisols (2.47%), Mollic Nitisols (2.05%), Dystirc Nitisols (1.54%), Eutric Nitisols (0.66%), and Alic Nitisols (0.11%).

Regosols

Regosols are young soils without subsurface diagnostic horizons. Most of these soils develop on the erosional surface of the highlands' slope and in some locations where their formation and development are retarded by the nature of the parent material. Profile development is limited to the formation of a thin A-C horizon sequence. They have shallow, medium-textured, and nondifferentiated mineral soils with low water-holding capacity and high permeability. These soils cover approximately 1.29% of the basin's total landmass, dominantly in the Tana, North Gojjam, Muger, Jemma, Dinder, Belles, and Rahad subbasins. The dominant soil units were Eutric Regosols (0.72%), Leptic Regosols (0.47%), and Dystric Regosols (0.1%).

Vertisols

The major processes in the formation of Vertisols are alternating swelling and shrinking of expandable clays resulting in deep cracks in the dry season. Shrink-swell behavior may also form gilgai microreliefs. Vertic horizons are clayey and when dry, they often have a hard to very hard consistency. Polished, shiny surfaces (slickenside), often at sharp angles, are distinctive. Vertisols of humid highlands (e.g., pockets of central and western highlands) exhibit shallower and narrower cracks, whereas, in semi-arid and subhumid zones, e.g., Across Dinder and Rahad subbasins Vertisols exhibit wider and deeper cracks.

They are commonly deep (> 100 cm) and they are typically characterized by their deep and wide cracks during the dry season. All Vertisols have high clay contents and they have angular to subangular blocky structures with high-water-holding capacities. Their horizons are characterized by wedge-shaped soil aggregates, slickensides, and shrink-swell cracks and they have hard to very hard consistency. They are commonly constrained by their imperfect drainage and hardness which impose workability problems. Their main features are critically limit agricultural production and management systems due to their heavy textural composition and shrinking properties. These soils covered approximately 11.67% of the basin landmass. The dominant soil units were Haplic Vertisols (6.18%), Pellic Vertisols (2.96%), Chromic Vertisols (1.15%), Leptic Vertisols (0.53%), Duric Vertisols (0.52%), Calcic Vertisols (0.32%), and Sodic Vertisols (0.01%).

Soil Physical Property Spatial Information

Soil Depth

The basin's soils have varying depths mainly attributed to the topography's diversity (Table 6.2). In the basin, the percentages of soils that are very shallow (less than 30 cm), sallow (30–50 cm), moderately deep (50–100 cm), deep (100–150 cm), and very deep (> 150 cm) are approximately 25.76, 1.34, 5.60, 1.40, and 54.29%, respectively (Fig. 6.5). Leptosols have very shallow soil depths and cover the biggest parts of the basin.

The eastern highlands of the Abbay basin, such as, areas around the North Gojjam, Beshilo, Welaka, Jemma, and some parts of the Muger and Guder subbasins are dominated by shallow soils. In the hilly regions of the Abbay basin, steeply sloping soils are very common. When a rugged toposequence is present, the soils are shallow in the upper positions and deeper in the lower parts within a very short distance (Sheleme 2017). The slope of the landscape, which affects soil formation and development by causing erosion and deposition as well as the infiltration and percolation of water deep into the soil at the lower topographic positions, is the cause of the variation in soil depth (Alemayehu and Sheleme 2013; Sheleme 2017).

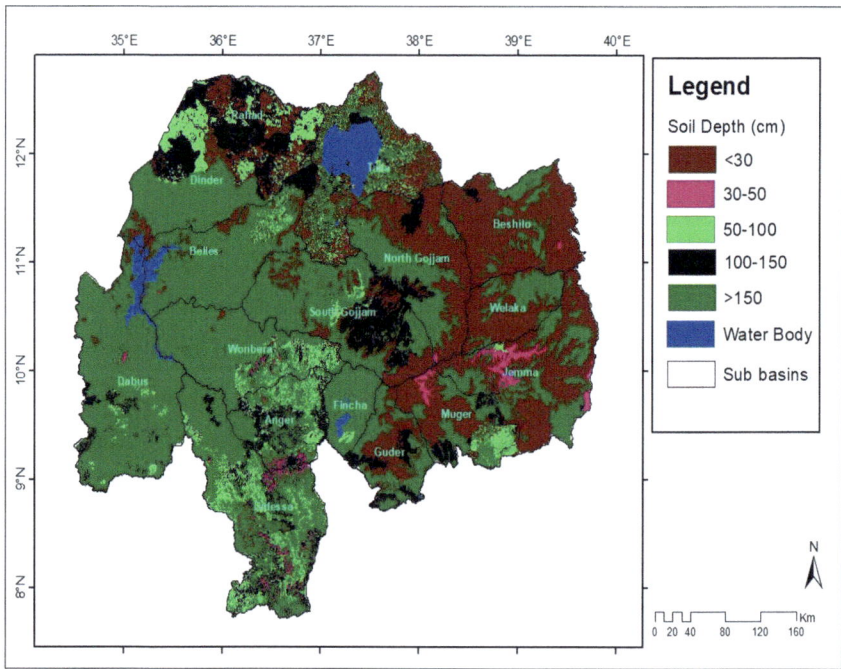

Fig. 6.5 Soil depth class map of the Abbay basin

Soil Texture

Soil texture is one of the essential and inherent characteristics of soils, which has an impact on practically all other soil characteristics. In particular, it affects soil characteristics that are crucial for agriculture (such as ease of cultivation, nutrient- and water-holding capacity and their transmission), the environment (e.g., susceptibility to pollution), and engineering purposes. Due to its static nature, tillage and other interventions have little impact on it unless they are dramatic (Chestworth 2008). The textural class refers to the overall textural classification of a soil based on the relative proportions of its sand, silt, and clay constituents. The USDA method classifies soils into three main textural groups: sandy (sand, loamy sand, and sandy loam), loamy (loam, silt loam, silt, sandy clay loam, clay loam, silty clay loam) and clayey (sandy clay, silty clay, and clay).

Ten of the 12 USDA textural class names are present in Abbay basin soils, with sand and silt missing textural class names. In terms of area coverage, clay (58.43%) and clay loam (20.03%), followed by sandy loam (5.21%) and sandy clay (5.14%), are the most dominant textural classes in the Abbay basin (Fig. 6.6). Generally, soils in the level lands part of the basin contain more clay, while those in the sloppy and summit areas are characterized by high sand content (Table 6.4).

Bulk Density

The mass (weight) of a dry soil per unit volume is known as bulk density. Solids and pores are both included in the volume. A soil's particle density is never greater than its bulk density. Compact soils have a higher weight per volume than loose soils, which have a lower weight per volume. As a dynamic soil attribute, it is influenced by a variety of factors, including soil depth, organic matter content, soil texture, structure, and management. Lower bulk densities are seen in soils with a greater pore space to solids ratio than in soils that are more compact and have fewer pores.

Therefore, bulk density will be impacted by any factor that alters soil pore space. According to Weil and Brady (2017), coarse-textured soils tend to have higher bulk density values (ranging from 1.4 to 1.9 Mg/m^3) than medium and fine-textured soils (0.9–1.4 Mg/m^3), while volcanic soils that are composed of special clay minerals that are amorphous have lower bulk density values (often less than 1.0 Mg/m^3). Higher organic matter content soils are anticipated to have lower bulk densities for the same textural class since organic matter weighs less per unit volume. Similarly, good and very good aggregated soils tend to have lower bulk density than compacted soils, because aggregation increases porosity. According to Regassa et al. (2023c), practical uses for soil bulk density data include determining the degree of compactness, which serves as an indirect indicator of soil structure, determining the mass of soil per unit area and depth, and converting soil water content and nutrient values from a mass to volumetric basis.

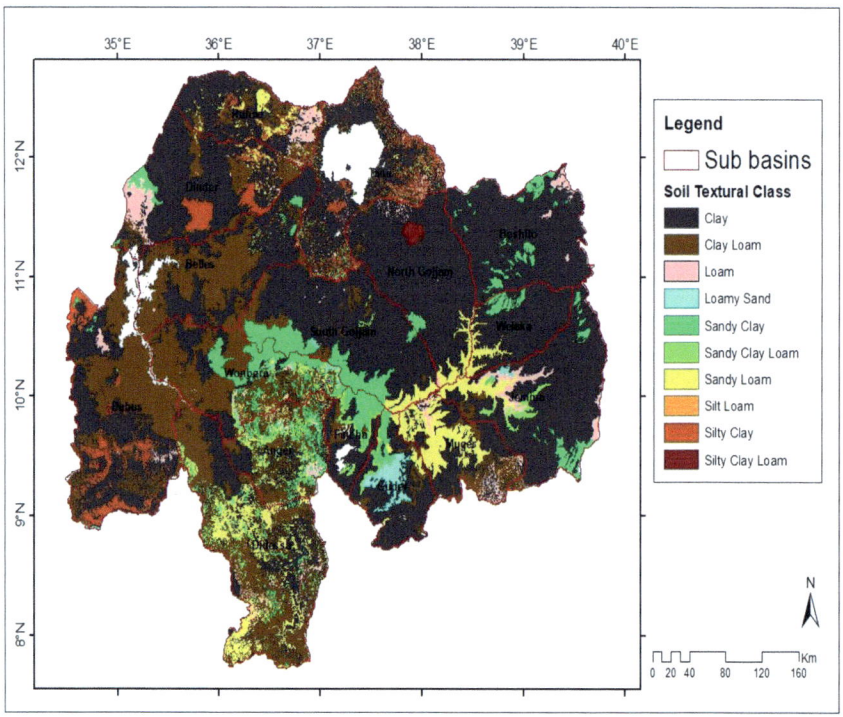

Fig. 6.6 Surface (0–30 cm) soil textural class map of Abbay basin

In the basin, approximately 16.5%, 30.6%, 52.4%, and 0.5% of the soils have a bulk density value of 0.8–1.1, 1.1–1.2, 1.2–1.4, and 1.4–1.5 g/cm^3, respectively (Fig. 6.7). In general, approximately 99.5% of the basin soil bulk density is between 0.8 and 1.4 g/cm^3, implying that the basin soils are cultivated and uncompacted and do not restrict root growth (Landon 2014). The bulk density along different watershed varies, which may be due to the variation in soil separation down the watershed (Table 6.4). In the basin, heavy equipment like tractors is not frequently used for tillage operations. For tillage or field preparation, most of farmers in this region use a tool known as Maresha. Due to the tiny size and low weight of this tool, severe compaction beneath the plow layer is unlikely to occur. Maresha-based tillage rarely causes compaction since it involves repeated plowing at the same depth (Regassa et al. 2023c).

Table 6.4 Status of texture and bulk density (BD) in different Abbay basin soils

Location	Sand (%)	Silt (%)	Clay (%)	BD (gm/cm^3)	Reference
Didessa Watershed	30–57	23–37	18–35	1.07–1.47	Deressa et al. (2020)
Jemma Watershed	11.4–43.4	6.0–34.8	21.8–77.8	–	Taye et al. (2018)
Bedele District	10–68	8–34	14–72	–	Sori et al. (2021)
Sibu Sire District	2.8–66.3	2.7–50.1	20.0–79.2	–	Takele and Iticha (2020)
Aba Gerima Catchment	5–66	13–41	11–80	–	Tiruneh et al. (2022)
Debre-Yakob Watershed	6–33	12–48	29–78	1.18–1.33	WLRC (2016b)
Atari-Mesk Watershed	15–57	12–46	14–65	0.9–1.45	
Andit-Tid Watershed	40–92	4–48	2–20	0.85–1.35	
Debre-Mewi Watershed	23–39	28–40	36–65	–	
Gerda Watershed	10–41	10–37	29–68	1.3–1.5	
Gosh Watershed	10–34	17–32	27–32	0.97–1.34	
Dembia Plain	2	18	80	–	Getahun and Selassie (2017)
Fogera Plain	3	16	80	–	
Kunzila Alefa Plain	2	13	85	–	
Mecha Plain	5	14	81	–	
Baso Tibe District	32–44	6–15	41–58	1.05–1.17	Dinssa and Elias (2021)
Kabe Watershed	42–54	27–30	18–29	1.3–1.4	Assefa et al. (2020)
Gerado Watershed	21–36	29–39	25–50	0.97–1.2	Asmamaw and Mohammed (2013)
TaraGedam Watershed	14–64	20–46	16–56	–	Menale (2014)

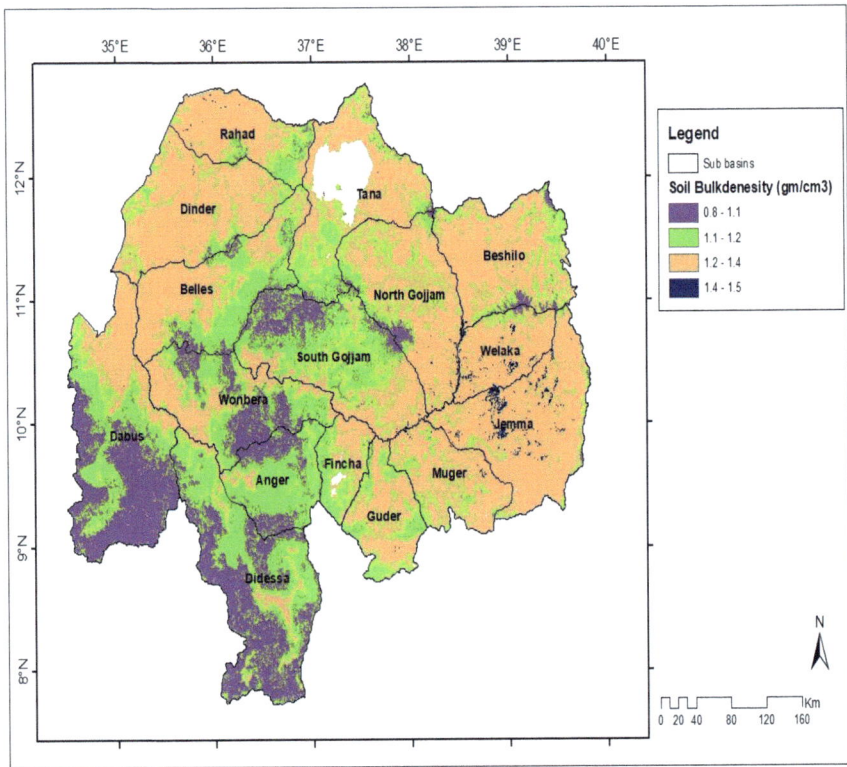

Fig. 6.7 Bulk density map of Abbay basin (Adapted from SoilGrids 2017. https://files.isric.org/soilgrids/former/2017-03-10/data/)

Soil Chemical Property Spatial Information

Soil Reaction (Acidity/Alkalinity)

Soil pH is a measurement of soil acidity or alkalinity, indicating the activity of hydrogen ions (H^+) and hydroxyl ions (OH^-) in a water solution. Soil pH defines the soil and can be used as a guide to the suitability of soils for a certain crop species. It is also an indicator of the chemical processes that occur in the soil, and is a guide to likely deficiencies and/or toxicities of a certain nutrients (Slattery et al. 1999). Furthermore, the availability of different nutrients, harmful substances, and chemical species to plant roots is also impacted by soil pH. According to Weil and Brady (2017), pH is a very good guide to some expected nutrient deficiencies and toxic effects. For instance, a higher pH tends to decrease the availability of Fe, Al, Zn, Mn, and B while increasing the availability of Mo. While Mg and P absorption is more challenging for plants when the pH is lower, but more Mn, Fe, and Al are absorbed.

The soil pH values of warm moist AEZs varied from 5.2 to 8.5 with a mean value of 6.5. The pH values of the warm moist AEZ soils of the bottom, flat, slope, and summit landforms varied from 5.3 to 9.2, 4.9 to 8.4, 5.2 to 8.4, and 5.2 to 7.9, respectively, indicating a wide range of variation from strongly acidic to strongly alkaline. The mean values were found to be 6.6, 6.4, 6.5, and 6.4 for bottom, flat, slope, and summit landforms of warm moist AEZs, respectively (Fig. 6.8; Table 6.5). The pH values of soils of the tepid moist, tepidsubhumid, warm subhumid, and cool submoist-humid AEZs varied from 4.8 to 8.6, 4.7 to 8.5, 5.0 to 7.5, and 4.8 to 8.0, respectively, indicating wide range of variation from strongly acidic to strongly alkaline. The mean values were found to be 6.2, 5.8, 6.5, and 6.3 for tepid moist, tepid subhumid, warm subhumid, and cool submoist-humid AEZs, respectively.

According to EthioSIS (2016), critical levels for soil reaction of approximately 7.3, 62.0, 27.5, and 3.1% of Abbay basin soils were found to be strongly acidic, moderately acidic, neutral and moderately alkaline, respectively (Fig. 6.8). Strongly acidic soil was mostly found in the tepid subhumid and tepid humid AEZs particularly in the highest altitude area of the South Gojjam, Tana, Fincha, Didessa, and Dabus subbasins. Moderately alkaline soils were mostly found in the lowlands of

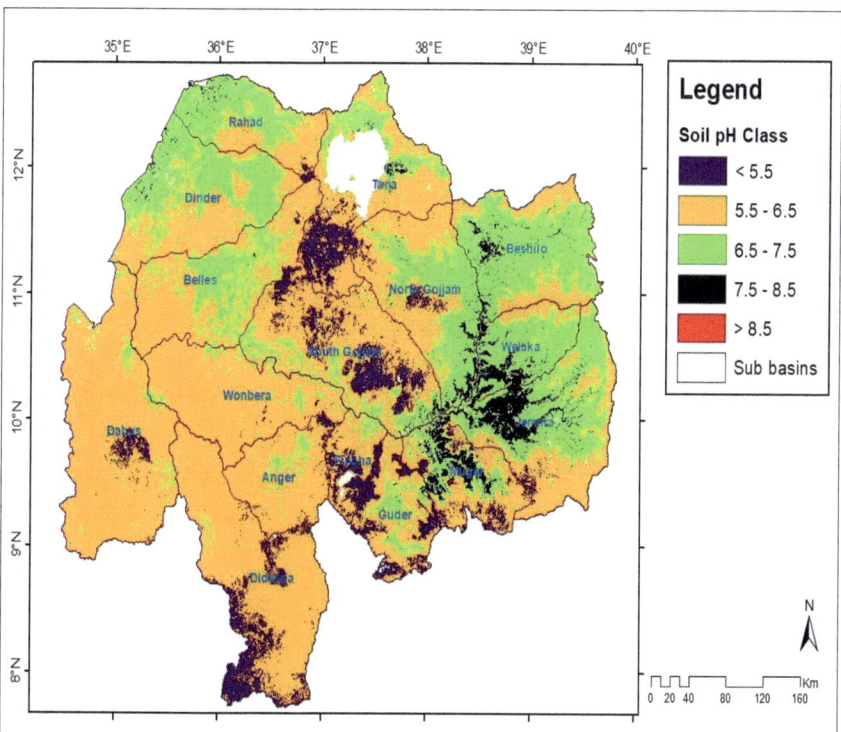

Fig. 6.8 Spatial distribution of surface (0–20 cm) soil pH in Abbay basin soils (Adapted from ATA 2016, 2018, 2019)

Table 6.5 Range values of selected surface (0–20 cm) soil chemical properties under different AEZs and geolandforms

No.	AEZ	GMK	pH	EC (dSm^{-1})	CEC cmol(+)kg^{-1}	Ca (mg/kg)	Mg (mg/kg)	K (mg/kg)	Na (mg/kg)	CaCO$_3$ (%)
1	Warm moist	Bottom	5.3–9.2	0.004–0.34	13.8–101.9	466–19,025	96–2892	30–877	13–509	0.1–38
		Flat	4.9–8.4	0.000–0.45	9.1–119.9	399–33,815	89–4117	20–1680	9–485	0.05–51
		Slope	5.2–8.4	0.000–0.37	10.9–105.9	250–34,781	37–3511	23–1582	1–400	0.09–58
		Summit	5.2–7.9	0.000–0.23	12.0–81.2	493–17,982	70–2222	47–1078	5–163	0.12–17
		Mean	5.2–8.5	0.001–0.35	11.4–102.2	402–26,401	73–3186	30–1304	7–389	0.09–41
2	Tepid-moist	Bottom	4.9–8.8	0.000–0.26	12.0–80.7	492–20,330	106–2531	49–1242	9–389	0.09–31
		Flat	4.7–8.6	0.000–0.36	10.7–108.1	200–28,247	29–3631	37–2651	1–794	0.06–45
		Slope	4.9–8.8	0.000–0.31	3.0–108.8	104–31,851	11–3767	31–1648	1–626	0.05–68
		Summit	4.8–8.3	0.000–0.26	10.7–95.5	94–21,278	9–3563	27–1873	1–522	0.07–42
		Mean	4.8–8.6	0.000–0.30	9.1–98.3	222–25,427	39–3373	36–1853	3–583	0.07–46
3	Tepid-subhumid	Bottom	4.8–8.8	0.000–0.27	11.8–67.8	154–11,684	18–2177	20–750	5–143	0.03–4.7
		Flat	4.8–8.5	0.000–0.24	7.9–82.6	92–15,032	9–2348	21–1634	1–489	0.02–20
		Slope	4.7–8.4	0.000–0.31	6.5–79.8	82–21,548	7–2803	15–1388	2–357	0.05–28
		Summit	4.7–8.3	0.000–0.27	5.4–68.4	49–18,386	10–2269	20–1413	2–850	0.07–25
		Mean	4.7–8.5	0.000–0.27	7.9–74.7	94–16,663	11–2399	19–1296	3–460	0.04–19
4	Warm-subhumid	Bottom	5.2–7.3	0.004–0.26	11.5–72.5	204–8192	29–2463	24–659	6–167	0.11–3.7
		Flat	4.9–7.9	0.008–0.27	8.9–79.7	33–13,461	6–2937	17–1280	5–461	0.1–11
		Slope	5.0–8.0	0.000–0.26	7.8–75.3	189–14,791	22–2434	19–1241	3–751	0.1–11
		Summit	4.9–7.0	0.002–0.21	12.0–51.1	185–4675	15–1035	21–1129	7–84	0.1–2
		Mean	5.0–7.5	0.004–0.25	10.0–69.6	153–10,280	18–2217	20–1077	5–366	0.1–6.9
5	Cool submoist-humid	Bottom	5.0–7.6	0.000–0.19	10.2–67.2	824–11,410	62–3142	82–628	28–186	0.17–2.2
		Flat	4.7–7.8	0.001–0.27	8.9–76.6	306–12,429	32–3182	60–2164	6–461	0.12–3.5

(continued)

Table 6.5 (continued)

No.	AEZ	GMK	pH	EC (dSm^{-1})	CEC cmol (+)kg^{-1}	Ca (mg/kg)	Mg (mg/kg)	K (mg/kg)	Na (mg/kg)	CaCO$_3$ (%)
		Slope	4.8–8.2	0.000–0.33	10.9–70.5	563–13,874	70–2814	58–1594	5–695	0.11–32
		Summit	4.7–8.3	0.000–0.26	9.5–69.4	349–14,189	47–3014	58–1913	6–541	0.08–12
		Mean	4.8–8.0	0.000–0.26	9.9–70.9	511–12,976	53–3038	65–1575	11–471	0.12–12
6	Hot-moist	Bottom	6.2–7.6	0.036–0.18	32.2–101.1	1465–9846	329–2555	30–689	69–85	0.1–2.1
		Flat	6.1–7.9	0.007–0.25	26.7–98.2	527–10,264	103–2683	13–1071	30–102	0.1–1.7
		Slope	6.0–7.5	0.005–0.18	25.0–94.9	1088–8925	202–3400	26–319	27–42	0.1–0.9
		Mean	6.1–7.7	0.016–0.20	27.9–98.1	1027–9678	211–2879	23–693	42–76	0.1–1.5
7	Tepid-humid	Bottom	4.9–6.2	0.028–0.26	13.9–36.2	689–3912	90–695	101–663	20–529	0.12–1.3
		Flat	4.6–6.3	0.017–0.24	10.7–45.1	272–6307	38–1252	69–844	14–436	0.12–2.2
		Slope	4.7–7.7	0.004–0.25	13.2–40.2	442–4377	73–1289	71–1040	9–619	0.1–3.1
		Summit	4.8–7.1	0.013–0.19	12.0–34.8	223–4824	32–723	53–1252	10–733	0.12–2.9
		Mean	4.8–6.8	0.016–0.23	12.4–39.1	407–4855	58–990	73–950	13–579	0.11–2.4
8	Tepid-submoist	Bottom	6.1–8.0	0.001–0.16	26.0–76.4	2707–10,098	575–2210	107–594	28–87	0.19–2.2
		Flat	5.7–8.2	0.001–0.21	17.8–76.1	2228–15,595	172–2299	120–1129	21–249	0.17–15
		Slope	5.7–8.2	0.000–0.20	22.5–74.3	1920–17,294	182–2359	81–1143	10–269	0.17–18
		Summit	5.8–7.6	0.000–0.20	18.9–78.5	2005–13,613	323–2359	148–1087	24–128	0.17–7.9
		Mean	5.8–8.0	0.001–0.19	21.3–76.3	2215–14,150	313–2307	114–988	21–183	0.18–11
9	Warm-submoist	Bottom	6.1–8.8	0.023–0.25	20.6–77.7	2194–15,171	250–2440	97–870	23–153	0.14–26
		Flat	6.1–8.1	0.000–0.24	29.7–78.3	2324–18,718	215–2399	99–664	29–119	0.07–36
		Slope	6.1–8.5	0.001–0.27	26.7–80.1	2114–18,858	198–2364	88–809	19–318	0.11–35
		Summit	6.6–7.3	0.049–0.16	34.9–63.4	2779–8463	364–1744	164–537	22–189	0.28–2.4
		Mean	6.2–8.2	0.018–0.23	27.9–74.9	2353–15,303	257–2237	112–720	23–195	0.15–25

(continued)

Table 6.5 (continued)

No.	AEZ	GMK	pH	EC (dSm^{-1})	CEC cmol (+)kg^{-1}	Ca (mg/kg)	Mg (mg/kg)	K (mg/kg)	Na (mg/kg)	CaCO$_3$ (%)
10	Warm-humid	Bottom	5.5–6.3	0.012–0.17	16.6–37.3	1250–5469	231–1082	101–632	20–49	0.12–0.9
		Flat	5.3–6.4	0.017–0.18	14.8–42.3	978–4996	219–1142	109–1110	16–212	0.12–0.9
		Slope	5.2–6.7	0.025–0.22	12.7–44.7	716–6152	208–954	99–1734	9–681	0.11–2
		Summit	5.3–6.0	0.05–0.17	16.5–29.7	776–3076	207–585	207–765	126–126	0.12–1.6
		Mean	5.3–6.4	0.026–0.19	15.2–38.5	930–4923	216–941	129–1060	43–267	0.12–1.3
11	Cold to very cold submoist-humid	Bottom	5.7–6.6	0.035–0.15	19.9–36.5	1276–4710	161–591	134–364	25–123	0.26–1.3
		Flat	5.8–6.2	0.076–0.13	21.0–25.3	1317–2571	185–309	147–357	23–112	0.76–1.4
		Slope	5.6–6.9	0.004–0.16	12.4–43.0	660–5340	83–990	70–463	18–93	0.14–8
		Summit	5.5–7.0	0.01–0.15	13.8–41.4	636–5501	84–1069	79–952	15–99	0.13–6
		Mean	5.7–6.7	0.031–0.15	16.8–36.5	972–4531	128.3–740	107–534	20–107	0.32–4.2

Source ATA (2016, 2018, 2019), overlay analysis by the authors

Muger, Jemma, Welaka, and Beshilo subbasins which could be the parent material's reflections and a concentration of basic cations in the surface layers because of minimal leaching.

Acrisols, Luvisols, and Nitisols are the soils with highly acidic reactions in Ethiopia (Agegnehu et al. 2019; Kidanemariam et al. 2013). In the Injibara area, Selassie (2002) noted pH values as low as 4.81. According to Elias (2016), the northcentral and southwestern highlands of Ethiopia are covered by 80% of the Nitisols and Luvisols reference soil groups, which are very strongly to strongly acidic with a pH of 4.5–5.5.

Soils with parent materials rich in silica (rhyolite and granite), large quantities of sand with low buffering capabilities, and areas with higher precipitation rates are all conducive condition to acidic soil development (Foster et al. 2016). Ammonia fertilizers, which contain nitrogen (N), are a source of acidification from a management standpoint (Fageria and Nascente 2014). Ammonia-based fertilizers (NH4), urea-based fertilizers ($CO(NH_2)_2$), and proteins (amino acids) are all forms of organic fertilizers that contain hydrogen. Acidity in the soil is produced when these types of N fertilizers are converted into nitrate (NO_3), which releases hydrogen ions (H^+).

In the western part of Ethiopia, in the overlaying Abbay basin, due to inadequate nutrient availability, poor soil microbial activity, and Al toxicity, acidic soil productivity is low and declines rapidly (Agegnehu et al., 2021). High levels of aluminum (Al) and manganese (Mn) and the associated lack of phosphorus (P), magnesium (Mg), calcium (Ca), and potassium (K) in locations with strong soil acidity (pH 5.5) limit crop growth (Tables 6.6 and 6.7) (Agegnehu et al. 2019; Mosissa 2018; Kidanemariam et al. 2013). Therefore, for both severely acidic and moderately acidic soils in the Abbay Basin, a suitable rate of lime must be applied, or cultivation of acid tolerant crops is advised for maximum crop yields. Moreover, the western and central portions of the Abbay basin should receive priority for integrated acid soil management activities.

Soil Salinity

The term "soil salinity" refers to the buildup of water-soluble salts, primarily composed of sodium but may also include potassium, calcium, and magnesium, which can be chlorides, sulfates, or carbonates forms. These can have a negative impact on plant growth, land use, and soil erosion. Most of saline soils are found in arid places where there is insufficient precipitation to remove the salts from the soil, which causes the salts to accumulate over time. Salinity levels are usually determined by measuring the electrical conductivity of soil/water suspensions.

The basin surface soil electrical conductivity (EC), across all AEZs ranged from trace (0.000) to 0.448 dSm^{-1} (Table 6.5). According to the EthioSIS EC rating (Ethiosis 2016), all the surface soils of the basin are salt free. In Ethiopia, the majority of the country's western, northern, and southern regions are salinity-free. However,

Table 6.6 Soil pH, exchangeable acidity, and CEC status across the Abbay basin soils

Location	pH	Ex Ac (cmol (+)/kg)	CEC (cmol(+)/kg)	Reference
Didessa Watershed	4.7–5.2	1.4–10.0	24–39	Deressa et al. (2020)
Choke Mountain	4.3–6.8	0.12–1.91	22–56	Yeneneh et al. (2022)
West Wollega	4.1–5.8	0.14–7.56	–	Deressa (2013)
East Wollega	4.2–6.4	0.10–10.9	–	
West Showa	4.9–5.9	0.07–1.44	–	
Yigossa Watershed	5.6–7.9	–	–	Selassie et al. (2014)
Bedele District	4.5–6.8	0.1–5.1	2.6–35.3	Sori et al. (2021)
Sibu Sire District	4.2–6.6	0.00–6.18	7.2–66.0	Takele and Iticha (2020)
Aba Gerima Catchment	4.7–6.0	–	–	Tiruneh et al. (2022)
Andit-Tid Watershed	4.8–6.3	–	16.9–72.3	WLRC (2016b)
Anjeni Watershed	4.8–6.0	–	15.3–34.5	
Atari-Mesk Watershed	5.7–6.7	–	39.1–55.6	
Debre-Yakob Watershed	5.2–6.1	–	23.4–45.1	
Debre-Mewi Watershed	6.3–7.9	–	12.2–64.2	
Gerda Watershed	4.7–6.4	–	8.6–51.9	
Gosh Watershed	4.9–6.5	–	12.3–53.5	
Dembia	5.5	–	44	Getahun and Selassie (2017)
Fogera	5.5	–	76	
Kunzila Alefa	5.7	–	58	
Mecha	4.7	–	47	
Bako Tibe District	5.2–6.6	–	25–51	Dinssa and Elias (2021)
Gerado Watershed	6.8–7.2	–	45.5–50.29	Asmamaw and Mohammed (2013)
TaraGedam Watershed	5.49–6.51	–	23.41–56.3	Menale (2014)

Table 6.7 Soil pH and CEC range values across Abbay subbasins

Sub basin	District	AEZ	pH	CEC (Cmol + kg^{-1})	Reference
Dabus	Ayira Guliso	Warm-subhumid	4.1–7.0	16–29	ORLEPB (2014a), OIDA (2018)
	Nejo	Tepid-subhumid	3.8–7.0	16–29	
Didessa	Amuru	Warm moist	5.0–6.7	12–45	ORLEPB (2014b), OIDA (2018)
	Bedele	Tepid-humid	4.0–5.9	17–49	
	Limu	Tepid moist	4.0–6.5	12–47	
	Limu Seka	Warm-humid	4.4–6.0	20–48	
Dinder	Alefa	Warm moist	5.4–7.5	8–48	WLRC (2016a)
	Dangur	Warm moist	6.3–8.8	13–55	
	Quara	Hot moist	5.7–8.8	18–57	
Guder	Ambo	Tepid moist	3.7–6.9	13–67	OIDA (2018), ORLEPB (2011)
	Dandi	Cool submoist-humid	3.8–6.9	13–67	
	Ginde Beret	Warm-subhumid	3.7–6.4	13–58	
	Jeldu	Tepid-subhumid	3.7–8.2	13–62	
Jemma	Angolala Tera	Cool submoist-humid	4.6–6.1	17–57	OIDA (2018), ORLEPB (2011)
	Dera	Warm moist	3.7–8.1	13–69	
	Kimbibit	Cool submoist-humid	4.6–6.1	17–57	
	Wuchale	Tepid moist	4.7–8.1	27–63	
Fincha	Abbay Chomen	Tepid moist	4.7–7.1	13–58	OIDA (2018), ORLEPB (2011)
	Guduru	Warm-subhumid	4.6–6.4	13–58	
	Jarti	Warm-moist	4.6–6.4	13–58	
	Jimma Horo	Tepid-subhumid	4.7–7.8	13–62	
Rahad	Alefa	Warm moist	5.3–7.9	8–53	WLRC (2016a)
	Chilga	Tepid moist	5–6.3	17–48	
	Metema	Hot moist	6.2–7.5	19–60	
	Quara	Warm moist	4.8–7.9	3–61	
Muger	Degem	Cool submoist-humid	5–7.8	13–69	ORLEPB (2011, 2018), OIDA (2018)
	Kuyu	Tepid moist	4.5–7.8	13–69	
	Were Jarso	Warm moist	3.7–8.1	13–69	

the development of large-scale irrigation projects in the basin without adequate irrigation water management and drainage systems for salinity control will have led to a rapid buildup of soil salinity and sodicity problems, which will further lead to the complete loss of land for crop cultivation.

Cation-Exchange Capacity and Exchangeable Bases

The soil's ability to store and exchange cations is known as its cation-exchange capacity. It acts as a buffer against variations in pH and the amount of available nutrients. Soil stability, nutrient availability for plant growth, soil pH, and the soil's response to fertilizers and other ameliorants are all significantly influenced by it (Hazelton and Murphy 2007). A low CEC indicates that the soil is less resistant to alterations in soil chemistry brought on by changes in land use. The amount of organic matter, soil pH, clay type, and clay percentage all affect a soil's CEC. Measurements of cation-exchange capacity are frequently performed as part of an overall evaluation of a soil's potential fertility and potential response to fertilizer application. In some cases, the CEC data can also be utilized as a general indicator of the kinds of clay minerals present (Landon 2014).

The CEC values of soils of warm moist AEZs varied from 11.4 to 102.2 cmol (+) kg^{-1} with a mean value of 40.7 cmol (+) kg^{-1} (ATA 2016, 2018, 2019) (Table 6.5). The CEC values of bottom, flat, slope, and summit landforms, of warm moist AEZs, varied from 44.7 to 101.9, 40.6 to 119.9, 10.9 to 105.9, and 12.0 to 81.2 cmol (+) kg^{-1}, respectively, indicating wide range of variation from low to very high (Landon 2014). The mean values were found to be 44.7, 40.6, 39.0, and 38.7 for bottom, flat, slope, and summit landforms of warm moist AEZs, respectively (Table 6.5). The mean CEC values in warm moist AEZs are within a high and very high range (Landon 2014).

For hot moist, tepid submoist, and warm submoist AEZs, the mean CEC values were found to be 65.6, 50.7, and 52.8 cmol (+) kg^{-1}, respectively, which are very high according to Landon (2014). High CEC values are mainly due to high clay content and the predominance of 2:1 clay mineral. Similarly, the mean CEC values were found to be 23.4, 23.5, 19.0, and 23.5 cmol (+) kg^{-1} for tepid subhumid, warm subhumid, tepid humid, and cold to very cold submoist-humid AEZ, respectively, which are medium according to Landon (2014). The CEC values of tepid subhumid, warm subhumid, tepid humid, and cold to very cold submoist-humid AEZs soils varied from 7.9 to 74.7, 10.0 to 69.6, 12.4 to 39.1, and 16.8 to 36.5 cmol (+) kg^{-1}, respectively, which are low to very high ranges according to Landon (2014). Acidic soils as Acrisols (Selassie 2002), Nitisols (Girma and Ravishankar 2004), and Plinthosols (Shimelis et al. 2007) have lower CEC.

Among the exchangeable bases, the mean exchangeable K values (melich-III extractable K) of the soils in the warm moist, tepid moist, tepid subhumid, warm subhumid, and hot moist AEZ soils varied from 30 to 1304, 36 to 1853, 19 to 1296, 20 to 1077 and 23 to 693 mg kg^{-1} (Table 6.5), respectively, indicating wide range from very low to very high according to EthioSIS rating (EthioSIS 2016). The mean exchangeable K values of the soils of tepid submoist, warm submoist and warm humid AEZs soils varied from 114 to 988, 112 to 720, and 129 to 1060 mg kg^{-1}, respectively, indicating wide range from low to very high according to EthioSIS rating (EthioSIS 2016). Exchangeable K in the western part of the basin (presumed to be high rainfall areas) is lower, due to leaching of basic cations, than the eastern counterparts which are characterized by low rainfall amounts.

According to EthioSIS critical levels (EthioSIS 2016) for Mehlich-III extractable exchangeable K, 2.87, 34.87, 61.95, 0.3, and 0.01% of the basin soils were found to be very low, low, optimum, high, and very high, respectively (Fig. 6.9). This indicates that 37.74% the soils of the Abbay basin are deficient in K. Low and very low exchangeable K values were found in most parts of the Rahad, Dinder, Belles, Wonbera, Dabus, and some parts of the southwestern Gojjam, Anger, and Didessa subbasins. Optimum and high exchangeable K values were found in most parts of the Tana, Welaka, Jemma, Muger, Guder, Fincha, and Didessa subbasins. Laekemariam et al. (2018) reported that 14.8% of Nitisols in southern Ethiopia were K-deficient based on Mehlich-III extractable exchangeable K rating. Hailu et al. (2015a, b) and Lelago et al. (2016) reported optimum amounts of exchangeable K, Ca, and Mg in Vertisols of central highlands.

The mean exchangeable Ca contents of soils in warm moist, tepid moist, tepid subhumid, warm subhumid and cool submoist humid AEZs were found to vary from 402 to 2601, 222 to 25,427, 94 to 16,663, 153 to 10,280, and 511 to 12,976 mg kg^{-1}, respectively, indicating wide ranges. The mean exchangeable Mg values were found to be 870, 865, 395, 438, and 868 mg kg^{-1}, for warm moist, tepid moist, tepid subhumid, warm subhumid, and cool submoist humid AEZs soils, respectively. Similarly, the mean exchangeable Na values were found to be 66, 60, 44, and 33 mg kg^{-1}, for warm moist, tepid moist, tepid subhumid, warm subhumid, and cool submoist humid AEZ soils, in the same order.

The contents of basic cations (Ca, Mg and Na) in all AEZs are high (Landon, 2014) indicating their accumulation in the surface soils of the basin. The mean relative abundances of basic cations in the exchange complex for all AEZ and geomorphic units were in the order of $Ca^{2+} > Mg^{2+} > K^+ > Na^+$ (Table 6.5) similar to the findings of Tadele et al. (2013) in Anjeni and Yihenew (2002) in Adet area soils. This could be related to the charge density where the divalent cations (Ca and Mg) have higher affinity toward the colloidal sites than monovalent cations (K and Na). High calcium carbonate contents affect soil physical and chemical properties (Cardone et al. 2020). Approximately, 93% of the basin lands mass surface soils contain a very low $CaCO_3$ level of less than 1% indicating no carbonate problem. However, approximately 7% of the basin surface soils in the Muger, Jemma, Welaka, Guder, Tana, and North Gojjam subbasins contained greater than 1% $CaCO_3$ (Table 6.5), which is also is confirmed by the occurrence of calcic and calcaric principal qualifiers in the lowland plains and highlands of the basin, respectively.

Exchangeable bases were higher in warm submoist, tepid submoist, and hot moist AEZs but lower in tepid humid, cold to very cold submoist, and tepid humid AEZs, possibly due to base leaching by high rainfall. Exchangeable bases were inconsistent

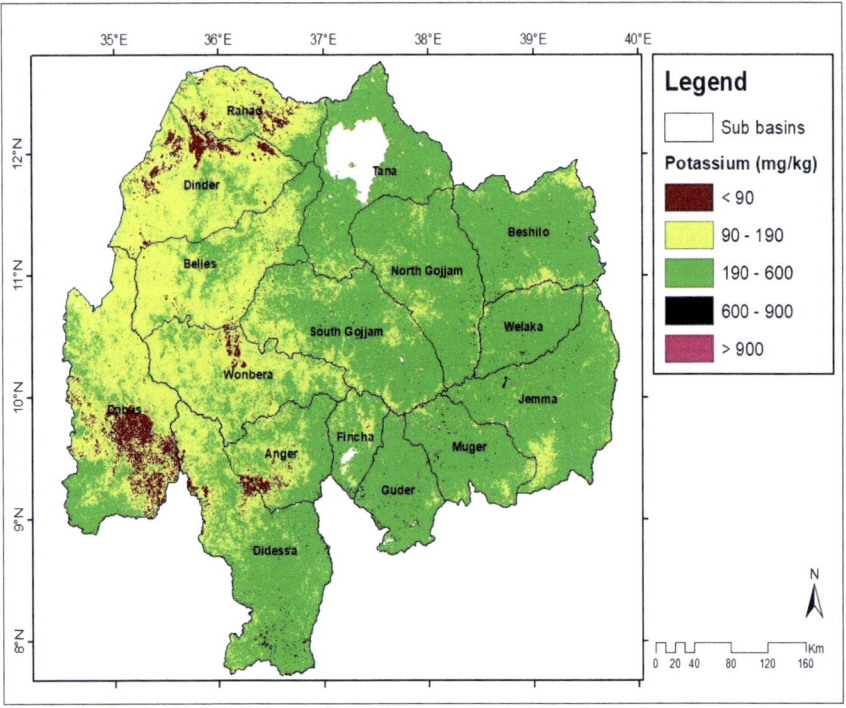

Fig. 6.9 Spatial distribution of exchangeable K in Abbay basin surface (0–20 cm) soils (Adapted from ATA 2016, 2018, 2019)

along the landforms. The studied soils had very low to medium Na, low to very high K, Ca, and Mg, following EthioSIS (2014) ratings. Previous studies in Ethiopia's agroecological settings found similar results (Tables 6.6 and 6.7) (Abate et al. 2014; Abu 2021; Ali et al. 2010; Bekele et al. 2021).

Organic Matter and Total Nitrogen

Soil organic matter (SOM), which is variable in composition, consists of the remains of soil organisms, plant roots, and plant material in varying states of decomposition and synthesis. Organic matter (OM), despite being present in relatively small proportions in soils, has a significant impact on soil aggregation, the availability of nutrient reserves, moisture retention, and biological activity. This dynamic soil component has a significant impact on a variety of soil physical, chemical, and biological characteristics as well as ecosystem functions. The majority of the soil's cation-exchange capacity (CEC) and water-holding capacity are provided by soil organic matter. It is a crucial element influencing soil biological activity and a major source of carbon for many species of soil bacteria.

According to EthioSIS critical levels for organic matter (OM), (EthioSIS 2016) 4.48, 27.59, 65.01, 1.21, and 1.72% of Abbay basin soils were found to be very low, low, optimum, high, and very high, respectively (Fig. 6.10). Soil organic matter was highly variable among AEZs and within each landform. The mean organic matter values of the soils in the warm moist, tepid moist, tepid subhumid, warm subhumid, and cool submoist humid AEZs were found vary from 0.51 to 8.34, 0.37 to 15.81, 0.63 to 18.62, and 1.37 to 12.11%, respectively, indicating wide ranges from very low to very high (Table 6.8). The only very high OM mean values (10.33%) were found in cold to very cold submoist-humid AEZs. Low OM mean values were found in the hot moist (2.73%), tepid submoist (2.4%), and warm submoist (2.23%) AEZs. Although, OM values vary along various geolandforms, they are low in slope and summit geolandforms.

Total nitrogen is a measure of all the nitrogen in the soil, a significant portion of which is stored in organic matter and not immediately available to plants. It might mineralize into available forms. The forms of nitrogen that are most accessible to

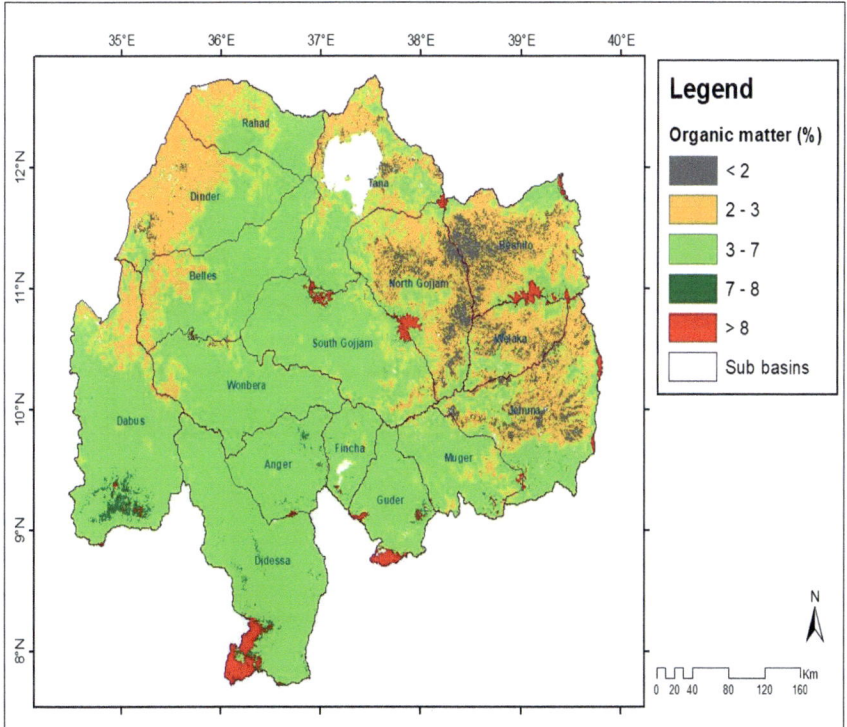

Fig. 6.10 Spatial distribution of organic matter in Abbay basin surface soils (Adapted from ATA 2016, 2018, 2019)

Table 6.8 Range values of selected soil fertility and micronutrient under different AEZs

No.	AEZ	GMK	OM	TN	Av P	Av S	Fe	Mn	Zn	Cu
			%		mg/kg					
1	Warm-moist	Bottom	0.5–7.8	0.05–0.3	0.05–90	0.05–30.3	17–446	19.5–319	0.05–5.1	0.54–8.7
		Flat	0.4–9.2	0.04–0.37	0.05–279	0.04–81.5	14.4–467	9.5–467	0.02–75.7	0.29–11.9
		Slope	0.2–9.2	0.01–0.51	0.1–241	0.05–51.3	8.5–462	5.4–496	0.02–12.3	0.32–12.1
		Summit	0.9–7.1	0.05–0.33	0.1–62	3.11–20.4	51.2–282	20.5–270	0.28–4.8	0.54–10.8
		Mean	0.5–8.3	0.04–0.38	0.08–168	0.81–45.9	22.8–414	13.7–388	0.09–24.5	0.42–10.9
2	Tepid moist	Bottom	0.4–13.9	0.04–0.35	0.05–234	0.25–50.2	19.4–499	19.5–310	0.18–15.6	0.77–8.8
		Flat	0.3–19.8	0.02–0.58	0.05–294.7	0.05–98.7	7.4–527	11.6–470	0.12–23	0.35–13.2
		Slope	0.4–13.9	0–0.56	0.05–298	0.04–114.4	14.8–483	9.8–367	0.08–14.3	0.34–28.2
		Summit	0.4–15.6	0.01–0.52	0.05–292	0.03–41.7	25.6–505	3.4–358	0.06–19.6	0.34–13.4
		Mean	0.4–15.8	0.02–0.5	0.05–279.7	0.09–76.3	16.8–503	11.1–376	0.11–18.1	0.45–15.9
3	Tepid-subhumid	Bottom	0.6–17.5	0.07–0.31	0.6–131.3	0.05–33.9	37.9–449	8.5–339	0.02–8.2	0.55–8.5
		Flat	0.6–18.5	0.05–0.49	0.05–133.6	0.06–38.5	30.4–516	8.8–535	0.01–13.3	0.31–8.3
		Slope	0.6–18.9	0.02–0.56	0.03–245	0.03–850	4–451	4.7–535	0.01–292	0.01–22.8
		Summit	0.8–19.5	0.05–0.58	0.05–212.8	0.03–44.7	32.2–489	3.8–375	0.01–35.4	0.14–9.4
		Mean	0.6–18.6	0.05–0.48	0.18–180.7	0.04–241.8	26.1–476	6.5–446	0.01–87.2	0.25–12.3
4	Warm-subhumid	Bottom	1.7–15.2	0.12–0.12	0.69–169	0.01–35.9	35.3–543	4.6–412	0.02–10	1.13–9.4
		Flat	0.9–12.7	0.07–0.34	0.05–118.4	0.02–34.1	30.2–504	6.8–384	0–12.4	0.2–11.3
		Slope	0.8–12.4	0.05–0.36	0.03–265	0–34.7	26.7–519	2.6–442	0–18.1	0.09–27.9
		Summit	2–8.2	0.03–0.48	0.05–113.3	0.04–28.8	30.6–405	20.9–392	0.03–7.4	0.53–7
		Mean	1.4–12.1	0.07–0.32	0.21–166.5	0.02–33.4	30.7–493	8.7–407	0.01–12	0.49–13.9

(continued)

Table 6.8 (continued)

No.	AEZ	GMK	OM	TN	Av P	Av S	Fe	Mn	Zn	Cu
			%		mg/kg					
5	Cool submoist-humid	Bottom	1.2–13.3	0.08–0.55	3–85	4.12–24.8	78.9–426	13.2–274	0.7–5.8	0.65–7.4
		Flat	0.6–18.3	0.05–0.53	0.94–249	0.03–29	66.1–457	13.2–276	0.38–17.2	0.57–10
		Slope	0.7–19.1	0.01–0.7	0.05–273.3	0.05–43.8	55.2–405	5.5–389	0.38–9.7	0.12–10.5
		Summit	0.6–19.3	0.05–0.59	0.12–288.9	0.03–37.3	60.1–498	8.3–295	0.45–15.3	0.34–11.7
		Mean	0.8–17.5	0.05–0.59	1.03–224	1.05–33.7	65.1–447	10–308	0.48–12	0.42–9.9
6	Hot-moist	Bottom	1.3–3.9	0.1–0.16	0.25–2.6	3.89–23.6	64.9–270	18.8–266	0.58–2.7	1.36–5.6
		Flat	1.8–4.2	0.07–0.26	0.1–35.2	4.07–24	47.5–286	11.4–266	0.48–3.1	0.78–7.3
		Slope	1.9–4.4	0.11–0.12	0.1–2.9	4.79–17.3	61.8–284	17.2–231	0.37–2.4	0.98–6.2
		Mean	1.6–4.1	0.09–0.18	0.15–13.6	4.25–21.6	58–280	15.8–254	0.48–2.7	1.04–6.4
7	Tepid-humid	Bottom	3.5–13.2	0.28–0.34	2.5–22	5.95–25.1	69.5–345	33.3–374	0.9–9.5	0.58–5.1
		Flat	3.5–13.3	0.16–0.42	0.8–26.2	5.13–39.5	60.2–335	35.2–373	0.51–15.5	0.48–5.5
		Slope	2.7–13.4	0.12–0.37	0.05–103.4	0.73–41	53.5–432	31.3–431	0.64–16.9	0.1–6.9
		Summit	3.8–17.3	0.19–0.52	0.56–69.5	0.29–30.2	54–383	18.1–424	0.49–18.5	0.17–8.3
		Mean	3.4–14.3	0.19–0.41	0.98–55.3	3.02–34	59.3–374	29.5–400	0.64–15.1	0.33–6.4
8	Tepid-submoist	Bottom	1.3–3.8	0.14–0.22	1.14–38	3.3–23.9	72.9–350	71.9–224	0.68–3.9	1.39–6.9
		Flat	0.3–5.9	0.05–0.21	0.78–276	3.44–24.1	27.3–468	51.3–272	0.7–8.5	0.68–10
		Slope	0.5–5.1	0.02–0.27	1–291	2.41–21.5	24–367	27.6–249	0.38–9.9	0.43–10.2
		Summit	1.2–6.2	0.1–0.18	1.7–141	3.01–23.6	76–325	38.2–220	0.71–5	0.79–8.1
		Mean	0.8–5.2	0.08–0.22	1.16–186.5	3.04–23.3	50–377	47.2–241	0.62–6.8	0.82–8.8

(continued)

Table 6.8 (continued)

No.	AEZ	GMK	OM	TN	Av P	Av S	Fe	Mn	Zn	Cu
			%		mg/kg					
9	Warm-submoist	Bottom	0.5–4.2	0.11–0.19	1.89–124	4.6–44.3	26.3–454	64.6–263	0.73–5.2	0.96–7
		Flat	0.8–4.7	0.08–0.16	1.7–146	4.44–20.1	17.2–353	40.1–280	0.63–4.7	1.02–6.9
		Slope	0.4–5.4	0.02–0.18	1.6–165	2.44–44.4	17.4–362	32.1–275	0.42–4.6	0.77–7.9
		Summit	1.3–2.9	0.05–0.14	3.56–33.4	5.33–17.6	76.8–191	96.4–201	0.91–2.8	1.31–5.8
		Mean	0.7–4.3	0.06–0.17	2.19–117.1	4.2–31.6	34.4–340	58.3–255	0.67–4.3	1.02–6.9
10	Warm-humid	Bottom	3.7–7.7	0.3–0.3	10.7–19	3.92–18.6	79.6–446	54.9–335	0.56–9	0.98–4.7
		Flat	3.3–7.7	0.22–0.37	0.25–66.5	0.38–28.8	77.5–402	51.4–343	0.38–10.1	0.51–4.6
		Slope	3.3–10.2	0.13–0.35	1–142.4	1.55–34.9	67.5–376	26–437	0.46–15	0.12–7.9
		Summit	4.4–6.6	0.19–0.35	2.6–2.6	4.28–20.7	74.4–210	93.2–270	1.16–6.3	1.24–3.6
		Mean	3.6–8	0.21–0.34	3.64–57.6	2.53–25.8	74.8–358	56.3–346	0.64–10.1	0.71–5.2
11	Cold to very cold submoist-humid	Bottom	4.2–10.4	0.21–0.5	5.93–16.7	8.34–14.9	106.1–212	13.3–68	1.21–2.6	0.97–2.9
		Flat	6.2–15.9	0.15–0.81	10.4–22.9	11.84–14.9	96.4–162	10.9–55	1.34–2.5	0.7–2.4
		Slope	3–20.2	0.58–0.58	6–83	5.57–25.6	68.1–239	8.3–124	0.59–3.6	0.37–3.5
		Summit	3.2–19.8	0.24–0.24	6–63	7.75–20.6	74.5–328	8.7–105	0.66–4	0.39–2.9
		Mean	4.2–16.6	0.3–0.53	7.08–46.4	8.38–19	86.3–235	10.3–88	0.95–3.2	0.61–2.9

Source ATA, Agricultural Transformation Agency (2016, 2018, 2019), adapted by the authors

plants are nitrate (NO^{3-}) and ammonia, which are produced when the nitrogen in the soil's organic matter pool is mineralized. Nitrate is easily leached or can be vulnerable to denitrification into nitrous oxide and nitrogen gas when there is a water logging condition. Nitrate and ammonium concentrations in soil are influenced by biological activity and change in environmental factors like temperature and moisture. When there is a lot of rain or irrigation, nitrate can easily be leached from the soil. According to EthioSIS critical levels for total nitrogen (TN) (EthioSIS 2016), 17.75, 31.71, 42.77, 6.96, and 0.81% of Abbay basin soil were found to be very low, low, medium, high, and very high, respectively (Fig. 6.11). The soil TN content was highly variable among AEZs and within each landform. The mean TN values of the soils in the warm moist, tepid moist, tepid subhumid, warm subhumid, and cool submoist humid AEZs were found to be vary from 0.037 to 0.377, 0.02 to 0.50, 0.046 to 0.48, and 0.06 to 0.32%, respectively, indicating wide ranges from very low to very high (Table 6.8).

Approximately, 0.31 and 0.42% of high TN mean values were found in the tepid humid and cold to very cold submoist-humid AEZs, respectively. In general, high and very high OM and TN values were found in the Dabus, Didessa, Wonbera, and Guder subbasins, and in some parts of the South Gojjam, North Gojjam, and Beshilo subbasins. The organic matter buildup in these soils is related to natural vegetation (forest), coffee plantation, cropping history, and temperature. Abegaz et al. (2021)

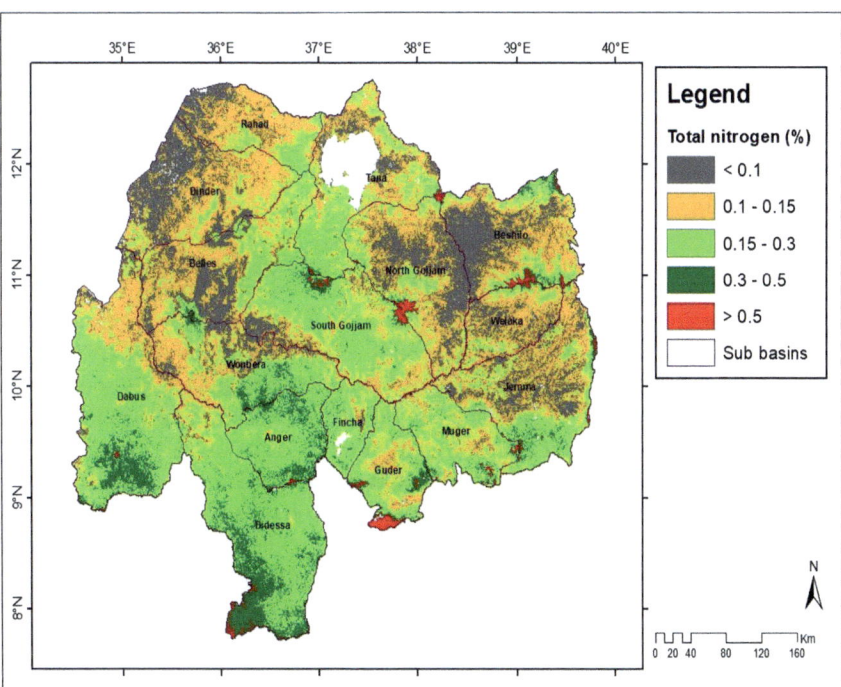

Fig. 6.11 Spatial distribution of total nitrogen in Abbay basin soils (Adapted from ATA 2016, 2018, 2019)

reported that the highest mean OM and TN were observed in alpine vegetation and forest areas probably due to the cool climate, lower soil pH and the dense vegetation cover that results in higher inputs of litter to the soil. Low TN mean values were found in the warm moist, hot moist, and warm submoist AEZs with percentages of 0.14, 0.18 and 0.12%, respectively. The soil TN values were inconsistent along the landforms and subabsins. Most parts of the Beshilo, Belles, and Dinder subbasins of Abbay contain very low and low OM and TN values. Generally, the soil TN content followed the soil OM distribution pattern. This is because most of the nitrogen is in organic form and therefore becomes part of the soil organic matter (Lelago 2016). Low organic matter content is particularly severe in Vertisols, Leptosols, and Cambisols, while Nitisols and Luvisols have moderately adequate levels of organic matter (Agegnehu and Amede 2017; Elias 2016).

The substantial loss of soil organic matter is one of Ethiopia's biggest problems in managing soil fertility. The total removal of crop residues from fields for use as livestock feed or for home energy production is the main cause of severe organic matter depletion. Animal waste is processed into dung "cake" and used as a source of energy for homes (Elias 2016). Due to low OM and TN content, most of Ethiopia's cultivated land soils are unable to sustain crop production for an extended period, necessitating the use of inorganic fertilizers and management of soil OM. To increase crop productivity, it is advised to use organic leftovers as a significant source of nutrients in these agricultural fields. Most of Ethiopian soils have low OM and TN contents, and crops respond well to N fertilizers in these regions, according to research findings from Elias (2017), Fekadu et al. (2018), Sebnie et al. (2021), Zewudie (2003) (Tables 6.9 and 6.10).

Available Phosphorus

One of the nutrients that plants need most is phosphorus, which is also a component of several chemicals used in biochemical processes including respiration and photosynthesis. Adenosine triphosphate (ATP) and adenosine diphosphate (ADP) both contain a significant amount of phosphorus. These serve as the energy source for a variety of metabolic processes in plants and animals. Phosphorus (P) is the second most important plant element after nitrogen in terms of its overall impact on both natural and agricultural ecosystems (Weil and Brady 2017). The mean available phosphorus (Melich-III extraction) content varied from 0.08 to 168.0, 0.05 to 279.7, 0.18 to 180.7, 0.21 to 166.5, and 1.03 to 224.0 mg kg^{-1} soils in the warm moist, tepid moist, tepid subhumid, warm subhumid, and cool submoist-humid AEZs, respectively. The mean value of available P was found to be 10.2, 13.4, 6.9, 11.8, and 15.6 mg kg^{-1} soil in warm moist, tepid moist, tepid subhumid, warm subhumid, and cool submoist humid AEZs, respectively. Based on the critical level adopted by EthioSIS (2016) for Mehlich-III extractable P, 91.79%, 7.59%, 0.63, 0.01, and 0.01% basin soils were found to be very low, low, optimum, high, and very high, respectively (Fig. 6.12). This indicates that 99.3% of the soils of the basin are P deficient. The

Table 6.9 Status of the available P and OM at different locations with the Abbay basin

Location	Av P (Olsen) (mg/kg)	OM (%)	Reference
Didessa Watershed	2.3–8.7	4.3–11.0	Deressa et al. (2020)
Choke Mountain	–	1.2–6.6	Yeneneh et al. (2022)
Yigossa Watershed	1.78–12.3	1.88–3.56	Selassie et al. (2014)
Bedele District	0.8–38.6	0.3–5.6	Sori et al. (2021)
Sibu Sire district	1.0–49.0	0.04–6.64	Takele and Iticha (2020)
Aba Gerima catchment	4.07–38	0.48–4.9	Tiruneh et al. (2022)
Andit-Tid Watershed	2.5–20.8	2.8–24	WLRC (2016a)
Anjeni Watershed	0.9–6.0	2.7–5.5	
Atari-Mesk watershed	3.8–11.8	1.7–10.0	
Debre-Yakob Watershed	0.7–14.0	1.2–4.9	
Debre-Mewi Watershed	2–48	0.36–3.39	
Gerda Watershed	0.20–92	3.0–7.5	
Gosh Watershed	3.9–155	2.4–7.3	
Dembia	2.4	2.76	WLRC (2016a)
Fogera	1.2	3.97	
Kunzila Alefa	3.5	2.41	
Mecha	6.4	3.45	
Baso Tibe District	6.0–7.9	3.6–5.2	Dinssa and Elias (2021)
Gerado watershed	6.69–23.15	2.18–5	Asmamaw and Mohammed (2013)
TaraGedam watershed	0.03–56.9	1.81–8.93	Menale (2014)

parent material, degree of P fixation, soil pH, and slope gradient variations may also play a role in the variation in accessible P levels among the soils of the basin. Most Ethiopian soils have low levels of available P (Tables 6.9 and 6.10), which is a common characteristic (Bellete 2014; Kibret et al. 2023; Laekemariam 2015; Lelago et al. 2016; Gebreselassie et al. 2015; Gebreselassie 2002; Mamo et al. 2002; Negassa and Gebrekidan 2003).

Available Sulfur

Sulfur (S), which is categorized as a macronutrient, has long been recognized as a crucial component for plant growth and development. Two amino acids, cysteine and methionine, which are necessary for the synthesis of proteins, contain sulfur as a significant component. Animals are unable to reduce sulfate, hence plants are crucial for supplying them with the required amino acids that include S. It is used by plants in amounts similar to those of phosphorus. However, S is often forgotten

Table 6.10 Range values of OM and available P in different subbasins of Abbay basin soils

Subbasin	District	AEZ	OM (%)	Av P (Olsen) (mg kg^{-1})	References
Dabus	Ayira Guliso	Warm-Subhumid	0.2–10.4	0.4–32.8	ORLEPB (2014a), OIDA (2018)
	Nejo	Tepid-Subhumid	0.2–4.4	1.0–8.8	
Didessa	Amuru	Warm Moist	2.4–10.4	0.1–18.1	ORLEPB (2014b), OIDA (2018)
	Bedele	Tepid-Humid	4.4–11.8	3.1–18.1	
	Limu	Tepid Moist	4.0–10.4	0.0–13.1	
	Limu Seka	Warm-Humid	2.2–11.8	3.0–42	
Dinder	Alefa	Warm Moist	2.2–4.9	0.1–2.0	ADSE (2011)
	Dangur	Warm Moist	2.6–6.0	0.1–1.1	
	Quara	Hot Moist	2.6–6.1	0.1–4.3	
Guder	Ambo	Tepid Moist	0.7–9.6	0.6–56.7	OIDA (2018), ORLEPB (2011)
	Dandi	Cool submoist-humid	0.9–10.8	0.6–56.7	
	Ginde Beret	Warm-Subhumid	1.0–9.5	0.4–27.8	
	Jeldu	Tepid-Subhumid	0.7–10.3	0.3–27.8	
Jemma	Angolala Tera	Cool submoist-humid	1.6–5.7	0.3–56.7	OIDA (2018), ORLEPB 2011
	Dera	Warm Moist	1.4–9.5	0.4–65.8	
	Kimbibit	Cool submoist-humid	1.6–5.7	0.3–56.7	
	Wuchale	Tepid Moist	1.0–5.7	0.3–57.5	
Fincha	Abbay Chomen	Tepid Moist	1.0–5.7	0.4–27.8	OIDA (2018), ORLEPB (2011)
	Guduru	Warm-Subhumid	1.0–9.5	0.4–27.8	
	Jarti	Warm Moist	1.0–9.5	0.4–27.8	
	Jimma Horo	Tepid-Subhumid	0.9–5.9	0.4–39.1	
Rahad	Alefa	Warm Moist	2.2–4.9	0–1.47	ADSE (2011)
	Chilga	Tepid Moist	3.3–4.4	0.2–1.0	
	Metema	Hot Moist	3.1–4.6	0.1–0.9	
	Quara	Warm Moist	2.2–6.0	0.1–1.5	
Muger	Degem	Cool submoist-humid	0.9–6.2	0.3–64.6	ORLEPB (2011, 2018), OIDA (2018)
	Kuyu	Tepid Moist	0.7–6.2	0.3–27.8	
	Were Jarso	Warm Moist	0.7–9.6	0.4–27.8	

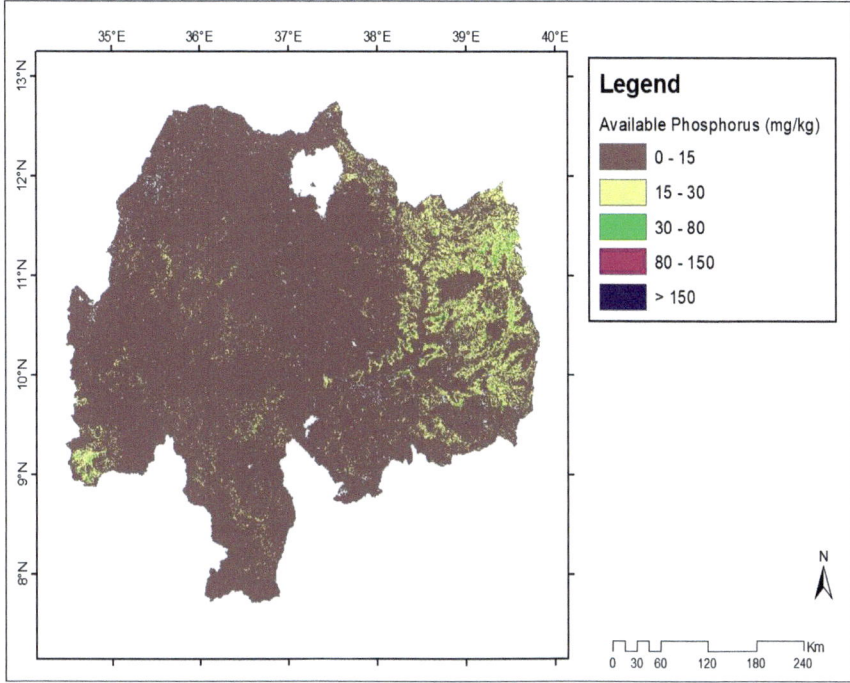

Fig. 6.12 Spatial distribution of available phosphorus in Abbay basin soils (Adapted from ATA 2016, 2018, 2019)

in discussions of soil fertility management which tend to focus on N, P, and K (Weil and Brady 2017). The mean available sulfur content (Mehlich-III extraction) varied from 0.81 to 45.9, 0.09 to 76.3, 0.04 to 241.8, 0.02 to 33.4, and 1.05 to 33.7 0 mg kg^{-1} soils in warm moist, tepid moist, tepid subhumid, warm subhumid, and cool submoist-humid AEZs. The mean value of available S was found in warm moist, tepid moist, tepid subhumid, warm subhumid, and cool submoist- humid AEZs. On the basis of the critical level adopted by EthioSIS (2014) for Mehlich-III extractable S, approximately 63.4, 35.9, and 0.7% of the Abbay basin soils were very low, low, and optimum, respectively (Fig. 6.13). This indicates that approximately 99.3% of the Abbay basin soils were S deficient.

In general, most of the agricultural soils of the Abbay basin were deficient in S (< 20 mg/Kg) and likely to respond to S fertilization. This might also be one of the factors that result in lower agricultural yields. Laekemariam (2015) and Lelago et al. (2016) reported that S was deficient in soils collected from the different parts of Ethiopia. Poor parent material that contains sulfur, land degradation, crop residue removal, crop mining, low soil organic matter (OM), usage of non-S fertilizers, and poor management techniques could all contribute to this. While most grain crops absorb sulfur in proportions between 10 and 30 kg/ha, similar to those of phosphorus, there has only been limited external addition of this nutrient. To increase agricultural

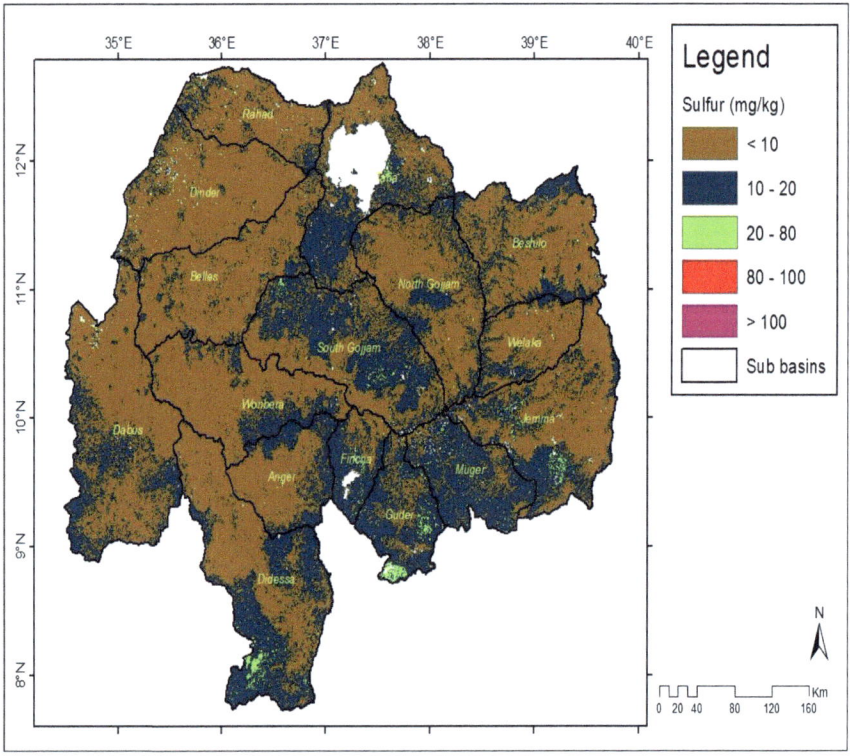

Fig. 6.13 Spatial distribution of available sulfur in Abbay basin soils (Adapted from ATA 2016, 2018, 2019)

production, S-containing fertilizers should be added to the fields in the Abbay basin as needed by a particular crop.

Available Micronutrients

Although plants only require trace amounts of micronutrients (Fe, Mn, Zn, and Cu), in recent years, their significance in crop productivity has grown. In plant nutrition, micronutrients play a variety of intricate roles. The specific roles played by the various micronutrients in plant and microbial growth processes vary widely, even though most micronutrients contribute to the operation of several enzyme systems. In many parts of the world, they are thought to be the micronutrients that have the most yield-limiting effects on crop production (Fageria 2009).

The mean available iron content (Melich-III extraction) varied from 22.8 to 414.4, 16.8 to 503.4, 26.1 to 476.0, 30.7 to 492.9, and 64.1 to 446.5 mg kg^{-1} soils in the warm moist, tepid moist, tepid subhumid, warm subhumid, and cool submoist humid AEZs, respectively, indicating wide variability within and among AEZ soils (Table 6.8). The mean values of available Fe were 126.5, 155.6, 134.0, 136.4, and 186.6 mg kg^{-1} soil in the warm moist, tepid moist, tepid subhumid, warm subhumid, and cool submoist humid AEZs, respectively. On the basis of the critical level adopted by EthioSIS (2016) for Mehlich-III extractable Fe, the mean Fe content of most the basin soils was found to be very high (> 100 mg/kg). Low Fe contents (< 25 mg/kg) were observed in flat and slope landforms of moist tepid and warm submoist AEZs.

As depicted in Table 6.8, Melich-III extracted mean available manganese (Mn) content varied from 13.7 to 388.0, 11.1 to 376.1, 6.5 to 446.1, 8.7 to 407.3, and 10.0 to 308.3 mg kg^{-1} soil in the warm moist, tepid moist, tepid subhumid, warm subhumid, and cool submoist-humid AEZs, respectively, indicating wide variability (from very low to very high range) within and among AEZs. The mean value of available Mn was found to be 114.2, 116.3, 106.0, 122.7, and 152.9 mg/kg soil in warm moist, tepid moist, tepid subhumid, hot moist, and tepid humid AEZs, respectively, which were optimum for crop production in accordance with ratings by EthioSIS (2016). On the other hand, the mean value of available Mn was found to be 93.3 and 88.7 mg/kg soils in warm subhumid and cool submoist-humid AEZs, respectively, which were low a (EthioSIS, 2016).

The Melich-III extracted mean available zinc (Zn) content varied from 0.09 to 24.47, 0.11 to 18.13, 0.01 to 87.22, 0.01 to 11.96, and 0.48 to 12.0 mg/kg soil in the warm moist, tepid moist, tepid subhumid, warm subhumid, and cool submoist-humid AEZs, respectively, indicating wide variability (from very low to very high range) within and among AEZs. The mean value of available Zn was found to be 1.69, 1.54, 2.06, 3.83 and 1.75 mg/kg soil in tepid moist, tepid subhumid, cool submoist-humid, tepid humid, and tepid submoist AEZs, respectively, which were optimum (EthioSIS, 2016) for crop production. On the other hand, the low mean value of available Zn was found to be 1.24, 1.17, and 1.09 mg/kg soils in warm moist, warm subhumid, and hot moist AEZs, respectively, indicating that it may be deficient and require Zn fertilizer application.

Based on the critical level adopted by EthioSIS (2014) for Mehlich-III extractable Zn, approximately 26.8, 35.2, 38.0, 0.01, and 0.0003% of Abbay basin soils were found to be very low, low, optimum, high, and very high, respectively (Fig. 6.14). This reveals that approximately 62% the basin soils were deficient in Zn. Low and very low available Zn values were found in most western parts of the basin such as in the Rahad, Dinder, Belles, Wonbera, Dabus, Anger, and Fincha subbasins. Most eastern parts of the basin contain optimum and high available Zn values, particularly in most parts of Welaka, Jemma, Muger, and Guder subbasins.

The mean Melich-III extracted available copper (Cu) content varied from 0.42 to 10.88, 0.45 to 15.88, 0.25 to 12.26, 0.49 to 13.87, and 0.42 to 9.90 mg/kg soil in the warm moist, tepid moist, tepid subhumid, warm subhumid, and cool submoist-humid AEZ, respectively, indicating wide variability (from very low to optimum range) within and among AEZs soils of the basin (Table 7.8). The mean value of available Mn was found to be 3.0, 3.47, 3.15, 2.77, and 3.04 mg/kg soil in the warm moist, tepid moist, tepid subhumid, warm subhumid, and cool submoist humid AEZ, respectively, which are optimum for crop production according to the EthioSIS (2016) rating.

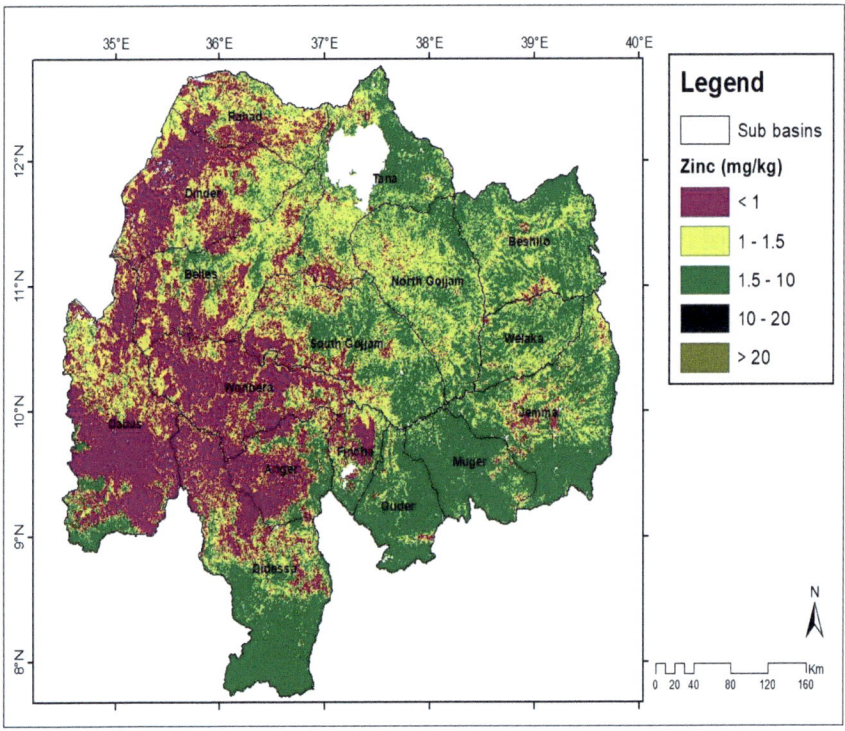

Fig. 6.14 Spatial distribution of available Zn in Abbay basin soils. (Adapted from ATA 2016, 2018, 2019)

Based on Mehlich-III extractable available Cu (EthioSIS, 2016), approximately 0.02, 0.58, and 99.4% of the Abbay basin soils were found to be very low, low, and optimum, respectively (Fig. 6.15). This shows that approximately 99.4% of the basin soils had optimum available Cu content. The remaining 0.6% of the basin soils was deficient.

The micronutrient content analysis depicts that approximately 62% and 0.6% of the Abbay basin soils are deficient in Zn and Cu, respectively. Micronutrient deficiencies have become a new issue for Ethiopia's crop productivity. Particularly, Zn and Cu deficiencies are pervasive throughout the nation. Future micronutrient demands from intensive cropping practices, adaptation of high-yielding cultivars that may have higher micronutrient demands, increased production of crops on marginal soils with low levels of essential nutrients, decreased use of animal manures, composts, and crop residues, and use of soils naturally deficient in micronutrient reserves are all expected to aggravate these issues (Fageria 2009).

Numerous investigations that have already been done have showed that Cu and Zn deficiency are widespread in many Ethiopian soils. As a case study, to gather national data on the status of micronutrients, Asgelil et al. (2007) argued that Cu and Zn are deficient in different agroecological zones of Abbay basin (Table 6.11). However,

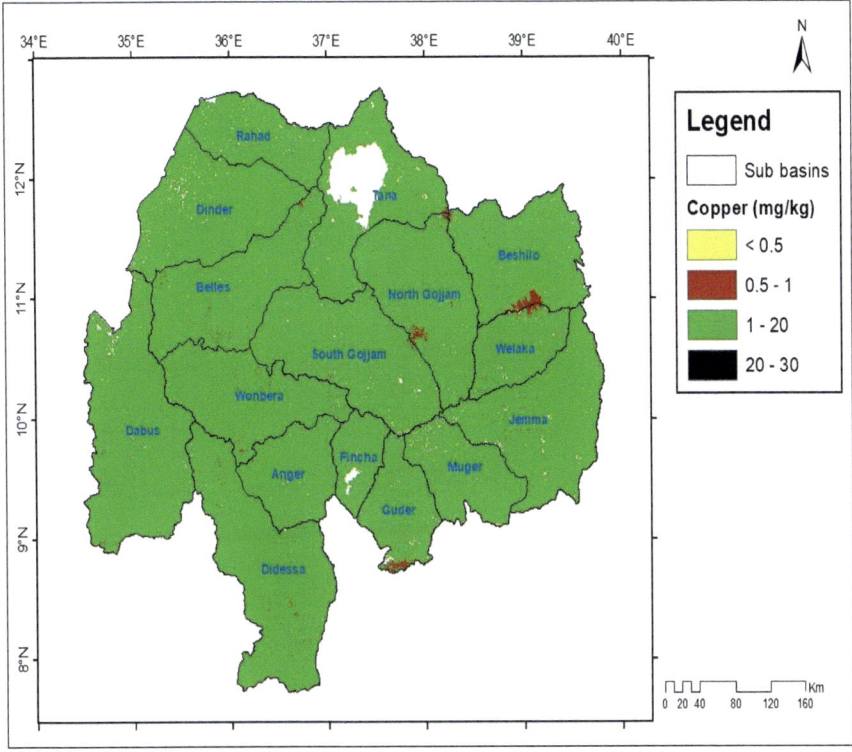

Fig. 6.15 Spatial distribution of available Cu in Abbay basin soils. (Adapted from ATA 2016, 2018, 2019)

it was discovered that Fe and Mn were beyond the acceptable level, and in some cases, Mn toxicity was identified. A study in Nitisols of western Ethiopia (Teklu et al. 2005), Nitisols of some highlands of Ethiopia (Elias 2017), Vertisols of central Ethiopia (Hillette et al. 2015), and the Fincha area (Fekadu 2020) indicated that Zn deficiency is most common (Table 6.11). Yifru and Mesfin (2013) also reported that the central Ethiopian highlands' vertisols lacked sufficient Fe and Zn. The addition of large doses of P fertilizers is thought to stress zinc deficiency.

Soil Biological Properties

All terrestrial ecosystems are built on top of soils, which are also home to a wide variety of bacteria, archaea, fungus, insects, annelids, and other invertebrates in addition to plants and algae. These organisms that reside in the soil supply food or nutrients to those that live above and below ground. According to Turmel et al. (2015), the soil is a living system made up of creatures that cycle nutrients, develop symbiotic

Table 6.11 DTPA extractable micronutrient (Fe, Mn, Zn and Cu) status in different Abbay basin soils

Location	Fe	Mn	Zn	Cu	References
	mg/kg				
West Shewa	17.3	39.7	1.1	0.9	Asgelil et al. (2007)
East Wellega	52.8	47.7	0.3	1.1	
Metekel	15.0	30.8	1.2	1.3	
West Gojam	8.9	26.7	0.9	1.6	
Awi	10.0	25.7	0.8	1.6	
East Gojam	7.9	40.7	1.1	2.9	
Jabi Tehnan (Cereal) North Central Highland	29–32	63–69	2–3	2	Elias (2017)
Muger sub-watershed (Wuchale district)	41.7–92.4	5.1–18.2	0.07–4.7	0.81–7.94	Asgelil et al. (2007)
Fincha Suger Estate (Melich extract)	101–190	94–130	1.4–1.5	3.2–4.9	Fekadu (2020)
Western Ethiopia (Nitisols)	4.56	–	0.32	0.34	Teklu et al. (2005)
Wolmera District	201.1	3.7	76.8	1.47	Asgelil et al. (2007)
Kabe Subwatershed	116–151	35–46	2.6–5.3	3.0–10.3	Hailu et al. (2015a, b)

connections with plant roots, control pests, weeds, and diseases, build soil aggregates, and aerate the soil. Soil ecology, which deals with microflora (fungi and bacteria), microfauna (such as protozoa and nematodes), mesofauna (such as collembolan and mites), and macrofauna (such as isopods and earthworms), is a young science in Ethiopia, and some of these sectors, particularly the fauna, have not been addressed in detail. A large diversity of the soil microbial community exists in Ethiopia due to the country's many altitudinal ranges, climatic conditions, vegetation, and crop varieties. The makeup of terrestrial microbial communities and their distribution, in general, and in the basin, have not been thoroughly studied.

Numerous papers have been written on various components of the soil's microflora, primarily rhizobial bacteria and mycorrhizal fungi. In Ethiopia, rhizobial biofertilizers are regarded as one of the best and most long-lasting soil fertility management strategies (Abere et al., 2016). The most effective indigenous and foreign rhizobial strains for the main grain legumes have been identified and characterized. The commercial strains that are currently on the market came from regional collections. According to Asfaw and Angaw (2006), the marketing of biofertilizers in Ethiopia only began in the 2000s. According to Fassil et al. (2018), numerous researches conducted in Ethiopia revealed that diverse species of rhizobia and rhizobacteria with various plant growth promoting (PGP) traits can be found in the nodules and rhizospheres of various legumes. The results of the research demonstrated that all grain legumes, including those in the basin, responded to rhizobial inoculation at various agroecologies throughout the nation.

Earthworms and termites are examples of soil macrofauna that are vital to the soil environment. In the savannah as well as in subtropical and tropical environments, termites play a critical role as decomposers of organic matter, nitrogen cycling, and soil structure improvement (Ayuke 2010). Despite the possibility of termites serving a positive purpose, 10% of them have been identified as pests of agriculture, forestry, residential buildings, and rangelands (Sileshi et al. 2008). Ten of the 62 termite species known in Ethiopia today—belonging to four families and 25 genera—are indigenous (Demisachew et al. 2018). Macrotermes subhyalinus (Rambur) and Macrotermes herus (Sjöstedt) are the two species that make up the genus Macrotermes (Abdurahman et al. 2010). The mounds of Macrotermes termites found in western Ethiopia, particularly the western portion of the basin, differ greatly in size and shape from those found in central Ethiopia, according to Abdurahman (1990), and Abdurahman et al. (2010). The management of rangeland/grassland and crop land is now threatened by termite invasion, a recent phenomenon in Ethiopia, especially the Abbay basin (Demisachew et al. 2018). The activity and diversity of soil fauna and flora have, however, generally decreased, mostly because of the depletion of organic matter and plant nutrients, soil acidity, and topsoil erosion (Regassa et al. 2023c).

Soil Degradation

According to Lal (2009), soil degradation is the rate at which soil properties deteriorate. It includes changes in the soil's physical, chemical, and biological characteristics. A serious ecological issue affects every country in the world and has consequences for future genetic resources, crop diseases, climate change, and agricultural resilience (Young et al. 2015). The main soil degradation processes include accelerated erosion, depletion of the soil's organic carbon pool, biodiversity loss, loss of soil fertility and elemental imbalance, acidification, and salinization (Lal 2015). The extent of these processes and their impact in the basin soils are presented in the following subsections.

Soil Fertility Decline

The soil chemical properties of the basin indicated that the soils are low in organic matter and total nitrogen, and deficient in P, S, and Zn. Furthermore, K and Cu deficiencies were reported in some parts of the basin. Currently, soils are degraded because of decades of repetitive tillage and cultivation, reduced or no fallow periods, and removal of crop residue. This gradual decrease in soil organic matter (SOM) and vital nutrients for plant growth is what is causing soil degradation. The soils have degraded and become less productive because of these not being replaced in recent decades through proper organic matter management (addition of compost and/or manure).

Important factors in the growth of plants include soil fertility and plant nutrition. To maximize soil productivity, nutrients must be supplied in sufficient and balanced proportions to support healthy plant development. Accordingly, standardizing laboratory procedures and determining the critical levels of macronutrients (K, P, and S) and micronutrients (Fe, Mn, Zn, and Cu) should be prioritized in light of the conflicting results that have been reported with various extractants and the fact that the critical levels that were used were not established in Ethiopia. Hence, correlation of extractants and calibration of soil test crop response experiments are needed to provide site-specific fertilizer recommendations across the Basin.

Soils with Depth Limitations

The Abbay basin soils with depth limitations of < 30 and 30–50 cm cover approximately 25.78 and 1.34% of the basin, respectively. Soil depth limitation is prominent in Leptosols. Because of the presence of abundant coarse earth fractions or continuous hard rock, these soils are relatively shallow. They are concentrated on steep side slopes, mountains, and hills, which are prominent in the Abbay basin, and they make up around 23.68% of the basin.

Soils with Drainage and Workability Limitations

Vertisols (12%), Gleysols (0.01%), and some reference soil groups with stagnic prefix qualifiers were identified in the Abbay basin. These soils have imperfect to poor drainage owing to the depressed topography, amount (> 60% clay) and type of clay mineral and associated flooding/ponding. In addition, most Vertisols have poor workability both when dry and wet since they are very hard to extremely hard when dry and very sticky and plastic when wet. This necessitates drainage improvement measures for successful crop production these soils. Furthermore, the following vertisols management techniques are practiced in different parts of Ethiopia: improved land preparation and drainage, drainage furrows, ridges, and furrows, broadbeds and furrows, green manure, decreased tillage, and post-rainy season planting techniques (Erkossa et al. 2009).

Soil Acidity

Approximately, 7.3 and 62.0% of basin soils were found to be strongly and moderately acidic, respectively. The basin's soils have significantly increased in acidity because of many natural events and poor farming practices (pH value of < 5.5). Among other factors, high rainfall leads to accelerated soil erosion and leaching which implies loss of nutrients and increasing soil acidification.

Due to inadequate nutrient availability, poor soil microbial activity, and Al toxicity, acidic soil productivity is low and declines quickly (Agegnehu et al. 2021). The basin should prioritize strategic research in managing acidic soils by combining agronomic management, better crop varieties, and soil and water management. These interventions could need to concentrate on the cultivation of crops that can tolerate acidic soil (such as millet, sorghum, sweet potatoes, potatoes, tomatoes, flax, tea, rye, and lupine), the enrichment of organic matter, the prevention of erosion, and the enhanced supply of cations and lime. Many advisory soil laboratories for lime requirement determination prefer the acid saturation method in conjunction with optimal soil acidity levels of different crops as a solution to the soil acidity problems, and it is advised to be used since it has been practically verified by numerous soil laboratories in Ethiopia.

Summary

Harmonized and updated basin-wide soil spatial information including selected soil forming factor spatial data was developed following a review, compilation, and synthesis of fragmented recent data sources. In terms of soil forming factor-topography, five physiographic units with their associated landforms were identified in the basin. Midlands and highlands with flat, slope, and summit geolandforms are the major topographic units of the basin. However, lowland plains with flat, slope, complex units with bottoms, mountains with slope and summit, and midlands with straight-flat geolandforms are common topographic features of the basin. Climate and agro-climatic zones showed spatial patterns across the basin including within the different subbasins. These mosaics of topographic and climatic features are among the key factors governing the spatial diversity of soil resources and properties across the basin.

The soil resource spatial information revealed that Nitisols, Leptosols, Cambisols, Vertisols, Acrisols, Luvisols, Fluvisols, Lixisols, Alisols, and Regosols are the dominant reference soil groups covering nearly 96% of the basin landmass. Soils with small area coverage identified in the basin are Arenosols, Ferralsols, Calcisols, Phaeozems, Gleysols, Andosols, and Umbrisols. About 27% of the basin soils have depth limitations most of which occur in sloping topographic positions requiring sound soil and land management interventions capable of averting soil erosion on these landscapes. Reference soil groups such as Vertisols, Gleysols, and some with stagnic properties depict soil drainage and workability problems requiring specific land management interventions.

The overall surface soil pH values indicate that approximately 69% of the basin land mass is covered by strongly to moderately acidic soils requiring the implementation of effective acid soil management strategies. The soil chemical property assessment indicated that most of the agricultural soils are low to medium in organic matter and total nitrogen content. In addition, significant soils are deficient in P, S, and Zn, while some areas are also deficient in Cu. However, since inconsistent results are reported with various soil nutrient assessment missions and lab extraction methods used, standardization of laboratory procedures and establishing the critical level of macronutrients (K, P, and S) and micronutrients (Fe, Mn, Zn, and Cu) for soils and crops in the basin should be given due emphasis.

Despite the attempts to provide soil spatial information in the basin, there are still issues with producing content full soil data and thus matching contemporary soil data requirements. Modern data demands, particularly for various modeling applications, require gridded quantitative soil data. However, most of the legacy soil data are generated by a conventional approach, and hence do not meet such modern data requirements. Furthermore, most of the environmental and crop modeling applications require detailed soil properties, such as hydraulic properties, than those commonly made available by legacy soil profiles. In addition, there have been limited and methodologically inconsistent efforts to derive critically missing soil properties using the easy-to-measure values using pedotransfer functions. Hence, the creation of future soil spatial information-making missions needed to be harnessed toward the development of national 3D gridded soil spatial information and site-specific pedo transfer functions for the basin.

Acknowledgements We deeply appreciate all of the institutions and platforms' diligent work in generating the soil data for the Abbay basin that formed the basis of the present study.

References

Abate N, Kibret K, Gebrekidan H, Esayas A (2014) Characterization and classification of soils along the toposequence at the Wadla Delanta Massif, north central highlands of Ethiopia. J Ecol Nat Environ 6:304–320

Abdurahman A (1990) Foraging activity and control of termites in western Ethiopia, PhD thesis, University of London. 277pp. https://spiral.imperial.ac.uk/bitstream/10044/1/47734/2/Abdulahi-A-1990-PhD-Thesis.pdf

Abdurahman A, Abraham T, Mohammed D (2010) Importance and management of termites in Ethiopia. Pest Manage J Ethiop 14:1–20

Abegaz A, Winowiecki LA, Vågen TG, Langan S, Smith JU (2016) Spatial and temporaldynamics of soil organic carbon in landscapes of the upper Blue Nile Basin of the Ethiopian Highlands. Agr Ecosyst Environ 218:190–208

Abegaz A, Tamene L, Abera W, Yaekob T, Hailu H, Nyawira SS, Da Silva M, Sommer R (2020) Soil organic carbon dynamics along chrono-sequence land-use systems in the highlands of Ethiopia. Agr Ecosyst Environ 300:106997

Abegaz A, Ali A, Tamene L, Abera W, Smith Jo U (2021) Modelling longterm attainable soil organic carbon sequestration across the highlands of Ethiopia. Environ Dev Sustain. https://doi.org/10.1007/s10668-021-01653-0

Abegaz A, Ashenafi A, Tamene L, Abera W, Smith JU (2022) Modeling long-term attainable soil organic carbon Sequestration across the highlands of Ethiopia. Environ Dev Sustain 24:131–5162

Abera W, Tamene L, Abegaz A, Hailu H, Piikki K, Söderström M, Girvetz E, Sommer R (2021) Estimating spatially distributed SOC sequestration potentials of sustainable landmanagement practices in Ethiopia. J Environ Manage 286:112191

Abere M, Yifru A, Getahun M (2016) Response of grain legumes to inorganic and biological fertilizers applications in Ethiopia: a review. Ethiop J Natl Resour 16(1):43–67

Abu RG (2021) Characterization of soils of Jello Chancho Watershed: the case of Liban District, East Shewa Zone Ethiopia. J Soil Sci Environ Manag 12(4):143–158

Adugna A, Abegaz A (2016) Effects of land use changes on the dynamics of selected soil properties in northeast Wellega, Ethiopia. Soil 2(1):63–70

Agegnehu G, Amede T (2017) Integrated soil fertility and plant nutrient management in tropical agro-ecosystems: a review. Pedosphere 27(4):662–680

Agegnehu G, Yirga C, Erkossa T (2019) Soil acidity management. Ethiopian Institute of Agricultural Research (EIAR), Addis Ababa

Agegnehu G, Amede T, Erkossa T, Yirga C, Henry C, Tyler R, Nosworthy MG, Beyene S, Sileshi GW (2021) Extent and management of acid soils for sustainable crop production system in the tropical agroecosystems: a review. Acta Agric Scand Sect B Soil Plant Sci. https://doi.org/10.1080/09064710.2021.1954239

Alemayehu K, Sheleme B (2013) Effects of different land-use systems on selected soil properties in southern Ethiopia. J Soil Sci Environ Manage 4:100–107

Alem HH, Kibebew K, Heluf G (2015) Characterization and classification of soils of Kabe Subwatershed in South Wollo Zone, Northeastern Ethiopia. Afr J Soil Sci 3(7):134–146. www.internationalscholarsjournals.org. ISSN 2375-088X

Ali A, Esayas A, Beyene S (2010) Characterizing soils of Delbo Wegene watershed, Wolaita Zone, Southern Ethiopia for planning appropriate land management. J Soil Sci Environ Manag 1:184–199

Ali A, Tamene L, Erkossa T (2020) Identifying, cataloguing, and mapping soil and agronomic data in Ethiopia. CIAT Publication No. 506. International Center for Tropical Agriculture (CIAT), Addis Ababa, Ethiopia, p 42

Ali A, Erkossa T, Gudeta K, Abera W, Mesfin E, Mekete T, Haile M, Haile W, Abegaz A, Tafesse D. Belay G (2022) Reference soil groups of Ethiopia based on legacy data and machine learning technique: EthioSoilGrids 1.0. EGUsphere, pp 1–40

Ali A, Erkossa T, Gudeta K, Abera W, Mesfin E, Mekete T, Haile M, Haile W, Abegaz A, Tafesse D, Belay G, Getahun M, Beyene S, Assen M, Regassa A, Selassie YG, Tadesse S, Abebe D, Wolde Y, Hussien N, Yirdaw A, Mera A, Admas T, Wakoya F, Legesse A, Tessema N, Abebe A, Gebremariam S, Aregaw Y, Abebaw B, Bekele D, Zewdie E, Schulz S, Tamene L, and Elias E (2024) Reference soil groups map of Ethiopia based on legacy data and machine learning-technique: EthioSoilGrids 1.0. Soil 10:189–209. https://doi.org/10.5194/soil-10-189-2024

Amare T, Hergarten C, Hurni H, Wolfgramm B, Yitaferu B, GSelassie Y (2013) Prediction of soil organic carbon for Ethiopian highlands using soil spectroscopy. ISRN Soil Sci 2013, Article ID 720589, 11 p. https://doi.org/10.1155/2013/720589

Asfaw HM, Angaw T (2006) Food and forage legumes of Ethiopia: progress and prospects. In: Ali K, Kenneni G, Ahmed S, Malhotra R, Beniwal S, Makkouk K (eds) Proceedings of the workshop on food and forage legume, ICARDA/EARO, 22–26 Sep 2003, Addis Ababa, Ethiopia, pp 172–176

Asgelil Debebe, Taye Bekele, Yesuf Assen (2007) The status of Micronutrients in Nitisols, Vertisols, Cambisols and Fluvisols in major maize, wheat, teff and citrus growing areas of Ethiopia. In: Proceedings of agricultural research fund research projects completion workshop held on 1–2 Feb 2007 at Ethiopian Institute of Agricultural Research, pp 77–96

Asmamaw LB, Mohammed AA (2013) Effects of slope gradient and changes in land use/cover on selected soil physico-biochemical properties of the Gerado catchment, northeastern Ethiopia. Int J Environ Stud 70(1):111–125. https://doi.org/10.1080/00207233.2012.751167

Assefa F, Elias E, Soromessa T, Ayele GT (2020) Effect of changes in land-use management practices on soil physicochemical properties in Kabe Watershed, Ethiopia. Air Soil Water Res 2020:13. https://doi.org/10.1177/1178622120939587

Aster Denekew, Seleshi Bekele (2009) Characterization and Atlas of the Blue Nile Basin and its Sub basins. International Water Management Institute

ATA, Agricultural Transformation Agency (2016) Soil fertility status and fertilizer recommendation Atlas of Amhara Regional State, Addis Ababa, Ethiopia

ATA, Agricultural Transformation Agency (2018) Soil fertility status and fertilizer recommendation Atlas of Benshangule national regional state, Addis Ababa, Ethiopia

ATA, Agricultural Transformation Agency (2019) Soil fertility status and fertilizer recommendation Atlas of Oromia national regional state, Addis Ababa, Ethiopia

Ayuke FO (2010) Soil macrofauna functional groups and their effects on soil structure, as related to agricultural management practices across agroecological zones of Sub-Saharan Africa. PhD thesis, Wageningen University, Wageningen, p 211

BCEOM (1998) Abbay River Basin integrated development master plan-phase 2—data collection—site investigation survey and analysis—volume VIII—land resources development—reconnaissance soil survey. Addis Ababa, Ethiopia

Bekele A, Lemma W, Samuel F (2021) Soil survey and characterization of soil of Argo-Gedilala Sub Watershed in Dugda District, Central Rift Valley of Ethiopia. J Soil Sci Environ Manag 12(3):94–106

Bellete T (2014) Fertility mapping of soils of Abbay–Chomen District, Western Oromia, Ethiopia. MSc Thesis, Haramaya University Ethiopia

Berhanu B, Melesse A, Seleshi Y (2013) GIS-based hydrological zones and soil geo-database of Ethiopia. CATENA 104:21–31

Chestworth W (ed) (2008) Encyclopedia of soil science. Springer, 3300 AA Dordrecht, The Netherlands

Debele M, Bedadi B, Beyene S, Mohammed M (2018) Characterization and classification of soils of Muger Sub-Watershed, Northern Oromia, Ethiopia. East Afr J Sci 12(1):11–28. ISSN 1993-8195 (Online)

Demisachew TA, Wondimu TA, Jaldesa DL, Abiyot Lelisa D, Amsalu TF (2018) Study on community perception of termite expansion and control in Borana plateau: case study of southern Oromia, Ethiopia. Int J Biodivers Conserv 10(9):365–371

Deressa A (2013) Evaluation of soil acidity in agricultural soils of smallholder farmers in South Western Ethiopia. Sci Technol Arts Res J 2(2):01–06

Deressa A, Yli-Halla M, Mohamed M, Wogi L (2018) Soil classification of humid Western Ethiopia: a transect study along a toposequence in Didessa watershed. CATENA 163:184–195

Deressa A, Yli-Halla M, Mohamed M, Wogi L (2020) Exchangeable aluminum as a measure of lime requirement of Ultisols and Alfisols in humid tropical Western Ethiopia Net J Agric Sci 8(3):46–58. ISSN: 2315-9766

Desta MK, Broadley MR, McGrath SP, Hernandez-Allica J, Hassall KL, Gameda S, Amede T, Haefele SM (2021) Plant available zinc is influenced by landscape position in the Amhara Region, Ethiopia. Plants 10:254. https://doi.org/10.3390/plants10020254

Dile YT, Tekleab S, Ayana EK, Gebrehiwot SG, Worqlul AW, Bayabil HK, Yimam YT, Tilahun SA, Daggupati P, Karlberg L, Srinivasan R (2018) Advances in water resources research in the Upper Blue Nile basin and the way forwards: a review. J Hydrol 560:407–423

Dinssa B, Elias E (2021) Characterization and classification of soils of Bako Tibe District, West Shewa, Ethiopia. Heliyon 7:e08279

Elias E (2016) Soils of the Ethiopian highlands: geomorphology and properties. CASCAPE Project, ALTERA, Wageningen University and Research Centre (Wageningen UR), The Netherlands, 385 pp

Elias E (2017) Characteristics of Nitisol profiles as affected by land use type and slope class in some Ethiopian highlands. Environ Syst Res 6:20

Elias E (2019) Selected chemical properties of agricultural soils in the Ethiopian highlands: a rapid assessment. South Afr J Plant Soil Sci 36:153–156

Erkossa T, Awulachew SB, Haileslassie A, Yilma AD (2009) Impacts of improving water management of smallholder agriculture in the upper Blue Nile Basin, CP 19 project workshop proceedings. IWMI, Addis Ababa. https://publications.iwmi.org/pdf/H042503.pdf

Fageria NK (2009) The Use of Nutrients in Crop Plants. CRC Press; Taylor & Francis Group. ISBN-13: 978-1-4200-7510-6

Fageria N, Nascente A (2014). Management of Soil Acidity of South American Soils for Sustainable Crop Production. https://doi.org/10.1016/B978-0-12-802139-2.00006-8

Fantaw A (2007) An overview of salt-affected soils and their management status in Ethiopia. A paper presented in the 3rd international workshop on water management project, Haramaya University, Ethiopia

FAO, IIASA (2023) Harmonized world soil database version 2.0. Rome and Laxenburg. https://doi.org/10.4060/cc3823en

Fassil A, Gemechu K, Negash D (2018) Overview of rhizobial inoculants research and biofertilizer production for increased yield of food legumes in Ethiopia. Ethiop J Crop Sci 6(Special Issue, 3). ISSN 2072-8506

Fekadu E, Kibret K, Bedadi B, Melese A (2018) Characterization and classification of soils of Yikalo Subwatershed in Lay Gayint District, Northwestern Highlands of Ethiopia. Eurasian J Soil Sci 7(2):151–166

Fekadu F (2020) Soil micronutrient status Assessment in sugarcane plantation of Ethiopia: Case of Fincha and Metahara. Int J Adv Res Biol Sci 7(11):156–162. ISSN: 2348-8069

Foster S, Urbanowitz S, Gatzke H, Schultz B (2016) Soil properties, Part 3 of 3: chemical characteristics, extension I University of Nevada, Reno, Fact Sheet FS-16-02

Gebrehiwot SG, Seibert J, Gärdenäs AI, Mellander PE, Bishop K (2013b) Hydrological change detection using modelling: half a century of runoff from four rivers in the Blue Nile Basin. Water Resour Res 49. https://doi.org/10.1002/wrcr.20319

Gebreselassie Y, Ayalew G, Elias E, Getahun M (2014) Soil characterization and land suitability evaluation to Cereal Crops in Yigossa Watershed, Northwestern Ethiopia. J Agric Sci 6(5):199–206

Gebreselassie Y, Anemut F, Addisu S (2015) The effects of land use types, management practices and slope classes on selected soil physico–chemical properties in Zikre watershed, North-Western Ethiopia. Environ Syst Res 4(3):1–7

Gebretsadik M (2014) Soil moisture prediction in an agricultural field of Gumara-Maksegnit watershed, North Gonder, Ethiopia. MSc thesis

Gessesse B, Ali A, Regassa A (2023) Land evaluation and land use planning. In: Beyene S, Regassa A, Mishra BB, Haile M (eds) The soils of Ethiopia. World Soils Book Series. Springer, Cham. https://doi.org/10.1007/978-3-031-170126_10

Getachew A, Tadesse T Gessesse B, Yigrem Y (2019) Soil moisture monitoring using remote sensing data and a stepwise-cluster prediction model: the case of Upper Blue Nile Basin, Ethiopia. Remote Sens 11(2):125.https://doi.org/10.3390/rs11020125

Getachew A, Tadesse T, Gessesse B, Yigrem Y, Melesse A (2020) Combined use of Sentinel-1 SAR and landsat sensors products for residual soil moisture retrieval over agricultural fields in the Upper Blue Nile Basin, Ethiopia. Sensors 20(11):3282. https://doi.org/10.3390/s20113282

Getahun M, Selassie YG (2017) Characterization, classification and mapping of soils of agricultural landscape in Tana Basin, Amhara National Regional State, Ethiopia. In: Stave K, Goshu G, Aynalem S (eds) Social and ecological system dynamics; characteristics, trends, and integration in the Lake Tana Basin, Ethiopia. AESS Interdisciplinary Environmental Studies and Sciences Series, Springer. https://doi.org/10.1007/978-3-319-45755-0

Han E, Ines AV, Koo J (2019) Development of a 10-km resolution global soil profile dataset for crop modelling applications. Environ Model Softw 119:70–83

Hailu H, Mamo T, Keskinen R et al (2015a) Soil fertility status and wheat nutrient content in Vertisol cropping systems of central highlands of Ethiopia. Agric Food Secur 4:19. https://doi.org/10.1186/s40066-015-0038-0

Hailu H, Mamo T, Keskinen R, Karltun E, Gebrekidan H, Bekele T (2015b) Soil fertility status and wheat nutrient content in Vertisol cropping systems of central highlands of Ethiopia. Agric Food Secur 4:19. https://doi.org/10.1186/s40066-015-0038-0

Hazelton PA, Murphy BW (2007) Interpreting soil test results: what do all the numbers mean? [2nd ed.]. Csiro Publishing, Australia

Jones JB (2003) Agronomic handbook: management of crops, soils, and their fertility (No. CRC Press.x, BOOK)

Jones A, Breuning-Madsen H, Brossard M, Dampha A, Deckers J, Dewitte O Zougmoré RB (2013) Soil atlas of Africa. European Commission

Kibret K, Beyene S, Erkossa T (2023) Soil Fertility and Soil Health. In: Beyene S, Regassa A, Mishra BB, Haile M (eds) The soils of Ethiopia. Springer International Publishing, Cham, pp 157–192

Kidanemariam A, Gebrekidan H, Mamo T, Fantaye KT (2013) Wheat crop response to liming materials and N and P fertilizers in acidic soils of Tsegede highlands, northern Ethiopia. Agric Forest Fish 2:126–135

Kumssa DB, Mossa AW, Amede T et al (2022) (2020) Cereal grain mineral micronutrient and soil chemistry data from GeoNutrition surveys in Ethiopia and Malawi. Sci Data 9:443. https://doi.org/10.1038/s41597-022-01500-5

Laekemariam F (2015) Soil spatial variability analysis, fertility mapping and soil plant nutrient relations in Wolaita zone. Haramaya University Ethiopia, Southern Ethiopia

Lal R (2009) Soil degradation as a reason for inadequate human nutrition. Food Secur 1:45–57

Lal R (2015) Restoring soil quality to mitigate soil degradation. Sustainability 7:5875–5895

Landon JR (2014) Booker tropical soil manual: a handbook for soil survey and agricultural land evaluation in the tropics and subtropics. Routledge

Leenaars JGB, Eyasu E, Wösten H, Ruiperez GM, Kempen B, Ashenafi A, Brouwer F (2016) Major soil-landscape resources of the cascape intervention woredas, Ethiopia: soil information in support to scaling up of evidence-based best practices in agricultural production (with dataset). CASCAPE working paper series No.OT_CP_2016_1. Cascape. Retrieved from https://edepot.wur.nl 428596

Leenaars JGB (2019) Geomorphic map of Ethiopia at 50m resolution. Bilateral Ethiopia Netherlands effort for food income and trade (BENEFIT) partnership-realise. ISRIC -World Soil Information, Wageningen University and Research Centre, SciLands Collaborator, The Netherlands

Leenaars JGB, Elias E, Wösten JHM, Ruiperez GM, Kempen B (2020a) Mapping the major soil landscape resources of the Ethiopian highlands using random forest. Geoderma. https://doi.org/10.1016/j.geoderma.2019.114067

Leenaars JGB, Ruiperez GM, Kempen B, Mantel S (2020b) Semidetailed soil resource survey and mapping of REALISE woredas in Ethiopia. Project report to the BENEFIT-REALISE programme. December 2020a, ISRIC–world soil information, Wageningen

Lelago A, Mamo T, Haile W et al (2016) Assessment and mapping of status and spatial distribution of soil macronutrients in kambata tembaro zone, southern Ethiopia. Adv Plants Agric Res 4(4):305–317. https://doi.org/10.15406/apar.2016.04.00144

Mamo T, Richter C, Heiligtag B (2002) Phosphorus availability studies on ten Ethiopian Vertisols. J Agric Rural Dev Trop Subtrop 103(2):177–183

Mitiku H, Herweg K, Stillhardt B (2006) Sustainable land management—a new approach to soil and water conservation in Ethiopia. Mekelle, Ethiopia: Land Resources Management and Environmental Protection Department, Mekelle University; Bern, Switzerland: Centre for Development and Environment (CDE), University of Bern, and Swiss National Centre of Competence in Research (NCCR) NorthSouth, 269 pp

Mosissa F (2018) Progress of soil acidity management research in Ethiopia. Adv Crop Sci Technol 6.https://doi.org/10.4172/2329-8863.1000377

Negassa W, Gebrekidan H (2003) Forms of phosphorus and status of available micronutrients under different land–use systems of Alfisols in Bako area of Ethiopia. Ethiop J Nat Resour 5(1):17–37

OIDA (Oromia Irrigation Development Authority) (2018) Oromia irrigation potential assessment project final report. Volume V: soil survey and land evaluation. Oromia National Regional State. Conducted by Oromia Water Works Design and Supervision Enterprise, Addis Ababa, Ethiopia

ORLEPB (Oromia Rural Land and Environmental Protection Bureau) (2011) Finfine surrounding special zone integrated land use planning study project final report. Section II: sectoral studies. Volume I: soils. Oromia National Regional State. Conducted by Oromia Water Works Design and Supervision Enterprise, Addis Ababa, Ethiopia

ORLEPB (Oromia Rural Land and Environmental Protection Bureau) (2014a) Didessa-Dabus integrated land use planning study project final report. Section II: sectoral studies. Volume I: Didessa Sub Basin soil survey. Oromia National Regional State. Conducted by Oromia Water Works Design and Supervision Enterprise, Addis Ababa, Ethiopia

ORLEPB (Oromia Rural Land and Environmental Protection Bureau) (2014b) Didessa-Dabus integrated land uses planning studies. Section II: sectoral study. Volume IV: Didessa Sub Basin erosion hazard and land degradation assessment. Oromia National Regional State, Addis Ababa

Poppe L, Frankl A, Poesen J, Admasu T, Dessie M, Adgo E, Deckers J, Nyssen J (2013) Geomorphology of the Lake Tana basin, Ethiopia. J Maps 9(3):431–437. https://doi.org/10.1080/17445647.2013.801000

Poggio L, de Sousa LM, Batjes NH, Heuvelink GBM, Kempen B, Ribeiro E, Rossiter D (2022) SoilGrids 2.0: producing soil information for the globe with quantified spatial uncertainty

Regassa A, Assen M, Ali A, Gessesse B (2023a) Major soil types. In: Beyene S, Regassa A, Mishra BB, Haile M (eds) The soils of Ethiopia. World Soils Book Series. Springer, Cham. https://doi.org/10.1007/978-3-031-17012-6_6

Regassa A, Ali A, Taye G (2023b) Soil classification. In: Beyene S, Regassa A, Mishra BB, Haile M (eds) The soils of Ethiopia. World Soils Book Series. Springer, Cham. https://doi.org/10.1007/978-3-031-17012-6_5

Regassa A, Kibret K, Selassie YG, Kiflu A, Tena W (2023c) Soil properties. In: Beyene S, Regassa A, Mishra BB, Haile M (eds) The soils of Ethiopia. World Soils Book Series. Springer, Cham. https://doi.org/10.1007/978-3-031-17012-6_7

Sebnie W, Adgo E, Kendie H (2021) Characterization and classification of soils of Zamra Irrigation Scheme, Northeastern Ethiopia. Air Soil Water Res 14:11786221211026576

Selassie Y (2002) Selected chemical and physical characteristics of soils of Adet research center and its testing sites in north–western Ethiopia. Ethiop J Nat Resour 4(2):199–215

Selassie YG, Ayalew G, Elias E, Getahun M (2014) Soil characterization and land suitability evaluation to cereal crops in Yigossa Watershed, Northwestern Ethiopia. J Agric Sci 6(5). ISSN 1916-9752 E-ISSN 1916-9760

Sheleme B (2017) Topographic positions and land use impacted soil properties along humbo larenaofa sere toposequence, southern Ethiopia. J Soil Sci Environ Manage 8:135–147. https://doi.org/10.5897/JSSEM2017.0643

Shimelis D, Mohammed A, Abbayneh E (2007) Characteristics and classification of the soils of tenocha-wenchacher microcatchment, southwest Shewa, Ethiopia. Ethiop J Natl Resour 9:37–62

Sileshi GW, Elias K, Patrick M, Philip ON (2008) Farmers' perceptions of tree mortality, pests and pest management practices in agroforestry in Malawi, Mozambique and Zambia. Agrofor Syst 72:87–101

Slattery WJ, Conyers MK, Aitken RL (1999) Soil pH, aluminium, manganese and lime requirement. In: Peverill KI, Sparrow LA, Reuter DJ (eds) Soil analysis: an interpretation manual, pp 103–128. (CSIRO Publishing: Melbourne.)

Sori1 G, Iticha B, Takele C (2021) Spatial prediction of soil acidity and nutrients for site-specific soil management in Bedele district, Southwestern Ethiopia. Sori et al. Agric Food Secur 10:59 https://doi.org/10.1186/s40066-021-00334-5

Takele C (2020) Iticha B (2020) Use of infrared spectroscopy and geospatial techniques for measurement and spatial prediction of soil properties. Heliyon 6:e05269

Taye M, Simane B, Selsssie YG, Zaitchik B, Setegn S (2018) Analysis of the spatial variability of soil texture in a tropical highland: the case of the Jema watershed, Northwestern highlands of Ethiopia. Int J Environ Res Public Health 15(9):1903. https://doi.org/10.3390/ijerph15091903

Teferi E (2015) Soil hydrological impacts and climatic controls of land use and land cover changes in the Upper Blue Nile (Abay) basin. PhD Dissertation, CRC Press

Teklu B, Amnat S, Yongyuth O, Sarobol Ed (2005) Status of B, Cu, Fe, Mo and Zn of soils of Ethiopia for Maize production: greenhouse assessment. Kasetsart J (Nat Sci) 39:357–367

Teshome Z, Kibret K (2015) Characterization of soil management groups of Metahara sugar estate in terms of their physical and hydraulic properties. Advanced Crop Sci Tech 3:159. https://doi.org/10.4172/2329-8863.1000159

Tibebe D, Teferi E, Bewket W, Zeleke G (2022) Climate induced water security risks on agriculture in the Abbay river basin: a review. Front Water 4:961948. https://doi.org/10.3389/frwa.2022.961948

Tibebe D, Mamo G, Zenebe A, Ali A (2023) Climate. In: Beyene S, Regassa A, Mishra BB, Haile M (eds) The soils of Ethiopia. World Soils Book Series. Springer, Cham. https://doi.org/10.1007/978-3-031-17012-6_3

Tiruneh GA, Meshesha DT, Adgo E, Tsunekawa A, Haregeweyn N, Fenta AA et al (2022) Use of soil spectral reflectance to estimate texture and fertility affected by land management practices in Ethiopian tropical highland. PLoS ONE 17(7):e0270629. https://doi.org/10.1371/journal.pone.0270629

Turmel MS, Speratti A, Baudron F, Verhulst N, Govaerts B (2015) Crop residue management and soil health: a systems analysis. Agric Syst 134(6–16). https://doi.org/10.1016/j.agsy.2014.05.009

Vågen TG, Winowiecki LA, Abegaz A, Hadgu KM (2013) Landsat-based approaches for mapping of land degradation prevalence and soil functional properties in Ethiopia. Remote Sens Environ 134:266–275

Weil RR, Brady NC (2017) The nature and properties of soils. Global Edition, 15th Edition, Pearson

WLRC (Water and Land Resource Center) (2016a) Soils of the Grand Ethiopian Renaissance Dam (GERD) Basin, volume II. Water and Land Resource Center (WLRC). Addis Ababa, Ethiopia

WLRC (Water and Land Resource Center) (2016b) Soils of the learning watershed (Andit-Tid, Anjeni, Atari-Mesk, Debre-Yakob, Debre-Mewi). Addis Ababa University, Ethiopia

WLRC (Water and Land Resource Center) (2019) Modelling hydrological processes and sediment yield in Ethiopia: the case of Abbay Basin. Water and Land Resource Centre, Addis Ababa University, Addis Ababa

Wonde M (2014) Spatial prediction and mapping of soil properties using geostatistics in the Ethiopian highlands. Ethiop J Natl Resour 14:17–37

Wösten JHM, Verzandvoort SJE, Leenaars JGB, Hoogland T, Wesseling JG (2013) Soil hydraulic information for river basin studies in semi-arid regions. Geoderma 195–196(79–86). https://doi.org/10.1016/j.geoderma.2012.11.021

Yeneneh N, Elias E, Legese GF (2022) Assessment of the spatial variability of selected soil chemical properties using geostatistical analysis in the northwestern highlands of Ethiopia. Acta Agric Scand Sect B Soil Plant Sci 72(1):1009–1019. https://doi.org/10.1080/09064710.2022.2142658

Yifru A, Mesfin Kebede (2013) Assessment on the Status of Some Micronutrients in Vertisols of the Central Highlands of Ethiopia. Int Res J Agric Sci Soil Sci. (ISSN: 2251-0044) 3(5):169–173

Young R, Orsini S, Fitzpatrick I (2015) Soil degradation: a major threat to humanity. Sustainable food trust, a global voice for sustainable food and health 38

Obsa Z (2022) Evaluation of micronutrient status in Nitisol, Vertisol and their relations under Wheat and Teff growing area of West Shewa Zone, Oromia. Am-Eurasian J Agron 15(1):13–19

Zewudie E (2003) Study on the physical, chemical and mineralogical characteristics of some Vertisols of Ethiopia. In: Chekol W, Mersha E (eds) Proceedings

Chapter 7
Forest Resources in the Abbay Basin

Aramde Fetene

Abstract The Abbay basin holds significant strategic importance in Ethiopia, spanning an extensive area composed of diverse landscapes and agroecological zones. These variations give rise to a wide range of plant communities, encompassing different types such as Combretum-Terminalia woodlands and wooded areas with grass, a diverse combination of dry evergreen Afromontane forests and grasslands, Afro-Alpine plant species, the Ericaceous Belt, moist evergreen Afromontane forests thriving in humid regions, as well as wetland habitats like freshwater marshes, floodplains, vegetation along lake shores, and plant life specifically adapted to open water environments. Riverine and bamboo forests also contribute to the basin's vegetation potential. The forests within the Abbay basin have a vital function in delivering essential ecological benefits, promoting sustainable development objectives, preserving the integrity of soil and water resources, and providing valuable resources for construction purposes and biomass energy production. However, studies indicate that deforestation and land degradation are widespread in most subbasins, with an expansion of cultivated land and settlements encroaching on forest areas. Anthropogenic factors specifically tree cutting, production of charcoal, and the transformation of natural vegetation into agricultural lands have been recognized as the primary catalysts behind the clearance of forests. In addition, challenges related to a lack of sense of ownership and institutional instability at the national level hinder forest resource development in the basin. In light of these challenges, there are favorable conditions that support the sustainable utilization and efficient governance of forest resources. There are promising opportunities for progress in the sustainable management of forest resources. The initiative designed to receive payments for environmental services, along with a supportive legal framework for forestry sector investment, and the active involvement of non-governmental actors in the forest and environmental sectors, contribute to these positive prospects. Hence, ensuring the protection and responsible stewardship of forest resources in Ethiopia, particularly in the Abbay basin, requires a crucial undertaking of further enhancing these favorable prospects.

A. Fetene (✉)
Department of Urban and Regional Planning, College of Technology and Built Environment, Addis Ababa University, Addis Ababa, Ethiopia
e-mail: aramde.fetene@aau.edu.et

Keywords Abbay basin · Deforestation · Ecological benefits · Forest resources · Land degradation · Sustainable development

Introduction

Overview of the Abbay Basin

Ethiopia is home to twelve significant river basins, each contributing to the country's water resources and ecological diversity. These basins include Abbay, Wabi-shebele, Dawa-Genale, Awash, Tekeze, Baro-Akobo, Omo-gibe, Ogaden, Denkele, Rift Valley Lakes, Mereb, and Ayisha (Fig. 7.1). Among these basins, eight are classified as river basins, namely Abbay, Baro-Akobo, Omo-Gibe, Tekeze, Genale-Dawa, Awash, Wabi Shebele, and Mereb. One basin, known as the Rift Valley, consists of lakes rather than rivers. The remaining three basins, Denakle, Ogaden, and Aysha, are classified as dry basins, lacking significant water flow (ABDO 2020).

The Abbay Basin is especially significant since it occupies a key strategic location in Ethiopia. It spans an extensive area of 199,812 square kilometers and is located in the North-Western part of the country. The basin's geographic coordinates range from approximately 7°45′ to 12°46′ latitude and 34°06′ to 40°00′ longitude. Its altitudinal

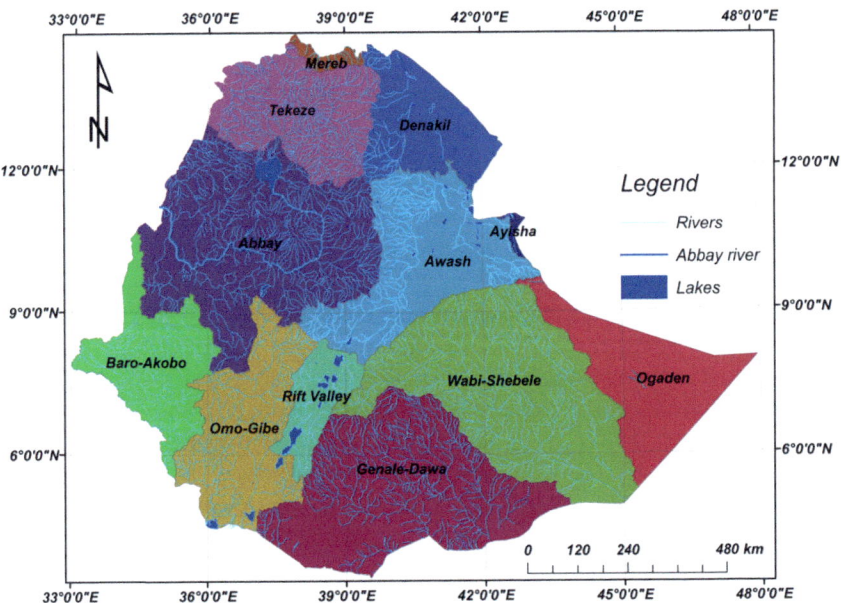

Fig. 7.1 Basin in Ethiopia (*Source of shapefile* Ethio-GIS, https://www.ethiogis-mapserver.org/)

range varies from 355 m above sea level (m.a.s.l.) at the Sudanese border to a towering 4235 m.a.s.l. at the summit of Mt Guna (Teferi et al. 2015).

The Abbay basin serves as a vital water source for the Abbay River, along with offering various demographic and ecological assets. It plays a significant role in supplying water to downstream countries such as Sudan and Egypt, contributing to the overall water flow of the Nile River.

To better understand the spatial distribution and characteristics of the Abbay basin, it is further divided into 16 subbasins (Fig. 7.2). This division is based on the major rivers and tributaries that traverse the basin, facilitating more effective study and management of its water resources and associated ecosystems (ABA 2016).

Ethiopia's river basins, including the Abbay basin, are crucial for the country's water supply and ecological well-being. The Abbay basin, in particular, stands out as a strategically important basin, covering a substantial area and providing a significant water supply to the Abbay River.

The Abbay basin in Ethiopia encompasses diverse landscapes and climatic zones, resulting in a rich variety of vegetation formations. This range of environments stretches from the lowlands near the Sudan border to the highest point of Mount Guna (Fig. 7.3).

In accordance with the traditional Ethiopian agroecological classification, the Abbay basin can be categorized into six distinct agroecological zones:

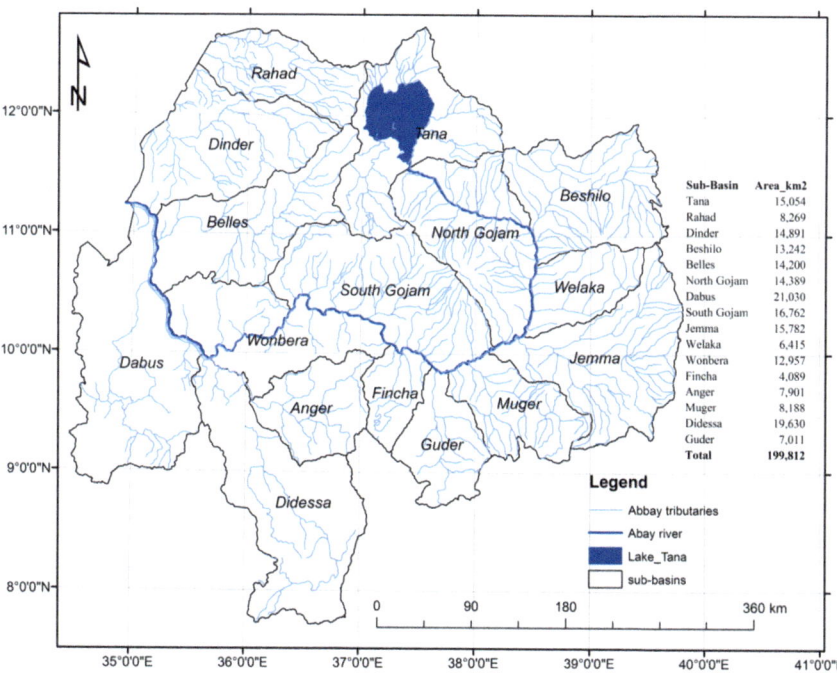

Fig. 7.2 Abbay basin subbasins

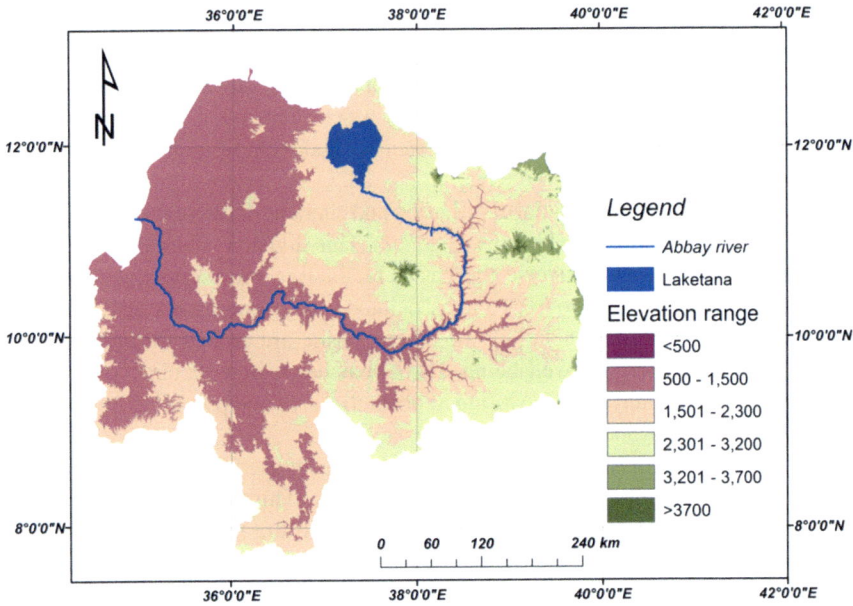

Fig. 7.3 Elevation range of the Abbay basin (*Data source* Ethio DEM 30 m)

1. *'Bereha'* represents the hot lowlands situated at an elevation of less than 500 m above sea level. Crop production in this arid eastern region is severely limited.
2. *'Kolla'* refers to the lowlands found between 500 and 1500 m above sea level. Dominant crops in this zone include sorghum, finger millet, sesame, cowpeas, and groundnuts.
3. *'Woina Dega'* designates the highlands ranging from 1500 to 2300 m. Major crops cultivated in this zone are wheat, teff, barley, maize, sorghum, and chickpeas.
4. *'Dega'* encompasses the highlands spanning 2300–3200 m above sea level. Barley, wheat, oilseeds, and pulses are the primary crops grown in this zone.
5. *'Wurch'* represents the highlands situated between 3200 and 3700 m above sea level, where barley cultivation is prevalent.
6. *'Kur'* denotes the highland areas located above 3700 m. These regions are predominantly utilized for animal grazing rather than crop production.

Generally, the Abbay basin exhibits a wide range of landscapes and climates, resulting in diverse vegetation formations. It encompasses all the above six agroecological zones, each characterized by specific elevations and supporting different crops and land use practices. The basin is primarily composed of Woina Dega, accounting for 41.64% of the area, followed by Kolla at 35.38% and Dega at 20.75%. There is a smaller portion of the area classified as arid and afroalpine regions (Table 7.1).

The Abbay Basin possesses a range of agroecologies with favorable conditions for diverse vegetation formations. However, the region is currently grappling with

Table 7.1 Proportions of agroecological zones and associated physical characteristics in the Abbay basin (traditional)

Agroecological zone	Elevation (m)	Climate	Mean annual temperature (°C)	Mean annual rainfall (mm)	Proportion (%)
Bereha	< 500	Hot arid	> 27.5	< 200	0.54
Kolla	500–1500	Warm semiarid	20.0–27.5	200–800	35.38
Woina Dega	1500–2300	Cool subhumid	17.5–20.0	800–1200	41.64
Dega	2300–3200	Cool and humid	11.5–17.5	1200–2200	20.75
Wurch	3200–3700	Cold and moist	< 11.5	Above 2200	1.5
Kur	> 3700	Cold and moist			0.18

Source (Own calculation based on pixel count; MoA 2000)
Note The table provides a breakdown of the agroecological zones within the Abbay basin based on traditional classification. Each zone is characterized by specific elevation ranges, climate conditions, average annual temperatures, average annual rainfall, and their respective proportions within the basin

significant environmental challenges, particularly deforestation and land degradation. Although several research endeavors have focused on tackling water management challenges within the basin, the exploration of forest resources has received scant consideration. Additionally, the available studies pertaining to land use and land cover offer limited insights and are confined to particular subbasins, failing to provide a holistic comprehension. Consequently, a significant dearth of comprehensive and detailed vegetation data exists in the Abbay basin, impeding the effectiveness of vegetation management planning and engagements for biodiversity conservation.

Forest Definition

The classification and understanding of forests can be a perplexing matter, subject to variations across different institutions and over time. Notably, the Food and Agriculture Organization (FAO) has introduced divergent definitions for forests over the years, as evidenced by the Forest Resources Assessment (FRA) spanning from 1948 to 2020. Nevertheless, commencing from 2015, the Food and Agriculture Organization (FAO) has embraced a distinct definition, whereby forests are defined as land areas that encompass over 0.5 ha, showcasing the existence of trees surpassing a height of 5 m, along with a canopy cover surpassing 10%. This definition excludes land primarily used for agricultural or urban purposes, and it continues to be in effect as of FAO's (2018) report.

In a separate development, Ethiopia introduced a new forest definition in February 2015, as outlined by the then Ministry of Environment and Forests (MEFCC 2017a). Based on this specific characterization, forests are defined as expansive territories exceeding 0.5 ha, containing trees (including bamboo) with a maximum width of 20 m. The height requirement for these trees is above 2 m, accompanied by a canopy cover exceeding 20%, or the potential to naturally achieve these criteria over a period of time. The adjustment made to the national forest definition aimed at more accurately reflecting the authentic primary condition of Ethiopia's forest vegetation.

State of Forest in the Abbay Basin

Forest Cover and Distribution Analysis

The presence and extent of forest resources within the Abbay Basin are significantly influenced by the region's climate and elevation. As a result, the lowland areas predominantly feature woodlands, while the highlands are characterized by natural montane forests, plantation forests, and limited Afro-Alpine vegetation. However, it is important to note that thus far, there has been a lack of comprehensive studies on forest resources conducted at the basin level. Instead, numerous studies have focused on land use and land cover analysis within specific sub-watershed areas.

To bridge this gap in knowledge, the current study aims to examine the forest resources across various subbasins within the Abbay Basin. By leveraging published articles and research papers pertaining to specific subbasins within the study area, an analysis of forest resources is undertaken. This approach allows for a better understanding of the forest cover and distribution in different subbasins, providing valuable insights into the state of forests within the Abbay Basin.

By considering the existing literature on specific subbasins, this study contributes to the assessment of forest resources within the Abbay Basin. It serves as a starting point for understanding the forest cover and distribution patterns at a subbasin level, shedding light on the diverse forest ecosystems present in the region.

Didessa River Subbasin (DRB)

The Didessa River, a significant tributary of the Abbay River, contributes the highest volume of water and spans a total area of 19,630 km^2. Based on Kabite et al. (2020), the primary land cover categories observed in the Downstream of the Abbay basin (DRB) from 1985 to 2017 comprised agriculture, human settlements, forests, bush and shrubland, grasslands, and water bodies. The study's results indicated a notable reduction in forest and shrubland areas over the course of the past three decades. In the upper Didessa region, Gebrehiwot et al. (2014) documented a decline in forest coverage, which decreased from 89% in 1957 to 45% in 2001.

In contrast, Wedajo et al. (2019) conducted a study spanning from 1983 to 2015 and observed spatiotemporally variable trends in the Normalized Difference Vegetation Index (NDVI) within the DRB. Generally, the NDVI trends indicated greening across various time scales. However, it is important to note that the increasing NDVI trend does not necessarily indicate an increase in vegetation cover within the study area. Instead, this trend could be attributed to the rise in dry season rainfall experienced during the study periods.

These studies collectively provide insights into the dynamics of land cover and vegetation changes within the Didessa Subbasin (DRB). The findings highlight the declining trend of forests and shrubland, as well as the spatiotemporal variability of NDVI, emphasizing the complex interplay between environmental factors and vegetation dynamics in the region.

South Gojjam Subbasin

The South Gojam subbasin is situated in the central region of the Abbay Basin, encompassing the Gojam and Awi zones within the Amhara National Regional State. It spans an estimated area of 16,762 km^2. Within this subbasin, notable tributaries include the Chemoga, Yeda, Gudla, Temcha, Leza, Lah, Dondor, Buchikis, Mesine, Abahim, Fettam, Talia, and Birr Rivers (ABDO 2020).

In the Chemoga watershed, a part of the South Gojam subbasin, Bewket (2002) observed an increasing trend in forest cover from 1957 to 1982. The forest area exhibited a growth rate of approximately 19% (equivalent to 7 ha/year) during this period. This expansion was largely attributed to the mass afforestation program implemented by the Derg regime, with a focus on preserving indigenous trees and forests within the subbasin. Presently, tree-planting initiatives at the household level, specifically using Eucalyptus trees, have become a common practice, particularly in the highland regions of the basin.

These findings shed light on the state of the South Gojam subbasin, highlighting the historical increase in forest cover within the Chemoga watershed. The afforestation efforts undertaken by the Derg regime contributed to this positive trend, emphasizing the importance of preserving indigenous trees and forests. Additionally, the prevalence of household-level tree planting, particularly with Eucalyptus trees, demonstrates the role of local communities in sustaining and enhancing the forest resources within the highland areas of the basin.

Guder Subbasin

The Guder River, which flows northward in central Ethiopia, covers a drainage area of about 7011 km^2 (Tana & Beles Integrated Water Resources Development, 2008). A comprehensive investigation conducted by Berihun et al. (2019) over a period of

35 years (1982–2017) unveiled significant alterations in land cover within the Guder River basin. Specifically, between 1982 and 2006, approximately 29.4 ha of bushland and 43.87 ha of forestland underwent conversion into cultivated areas.

Additionally, Muleta and Biru (2019) reported a decline of 48.5% in shrubland and 37.5% in forestland between 1973 and 2015. These findings highlight the ongoing transformation of natural vegetation in the Guder River basin, with substantial losses observed over the decades. Consequently, the reduction in vegetation cover and the expansion of barren land have had significant impacts on the Land Surface Temperature (LST) in the study area. Moisa et al. (2022) demonstrated that the LST in the region increased by 11.3 °C from 1990 to 2020. This rise can be attributed to the decrease in vegetation cover and the subsequent expansion of unproductive land.

The aforementioned studies collectively emphasize the alarming land cover changes taking place within the Guder River basin. The conversion of bushland and forestland into cultivated areas, along with the loss of shrubland and forestland, has led to higher land surface temperatures and the expansion of unproductive land. Understanding these changes is crucial for informed decision-making and effective land management strategies in the Guder River basin.

Anger Subbasin

The Anger subbasin, which covers an area of 7902 km^2 (ABDO 2020), is located north of Nekemte town. It stretches from the elevated plateau region to the Abbay River. Over the past three decades, significant transformations have taken place in the land cover of the Anger catchment. One notable change has been the decrease in vegetation, primarily attributed to the expansion of agricultural land. A study conducted by Gebre et al. (2022) has provided compelling evidence that cropland has expanded from 38.1% in 1984 to 78% in 2000. In contrast, forest coverage has experienced a decline from 10.8% in 1984 to 3.26% in 2000. Consequently, the conversion of forested areas into farmland has led to increased surface runoff due to reduced infiltration rates, resulting in higher stream flow and elevated risk of downstream flooding during the wet season. Additional findings by Galata (2020) support these observations, highlighting a 32% reduction in forest cover and a corresponding 28% increase in cultivated land within the study area from 1987 to 2017. Such alterations in land use patterns not only contribute to heightened sediment yield in the Abby River but also impact the functioning of the Grand Ethiopian Renaissance Dam (GERD).

Lake Tana Subbasin

According to the Nile Basin Initiative report, the Tana subbasin encompasses an area of 15,054 square kilometers (NBI 2007). The major rivers contributing to this subbasin are Gilgel Abbay, Birr, Gumera, Ribb, Megech, Mizewa, Debre Mawi, and Enchilala. Within the Lake Tana basin, a region known for its diverse ecosystems, the dry evergreen Afromontane forest type is abundant, hosting numerous plant species that are endemic to the area (Song et al. 2018). Unfortunately, the expansion of cropland, fuelwood collection, and charcoal production has resulted in the depletion of forested areas surrounding the church, leaving only small remnants.

A study conducted by Asres et al. (2015) in the Mizewa, Debre Mawi, and Enchilala watersheds from 1973 to 2013 revealed a significant expansion of cultivated land at the expense of shrub/bush, forest, and grazing land. Over this period, cultivated land increased by 60 ha, causing a reduction of 22 ha in grazing land, 16 hectares in shrub/bush areas, and 22 ha in forested regions across the three watersheds. These changes resulted in a decrease in evapotranspiration and an increase in overland and subsurface flows, leading to intensified soil erosion.

In another study by Asimamaw (2019) focusing on the upper Ribb watershed, it was observed that cultivated land expanded by 29.947% between 1973 and 2016. This expansion of cultivated land corresponded to a 6.143 m^3 per second increase in stream flow and a sediment yield of 343.25 tons per square kilometer per year.

When examining changes in forest cover in the Gilgel Abbay and Birr Rivers, which are major tributaries of Lake Tana, Gebrehiwot et al. (2014) discovered an upward trend in forest cover in Gilgel Abbay (from 10% in 1957 to 22% in 2001), primarily due to the expansion of eucalyptus plantations. However, during the same period, forest cover in Birr declined from 29 to 22%. The reduction in natural forest cover in Gilgel Abbay has resulted in increased surface run-off and decreased groundwater levels (Woldesenbet et al. 2017).

According to Song et al. (2018), the Tana basin contains two types of natural forests: dry evergreen Afromontane forest and riverine forest. Key species in the upper story of these forests include *Podocarpus falcatus* and *Juniperus procera*, while the understorey is dominated by *Croton macrostachyus, Ficus spp., Olea europaea subsp. cuspidata, Trema orientalis,* and *Maesa lanceolata*. Riverine forests along riversides and lakeshores feature species such as *Diospyros mespiliformis, Mimusops kummel,* and *Syzygium guineense*. Additionally, the Lake Tana basin comprises two types of woodland communities: Combretum-Terminalia and Acacia-Commiphora woodlands (IBC 2005).

The Combretum-Terminalia woodland community, as reported by Friis et al. (2011), showcases a diverse range of key plant species that contribute to its composition and ecological significance. These species include *Combretum spp., Terminalia spp., Oxytenanthera abyssinica, Anogeissus leiocarpa, Stereospermum kunthianum, Pterocarpus lucens, Lonchocarpus laxiflorus, Lannea welwitschii, Albizia malacophylla,* and *Entada africana*. Combretum species are prominent members of this community, characterized by their unique features and ecological roles.

These trees exhibit distinct foliage and often highlight vibrant, attractive flowers. They contribute to the canopy structure and offer valuable shade and shelter for various organisms residing in the woodland. Terminalia species also play a crucial role within the Combretum-Terminalia community. These trees are known for their impressive height, expansive canopies, and prized timber. They serve as important habitat providers for a diverse range of wildlife, including birds, mammals, and insects. The interactions and dynamics among these key plant species under the

Combretum-Terminalia community form a complex network, influencing essential ecological processes such as nutrient cycling, water regulation, and habitat provision for numerous organisms. Thus, it is crucial to prioritize the conservation and preservation of this community to ensure the long-term health and sustainability of the Lake Tana basin ecosystem.

The Acacia-Commiphora communities found in the Lake Tana basin are characterized by the presence of certain dominant species, which contribute significantly to the structure and ecological dynamics of the region. These species, including *Acacia tortillas, A. millifera, Balanites aegyptiaca, Acalypha spp., Boswellia papyrifera*, and *Commiphora spp.*, hold key roles in shaping the vegetation and maintaining the ecological balance in this area. Their abundance and distribution influence the overall composition and functioning of the ecosystem within the basin. In addition to the Acacia-Commiphora communities, the Lake Tana basin also features scrub land, which is particularly prominent in areas with steep slopes, hills, escarpments, mountains, and gorges. The scrub land in this area displays a wide variety of dominant species, playing a crucial role in enhancing the overall biodiversity of the region. Among the notable species found within the scrub land are *Maytenus senegalensis, Carissa spinarum, Clausena anisata, Clerodendrum myricoides, Grewia ferruginea, Caesalpinia decapetala, Euclea divinorum, Ficus verruculosa, Calpurnia aurea, Erica arborea, Hypericum revolutum, Vernonia spp., Senna spp., Cordia spp.*, and *Acacia spp*. These dominant species within the scrub land play crucial roles in the ecosystem through providing habitats, food sources, and shelter for a variety of organisms, including animals, insects, and birds. The presence of these species contributes to the biodiversity and ecological stability of the Lake Tana basin. Moreover, the Lake Tana basin is one of the areas in the Abbay basin that has undergone restoration through plantation forests. The key species chosen for plantation efforts in the Lake Tana basin primarily consist of Eucalyptus species, specifically *Eucalyptus globulus* and *Eucalyptus camaldulensis*, along with *Cuperusus lusitanica* and *Acacia mearnsii* (Yitaferu et al. 2013; Wassie 2017). Therefore, conservation efforts aimed at protecting these vegetation formations and their associated habitats will contribute to the preservation of biodiversity and the overall health of the ecosystem in the basin.

North Gojjam Subbasin

The North Gojam subbasin extends between the Choke and Guna mountains and consists of six perennial rivers: Suha, Wenka, Tigder, Muga, Sedie, and Andasa Rivers (Jembere 2017). This subbasin encompasses the upper Abbay River catchment, covering an estimated area of 14,390 km^2 with elevations ranging from 2800 to 4090 masl. Within this subbasin, *E. globulus* is the dominant plantation species among introduced trees, while natural forest cover is limited and primarily found along riverbanks, hillsides, and church yards (Ewnetu et al. 2021a).

A study conducted by Gashaw et al. (2018) in the Andassa watershed uncovered substantial growth in cultivated land, expanding from 36,820 ha in 1985 to 45,108 ha in 2015. Alongside this expansion, the built-up area also increased from

35 hectares in 1985 to 672 ha in 2015, resulting in a decrease in forest cover from 2068 hectares in 1985 to 1138 ha in 2015. Similarly, in the Muga watershed, an upward trend was observed in cultivated land and urban areas, with increases of 12% and 270%, respectively, between 1985 and 2017, while forestland experienced a decline of approximately 40% (Belay and Mengistu 2019). The depletion of forest resources in the North Gojam basin can be attributed to various human activities such as wood cutting for household fuel, charcoal production, construction, timber trade, furniture making, and the manufacturing of agricultural tools (Ewnetu et al. 2021b). Consequently, the natural vegetation is now mostly limited to church areas, riverbanks, and hillsides. However, there has been an increase in plantation forests, particularly with *E. globulus*, as farmers utilize their land commercially due to declining productivity of croplands caused by land degradation.

Dabus Subbasin

The Dabus subbasin, situated in South-Western part of the Abbay basin, encompasses a drainage area of 21,030 km^2 (ABDO 2020). Flowing northward, the primary river within this basin is the Dabus River, which serves as a tributary to the Abbay River. Similar to other regions in the Abbay basin, there has been a noticeable decrease in vegetation cover in the Dabus subcatchment. Research conducted between 1986 and 2019 on land use and land cover change in this area demonstrated specific transformations. Notably, between 1986 and 2019, there was a noteworthy expansion of agricultural land from 442.6 to 512.4 km^2. Concurrently, the built-up area also experienced substantial growth, increasing from 222.0 to 971.7 km^2. Conversely, forestland witnessed a decline from 94.7 km^2 in 1986 to 86.2 km^2 in 2019, representing an annual decrease of 0.4%, equivalent to 8.5 km^2, as reported by Akalu et al. (2019). This shift can be attributed to the cultivation of crops, which has led to the degradation of forests due to heavy reliance on irrigation methods. Consequently, this has resulted in a decline in water bodies. Moreover, the reduction in forest cover may have adverse effects on biodiversity, the provision of ecosystem services, and the local climate.

Beshilo Subbasin

The Beshilo catchment, situated within the Abbay basin, serves as one of its significant tributaries, encompassing an estimated area of 13,243 km^2. This catchment spans across portions of the Semien Gondar, Semien Wollo, and Debub Wollo Zones, as documented by Yesuph and Dagnew (2019). However, it faces a pressing issue in the form of land degradation within the broader context of the Abbay Basin.

Conducting an analysis of land cover change, it was observed that the Beshilo catchment has undergone notable transformations between the years 1973 and 2017. The results indicate a considerable reduction in forestlands, shrub lands, and grasslands, accompanied by an expansion of agricultural land. This shift in land use has had unintended negative socio-ecological consequences on the watershed (Yesuph 2020). The same source revealed that the transition matrix was approximately 21.25% of Afro/subalpine landscapes, 17.59% of grasslands, and 8% of shrublands have been converted into cultivated areas or settlements. Such conversions disrupt the natural balance and functioning of the ecosystem, leading to adverse effects on both the society and the environment in the Beshilo catchment.

Fincha'a Subbasin

The Fincha'a subbasin, situated in the Oromia region within the Horro Guduru Wollega zone, forms an integral part of the Abbay basin. Covering a catchment area of 4089 km^2, it serves as a vital water source for both irrigation and hydroelectric power generation. A comprehensive study conducted by Kenea et al. (2021) examined the dynamics of forest cover in the Fincha'a subbasin. The research findings revealed a decline in forest cover by 20.0% between 1994 and 2004 and a subsequent decrease of 11.8% between 2004 and 2018. These changes were predominantly attributed to the expansion of cultivated land, which experienced an increase of 16.4% and 10.81% in the respective time periods. Notably, the study documented a rapid growth in cultivated land within the subbasin, with its proportion expanding from 42.8% of the total area in 1994 to 70.1% in 2018.

Similar studies conducted in the Fincha'a watershed by Tolessa et al. (2021) covering the period from 1987 to 2019 highlighted the significant reduction in forest cover due to the expansion of cultivated areas, settlements, water bodies, and sugarcane plantations. The rise in water bodies can be attributed to the establishment of the Fincha'a Hydroelectric Dam, which led to the submergence of approximately 120 km^2 of swamp, 100 km^2 of grazing land, 18 km^2 of cropland, and 1.2 km^2 of forest, as reported by Tessema and Simane (2019).

Muger Subbasin

The Muger subbasin, which serves as a north-flowing tributary of the Abbay River, encompasses an area of approximately 8188 km^2, as documented by Amare and Simane (2018). Within the Muger subbasin, notable tributaries include the Labbu, Aleltu, Sibilu, Roba, and Gerbi Rivers, contributing to its hydrological network.

Like other watersheds within the Abbay basin, the Muger subbasin has experienced a decline in forest cover over time. A study conducted by Teshome et al. (2022) revealed a reduction in forest cover from 11.80% of the subbasin's total area in 1986 to 5.90% in 2020. In contrast, the proportion of cultivated land exhibited an increase from 68.86 to 70.44% during the same period.

The decline in forest cover and the concurrent expansion of cultivated land in the Muger subbasin signify a significant change in land use patterns. These transformations have implications for the ecological balance, water resources, and overall sustainability within the subbasin. Preserving and managing the remaining forested areas while addressing the drivers behind the expansion of cultivated land are crucial for maintaining the ecological integrity and long-term health of the Muger subbasin.

Jemma Subbasin

The Jemma subbasin is situated in the South-Eastern portion of the Upper Abbay Basin, resembling the shape of Ethiopia as depicted in Fig. 7.2.

It covers a substantial area of approximately 15,782 km^2. A study conducted by Worku et al. (2021) revealed that agricultural land occupies the majority land cover, accounting for 56.72% of the total area. Grazing land, bare land, shrubland, woodland, and forestland follow, comprising 14.64%, 10.45%, 6.44%, 6.05%, and 1.20% of the subbasin, respectively. Additionally, smaller areas are covered by Afro-Alpine vegetation, eucalyptus plantations, and water bodies. The Jemma River, the primary river in the basin, is fed by several smaller tributaries, including the Beressa, Wizer, Chacha, Shay, Aleltu, and Robi Rivers.

In the Jemma Basin, land cover change is an on-going phenomenon driven by intensive agricultural activities and the expansion of settlements within the subbasin. Consequently, there has been a significant decline in forestland over time. A recent study by Bishaw (2022) indicates that forestland accounted for 33.7% of the area in 1985, decreasing to 30.7% in 2021. On the other hand, agricultural land increased from 32.6% in 1985 to 35% in 2021. The proportion of residential and settlement land also experienced an increase from 15.6% in 1985 to 17.8% in 2021. Furthermore, bush and shrub land witnessed a slight increase from 18.1% in 1985 to 18.5% in 2021.

Intensive land evaluation conducted by Taye et al. (2019) sheds light on the categorization of land within the Jemma watershed. Approximately 1707.7 ha of land fall under the classification of no arable land, rendering it unsuitable for annual crop cultivation at any level of intensity. Surprisingly, around 3.5% of this no arable land is still utilized for agricultural purposes, leading to increased soil erosion in the upstream regions and heightened risks of siltation and flooding in the downstream areas.

Welaka Subbasin

The Welaka subbasin, covering an area of approximately 6415 km^2, is characterized by the Selgi and Lega Kora Rivers as its tributaries. A recent study by Shobary et al. (2021) analyzed a Landsat 8 image from 2020 to assess the land cover composition within the subbasin. The results indicate that water bodies occupy 20.67% of the area, while urban areas account for 32.65%. Barren land comprises 28.96%, forestland only represents 2.32%, grassland covers 5.29%, and crop land accounts for 10.11% of the subbasin.

Importantly, the study highlights that forestland has the smallest spatial extent compared to other land cover categories in the Welaka subbasin. This suggests significant changes in land use, with a notable reduction in the size of forests.

Wombera Subbasin

The Wenbera subbasin spans an area of 12,957 km^2. A study conducted by Shobary et al. (2021) revealed that the dominant land cover type within the subbasin is barren land, comprising 86.4% of the total area. Furthermore, an analysis of the Landsat 8 image from 2020, the same study reported that forest cover accounts for 7.9% of the subbasin. Specifically, the Wenbera Mountains in the subbasin host a moisture-loving mountain forest characterized by the presence of *Hagenia abyssinica* and *Podocarpus flactus*. These natural vegetation types contribute to the forest cover within the area. Additionally, it is common to find Eucalyptus plantations near villages in the Wenbera subbasin, which are primarily utilized for construction purposes and as a source of fuelwood (Nyssen et al. 2018).

Beles Subbasin

The Beles subbasin, spanning an area of 14,200 km^2, boasts a diverse array of indigenous vegetation types, including *Cordia africans, Acacia spp., P. flactus, Ficus sycomorus, O. europaea,* and *O. abyssinica* (Nyssen et al. 2018). However, the subbasin has witnessed a gradual reduction in vegetation cover over time, primarily attributed to activities such as charcoal making, fires, and other means of livelihood.

The decline in woodland and the expansion of cultivated land have emerged as major factors contributing to increased surface run-off and water yield (Woldesenbet et al. 2016).

Research conducted by Semeneh Bessie Desta (2014) delved into the estimation of deforestation patterns and magnitudes using NDVI data obtained from two distinct periods, namely 2001 and 2012. The study highlighted a significant decrease in forest cover, which can be attributed to various factors including high population pressure, expansion of agricultural land, large-scale investments, uncontrolled wildfires, illegal logging, firewood and charcoal production, as well as forced resettlement programs. The bamboo forest in the subbasin, which holds significant value for supporting livelihood development, has experienced changes between 1985 and 2019. Initially accounting for 5.1% of the subbasin, the bamboo coverage witnessed a decline of 0.8% during the study period (Abebe et al. 2021).

Rahad Subbasin

The Rahad River originates from the western regions of Ethiopia and stretches all the way to Sudan. The Ethiopian portion of the Rahad subbasin covers an estimated area of 8269 km^2. The hydrology of the Rahad River basins is characterized by its complexity, influenced by diverse factors such as climate, topography, soil composition, vegetation, and geology (Hassaballah et al. 2016).

Research conducted by Hassaballah et al. (2016) revealed significant changes in the land use within the Rahad subbasin. Over the period of 1972 to 2011, there has been a noteworthy decline in woodland coverage, decreasing from 35 to 14%. Conversely, cropland has experienced substantial growth, increasing from 18 to 68% in the same subbasin.

Dinder Subbasin

Originating from the western part of Ethiopia and extending into Sudan, the Dinder River plays a crucial role in the Dinder subbasin, which spans an estimated area of 14,891 km^2. The river operates as a seasonal water source, primarily active from July to November, supplying water to the catchment area. Unlike the Rahad River, the Dinder River exhibits a limited expansion of cultivated land in its catchment, resulting in a reduced impact of land use and land cover change on stream flow within the Dinder subbasin (Hassaballah et al. 2017).

The vegetation types identified within the Dinder subbasin include wooded grassland, open grassland, woodland, and riverine forest. Dominant species found in the basin consist of *Acacia seyal, B. aegyptiaca, Combretum spp., Anogeissus leocarpa*, and *Dichrostachys cinerea* (Dasmann 1972; Hassaballah et al. 2016). Notably, a significant degradation of woodland has been observed in the Dinder basin, with the woodland coverage declining from 42% in 1972 to 14% in 2011 (Hassaballah et al. 2017).

Forest Resource Types

The forest resource types in the Abbay Basin are determined by analyzing the land use and land cover in the region. Spanning a vast area of 199,812 km^2 in Ethiopia, the Abbay subbasin consists of various land cover categories. These include cultivated land (34%), along with different types of forests such as tree crops (0.1%), plantation forests (0.3%), Afro-Alpine vegetation (0.6%), disturbed forests (1.1%), and bamboo forests (3.7%). Collectively, the forested areas make up approximately 5.8% of the entire basin. Additionally, woodland, bushland, and shrublands cover 30.2% of the area, while grasslands encompass 23.1%. Wetlands account for 1.2% of the land cover, water bodies occupy 1.6%, and rock and urban areas make up 4% (ENTRO 2006; Guchie 2007).

The primary land cover types dominating the Abbay Basin are cultivation, grassland, and woodland (MoWR 1998). Dense woodlands are mainly concentrated on the lower western slopes in Wellega, North Gondar, and Benishangul-Gumuz. Conversely, open woodlands are notably present in Wellega, Illubabor, and Jimma, often accompanied by grass understory or grassed areas amidst regions with sparser woody vegetation (Teferi et al. 2015). In the highland areas of the basin, Eucalyptus plantations are widespread, serving as a common source of construction materials and fuelwood for nearby villages (WBISPP 2002).

Forest Characteristics

The forest features seen in the Abbay Basin were analysed and classified into various vegetation types (Friis et al. 2011). As a result, they identified six distinct vegetation types in this area. These include Combretum-Terminalia woodland and wooded grassland, dry evergreen Afromontane Forest and Grassland complex, Afro-Alpine vegetation, Ericaceous Belt, moist evergreen Afromontane Forest, and freshwater marshes, flood plains, lake shore vegetation, and open water vegetation (Fig. 7.4).

In the western and North-Western parts of the basin, the presence of Combretum-Terminalia woodland and wooded grassland areas is notable. These areas are characterized by the abundance of various plant species, including *Combretum spp., Terminalia spp., O. abyssinica, B. papyrifera, A. leiocarpa, S. kunthianum, P. lucens, L. laxiflorus, Lannea spp., A. malacophylla,* and *E. africana.*

A separate research conducted by Awas et al. (2009) focused on the Banishangul Gumuz region and identified five distinct plant communities. These communities are Hyphaene thebaica-Pterocarpus lucens, Boswellia papyrifera-Pterocarpus lucens, Securidaca longepedunculata-Albizia malacophylla, Croton macrostachyus-Albizia

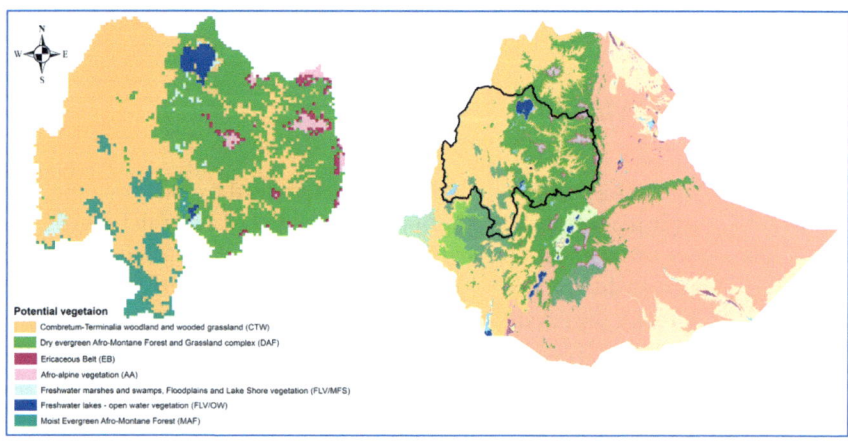

Fig. 7.4 Potential vegetation map of the Abbay basin (*Shapefile source* Friis et al. 2011)

malacophylla, and Breonadia salicina-Phoenix reclinata. Afromontane forests are found in the eastern, North-Eastern, and central highlands of the Abbay Basin.

The dominant species in this category include *P. falcatus* and *J. procera* in the upper canopy, and *C. macrostachyus, Ficus spp., O. europaea subsp. cuspidata, T. orientalis,* and *M. lanceolata* in the understorey. These Afromontane forests have been heavily exploited and fragmented in the basin. Riverine forests, predominantly located near lakes and rivers, are characterized by dominant species such as *D. mespiliformis, M. kummel,* and *S. guineense,* along with the Combretum-Terminalia woodlands (Wassie 2017; Song et al. 2018).

Bamboo forests are other significant forest resources in the Abbay Basin. Ethiopia is home to two indigenous bamboo species, namely *Yushane alpine* and *Oxytenanthera abyssinica*, which are endemic to Africa. *Yushane alpine* and *O. abyssinica*, commonly known as highland and lowland bamboo, respectively, are found in various parts of the Amhara and Benishangul-Gumuz regions (Embaye 2000). According to Wassie (2017), highland bamboo (*Y. alpine*) is predominantly distributed in the Amhara Region, including areas such as Awi zone, West Gojam zone, East Gojam, South Gondar, and North Shewa. Lowland bamboo (*Oxytenanthera abyssinica*) is mainly found in the Awi Zone of the Amhara National Regional State, as well as in Assosa Zone, Metekel Zone, Kemash Zone, Mao-Komo in the Benishangul-Gumuz region, and various zones in the Oromia region, such as West Wollega, East Wollega, and Buno Bedele (Anjulo et al. 2022).

The presence of plantation forests in the Abbay Basin is documented based on ownership, including state-owned plantations and community and individual woodlots (Wassie 2017). These plantations primarily consist of Eucalyptus species such as *E. globulus* and *E. camaldulensis*, as well as *C. lusitanica* and *Pine species* in certain areas. However, there is a lack of sufficient data at the basin level to determine the extent of forest disturbance and recovery over time. It is worth noting that Eucalyptus species have faced ecological criticism from various sources, as highlighted by Yitaferu et al. (2013). Despite this criticism, the use of Eucalyptus has significantly expanded in the central highlands, potentially contributing to a notable increase in forest cover in certain parts of the basin.

Designated Major Functions of Forests

Forest Contributes to SDGs

Forests have a significant impact on the achievement of the Sustainable Development Goals (SDGs) as they offer fundamental ecosystem services that are crucial for human welfare. The development of the Grand Ethiopian Renaissance Dam (GERD) in Ethiopia encompasses more than just energy production, as it also addresses global issues like carbon sequestration and efforts to mitigate climate change. To ensure the success of these efforts, it is necessary for downstream countries like Sudan and Egypt to collaborate and actively participate in afforestation programs. These initiatives aim to preserve and conserve clean water resources, while also mitigating the risks of flooding and sedimentation. Therefore, allocating resources toward the sustainable development of forest resources holds the capacity to make a positive impact on the SDGs, incorporating a diverse set of socio-economic and environmental goals.

Managing Forest for Water

The development of forest resources plays a crucial role in determining the availability and quality of water resources. Forests serve as natural protectors of water bodies and watercourses by effectively capturing sediments and pollutants. Furthermore, they contribute to the replenishment of water sources such as groundwater, surface watercourses, and stormwater, thereby increasing the overall water supply (FAO 2008).

During the last four decades, noteworthy transformations in the forested areas of the Abbay basin have been evident, and this region bears great strategic significance due to its association with the Abbay River. These alterations have exhibited a strong correlation with climate change, leading to notable repercussions on the hydrological patterns and water reserves of the area. As a result, the basin has witnessed a rise in calamitous occurrences such as floods and droughts (Bergkamp et al. 2003), along with a scarcity of water during the dry season (Gebrehiwot 2012).

In light of these challenges, the restoration of degraded areas and the enhancement of forest ecosystems emerge as effective strategies for mitigating the adverse effects of climate change. By doing so, the occurrence of water scarcity within the basin can be minimized (Bruijnzeel 2004). Through these measures, the restoration and preservation of forested areas can help alleviate the impacts of climate change, foster sustainable water resources, and promote a more resilient hydrological system in the Abbay basin.

Managing Forest for Soil

Forests play a vital role in the preservation of soil organic matter and serve as significant repositories of carbon (C) and nutrients, mainly through the supply of litterfall to the forest floor (Friggens 2020 Sayer 2021). Consequently, forest ecosystems are widely recognized for their positive contributions to sustainable land management practices. They offer a range of services that include enhancing soil fertility, safeguarding soil and water resources, rehabilitating degraded lands, and combating desertification (Bredemeier 2011). One of the notable benefits of forests is their capacity to enhance soil infiltration, leading to increased groundwater recharge. Forests also serve as an effective solution for the conservation of riparian areas. Riparian forests, in particular, possess unique ecosystem functions as they act as a transitional zone between upland and aquatic habitats. The roots of trees and understory shrubs within these forests stabilize riverbanks, preventing erosion, and exerting a significant influence on the diversity and communities associated with aquatic ecosystems (Ruzicka et al. 2014).

The Abbay basin is highly susceptible to severe soil erosion (Bewket and Teferi 2009) due to factors such as extensive deforestation, rugged terrain, intensive farming on steep slopes, and inadequate land management practices (Reusing et al. 2000). Consequently, prioritizing soil and water conservation efforts is essential in the Abbay basin (Zeleke 2014). To achieve this, a comprehensive approach is needed, which includes intensive rehabilitation activities, thoughtful species selection, and active participation from the public. By implementing these measures, the sub-watersheds within the basin can be restored to a sustainable condition, ensuring the long-term preservation of soil and water resources.

Managing Forest for Biomass Energy

In Ethiopia, particularly in the Abbay basin, biomass energy represents a significant portion, over 85%, of the total energy consumed (Yalew, 2022). Consequently, the efficient production of energy from forest biomass necessitates the formulation of interdisciplinary strategies in conjunction with exploring alternative renewable energy sources (Hailu and Kumsa, 2020). In order to enhance Ethiopia's renewable energy output, significant endeavors such as the Great Ethiopian Renaissance Dam (GERD) are being undertaken along the Abbay River. To maintain a sustainable energy supply, forest resources play a vital role by not only serving as a source of fuel but also aiding in the replenishment of water in streams, as well as reducing the occurrence of flooding and sedimentation in the newly constructed dams.

In this context, when rehabilitating degraded lands, the selection of tree species should take into account the demand for fuelwood as well as interventions for soil and water conservation. Furthermore, the Abbay basin plays a vital role in supporting the livelihoods of local communities by providing non-timber forest products. These include various resources such as bamboo products, medicinal plants, gum and incense products, honey bee production, and cultural values, which contribute to the economic and cultural well-being of the people in the region.

Challenges and Opportunities

Challenges

Deforestation and Forest Degradation in the Abbay Basin

Deforestation and forest degradation pose significant challenges in the Abbay basin, although the extent of forest loss and gain varies across different sub-watersheds and over time (Bewket 2002; FAO 2010).

In the Chemoga watershed, e.g., Bewket (2002) observed a yearly increase of approximately 1% in forested land between 1957 and 1998. On the other hand, in a distinct study by Gebrehiwot et al. (2010), the focus was placed on the detrimental effects of deforestation on soil and water degradation in the Abbay basin. Through their comprehensive examination of the Koga watershed, situated within the Abbay basin, they observed a significant decline in the extent of natural forest, occurring at a rate of 3.6% annually between 1957 and 1982. In a separate investigation by Gashaw et al. (2020), it was similarly noted that the Gilgel Abbay watershed predominantly consists of cultivated land, grassland, and shrubland, with a discernible decline in forested areas. Consequently, these alterations have contributed to an elevated average yearly erosion of soil within the basin and the deposition of sediment in Lake Tana.

Drivers of Deforestation in the Abbay Basin

There are two primary processes that contribute to the reduction of forest area: deforestation and natural disasters. The main drivers of forest clearance are closely linked to human activities, including tree cutting, charcoal production, timber harvesting for construction materials, and the conversion of natural vegetation into agricultural land (Zeleke and Hurni 2001; Gebrehiwot et al. 2014, Belay and Mengistu 2019). Additionally, natural disasters such as wildfires, droughts associated with climate change, floods, and outbreaks of pests and diseases can also lead to the destruction of forests. For instance, in the lowland areas of the basin, forest fires are a recurring phenomenon.

These fires are intentionally ignited by farmers and pastoralists each year as a means of preparing land for cultivation and promoting the regrowth of pasture grasses. However, these fires often cause severe damage to natural forests, resulting in significant changes in the composition of plant species and potential impacts on biodiversity. Large-scale investment activities also contribute significantly to deforestation and forest degradation within the basin. In the Benishangul-Gumuz Region (BGR), e.g., Tsegaye (2013) identified a considerable amount of land that has been allocated to domestic and foreign investors for agricultural purposes without proper assessment of existing land uses. This phenomenon, commonly referred to as 'land grabbing' by Dessalegn (2011), has detrimental effects on both land resources and local communities. The intervention of such investments has accelerated the degradation of forest resources and posed a threat to the livelihood security of rural communities. These communities heavily rely on forest resources for their sustenance, resulting in additional pressures on already strained common resources.

Ownership and Management Rights

One of the most significant obstacles to achieving sustainable forest management is the issue of insecure land tenure and unequal distribution of benefits among forest users (Andersson et al. 2018). A considerable portion of the woody biomass in the Abbay basin, especially in lowland areas, is considered 'common land,' which is often exploited but not effectively managed. While the Forest Development, Conservation, and Utilization Proclamation No. 1065/2018 (FDRE 2018) has categorized forests into four types—private forest, community forest, association forest, and state forest—limited attention has been given to the first three types of ownership. In recent years, notable efforts have been made to mobilize communities for restoration work and minimize forest loss. Although, the Ethiopian government has garnered global attention through its ambitious tree-planting targets, initiated under the Green Legacy Initiative (GLI) in 2019, these efforts have not been supported by competent leadership in areas such as species and site selection, planting techniques, and the sustainability of management interventions. Furthermore, promoting farm forestry activities and implementing agroforestry interventions on private land within the Abbay basin can contribute to increased productivity and the conservation of nature within the basin and beyond.

Institutional Barriers

Despite Ethiopia's strong commitment to increasing forest cover and enhancing the role of forests in the country, a major challenge lies in the institutional changes required for effective forest resource conservation (Fetene et al. 2015). The shift in government focus from production-oriented approaches to multifunctional forests and broader environmental conservation concepts occurred after the devastating drought and famine experienced during the 'Derg regime' in 1984–1985 (Ayana et al. 2013). However, with the overthrow of the 'Derg regime' and the subsequent establishment of the EPRDF in 1991, the forestry and environmental conservation sectors took a backseat to the priority of achieving rapid economic growth through agricultural intensification.

Consequently, the forestry sector gradually lost its autonomy and presence within the national agenda, without alternative institutions of comparable strength and function being established (Yemishaw 2001; Bekele 2008; Mengesha 2009). These changes and institutional instability at the national level have had detrimental effects on forest resources, leading to land degradation in various parts of the country, including the Abbay basin.

Opportunities

Progress Toward Sustainable Forest Management

The Ethiopian government has shown a strong determination to improve the sustainable governance of previously state-controlled forests and woodlands. This commitment is clearly evident through the utilization of the Participatory Forest Management (PFM) approach, which was integrated into the Climate Resilient Green Economy (CRGE) Strategy in 2011. Additionally, in 2016, the government introduced the African Forest Landscape Restoration (FLR) commitment as a further testament to its dedication. It is worth noting that Ethiopia's government has recently updated its nationally determined contribution (NDC) with an ambitious plan to achieve carbon neutrality and attain middle-income status by 2030 (Kassa et al. 2022).These various initiatives hold great promise for the conservation of natural resources, particularly forest resources, at a larger scale encompassing the Abbay basin.

Payment for Environmental Services

Ethiopia has dedicated significant efforts to enhance the livelihoods of farmers through the implementation of the REDD+ investment program.

This program incentivizes farmers to participate in the conservation and protection of high forests, woodlands, traditional agroforestry systems, and restoration activities such as area closures, soil and water conservation, reforestation/afforestation, and biosphere reserves. In 2015, the Ethiopian government submitted its Intended Nationally Determined Contribution (INDC) to the UNFCCC, which later became Ethiopia's first NDC upon ratifying the Paris Agreement in 2017. However, the updated NDC in 2021 demonstrates Ethiopia's increased ambition by proposing a more substantial emission reduction target of 68.8% (-277.7 Mt CO2eq) by 2030, surpassing its initial NDC target of 64% (FDRE 2021).

Ethiopia's commitment to rehabilitating degraded land through a large-scale afforestation program aligns with its NDC objectives and presents an opportunity to receive payments for environmental services (PES). This initiative aims to establish a willingness to compensate (WTC) for improved land and water management practices throughout the country, with a specific focus on the Abbay basin (Alemayehu et al. 2018). By ensuring the equitable sharing of costs and benefits among all stakeholders, Ethiopia fosters the potential for sustainable environmental management in the Abbay basin and beyond.

Establishment of Forest Enterprises in the Basin

The forestry sector in Ethiopia currently presents promising investment opportunities. The escalating demand for wood, driven by economic growth, urbanization, and infrastructure development, creates a favorable environment for forest enterprises. Notably, the Amhara and Oromia regions have already seen effective operations of forest enterprises, indicating significant potential for job creation and income generation in both rural and urban communities. This potential extends to diverse demographic groups, including women and youth.

Moreover, Ethiopia has established a supportive legal framework that encourages investment in the forestry sector. The Forest Proclamation, numbered 1065/2018, serves as a facilitator for such investments (FDRE 2018). Additionally, the government has made commitments to bolster investment in the sector through initiatives like the National Forest Sector Development Program (MEFCC 2017a, b), the REDD + Strategy (MEFCC 2018), and the Bamboo Development Strategy and Action Plan (EFCCC 2020). To maximize the benefits of these supportive policies and incentives, it is crucial to raise awareness at the regional and subbasin levels, ensuring potential investors are well-informed about the opportunities available.

NGOs Involvement in Watershed Management

The involvement of non-governmental organizations (NGOs) in forest management and environmental sectors has witnessed significant growth. These NGOs have embraced the approach of participatory forest management (PFM), which has demonstrated better environmental, social, and economic outcomes compared to traditional state-led forest management methods (Ayana 2014). Various governments and NGOs are actively supporting the restoration and sustainable utilization of natural resources in the Abbay basin region.

Numerous organizations are actively participating in these endeavors. For example, the Nile-SEC (NBI Secretariat), which is an intergovernmental partnership comprising ten Nile Basin countries including Burundi, DR Congo, Egypt, Ethiopia, Kenya, Rwanda, South Sudan, Sudan, Tanzania, and Uganda, plays a vital role. Similarly, the Eastern Nile Technical Regional Office (ENTRO) serves as the executive and technical body of the Eastern Nile Subsidiary Action Program (ENSAP) and is jointly owned by Egypt, Ethiopia, South Sudan, and Sudan.

The International Water Management Institute (IWMI) focuses on improving water and land management practices, while the Organization for Rehabilitation and Development in Amhara (ORDA) places significant emphasis on water and land resource management programs as crucial components of the watershed management strategy. Lastly, the Nature and Biodiversity Conservation Union (NABU) actively engages in the rehabilitation of the Lake Tana Biosphere Reserve and its surrounding watershed, aiming for long-term sustainability.

The Abbay basin offers a valuable opportunity for development organizations to participate in the management of natural resources, particularly through tree-planting initiatives aimed at ensuring a sustainable water supply to the Grand Ethiopian Renaissance Dam (GERD) and downstream countries. Consequently, NGOs operating in Sudan and Egypt can enhance collaboration with Ethiopian institutions involved in the forestry sector, working toward shared benefits for the sustainable conservation and utilization of the Nile basin.

Conclusions

Numerous research investigations have been carried out within the Abbay basin; however, these studies have predominantly concentrated on specific subbasins or micro-watersheds, lacking a comprehensive analysis of the entire basin. Human actions, particularly the growth of farming and residential areas, stand as the primary catalysts behind deforestation and the deterioration of land. In most of the study basins, cultivated land has expanded while forestland has decreased. The Tana watershed stands out as an exception, where the expansion of Eucalyptus plantations has led to an increase in forestland, benefiting the local community as a crucial source of livelihood. A significant challenge in the Abbay basin is the absence of a strong sense of ownership regarding forest resources. This has resulted in the neglect and unregulated exploitation of much of the existing woody biomass, particularly in lowland areas. The forestry sector in Ethiopia has experienced multiple restructurings over the past five decades, which, combined with institutional instability at the national level, have negatively impacted forest resources and contributed to land degradation throughout the country, including the Abbay basin. Despite these challenges, there are positive indications pointing toward the sustainable advancement and administration of forest reserves. The Ethiopian administration has established ambitious objectives, striving to realize a carbon–neutral economy and attain middle-income status by 2030. This initiative holds significant potential for conserving natural resources, especially forest resources, at a broader basin level, including the Abbay basin. Ethiopia's commitment to rehabilitating degraded land through a large-scale afforestation program aligns with its nationally determined contribution (NDC), creates opportunities for receiving payments for environmental services (PES), which serve as incentives for implementing improved land and water management practices nationwide. Moreover, a supportive legal framework, such as the Forest Proclamation (no 1065/2018), facilitates investment in the forestry sector.

Moreover, there has been a significant increase in the engagement of nongovernmental entities within the realms of forestry and environmental sectors. These actors have embraced participatory forest management (PFM) methodologies, showing the potential for attaining more favorable environmental, social, and economic results compared to conventional state-led forest management approaches.

References

ABA (2016) Atlas of Abbay basin, magazine. Abbay Basin Authority, High Council Secretariat and International Relations Directorate. Bahir Dar, Ethiopia

ABDO (2020) Water Atlas of Abbay Basin. Abbay Basin Development Office, Hydrometry and water quality directorate

Abebe S, Minale AS, Teketay D (2021) Spatiotemporal bamboo forest dynamics in the Lower Beles River Basin, North-Western Ethiopia. Remote Sens Appl Soc Environ 23:100538. https://doi.org/10.1016/j.rsase.2021.100538

Akalu F, Raude JM, Sintayehu EG, Kiptala J (2019) Evaluation of land use and land cover change (1986–2019) using remote sensing and GIS in Dabus Sub-Catchment, South-Western Ethiopia. J Sustain Resear Eng 5(2):91–100

Alemayehu B, Hagos F, Haileslassie A, Mapedza E, Awulachew SB, Peden D, Tafesse T (2018) Prospect of payments for environmental services in the Abbay Basin: examples from Koga and Gumera Watersheds, Ethiopia. CP 19 Project workshop proceedings, pp 254–280

Amare A, Simane B (2018) Does adaptation to climate change and variability provide household food security? Evidence from Muger subbasin of the upper Blue-Nile, Ethiopia. Ecol Process 7:13. https://doi.org/10.1186/s13717-018-0124-x

Andersson KP, Smith SM, Alston LJ, Duchelle AE, Mwangi E, Larson AM, Wong GY (2018) Wealth and the distribution of benefits from tropical forests: Implications for REDD+. Land Use Policy 72:510–522. https://doi.org/10.1016/j.landusepol.2018.01.012

Anjulo A, Mulatu Y, Kidane B, Reza S, Getahun A, Mulat S, Abere M, Teshome U (2022) *Oxytenanthera abyssinica* A. Rich. Munro species-site suitability matching in Ethiopia. Adv Bamboo Sci 1:100001. https://doi.org/10.1016/j.bamboo.2022.100001

Asimamaw N (2019) Impact of land use/land cover change on hydrology of the catchment: the case of upper ribb catchment, Lake Tana Sub Basin, Ethiopia. J Environ Earth Sci 9(6):13–29

Asrat T (2020) Land use land cover change and its effect on soil erosion using remote sensing data in Yewoll Watershed, Abbay Basin. Thesis, Bahir Dar University, Ethiopia, Ethiopia. MSc

Asres RS, Tilahun SA, Ayele GT, Melesse AM (2015) Analyses of land use/land cover change dynamics in the upland watersheds of Upper Abbay Basin. In: Melesse AM, Abtew W (eds) Landscape dynamics, soils and hydrological processes in varied climates, pp 73–91. https://doi.org/10.1007/978-3-319-18787-7_5

Awas T, Nordal I, Demissew S (2009) Plant communities of the woodland vegetation of Benishangul Gumuz National Regional State, Western Ethiopia. Ethiop J Nat Resour 11(2):169–193

Ayana AN (2014) Forest governance dynamics in Ethiopia: histories, arrangements, and practices. Wageningen University, Wageningen

Ayana AN, Arts B, Wiersum KF (2013) Historical development of forest policy in Ethiopia: trends of institutionalization and deinstitutionalization. Land Use Policy 32:186–196. https://doi.org/10.1016/j.landusepol.2012.10

Bekele M (2008) Ethiopia's environmental policies strategies and programs. In: Assefa T (ed) Digest of Ethiopia's national policies, strategies and programs. Forum for social studies, Addis Ababa, pp 337–369

Belay T, Mengistu DA (2019) Land use and land cover dynamics and drivers in the Muga watershed, Upper Abbay basin, Ethiopia. Remote Sens Appl Soc Environ. https://doi.org/10.1016/j.rsase.2019.100249

Bergkamp G, Orlando B, Burton I (2003) Change: adaptation of water resources management to climate change. Gland, Switzerland, IUCN

Berihun ML, Tsunekawa A, Haregeweyn N, Meshesha DT, Adgo E, Tsubo M et al (2019) Exploring land use/land cover changes, drivers and their implications in contrasting agro-ecological environments of Ethiopia. Land Use Policy 87:104052. https://doi.org/10.1016/j.landusepol.2019.104

Bewket W (2002) Land Cover Dynamics Since the 1950s in Chemoga Watershed, Abbay Basin, Ethiopia. Mt Resear Develop 22(3):263–269. https://doi.org/10.1659/0276-4741(2002)022[0263:lcdsti]2.0.co;2

Bewket W (2003) Household level tree planting and its implications for environmental management in the North-Western highlands of Ethiopia: a case study in the Chemoga watershed, Abbay basin. Land Degrad Dev 14:377–388

Bewket W, Teferi E (2009) Assessment of soil erosion hazard and prioritization for treatment at the watershed level: case study in the Chemoga watershed, Abbay basin, Ethiopia. Land Degrad Dev 20:609–622

Biru, M.K., Minale, A.S., Debay, A.B. (2015) Multitemporal land use land cover change and dynamics of Abbay Basin by using GIS and remote sensing techniques, North-Western Ethiopia. Int J Environ Sci 4(2):81–88

Bishaw DA (2022) Assessment of land use land cover change detection using remotely sensed data in Lower Jemma Sub-Basin of Upper Abbay, Central Ethiopia. Ethiop J Environ Stud Manag 15(4):546–556. ISSN: 1998-0507. https://ejesm.org/doi/v15i4.11

Bredemeier M (2011) Forest, climate and water issues in Europe. Ecohydrology 4(2):159–167. https://doi.org/10.1002/eco.203

Bruijnzeel LA (2004) Hydrological functions of tropical forests: not seeing the soil for the trees? Agr Ecosyst Environ 104:185–228

Dasmann W (1972) Development and management of the Dinder N. Park and its wildlife: A report to the Government of Sudan, 61p FAO No TA 31 1 3, Rome

Dessalegn R (2011) Land to investors: large-scale land transfers in Ethiopia, forum for social studies, Addis Ababa

Eastern Nile Technical Regional Office (ENTRO) (2006) Cooperative regional assessments for watershed management. Transboundary analysis of the Abbay–Blue–Nile subbasin. Nile Basin Initiative, Eastern Nile Technical Regional Office (ENTRO), Addis Ababa, Ethiopia

EFCCC (2020) Ethiopian Bamboo Development Strategy Action Plan (2019–2030). Environment, Forest and Climate Change Commission of Ethiopia & International Bamboo and Rattan Organization (INBAR), Addis Ababa, Ethiopia

Embaye K (2000) The Indigenous Bamboo forests of Ethiopia: an overview. AMBIO A J Hum Environ 29(8):518–521. https://doi.org/10.1579/0044-7447-29.8.518

Ewunetu A, Simane B, Teferi E, Zaitchik BF (2021a) Land cover change in the Abbay River headwaters: farmers' perceptions, pressures, and satellite-based mapping. Land 10(1):68. https://doi.org/10.3390/land10010068

Ewunetu A, Simane B, Teferi E, Zaitchik BF (2021b) Mapping and quantifying comprehensive land degradation status using spatial multicriteria evaluation technique in the headwaters area of Upper Abbay River. Sustainability 13:2244. https://doi.org/10.3390/su13042244

FAO (2008) Forests and water. A thematic study prepared in the framework of the Global Forest Resources Assessment (2005) FAO Forestry Paper 155. Italy, Rome

FAO (2010) Global forest resources assessment 2010. Main report. FAO forestry paper 163. Food and Agriculture Organization of the United Nations, Rome

FAO (2018) 1948–2018 Seventy years of FAO's Global forest resources assessment historical overview and future prospects. Food and Agriculture Organization of the United Nations, Rome, Italy

FDRE (2018) Proclamation No. 1065/2018 Forest Development, Conservation and Utilization Proclamation, Negarit Gazette, 24th Year, No. 21, Addis Ababa, 23rd Jan 2018

FDRE (2021) Updated nationally determined contribution (NDC). Federal Democratic Republic of Ethiopia. 48pp

Fetene A, Hilker T, Yeshitela K, Prasse R, Cohen W, Yang Z (2015) Detecting trends in landuse and landcover change of Nech Sar National Park, Ethiopia. Environ Manag 57(1):137–147. https://doi.org/10.1007/s00267-015-0603-0

Friggens NL, Aspray TJ, Parker TC, Subke JA, Wooke PA, Sangil LL, Rodtassana C (2020) Spatial patterns in soil organic matter dynamics are shaped by mycorrhizosphere interactions in a treeline forest. Plant Soil 447:521–535. https://doi.org/10.1007/s11104-019-04398-y

Friis I, Sebsebe D, van Paulo B (2011) Atlas of the Potential Vegetation of Ethiopia, Addis Ababa, Addis Ababa University Press & Shama Books

Galata AW (2020) Analysis of land use/land cover changes and their causes using landsat data in Anger watershed, Abbay basin, Ethiopia. J Sed Environ 5(4):415–423. https://doi.org/10.1007/s43217-020-00025-4

Gashaw T, Tulu T, Argaw M, Worqlul AW, Tolessa T, Kindu M (2018) Estimating the impacts of land use/land cover changes on Ecosystem Service Values: the case of the Andassa watershed in the Upper Abbay basin of Ethiopia. Ecosyst Serv 31:219–228. https://doi.org/10.1016/j.ecoser.2018.05.001

Gashaw T, Worqlul AW, Dile YT, Addisu S, Bantider A, Zeleke G (2020) Evaluating potential impacts of land management practices on soil erosion in the Gilgel Abbay watershed, upper Abbay basin. Heliyon 6(8):e04777. https://doi.org/10.1016/j.heliyon.2020.e04777

Gebre SL, Getahun D, Fufa F, Diriba OH, Nigussie A, Alemayehu E (2022) Hydrological responses of land use land cover change on the anger catchment, Abbay Basin. Ethiop J Sci Technol 13(1):33–44

Gebrehiwot SG (2012) Hydrology and forests in the Abbay Basin. What can be learned from half a century of observations and community perception for water management? Doctoral Thesis Swedish University of Agricultural Sciences, Uppsala

Gebrehiwot SG, Taye A, Bishop K (2010) Forest cover and stream flow in a headwater of the Blue Nile: complementing observational data analysis with community perception. AMBIO 39:284–294. https://doi.org/10.1007/s13280-010-0047-y

Gebrehiwot SG, Bewket W, Gärdenäs AI, Bishop K (2014) Forest cover change over four decades in the Abbay Basin, Ethiopia: comparison of three watersheds. Reg Environ Change 14(1):253–266. https://doi.org/10.1007/s10113-013-0483-x

Guchie A (2007) Synthesis of environmental assessment of Eastern Nile Sub Basins. Eastern Nile Technical Regional Office (ENTRO), Addis Ababa, Ethiopia

Hailu AD, Kumsa DK (2020) Ethiopia renewable energy potentials and current state. AIMS Energy 9(1):1–14. https://doi.org/10.3934/energy.2021001

Hassaballah K, Mohamed YA, Uhlenbrook S (2016) The Mayas wetlands of the Dinder and Rahad: tributaries of the Abbay Basin (Sudan). In: Finlayson CM, Milton GR, Prentice RC, Davidson NC (eds) The wetland book: II: distribution, description and conservation. Springer, Netherlands, pp 1–13

Hassaballah K, Mohamed Y, Uhlenbrook S, Biro K (2017) Analysis of streamflow response to land use and land cover changes using satellite data and hydrological modelling: case study of Dinder and Rahad tributaries of the Abbay (Ethiopia–Sudan). Hydrol Earth Syst Sci 21(10):5217–5242. https://doi.org/10.5194/hess-21-5217-2017

IBC (2005) Ethiopian Biodiversity Strategy and Action Plan, available at: https://www.cbd.int/doc/world/et/et-nbsap-01-en.pdf. Last access: 11 Nov 2022

Jembere TK (2017) GIS Based Surface Irrigation Potential Assessment in North Gojjam Basin, Ethiopia. M.Sc Thesis, Arba Minch University, Ethiopia

Kabite G, Muleta MK, Gessesse B (2020) Spatiotemporal land cover dynamics and drivers for Didessa River Basin (DRB), Ethiopia. Modelling Earth Systems and Environment. https://doi.org/10.1007/s40808-020-00743-8

Kassa H, Abiyu A, Hagazi N, Mokria M, Kassawmar T, Gitz V (2022) Forestlandscape restoration in Ethiopia: progress and challenges. Front Forests Glob Change, 06 Oct 2022. https://doi.org/10.3389/ffgc.2022.796106

Kenea U, Adeba D, Regasa MS, Nones M (2021) Hydrological responses to land use land cover changes in the Fincha'a Watershed, Ethiopia. Land 10:916. https://doi.org/10.3390/land10090916

MEFCC (2017a) Ethiopia's forest reference level submission to the UNFCCC. Environment, Forest and Climate Change Commission of Ethiopia, Addis Ababa

MEFCC (2017b) National forest sector development program, Ethiopia, Volume I situation analysis. Addis Ababa, Ethiopia

MEFCC (2018) National REDD+ Strategy (2016–2030). Final draft, version 1.2. Addis Ababa, Ethiopia

Mengesha B (2009) Ethiopian forest conservation, development and utilization: past, present and future. In: Heckett T, Aklilu N (eds) Proceedings of Ethiopian forestry at crossroads: the need for a strong institution. Occasional report No.1, forum for environment, Addis Ababa, Ethiopia

MoA (2000) Agro-ecological Zonations of Ethiopia. The Ministry of Agriculture, Addis Ababa, Ethiopia

Moisa MB, Dejene IN, Gemeda DO (2022) Geospatial technology–based analysis of land use land cover dynamics and its effects on land surface temperature in Guder River subbasin, Abbay Basin, Ethiopia. Appl Geomat 14:451–463

MoWR (1998) Abbay River basin integrated development master plan project. Ministry of Water resources, Addis Ababa, Ethiopia

Muleta TT, Biru MK (2019) Human modified landscape structure and its implication on ecosystem services at Guder watershed in Ethiopia. Environ Monit Assess 191(5). https://doi.org/10.1007/s10661-019-7403-6

NBI (2007) Trans-boundary Analysis Abbay-Abbay Sub-Basin. Eastern Nile watershed management project. Nile Basin Initiative, Addis Ababa, Ethiopia

Nyssen J, Fetene F, Dessie M, Alemayehu G, Sewnet A, Wassie A et al (2018) Persistence and changes in the peripheral Beles basin of Ethiopia. Reg Environ Change 18(7):2089–2104

Reusing M, Schneider T, Ammer U (2000) Modelling soil loss rates in the Ethiopian highlands by integration of high resolution MOMS-02/D2-stereo-data in a GIS. Int J Remote Sens 21(9):1885–1896. https://doi.org/10.1080/014311600209797

Ruzicka KJ, Puettmann KJ, Olson DH (2014) Management of Riparian buffers: upslope thinning with downslope impacts. Forest Sci 60(5):881–892. https://doi.org/10.5849/forsci.13-107

Sayer EJ, Baxendale C, Birkett AJ, Bre´chet LM, Castro B, Byrne BDK (2021) Altered litter inputs modify carbon and nitrogen storage in soil organic matter in a lowland tropical forest. Biogeochemistry 156:115–130. https://doi.org/10.1007/s10533-020-00747-7

Semeneh Bessie Desta SB (2014) Deforestation and a strategy for rehabilitation in Beles Sub-Basin, Ethiopia. J Econ Sustain Develop 5(15):285–295

Song C, Nigatu L, Beneye Y, Abdulahi A, Zhang L, Wu D (2018) Mapping the vegetation of the Lake Tana basin, Ethiopia, using Google Earth images. Earth Syst Sci Data 10:2033–2041. https://doi.org/10.5194/essd-10-2033-2018

Shobary AS, Elsharkawy A, El-Hanafy HEM, Moussa OM (2021) Updating land use and land cover classes of Blue Nile Basin using landsat-8 images. IOP Conf Ser Mater Sci Eng 1172:012014. https://doi.org/10.1088/1757-899X/1172/1/012014

Tamirat S, Paulos L (2022) Land use land cover analysis of the Great Ethiopian Renaissance Dam (GERD) catchment using remote sensing and GIS techniques. Geol Ecol Landscapes.https://doi.org/10.1080/24749508.2022.2138027

Tana & Beles Integrated Water Resources Development (2008) Project Appraisal Document (PAD), vol. 1, World Bank document. Accessed 11 Nov 2022

Taye M, Simane B, Zaitchik F., B., G. Selassie, Y., & Setegn, S. (2019) Land Use Evaluation over the Jema Watershed, in the Upper Abbay River Basin, North-Western Highlands of Ethiopia. Land 8(3):50. https://doi.org/10.3390/land8030050

Teferi E, Bewket W, Uhlenbrook S, Wenninger J (2013) Understanding recent land use and land cover dynamics in the source region of the Upper Abbay, Ethiopia: spatially explicit statistical modelling of systematic transitions. Agr Ecosyst Environ 165:98–117. https://doi.org/10.1016/j.agee.2012.11.007

Teferi E, Uhlenbrook S, Bewket W (2015) Interannual and seasonal trends of vegetation condition in the Upper Abbay (Abbay) Basin: dual-scale time series analysis. Earth Syst Dyn 6(2):617–636. https://doi.org/10.5194/esd-6-617-2015

Teshome DS, Taddese H, Tolessa T, Kidane M, You S (2022) Drivers and implications of land cover dynamics in Muger Sub-Basin, Abbay Basin, Ethiopia. Sustainability 14(18):11241. https://doi.org/10.3390/su141811241

Tessema I, Simane B (2019) Vulnerability analysis of smallholder farmers to climate variability and change: an agro-ecological system-based approach in the Fincha'a subbasin of the upper Abbay Basin of Ethiopia. Ecol Process 8(1). https://doi.org/10.1186/s13717-019-0159-7

Tolessa T, Kidane M, Bezie A (2021) Assessment of the linkages between ecosystem service provision and land use/land cover change in Fincha watershed, North-Western Ethiopia. Heliyon 7(7):e07673. https://doi.org/10.1016/j.heliyon.2021.e07673

Tsegaye M (2013) Postponed local concerns? Implications of land acquisitions for indigenous local communities in Benishangul-Gumuz Regional state, Ethiopia. LDPI working paper 13

Wassie A (2017) Forest resources in Amhara: brief description, distribution and status. Soc Ecol Syst Dyn. https://doi.org/10.1007/978-3-319-45755-0_15

WBISPP (2002) Amhara regional state: a strategic plan for the sustainable development, conservation, and management of the woody biomass resources, Final Report

Wedajo GK, Muleta Mk, Gessesse B, Koriche SA (2019) Spatiotemporal climate and vegetation greenness changes and their nexus for Didessa River Basin, Ethiopia. Environ Syst Res (2019) 8:31. https://doi.org/10.1186/s40068-019-0159-8

Woldesenbet TA, Elagib NA, Ribbe L, Heinrich J (2017) Hydrological responses to land use/cover changes in the source region of the Upper Abbay Basin, Ethiopia. Sci Total Environ 575:724–741. https://doi.org/10.1016/j.scitotenv.2016.09.124

Worku G, Teferi E, Bantider A, Dile YT (2021) Modelling hydrological processes under climate change scenarios in the Jemma subbasin of upper Abbay Basin, Ethiopia. Clim Risk Manag 31:100272. https://doi.org/10.1016/j.crm.2021.100272

Yalew AW (2022) Environmental and economic accounting for biomass energy in Ethiopia. Energ Sustain Soc 12:30. https://doi.org/10.1186/s13705-022-00356-2

Yalew S, Mul M, van Griensven A, Teferi E, Priess J, Schweitzer C, van Der Zaag P (2016) Land-use change modelling in the Upper Abbay Basin. Environments 3(4):21. https://doi.org/10.3390/environments3030021

Yemishaw Y (2001) Status and prospects of forest policy in Ethiopia. In: Biological society of Ethiopia: imperative problems associated with forestry in Ethiopia. In: Proceedings of a workshop, Addis Ababa University, Addis Ababa, pp 9–30

Yesuph AY, Dagnew AB (2019) Land use/cover spatiotemporal dynamics, driving forces and implications at the Beshillo catchment of the Abbay Basin, North-Eastern Highlands of Ethiopia. Environ Syst Res 8(1). https://doi.org/10.1186/s40068-019-0148-y

Yesuph AY (2020) Prospects of sustainable land management amidst interlocking challenges in the Upper Beshillo Catchments, North-Eastern Highlands of Ethiopia. PhD dissertation, University of South Africa

Yilma AD, Awulachew SB (2009) Characterization and Atlas of the Abbay Basin and its Sub basins. International Water Management Institute, 253p

Yitaferu B, Abewa A, Amare T (2013) Expansion of Eucalyptus woodlots in the fertile soils of the highlands of Ethiopia: could it be a treat on future cropland use? J Agric Sci 5(8). https://doi.org/10.5539/jas.v5n8p97

Zeleke G (2014) Exit strategy and performance assessment for watershed management: a guideline for sustainability. Addis Ababa, Ethiopia

Zeleke G, Hurni H (2001) Implications of land use and land cover dynamics for mountain resource degradation in the Northwestern Ethiopian highlands. Mt Res Dev 21(2):184–191. https://doi.org/10.1659/0276-4741(2001)021[0184:IOLUAL]2.0.CO;2

Chapter 8
Hydrology of the Abbay River Basin, Ethiopia

Getachew Tegegne and Assefa M. Melesse

Abstract The Abbay River basin has been identified as the main socioeconomic growth corridor in Ethiopia due to its significant water resource capacity for agricultural and hydropower production. Sustainable water resources planning and management require detail study of the hydrological processes of the basin. Several hydrological studies have been reported in the Abbay River basin; however, water resources planners are often faced with the large uncertainty in the previous hydrological studies results. Thus, this study explores the advancement in the hydrological studies in the Abbay River basin. This review covers the main hydrological studies, such as spatial and meteorological data analysis, water resources potential and development, application of hydrological models, ungauged hydrological modelling, and climate change impact on hydrology. The review found that the majority of previous studies reported unreliable hydrological modelling results in some of the Abbay River subbasins. The sources of uncertainty for hydrologic modelling in the Abbay River basin might be associated with water diversion for small-scale irrigation, infrastructure development in the headwater catchments, the existence of artificial open reservoirs, upstream water supply abstraction, poor watershed management, and hydro-meteorological variables measurement errors. Moreover, the lack of reliable spatiotemporal coverage and a disparate distribution of hydrometric gauging locations results in considerable uncertainty in the Abbay River basin hydrological prediction. Thus, one of the important areas for possible research in the Abbay River basin is reducing the hydrological processes modelling uncertainty. This can be achieved by utilizing emerging technologies for hydro-meteorological data observation. The application of merged satellite products and ground rain-gauge observations can enhance the reliability of hydrological modelling results in the Abbay River basin. Furthermore, the majority of prior hydrological studies in the Abbay River basin have been relied on secondary hydro-meteorological data; however, primary data (field

---

G. Tegegne (✉)
Department of Civil Engineering, Sustainable Energy Center of Excellence, Addis Ababa Science and Technology University, Addis Ababa, Ethiopia
e-mail: getachewtegegne21@gmail.com; getachew.tegegne@aastu.edu.et

A. M. Melesse
Department of Earth and Environment, Institute of Environment, Florida International University, Miami, FL, USA

data collection) could help to better understand hydrological processes. Integrating primary and secondary hydro-meteorological data can help to obtain improved hydrological process modelling results. The lack of continuous ground monitoring in high-elevation zones is one of the most significant challenges in the hydrological process modelling in the Abbay River basin. The satellite meteorological products and citizen science ground-based meteorological data can provide more reliable information on the spatiotemporal variability of Abbay River basin hydrology. In general, more reliable hydrological modelling results can be achieved by citizen science-based data collection, frequent monitoring and maintenance of hydrometric gauging stations, merging satellite products and ground rain-gauge meteorological observations, and advances in the simplified representation of hydrological processes in the physical hydrological models.

Keywords Hydrological modelling · Water resources · Climate change · Abbay River basin · Ethiopia

Introduction

The Abbay basin, one of the primary sources of the Nile River, has been considered the main region for socioeconomic growth in Ethiopia for irrigation and hydropower development (MoFED 2005) due to its substantial water resource development capacity. Numerous water resource projects are ongoing and proposed in the region for food and energy security as well as for economic growth of Ethiopia. The Abbay River basin (see Fig. 8.1) covers approximately 176,000 km^2 natural watershed area and geographically extends between latitudes 7°45′ to 12°45′N and 34°05′ to 39°45′E with elevations ranging between 500 and 4160 m above mean sea level. Water resource projects development in the Abbay River basin (e.g. Grand Ethiopian Renaissance Dam hydropower project, Koga irrigation project, Tana-Beles hydropower projects, among others) has significantly contributed to the food and energy security and socioeconomic growth of the country. Moreover, water resource projects can significantly alter the natural flow characteristics (the magnitude and frequency of natural flow), which adversely affects the hydrological model applications as well as the river quality and habitats health in the Abbay River. Therefore, a systematic review of previous hydrological processes modelling researches in the Abbay River basin is often a prerequisite for reliable water resource assessment and management in the region.

The hydrology, flow variability, rainfall trends of the Nile River basin, and other similar basin have been studied my many researchers (Chebud and Melesse 2009a, b; 2013; Dessu and Melesse 2012, 2013; Dessu et al. 2014). The Abbay River, which springs in the Ethiopian highlands, delivers 60% of the Nile River flow at the Aswan dam in Egypt (Conway and Hulme 1993; Sutcliffe and Parks 1999). The Abbay River water is crucial for water resource projects development in Ethiopia, Egypt, and Sudan, with competing demands for hydropower, irrigation, water supply, industry,

Fig. 8.1 Geographical study map of the Abbay River basin

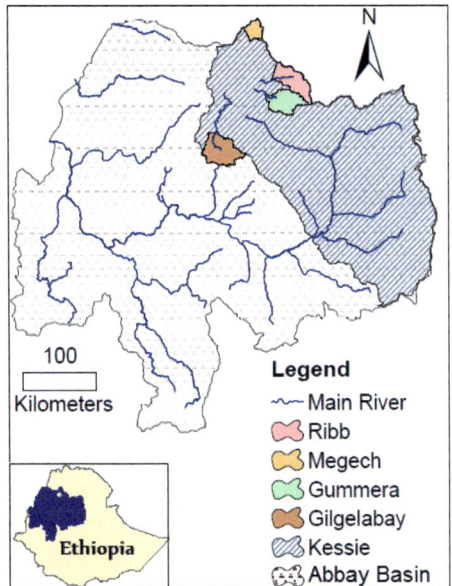

and other ecosystem services. It is important to know that the water resource development in Sudan, Ethiopia, and Egypt is severely affected by soil erosion and reservoirs and irrigation canals sedimentation. Moreover, climate variability and improper watershed management provide further obstacles to the basin's water resources development (Hurni et al. 2005). The long-term average flow of the Abbay River measured at the basin outlet in Sudan (Eldiem station) is approximately 48.9 Gm^3/year (Tesemma et al. 2010).

The United States Bureau of Reclamation reported the Abbay River basin's first comprehensive water resources project development master plan study (USBR 1964). A master plan for water projects development in the study region was also developed by a French engineering consultant, BCEOM (1999). There has been growing research interest in the investigation of the hydrology and water resources potential of the basin in the recent past. Several studies have assessed the implications of climate change on hydrological ecosystem services in the Abbay River basin (Tegegne et al. 2020; Tegegne and Melesse 2021; Beyene et al. 2010; Kim and Kaluarachchi 2009; Elshamy et al. 2009; Taye et al. 2011; Worqlul et al. 2018; Gebere et al, 2015; Bekele et al. 2017; Berhanu et al. 2015). Taye et al. (2011) reported that there were discrepancies in the hydrological ecosystem services under climate change in the region of interest. Moreover, previous hydrological processes modelling studies have assessed the hydrology of the Abbay River basin using various hydrological models spanning from lumped to fully distributed physical models. However, most of the previous research findings have not been documented and organized well, indicating the difficulty of identifying the hydrological research gap in the Abbay River basin. Thus, a systematic literature review could help identify research gaps, advance hydrological research, and suggest future possible studies in the Abbay River basin. The primary goals of this review are to understand the advances in hydrology research, identify the research gaps, and make recommendations for potential future study.

Spatial Data for Hydrological Modelling

Soil and water assessment modelling using physical-based models requires spatial data including the digital elevation models, land and soil uses. Without good knowledge of the spatial data (topography, land and soil uses) of the study watershed and subbasin characteristics, it is impossible to understand the basin's hydrological process (Sutcliffe and Parks 1999). Thus, it is crucial to assess and process the spatial data of the Abbay River basin for physically based hydrological processes modelling. Figure 8.2 depicts the land use map of the Abbay River basin land use map, which is mainly characterized by wooded grassland, agro-pastoral, agriculture, and sylvo-pastoral areas. Mostly, the land use/land cover in the Abbay River basin is significantly shifting to the cultivated land; the woody vegetation deforestation and agricultural expansion are rapidly increasing in the Abbay River basin (Tegegne et al. 2020). Thus, the significant change in the land use/cover through time can influence the hydrological processes modelling in the study region. For instance, the conversion of forest into farm land will increase the runoff and soil erosion. Thus, it is crucial to regularly update the land use/cover change impact on the surface runoff and watershed erosion in the study region.

The soils of the Abbay River basin are highly spatially variable. The spatial change in soil characteristics within a short distance might affect agricultural planning and development in the study region. Figure 8.3 depicts the spatial distribution of soil use in the Abbay River basin, which evidenced that the soil types of Nitisols, Leptosols, Luvisols, Vertisols, and Cambisols cover approximately 85% of the Abbay River basin. These soil groups are often characterized by acidity, depth, and permeability, which adversely affect the productivity of the region of interest. The soil use analysis indicates that 39.08% of the Abbay River basin area (199,812 km^2) is covered by Nitisols, 23.68% by Leptosols, 14.20% by Luvisols, 12.05% by Vertisols, and 5.07% by Cambisols.

The landscape of Abbay River basin is complex with elevation ranging between 502 m above mean sea level (m.a.s.l) along the Ethiopia-Sudan border and 4133 m.a.s.l. in the middle highlands. The study basin covers an area of about 199,812 km^2 (including Rahad and Dinder). A similar topographic layout can be seen in the small-scale watersheds (the Gummara, Gilgel Abay, Megech and Ribb small-scale watersheds); while the medium-scale basin (Kessie river basin) and large-scale basin (Abbay River basin) show diverse slope patterns. The diverse topography can influence the responses of hydrological process modelling; thus, it is essential to choose physical models that suit such landscape dynamics. The topography of the Abbay River basin is diversified with mountains, floodplains, lakes, wetlands, and gorges. The majority of the basin area is located in low land ranges (western parts

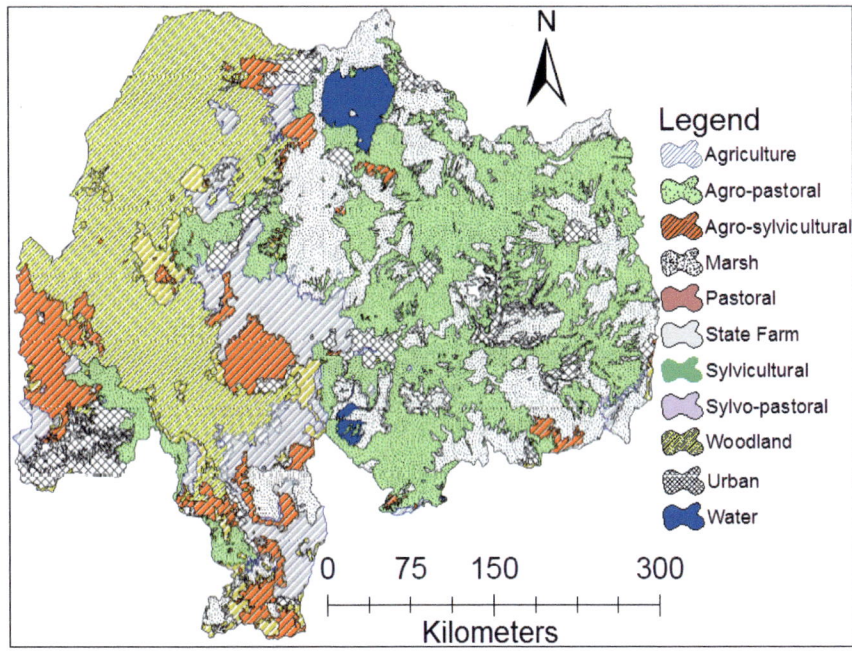

Fig. 8.2 Map shows the spatial distribution of land use in the Abbay River basin

of the basin) with elevations less than 1500 (m.a.s.l). The Abbay River basin eastern part, which primarily highlands, is the main source of rainwater. In general, topographic analysis in the Abbay River basin is the backbone for hydrological processes modelling by delineating the basin and investigating the smallest hydrologic response unit (Fig. 8.4).

Meteorological Data for Hydrological Modelling

The hydrology in the Abbay River basin is heavily reliant on rainfall intensity, frequency, and duration of rainfall. Thus, analysis of the spatiotemporal variability of rainfall is crucial to understand the hydrological processes of the study basin. The spatiotemporal variability of rainfall can greatly influence the basin hydrological response, such as low flow, peak flow, runoff volume, and timing of the hydrograph. The Abbay River basin has three seasons: dry (from October to the end of February), moderate wet (from March to May), and the major rainy (from June to September) (Tesemma et al. 2010). The temperature and rainfall variability in the Abbay River basin is investigated using the daily meteorological data collected from a21 gauging stations and recorded in the 1985–2014 period. The Ethiopian National Metrological Agency provided the observed meteorological data from 21 gauging stations. The

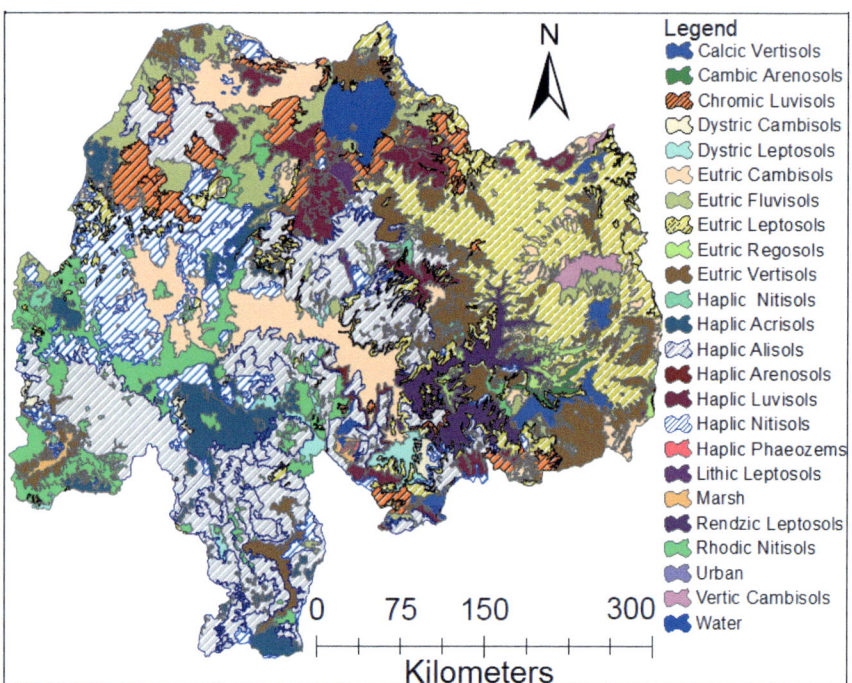

Fig. 8.3 Map shows the soil use spatial distribution in the Abbay River basin

mean annual rainfall over the Abbay River basin is between 2056 and 961 mm. The analysis showed that the rainfall is extremely variable over both space and time (see Fig. 8.5). The southern parts (mostly in the Nekemte, Arjo, Bedelle, and Jimma areas) receive more rainfall, while the majority of eastern and some of the western basin regions have the lowest rainfall amounts. The rainfall temporal variability calculated based on the coefficient of variation (CV) was found to vary between 0.12 and 0.48. There is no significantly extreme annual rainfall temporal variability in the basin. Most of the Abbay River basin showed almost similar year to year temporal variability patterns (see Fig. 8.5). The Lake Tana basin eastern part (Gummara and Ribb catchments) showed a relatively high temporal variability, with a CV of 0.48.

The spatial and temporal dynamics of maximum and minimum temperature was assessed using the recorded data from 21 gauging stations measured from 1985 to 2014. The maximum temperature was observed to spatially vary between 18.13 °C (eastern parts) and 32.91 °C (western parts). The temporal variability of maximum temperature was found to vary between 0.09 and 0.14 (see Fig. 8.6).

The minimum temperature was also calculated to spatially vary between 6.54 and 20.39 °C (see Fig. 8.7). The minimum temperature was observed to be higher over the western region, while the eastern parts showed relatively lower minimum temperatures. The temporal variability of minimum temperature was found to vary between 0.11 (over the western region) and 0.44 (over the eastern region). In general,

8 Hydrology of the Abbay River Basin, Ethiopia

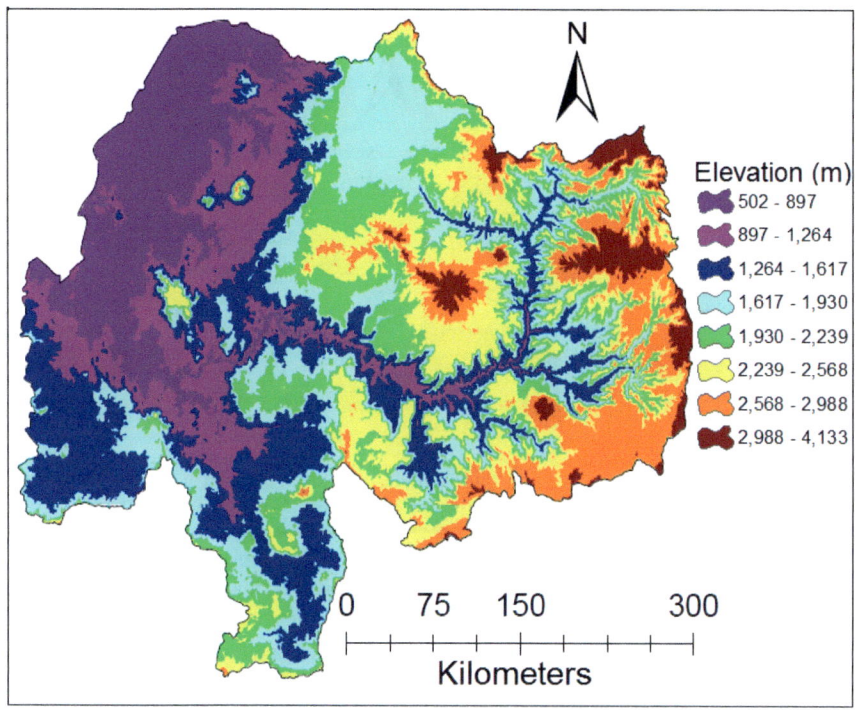

Fig. 8.4 Topography of the Abbay River basin

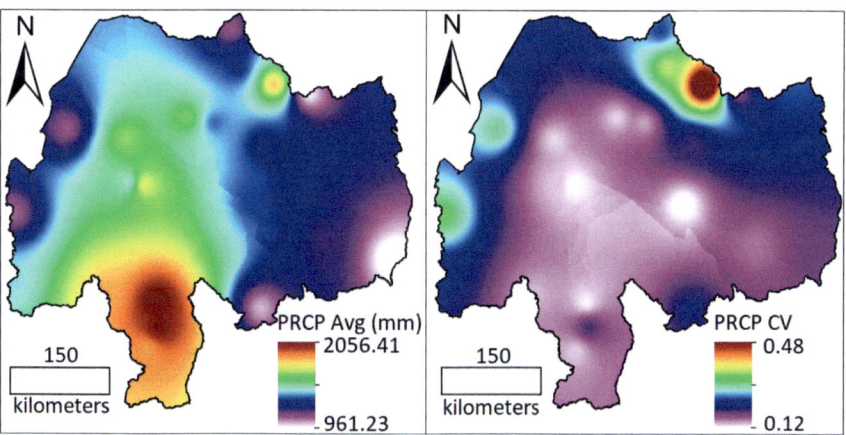

Fig. 8.5 The Abbay River basin spatial (left) and temporal (right) variability of the mean annual rainfall calculated based on the data recorded between 1985 and 2014

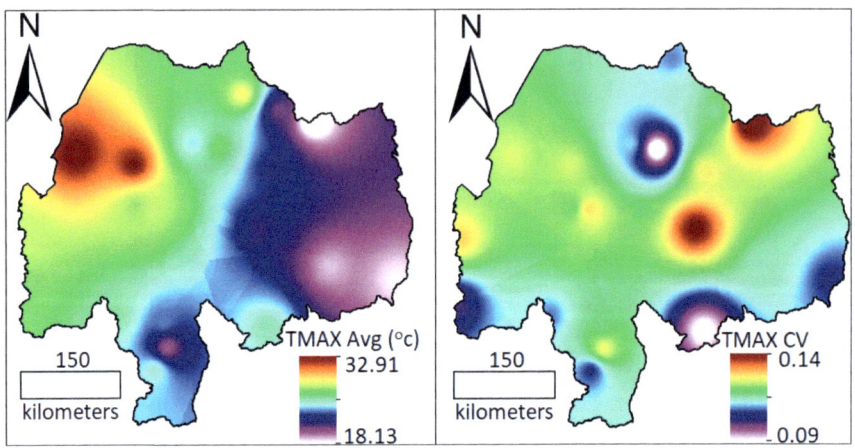

Fig. 8.6 Spatial (left) and temporal (right) variability of the long-term average maximum temperature calculated based on the analysis period 1985–2014 in the Abbay River basin

the western parts of the Abbay River basin had greater maximum and minimum temperatures, whereas the eastern parts had lower values.

The statistical trend analysis of the hydro-meteorological data has been investigated by various studies in the Abbay River basin (Tekleab et al. 2013; Tesemma et al. 2010; Gebremicael et al. 2013; among others). For instance, Tekleab et al. (2013) employed Mann–Kendall and Pettit tests to investigate the trends of streamflow, temperature, and rainfall in the Abbay River basin. They discovered a statistically significant increasing trend in temperature, but no significant trends in mean annual

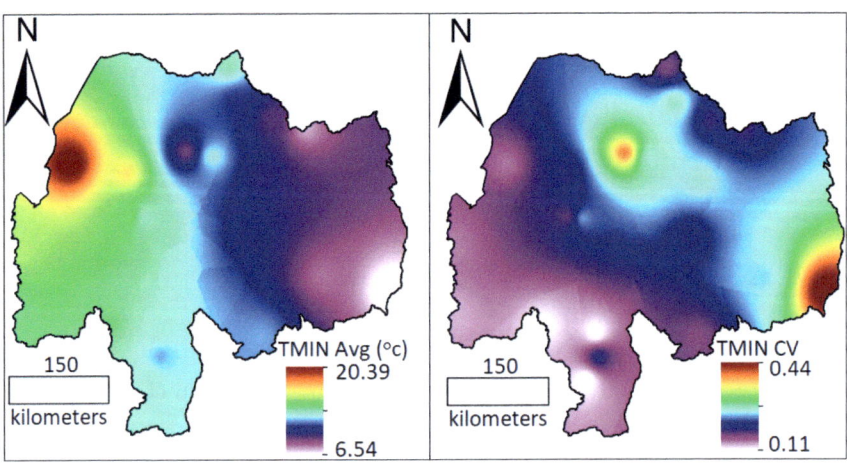

Fig. 8.7 Spatial (left) and temporal (right) fluctuation of the long-term average minimum temperature in the Abbay River basin calculated from 1985 to 2014

and seasonal rainfall, and a heterogeneous trend in streamflow (no decreasing or increasing trend).

Water Resource Development in the Abbay River Basin

The Ethiopian government has designated the Abbay River basin as a region for irrigation and hydropower development (Tegegne and Kim 2020), both of which are critical for Ethiopia's food security and economic growth (MoFED 2005). The Abbay (Blue Nile) basin is the primary source of the Nile River, accounting for more than 60% of the Nile water at Aswan reservoir (Mulat and Moges 2014), with a long-term annual average streamflow of 49.4 billion meter cubic (BCM) at the Sudan border (BCEOM 1999). Note that the annual average streamflow at the Lake Tana outlet (~3.5 BCM) contributes approximately 7% of the Abbay streamflow at the Sudan border (BCEOM 1999). Large-scale water resources projects (hydropower and irrigation) have been implemented and planned in the Abbay River basin due to its substantial water resources potential. The Abbay basin's potential irrigable area is estimated to be approximately 760,000 ha (FDREMW 2002). The Grand Ethiopian Renaissance Dam (GERD) hydropower project, which is near completion of construction, is one of the main mega water resources projects in the Abbay River basin. In addition, there is a development plan for the cascade reservoir system upstream of GERD. Most of the water used by Sudan and Egypt is blue water, which originates from the Abbay River basin. Conversely, the agricultural sectors in Ethiopia are based on the rainfed system. Climate change, population growth, and industrial development are the main driving factors for Ethiopia to plan and implement new water resource development projects towards food and energy security. The planned and ongoing water resources projects are likely to change the spatial and temporal dynamics of the hydrological components in the Abbay River system. Hydrological modelling, for example, may be challenging when utilizing regulated flow characteristics since natural flow conditions would be regulated by water infrastructures. As a result, it is critical to create a hydrological model that accurately depicts the hydrological processes in the Abbay River basin utilizing undisturbed flow characteristics.

Advances in Hydrology Research in the Abbay River Basin

Application of Hydrological Models

Quantification of the water balance components, with a proper hydrological model, in the Abbay basin is crucial for sustainable water resource development. Several hydrological modelling studies in the Abbay River basin have been published (Conway 1997; Wale et al. 2009; Tegegne et al. 2017; Tegegne and Kim 2018; Tegegne et al.

2019; Easton et al. 2010; Kim and Kaluarachchi 2008; Steenhuis et al. 2009; Uhlenbrook et al. 2010; Haile et al. 2017; Hurni et al. 2005; Gebremicael et al. 2013; Gebeyehu et al. 2023a; among others). The spatiotemporal data availability and computer processing capacity allows the use more complex physically based hydrological models, such as the Variable Infiltration (VIC), Soil and Water Assessment Tool (SWAT), and Hydrologiska Byråns Vattenbalansavdelning (HBV), in the Abbay River basin. Conway (1997) created a basic hydrological model for the Abbay basin to investigate runoff sensitivity to rainfall and potential evaporation changes and concluded that streamflow sensitivity to rainfall. Uhlenbrook et al. (2010) used the HBV in the Abbay basin with three simplified representation modes of the watershed: lumped, lumped with multiple vegetation zones, and semi-distributed with multiple vegetation and elevation zones. The SWAT is a commonly and popularly used water balance model in the Abbay River basin (Tegegne et al. 2017, 2019; Tegegne and Kim 2018; Dile and Srinivasan 2014). The majority of the previous studies reported the applicability of the SWAT to quantify the water availability in the Abbay River basin. For instance, the majority of prior studies that applied the SWAT model in the Abbay River basin were critically reviewed by van Griensven et al. (2012) and indicated that the majority of SWAT applications in the Abbay River basin resulted in satisfactory performances. Tegegne et al. (2017) employed simple conceptual and semi-distributed hydrological models to investigate the water availability in the sources of the Abbay River (Lake Tana basin). They reported that the lumped models performed best in small-scale catchments in reproducing the observed flow, while the complex semi-distributed model was found to be best for the largest watersheds with diverse spatial characteristics, indicating the applicability of distributed hydrological models for the large-scale watershed because of their diversified hydro-geographical characteristics. Their finding revealed that the simple conceptual and physically based semi-distributed models comparably reproduced the observed streamflow in the selected gauging stations. They also indicate that integrating the simple conceptual and physically based distributed models with the artificial neural network improved the overall accuracy of the individual models. Moreover, Lemann et al. (2018) employed SWAT for water balance modelling in the Abbay basin by dividing the entire basin into eight subbasins. They modelled the seven subbasins individually using SWAT to investigate the hydrological processes in the Abbay basin and reported the satisfactory performance of the model for flow simulation. Furthermore, Haile et al. (2017) employed the HBV to investigate the water availability fluctuation under climate changes in the Abbay River basin; they concluded that the model satisfactorily reproduced the observed flow in the 20 selected typical watersheds in the Abbay basin. Kebede et al. (2006) applied the SWAT model to model the Lake Tana basin; they reported that the major gauged catchments accounts for about 93% of the total surface flow into Lake Tana. According to SMEC (2007) and Wale et al. (2009), the primary gauged catchments of the Lake Tana basin provided 71% and 58%, respectively, of the total surface flow into Lake Tana.

Ungauged Hydrology Modelling

Despite advances in the development of hydrological models for hydrological process modelling in gauged catchments, modelling ungauged catchments in the Abbay River basin is still a major obstacle for hydrological modelling at the subcatchment level. Numerous parameter transfer approaches have been developed to model ungauged hydrology. The most commonly used regionalization approaches include the arithmetic mean (Merz and Blöschl 2004; Oudin et al. 2008), physical similarity (Oudin et al. 2008; Samuel et al. 2011; Tegegne and Kim 2018; Gebeyehu et al. 2023b), spatial proximity (Merz and Blöschl 2004; Oudin et al. 2008; Parajka et al. 2005; Tegegne and Kim 2018; Gebeyehu et al. 2023b), and regression (Merz and Blöschl 2004; Oudin et al. 2008; Parajka et al. 2005; Young 2006). Tegegne and Kim (2018) compared the commonly used regionalization approaches in the Lake Tana basin. They concluded that the geographical proximity and physical similarity approaches performed similarly. They also developed a new regionalization approach for the Lake Tana basin termed catchment runoff response similarity (CRRS). They found that the CRRS regionalization technique outperformed the conventional approaches in the Lake Tana basin.

Climate Change Impact on Hydrology

Several researches have assessed the impact of climate change on the hydrology and water resources availability in the Abbay basin (Beyene et al. 2010; Kim and Kaluarachchi 2009; Elshamy et al. 2009; Dile et al. 2013; Taye et al. 2011; Tegegne et al. 2021; Tegegne and Melesse 2021). Beyene et al. (2010) investigated the potential impact of climate change on the hydrological components in the Abbay River basin using the VIC model forced with climate data from 2 emission scenarios and 11 global climate models (GCMs). Kim and Kaluarachchi (2009) employed climate data from 6 GCMs with one emission scenario to assess the influence of climate change on the water availability in the Abbay basin using the two-layer water balance model. Dile et al. (2013) employed the SWAT, while Haile et al. (2017) used the HBV models to investigate the climate change potential impact on the water balance components in the Abbay River basin. Taye et al. (2011) examined the climate change impact on hydro-meteorological extremes in the Lake Tana basin using two simple conceptual water balance models. Taye et al. (2011) examined climatic data from 15 and 17, respectively, GCMs for climate change impact analysis in the study basin. Based on the dynamically downscaled outputs of six global climate models, Haile et al. (2017) found that global warming will likely intensify hydro-climatic extreme changes in the Abbay River basin. They demonstrated that the Abbay River basin annual rainfall projected to be changed by − 2.8 to 2.7% with a projected rise in annual potential evapotranspiration in the middle future period under the balanced representative concentration pathway. In general, Taye et al. (2015) systematically reviewed the potential impact of climate change on the hydro-meteorological extremes in the Abbay River basin and reported that most of the previous climate change studies produced notably heterogeneous results. However, the reasons for these disparities remain unknown because each study investigated the climate change potential impact on the hydrology of the Abbay River basin using a different number and type of GCMs, emission scenarios, downscaling techniques, and/or hydrological models. Tegegne and Melesse (2021) compared the performances of three bias correction and climate downscaling algorithms in the Lake Tana basin: quantile mapping (QM), detrended QM (DQM), and quantile delta mapping (QDM). They concluded that the conventional QM algorithm exhibits significant quantile inflation, but the QDM and DQM algorithms fared best in replicating all rainfall extremes in the study basin. This indicates the suitability of the DQM and QDM algorithms for assessing the influence climate change on the hydro-climatic extremes of the Abbay River basin.

Conclusions and Future Research Direction

The results of hydrological process modelling in some of the Abbay subbasins were unreliable. In most of the previous studies, for example, the SWAT performed poorly in the Megech and Koga watersheds (headwater catchments of the Abbay River). This could be attributed to SWAT's failure to reproduce streamflow in hilly places where credible data sets are lacking due to spatially insufficient rain-gauge distributions (Tegegne et al. 2017). Moreover, there are considerable sources of uncertainty in hydrologic modelling in the Abbay River basin, such as water diversion for small-scale irrigation, infrastructure development in the basin, the presence of open reservoirs, upstream water supply abstraction, and measurement errors in the hydro-meteorological variables. The poor watershed management practices (deforestation and excessive erosion) in the study region can be considered the primary attribute for high spatiotemporal dynamics of the hydrological components. For instance, Tegegne and Kim (2018) stated that the SWAT flow parameters do not account for the soil erosion effect on flow simulation, which might result in high hydrological model prediction uncertainty. The Abbay River subbasins have considerable hydrological processes modelling uncertainty due to inadequate geographic coverage and unequal distribution of hydrometric gauging stations. Zeng et al. (2018), for example, observed that improvements in data quality and spatial coverage resulted in reduced hydrologic modelling uncertainty. The most important research topic for future study in the Abbay River basin is thus how to reduce the predicted uncertainty in hydrologic models in the gauged and ungauged basins.

One of the hydrologic prediction uncertainty improvement approaches can be by improving the quality and quantity of hydro-meteorological gauging stations. Hydro-meteorological data measurement often requires monitoring time and financial resources (Hrachowitz et al. 2013), resulting in little progress in improving the data quality and quantity in the Abbay River basin (Dile et al. 2018). It is known that hydro-meteorological data quality and spatial coverage form the backbone for watershed hydrological process modelling. In this direction, there have been significant global efforts to advance hydro-meteorological data observation technologies and strategies. Emerging technologies for hydro-meteorological observation, such as radar and satellite technologies, have become alternative data sources for hydrological process modelling. Faridzad et al. (2018) found that the use of satellite rainfall products improved hydrological process prediction in data-scarce basins. Some studies have evaluated the satellite rainfall data application for hydrological process modelling in the Abbay River basin (Romilly and Gebremichael 2011; Duan and Bastiaanssen 2013; Worqlul et al. 2018); they reported improvements in the hydrological modelling results. Despite the advancement in weather observation technologies, future research in the Abbay River basin should focus on investigating reliable satellite products and integrating them with ground-based measured data to improve hydrological process modelling results.

The application of merging remotely sensing data and ground rain-gauge observations has not been well investigated in the Abbay River basin. Various merging algorithms have been proposed in recent decades, including Bayesian model averaging (Duan and Phillips 2010), random forest-based approaches (Baez-Villanueva et al. 2020), and geostatistical methods (Verdin et al. 2016). Hu et al. (2019) and Belay et al. (2022) reported a detailed review of satellite products and ground-based measurement merging approaches.

Most of the prior hydrological processes modelling studies in the Abbay River basin have relied on secondary information. However, primary data (field data collection) could advance the hydrological modelling results in the Abbay River basin (Dile et al. 2018). Thus, future research should also focus on collecting detailed field information from representative watersheds by means of emerging ground-based measurement devices and citizen science data collection approaches. Integrating primary and secondary hydro-meteorological data can help to obtain improved hydrological process modelling results.

Acknowledgements This research was financially supported by the Ethiopian Water Technology Institute and Addis Ababa Science and Technology University under project number EG-60/12-1/23.

Conflicts of Interest The authors state that there are no conflicts of interest for the publication of this research.

References

Baez-Villanueva OM, Zambrano-Bigiarini M, Beck HE, McNamara I, Ribbe L, Nauditt A et al (2020) RF-MEP: a novel Random Forest method for merging gridded precipitation products and ground-based measurements. Remote Sens Environ 239:111606

BCEOM (Egis Bceom International) (1999) Abbay River Basin integrated development master plan project and associates. Report to the Ministry of Water Resources, Ethiopia

Bekele D, Alamirew T, Kebede A, Zeleke G, Melese AM (2017) Analysis of rainfall trend and variability for agricultural water management in Awash River Basin, Ethiopia. J Water Clim Change 8(1):127–141

Belay H, Melesse AM, Tegegne G (2022) Merging satellite products and rain-gauge observations to improve hydrological simulation: a review. Earth 3(4):1275–1289

Berhanu B, Seleshi Y, Demisse SS, Melesse AM (2015) Flow regime classification and hydrological characterization: a case study of Ethiopian rivers. Water 7(6):3149–3165

Beyene T, Lettenmaier DP, Kabat P (2010) Hydrologic impacts of climate change on the Nile River Basin: implications of the 2007 IPCC scenarios. Clim Change 100(3–4):433–461

Chebud YA, Melesse AM (2009a) Numerical modeling of the groundwater flow system of the Gumera Sub-Basin in Lake Tana Basin, Ethiopia. Hydrol Process Special Issue: Nile Hydrol 23(26):3694–3704

Chebud YA, Melesse AM (2009b) Modeling Lake stage and water balance of Lake Tana, Ethiopia. Hydrol Process 23(25):3534–3544

Chebud Y, Melesse AM (2013) Stage level, volume, and time-frequency change information content of Lake Tana using stochastic approaches. Hydrol Process 27(10):1475–1483. https://doi.org/10.1002/hyp.9291

Conway D (1997) A water balance model of the Upper Blue Nile in Ethiopia. Hydrol Sci J 42(2):265–286

Conway D, Hulme M (1993) Recent fluctuations in precipitation and runoff over the Nile subbasins and their impact on main Nile discharge. Clim Change 25:127–151

Dessu SB, Melesse AM (2012) Modeling the rainfall-runoff process of the Mara River Basin using SWAT. Hydrol Process 26(26):4038–4049

Dessu SB, Melesse AM (2013) Impact and uncertainties of climate change on the hydrology of the Mara River Basin. Hydrol Process 27(20):2973–2986

Dessu SB, Melesse AM, Bhat M, McClain M (2014) Assessment of water resources availability and demand in the Mara River Basin. CATENA 115:104–114

Dile YT, Srinivasan R (2014) Evaluation of CFSR climate data for hydrologic prediction in data-scarce watersheds: an application in the Blue Nile River Basin. JAWRA J Am Water Resour Assoc 50(5):1226–1241

Dile YT, Berndtsson R, Setegn SG (2013) Hydrological response to climate change for Gilgel Abay river, in the lake tana basin-upper blue Nile basin of Ethiopia. PLoS ONE 8(10):e79296

Dile YT, Tekleab S, Ayana EK, Gebrehiwot SG, Worqlul AW, Bayabil HK et al (2018) Advances in water resources research in the Upper Blue Nile basin and the way forward: a review. J Hydrol 560:407–423

Duan Q, Phillips TJ (2010) Bayesian estimation of local signal and noise in multimodel simulations of climate change. J Geophys Res Atmosp 115(D18)

Duan Z, Bastiaanssen WGM (2013) First results from Version 7 TRMM 3B43 precipitation product in combination with a new downscaling–calibration procedure. Remote Sens Environ 131:1–13

Easton ZM, Fuka DR, White ED, Collick AS, Biruk Ashagre B, McCartney M et al (2010) A multi basin SWAT model analysis of runoff and sedimentation in the Blue Nile, Ethiopia. Hydrol Earth Syst Sci 14(10):1827–1841

Elshamy ME, Seierstad IA, Sorteberg A (2009) Impacts of climate change on Blue Nile flows using bias-corrected GCM scenarios. Hydrol Earth Syst Sci 13(5):551–565

Faridzad M, Yang T, Hsu K, Sorooshian S, Xiao C (2018) Rainfall frequency analysis for ungauged regions using remotely sensed precipitation information. J Hydrol 563:123–142

FDREMW, Federal Democratic Republic of Ethiopia, Ministry of Water Resources (2002) Water sector development program. Addis Ababa

Gebere SB, Alamirew T, Merkel BJ, Melesse AM (2015) Performance of high-resolution satellite rainfall products over data scarce parts of Eastern Ethiopia. Remote Sens 7(9):11639–11663

Gebeyehu BM, Jabir AK, Tegegne G, Melesse AM (2023a) Subbasin spatial scale effects on hydrological model prediction uncertainty of extreme stream flows in the Omo Gibe River Basin, Ethiopia. Remote Sens 15(3):611

Gebeyehu BM, Tegegne G, Melesse AM (2023b) Reliability-weighted approach for streamflow prediction at ungauged catchments. J Hydrol 624:129935

Gebremicael TG, Mohamed YA, Betrie GD, Van der Zaag P, Teferi E (2013) Trend analysis of runoff and sediment fluxes in the Upper Blue Nile basin: a combined analysis of statistical tests, physically based models and landuse maps. J Hydrol 482:57–68

Haile AT, Akawka AL, Berhanu B, Rientjes T (2017) Changes in water availability in the Upper Blue Nile basin under the representative concentration pathways scenario. Hydrol Sci J 62(13):2139–2149

Hrachowitz M, Savenije HHG, Blöschl G, McDonnell JJ, Sivapalan M, Pomeroy JW et al (2013) A decade of predictions in ungauged basins (PUB)—a review. Hydrol Sci J 58(6):1198–1255

Hu Q, Li Z, Wang L, Huang Y, Wang Y, Li L (2019) Rainfall spatial estimations: A review from spatial interpolation to multisource data merging. Water 11(3):579

Hurni H, Tato K, Zeleke G (2005) The implication of changes in population, Land use and land management for surface runoff in the upper Nile basin area of Ethiopia. Mountain Res Develop 25(2):147–154

Kebede S, Travi Y, Alemayehu T, Marc VJJOH (2006) Water balance of Lake Tana and its sensitivity to fluctuations in rainfall, Blue Nile basin, Ethiopia. J Hydrol 316(1–4):233–247

Kim U, Kaluarachchi JJ (2008) Application of parameter estimation and regionalization methodologies to ungauged basins of the Upper Blue Nile River Basin, Ethiopia. J Hydrol 362(1–2):39–56

Kim U, Kaluarachchi JJ (2009) Climate change impacts on water resources in the upper blue Nile River Basin, Ethiopia 1. JAWRA J Am Water Resour Assoc 45(6):1361–1378

Lemann T, Roth V, Zeleke G, Subhatu A, Kassawmar T, Hurni H (2018) Spatial and temporal variability in hydrological responses of the Upper Blue Nile basin, Ethiopia. Water 11(1):21

Merz R, Blöschl G (2004) Regionalization of catchment model parameters. J Hydrol 287(1–4):95–123

MoFED (Ministry of Finance and Economic Development) (2005) A plan for accelerated and sustained development to end poverty, (2005/06–2009/10 Vol. I MoFED, Addis Ababa, Ethiopia

Mulat AG, Moges SA (2014) Assessment of the impact of the grand Ethiopian Renaissance Dam on the performance of the high Aswan Dam. J Water Resour Protection

Oudin L, Andréassian V, Perrin C, Michel C, Le Moine N (2008) Spatial proximity, physical similarity, regression and ungaged catchments: a comparison of regionalization approaches based on 913 French catchments. Water Resour Res 44(3)

Parajka J, Merz R, Blöschl G (2005) A comparison of regionalization methods for catchment model parameters. Hydrol Earth Syst Sci 9(3):157–171

Romilly TG, Gebremichael M (2011) Evaluation of satellite rainfall estimates over Ethiopian river basins. Hydrol Earth Syst Sci 15(5):1505–1514

Samuel J, Coulibaly P, Metcalfe RA (2011) Estimation of continuous streamflow in Ontario ungauged basins: comparison of regionalization methods. J Hydrol Eng 16(5):447–459

SMEC (Snowy Mountains Engineering Corporation) (2007) Hydrological study of the Tana–Beles Subbasins, main report. Ministry of Water Resources, Addis Ababa, Ethiopia

Steenhuis TS, Collick AS, Easton ZM, Leggesse ES, Bayabil HK, White ED et al (2009) Predicting discharge and sediment for the Abay (Blue Nile) with a simple model. Hydrol Process Int J 23(26):3728–3737

Sutcliffe JV, Parks YP (1999) The hydrology of the Nile, IAHS special publication no. 5, IAHS Press, Institute of Hydrology, Wallingford, Oxfordshire OX10 8BB, UK

Taye MT, Ntegeka V, Ogiramoi NP, Willems P (2011) Assessment of climate change impact on hydrological extremes in two source regions of the Nile River Basin. Hydrol Earth Syst Sci 15(1):209–222

Taye MT, Willems P, Block P (2015) Implications of climate change on hydrological extremes in the Blue Nile basin: a review. J Hydrol Reg Stud 4:280–293

Tegegne G, Kim YO (2018) Modelling ungauged catchments using the catchment runoff response similarity. J Hydrol 564:452–466

Tegegne G, Park DK, Kim YO (2017) Comparison of hydrological models for the assessment of water resources in a data-scarce region, the Upper Blue Nile River Basin. J Hydrol Regional Stud 14:49–66

Tegegne G, Kim YO, Seo SB, Kim Y (2019) Hydrological modelling uncertainty analysis for different flow quantiles: a case study in two hydrogeographically different watersheds. Hydrol Sci J 64(4):473–489

Tegegne G, Kim YO (2020) Representing inflow uncertainty for the development of monthly reservoir operations using genetic algorithms. J Hydrol 586:124876

Tegegne G, Melesse AM, Asfaw DH, Worqlul AW (2020) Flood frequency analyses over different basin scales in the Blue Nile River basin, Ethiopia. Hydrology 7(3):44

Tegegne G, Melesse AM (2021) Comparison of trend preserving statistical downscaling algorithms toward an improved precipitation extremes projection in the headwaters of blue Nile river in Ethiopia. Environ Process 8:59–75

Tegegne G, Melesse AM, Alamirew T (2021) Projected changes in extreme precipitation indices from CORDEX simulations over Ethiopia, East Africa. Atmospheric Res 247:105156

Tekleab S, Mohamed Y, Uhlenbrook S (2013) Hydroclimatic trends in the Abay/upper Blue Nile basin, Ethiopia. Phys Chem Earth, Parts a/b/c 61:32–42

Tesemma ZK, Mohamed YA, Steenhuis TS (2010) Trends in rainfall and runoff in the Blue Nile Basin: 1964–2003. Hydrol Process.https://doi.org/10.1002/hyp.7893

Uhlenbrook S, Mohamed Y, Gragne AS (2010) Analysing catchment behavior through catchment modelling in the Gilgel Abay, upper Blue Nile River basin, Ethiopia. Hydrol Earth Syst Sci 14(10):2153–2165

USBR (US Bureau of Reclamation) (1964) Land and water resources of the Blue Nile Basin. Main report, United States Department of Interior Bureau of Reclamation, Washington, DC

Van Griensven A, Ndomba P, Yalew S, Kilonzo F (2012) Critical review of SWAT applications in the upper Nile basin countries. Hydrol Earth Syst Sci 16(9):3371–3381

Verdin A, Funk C, Rajagopalan B, Kleiber W (2016) Kriging and local polynomial methods for blending satellite-derived and gauge precipitation estimates to support hydrologic early warning systems. IEEE Trans Geosci Remote Sens 54(5):2552–2562

Wale A, Rientjes THM, Gieske ASM, Getachew HA (2009) Ungauged catchment contributions to Lake Tana's water balance. Hydrol Process Int J 23(26):3682–3693

Worqlul AW, Dile YT, Ayana EK, Jeong J, Adem AA, Gerik T (2018) Impact of climate change on streamflow hydrology in headwater catchments of the Upper Blue Nile Basin, Ethiopia. Water 10(2):120

Young AR (2006) Stream flow simulation within UK ungauged catchments using a daily rainfall-runoff model. J Hydrol 320(1–2):155–172

Zeng Q, Chen H, Xu CY, Jie MX, Chen J, Guo SL, Liu J (2018) The effect of rain gauge density and distribution on runoff simulation using a lumped hydrological modelling approach. J Hydrol 563:106–122

Chapter 9
Abbay Basin's Regional Groundwater Flow System

Mebruk Mohammed Nurhusein

Abstract Regional groundwater flows generally provide a constant source of water because they are commonly less affected by seasonal recharge variation. Such a regional groundwater system in the Abbay Basin's aquifer can be an important source of water. The focus of this study is, thus, to describe the state of the regional groundwater flow components of the Abbay Basin. Two conceptual models, larger northwestern groundwater basin and Abbay basin, were evaluated to test the conceptualization of inter- and intra- boundary regional groundwater flow for the basin. The northwestern groundwater basin that merges Abbay, Tekeze, Mereb and Baro Akobo surface water basins was found to better describe Abbay Basin's regional groundwater system. A regional groundwater flow direction, access depths and potentiometric surface maps were estimated by use of output from this model. The model was calibrated using 11,474 water points inventoried within this larger basin. The model has shown that there is effective recharge of 15.385 BCM per year in the Abbay Basin. The study reveals, three quarter (by area) of the basin's regional groundwater can be accessed by drilling up to 500 m depth. Shallower regional groundwater is observed closer to the main water bodies (main river course and Lakes) and in the lowlands of Abbay basin. The recharge and potentiometric maps together reveal that at high recharge locations, regional groundwater flow paths extend from high to low regional potentiometric areas over distances of greater than tens of kilometres.

Keywords Abbay Basin · Groundwater flow system · Northwestern groundwater basin · Potentiometric surface access depth · Recharge

---

M. M. Nurhusein (✉)
School of Civil and Environmental Engineering, Addis Ababa Institute of Technology, Addis Ababa University, Addis Ababa, Ethiopia
e-mail: mebruk.mohammed@aait.edu.et

Introduction

In Ethiopia, there is increasing water demand for agricultural, industrial and municipal uses. These demands require more surface and groundwater abstraction and/or changes in water resources management approach being pursued. Ongoing activities in Abbay Basin, such as irrigable land expansion, economic development and industrial development, create larger demands and consumption of water. In addition to dam construction, exploring the possibility of groundwater as additional water resources has been the focus of the country. An understanding of the groundwater flow system is thus needed so that the water-resource management strategies in the basin are properly designed and employed.

According to Tóth (1963), an aquifer system can comprise regional, intermediate and local groundwater flow systems. Recharge and discharge areas are adjacent to each other in the local groundwater flow system while they are separated by one or more topographic highs and lows in an intermediate flow system. Groundwater divides and bottom of major surface water bodies represent recharge and discharge areas, respectively, in a regional groundwater flow system. Regional flow systems are more of steady-state situation than intermediate and local flow systems (Tóth 1963). Researchers have studied groundwater flow, variability, recharge and interaction with surface water extensively (Stiefel et al. 2009; Chebud and Melesse 2009, 2011).

Regional groundwater flows generally provide a constant source of water because they are commonly less affected by seasonal recharge variation. Among the three aquifer systems, the regional groundwater system is thus more important source of water in the Abbay Basin's. The overall objective of this study is thus to provide a description of the regional groundwater flow system in the Abbay Basin. "Regional groundwater flow system" here describes flow systems that provide a sustained source of groundwater, which is less affected by seasonal variations in recharge. The conceptual model of the flow system and the results of the study can be used to support actions taken by concerned establishments with respect to groundwater availability in the basin. To satisfy this main objective, the study deals with estimating aquifer characteristics, including groundwater flow direction, possible access depths below the ground surface, and recharge distribution.

The goal of this study is to develop a State of the Basin report that synthesizes the current state of regional groundwater in Abbay Basin, as is possible with existing information. Numerical modelling approach was used to accomplish this goal. The purpose of the modelling effort here is to estimate the regional groundwater flow system with reasonable accuracy (as much as the input data prevail). Thus, the model was constructed for system interpretation rather than prediction of consequences of some development interventions. The model development includes data on the geologic system and hydraulic properties of the hydrogeologic characters (including boundary conditions) constituting the aquifer system. Here, groundwater inventory data were used to offer a better insight in the reliability of the modelling effort in describing the current situation of groundwater resources of the basin.

Abbay Basin's Groundwater Review

Geomorphology

Geophysical and climatic conditions of an area influences the occurrence and behaviour of the groundwater flow system. Alemayehu (2006) reveals that because of the wide variation and heterogeneity of geology, topography and environmental conditions, locating productive aquifers in Abbay Basin has been a challenge. Volcanic rocks, ancient crystalline basement rocks and sediments of various ages comprise the geology of the basin. Because of their stratigraphic position, volcanic rocks associated with the east African rift system, form the most accessible aquifers in the highlands of the basin. In locations, like Ambo-Guder valley and the Didessa valley, where volcanic caps either are exhumed by erosion or did not exist originally, sedimentary rock forms the aquifers (Kebede 2013). All the younger formations were deposited over the Precambrian rocks, with ages of over 600 million years. Along the western border of the basin where the younger cover rocks have been eroded, oldest rocks are seen to be exposed (Fig. 9.1).

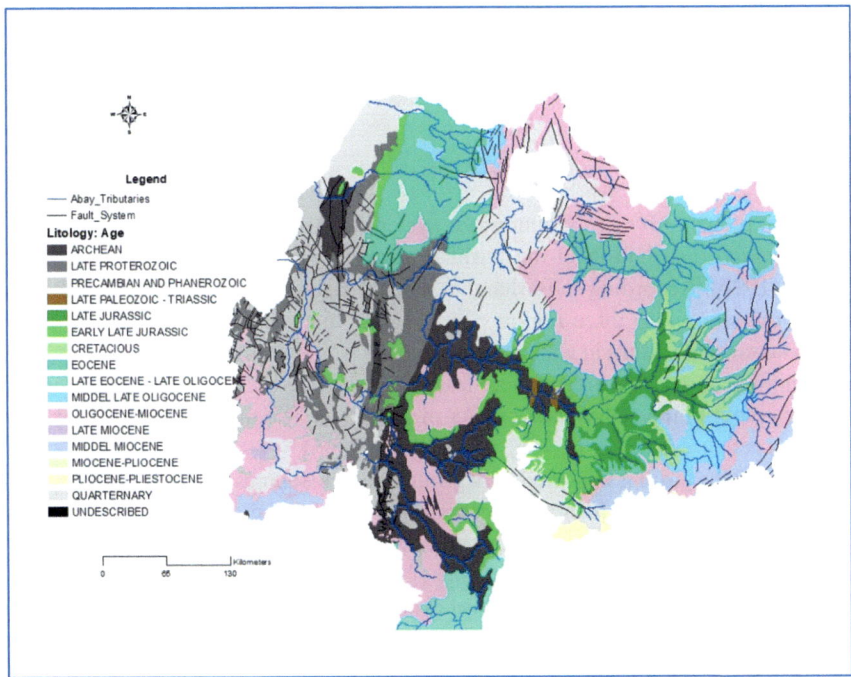

Fig. 9.1 Formation in Abbay Basin with geologic timescale. Modified from Geological Map of Ethiopia

Volcanic rocks, mainly of Tertiary and younger age, forms the highlands of the basin flanking the Rift Valley. Although major displacement along the fault systems did not occur until later in the Tertiary, extensive fracturing occurred in the earliest rocks of the Cenozoic, which are dated at 65 million years. Volcanic activity together with faulting, which are partially related, have formed the landscapes in the Abbay Basin. Basaltic lava alternated with ash and coarser fragmented materials from the volcanic eruption forms the Trap Series in the basin. Around the eastern edge of the Lake Tana depression, shield volcanoes consisting of alkali basalts and fragmental material, developed later. In the basin, the more recent volcanic activities are linked with the development of the Rift Valley along the edge of the adjoining Abbay Plateau.

Abbay terrain, bounded from northeast by the East African Rift Valley and Baro Akobo Basin in the south and Tekeze Basin, in the north consists largely of mountains and plateaus with elevations above 1500 m above sea level (m asl). The western boundary is along the Ethio-Sudanese border. For the past 30 million years, the elevated terrain of the basin has been undergoing erosion at 0.029–0.185 mm/yr rate (Pik et al. 2003). The prominent topographic features (Rugged terrain, plateaus, deep gorges, canyons, mountains, buttes and mesas) of the basin have been affecting the water resource distribution, (rainfall pattern, and groundwater occurrence and movement) in the basin. Elevations in the basin vary from highest point at Ras Dashen (4620 m) to the lowest along the Ethio-Sudanese Border (430 m).

Hydrogeology

The main recharge for the groundwater system in Abbay basin is the high rainfall whose spatial distribution in the basin is considerably influenced by topography (NMSA 1996). According to Berhanu et al. (2014), the rainfall regimes of Ethiopia can be classified into three regimes; the regime that includes Abbay basin has a unimodal rainfall pattern in which rainfall occurs from February to November. Quick infiltration occurs at locations where fractured volcanic rocks are prominent and to a lesser extent in the sedimentary rocks of the basin. The major rivers and lakes in the basin act as a discharging zone (Alemayehu 2006).

Although the basin is endowed with groundwater resources, its occurrence (spatial and temporal) were not quantified accurately due to lack of sufficient hydrogeological data, rendering its development and management (Nyagwambo 2006). Local groundwater closer to the highlands and plateau of the basin flows at a relatively shallow depth. In the upper reaches of Abbay Basin, the rivers are hydraulically linked to the aquifers. The highland or recharge areas have shallow groundwater that sustains only surface water resources and springs (Asfaw et al. 2001). Near-surface conditions affect groundwater recharge in semiarid areas than that in humid areas of the basin (Nyagwambo 2006). Ayenew and Alemayehu (2001) and BGS (2016) estimate that the recharge in Ethiopia varies between 50 mm in the western lowlands and 250 mm in the highlands of Abbay Basin.

Aquifers of high, moderate or low productivity in the basin consist of considerable groundwater potentials. Relatively lower fracture and larger amounts of clay filling in the highland rocks of older ages made them form moderate to low productive aquifers. Although phreatic groundwater is common in these rocks, semiconfined and some flowing wells are known in Ambo area of Guder sub-basin. Generally, because of their interlocking crystals, Precambrian intrusive igneous rocks and metamorphic rocks have very small (less than 1%) primary porosities. In comparison with intrusive rocks, metamorphic rocks have low storage capacity; as a result, metamorphic rocks of the basin have very low productivity and are commonly recognized as aquicludes. The yield is very low in un-weathered basement rocks and highest (~ 0.1 L/s) where weathering and regolith are most developed; however, the groundwater quality in these aquicludes is very good (BGS 2016). The recharge according to the BGS (2016) varies from 10 to 250 mm/year depending on the rainfall regime.

At locations where the climate situation creates larger recharges, Paleozoic and Mesozoic sedimentary rocks in the basin bears larger amount of groundwater. Paleozoic sediments in Abbay Basin overlie impervious basement rocks. These rocks consists particularly of channel fills and within classic sediment horizons, which give them a good opportunity of storing water. The conglomeratic and weathered zones of Paleozoic sediments mainly holds the groundwater. Aquifers of the Mesozoic formations can contain ample amount of groundwater under favourable climatic conditions. These Mesozoic aquifers very often are fossiliferous and consist of calcareous-dolomitic layers, conducive for karst aquifer formation.

Water-bearing potential varies with structure, mineralogy and texture of volcanic rock. Groundwater flow (velocity and storage) in volcanic rocks depends on the rocks characters (pore spaces) produced during and after rock formation. Primary pore spaces may not necessarily represent permeability, because the interconnection between pore spaces during rock formation may not lend itself for possible water flow in the rock. However, natural processes (weathering, faulting, fracturing, etc.) that later induce interconnection result in secondary permeability.

Tertiary and quaternary formations of the basin, groundwater bearing and transmitting formations can be located in channels and valleys filled by younger sediments. Coarse sandstone and conglomeratic layers interbedded with shale, interstratification of basaltic lava flows with loose sediments and/or alluvial sediments on flood plains are also groundwater-bearing structures in the basin. Quaternary sediments in the basin are the main sources of shallow groundwater. These sediments are made of unconsolidated mixtures of sand and coarse gravel. These sediments are productive and located in topographically depressed areas along river valleys. The transmissivity and storage characteristics of sands and gravel are the highest of any geologic material. Good groundwater quality is common in these sediments. Generally, these quaternary sediment aquifers in the basin are important formations, because they usually receive direct recharge from perennial rivers and rainfall.

Groundwater Flow System Characterization

Several approaches have been used to explore temporal and spatial variation of groundwater flow systems. Researchers and practitioners have proposed different methods. Rushton and Rathod (1985) use the information from head distribution to determine groundwater velocity variation. Using digital simulation models, Sondhi et al. (1989) have shown the possibility of computing recharge distribution coefficients. Serrano and Unny (1987), considering the uncertainty related to environmental fluctuations and measurement errors, has developed innovative mathematical model to solve groundwater forecasting problems. Chiew and McMahon (1990) adopted a watershed modelling approach for recharge estimation. Boonstra and Bhutta (1996) has developed a numerical modelling approach to estimate seasonal net recharge using historical water table and spatiotemporal recharge variations as input. A similar attempt was made by Mohan and Ramani (2010) to estimate the groundwater recharge at river basin scale.

Based on the objectives of this study, the following three-step method is used to understand the regional groundwater flow system.

(i) The first steps creates the conceptual hydrogeologic model of the basin, which includes

- Compiling, documenting, and assessing available data (well inventory, geology, surface water, rainfall, etc.),
- Initial data collection in relation to hydraulic and physical boundary conditions in the basin.

(ii) The second step involves data collection for the regional groundwater flow system work elements of the numerical groundwater model. The above two phases of study together provide data and information needed to build the conceptual model necessary for groundwater flow system analysis.

(iii) The third step, involves constructing a three-dimensional, steady-state regional-scale numerical model of the groundwater flow system.

The model describes the general groundwater flow direction, water table/potentiometric surface depth below the ground surface and the possible recharge and hydrogeologic conditions throughout the basin. Data needed for the modelling are related to well and spring inventory data collected and monitored for a long period of time at different but predetermined monitoring sites. However, such data are scarcely available in Ethiopia. Besides, the few data that exists are not collected with the intention of estimating the aforementioned objectives. Thus, with the available information on well and spring data in the basin, a modelling effort is made to estimate the groundwater flow direction and possible water table/potentiometric surface level. For such analysis, a multidimensional groundwater flow equation at the regional scale needs to be solved; in which case, using a numerical method is indispensable.

The model uses discretization, which is appropriate to capture the existing conditions of the basin (its hydrogeology, well, boundaries and recharge locations). The main river course, lakes and their major tributaries are expected to be included in the model as either seepage faces or stream flow-routing cells. Top surface recharge is estimated by trial similar to the hydrogeologic parameter estimation.

The model was calibrated to the steady-state conditions based on the available well and spring inventory data. Traditional trial-and-error method of calibration was used. At well and spring observation points, the goodness of fit indicators computed include, the average residual (difference between simulated and measured hydraulic heads), root-mean-square error divided by the range (total difference in water levels between observations in the analysis area) and both unweighted and weighted residuals. Once the model was calibrated, a water table/potentiometric surface map was constructed based on the hydraulic head computed from the numerical model for the selected conceptual model.

Conceptual Model

A groundwater conceptual model is a simplified depiction of the groundwater flow system with descriptive plan and cross-section of the aquifer. In this plan, clear pictures of how and where groundwater originates, flows and exits in the system are critical for the development of accurate numerical model. This pictorial representation of the flow system is needed to show the extent of the flow domain to be analysed and is expected to include natural physical features that can be represented as model boundaries (Thomas 2001). According to USACE (1999), features in a conceptual models need to describe the relationship and extent of hydrogeologic units, aquifer material properties (transmissivity), potentiometric surfaces, water budget, boundary locations (depth to bedrock, impermeable layer boundaries, etc.), boundary conditions (fluxes, heads, natural water bodies) and system stresses (withdrawal wells, infiltration trenches, etc.).

The spatial scale of analysis adopted in this study (being basin scale) warrants the adoption of regional groundwater flow analysis, which predominantly contains fractured and basement rocks. Thus, the regional geology forms the framework of the conceptual model. The fundamental characteristic of fractured rock aquifers is extreme spatial variability in transmissivity and hence groundwater flow rate. To conceptualize the movement of groundwater through the system including the existing hydrologic information and data on precipitation, evaporation, river runoff, as well as river stage data are important.

Two different sets of model areas are selected for the regional flow system analysis.

(i) Model area that coincides with the surface water divide of the Abbay Basin (in which the surface water divide of Abbay Basin is assumed to coincide with the groundwater divide) and
(ii) Model area (Northwestern groundwater basin) that integrates adjacent major river basins that may contribute to the groundwater flow system in the Abbay basin.

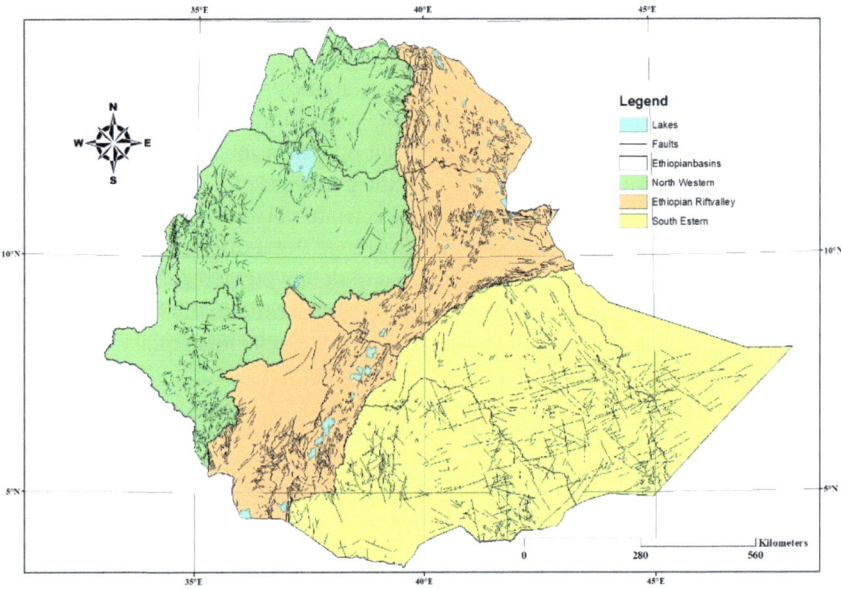

Fig. 9.2 Fault system and the three groundwater basins of Ethiopia

The second model area conceptualization is set as shown in Fig. 9.2. Three larger groundwater basin conceptualizations for the country are set: viz. the northwestern groundwater basin (which merges Abbay, Tekeze, Mereb and Baro Akobo basins), the Ethiopian Rift Valley groundwater basin (which merges Omo-Gibe, Rift Valley Lakes, Denakil and Aisha Basins) and the southeastern groundwater basin, which consists of Genale-Dawa, Wabe-Shebelle and Ogaden basins. In addition to fault formation and orientation and the physical surface water hydrology (boundaries), the general rainfall regime in the country (as discussed in Berhanu et al. (2014) was considered for these three larger basin conceptualization.

Hydrogeology Conceptualization

The hydrogeological conceptual model consists of static and dynamic components. The static components include aquifer matrix descriptions, viz. rocks and soils (lithology, geological structures, etc.), while the dynamic component includes descriptions related to water (hydrology, hydrochemistry, isotope hydrology, etc.). A fractured system hydrogeologic conceptualization detailed description of the aquifer properties controlling flow, which entails costly exploration works, thus it is common to embark on the cheaper approach that involves gross simplification. Fractured systems are typically modelled using one or more of the following conceptual approaches: (1) equivalent porous medium (EPM); (2) discrete fractures; and (3) dual porosity. Among these approaches, the EPM approach adequately represents

Table 9.1 Basin discretization information and groundwater inventory

Basin/Model	Area (km^2)	Element side length (m)	No. of elements	No. of nodes
Northwestern GWB	371,147	2200	175,262	88,708
Abbay	201,346	1500	207,145	104,779

the behaviour of a regional flow system (Long and Billaux 1987; Cacas et al. 1990). Accordingly, in this study, the EPM approach was used to model the regional groundwater flow system. In the EPM approach, fractured material is represented by equivalent porous material having so-called equivalent or effective hydraulic properties. The effective hydraulic properties were selected so that the flow pattern in the EPM and the real fractured system are similar. In EPM approach, it is assumed that the fractured medium can be discretized into sizes of materials represented by effective hydraulic parameters.

The concept of a hydrostratigraphic unit is most useful for simulating geologic systems at a regional scale (Anderson and Woessner 2002). The concept of the hydrostratigraphic unit, which was first presented by Maxey (1964) and later reconsidered by Seaber (1988), can be used to define aquifers and confining units in modelling regional flow systems. Hydrostratigraphic units contain geologic units of similar hydrogeologic characters. A hydrostratigraphic unit may combine several geologic formations, or a geologic formation may be subdivided into several hydrostratigraphic units. Geologists rely on stratigraphic information and an understanding of the depositional history to help reconstruct the depositional environment. According to the EPM approach, simulation of the groundwater flow requires definition of effective values of transmissivity and specific yield (for transient analysis) for each hydrostratigraphic unit defined in the study area.

Since the model area, the study opts to consider is hundreds of thousands of square kilometres (Table 9.1), an EPM approach was selected. The approach uses the concept of a two-dimensional areal model for which a single model parameter that lumps a number of hydrostratigraphic units across the depth is used. Since regional flow systems are less transient (Tóth 1963) and no time-dependent observations on groundwater level and spring yield are obtained, a steady-state model is used to calibrate the observed data.

Recharge

A groundwater flow model can be used to estimate recharge. With actual measurements of groundwater discharges and head at hand, estimates of recharge can be obtained through groundwater model calibration. In the calibration process, trial values diffuse recharge are used as input to the groundwater model to bring the simulated heads and fluxes into agreement with the measured values. A uniform value of diffuse recharge can be assigned over the top surface of the model on an element-by-element basis. The top surface of the model domain can be divided into a number of polygons (e.g. Thiessen polygons) having uniform recharge values (Cooley 1979). Recharge rates that produce the best model results, relative to the measured values, will be selected as the best estimates. Many studies, including Cooley (1979), Boonstra and Bhutta (1996), Sanford et al. (2004) and Dripps et al. (2006), have adopted the same approach.

During calibration, a trial fraction of the total rainfall over a surface was used to represent the recharges. Porous soils in humid areas as large as 25% of annual rainfall may recharge; in contrast, less than 1% of rainfall recharges in desert regions (Moore and Bell 2002). Accordingly, while calibrating, the recharge fraction in the model was made to vary between 1 and 25% of the total annual rainfall. The annual rainfall distribution was estimated by adopting the Thiessen polygon method of interpolation (as described in Ragunath 2006). 482 metrological stations in and around the study area were first identified. The monthly rainfall data of these selected stations were checked for data quality and consistency, which finally brought about the annual rainfall amount. Figure 9.3 shows the annual rainfall (in mm) distribution for the northwestern groundwater basin.

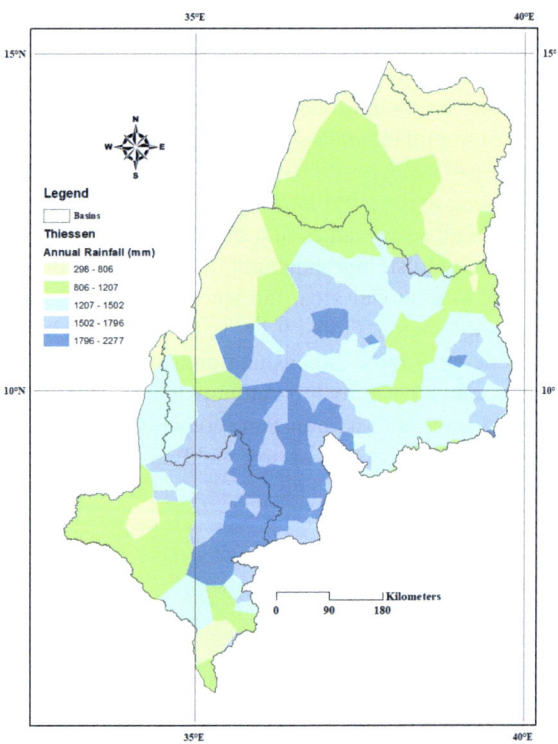

Fig. 9.3 Annual rainfall distribution in mm

Boundary Conditions

Groundwater flow equations derived for porous medium are categorized as boundary value problems. Information about the physical state of the groundwater flow, described as boundary and initial conditions, is needed for specific solution. For steady-state problems, only boundary conditions are required. In groundwater flow analysis, the two basic types of boundary conditions are the constant head boundaries and no-flow boundaries. Lakes or reservoirs represent constant head boundaries while interfaces with lower hydraulic conductivity represent the no-flow boundaries.

Diffuse and focused recharge are simulated differently in groundwater flow models. Constant flux rate, i.e. volumetric flow per unit horizontal surface area is usually assigned for diffuse recharge. Focused recharge from a river or lake is typically simulated with fixed head or head-dependent boundaries (Healy 2010); calculated heads are used to estimate these recharges and therefore are not known a priori.

In Abbay basin regional ground water flow model, a constant head boundary condition was set for the major rivers where the flow at the location is perennial (Fig. 9.4). The nodes representing lakes and their peripheries were also treated as constant head boundaries. The perimeter of the study area to the east extends to the watershed divide that exists between the model area and adjacent major basins of Ethiopia. These boundaries are considered as no-flow boundaries. The ground surface is considered as recharge boundary, the value of which is taken as a fraction of the annual rainfall.

Numerical Groundwater Modelling

Regional groundwater modelling that deals with fractured rock aquifers with larger scale of analysis, employs an EPM modelling approach. EPM assumes modelling results are only valid if discretization of the model area is done at larger than representative elementary volume (REV). At REV scale, a material characterized by equivalent hydraulic parameters can be set for the discretized domain (Long et al. 1982). At REV scale, it is assumed that the geologic structures (rock matrix, faults and fractures) with which groundwater interacts are small and densely distributed enough to act as porous media. Other assumptions include:

- Transmissivity and small discharge boundaries (springs and rivers), are adequately refined at the selected element sizes (0.1–1 km^2);
- Recharge is sufficiently distributed at the selected nodal distribution (1500–2200 m);

In this study, a finite element-based groundwater flow modelling software called TAGSAC was used. TAGSAC has been applied in a number of researches and studies across the globe and proven to have performed well (Mohammed et al. 2010). TAGSAC requires the model domain is discretized into a network of two-dimensional

Fig. 9.4 Major rivers and lakes taken in boundary condition assignment

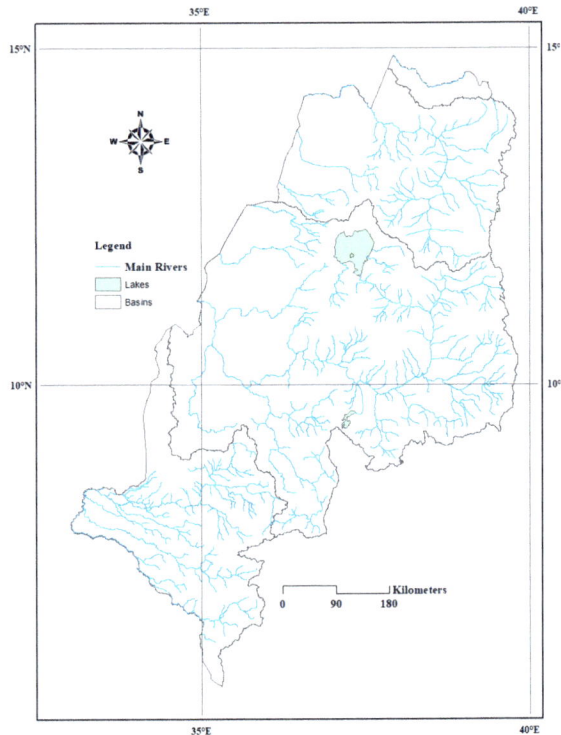

or three-dimensional finite elements. TAGSAC uses triangles and trapezoids for two-dimensional and triangular and trapezoidal prisms for three-dimensional flow analysis. In this study, a two-dimensional areal flow analysis that adopts triangular elements was selected. The hydrogeologic parameter, here transmissivity, is specified for each discretized triangular element. Nodal spacing is governed by the overall size of the model domain. To minimize computation time and data handling, a grid with a small number of nodes is preferred; however, a large number of nodes are essential to represent the system precisely. A compromise between precision and practicality is thus necessary. Considering the study area extent and the computer in hand, the finite element size was made to vary (Table 9.1) for the two conceptual models.

Model Data Inventory

The data collected were mainly focused on regional and local geology, including hydrostratigraphy, alteration and/or structural features (mainly shear zones and faults). Water-level and spring discharge measurements including their location are the principal sources of information. Long-term, systematic measurements of water levels and spring discharges provide essential data needed to evaluate changes in resources over time, to develop groundwater models and forecast trends and to design, implement and monitor the effectiveness of groundwater management (Anderson and Woessner 2002). However, continuous water point data monitoring seldom exists in the country. Water point data collected included the well and spring location (latitude and longitude), elevation and water depth below the surface.

Due to the difficulty of identifying whether a given well and/or spring represents the regional groundwater flow, most inventories with water-level measurements were not used for model calibration. Water points were, thus, selected based on (i) springs, hand dug and shallow wells closer to each other shall have acceptable variation in their corresponding measured data, (ii) all deep boreholes and (iii) wells that were located along general hydraulic gradients in the basin. Accordingly, among the 15,351 well and spring inventory data, 11,474 data points were selected for further analysis. Table 9.2 gives the number, and Fig. 9.6 shows the distribution of groundwater inventories made in the study area.

Water levels at 11,474 wells (Table 9.2) were taken as steady-state water-level observations. Almost all of the spring data lack discharge data; however, the elevation (altitude) at the location of the spring was taken as the hydraulic head. In addition, most of the hand dug wells recorded do not have the actual groundwater table depth information, and the average water level of a hand dug well in Ethiopia (some 2–15 m) was assigned and used in the model calibration. The calibration target at each well is the average water-level measurements taken in the well.

The quality of data, even the basic borehole information obtained, was difficult to assess. For instance, geographic locations are uncertain because coordinate systems are neither consistent nor identified or the coordinates do not fall within the region the well is identified as being in. Locations of wells in each basin are suspect because it is a simple compilation with little quality assurance. Because of this, it was difficult to construct credible water-level surface maps. In addition, the water-level information gathered is mainly concentrated in the Mereb and Tekeze Basins, while basins such as Baro Akobo have very few data (Table 9.2).

These selected water point and spring inventories have many sources of inaccuracies that should somehow be recognized in determining residuals used for model calibration. These inaccuracies can be due to (i) errors in the measurement of water-level altitude (V_1); (ii) errors in locating the well (V_2); and (iii) fluctuations caused by variations in non-simulated transient stress (e.g. anthropogenic and seasonal stresses) (V_3). Following the approach discussed by San Juan et al. (2004), these errors were estimated from the inventories and were used to compute the variances of each observation. In addition, model discretization errors, resulting from inaccuracies in the

Table 9.2 Number of water point inventories with relevant information

Basin	Abbay	Tekeze	Mereb	Baro Akobo	Total
Number of inventories	2495	10,806	1944	106	15,351
Inventories used for analysis	1071	8474	1823	106	11,474

Table 9.3 Variance and standard deviation of water-level observations

Basin	Range	V_1	V_2	V_3	V_4	Sd_T
Northwestern GWB	3483.1	2.824	15.20	5.75	112.245	113.45
Abbay	3174.8	2.824	15.20	5.75	76.531	78.288

geometric representation of hydrostratigraphy and major structural features (shear zones and faults) in the model (V_4), were evaluated using the approach discussed in Faunt et al. (2004). Table 9.3 gives the results of these variances. Based on the four potential variances, the standard deviation of each water-level observation (Sd_T) was computed by the equation:

$$Sd_T = \sqrt{(V_1 + V_2 + V_3 + V_4)} \tag{9.1}$$

Groundwater Flow System in Abbay Basin

Model Evaluation

Model calibration is important to demonstrate that models are realistic. Measurements of aquifer response shall be compared with the corresponding model outputs, with the general objective of reducing their difference. In this study, manual trial-and-error calibration process was followed. The two calibrated regional groundwater models (Abbay Basin and Northwest Groundwater basin) were evaluated to assess which conceptualization better represents the groundwater flow system in the Abbay Basin. As part of the model evaluation, hydraulic heads at observation points (location of water point inventories) were evaluated. The results of this evaluation (shown in the following sections) were used to select a reasonable representation of the regional groundwater system in the Abbay Basin.

Both unweighted and weighted residuals were used to evaluate model fit to observations. Unweighted residuals have the same dimensions as hydraulic heads and are prone to interpretation errors, because different observations commonly have different accuracies, which entails the proposition that unweighted residuals of equal magnitude may not represent comparable model fit. As the two conceptual regional models are too large, calibration attempts were focused on reducing unweighted residuals to two standard deviations of the measured water point observations. Average weighted residuals (AWR) on the other hand are dimensionless quantities. According to Hill and Tiedeman (2007), AWR reflects model fit in the context of the expected accuracy of the observations. A weighted residual of 2.0 shows that the unweighted residual is twice the observation error or the observation error is twice that of the standard deviation. Hill and Tiedeman (2007) define standard error (SE) as the sum of square weighted residuals divided by the number of observations. An SE value of 1.0 indicates a consistent observation, error evaluation used to determine the weighting (Hill and Tiedeman 2007).

The primary calibration target in this study is the hydraulic head. Steady-state calibration was made using static water-level observations of 1071 (wells/springs inventories in the Abbay Basin) and 11,474 (inventories in Northwest groundwater basin). The maximum acceptable value of the calibration criterion depends on the magnitude of the change in heads over the problem domain. If the ratio of the root-mean-square error to the total head loss (THL), here defined as ARPR, in the system is small, the errors are only a small part of the overall model response (Anderson and Woessner 2002). The THL is estimated to be the range of the observed values. The goodness of fit indicator values can be seen in Table 9.4.

For clear evaluation of the conceptual models' performance, the larger Northwest groundwater basin against the sole Abbay basin model was compared. The comparison was also made on the basis of the results obtained by the Northwest groundwater basin model results of the groundwater inventories made in Abbay River basin.

An in-depth comparison of the results for other basins (Mereb and Tekeze) in the northwestern groundwater basin is also shown in Fig. 9.5. From Table 9.4 and Fig. 9.5, it is evident that the northwestern groundwater basin's (LGWE in Fig. 9.5) model outperforms the individual basins (Basins in Fig. 9.5) on the basis of weighted residual analysis (AWR, SE) values. Similar performance results were obtained in ARPR values for the two conceptualized models. The two models being in large-scale domain, calibration shall be acknowledged on the basis of weighted residuals rather than unweighted residuals. Accordingly, the northwestern groundwater basin results were selected and used for Abbay Basin groundwater flow system interpretation. The root-mean-square error to the range of observations ratio for all observations in the northwestern groundwater basin (ARPR) was 0.06, where the range equals the difference between the highest and lowest observed hydraulic head. These values indicate that the errors made in the northwestern groundwater basin are only 6% of the corresponding overall model response.

The magnitude of the residue between the modelled and observed data can be seen in Fig. 9.6. In the figure, out of 11,474 water points, the residual for 8102 (> 70%) was found to be within one standard deviation. For Abbay Basin, out of 1071 water points, 857 (> 80%) residuals were found to be less than one standard deviation. This indicates the acceptability of the results. In addition, the results with larger errors were distributed across the modelled area, which implies that these errors might be due to the non-representativeness of the wells/springs of the regional system. These

Table 9.4 Measures of model fit for the two regional models

Basin	AWR	SE	THL	ARPR
Northwestern GWB	− 0.92/− 0.75*	1.8/1.7*	3483	0.06
Abbay Basin	− 1.03	2.4	3175	0.06

*Northwestern GWB model result for inventories in Abbay Basin only

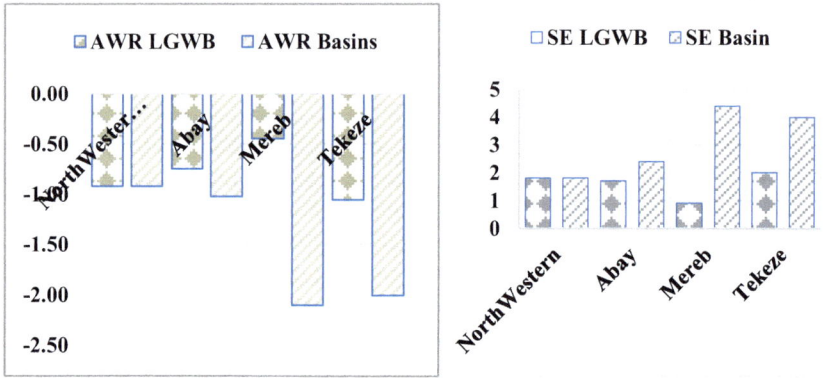

Fig. 9.5 AWR and SE results Northwest GWB and other basins model

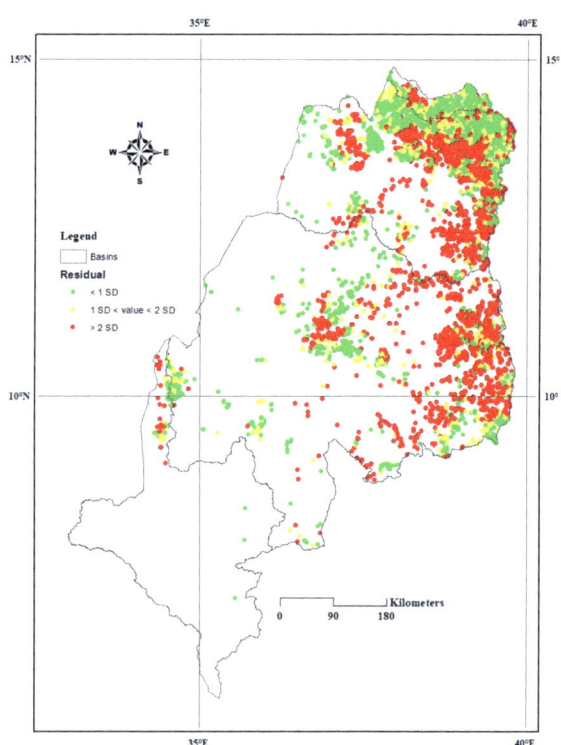

Fig. 9.6 Magnitude of residual for the 11,474 wells/spring inventories

wells might be tapping the local groundwater flow system rather than the regional aquifer.

For calibrated transmissivity and recharge values, the modelled and measured values comparison can be seen in Fig. 9.7 for the northwestern groundwater basin.

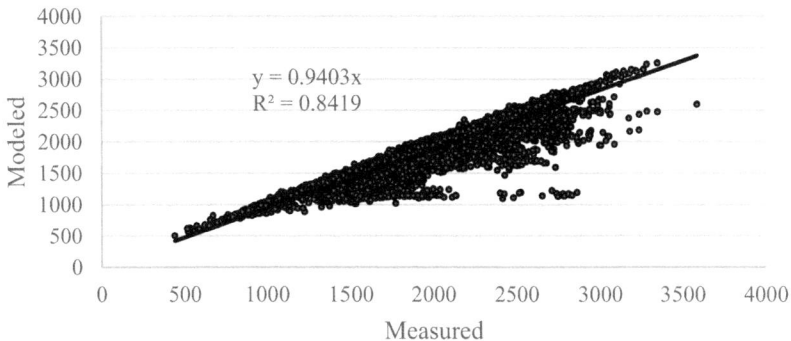

Fig. 9.7 Simulated versus measured head comparison for the Northwestern groundwater basin

In the figure, the best fit linear equation (with intercept zero) between the modelled head (y-axis) and measured head (x-axis) is also displayed. In an ideal calibration, the points fall on a straight line with a 1V:1H slope, meaning that both computed and measured values are equal. The degree of distribution about this theoretical line can be a measure of overall calibration quality (Anderson and Woessner 2002). The slope value (0.9403) in the figure being closer to one suggests the adequacy of the modelling results. The coefficient of determination (R^2) ranges in value from zero to one and relates estimated and measured head. R^2 being one represents a perfect correlation between measured and modelled values. At the other extreme, R^2 being zero denotes unacceptable modelling effort. The values seen in these figures (0.8419) again describe a good performance of the calibrated transmissivity and recharge values.

Hydrogeologic Setting and Recharge Distribution

In each of the two conceptual models, hydraulic head was computed for each basin by varying the transmissivity and recharge values for the study area. After a number of trial transmissivity for the hydrostratigraphic units (Fig. 9.1) and surface recharge (Fig. 9.3), the distribution maps shown in Figs. 9.8 and 9.9 were selected as the best. The recharge distribution across the model area is found to be between 0.006 and 0.568 mm/day. The percentage of the annual rainfall that seeps as recharge is found to be in the range of 1–15%, which is in globally recognized ranges. For hydrostratigraphic units, transmissivity estimates of 10–10,200 m²/day were obtained, which is the common range reported by Anderson and Woessner (2002) for fractured aquifers. The goodness of fit indicators given in Table 9.4 are for these selected parameters. The final parameter estimates of the northwestern groundwater basin model are considered reasonable for the lithology and conditions found in the Abbay Basin.

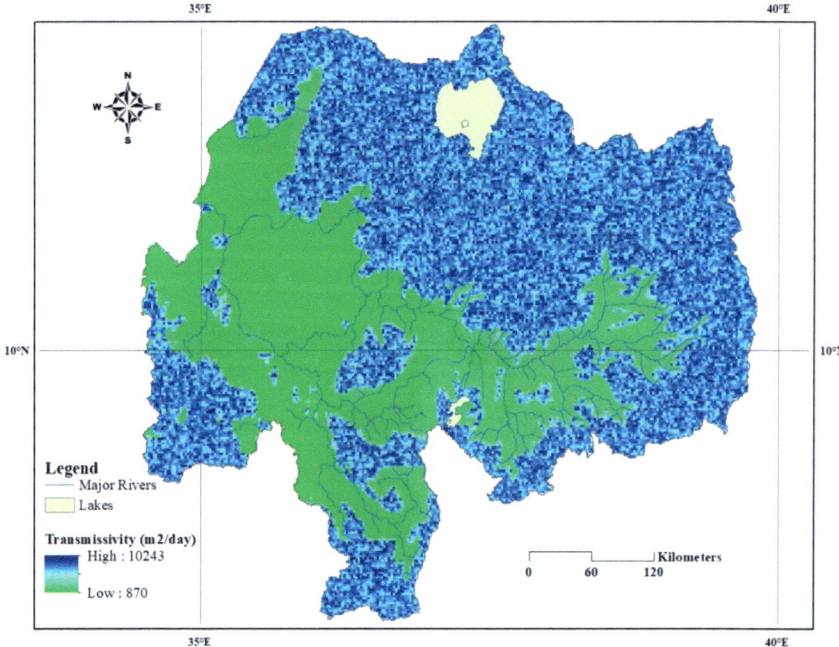

Fig. 9.8 Transmissivity (m^2/day) distribution

These figures show the results of the northwestern groundwater model clipped for the Abbay Basin.

In general, the lower estimated transmissivity was for the older Precambrian intrusive rocks. Higher transmissivity values have been associated with the prevalence of younger rock formations in the basin. Within a mapped area, transmissivity usually varies by an order of magnitude; locally the estimated transmissivity values may be either greater or less than the range indicated. Therefore, for potential local uses of groundwater resources in the Abbay Basin, site-specific hydrogeology investigations are necessary. Values shown in the mapping are subjected to the limitations of modelling and data quality previously discussed. Due to the maximum depth consideration in this study, the transmissivity values may not be exceeded. In this mapping, the lower transmissivity values were thus emphasized because they are believed to be close to the true transmissivity for local level studies.

Abbay Basin's water divides in the south and most of the highlands in the north have higher values. Average simulated recharge estimates range spatially from essentially zero to a maximum of 0.568 mm/day (~ 200 mm/year). The effective recharge across the Abbay Basin depends on surface geology and rainfall behaviour. The northern section of the basin recharge is related to the surface geologic nature (being more permeable) than the rainfall amount, while the southern eastern parts of the basin recharge mainly come from higher amounts of rainfall. Generally, under favourable geological and topographic conditions, areas with high precipitation receive large

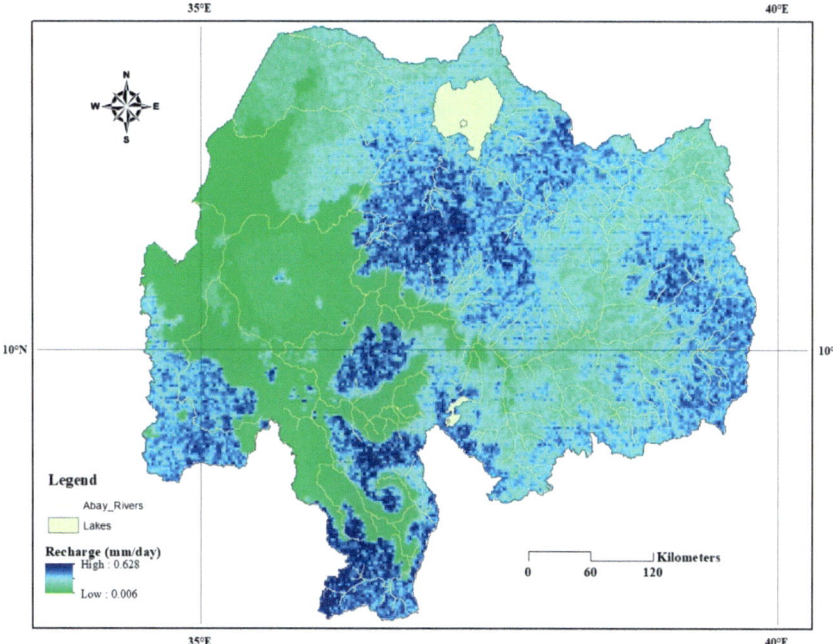

Fig. 9.9 Recharge (mm/day) distribution

recharge. The elevated mountain ranges and plateau of the basin are the most important recharge zones. The model has shown that there is a basin-level effective recharge of 15.385 billion cubic metres per year. This budget does not account for localized groundwater flow, which is an important part of the overall groundwater flow budget within the regional aquifer system.

The perennial springs emanating at the base of larger mountains are evidence of sustained water supply from the permeable volcanic rocks to the intermediate and shallow aquifer systems. The existence of high discharge perennial rivers in the basin show that the recharge from high rainfall areas of the basin are also contributing to sustained water supply to the lowland plains. Although the precipitation seems to be not very high, the existence of highly fractured basalts with underlying fractured basement complex rocks favours groundwater recharge north western of the basin, including part of the Tana Basin.

Occurrence of springs and an increase in river flow during the dry season are evidence of discharges of the groundwater. When the groundwater flow (pathlines) intersects the land surface, natural groundwater discharge develops. Groundwater discharge areas are present in the lowlands, local and regional depressions, and along the banks of rivers (where the rivers crossover relatively old geologic formations). The groundwater flow directions are found to converge towards the major rivers and lakes in the basin. The regional groundwater flow follows the topographic gradient. Many of the springs are localized along the bottom of the ridges and along the surface water bodies of the basin. In contrast, in the lowlands, groundwater recharge from rivers is dominant. In general, in the highlands of the basin, contribution of groundwater to rivers is higher while at the lowlands substantial amounts of river water join the groundwater system through permeable fractured rocks and quaternary sediments.

Groundwater Table/Potentiometric Surface and Flow Direction

From the calibrated model, the groundwater table/potentiometric surface obtained is as shown in Fig. 9.10. In the same figure, the groundwater flow gradient can be seen. From Fig. 9.10, it is evident that the main surface water bodies are the regional groundwater discharge areas in the basin. This could be explained by the flatter hydraulic gradient obtained in these areas than in the rest (highlands) of the basin.

The groundwater potentiometric surface below the ground (groundwater access depth) for Abbay Basin is shown in Fig. 9.11. From the figure, it is evident that groundwater is more accessible closer to the rivers than away from them. In general, nearly 80% (by area) of the Abbay Basin groundwater could be accessed within 500 m drilling. The access depth can have an additional ± 52 m error because of the assumptions made in the model generation and inventory data quality.

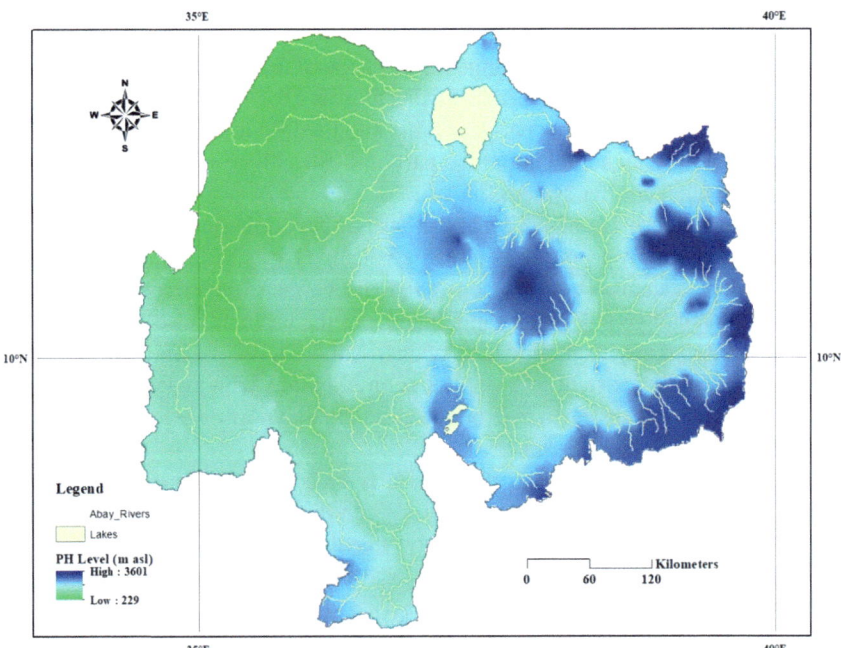

Fig. 9.10 Groundwater table/potentiometric surface in m asl

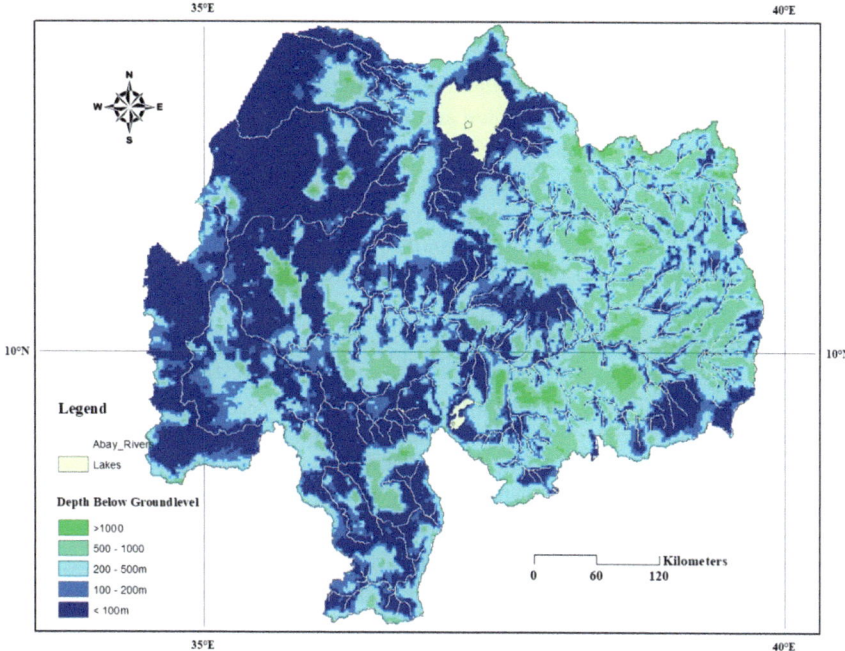

Fig. 9.11 Groundwater potentiometric surface in m below ground

Summary and Conclusions

The focus of this study is the regional groundwater flow components of the aquifers within the Abbay Basin. The regional groundwater flow system describes flow systems that provide a sustained source of groundwater, which is less affected by seasonal variations in recharge. The practicality of collecting and analysing data for such a large area and the resulting coarseness of the regional flow model grid limited this study to a description of regional groundwater flow. Consequently, the groundwater flow simulations were not designed to address shallow, local components of groundwater flow paths.

Two conceptual models (Northwestern groundwater and Abbay basins) for regional groundwater flow simulations were evaluated to test the conceptual model of the regional aquifer system in Abbay Basin, specifically, to test the conceptualization of external and internal boundaries of groundwater flow. The northwestern groundwater basin encompasses approximately 371,147 square kilometres (km^2) and merges Abbay, Tekeze, Mereb and Baro Akobo basins. These conceptual regional groundwater flow models were constructed to (i) assess and advance the conceptual model of regional groundwater flow, (ii) characterize regional aquifers, (iii) determine recharge and (iv) identify groundwater flow patterns within the Abbay Basin.

The steady-state regional groundwater flow models were designed to simulate flow conditions within each of the two groundwater basins without explicitly representing local flow systems that are too small to be adequately simulated with a large-scale model. A comparison of the simulated and measured hydraulic heads indicates that the larger northwestern groundwater basin conceptualization has achieved better results than that of the surface water basin (Abbay Basin only) regional groundwater flow model. The northwestern groundwater basin has thus been developed as the primary means for the assessment of groundwater availability in the Abbay Basin. The study area was discretized horizontally by 2200 m length triangular elements. A finite element-based model called TAGSAC was used to approximate the solution of the equations governing two-dimensional groundwater flow. These models simulate 11,474 spring and well inventories made across these four surface water basins.

A regional groundwater flow direction, groundwater access depths and groundwater table (potentiometric surface) maps were estimated using output from the larger groundwater basin's calibrated groundwater flow model. The model has shown that there is a basin-level effective recharge of 15.385 BCM per year. This budget does not account for localized groundwater flow, which is an important part of the overall groundwater flow budget within the regional aquifer system. More than three fourth of the Abbay Basin's locations regional groundwater could be accessed within 500 m drilling. The recharge and groundwater potentiometric maps together reveal that at high recharge locations, regional groundwater flow paths extend from high to low regional potentiometric areas over distances greater than tens of kilometres; instead, here, flow is dominated by shorter flow paths/in high recharge areas, the discharge locations are closer to each other.

References

Alemayehu T (2006) Groundwater occurrence in Ethiopia. Addis Ababa University, Addis Ababa

Anderson MP, Woessner WW (2002) Applied groundwater modelling: simulation of flow and advective transport. Academic press, San Diego, pp 1–143 and 295–314

Asfaw B, Abaire B, Tefera G (2001) Hydrogeological report of Gore area (NC36-16). Ethiopian Institute of Geological Survey, Addis Ababa

Ayenew T, Alemayehu T (2001) Principles of hydrogeology. Addis Ababa University, Addis Ababa, p 125

Berhanu B, Seleshi Y, Melesse AM (2014) Surface water and groundwater resources of Ethiopia: potentials and challenges of water resources development. In: Melesse A, Abtew W, Setegn S (eds) Nile River Basin: ecohydrological challenges, climate change and hydropolitics. Springer, New York

BGS, British Geological Survey (2016) Africa groundwater atlas: hydrogeology of Ethiopia. https://earthwise.bgs.ac.uk/index.php/Hydrogeology_of_Ethiopia. Accessed date April 2023

Boonstra J, Bhutta MN (1996) Groundwater recharge in irrigated agriculture: the theory and practice of inverse modelling. J Hydrol 174(3–4):357–374

Cacas MC, Ledoux E, de Marsily G, Tillie B, Barbreau A, Durand E, Feuga B, Peaudecerf P (1990) Modelling fracture flow with a stochastic discrete fracture network: calibration and validation. 2. The transport model. Wat Resour Res 26(3):491–500

Chebud YA, Melesse AM (2009) Numerical modeling of the groundwater flow system of the Gumera sub-basin in Lake Tana Basin, Ethiopia. Hydrol Process 23(26):3694–3704

Chebud YA, Melesse AM (2011) Operational prediction of groundwater fluctuation in South Florida using sequence based Markovian stochastic model. Water Resour Manage 25(9):2279–2294

Chiew FHS, McMahon TA (1990) Estimating groundwater recharge using a surface watershed modelling approach. J Hydrol 114(3–4):285–304

Cooley RL (1979) A method of estimating parameters and assessing reliability for models of steady state groundwater flow. Application of statistical analysis. Wat Resour Res 15(3):603–617

Dripps WR, Hunt RJ, Anderson MP (2006) Estimating recharge rates with analytic element models and parameter estimation. Ground Water 44(1):47–55

Faunt CC, Sweekind DS, Belcher WR (2004) Three-dimensional hydrogeological framework model, Chapter E of death valley regional groundwater flow system, Nevada and California-hydrogeological framework and transient groundwater flow model. U.S. Geological Survey. Scientific Investigation Report 2004e5205

Healy WR (2010) Estimating groundwater recharge. Cambridge University Press, New York

Hill MC, Tiedeman CR (2007) Effective groundwater model calibration: with analysis of data, sensitivities, predictions and uncertainty. John Wiley and Sons Inc., Hoboken, N.J., p 455

Kebede S (2013) Groundwater in Ethiopia. Springer-Verlag Hydrogeology, Berlin Heidelberg. https://doi.org/10.1007/978-3-642-30391-3_2

Long JC, Billaux DM (1987) From field data to fracture network modelling: an example incorporating spatial structure. Water Resour Res 23(7):1201–1216

Long JCS, Remer JS, Wilson CR, Witherspoon PA (1982) Porous media equivalents for networks of discontinuous fractures. Water Resour Res 18(3):645–658

Maxey GB (1964) Hydrostratigraphic units. J Hydrol 2:124–129

Mohammed M, Watanabe K, Takeuchi S (2010) Grey model for prediction of pore pressure change. Environ Earth Sci 60:1523–1534

Mohan S, Ramani BV (2010) Groundwater potential estimation—a comparative analysis, Ramachandra, Raj Murthy and Ahalya, Lake 2000, Bangalore

Moore RJ, Bell VA (2002) Incorporation of groundwater losses and well level data in rainfall-runoff models illustrated using the PDM. Hydrol Earth Syst Sci 6:25–38

NMSA, National meteorological services agency of Ethiopia (1996) Climatic and agro-climatic resources of Ethiopia. Meteorol Res Rep Ser 1:1–137

Nyagwambo NL (2006) Groundwater recharge estimation and water resources assessment in a tropical crystalline basement aquifer. PhD thesis, UNESCO-IHE, Delft, The Netherlands

Pik R, Marty B, Carignan J, Lave J (2003) Stability of upper Nile drainage network (Ethiopia) deduces from (U–Th)/He thermochronometry: implication of uplift and erosion of the Afar plume dome. Earth Planetary Sci Lett 215:73–88. https://doi.org/10.1016/S0012-821X(03)00457-6

Ragunath HM (2006) Hydrology: principles analysis and design. New Age International (P) Limited Publishers, New Delhi

Rushton KR, Rathod KS (1985) Horizontal and vertical components of flow deduced from groundwater heads. J Hydrol 79(1):261–278

San Juan CA, Belcher WR, Laczniak RJ, Putnam HM (2004) Hydrologic components for model development. In: Belcher WR (ed) Death Valley regional groundwater flow system, Nevada and California—Hydrologic framework and transient groundwater flow model. USGS Scientific Investigations Report, pp. 103–136

Sanford WE, Plummer LN, McAda DP, Bexfield LM, Anderholm SK (2004) Hydrochemical tracers in the Middle Rio Grande Basin, USA: calibration of a groundwater flow model. Hydrogeol J 12:389–407

Seaber PR (1988) Hydrostratigraphic units in hydrogeology. In: Back W, Rosenshein JS, Seaber PR (eds) The geology of North America. Geological Society of America

Serrano SE, Unny TE (1987) Predicting groundwater flow in a phreatic aquifer. J Hydrol 95(3–4):241–268

Sondhi S, Rao N, Sarma P (1989) Assessment of groundwater potential for conjunctive water use in a large irrigation project in India. J Hydrol 107(1–4):283–295

Stiefel J, Melesse A, McClain M, Price RM, Anderson AP, Chauhan NK (2009) Effects of rainwater harvesting induced artificial recharge on the groundwater of wells in Rajasthan, India. Hydrogeol J 17(8):2061–2073

Thomas G (2001) Designing successful groundwater banking programs in the Central Valley: lessons from experience. Natural Heritage Institute, Berkeley

Tóth J (1963) A theoretical analysis of groundwater flow in small drainage basins. J Geophys Res 68:4795–4812. https://doi.org/10.1029/JZ068i016p04795

USACE (U.S. Army Corps of Engineers) (1999) Engineering and design, groundwater hydrology, Washington DC

Part IV
Natural Resources Degradation in the Basin

Chapter 10
Land Use/Land Cover Changes, Drivers, and Implications

Gizachew Kabite Wedajo

Abstract Land use/land cover (LULC) is critical as an environmental variable that influences land surface processes and natural resources availability. Land use/land cover changes (LULCC) driven by biophysical and anthropogenic factors are the major components of global environmental changes. Human-induced LULCC greatly alter the functioning of ecosystem services. The Abbay basin, which is characterized by diverse topography, climate and natural resources, experienced substantial land use/land cover changes for the past several years. Therefore, up-to-date information on the trends, drivers and implications of LULCC has paramount importance in designing appropriate land use policies and, thus minimize the negative consequences of the change. This book chapter assesses the trends, drivers and consequences of LULCC in the Abbay basin. Accordingly, peer-reviewed articles were selected, and a systematic literature review was performed. Moreover, LULCC and its responses on the hydrologic processes, sediment yield and soil erosion were examined as case studies. The results showed considerable spatiotemporal changes in land use/land cover for the past couples of decades in the Abbay basin. Most of the changes include increasing cultivated land, built-up areas and waterbodies while decreasing forestland, bush/shrubland and grassland. The major drivers include population growth, resettlement programs, unwise utilization of land resources and development projects like large-scale agricultural investment and construction of dams. The substantial human-induced LULCC observed in the Abbay basin aggravated soil erosion, increased sediment yield, altered hydrologic processes, decreased ecosystem services and affected the socio-economy of the local community and the country at large. Therefore, proper land use planning should be implemented to minimize the negative consequences of LULCC on the socio-economy and the environment of the area and the surroundings.

Keywords Land use/land cover change · Drivers · Implications · Abbay basin

---

G. K. Wedajo (✉)
Space Science and Geospatial Institute, Addis Ababa, Ethiopia
e-mail: kabiteg@gmail.com

© The Author(s), under exclusive license to Springer Nature Switzerland AG 2025
A. Melesse et al. (eds.), *Abbay River Basin*, Springer Geography,
https://doi.org/10.1007/978-3-031-65241-7_10

Introduction

The terms land use and land cover connote different meanings and concepts but are commonly used together and interchangeably. Land use refers to the socioeconomic use of land, while land cover is a biophysical characteristic of the land surface (Mariye et al. 2020). However, the term land use/land cover is commonly used interchangeably. In this book chapter, land use/land cover (LULC) is used to describe both concepts. The LULC is the major environmental variable that influences the energy and material fluxes between the atmosphere and the terrestrial land surface (Hu et al. 2019). As such, it is directly related to the fundamental processes of the Earth, such as land productivity, biodiversity, land degradation, hydrologic cycle and other environmental processes (Oliver and Morecroft 2014; Msofe et al. 2019). Studies used land cover mapping tools and methods to understand land use changes, inventory of forest and natural resources as well as understand the changes in the hydrologic behavior of watersheds (Getachew and Melesse 2012; Mango et al. 2011a, b; Wondie et al. 2011, 2012; Melesse and Jordan 2002, 2003; Melesse et al. 2007; Mohammed et al. 2013).

As a result, it has been a hot topic for the global environmental changes and land resource management studies. In particular, human-induced LULCC has been one of the major factors leading to climate changes (Turner et al. 1993). Changes in land use/land cover, e.g., reduce ecosystem services at multiple spatial scales and disturb the capability of natural systems in supporting human wellbeing (Lin et al. 2018; Alemayehu et al. 2019). For instance, the LULCC is the major driver of land degradation by aggravating soil erosion (Aneseyee et al. 2020) and affects land productivity and climate change (Lambin et al. 2001), biodiversity (Lambin et al. 2003), food security (Lambin et al. 2003) and public health (Shi et al. 2018). Generally, LULCC greatly threatens the environment, economic development and food security (Ewers 2006). Therefore, to minimize the negative consequences of LULCC, understanding the drivers and extent of the change is very important.

The drivers of LULCC vary from location to location and community to community. However, the common drivers of the change include socioeconomic (Xie et al. 2005), demographics (Shi et al. 2010), biophysical (Serra et al. 2008), technology (Hasselmann et al. 2010) and industrial factors (Xiao et al. 2006), politics (Kanianska et al. 2014). In Ethiopia, the main drivers of LULCC include population pressure (e.g., Betru et al. 2019; Ewunetu et al. 2021; Mathewos et al. 2022; Mathewos et al. 2022), land tenure and property rights (e.g., Ewunetu et al. 2021;), persistent poverty (e.g., Ewunetu et al. 2021), resettlement programs (e.g., Dibaba et al. 2020; Betru et al. 2019), climate change (e.g., Kabite et al. 2020; Amini et al. 2022), and fuel wood extraction (e.g., Mathewos et al. 2022), agricultural expansion (e.g., Mathewos et al. 2022; Dibaba et al. 2020). Of all the driving factors, population pressure is the major driver of LULCC in Ethiopia, as the majority of the population depends on natural resources as their livelihood. Natural landscapes like forest, bush/shrubland, wetland, and grassland were converted to agricultural and urban land to meet the increasing demands of food and material for the ever-increasing population (Amini et al. 2022), which leads to land degradation and declined ecological services.

Therefore, analyzing LULCC is important for understanding its spatiotemporal dynamics, impacts, and the drivers of the change (Haregeweyn et al. 2014). Such information is important for policymakers and development practitioners, decision-makers, and global environmental change and sustainability studies (Hu et al. 2019). Generally, up-to-date and reliable LULCC information is important for understanding the conditions of the land (Anderson 1976), food and energy security for the growing population (Erb et al. 2017), assessing its impacts on the environment and ecosystem services (Anderson 1976), land management considerations (Erb et al. 2017), change detection analysis (Ran et al. 2012), and monitoring natural resource status and input for policymakers. As such, several studies have been conducted on spatiotemporal LULC changes at local, regional and global scales, and the topic remains a hot research topic (Hu et al. 2018).

Study showed significant changes in land use/land cover during the last few decades in Ethiopia (e.g., Teferi et al. 2013; Betru et al. 2019; Berihun et al. 2019; Dibaba et al. 2020; Regasa et al. 2021). In most basins of Ethiopia, agricultural land, built-up areas, and bare land were increased, while forestland, grassland, and shrubland have decreased during the last few decades (Betru et al. 2019; Regasa et al. 2021; Ewunetu et al. 2021). Changes in LULC significantly affect soil erosion intensities (e.g., Kidane et al. 2019; Berihun et al. 2019; Aneseyee et al. 2020), hydrological processes (Kabite et al. 2022), and ecological services (Tolessa et al. 2017; Assefa et al. 2021). The Abbay basin is one of the basins in Ethiopia that experienced significant changes in LULCC during the past several years.

This book chapter is intended to assess LULCC, identify drivers of the change, and assess implications of the change. This chapter is divided into five sections. The first part is about background information, the second part is about the LULCC in the Abbay basin, part three describes about the drivers of the land use/land cover change, part four describes implication of the changes and the fifth part describes two case studies on the impacts of LULCC on hydrologic processes and soil erosion and sediment yield.

Land Use/Land Cover Changes in the Abbay Basin

It is challenging to generalize LULCC trends within a given area, as it is influenced by human activities (Zeleke and Hurni 2001) and the landscape that varies across a given area. As such, most previous studies in the Abbay basin focused on assessing LULCC, focusing only on specific watersheds (Betru et al. 2019; Berihun et al. 2019; Mathewos et al. 2022; Ewunetu et al. 2021).

The Abbay basin, in general, is characterized by diverse biophysical and socioeconomic environments and endowed with diverse natural resources that includes land, water, vegetation, and genetic diversity (Hurni et al. 2005). However, a study showed that immense LULCC was observed in the basin. For example, significant land use/

land cover changes were observed during the last 50 years in the Jedeb watershed, Abbay Basin (Teferi et al. 2013). Study generally showed that cultivated land significantly increased at the expense of natural vegetation. According to Dibaba et al. (2020), the Fincha'a watershed, which is one of the tributaries of the Abbay River, experienced increasing agricultural land, wetland and urban land while declining forestland, bush/shrubland, and grassland. According to Ewunetu et al. (2021), agricultural land and plantation forest increased, while natural forestland, shrubland, bushland, grassland, and bare land declined in the headwater of the Abbay River basin from 1986 to 2017. In addition, Betru et al. (2019) reported that woodland declined by 40%, while agricultural land increased by 39% during the last 38 years in Ethiopia.

Most LULCC studies done in the Abbay basin showed increasing agricultural land and built-up areas and declining forestland, grassland, and bush/shrubland. Moreover, the major drivers are mainly anthropogenic factors (Table 10.1). Anthropogenic factors include population pressure, resettlement and unwise utilization of land resources (Ewunetu et al. 2021), agricultural expansion, urbanization and infrastructural development, and weak environmental considerations, and timber and woodworks (Dibaba et al. 2020). Such a human-induced LULCC greatly affects the environment and natural resource availability. In particular, the transformation of natural landscapes to human-dominated landscapes accelerates soil erosion and sediment yield, leading to land degradation (Moges and Bhat 2017; Aneseyee et al. 2020; Wassie 2020; Moisa et al. 2021). As such, the Abbay basin is one of the most degraded lands in Africa (Fenta et al. 2021), which is aggravated due to rapid LULCC and high population pressure coupled with roughened topography (Hurni et al. 2015; Kassawmar et al. 2018; Berihun et al. 2019). The increased soil erosion in response to land use/land cover change resulted in siltation of reservoirs, affecting water resource development projects such as irrigation and hydropower (e.g., Megech Dam in the Abbay basin) (Lemma et al. 2018; Abebe and Gebremariam 2019) and hydropower generation (e.g., Grand Ethiopian Renaissance Dam in the Abbay basin) (Welde 2016; Haregeweyn et al. 2017). Table 10.1 summarizes the LULCC trends, drivers and their effects in the Abbay basin.

Drivers of Land Use/Land Cover Changes in the Abbay Basin

Up-to-date and reliable information on the drivers of LULCC is crucial to design sustainable land use planning and minimize the negative consequences of LULCC. Several studies were done in various parts of the Abbay basin and identified major drivers of LULCC. Most of the studies depend on socioeconomic surveys to assess drivers of LULCC. For instance, focus group discussions (FGD) and key informant interviews (KII) conducted in the Bechet watershed of the Abbay basin showed that farmland expansion and deforestation are the major drivers of LULCC (Sisay et al.

Table 10.1 Summaries of LULCC, drivers, and its effects in the Abbay basin

S. No	Land use/land cover change		Drivers	Effects	Subbasins	Study periods	References
	Increasing	Decreasing					
1	Cultivated land	Forest, grassland, and bushland	Population growth and changing farming system	Increasing soil erosion and surface runoff	Aba Gerima and Debatie	1982–2017	Berihun et al. (2019)
2	Cultivated land and plantation forest	Forest, grassland, and bushland	Population growth and changing farming system	Increasing soil erosion and surface runoff	Gudar	1982–2017	Berihun et al. (2019)
3	Cultivated land, bare land, built-up	Forest, grassland, and shrubland	Population growth	Modified hydrological processes of the watersheds (runoff and sediment yield)	Jedeb and Chemoga	1990–2018	Birhanu et al. (2022)
4	Cultivated land, commercial farm, built-up, and waterbody	Forestland, grassland, bush/shrub and swampy areas	Biophysical, socioeconomic and demographic factors	Declined agricultural yield, loss of biodiversity, and decline of water resource availability	Fincha'a	1987–2017	Dibaba et al. (2020)
5	Increased cultivated land	Decreased forestland	Population growth, resettlement program, unwise resource utilization	–	Western part of Abbay basin	1978–2016	Betru et al. (2019)
6	Cultivated land, shrubland, bare land and built-up	Forestland		Declining ecosystem services (e.g., nutrient cycling, provision of raw material, and erosion control	Chilimo forest (central highlands of Abbay basin)	1973–2015	Tolessa et al. (2017)
7	Cultivated land	Forestland and grassland	Population pressure and regime change	–	Tana	1973–2015	Fikirte et al. (2017)
8	Cultivated land and built-up areas	Grassland, forestland, and bush/shrubland	Population growth, deforestation for fuelwood and construction purpose	Land degradation	Muga	1985–2025	Belay and Mengistu (2019)

(continued)

Table 10.1 (continued)

S. No	Land use/land cover change		Drivers	Effects	Subbasins	Study periods	References
	Increasing	Decreasing					
9	Built-up, and cultivated land	Wetlands and water bodies	Population growth	Declining ecosystem services	Tana	1984–2019	Assefa et al. (2021)
10	Cultivated land and built-up area	Bush/shrubland, forestland and grassland	Population growth	Alter hydrological processes (e.g., increased surface runoff, but decreased baseflow, water yield and ET)	Birr	1997–2018	Malede et al. (2023)
11	Cultivated land	Forestland, grassland and wetland	–	Increased soil erosion and sediment yield	Muger	1986–2020	Teshome et al. (2022)
12	Cultivated land and built-up areas	Forestland and bush/shrubland	Population growth, resettlement program and agricultural investment	Altered hydrologic processes	Dhidhessa River basin	1986–2018	Kabite et al. (2022)
13	Cultivated land and built-up areas	Forestland, shrubland, and grassland	–	Declining ecosystem services	Andassa	1985–2015	Gashaw et al. (2018)
14	Cultivated land and plantation forest	Forestland, grassland, bush/shrubland, waterbody, and bare land	Population growth, land tenure, poverty, lack of awareness, and poor infrastructure	–	North Gojjam	1986–2017	Ewunetu et al. (2021)

(continued)

Table 10.1 (continued)

S. No	Land use/land cover change		Drivers	Effects	Subbasins	Study periods	References
	Increasing	Decreasing					
15	Increased cultivated land	Decreased forestland	Expansion of cultivated land and deforestation due to increasing population	Deforestation and land degradation	Bechet	1984–2020	Sisay et al. (2020)
16	Increased cultivated land and built-up areas	Declined forestland, andbush/shrub land	–	Population pressure, lack of land use policy, unwise utilization of resources	Asosa, western Abbay basin	1978–2016	Betru et al. (2019)

2021). Moreover, population pressure and poor farming systems are the main drivers of LULCC in the Abbay basin (Berihun et al. 2019). According to Shiferaw et al. (2019), climate change, invasive species, and weak natural resource management policies were the major drivers of LULCC during the last 40 years in the Afar region. Expansion of agricultural land, urban land and infrastructural development, uncontrolled cutting of trees and grazing, resettlement, and weak environmental considerations during development activities is the major drivers of LULCC in the Abbay basin (Dibaba et al. 2020). Population pressure, resettlement, and unwise utilization of land resources are the major drivers of LULCC in Ethiopia (Regasa et al. 2021). Furthermore, Mathewos et al. (2022) reported that the expansion of agricultural, increasing population, and deforestation are the main drivers of LULCC in the Rift Valley basin of Ethiopia, which is adjacent to the Abbay basin.

The main drivers of LULCC in western Ethiopia include the expansion of agricultural land, unsustainable exploitation of forest products, increasing population, and implementation of resettlement programs (Betru et al. 2019). According to Ewunetu et al. (2021), the main drivers of LULCC in the Abbay basin include population growth, land tenure, poverty, weak enforcement of rules, lack of community awareness, and poor infrastructure.

Land use/land cover change is greatly affected by economic (e.g., land price), political (e.g., war), demographic (e.g., population growth), development activities (e.g., construction of infrastructures and large-scale agricultural investments), and climate change (e.g., severe drought and flooding) factors (Rabiei-Dastjerdi et al. 2022). For example, Kuma et al. (2022) reported that population pressure, inappropriate land management, and farming slopy lands are the major drivers of LULCC in southern Ethiopia. According to Bekele et al. (2019), the major drivers of LULCC in the Rift Valley basin of Ethiopia includes population, deforestation due to fuelwood extraction, expansion of agricultural land, climate change (e.g., recurrent drought) and overgrazing.

Generally, the major drivers of LULCC in Ethiopia including the Abbay Basin in particular are human factors such as population pressure, unwise land utilization, lack of awareness, government-sponsored resettlement programs, and socioeconomic activities like large-scale agricultural investment and the construction of reservoirs (Table 10.1).

Implications of Land Use/Land Cover Changes

Soil Erosion

Land use/land cover change alters the ecosystem services of the landscape. For instance, the transformation of vegetated lands to cultivated lands aggravates gully formation and density (Berihun et al. 2019). Moreover, gully erosion is aggravated by inappropriate agricultural activities, road construction, and overgrazing (Valentin et al. 2005). Therefore, the continual increasing of cultivated land combined with population growth in Ethiopia, particularly in the Abbay basin, has aggravated the on-site and off-site effects of soil erosion. Conversion of natural landscapes such as vegetated land to human-dominated landscapes such as agricultural land with poor land management accelerates soil erosion compared to other landscape types (Wassie 2020; Han et al. 2020). Fenta et al. (2021) described that soil loss from cropland was 2–15 times that of noncropland cover types in most basins of Ethiopia. Moreover, Hurni (1988) reported that the mean soil loss rates of cropland were 42 tonnes per hectare per year (t ha^{-1} yr^{-1}), whereas they were 5 t ha^{-1} yr^{-1} and 1 t ha^{-1} yr^{-1} for woodland and bushland and forestland, respectively. As such, the conversion of land from natural vegetation to cropland would aggravate land degradation by soil erosion (Fenta et al. 2021).

Land use/land cover changes contributed to the increasing soil erosion in the Yezat (Tadesse et al. 2017) and Rib (Moges and Bhat 2017) watersheds of Abbay basin. Belay and Mengistu (2021) also reported increasing soil erosion in the Muga watershed of the Abbay basin due to the increasing of agricultural land at the expense of natural vegetation. Aggravated soil erosion response to LULCC in the Abbay basin leads to soil fertility decline and subsequently reduced production and downstream reservoir siltation (Hurni et al. 2015; Haregeweyn et al. 2017; Lemma et al. 2018). Furthermore, Teshome et al. (2022) reported an increasing average soil loss rate from 53.2 t ha^{-1} year^{-1} in 1986 to 64 t ha^{-1} yr^{-1} in 2020 and correspondingly increasing mean sediment yield from 7.8 t ha^{-1} year^{-1} in 1986 to 10.2 t ha^{-1} yr^{-1} in 2020 in response to increasing agricultural land at the expense of vegetated land in the Mugar watershed of the Abbay basin. Accordingly, the Abbay basin is the second most eroded basin in Ethiopia next to the Tekeze basin due to its high R-factor and cropland-dominated land use (i.e., half of the basin is covered by cropland) (Fenta et al. 2021). Even the annual sediment yield of the Abbay basin is the highest in the country.

The increased soil erosion and sediment yield due to land use/land cover change not only affect soil fertility but also affect downstream reservoir functionality. The main on-site effects of soil erosion are the decline in soil fertility and the subsequent reduction in the production functions of the land, while the off-site effect includes siltation of downstream reservoirs. The high soil erosion and sediment yield reported in the different watersheds of the Abbay basin have negative consequences on water resource development projects in the Abbay basin (e.g., Grand Ethiopian Renascences Dam; GERD and other hydroelectric and irrigation structures). The government of Ethiopia adopted a strategic plan to reduce the negative consequences of soil erosion and build a climate-resilient green economy by 2025 (FDRE 2011). The government has been trying to implement this strategy through community-based land management practices that could reduce the ongoing land degradation by soil erosion. However, little is known about the effectiveness and the potentials of these practices in reducing soil loss at the national and basin scales (Hurni et al. 2015).

Surface Runoff

Changes in land use/land cover modify the runoff coefficient of the landscape, and thus alter the hydrologic processes of a watershed. A study conducted in one of the watersheds of the Abbay basin showed that average runoff coefficient of cultivated land is 21% higher than that of the vegetated land. The increased runoff coefficient resulted in increasing surface runoff, which in turn increased the on-site and off-site impacts of soil erosion, such as declining soil moisture and increasing soil erosion and siltation of the downstream reservoirs (Berihun et al. 2019). Moreover, studies showed that the runoff rate is strongly influenced by the land cover types (Hurni et al. 2005; Haregeweyn et al. 2016; Sultan et al. 2017).

According to Getachew and Manjunatha (2022), the increasing of agricultural and built-up lands at the expense of shrublands and grasslands in the Tana subbasin of the Abbay basin experienced increasing surface runoff and streamflow while declining lateral flow and evapotranspiration. Moreover, Kabite et al. (2022) reported increasing runoff and wet season streamflow and declining groundwater recharge in the Dhidhessa River basin of the Abbay basin in response to increasing cultivated land and declining forestland. The transformation of vegetated land to cultivated land increased surface runoff by 20% in the Endorheic Hayk Lake of the Abbay basin (Mewded et al. 2021). Expansion of agricultural land and built-up areas while declining forestland, bushland, and grassland resulted in increasing surface runoff but declining water yield and baseflow in the Birr watershed, Abbay basin (Malede et al. 2023). Moreover, increased cultivated land at the expense of wood land in the Tana subbasin of the Abbay basin resulted in increasing surface runoff and declining groundwater recharge. However, increasing cultivated land while decreasing woodland increased surface runoff and wet season water yield (Woldesenbet et al. 2017).

A systematic review showed that increased cultivated land and urban lands at the expense of forestland resulted in increased streamflow, surface runoff, and sediment yield, while evapotranspiration and groundwater recharge declined in tropical regions during the last few decades (Kayitesi et al. 2022). Therefore, proper land management and utilization are highly required to conserve both water and soil, which improves soil fertility and water resource availability in the Abbay basin.

Socioeconomic and Environmental Implications

Depending on the type of change, LULCC could have positive and negative socioeconomic and environmental implications (Belete 2015; Desalegn et al. 2014; Berihun et al. 2019). For instance, plantations have both socioeconomic and environmental benefits. Belete (2015) reported that farmers benefited more from the sale of *A. decurrents* charcoal, which gained an average net return of 400% compared to the sale of annual crops, as it reduced labor and fertilizer costs. Moreover, the plantation of

Acacia. decurrens mitigated land degradation by modifying the physical, biological and chemical properties of soil (Belete 2015). According to the study result, the pH of soil under *A. decurrens* plantations was 2% lower than that of soil under cultivated land. The available phosphorus in soil under the *A. decurrens* plantation was 1.25 mg/kg lower than that in soil under cultivation, whereas the total nitrogen of soil under the *A. decurrens* plantation was 43.5% higher compared to the soil under cultivation. However, the transformation of grazing land to plantations reduced the number of livestock by reducing free grazing land and livestock fodder (Desalegn et al. 2014). Moreover, the conversion of forestland to cultivated land could affect the economy of the community that depends on forest products for their livelihood.

Ecosystem Service Decline

A study conducted in the Afar Region indicated that the expansion of *Prosopis juliflora*, which invade over 1 million ha resulted in the loss of about US$602 million on average (Shiferaw et al. 2019). Moreover, due to the declining of forest cover from 1973 to 2015 in the central highlands of Ethiopia, about US$3.69 million of ecosystem service values (ESV) was lost (Tolessa et al. 2017), which includes soil erosion control, nutrient cycling, and provision of raw material. A study conducted in the Andassa watershed of the Abbay basin showed that the increasing cultivated land and built-up areas while declining of vegetation lands (e.g., grassland, forest, and shrubland) from 1985 to 2015 leads to the loss of total ESV of approximately US$6 million. Assefa et al. (2021) reported that declining wetlands and waterbodies while increasing cultivated land and built-up areas resulted in declining ecosystem services in the Tana subbasin of the Abbay basin. The study further reported that from 1984 to 2019, about US$8.9 * 10^6 ecosystem service values, which includes water regulation, biodiversity habitat, and waste water treatment were lost. Moreover, due to the increasing agricultural land and built-up areas while declining natural vegetation in the afro-alpine area of Guna Mountain located in the Abbay basin, ecosystem service values decreased by approximately US$9.78 * 10^6 from 1995 to 2020 alone. According to the study, of this ecosystem services value loss, 42% accounts for regulation service values, while 29% and 13% account for provision and supporting service values, respectively.

Case Studies

Land Use/Land Cover Changes and Their Impacts on Hydrological Responses in the Dhidhessa River Basin

Kabite et al. (2020) conducted LULCC analysis from 1985 to 2017 for the Dhidhessa River basin (DRB), which is one of the magnificent tributaries of the Abbay River that contributes more than 25% of discharge during the dry season. According to the study, agricultural land, built-up areas, forestland, grassland, bush/shrubland, and waterbodies are the major land use/land cover types of the DRB. The results showed that forestland and bush/shrubland were the major land use/land cover classes covering approximately 67% and 13% of the basin, respectively, during 1985. During the 2001 study period, however, bush/shrubland and agricultural land were the dominant classes, covering approximately 40% and 26%, respectively, whereas agricultural land and bush/shrubland were the major land use/land cover types, with areal coverages of 42% and 35%, respectively, during the 2017 study period (Figs. 10.1 and 10.2).

Generally, the Dhidhessa River basin experienced declining bush/shrubland and forestland and increasing agricultural land, grassland, built-up areas and waterbodies from 1985 to 2017. Accordingly, bush/shrublands decreased by 40% (7655 km^2) from 1985 to 2001 and then by 12% (1330 km^2) from 2001 to 2017. Similarly, forestland decreased by 547 km^2 (15%) from 1985 to 2001 and by 195 km^2 (6%) from 2001 to 2017. However, agricultural land increased by 4190 km^2 (101%) from 1985 to 2001 and by 4441 km^2 (60%) from 2001 to 2017. Increasing agricultural land while

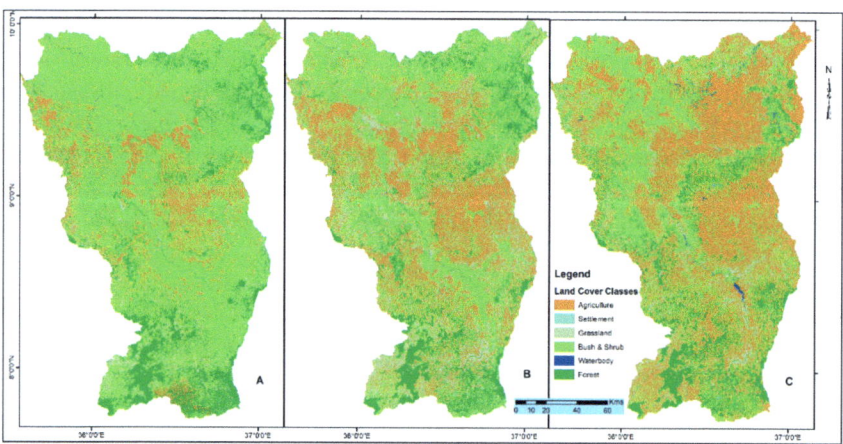

Fig. 10.1 Land use/land cover map of 1985 (**a**), 2001 (**b**) and 2017 (**c**) for the DRB. *Source* Kabite et al. (2020)

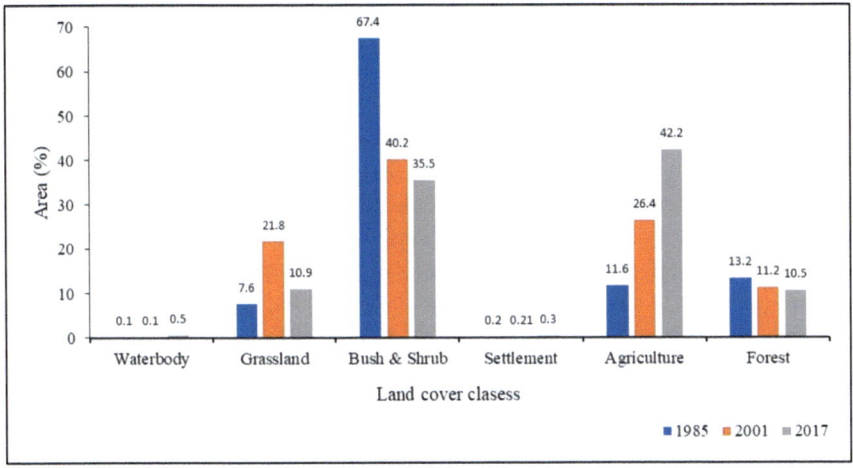

Fig. 10.2 Land use/land cover area coverage for 1985, 2001 and 2017. *Source* Kabite et al. (2020)

decreasing bush/shrub land and forestland was mainly due to the increasing population in the basin, which was mainly attributed to the 1984/1985 and 2002/2003 government-sponsored implementation of resettlement programs. In response to the increased population, built-up areas also expanded from 43 km^2 in 1985 to 89 km^2 in 2017. In addition, the coverage of waterbodies increased by 525% (119.6 km^2) from 1985 to 2017 in the basin. The increased waterbodies in the basin could be due to the construction of reservoirs such as the Sorga, Dega, and Arjo-Dhidhessa dams during the study periods. However, grassland increased by 186% (4000 km^2) from 1985 to 2001 but decreased by 50% (3070 km^2) from 2001 to 2017. This implies that grassland increased from 1985 to 2017 by a net of 43% (930 km^2). The increasing grassland during 1985–2001 was attributed to bush/shrubland clearance in response to the settlement of a large number of people from the drought-affected areas of the northern parts of the country during the 1984/85 resettlement program. On the other hand, the decrease in grassland from 2001 to 2017 could be attributed to the conversion of grassland into agriculture land in response to population growth. Generally, approximately 8985 km^2 (2.07%/year) of bush/shrubland was lost, whereas approximately 8632 km^2 (4.17%/year) of the basin was converted to agriculture from 1985 to 2017 in the basin (Fig. 10.2).

The study further showed that approximately 42% of the Dhidhessa River basin experienced land use/land cover conversion mainly from bush/shrubland to agriculture (16%) and grassland (14%) from 1985 to 2001. Similarly, approximately 47% of the study area experienced land use/land cover transformation during the 2001–2017 study periods, of which 14% of the change represented bush/shrubland transformation into agricultural land and 9% of the change was shared by the transformation of grassland to agricultural land. The result confirms that agricultural land was expanded toward the bush/shrubland and forestlands (Kabite et al. 2020). The results of this study agree with study conducted by Alemayehu et al. (2019), who reported that approximately 51% of the Somoda watershed, which is located in the southwestern part of Ethiopia, changed from at least one type to another.

The major drivers of land use/land cover changes observed in the Dhidhessa River basin from 1985 to 2001 include (i) 1984/5 government-sponsored resettlement program, (ii) illegal and informal settlement during the 1991 regime change, (iii) policy of the government in expanding agricultural land to meet the increasing demands of food, and (iv) the return of some of the 1984/5 settlers to their original areas. For the 2001–2017 land cover dynamics, the drivers include (i) the 2002/3 government-induced resettlement program, (ii) the continuous illegal and informal influxes of people to the area in search of employment opportunities and agricultural land and (iii) socioeconomic development (e.g., construction of Dams, urbanization, and large-scale agricultural investment). The study also revealed that natural drivers like climate and topography, and human factors like population growth are the major drivers of land use/land cover changes in the Dhidhessa River basin. Generally, the study showed that the Dhidhessa River basin experienced significant changes in land use/land cover that vary in time and space during the past 30 years. The change could have negative consequences on the socio-economy, the environment and natural resource availability. Altering hydrological processes through disturbing natural environmental and ecological conditions is one of the negative consequences of human-induced land use/land cover changes.

The impacts of LULCC on the hydrological processes of the Dhidhessa River basin were quantified by Kabite et al. (2022). The study showed increasing surface runoff, actual evapotranspiration, and streamflow by 40%, 5%, and 4%, respectively, while decreasing groundwater by 3% during the last three decades attributed to the substantial changes in land use/land cover and climate change. Conversion of natural vegetation (i.e., bush/shrubland and forest) to cultivated land during the study periods was the main factor that contributed to the increasing surface runoff and declining groundwater recharge. In addition, streamflow increased during the wet season and decreased during the dry season, which negatively affected agricultural production and water resource availability (Table 10.2).

Land Use/Land Cover Change and Its Impacts on Soil Erosion and Sediment Yield in the fincha'a Watershed

The Fincha'a watershed is one of the significant watersheds of the Abbay basin that contributed a significant portion of streamflow owing to the Chomen wetland and Fincha'a reservoirs. Study showed that the watershed experienced significant changes in land use/land cover during the last couple of decades driven by population pressure and developmental activities such as Fincha'a Sugarcane farm and construction of reservoirs. Recent LULCC analysis was performed in the watershed from 1991 to 2021. According to the study result, cultivated land was the dominant land use covering approximately 28%, 50% and 53% during 1991, 2006 and

Table 10.2 Relative contributions of separate and combined effects of land use/land cover and climate changes

Hydrologic variables	1982–2001			2001–2018			1982–2018		
	$\Delta H_{sep,cc}$ (%)	$\Delta H_{sep,Lc}$ (%)	ΔH_{comb} (%)	$\Delta H_{sep,cc}$ (%)	$\Delta H_{sep,Lc}$ (%)	ΔH_{comb} (%)	$\Delta H_{sep,cc}$ (%)	$\Delta H_{sep,Lc}$ (%)	ΔH_{comb} (%)
AET	2.4	−0.9	1.4	2.8	0.4	3.4	5.3	−0.3	4.8
Runoff	9.4	1.2	10.4	−8.5	38.7	26.4	2.1	36.5	39.6
Recharge	8.0	−7.4	1.2	−4.8	−1.2	−4.2	3.7	−7.4	−3.1
Yield	7.0	0.5	7.5	−2.3	−0.5	−2.9	4.7	−0.1	4.4

Source Kabite et al. (2022)

$\Delta H_{sep,cc}$ = separate impacts of climate change; $\Delta H_{sep,Lc}$ = separate impacts of land use/land cover change; ΔH_{comb} = combined impacts of land use/land cover and climate change

2021, respectively. On the other hand, the areal coverage of forestland, grassland, and shrubland was 25%, 20%, and 7.1% during 1991, 16.8%, 14.5%, and 7.8% during 2006 and 13.8%, 14.2%, and 7.1% during 2021, respectively. Likewise, waterbodies covered approximately 4.8%, 5%, and 6.1% during 1991, 2006, and 2021, respectively, which showed increasing trends due to the construction of dams (e.g., Amerti and Neshe Dams) (Table 10.3). Generally, cultivated land, settlement, and waterbodies showed increasing trends, while forestland, grassland, and shrubland showed decreasing trends during the last 30 years in the Fincha'a watershed (Fig. 10.3 and Table 10.3). The results of this study are in agreement with the previous studies done in the Fincha'a watershed (e.g., Regasa and Nones 2022; Moisa et al. 2021).

The main drivers of LULCC in the study area include increasing population pressure and development activities such as the construction of Fincha'a sugarcane factories and farms, and the construction of Dams in the watershed (e.g., Fincha'a, Amerti and Neshe Dams). For example, the increase in waterbodies from 4.8% in 1991 to 6% in 2021 was mainly due to the construction of the Amerti and Neshe Dams. Moreover, the construction of these reservoirs and expansion of sugarcane farms drove the local communities to expand their farms toward marginal and bush/shrublands areas.

The results generally showed that cultivated land increased by 842 km^2 (22%) from 1991 to 2006 and further increased by 949 km^2 (24.8%) from 1991 to 2021. However, forestland, grassland, and shrubland declined by 11.3%, 5.7%, and 10.0% from 1991 to 2021, respectively. Nevertheless, built-up areas, waterbodies, and swampy areas slightly increased during the past 30 years in the Fincha'a watershed. From 1991 to 2021, increasing agricultural land, built-up areas, swamp area, and waterbodies by 6323.5 ha/year (1.7%), 131.8 ha/year (0.034%), 106.6 (0.03),

Table 10.3 Land use/land cover area coverage and changes from 1991 to 2021

LULC type	1991		2006		2021		1991–2006	2006–2021	1991–2021
	Area (km^2)	(%)	Area (km^2)	(%)	Area (km^2)	(%)	Area (km^2/yr)	Area (km^2/yr)	Area (km^2/yr)
Cultivated land	1071.1	28.0	1912.6	49.9	2019.6	52.7	56.1	7.1	63.2
Forestland	962.5	25.1	643.5	16.8	530.5	13.8	− 21.3	− 7.5	− 28.8
Grass land	764.5	20.0	554.6	14.5	545.1	14.2	− 14.0	− 0.6	− 14.6
Settlement	78.1	2.0	95.0	2.5	97.9	2.6	1.1	0.2	1.3
Shrub land	656.4	17.1	279.8	7.3	272.8	7.1	− 25.1	− -0.5	− 25.6
Swamp area	115.4	3.0	154.0	4.0	131.4	3.4	2.6	− 1.5	1.1
Water body	183.4	4.8	192.1	5.0	234.2	6.1	0.6	2.8	3.4
Total	3831.4	100	3831.4	100	3831.4	100			

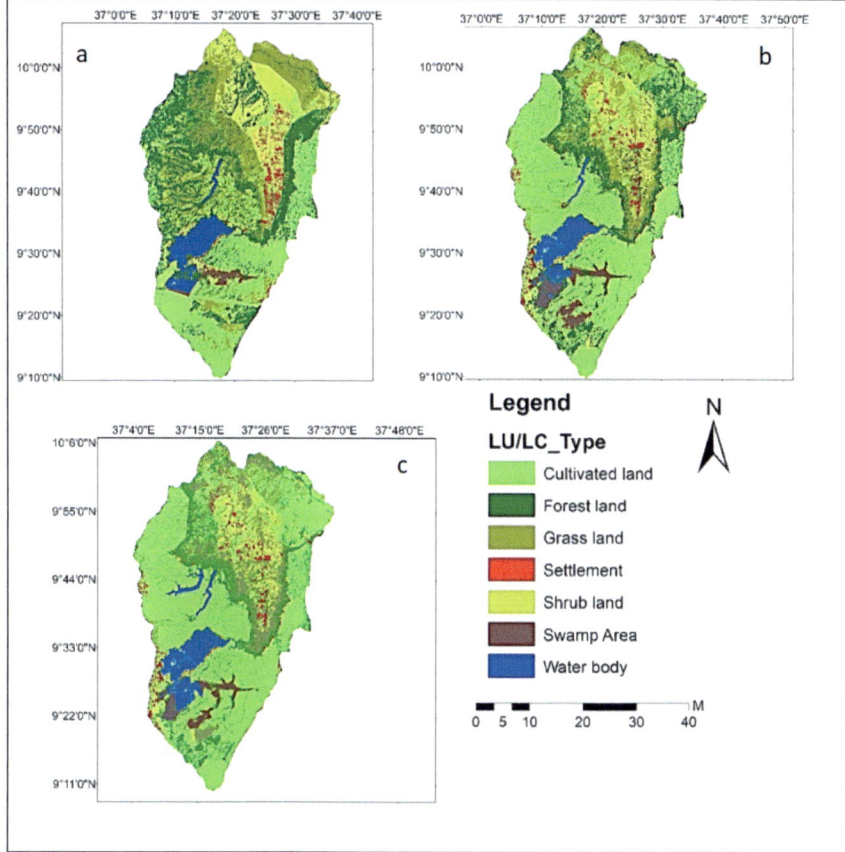

Fig. 10.3 Land use/land cover maps of 1991 (**a**), 2006 (**b**), and 2021 (**c**)

and 338.7 ha/year (0.1%), respectively, were observed. However, forestland, grassland, and shrubland declined by 2880.4 ha/year (0.8%), 1463.0 ha/year (0.4%), and 2557.2 ha/year (0.7%), respectively, from 1991 to 2021. The study results indicated that agricultural land significantly increased at the expense of natural vegetation in the Fincha'a watershed during the last 30 years, mainly due to increasing population. The observed land use/land cover changes during the past three decades in the Fincha'a watershed could lead to environment degradation. In particular, the conversion of natural vegetation to human-modified landscapes such as cultivated land aggravated soil erosion and sedimentation, which could result in significant on-site and off-site negative effects.

To assess the impact of LULCC on soil erosion in the Fincha'a watershed, the Revised Universal Soil Loss Equation (RUSLE) model was run for 1991, 2006, and 2021 separately by changing the C-factor and P-factor (estimated from the corresponding land use/land cover maps of the study periods) and keeping the other factors the same. Accordingly, the results show that the estimated annual soil loss of the Fincha'a watershed ranges from 0.0 t ha^{-1} year^{-1} to 308.7 t ha^{-1} year^{-1} for 1991, from 0.0 t ha^{-1} year^{-1} to 410.9 t ha^{-1} year^{-1} for 2006, and from 0.0 to 492.24 t ha^{-1} year^{-1} for 2021 (Fig. 10.4 and Table 10.4).

The estimated average annual soil loss from the whole study area was 34.5 t ha^{-1} yr^{-1}, 46.4 t ha^{-1} yr^{-1}, and 58.7 t ha^{-1} year^{-1} during 1991, 2006, and 2021, respectively. The results indicated increasing mean annual soil loss from 1991 to 2021 in response to the observed land use/land cover changes during the study period in the Fincha'a watershed. In particular, the conversion of forestland, grassland, and shrubland to cultivated land contributed for the increased soil erosion in the study area.

The results further showed that approximately 291.80 km^2 (7.6%), 400.95 km^2 (10.5%), and 409.69 km^2 (10.7%) of the Fincha'a watershed was categorized as very

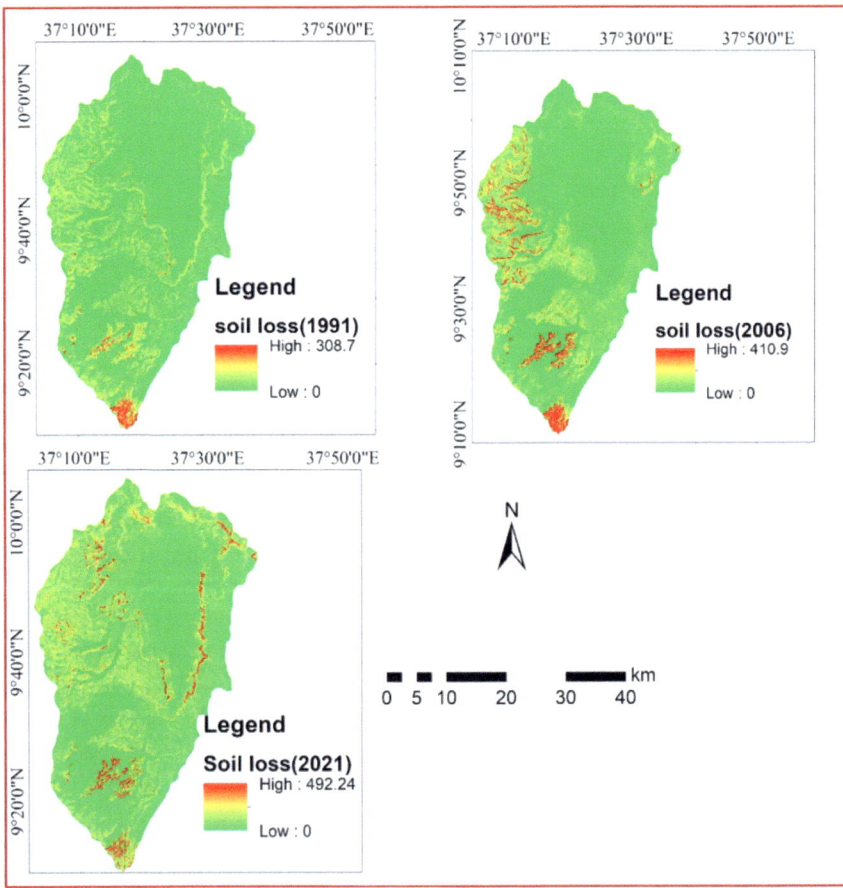

Fig. 10.4 Soil loss map of the 1991, 2006 and 2021 Fincha'a watersheds

Table 10.4 Mean soil loss, areal coverage, and soil erosion severity class for 1991, 2006, and 2021

Severity ranges (t ha^{-1} yr^{-1})	Severity class	Priority class	1991		2006		2021	
			Area (km^2)	Area (%)	Area (km^2)	Area (%)	Area (km^2)	Area (%)
< 10	Low	IV	2040.8	53.3	1854.3	48.4	1774.9	46.3
10–50	Moderate	III	1101.2	28.7	1160.9	30.3	1206.2	31.5
50–100	High	II	397.7	10.4	415.3	10.8	440.7	11.5
> 100	Very high	I	291.8	7.6	400.9	10.5	409.7	10.7
Mean annual soil loss (t ha^{-1} yr^{-1})			34.5		46.4		58.7	

highly eroded during 1991, 2006, and 2021, respectively (Fig. 10.5 and Table 10.4). The results showed that the coverage of very highly eroded areas increased from 1991 to 2021 mainly in response to declining natural vegetation coverage and increasing cultivated land in the Fincha'a watershed.

The study further showed that high soil loss was observed in cultivated land, with mean annual soil losses of 76.0 t ha^{-1} yr^{-1}, 89.0 t ha^{-1} yr^{-1}, and 96.0 t ha^{-1} yr^{-1} during 1991, 2006, and 2021, respectively (Table 10.5). This is because cultivated land is more susceptible to soil erosion relative to the other land cover types, as agricultural activities expose the soil to raindrops and surface runoff.

Likewise, the sediment yield of the Fincha'a watershed was estimated for each LULC map. Accordingly, the results showed that the annual sediment yield of the Fincha'a watershed ranged from 0.0 t ha^{-1} yr^{-1} to 65 t ha^{-1} yr^{-1} during 1991, from 0.0 t ha^{-1} year^{-1} to 89 t ha^{-1} yr^{-1} during 2006, and from 0.0 to 97.6 t ha^{-1} yr^{-1} during 2021. Accordingly, the mean annual sediment yield of the entire watershed was 6.7 t ha^{-1} yr^{-1}, 8.5 t ha^{-1} yr^{-1}, and 10.3 t ha^{-1} yr^{-1} during 1991, 2006, and 2021, respectively. The results indicated that sediment yield increased from 1991 to 2021 in line with increasing agricultural lands and declining natural vegetation. The increased sediment yield in response to the substantial LULCC could negatively affect the downstream reservoirs and irrigation infrastructure and croplands.

Conclusions

Land use/land cover is an important environmental variable that determines the flow of energy and material between the terrestrial land surface and the atmosphere. As such, the change in LULC influences climate system, ecosystem services, and natural resources availability. The drivers of LULCC are driven by natural and anthropogenic factors. Therefore, assessing the changes, drivers and implications of LULCC is vital for designing appropriate land management strategies and minimizing its negative consequences. The Abbay River is one of the transboundary rivers in Ethiopia,

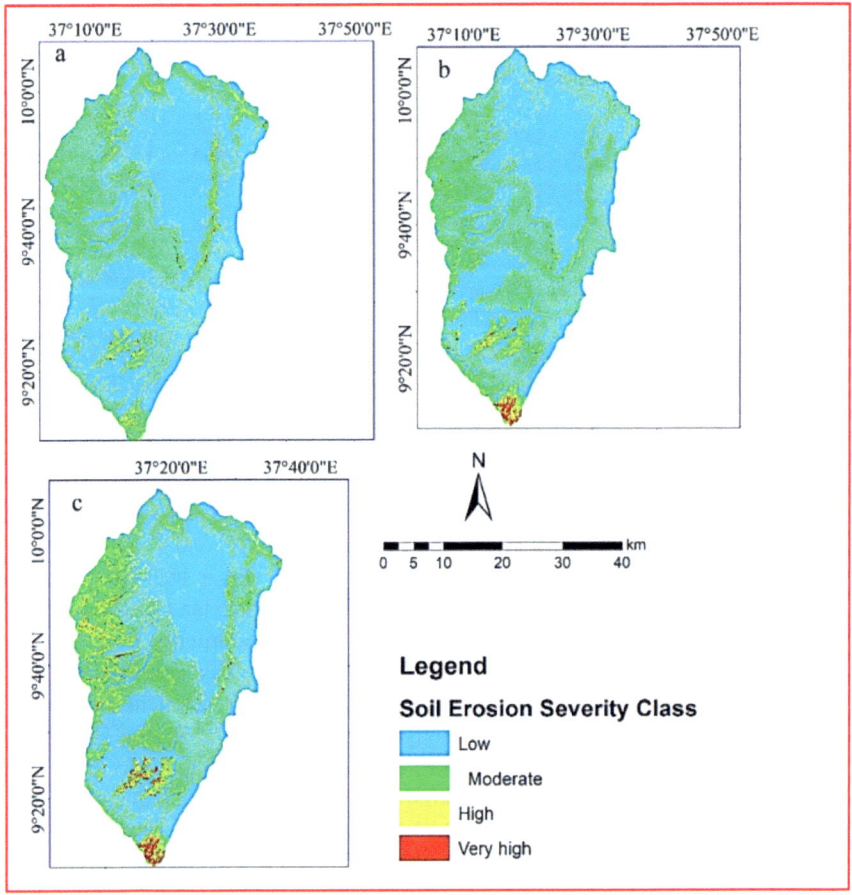

Fig. 10.5 Soil loss severity classes for 1991 (**a**), 2006 (**b**), and 2021 (**c**)

Table 10.5 Mean soil erosion rates across different LULC classes in the Fincha'a watershed

LU/LC type	1991		2006		2021	
	Area (km^2)	Mean soil loss t ha^{-1} yr^{-1}	Area (km^2)	Mean soil loss t ha^{-1} yr^{-1}	Area (km^2)	Mean soil loss t ha^{-1} yr^{-1}
Cultivated land	1071.1	76	1912.6	89	2019.6	96
Settlement	78.1	16.9	95.0	18	97.9	33.2
Shrubland	656.4	15.2	280.0	17.3	272.8	29.1
Grassland	764.53	16.2	554.6	20.2	545.1	25
Forestland	962.6	6.7	643.5	7.3	530.5	9
Swamp area	115.4	0.9	154.0	0.3	131.4	1.3
Water body	183.41	0	192.1	0.2	234.2	0.1

supports millions of people, is characterized by diverse biophysical and socioeconomic environments and reaches with diverse natural resources such as land, vegetation, genetic diversity and water. However, the basin has been experiencing substantial spatiotemporal LULCC during the past several years. Most of the LULCC studies conducted in the basin showed increasing agricultural land, bare land, built-up areas, and waterbodies and decreasing forestland, bush/shrubland, and grassland during the last couple of decades. The major drivers of the change are anthropogenic factors such as population pressure, unwise land utilization, lack of awareness, government-sponsored resettlement programs, weak land use policy, and socioeconomic activities such as large-scale agricultural investment and construction of reservoirs.

The human-induced LULCC observed in the Abbay basin over several decades has resulted in negative consequences on the environment and natural resource availability. Several studies have shown that LULCC, particularly the conversion of natural landscapes to human-dominated landscapes, aggregate soil erosion, and sediment yield, leading to land degradation and siltation of downstream reservoirs. Moreover, human-induced LULCC alters hydrological processes, decreases ecosystem services, and affects the socioeconomic activities of the local community and the country at large. Generally, the observed LULCC in the Abbay basin influences the environmental, social, and economic factors of the basin and the country in general. Therefore, proper land use planning, wise land resource utilization, and community-based soil and water conservation could minimize LULCC and their negative consequences.

Acknowledgements We would like to acknowledge Space Science and Geospatial Institute for creating conducive working environment for writing this book chapter. We are also grateful to reviewers and editor in chief for their constructive and useful comments.

References

Abebe T, Gebremariam B (2019) Modelling runoff and sediment yield of Kesem dam watershed, Awash basin, Ethiopia. SN Appl Sci 1:44

Alemayehu F, Tolera M, Tesfaye G (2019) Land use land cover change trend and its drivers in Somodo watershed south western, Ethiopia. Afr J Agric Res 14:102–117

Amini S, Saber M, Rabiei-Dastjerdi H, Homayouni S (2022) Urban land use and land cover change analysis using random forest classification of landsat time series. Remote Sens 14:2654. https://doi.org/10.3390/rs14112654

Anderson JR (1976) A land use and land cover classification system for use with remote sensor data. U.S. Government Printing Office, Washington, DC, USA

Aneseyee AB, Elias E, Soromessa T, Feyisa GL (2020) Land use/land cover change effect on soil erosion and sediment delivery in the Winike watershed, Omo Gibe Basin, Ethiopia. Sci Total Environ. https://doi.org/10.1016/j.scitotenv.2020.138776

Assefa WW, Eneyew BG, Wondie A (2021) The impacts of land-use and land-cover change on wetland ecosystem service values in peri-urban and urban area of Bahar Dar City, upper Blue Nile Basin, North western Ethiopia. Ecol Process 10:39

Bekele B, Wu W, Yirsaw E (2019) Drivers of land use-land cover changes in the Central Rift Valley of Ethiopia. Sains Malaysiana 48(7):1333–1345

Belay T, Mengistu DA (2019) Land use and land cover dynamics and drivers in the Muga watershed, Upper Blue Nile basin, Ethiopia. Remote Sens Appl: Soc Environ.https://doi.org/10.1016/j.rsase.2019.100249

Belay T, Mengistu DA (2021) Impacts of land use/land cover and climate changes on soil erosion in Muga watershed, Upper Blue Nile basin (Abay), Ethiopia. Ecol Process 10:68. https://doi.org/10.1186/s13717-021-00339-9

Belete AK (2015) Integration of Acacia decurrens (J.C. Wendl.) Willd. into the farming system, its effects on soil fertility and comparative economic advantages in North-Western Ethiopia. MSc Thesis. Bahir Dar Univ., Ethiopia

Berihun ML, Tsunekawa A, Haregeweyn N, Meshesha DT, Adgo E, Tsubo M, Masunaga T, Fenta AA, Sultan D, Yibeltal M (2019) Exploring land use/land cover changes, drivers and their implications in contrasting agro-ecological environments of Ethiopia. Land Use Policy 87:104052

Betru T, Tolera M, Sahle K, Kassa H (2019) Trends and drivers of land use/land cover change in Western Ethiopia. Appl Geogr 104:83–93

Birhanu SY, Moges MA, Sinshaw BG, Tefera AK, Atinkut HB, Fenta HM, Berihun ML (2022) Hydrological modelling, impact of land-use and land-cover change on hydrological process and sediment yield: case study in Jedeb and Chemoga watersheds. Energy Nexus 5:100051

Desalegn T, Cruz F, Kindu M, Turrión MB, Gonzalo J (2014) Land-use/land-cover (LULC) change and socioeconomic conditions of local community in the central highlands of Ethiopia. Int J Sustain Dev World Ecol 21:406–413

Dibaba WT, Demissie TA, Miegel K (2020) Drivers and implications of land use/land cover dynamics in Finchaa Catchment, Nortwestern Ethiopia. Land 9:113

Erb K-H, Luyssaert S, Meyfroidt P, Pongratz J, Don A, Kloster S, Kuemmerle T, Fetzel T, Fuchs R, Herold M et al (2017) Land management: data availability and process understanding for global change studies. Glob Chang Biol 23:512–533

Ewers RM (2006) Interaction effects between economic development and forest cover determine deforestation rates. Glob Environ Chang 16:161–169

Ewunetu A, Simane B, Teferi E, Zaitchik BF (2021) Land cover change in the Blue Nile River headwaters: farmers' perceptions, pressures, and satellite-based mapping. Land 10:68

FDRE (2011) Ethiopia's climate-resilient green economy strategy document. Federal Democratic Republic of Ethiopia, Addis Ababa, p 200

Fenta AA, Tsunekawa A, Haregeweyn N, Tsubo M, Yasuda H, Kawai T, Ebabu K, Berihun ML, Belay AS, Sultan D (2021) Agroecology-based soil erosion assessment for better conservation planning in Ethiopian river basins. Environ Res 195:110786

Fikirte D, Kumelachew Y, Mengistie K, Thomas S (2017) Land use/Land cover changes and their causes in Libokemkem District of South Gonder, Ethiopia. Remote Sens Appl: Soc Environ 8:224–230

Gashaw T, Tulu T, Argaw M, Worqlul AW, Tolessa T, Kindu M (2018) Estimating the impacts of land use/land cover changes on ecosystem service values: the case of the Andassa watershed in the upper Blue Nile basin of Ethiopia. Ecosyst Serv 31:219–228

Getachew B, Manjunatha BR (2022) Impacts of land use change on the hydrology of lake Tana basin, upper Blue River basin, Ethiopia. Global Challenges 6:2200041

Getachew HE, Melesse AM (2012) Impact of land use /land cover change on the hydrology of Angereb Watershed, Ethiopia. Int J Water Sci 1(4):1–7. https://doi.org/10.5772/56266

Han J, Ge W, Hei Z, Cong C, Ma C, Xie M, Liu B, Feng W, Wang F, Jiao J (2020) Agricultural land use and management weaken the soil erosion induced by extreme rainstorms. Agric Ecosyst Environ 301:107047

Haregeweyn N, Tesfaye S, Tsunekawa A, Tsubo M, Meshesha DT, Adgo E, Elias A (2014) Dynamics of land use and land cover and its effects on hydrologic responses: case study of the Gilgel Tekeze catchment in the highlands of Northern Ethiopia. Environ Monit Assess 187:1–14

Haregeweyn N, Tsunekawa A, Tsubo M, Meshesha D, Adgo E, Poesen J, Schütt B (2016) Analysing the hydrologic effects of region-wide land and water development interventions: a case study of the Upper Blue Nile basin. Reg Environ Change 16:951–966

Haregeweyn N, Tsunekawa A, Poesen J, Tsubo M, Meshesha DT, Fenta AA, Nyssen J, Adgo E (2017) Comprehensive assessment of soil erosion risk for better land use planning in river basins: case study of the upper Blue Nile River. Sci Total Environ, 574:95–108. https://doi.org/10.1016/j.scitotenv.2016.09.019

Hasselmann F, Csaplovics E, Falconer I, Bürgi M, Hersperger AM (2010) Technological driving forces of LUCC: conceptualization, quantification, and the example of urban power distribution networks. Land Use Policy 27(2):628–637

Hu Y, Dong Y, Batunacun (2018) An automatic approach for land-change detection and land updates based on integrated NDVI timing analysis and the CVAPS method with GEE support. ISPRS J Photogramm Remote Sens 146:347–359

Hu Y, Batunacun, Zhen L, Zhuang D (2019) Assessment of land use and land cover change in Guangxi, China. Sci Rep 9:2189.https://doi.org/10.1038/s41598-019-38487w

Hurni H (1988) Degradation and conservation of the resources in the Ethiopian highlands. Mt Res Dev 8:123–130

Hurni H, Tato K, Zeleke G (2005) The implications of changes in population, land use, and land management for surface runoff in the Upper Nile Basin Area of Ethiopia. Mt Res Dev 25:147–154

Hurni K, Zeleke G, Kassie M, Tegegne B, Kassawmar T, Teferi E, Moges A, Tadesse D, Ahmed M, Degu Y, Kebebew Z, Hodel E, Amdihun A, Mekuriaw A, Debele B, Deichert G, Hurni H (2015) Soil degradation and sustainable land management in the rainfed agricultural areas of Ethiopia: an assessment of the economic implications. In: Economics of Land Degradation (ELD) Ethiopia case study report for the economics of land degradation initiative, p 94

Kabite G, Muleta MK, Gessesse B (2020) Spatiotemporal land cover dynamics and drivers for the Dhidhessa River basin (DRB), Ethiopia. Model Earth Syst Environ. https://doi.org/10.1007/s40808-020-00743-8

Kabite GW, Muleta MK, Gessesse BA (2022) Impacts of combined and separate land cover and climate changes on hydrologic responses of Dhidhessa River basin, Ethiopia. Int J River Basin Manage. https://doi.org/10.1080/15715124.2022.2101464

Kanianska R, Kizekova M, Novacek J, Zeman M (2014) Land-use and land-cover changes in rular areas during different political systems: a case study of Slovakia from 1782 to 2006. Land Use Policy 36:554–566

Kassawmar T, Zeleke G, Bantider A, Gessesse GD, Abraha L (2018) A synoptic land change assessment of Ethiopia's Rainfed Agricultural Area for evidence-based agricultural ecosystem management. Heliyon 4:e00914

Kayitesi NM, Guzha AG, Mariethoz G (2022) Impacts of land use land cover change and climate change on river hydromorphology—a review of research studies in tropical regions. J Hydroology 615:128702

Kidane M, Bezie A, Kesete N, Tolessa T (2019). The impact of land use and land cover (LULC) dynamics on soil erosion and sediment yield in Ethiopia, Heliyon 5(12): e02981.https://doi.org/10.1016/j.heliyon.2019.e02981

Lambin EF et al (2001) The causes of land-use and land-cover change: moving beyond myths. Glob Environ Chang 11:261–269

Lambin EF, Geist HJ, Lepers E (2003) Dynamics of land-use and land-cover change in tropical regions. Annu Rev Environ Resour 28:205–241

Lemma H, Admasu T, Dessie M, Fentie D, Deckers J, Frankl A, Poesen J, Adgo E, Nyssen J (2018) Revisiting lake sediment budgets: how the calculation of Environ. 574:95–108

Lin X, Xu M, Cao C, Singh RP, Chen W (2018) Land-use/land-cover changes and their influence on the ecosystem in Chengdu City, China during the period of 1992–2018. Sustainability 10:3580

Malede DA, Alamirew T, Andualem TG (2023) Integrated and individual impacts of land use land cover and climate changes on hydrological flows over Birr River Watershed, Abbay Basin, Ethiopia. Water 15:166. https://doi.org/10.3390/w15010166

Mango L, Melesse AM, McClain ME, Gann D, Setegn SG (2011a) Land use and climate change impacts on the hydrology of the upper Mara River Basin, Kenya: results of a modeling study to support better resource management. Special Issue: Climate, weather and hydrology of East African Highlands, Hydrol. Earth Syst Sci 15:2245–2258. https://doi.org/10.5194/hess-15-2245-2011

Mango L, Melesse AM, McClain ME, Gann D, Setegn SG (2011b) Hydro-meteorology and water budget of Mara River basin, Kenya: a land use change scenarios analysis, In: Melesse A (ed) Nile River Basin: hydrology, climate and water use. Springer Science Publisher, pp 39–68. https://doi.org/10.1007/978-94-007-0689-7_2

Mariye M, Mariyo M, Changming Y, Teffera ZL, Weldegebrial B (2020) Effects of land use and land cover change on soil erosion potential in Berhe district: a case study of Legedadi watershed, Ethiopia. Int J River Basin Manag 1–13

Mathewos M, Lencha SM, Tsegaye M (2022) Land use and land cover change assessment and future predictions in the Matenchose Watershed, Rift Valley Basin, using CA-Markov simulation. Land 11:1632. https://doi.org/10.3390/land11101632

Melesse AM, Jordan JD (2002) A comparison of fuzzy vs augmented-ISODATA classification algorithm for cloud and cloud-shadow discrimination in landsat imagery. Photogram Eng Remote Sens 68(9):905–911

Melesse AM, Jordan JD (2003) Spatially distributed watershed mapping and modeling: land cover and microclimate mapping using landsat imagery part 1. J Spat Hydrol (e-journal) 3(2)

Melesse A, Weng Q, Thenkabail P, Senay G (2007) Remote sensing sensors and applications in environmental resources mapping and modeling. Sensors 7:3209–3241

Mewded M, Abebe A, Tilahun S, Agide Z (2021) Impact of land use and land cover change on the magnitude of surface runoff in the endorheic Hayk Lake basin, Ethiopia. SN Appl Sci 3:742. https://doi.org/10.1007/s42452-021-04725-y

Moges DM, Bhat HG (2017) Integration of geospatial technologies with RUSLE for analysis of land use/cover change impact on soil erosion: case study in Rib watershed, north western highland Ethiopia. Environ Earth Sci 76:765

Mohammed H, Alamirew A, Assen M, Melesse A (2013) Spatiotemporal mapping of land cover in Lake Hardibo Drainage Basin, Northeast Ethiopia: 1957–2007. In: Water conservation: practices, challenges and future implications. Nova Publishers, pp 147–164

Moisa MB, Negash DA, Merga BB, Gemeda DO (2021) Impact of land-use and land-cover change on soil erosion using the RUSLE model and the geographic information system: a case of Temeji watershed, Western Ethiopia geographic information systems. J Water Clim Change 12:3404–3420

Msofe NK, Sheng L, Lyimo J (2019) Land use change trends and their driving forces in the Kilombero Valley Floodplain, Southeastern Tanzania. Sustainability 11:505

Oliver TH, Morecroft MD (2014) Interactions between climate change and land use change on biodiversity: attribution problems, risks, and opportunities. Wiley Interdiscip Rev Clim Chang 5:317–335

Rabiei-Dastjerdi H, Amini S, McArdle G, Homayouni S (2022) City-region or city? That is the question: modelling sprawl in Isfahan using geospatial data and technology. GeoJournal 1–21

Ran YH, Li X, Lu L, Li ZY (2012) Large-scale land cover mapping with the integration of multi-source information based on the Dempster-Shafer theory. Int J Geogr Inf Sci 26:169–191

Regasa MS, Nones M (2022) Past and future land use/land cover changes in the Ethiopian Fincha SubtoBasin. Land 11(8):1239

Regasa MS, Nones M, Adeba D (2021) A review on land use and land cover change in Ethiopian Basin. Land 10:585. https://doi.org/10.3390/land10060585

Serra P, Pons X, Saur D (2008) Land-cover and land-use change in a Mediterranean landscape: a spatial analysis of driving forces integrating biophysical and human factors. Appl Geogr 28(3):189–209

Shi Y, Wang R, Fan L, Li J, Yang D (2010) Analysis on land use change and its demographic factors in the original-stream watershed of Tarim River based on GIS and statistic. Procedia Environ Sci 2:175–184

Shi G, Jiang N, Yao L (2018) Land use and cover change during the rapid economic growth period from 1990 to 2010: a case study of Shanghai. Sustainability 10:426

Shiferaw H, Bewket W, Alamirew T, Zeleke G, Teketay D, Bekele K, Schaffner U, Eckert S (2019) Implications of land use/land cover dynamics and Prosopis invasion on ecosystem service values in Afar Region, Ethiopia. Sci Total Environ 675:354–366

Sisay G, Gitima G, Mersha M, Alemu WG (2021) Assessment of land use land cover dynamics and its drivers in Bechet Watershed upper Blue Nile basin, Ethiopia. Remote Sens Appl: Soc Environ 24(2021):100648

Sultan D, Tsunekawa A, Haregeweyn N, Adgo E, Tsubo M, Meshesha DT, Masunaga T, Aklog D, Ebabu K (2017) Analysing the runoff response to soil and water conservation measures in a tropical humid Ethiopian highland. Phys Geogr 38:423–447

Tadesse L, Suryabhagavan KV, Sridhar G, Legesse G (2017) Land use and land cover changes and Soil erosion in Yezat Watershed, North Western Ethiopia. Int Soil Water Conserv Res 5(2):85–94

Teferi E, Bewket W, Uhlenbrook S, Wenninger J (2013) Understanding recent land use and land cover dynamics in the source region of the Upper Blue Nile, Ethiopia: spatially explicit statistical modelling of systematic transitions. Agr Ecosyst Environ 165:98–117

Teshome DS, Moisa MB, Gemeda DO, You S (2022) Effect of land use-land cover change on soil erosion and sediment yield in Muger Sub-Basin, Upper Blue Nile Basin, Ethiopia. Land 11:2173. https://doi.org/10.3390/land11122173

Tolessa T, Senbeta F, Kidane M (2017) The impact land use/land cover change on ecosystem services in the central highlands of Ethiopia. Ecosyst Serv 23:47–54

Turner BL, Moss RH, Skole DL (eds) (1993) Relating land use and global land-cover change: a proposal for an IGBP-HDP core project. Report from the IGBP-HDP working group on land-use/land-cover change. Joint publication of the International Geosphere-Biosphere Programme (Report No. 24) and the Human Dimensions of Global Environmental Change Programme (Report No. 5). International Geosphere-Biosphere Program (IGBP) Secretariat, Royal Swedish Academy of Sciences, Stockholm

Valentin C, Poesen J, Li Y (2005) Gully erosion: impacts, factors and control. CATENA 63:132–153

Wassie SB (2020) Natural resource degradation tendencies in Ethiopia: a review. Environ Syst Res 9:33

Welde K (2016) Identification and prioritization of subwatersheds for land and water management in Tekeze dam watershed, Northern Ethiopia. Int Soil Water Conserv Res 4:30

Woldesenbet TA, Elagib NA, Ribbe L, Heinrich J (2017) Hydrological responses to land use/land cover changes in the source region of the upper Blue Nile Basin Ethiopia. Sci Total Environ 575:724–741

Xiao J, Shen Y, Ge J, Tateishia R, Tanga C, Liang Y, Huang Z (2006) Evaluating urban expansion and land-use change in Shijiazhuang, China by using GIS and remote sensing. Landscape Urban Plan 75(1–2):69–80

Xie Y, Mei Y, Guangjin T, Xuerong X (2005) Socioeconomic driving forces of arable land conversion: a case study of Wuxian City China. Glob Environ Change 15(3):238–252

Wondie M, Schneider W, Melesse AM, Teketay D (2011) Spatial and temporal land cover changes in the Simen Mountains National Park, a world heritage site in Northwestern Ethiopia. Remote Sens 3:752–766. https://doi.org/10.3390/rs3040752

Wondie M, Teketay D, Melesse AM, Schneider W (2012) Relationship between topographic variables and land cover in the Simen Mountains National Park, a World Heritage Site in northern Ethiopia. Int J Remote Sens Appl 2(2):36–43

Zeleke G, Hurni H (2001) Implications of land use and land cover dynamics for mountain resource degradation in the northwestern Ethiopian highlands. Mt Res Dev 21:184–191

Chapter 11
Soil Erosion and Sediment Yield Status of the Abbay Basin

Gizachew Kabite Wedajo, Berhan Gessesse, Worku Zewdie, Wubetu Anley, and Seyoum Eshetie

Abstract With a total area of about 173,000 km^2 and contributing roughly 60% of streamflow to the Nile basin, the Abbay Basin is one of the significant transboundary rivers for both upstream and downstream countries. However, rugged topography, poor land management practices and high erosive rainfall, and an agricultural dominated land use system made the basin highly vulnerable to surface runoff and soil erosion that negatively affect agricultural production and water resources availability. Land management strategies have been planned and implemented by the Ethiopian government in the upstream part of the Abbay Basin for the last couple of decades to minimize the effects of soil erosion. Nevertheless, the effectiveness of these approaches is not precisely known and quantified. Therefore, this review was intended to assess approaches of soil erosion and sediment yield estimation, the status and extent of soil erosion and sediment yield, and the effectiveness of land management strategies being implemented in the Abbay Basin. The results highlight that empirical models like the Revised Universal Soil Loss Equation (RUSLE), Sediment Delivery ration (SDR), and Soil and Water Assessment Tools (SWATs) are the widely used models for estimating soil loss and sediment yield in the Abbay Basin. Furthermore, the results of the studies showed that soil erosion remains severe in the basin, despite several government-sponsored land management strategies. This could be partly attributed to inappropriate design of the soil and water conservation (SWC) practices, lack of considering local conditions, lack of continuous monitoring, and top-down approaches. Therefore, for the effectiveness of the land management practices, proper planning and design, continuous maintenance, community-based sustainable land management, and integration of different soil and water conservation (SWC) practices are highly required. Moreover, the soil erosion and sediment yield status of the basin should be estimated more precisely using improved state-of-the-art machine learning algorithms.

Keywords Soil erosion · Sediment yield · Land management · Abbay Basin · Ethiopia

---

G. K. Wedajo (✉) · B. Gessesse · W. Zewdie · W. Anley · S. Eshetie
Remote Sensing Department, Ethiopian Space Science and Geospatial Institute, Addis Ababa, Ethiopia
e-mail: kabiteg@gmail.com

Introduction

Abbay Basin is the major tributary of the Nile River with a total drainage area of about 173,000 km^2 (Conway, 2000). The basin supports more than 17 million people and contributes about 60% of the Nile's flow at Khartoum, Sudan (Sutcliffe and Parks 1999; Conway 2000). The basin has ample arable land and multiple water resources development projects designed for irrigation and hydroelectric power (e.g., Tana Beles and Arjo-Dhidhessa, Grand Ethiopian Reminiscence Dam (GERD)) (Berihun et al. 2022). The GERD will be the largest hydroelectric Dam in Africa when it is completed. Moreover, many dams are currently operational in downstream countries like Sudan and Egypt. As such, the Abbay River is a crucial river for upstream and downstream users for irrigation, domestic use, industry, hydropower, and ecosystem services (Dile et al. 2018). However, the competing water use, climate change, poor land management, and increasing population pressure challenge water resources availability in the basin (Hurni et al. 2005). Particularly, poor land management and human-induced land use/land cover changes (LULCC) accelerated soil erosion and sediment yield productions resulting in negative consequences on agricultural production and the water resources availability as on-and off-site effects (Berihun et al. 2022).

Soil erosion by water is one of the most common environmental challenges that lead to soil loss, declining land productivity, agricultural land loss, and declining water resources quality and quantity (Tilahun et al. 2014). Studies used different approaches to estimate soil erosion and sediment dynamics in various regions (Maalim and Melesse 2013; Maalim et al. 2013; Setegn et al. 2010; Setegn et al. 2009; Mohammed et al. 2015).

The Ethiopian highlands like the Abbay Basin are highly vulnerable to soil erosion due to rugged topography, erosive rainfall, high population pressure, poor land management, deforestation, and agriculture-dominated land use types (Hurni et al. 2005; Haregeweyn et al. 2017). A sediment yield study performed in the basin showed that about 95% of the sediment that the Nile River receives comes from the highlands of Ethiopia (Ahmed and Ismail 2008). Moreover, agricultural lands of the highland areas, where Abbay Basin is located, experienced soil loss that is beyond the tolerable soil loss (TSL) rate, which ranges from 5 to 11 tons per hectares per year (t ha^{-1} yr^{-1}) (Moges and Bhat 2017; Yesuph and Dagnew 2019; Ebabu et al. 2019). The soil loss rate of the Abbay Basin ranges from 37 t ha^{-1} yr^{-1} to 246 t ha^{-1} yr^{-1} (Ebabu et al. 2019). Similarly, Yesuph and Dagnew (2019) reported a mean annual soil loss rate of about 37 t ha^{-1} yr^{-1} in the Beshillo catchment, which is more than three times the maximum TSL value. On the other hand, the mean soil loss rates of the Abbay Basin were 27.5 t ha^{-1} yr^{-1} and the soil loss rates of about 39% of the basin were > 30 t ha^{-1} yr^{-1}, indicating that the basin experienced severe to very severe soil erosion risks (Haregeweyn et al. 2017). The mean annual soil loss from Rib watershed of the Abbay Basin was about 26 t ha^{-1} yr^{-1}, which is beyond the TSL (Sinshaw et al. 2021). Moreover, 90 t ha^{-1} yr^{-1} soil erosion rates were reported for the Chemoga watershed (Bewket and Teferi 2009) and 47 t ha^{-1} yr^{-1} for the

Koga watersheds (Gelagay and Minale 2016). To reduce soil erosion and sediment yield, land management strategies have been implemented in different parts of the country and Abbay Basin in particular for the last couple of decades.

The government of Ethiopia has been designing various soil and water conservation and management strategies at different times. For example, the government implemented a 15-year strategy to combat land degradation and build a climate-resilient green economy by 2025 (FDRE 2011). The strategy includes implementing participatory soil and water conservation practices to minimize soil erosion and sedimentation. Accordingly, soil and water conservation measures have been practiced in different parts of the country including in the Abbay Basin, particularly since the implementation of Sustainable Land Management Project (SLMP) in 2008 that targeted about 135 watersheds (SLMP 2013). However, this project has not been effectively implemented in the Abbay Basin partly since the watersheds were not selected based on prioritized conservation needs but rather based on their size (Haregeweyn et al. 2017). In addition, the National Green Legacy program was launched in 2019 as means of climate change adaptation strategies and combat environmental degradation (Woldegiorgis 2020). In this program, it was planned to cover 1.5 million hectares by planting more than 4 billion trees. However, the effectiveness and the impacts of these management programs on soil erosion loss are not well known in the country in general and in the Abbay Basin in particular except for a few studies at small watershed and plot levels.

Studies reported the effectiveness of land management practices in minimizing soil loss and sediment yield (Yibeltal et al. 2019; Gashaw et al. 2020; Gizaw et al. 2021; Berihun et al. 2022). For instance, in a study done in six catchments located in the Abbay Basin with contrasting agro-ecology, catchments with soil and water conservation practices experienced lower gully densities compared with catchments with no conservation activities (Yibeltal et al. 2019). A systematic literature review in Ethiopia showed that land management played a positive role in minimizing soil loss and increasing crop yield (Gizaw et al. 2021). Berihun et al. (2022) reported integrating suitable land use and land management practices which reduced runoff and sediment yield by 72% and 95%, respectively, in the Laguna watershed of the Abbay Basin. In addition, a study done in Jamma watershed indicated that soil and water conservation practices implemented at watershed and plot levels minimized soil erosion and sediment yield. Similarly, Lemann et al. (2016) reported that the implementation of SWC in the Gerda catchment of the Abbay Basin reduced sediment yield from 37 t ha^{-1} to 17 t ha^{-1} (30%). Moreover, Melaku et al. (2018) reported a 28–38% reduction in sediment yield due to SWC practices in the Gumara-Maksegnit watershed of the Abbay Basin.

However, the study indicated that the effectiveness of the conservation practices declines with time due to lack of maintenance and poor design (Yaekob et al. 2022). Moreover, Ebabu et al. (2023) reported that seasonal soil loss from watersheds with soil and water conservation in the highlands of Ethiopia in the Abbay Basin remains higher than the TSL rates. The result indicated that local conditions like land use and climatic conditions should be carefully considered for designing soil and water conservation practices in a given area. In addition, proper planning and design are vital for the effectiveness of land management in controlling soil loss. Belayneh et al. (2020) reported that graded soil bunds significantly minimized soil loss and sediment yield, but the effectiveness of the structure has been highly reduced with time. However, soil loss from the treated plots in this area is still higher than the TSL (> 11 t ha^{-1} yr^{-1}).

Similarly, despite the implementation of various soil and water conservation practices in the upstream catchments, most reservoirs in the downstream areas are highly silted (Haregeweyn et al. 2015). This is because the implementation of soil and water conservation measures in most watersheds was without proper design and without considering existing biophysical conditions like land cover and topography. For example, the nine years of an experimental study conducted in the Debre Mawi watershed indicated that the government-sponsored SWC reduced soil loss and runoff, but it increased risks of gully formation in the downstream areas due to greater saturation and soil weakening (Mhiret et al. 2020). If soil and water conservation practices are properly designed and maintained, it would reduce surface runoff by 50% and sediment yield by 85% (Schmidt and Zemadim 2015). Therefore, for the effectiveness of the land management measures, proper planning and design, continuous maintenance and integration of different SWC practices are highly required. However, the implementation and continuous maintenance of the practices are labor-intensive and negatively affect other agricultural activities. Moreover, a study done in three catchments of the Abbay basin showed that livelihood diversification negatively affected the adoption of the large number of SLM. As such, the livelihood portfolios of rural households should be considered for the effective implementation of SLM (Abeje et al. 2019). As such, policy measures are required for incentivizing the implementation and maintenance of sustainable land management measures. Hence, this review evaluates the effectiveness of various land management practices in the basin for reducing soil erosion and sediment yield.

Methods

Literature Search

Up to date literatures used for this review book chapter were collected from an online resource using "keywords" from major databases such as Web of sciences, Scopus, and Google scholar and major publishers that provide access to scientific research like ScienceDirect, Springer, and MDPI. During the search, we used keywords like 'soil erosion', 'sediment yield', 'upper Blue Nile basin', 'Abay basin', 'Abbay Basin', 'on-site and off-site soil erosion effect', 'land management', 'effectiveness of land management', 'soil and water conservation'. The search mainly focused on soil erosion and sediment estimation techniques, the on-site and off-site effects of soil erosion, erosion and sediment yield status of the Abbay Basin, and the effectiveness of land management practices in the basin. In this search, more than 300 articles were identified. Further screening was made after the first initial search results based on the following criteria: (1) study conducted in the Abbay Basin; (2) study conducted in the last two decades (from 2010 to 2023); (3) articles published outside the Abbay Basin but that deals with techniques of soil erosion and sediment yield estimations, and (4) peer-reviewed articles. Accordingly, duplicated articles were excluded, and finally, about 140 articles were collected for this review chapter.

Techniques of Soil Erosion Estimation

Soil erosion, which is accelerated by climate change and human activity, is resulted in negative consequences on the environment and socio-economy of a country. As such, accurately estimating soil loss is important for designing appropriate soil erosion control strategies. Soil erosion loss is widely estimated for the Abbay Basin using various techniques like Modified Universal Soil Loss Equation (MUSLE) and the Universal Soil Loss Equation, Revised Universal Soil Loss Equation (RUSLE) (Tsige et al. 2022).

The Universal Soil Loss Equation (USLE)

The USLE was developed by Wischmeier and Smith (1965) and modified with data from rainfall simulators, runoff plots, and field experiences. The USLE model is one of the most commonly used techniques for estimating soil erosion from agriculture-dominated watersheds that have been used for planning erosion control practices. The model estimates long-term average soil loss from sheet and rill erosions from a specified land conditions by considering multiple factors like soil erodibility, rainfall erosivity, slope steepness, slope length, support practice, and cover management. However, this model is restricted to estimate the mean annual soil loss for a given area and does not show the processes of soil erosion (Gelagay and Minale 2016). Moreover, the model does not consider gully erosion and deposition. Despite these limitations, few studies employed the model for estimating soil erosion for the Abbay Basin (Girmay et al. 2020; Duguma 2022). For example, Girmay et al. (2020) estimated soil erosion loss in Agewmariam watershed, Abbay Basin, using USLE and reported mean annual soil loss of about 25 t ha^{-1} yr^{-1} and more than 33% of the study areas with above TSL (> 11 t ha^{-1} yr^{-1}).

Revised Universal Soil Loss Equation (RUSLE)

The RUSLE model is the modified version of the USLE model. The modification includes an improved theory describing the fundamental hydrologic and erosion processes, improved determination of R-factor, consideration of rock outcrops and considering three separate slope length relationships (Kenneth et al. 1991). As such, RUSLE model is the most widely used technique for estimating long-term rates of soil loss due to sheet and rill erosion. The model estimated long-term soil loss by considering six primary parameters, namely soil erodibility (K), rainfall erosivity (R), cover management (C), slope length and steepness (LS), and support practice (P). The RUSLE equation is described as Eq. 11.1 (Renard and Ferreira 1993).

$$A = R * K * LS * C * P, \tag{11.1}$$

where A is mean annual soil loss (t ha^{-1} yr^{-1}), R is rainfall and runoff erosivity factor (MJ mm ha^{-1} yr^{-1}), K is soil erodibility factor, LS is topographic factor, C is crop management factor, and P is supporting conservation practice factor.

Each factor can be calculated using an empirical formula by considering local factors. Hurni (2005) customized each empirical formula for the Ethiopian condition. Accordingly, the RUSLE model has been widely employed for estimating soil erosion in Ethiopia in general and the Abbay Basin in particular (Endalamaw et al. 2021; Legese et al. 2022). For example, Legese et al. (2022) quantified soil erosion in the Megech catchment, which is located in the Abbay Basin by integrating the geospatial technologies and RUSLE model. Moreover, Endalamaw et al. (2021) estimated the mean annual soil loss rate for the Gilgel Belles watershed by integrating the RUSLE model and Multi-Criteria Evaluation (MCE). The simplicity, low data requirement, and ability to be customized anywhere nature of the model attributed to the wide applicability of the model everywhere in the world including in Ethiopia and the Abbay Basin.

Modified Universal Soil Loss Equation (MUSLE)

The MUSLE model is the modified version of the USLE in which the rainfall energy factor was substituted with the runoff factor (Williams 1975) (Eq. 11.2). The model assumed that the potential energy of runoff is directly proportional to the shear stress for sediment transportation from a slope field while the kinetic energy of the runoff at the bottom of the slope field for gully formation (Tsige et al. 2022). It is a sediment yield model that can be applied to individual storms, and it does not require a separate estimation of the sediment yield ratio. The MUSLE can be integrated with a hydrological model and simulates the sediment yield of a watershed at daily, monthly, and annual timescales. MUSLE model is represented as:

$$y = a(Q_q)^b KLSCP, \tag{11.2}$$

where y represents the sediment yield in metric tons, a represents the coefficient, b represents the exponent ($a = 11.8, b = 0.56$ for USA, where the MUSLE was originally developed), Q represents the runoff volume in m^3, q represents the peak runoff rate in m^3 s^{-1}, K is the soil erodibility factor, L represents the slope length factor, S represents the slope steepness factors, C is the cover factor, and P is the soil conservation practices factor.

To apply to larger watersheds and in different location, integrating the MUSLE model with hydrological models like SWAT is recommended (Tsige et al. 2022).

Few studies used this model in Ethiopia conditions (e.g., Muche et al. 2013; Tsige et al. 2022). For instance, Tsige et al. (2022) tested the performance of MUSLE in four catchments of Ethiopia (i.e., two catchments in the upper Awash basin, one catchment in the Abbay Basin, and one watershed in the Omo-Gibe basin) and reported better performance (> 80%) for all the four watersheds. Moreover, Muche et al. (2013) compared the USLE and MUSLE models at Choke Mountain of the Abbay Basin and reported that MUSLE model better-estimated soil erosion compared to the USLE model. The better performance of the MUSLE model is attributed to its consideration of runoff instead of rainfall, which is the major agent of soil erosion.

Techniques of Sediment Yield Estimation

Sediment yield estimation is a critical aspect of sediment control and watershed management, especially in regions where soil erosion is very common. Moreover, estimating sediment yield is a crucial task in understanding and managing erosion in a watershed. The techniques of sediment yield estimation vary widely, and they have evolved over the years to become more accurate, efficient, and cost-effective (Ayele et al. 2017). The common sediment estimation techniques include empirical, physical, sediment monitoring and remote sensing techniques.

Empirical Techniques

The empirical techniques use sediment rating curves and regression models to estimate sediment yield based on streamflow measurements and sediment samples collected at different locations in a watershed. This technique is primarily used to estimate soil loss from which sediment yield can be estimated. The Universal Soil Loss Equation (USLE), the Modified Universal Soil Loss Equation (MUSLE), and Revised Universal Soil Loss Equation (RUSLE) are the widely used sediment yield estimation empirical techniques (Wischmeier and Smith 1978). These techniques have been widely used for estimating soil loss in the Abbay Basin. The sediment yield can be estimated from the soil loss using several empirical methods (e.g., Sediment Delivery ratio) by establishing a relationship between sediment loss and other variables such as precipitation, drainage area, land use, and soil type.

Several studies have been conducted for evaluating the performance of the empirical methods in the Abbay Basin (Setegn et al. 2010; Ayele et al. 2017). For instance, Setegn et al. (2010) used the Sediment Delivery ratio (SDR) technique to estimate the sediment yield in the Lake Tana basin. This technique is based on the theory that the amount of sediment delivered to the outlet of a watershed is proportional to the total amount of sediment generated in the watershed. Moreover, according to Ayele et al. (2017), the MUSLE equation has been used to estimate sediment yield in the basin and has shown reasonable accuracy.

Physically Based Techniques

Physically based techniques use mathematical models to simulate sediment transport and erosion processes in the watershed. These techniques require detailed information on the watershed topography, soil properties, land use, and hydrological processes. The Soil and Water Assessment Tool (SWAT) (Arnold et al. 1998) and the Water Erosion Prediction Project (WEPP) (Laflen and Flanagan 2013) are the two common physically based techniques, which have been widely applied in the Abbay Basin for estimating sediment yield. Several studies have used SWAT to estimate sediment yield in the basin (Betrie et al. 2011; Teklehaimanot et al. 2014; Sinshaw et al. 2021; Nigus et al. 2018; Dagnaw and Debele 2019; Gizaw and Leta 2019).

For example, Teklehaimanot et al. (2014) used SWAT to simulate sediment yield in the Tana Beles sub-basin of the Abbay Basin. They found that the model had good predictive ability and can be used for estimating sediment yield. Similarly, Gizaw and Leta (2019) used the SWAT model for estimating sediment yield in the Gilgel Abbay watershed, a sub-watershed in the Abbay Basin. The study highlighted the effectiveness of the method in estimating the potential impacts of soil and water conservation practices on sediment generation. Moreover, Betrie et al. (2011) used the SWAT model for the estimation of sediment yield in the Abbay Basin and reported that the model is effective for estimating sediment yield. Generally, SWAT-based techniques are reported to be effective for sediment yield estimation in the Abbay Basin. The easy availability of high-resolution spatial data has made the application of SWAT model an effective tool for estimating sediment yield globally including in the Abbay Basin.

Similarly, several studies have applied the WEPP model to estimate sediment yield in the Abbay Basin (Admas et al. 2022; Nigus et al. 2018). For instance, Admas et al. (2022) used the WEPP model to estimate sediment yield in the basin and reported the effectiveness of the model in predicting sediment yield for the Abbay Basin. The study showed that the topography of the basin played a significant role in producing high sediment yields where the steeper the slopes, the higher the sediment yields. Moreover, Nigus et al. (2018) investigated sediment yield estimation using the WEPP model in the Abbay Basin and reported that the model is effective in estimating soil erosion and sediment yield in the basin. The study also showed that land use changes significantly affect sediment yield, and the model can be used as a valuable tool for land use planning and soil conservation in the basin. In addition, Dagnaw and Debele (2019) used the WEPP model to predict sediment yield in the source catchments of the Abbay Basin. The study reported that the WEPP model can accurately estimate sediment yield in the catchments and that conservation practices such as terracing and vegetative cover can effectively reduce sediment yield.

Sediment Monitoring Techniques

Sediment monitoring techniques are the other types of sediment yield estimation techniques that involve the measurement of sediment concentration and sediment load in streams and rivers. Generally, sediment monitoring techniques are important for estimating sediment yield in river systems. These techniques use sediment samplers, acoustic instruments, and other devices to collect sediment samples and measure sediment transport rates. The suspended sediment concentration method, the bedload sampling method, and the acoustic Doppler current profiler (ADCP) method are common types of sediment monitoring techniques (Rice et al. 2012).

Several studies estimated sediment yield using sediment monitoring techniques in the Abbay Basin (Moges et al. 2016; Gebiyaw et al. 2017; Teferi et al. 2019). For instance, Moges et al. (2016) estimated sediment yield for the Abbay Basin using sediment rating curves, which relate suspended sediment concentration to discharge. Gebiyaw et al. (2017) used sediment samples from streams to analyze the spatial and temporal variability of suspended sediment in the Abbay Basin and reported that sediment concentration varied significantly between different sub-basins. Moreover, Teferi et al. (2019) used sediment samples to analyze the effect of land use change on sediment yield in the Abbay Basin and reported that land use changes significantly affect sediment yield in the basin.

The sediment rating curve technique is easy to be used and provides accurate results (Moges et al. 2016). However, they are limited by the availability of discharge data and detailed sediment concentration data. Moreover, collecting detailed sediment samples at high spatiotemporal resolution is limited and costly.

Remote Sensing Techniques

Remote sensing techniques have emerged as powerful techniques for monitoring and estimating sediment yield from watersheds. Remote sensing provides an efficient and cost-effective means of monitoring and modeling sediment transport and deposition. Recently, sediment yield estimation techniques have been developed using satellite and airborne sensors based on the reflectance properties of the watershed. Remote sensing can provide high-resolution data on land use, vegetation cover, and soil moisture content, which can be used as an input to estimate sediment yield. For example, indices such as the Soil Adjusted Vegetation Index (SAVI), the Normalized Difference Vegetation Index (NDVI), and the Radar Remote Sensing (RRS) technique are commonly used to estimate sediment yield (Yang et al. 2017).

Studies used remote sensing techniques for sediment yield estimation in the Abbay Basin (Mengesha et al. 2018; Veera et al. 2020; Sinshaw et al. 2021). For instance, Veera et al. (2020) integrated remote sensing data with Soil and Water Assessment Tool (SWAT) model to estimate sediment yield for the Abbay Basin. The study reported that remote sensing-derived indices like the NDVI and Enhanced Vegetation Index (EVI) are effective to estimate vegetation cover and identifying land use changes that affect sediment yield. In addition, Sinshaw et al. (2021) also integrated remote sensing and the SWAT model for estimating sediment yield in the Abbay Basin, and reported the effectiveness of remote sensing data particularly Landsat for identifying and monitoring land use changes, which significantly influence sediment yield. Mengesha et al. (2018) applied remote sensing techniques to estimate sediment yield in the Abbay Basin and reported that remote sensing data could provide accurate and efficient means of monitoring land use changes and their impact on sediment yield in the region.

Generally, the Abbay Basin is a critical region for studying sediment yield due to its high vulnerability to soil erosion and sedimentation. As such, sediment yield estimation is a critical aspect of watershed management and sediment control. However, accurately measuring sediment yield is challenging. Therefore, sediment yield can be estimated using various techniques. Each of these techniques has its strengths and weakness. The choice of sediment yield estimation technique depends on several factors such as the size of the watershed, the availability of data, and the required accuracy. Overall, further research should be conducted to select suitable and best sediment yield estimation techniques for the basin.

Magnitude, Causes, and Effects of Soil Erosion

Magnitude and Status of Soil Erosion

Because of the geographical location and altitudinal variation advantage, river basins in Ethiopia have experienced both temperate and tropical climates, which result in abundant biophysical resources like fertile soils, rich biodiversity, and plentiful freshwater resources. These resources have supported agricultural growth and the provision of necessities for the nation's millions of residents for millennia (Zeleke et al. 2006). The Ethiopian highlands (above 1500 m a.s.l.) accounted for more than 52% of the total areas of the country and are the home of almost 90% of the population of the country and 60% of the county's livestock (Hurni et al. 2010).

Ethiopia's river basins found both in the highlands and lowlands are rich in a variety of natural resources that can be used to benefit society in some ways with economic, social, and ecological services and functions. However, river basins in Ethiopia have been experiencing severe soil erosion challenges, and these problems are the critical impediments affecting the sustainable development program of the country. Besides, environmental deterioration, particularly soil erosion, has been the main issue in various river basins, and currently, the functions and services of river basins in Ethiopia have been at risk due to soil erosion issues (Hurni et al. 2010). Consequently, the majority of the river basins in Ethiopia are the world's major soil degradation hotspots these days (Hurni 1993, 2010; Gessesse et al. 2015; Haregeweyn et al. 2017). However, the amount of surface soil erosion and runoff in each basin varies concerning its spatial position, which is a reflection of topographic diversity, land use types, and soil characteristics both within and between basins. In this regard, Kidane and Alemu (2015) noted that nutrient depletion due to soil erosion has been one of the main environmental challenges in the Abbay Basin. Specifically, soil erosion in the form of gully, rill, and sheet erosion in the basin is a life-threatening environmental issue (Haregeweyn et al. 2017).

Subsequently, billions of tons of topsoil have been eroded from mainly the upstream parts of the basins every year (FAO 1986; Zeleke and Hurni 2001). Accordingly, since a long-term research network was initiated by the Soil Conservation Research Program (SCRP) in 1981, several efforts were made to monitor the magnitude of soil erosion processes as well as estimate the amount of soil erosion in the country (Hurni 1993). The long-standing soil erosion status analysis revealed that the amount of soil loss on cultivated slopes ranges from a few $t\ ha^{-1}\ yr^{-1}$ to more than $300\ t\ ha^{-1}\ yr^{-1}$ (SCRP 1996). In an extreme case, Hurni (1993) and Berry (2003) reported that the soil loss rate reaches ~ $300\ t\ ha^{-1}\ yr^{-1}$ from cultivated land.

Besides, the observed rates of soil loss due to water erosion in the highlands of Ethiopia range from 3.4 to $84.5\ t\ ha^{-1}\ yr^{-1}$, with mean value of $32.0\ t^{-1}\ h^{-1}\ yr^{-1}$ from the entire lands and 42 Megaton per year ($Mt\ yr^{-1}$) from agricultural fields (Hurni 1993). Sonneveld (2002) also produced a rough nationwide map that shows the mean annual soil loss, in which soil loss varied noticeably, from nil in eastern and southeastern Ethiopia to $100\ t\ ha^{-1}\ yr^{-1}$ in the north and northwestern regions of the country including the Abbay Basin. Accordingly, for example, the mean values of $40\ t\ h^{-1}\ yr^{-1}$ of soil loss were estimated from cropland plots, while much lower value was quantified from grassland and forestland plots (Hurni 1993; Hurni et al. 2010).

On the other hand, Berry (2003) and Hurni et al. (2010) warned that since the mid of 1980s, about 50% (27 million hectares) of the highland region had been affected by substantial erosion, 14 million hectares (ha) seriously damaged and nearly 2 million ha of farmland are deemed to be beyond the point of regeneration capacity. More than 2 million hectares of land have already been seriously harmed, and each year water erosion damages roughly 30,000 ha of land (Berry 2003). The same author reported that soil erosion rates were estimated to be $130\ t\ ha^{-1}\ yr^{-1}$ from agricultural fields and $35\ t\ ha^{-1}\ yr^{-1}$ on average for all land in the highlands.

Besides, the Abbay Basin is among the high soil erosion-prone basins with ranges from zero in water bodies to $200\ t\ ha^{-1}\ yr^{-1}$ on degraded slopes and with a mean value of $27.5\ t\ ha^{-1}\ yr^{-1}$. Accordingly, about 39% of the total basin area is experiencing severe to very severe soil erosion risk ($> 30\ t\ ha^{-1}\ yr^{-1}$) (Haregeweyn et al. 2017). In addition, Hurni (1988) argued that the potential soil loss for the Abbay Basin, which covers ~ 16% of the nation's area, accounts for ~ 31% of the national gross soil loss amount. The spatial variability of soil erosion in the basin ranges from 4 to 49 t ha^{-1} yr^{-1} (Balthazar et al. 2013). More specifically, the mean annual soil loss due to soil erosion by water in the Chemoga watershed of the Abbay Basin was estimated at 93 Mt yr^{-1} (Bewket and Sterk 2003). Moreover, Shiferaw and Holden (1999) estimated soil loss in the Borena district using RUSLE model and reported annual soil loss that ranges from no loss in the flat plain areas and waterbodies to over 154 Mt yr^{-1} in steeper gradient areas.

In contrast, several case studies at plot, sub-watershed, and sub-basin levels in the Abbay Basin indicated that average soil erosion loss ranged from 0 to 170 Mt yr^{-1} (Zeleke and Hurni 2001; Hurni et al. 2005; Setegn et al. 2010; Tibebe and Bewket 2010; Setegn et al. 2010; Haregeweyn et al. 2017; Elnashar et al. 2021; Kabite et al. 2022). In general, the actual soil loss from the whole basin is ~ 473 Mt yr^{-1}, of which ~ 10% are due to gullies and 26.7% of sediments leaves the country. Besides, the time series analysis of the estimated amounts of sediment leaving the gauging station at El Deim, just near the Ethio-Sudan border, is ranging from 111 to 140 Mt y^{-1} (Easton et al. 2010; Betrie et al. 2011).

Causes of Soil Erosion

In general, land degradation attributed to soil erosion is a decline or loss of ecosystem services of the land resource bases. It can be triggered by natural and socioeconomic causes resulting in a change in land resources affecting the ecosystem services (Abu Hammad and Tumeizi 2010). In Ethiopia, most of the farming land is tilled using an ox-plow system, which exposes the soil to rain and surface runoff, especially during the rainy seasons. The process of soil erosion in the basin is further exacerbated by this type of farming operations on steep slopes combined with a lack of proper and acceptable land management methods (Hurni et al., 2010). In Ethiopia's undulating topography, soil erosion by water is a complicated process, and evidence confirmed that the agricultural landscapes of the nation are the primary hotspots for land degradation, mostly because of rapid soil erosion by water (Hurni 1988, 1993, 2010; Gessesse et al. 2015). In Ethiopia, the proximate and underlying drivers of soil erosion are known as the major source of land degradation (Taddese 2001). The high topographic characteristics (i.e., slope length and steepness), rainfall intensity, erodibility of soils, framing practices, LULC, conditions, and land management practices are the major proximate governing factors for soil erosion, surface runoff generation, and sediment production in the Abbay Basin in particular and Ethiopia in general (Hurni et al. 2005; Tibebe and Bewket 2010; Gessesse et al. 2015).

Although soil erosion in Ethiopia is influenced by biophysical factors like topography, land cover, and climate, the process is highly influenced by anthropogenic or underlying driving forces such as weak policy and regulatory institution, low empowerment of local communities, infrastructure development, demographic growth, unclear user rights, the prevalence of resource-poor farmers, and scarcity of farmlands (Tedla and Lemma 1998; Taddese 2001; Wassie 2020). Besides, lack of effective, integrated, and participatory watershed management practices, encroachments of agricultural activities on marginal lands without adequate land management investments, intensive land use/land cover changes, and malfunctioning land management systems are additional intertwined demographic and socioeconomic factors, which aggravate soil erosion in the basin (Hurni 1993; Hurni et al. 2005; Dibaba et al. 2020).

Furthermore, socioeconomic and institutional factors are the major underlying causes that affect soil erosion by influencing the decision of the farmer concerning land management and land use practices (Tefera et al. 2002). In addition, Desta et al. (2000) reported that limited access to agricultural inputs and credit facilities, high population pressure, the absence of comprehensive land use and administration policies, top-down land management planning systems, lack of land management policies and strategies implementation strategies, lack of sustainable land management practices as well as frequent organizational restructuring are factors that contribute to soil erosion in the Abbay Basin.

On-Site and Off-Site Effects of Soil Erosion

The Abbay Basin is a section of Ethiopia's southwest highlands and western lowlands, where soil erosion has reached extremely serious levels of land degradation, resulting in negative consequences on local ecosystem services and processes like soil fertility loss and hydrologic regime. In most cases, land degradation due to soil erosion triggers the loss of biodiversity and fertile topsoil, which is vital to agricultural productivity to fulfill the food security of the country.

Both on-site and off-site effects of soil erosion are observed in the Abbay Basin. Some of the main on-site impacts are a decrease in topsoil depth, nutrient depletion, and a subsequent decrease in crop yields (Zeleke et al. 2006; Hurni et al. 2005). On the other hand, sedimentation, flooding, and decreased productivity of irrigated land are some of the main adverse downstream repercussions of soil erosion in the Abbay Basin (Zeleke et al. 2006).

According to Berry et al. (2003), the costs of nutrients lost due to removal of topsoil by soil erosion, the capital costs needed to replace these nutrients using artificial fertilizers, the production and productivity costs lost due to nutrient and soil losses, declining of livestock carrying capacity, and the decrease in the cropped area are the direct costs of soil erosion in Ethiopia. However, the same author affirmed that the major indirect costs of soil erosion include the siltation of reservoirs, loss of environmental services, fluctuation of streamflow, the reduction of groundwater reserves, and flooding as well as other social and community related costs like malnutrition, poverty, and migration.

Besides, soil erosion by water greatly influences ecosystem services (Gebrehiwot et al. 2014), flood inundation and siltation of the downstream areas (Haregeweyn et al. 2017), crop production and productivity (Hurni et al. 2015), and economic costs (Zeleke et al. 2006; Hurni et al. 2015). Studies on soil erosion at the hillslope scale have been also conducted for decades (Hurni 1985; Taye et al. 2013) and in small watersheds (Gashaw et al. 2020). Nevertheless, very few estimates are available for overall rates of total soil loss, which is mainly attributed to sheet and rill erosion at regional or national scales (FAO 1986; Hurni 1988; Sonneveld et al. 2011).

Accordingly, Haregeweyn et al. (2015) reported that excessive soil loss in Abbay Basin in particular and in Ethiopia in general is severely impeding rural development and land productivity, the operation and functionality of water development infrastructure, and the goods and services that support livelihoods. The farming system is thus significantly challenged, which hurts farmer employment, revenue, and profitability. Subsequently, soil erosion endangers the livelihoods of rural farmers and the ability of a country in producing crops, livestock, and other natural resource products. On top of that, soil erosion poses a risk to reservoirs and dams in the Abbay Basin's downstream region, including the Grand Ethiopian Renaissance Dam (GERD). This is because siltation, sedimentation, and nutrient influx is coming from the highlands of the Abbay Basin and ends up at the dam and it will fill the dam quickly. As a result, the functionality and the life span of the hydraulic structures of the dam will be reduced because of the siltation. Additionally, the effective operation of the dam will be hampered by silt accumulation. Accordingly, the research findings in the Abbay Basin commented that surface runoff, soil erosion, siltation, and sedimentation have been a threat to both upstream (on-site) and downstream (off-site) communities for centuries due to nutrient loss, decreased agricultural production, dam siltation, flooding, obstruction of irrigation canals, and groundwater depletion.

Sediment Yield Status of the Abbay Basin

Deforestation, the continuous removal of vegetation cover, exposes the topsoil to soil erosion that ends in transporting sediments during the rainy season. In addition, unwise land management also exposes the cultivated land to losing soil due to erosion and flooding. Numerous studies indicated that the cost of land degradation by far exceeds the cost of actions to prevent it (Nkonya and Mirzabaev 2016). These authors also estimated the annual cost of land degradation due to soil erosion that resulted from LULCC change in Ethiopia to be about $4.3 billion. However, in recent years there are several efforts from the international community to pause and reverse land degradation incorporated in the agenda 2030 focusing on preventing and reversing soil erosion-induced biodiversity loss and land degradation. Ethiopia has also made various activities to halt soil erosion-induced land degradation through the implementation of soil and water conservation practices and tree-planting activities. However, the loss of soil and sediment transportation is still challenging the country.

A number of approaches were developed to monitor and quantify land degradation and sediment yield that results from removing soils from the watersheds. Preetha and Al-Hamdan (2022) employed the crop management (C) and soil erodibility (K) factors of the USLE in the SWAT model and satellite data to predict sediment yield in the Southeastern USA. Their result showed that the estimation of sediment yield using dynamic assessment of multiple factors is comparable with the measured data.

Betrie et al. (2011) also examined the impact of catchment management intervention in reducing soil erosion and sediment transport by adopting scenarios (business as usual, stone bunds, filter strips, and reforestation) using the SWAT model. In their study, they indicated that the measured mean sediment yield at the outlet of the Abbay Basin was 131×10^6 t yr^{-1}. On the other hand, the SWAT model predicted a sediment yield of 117×10^6 t yr^{-1} if there is no action taken to manage the land use in the basin. The remaining three simulation scenarios (filter strips, stone bunds, and reforestation) resulted in a sediment yield reduction from the current condition by 44%, 41%, and 11%, respectively.

Similarly, limited studies have estimated mean annual sediment yield rates, which show the amounts of sediment leaving from the basin. Study showed that the Abbay River alone transports 131 million tons of sediment annually, compared to the 1.3 billion tons transported annually by Ethiopia's international rivers that originate in the highland parts of the basin (Hurni et al. 2010; Betrie et al. 2011). Moreover, Ali et al. (2014) employed the rating curves for the areas located along the rivers network in the Abbay Basin and estimated sediment load that ranges from 130 to 170 Mt yr^{-1}. In addition, the estimated potential annual sediment yield budget of the Blue Nile (Abbay) river in Ethiopia and Sudan measured at Khartoum was about 140 ± 20 Mt yr^{-1} (Garzanti et al. 2006).

Sediment yield of a watershed could be reduced through proper land management. For example, a lower reduction in sediment transport is obtained from the reforestation scenario which is attributed to the smaller implementation area compared to the other scenarios (Betrie et al. 2011). Likewise, Lemann et al. (2016) presented the contribution of soil and water conservation practices on sediment yield of the Gerda catchment in the Abbay Basin of northwestern Ethiopia using two scenarios. They considered scenario 1 with full soil and water conservation (SWC) and scenario 2 with no soil and water conservation practices. Their result showed that through the implementation of SWC at the catchment, there is a declining of sediment loss by 30%. The assessment made by Asres and Awulachew (2010) for evaluating runoff and sediment yield in the Gumera watershed also indicated that the level of runoff and sediment yield varies among the sub-watersheds. Accordingly, among the 30 sub-watersheds, 18 of the sub-watersheds located in the upstream parts of the watershed produced a mean annual sediment yield of 11 to 22 t ha^{-1} yr^{-1}, while the lower and wetland parts of the watershed produced a lower sediment yield per year indicating watershed characteristics control sediment yield of the watershed.

Ayele et al. (2017) assessed the spatial variability of sediment yield of Koga catchment from 2002 to 2007. The analysis showed that 17 out of the 22 sub-catchments produced a variable amount of sediment yield ranging from 0 to 3.55 t ha^{-1} yr^{-1} (0–2 t ha^{-1} yr^{-1} in highlands and 2–3.55 t ha^{-1} yr^{-1} in lowlands). The variation in sediment yield differs with the type of land use and soil types at hydrological response units. However, Abebe et al. (2022) estimated sediment yield using the SWAT model in the Andassa watershed for the years 1992–2012. Accordingly, their study showed that the annual mean sediment yield ranges between 5 and 28 t ha^{-1} yr^{-1}.

As mentioned above, most of the studies done in the basin showed a significant variation in land use management practices and a huge amount of transportation of sediments. This significantly affects the agricultural activity and the life span of several dams built and planned within the basin. Most studies revealed the accumulation of sediments from the upstream watersheds endangering the storage capacity and life span of downstream reservoirs (Tsegaye and Bharti 2021). The removal of top soils causes land unsuitable for agriculture impacting food production and food security of the country. In addition, the degraded land grows fewer plants that can take in carbon dioxide resulting in climate change. Hence, proper land management activities should be practiced in the basin including SWC and afforestation, which support the growth of more trees that sequester more carbon dioxide, reduce soil erosion, and trap sediment.

Sustainable Land Management Practices

Sustainable land management (SLM) contributes to the protection of the environment from soil erosion- and deforestation-induced land degradation. Land provides agricultural production and sinks for greenhouse gases, recycling of nutrients, and maintaining the hydrological cycle. SLM can be defined by the UN Earth Summit in 1992 as the use of land resources like water, soils, animals, and plants for the production of goods and services to meet the increasing human needs while at the same time ensure the sustainability and management of these resources. Smyth and Dumanski (1993) also described SLM aim as harmonizing the complementary goals of delivering economic, environmental, and social opportunities to support the current and future generations by improving the quality of the land. SLM is regarded as an attempt to avoid and reverse land degradation and act as a climate change adaptation strategy.

Ethiopia as a signatory of several conventions to combat land degradation and promoted SLM through implementing appropriate soil and water conservation measures. This is due to the fact the SLM is a prominent activity for protecting the health of an environment, which is drought-prone and vulnerable to natural disasters. Accordingly, Ethiopia adopted several measures in response to climate change and land degradation through ratifying the United Nations Framework Convention on Climate Change (UNFCC) in 1994. The country also endorsed a Climate Resilient Green Economy (CRGE) strategy in 2011 to build a green and resilient economy. It developed the CRGE strategy to maintain the health of its environment and also minimize the carbon emission from agriculture and the unwise use of natural resources. Among the four pillars of CRGE deforestation and agriculture are incorporated to be part of reducing the emission of greenhouse gas and loss of vegetation cover of the environment (FDRE 2011). The agriculture sector promised to boost livestock and crop production for increasing food security while reducing Green House Gas (GHG) emissions. On the other hand, the forestry sector also aims at avoiding deforestation and promoting reforestation to provide economic and ecosystem services. Ethiopia is also involved in another program to deal with deforestation to build a green and climate-resilient country through planting trees. In this regard, the country began Green Legacy Initiative (GLI) in 2019 for planting 4 billion various types of seedlings.

This is because forests significantly contribute for mitigating and adapting to climate change, and minimize soil erosion and sedimentation. They trap the carbon dioxide from the atmosphere to prepare their food and have an important role to maintain the soil by preventing soil erosion and the addition of litter and other organic matter to the soil. In this regard, the tree-planting activity of Ethiopia is aimed at biofuel production, carbon trade, and soil and water conservation. Even if all these efforts are undertaken in the country, the loss of forests due to deforestation and forest degradation as well as soil loss is immense.

Different sources indicate that the annual forest loss in the country is continuing through the years (Tadesse et al. 2014). The global forest resources assessment study indicated 0.42% forest loss between 2010 and 2020 alone (FAO 2020). Forest loss is linked to high dependency on fuelwood, encroachment of agriculture to forestland, changes in settlement, population growth, and forest fire (Tadesse et al. 2014). Dibaba et al. (2020) evaluated the LULC change in the Fincha'a catchment using Landsat images of three periods (1987, 2002, and 2007), a digital elevation model and field data. The result indicated that there is a significant increase in agricultural land with a sharp decline in forest land. The major drivers of LULC change include expansion of agricultural land and urban land, infrastructural development, deforestation, resettlement, and poor grazing land mismanagement. The study also reported significant land and soil degradation and a decline in soil fertility.

The LULC dynamics assessment made by Terefe et al. (2019) at the Dhidhessa sub-basin for the period 1974–2014 demonstrated that there is a rapid decline in the size of the natural environment including wetlands, grassland, forestland, and shrubland. In contrast, settlement and agricultural land showed a sharp increase during the same period. This study showed a decrease in agricultural productivity linked with land degradation and population pressures. Liyew et al. (2019) evaluated the trends, driving factors and implications of LULC change in three watersheds (Aba Gerima, Debatie, and Guder) for the period 1982–2017. The LULC change analysis was performed by combining satellite images and field observation. The assessment displayed that in all the three watersheds, forestland, grazing land, and bushland declined while cultivated land significantly increased in coverage. The main drivers of change are attributed to farming practices and population growth. The study also indicated an increase in gully erosion and surface runoff potential.

Gashaw et al. (2017) also analyzed LULC dynamics in the Andassa watershed using five land use classes, namely cultivated land, shrubland, forest, grassland, and built-up. The LULC change investigation was performed for the year between 1985 and 2015. In their analysis, they indicated that cultivated land significantly increased in size at the expense of other land use types. On the other hand, forest, grassland and shrubland showed a sharp decline in the study period. The main driver of the change is indicated as increasing population, which leads to higher demand for cultivated land and wood for different purposes. This study also showed that the current dynamics in land use will lead to soil loss and impact the hydrology of the watershed.

Sewnet and Abebe (2017) also assessed the impact of LULC change on the Koga watershed degradation during the period from 1973 to 2011. According to their analysis, cultivated land and settlement significantly increased in the study period while bushland, grassland and wetland showed a declining trend. This study also demonstrated that continuous plowing without proper conservation measures facilitated erosion to move sediment as runoff in the catchment. Moreover, Badasa et al. (2022) also evaluated the LULC dynamics over the Geba watershed for the period 1990–2020. In their assessment, they indicated that there is a significant amount of land use dynamics in the watershed. In the study period, forestland showed a significant decline, while agricultural land and settlement showed a sharp increase. They also revealed the decline in forest cover and accelerated soil erosion in the watershed area which may lead to a loss in agricultural productivity.

Malede et al. (2023) made a LULC dynamics analysis for the period from 1986 to 2018 over the Birr River watershed. The assessment showed increasing agricultural land and settlement while declining bushland, forest, and grassland demonstrated. This also confirms the transition of the natural environment to human-manipulated land use types exposing the soil of the watershed. Their evaluation also revealed that at the local level, the land use transformation affected the runoff, soil erosion, and sediment load of the watershed.

Yalew et al. (2016) have also made an assessment of land use change modeling and trajectories for the year 1986–2009 and a prediction of change for 2009–2025 at Jedeb catchment in the southwest part of Mount Choke. The analysis showed a decline in woody vegetation and grassland while cultivated land and plantation forests showed a significant amount of increase in size during the study period. The model prediction also showed a similar trend with an increase in cultivated land and plantation forests till 2025. Adgo et al. (2013) evaluated the contribution of soil and water conservation (SWC) practices in the Anjenie watershed for improving agricultural productivity. Accordingly, their assessment indicated that SWC enhanced crop productivity. In addition, the soil erosion in the watershed was radically reduced with improved household income and food security. A study made in the Gumara watershed regarding SWC also indicated an improvement in soil condition by reducing runoff and sediment yield from the watershed (Mengie et al. 2019).

Mekuria et al. (2018) examined variations in vegetation composition, biomass accumulation, and soil properties by developing exclosure in degraded lands of the Aba Gerima watershed. Several activities were performed in the closed land including the construction of physical structures for SWC, planting trees, and seeding of grass and tree. The analysis of soil properties and vegetation composition in the area confirmed that enclosures improved soil properties and restored degraded native vegetation. The soil analysis also indicated the availability of a higher amount of organic matter, nitrogen, and phosphorus in the 0–15 cm soil depth.

Admas et al. (2023) examined the contribution of best management practices in reducing soil loss from catchments and sediment load into dam reservoirs in the Megech watershed. Their investigations indicated that implementing these practices impacted runoff, soil loss, and sediment yield compared to conventional agricultural practices. Among the tested practices, agroforestry with five-year perennial trees significantly reduced runoff, soil loss, and sediment yield in the catchment. Erkossa et al. (2020) also evaluated infiltration trenches and reseeding contribution to restoring degraded grazing land found near the Dhidhessa River, Abbay Basin. The analysis showed an enhanced soil moisture content and a prolonged growing period with higher biomass yield. They also indicated that the improved soil properties benefited the farmers to increase their agricultural productivity.

An improved land management and afforestation/reforestation activity in the basin enhance the soil quality to hold more water through infiltration. The organic matter from tree roots and leaf litter enhances soil structure, which increases soil infiltration rates and reduces soil erosion. The vegetation cover protects runoff and hence stops the transportation of sediments from the surface of the land. The trees also intercept rainfall during heavy rainstorms to prevent splash erosion from the surface. For example, Naomi (2020) examined the impacts of afforestation of *Acacia decurrens* plantations on runoff and sediment transportation in Fagita Lekoma District within the Abbay Basin. The model analysis indicated a significant reduction in surface runoff and subsequently soil erosion. Accordingly, the afforestation activity and the implementation of SWC practices reduced sediment yields from 26.5 t ha^{-1} yr^{-1} to 15.5 t ha^{-1} yr^{-1} within the watershed.

Mersha et al. (2022) assessed the effectiveness of SLM (e.g., soil bund and contour trenches) in improving water security through five-year catchment restoration efforts in the Aba Gerima catchment, Abbay Basin. A comparison was made between two parts of the catchment, one treated with SLM activities and one a control without any SLM activities. Their evaluation indicated a significant reduction in runoff volume and the number of events and an increase in soil moisture in the treated part of the catchment that leads to the recovery of the hydrologic functionality of the landscape. Moreover, Ebabu et al. (2019) examined the effects of SLM practice on runoff and soil loss in the Abbay Basin. They made a comparison between two land use types such as croplands and none croplands. They used four treatments in cropland (control, soil bund, Fanyajuu, and soil bund reinforced with grass) and three treatments in non-cropland (control, exclosure, and enclosure with trenches). According to their assessment, runoff and sediment loss were significantly reduced in SLM plots than in the control plots.

The government of Ethiopia has implemented several land management practices to minimize soil erosion and maintain land productivity in different watersheds of the country including in the Abbay Basin. The SWC activities had begun in Ethiopia in the mid-1970s and had great attention to the conservation of catchments (Mersha et al. 2022). Accordingly, Food for Work (1973–2002), Community Mobilization through free-labor days (1998-present), Managing Environmental Resources to Enable Transition to more sustainable livelihoods (MERET) (since 2003), Productive Safety Net Programs (PSNP) (since 2005), and National Sustainable Land Management Project (SLMP) (since 2008) (Andersson et al. 2011; Adgo et al. 2013; Adimassu et al. 2017). Through these programs, several SWC and tree-planting activities were implemented in different watersheds of the country including Abbay Basin. All these activities played a pivotal role in minimizing soil erosion and sediment yield. However, even if there are progresses in protecting the vegetation cover and soil resources of the country, there is still a lot to be done to maintain the natural resources to cope with the current climate change and anthropogenic impacts. Human activities are impacting the physical environment by transforming to other land use types.

Opportunities and Challenges of Estimating Soil Erosion and Sediment Yield

Opportunities

The advancement of geospatial technologies and machine learning algorithms created an opportunity in estimating soil erosion and sediment yield more accurately than it was possible before. For instance, the integration of Earth observation and geographic information system (GIS) provide huge potential for analyzing and mapping soil erosion estimation factors quantitatively and qualitatively. Remote sensing provides climate, topography, land use/land cover, and topography datasets, whereas GIS combines the factors and model soil erosion, transport, and sedimentation. Moreover, the resolution, accuracy, and types of remote sensing products have been improving on the one hand and the algorithms to analyze the data have been advancing on the other hand. As such, nowadays, machine learning (deep learning) and cloud computing and high-resolution satellite images are available that improved soil erosion and sediment estimation accuracies, which are determinants for reducing the negative consequences of soil erosion and siltation in a given basin.

Recently, cloud computing and machine learning algorithms have been used to estimate soil erosion and sediment yield more accurately. Deep learning, the new types of machine learning (ML) algorithms have greater flexibility and better predictive performance compared to traditional ML models (Ghorbanzadeh et al. 2019). Deep learning algorithms such as Convolution Neural Network (CNN), Long Short-Term Memory (LSTM), and Recurrent Neural Network (RNN) are the widely used algorithms used for flood modeling (Khosravi et al. 2020), landslides susceptibility assessment (Le et al. 2021), and soil erosion susceptibility modeling (Khosravi et al. 2023). For example, Titti et al. (2022) developed cloud-based interactive susceptibility modeling of gully erosion in Google Earth Engine. Moreover, Khosravi et al. (2023) tested three deep learning algorithms [i.e., Convolutional Neural Network (CNN), Recurrent Neural Network (RNN), and Long Short-Term Memory (LSTM)] for soil erosion susceptibility assessment and reported that all the models predicted well with relatively better results of the CNN. However, soil erosion and sediment yield estimation using such state-of-the-art-algorithm is missing in Ethiopia and Abbay Basin in particular.

Challenges

It is difficult to measure accurately soil loss and sediment yield by water at various spatiotemporal scales. As such, soil loss is commonly quantified using soil erosion models. There are three major approaches for quantifying soil erosion and sediment yield, namely empirical, physically based, and machine learning models, which is a data-driven approaches (Raza et al. 2021). All the techniques used for estimating soil erosion and sediment yield exhibit limitations and strengths. The empirical models such as USLE, RUSLE, and MUSLE estimate parameters that influence soil erosion and sediment yield. These models depend on a mathematical algorithm that represents the relationship between the parameters and soil erosion rates and sediment yield (Khosravi et al. 2023). The performance of these models is well for conditions for which they are calibrated while poorly performed in conditions outside those used in calibration (Tan et al. 2018).

The RUSLE model has been widely used due to its moderate data demand and its simplicity (e.g., Haregeweyn et al. 2017; Panagos et al. 2017; Fenta et al. 2020). However, the RUSEL model output is associated with uncertainties (Panagos et al. 2015) as the model was developed in the USA based on standardized soil loss measurements (Renard et al. 1997). The uncertainties have mainly resulted from the C- and P-factors, which are influenced by local variables like LULC, land management practices, soil type, climate, topography (Taye et al. 2017). Therefore, accurate representation of these factors by considering the local factors improves the accuracy of soil loss estimation (Kuok et al. 2013; Panagos et al. 2015; Taye et al. 2017). However, C-and P-factors are less studied and are roughly estimated in most cases.

For the Abbay Basin and elsewhere in Ethiopia, the *C*- and *P*-factor values were determined for a specific watershed by considering land use and management practices, respectively (Hurni 1985; Nyssen et al. 2009; Taye et al. 2017). However, *C*- and *P*-factors have not been estimated for the various land uses and land management structures for the Abbay Basin for the various agro-ecologies. In most cases, *C*- and *P*-factors are determined based on whether they are cultivated and non-cultivated land. Moreover, land management practices' data for the basin under study are lacking in most cases. Furthermore, no model considers the storm kinetic energy of rainfall in the Abbay Basin and calculates an accurate R-factor value. Therefore, determining accurate values of *C*-, *P*-, and *R*-factor values reduces uncertainties in estimating soil erosion loss (Kebede et al. 1991).

For instance, Kebede et al. (2020) estimated the *C*- and *P*-factors for the different land management practices and cover conditions for the different agro-ecologic zones of the Abbay Basin. Accordingly, the smallest *P* values were observed for areas with soil bund integrated with grass in cropland and pasture land with the trench in the non-cropland plots. Similarly, a lower average *C* value (0.1) was observed for the non-cropland plots compared with the cropland with a *C* value of 0.24. The study concluded that accurate estimation of *C*- and *P*-factors and *R*-factor values with an appropriate model will minimize uncertainties in the processes of soil loss estimation.

For instance, Tian et al. (2021) also describe that accurate determination of *P*-factor would improve the accuracy of soil erosion estimation.

The physically based models like Watershed Erosion Prediction Project (WEPP) and KINematic runoff and EROSion (KINEROS) estimate soil erosion by mathematically representing the physics behind the processes of soil detachment, transportation, and deposition. However, the models require several variables with detailed spatiotemporal watershed dataset for building, calibrating, and validating the models. As such, it is challenging to use the models as such types of data are rarely available in developing countries like Ethiopia and it is very costly for larger basins (Conoscenti et al. 2008; Khosravi et al. 2023).

The development of machine learning (ML) approaches advanced soil erosion susceptibility modeling (Vu et al. 2020; Mosavi et al. 2020). The ML approaches established nonlinear robust relationship between input and output parameters and recognize patterns within the data. ML-based approach is more effective in areas with scarce data or where physically based models cannot be applied (Khosravi et al. 2023). However, the traditional ML algorithms are characterized with slow computational speed during the training procedure, generalization power, high uncertainty in the modeling phase, and a low convergence (Kisi et al. 2012). For example, Kisi et al. (2016) reported poor prediction power of Artificial Neural Network (ANN) when different ranges of test datasets and training data are used. As such, Tien et al. (2012) reported superior performance of tree-based models (e.g., RF), but they are less suitable for scarce data catchments as they are sensitive to noisy data.

Concluding Remarks

Abbay Basin is one of the significant transboundary rivers that support the well-being of millions of people in the Ethiopia and in the downstream countries. The basin is a potential area for agricultural production and developing water resource projects. As a result, roughly 50% of the basin is used for mixed farming practices, and there are several dams intended for irrigation and hydropower generation including the GERD, which will be the biggest dam when completed. However, the basin is highly vulnerable to severe soil erosion. This chapter reviewed previous studies on the techniques on estimating soil erosion and sediment yield; magnitude, causes, and impacts of soil erosion; sediment yield status; sustainable land management for controlling soil erosion; opportunities and challenges in assessing soil erosion and sediment yield in the Abbay Basin; and intended to draw the implications on the policies, programs, and future research on sustainable land management (SLM) practices.

Severe soil erosion and sedimentation in the Abbay Basin have resulted from the rugged topography, erosive rainfall, poor land management, deforestation, high population pressure, and low soil fertility due to prolonged farming without proper land management. The severe soil erosion and the resulted sediment yield negatively affect agricultural production and water resources availability resulting in socioeconomic and environmental problems. In this connection, the various research findings conducted in the Abbay Basin confirmed surface runoff and soil erosion, and sedimentation had been threatening both the upstream (on-site) and downstream (off-site) parts of the basin for centuries. The effects have been observed through nutrient loss, reduced agricultural production, siltation of dams, flooding, blockage of irrigation canals, and depletion of groundwater. To that end, it is crucial to decrease sediment and nutrient influx using various management techniques. Moreover, the use of efficient and sustainable watershed management techniques might increase the life span of the basins and improve the ecosystem services that land offers.

Subsequently, the government of Ethiopia has implemented several land management programs in different parts of the country including in the Abbay Basin during the last couple of decades. The program includes Food for work (1973–2002), Managing Environmental Resources to Enable Transition to more sustainable livelihoods (MERET) (since 2003), Community Mobilization through free-labor days (1998 to present), National Sustainable Land Management Project (SLMP) (since 2008), Productive Safety Net Programs (PSNP) (since 2005), and Green Legacy Initiatives (since 2019). Through these programs, multiple SWC and tree-planting activities were performed in different basins of the country including the Abbay Basin. Particularly, SWC measures were implemented in different parts of the country in general and Abbay Basin in particular. Studies showed the effectiveness of these programs in reducing soil erosion and sediment yield in some areas.

However, the study also showed that soil erosion and sedimentation of reservoirs and irrigation structures remained significant problems in the Abbay Basin, indicating that the effectiveness of land management has not been as expected. This could be due to (i) lack of maintenance and monitoring, (ii) the approach is not fully participatory, (iii) the farmers do not exactly know the cost and benefits of the conservation measures, (iv) lack of proper plan and design, (v) local factors are not fully considered, (vi) complexities of topography, climate, soil and land use and management practices in the highlands of Ethiopia for prioritizing and designing appropriate measures, and (vi) benefits of SWC do not outweigh costs of construction and maintenance at household level (incentives are required). Therefore, integrated and location-specific land management measures should be designed and implemented in the basin to reduce soil erosion and siltation of the downstream reservoirs (e.g., GERD) and ensure sustainable land resource availability. Moreover, hot spots and prioritized watersheds should be identified through accurate quantification of soil loss and sediment yield estimation using state-of-the-art machine learning algorithms.

Acknowledgements We would like to acknowledge Space Science and Geospatial Institute for creating conducive working environment for writing this book chapter. We are also grateful to reviewers and editor in chief for their constructive and useful comments.

References

Abebe BK, Zimale FA, Gelaye KK, Gashaw T, Dagnaw EG, Adem AA (2022) Application of hydrological and sediment modeling with limited data in the Abbay (Upper Blue Nile) Basin, Ethiopia. Hydrology 9(10):1–21

Abeje MT, Tsunekawa A, Adgo E, Haregeweyn N, Nigussie Z, Ayalew Z, Elias A, Molla D, Berihun (2019) Exploring drivers of livelihood diversification and its effects on adoption of sustainable land management practices in the upper Blue Nile basin, Ethiopia. Sustainable 11:1991

Abu Hammad A, Tumeizi A (2010) Land degradation: socioeconomic and environmental causes and consequences in the eastern mediterranean. Land Degrad Dev 23(3):216–226

Adgo E, Teshome A, Mati B (2013) Impacts of long-term soil and water conservation on agricultural productivity: the case of Anjenie watershed, Ethiopia. Agric Water Manag 117:55–61

Adimassu Z, Langan S, Johnston R, Mekuria W, Amede T (2017) Impacts of soil and water conservation practices on crop yield, run-off, soil loss and nutrient loss in Ethiopia: review and synthesis. Environ Manage 59(1):87–101

Ahmed AA, Ismail UHAE (2008) Sediment in the Nile River system: UNESCO-international hydrological program international sediment initiative. UNESCO. Available: http://isi.irtces.org/isi/rootfiles/2017/07/07/1487239390353757-1498713528334367.pdf

Ali YSA, Crosato A, Mohamed YA, Abdalla SH, Wright NG (2014) Sediment balances in the Blue Nile River Basin. Int J Sedim Res 29:316–328

Andersson C, Mekonnen A, Stage J (2011) Impacts of the productive safety net program in Ethiopia on livestock and tree holdings of rural households. J Dev Econ 94(1):119–126

Arnold JG, Srinivasan R, Muttiah RS, Williams JR (1998) Large area hydrologic modeling and assessment part I: model development. J Am Water Resour Assoc 34(1):73–89

Asres MT, Awulachew SB (2010) SWAT based runoff and sediment yield modeling: a case study of the Gumera watershed in the Blue Nile basin. Ecohydrol Hydrobiol 10(2–4):191–199

Ayele GT, Teshale EZ, Yu B, Rutherfurd ID, Jeong J (2017) Streamflow and sediment yield prediction for watershed prioritization in the upper Blue Nile river basin, Ethiopia. Water 9(10)

Badasa M, Indale M, Dejene N, Busha L, Dessalegn H, Gemeda O (2022) Land use/land cover change analysis using geospatial techniques : a case of Geba watershed, western Ethiopia. SN Appl Sci 4:187

Balthazar V, Vanacker V, Girma A, Poesen J, Golla G (2013) Human impact on sediment fluxes within the Blue Nile and Atbara River basins. Geomorphology 180–181:231–241

Belayneh M, Yirgu T, Tsegaye D (2020) Runoff and soil loss responses of cultivated land managed with graded soil bunds of different ages in the upper Blue Nile basin, Ethiopia. Ecol Process 9:66

Berihun ML, Tsunekawa A, Haregeweyn N, Tsubo M, Fenta AA, Ebabu K, Sultan D, Dile YT (2022) Reduced runoff and sediment loss under alternative land capability-based land use and management options in a sub-humid watershed of Ethiopia. J Hydrol: Reg Stud 40:100998

Berry L (2003) Land degradation in Ethiopia: its extent and impact. Commissioned by the GM with WB Support Group

Betrie GD, Mohamed YA, van Griensven A, Srinivasan R (2011) Sediment management modelling in the Blue Nile Basin using SWAT model. Hydrol Earth Syst Sci 15:807–818

Bewket W, Sterk G (2003) Assessment of soil erosion in cultivated fields using a survey methodology for rills in the Chemoga watershed, Ethiopia. Agric Ecosyst Environ 97:81–93

Bewket W, Teferi E (2009) Assessment of soil erosion hazard and prioritization for treatment at the watershed level: case study in the Chemoga watershed, Blue Nile basin Ethiopia. Land Degrad Dev 20(6):609–622

Conoscenti C, Di Maggio C, Rotigliano E (2008) Soil erosion susceptibility assessment and validation using a geostatistical multivariate approach: a test in Southern Sicily. Nat Hazards 46:287–305

Conway D (2000) The climate and hydrology of the Upper Blue Nile River. Geogr J 166:49–62

Dagnaw DC, Debele B (2019) Spatial prediction of soil erosion and sediment yield using WEPP model in the headwater catchments of the Blue Nile Basin, Ethiopia. Model Earth Syst Environ 5(3):1159–1173

Desta L, Kassie M, Benin S, Pender J (2000) Land degradation in the highlands of Amhara Region and strategies for sustainable land management. Livestock policy analysis program. Working paper No. 32. International Livestock Research Institute (ILRI), Nairobi

Dibaba WT, Demissie TA, Miegel K (2020) Drivers and implications of land use/land cover dynamics in Finchaa catchment, Northwestern Ethiopia. Land 9(4):1–22

Dile YT, Tekleab S, Ayana EK, Gebrehiwot SG, Worqlul AW, Bayabil HK, Yimam YT, Tilahun SA, Daggupati P, Karlberg L, Srinivasan R (2018) Advances in water resources research in the Upper Blue Nile basin and the way forward: a review. J Hydrol 560:407–423

Duguma TA (2022) Soil erosion risk assessment and treatment priority classification: a case study on Guder watersheds, Abbay river basin, Oromia, Ethiopia. Heliyon 8(2022):e10183

Easton Z, Fuka D, White E, Collick A, Ashagre BB, McCartney M, Awulachew S, Ahmed A, Steenhuis T (2010) A multibasin SWAT model analysis of runoff and sedimentation in the Blue Nile, Ethiopia. Hydrol Earth Syst Sci 14:1827–1841

Ebabu K, Taye G, Tsunekawa A, Haregeweyn N, Adgo E, Tsubo M, Fenta AA, Meshesha DT, Sultan D, Aklog D, Admasu T, van Wesemael B, Poesen J (2023) Land use, management and climate effects on runoff and soil loss responses in the highlands of Ethiopia. J Environ Manage 326(2023):116707

Ebabu K, Tsunekawa A, Haregeweyn N, Adgo E, Tsegaye D (2019) Effects of land use and sustainable land management practices on runoff and soil loss in the Upper Blue Nile basin, Ethiopia. 648(August 2018):1462–1475

Elnashar A, Zeng H, Wu B, Fenta AA, Nabil M, Duerlerv R (2021) Soil erosion assessment in the Blue Nile basin driven by a novel RUSLE-GEE framework. Sci Total Environ 793:1–17

Endalamaw NT, Moges MA, Kebede YS, Alehegn BM, Sinshaw BG (2021) Potential soil loss estimation for conservation planning, upper Blue Nile basin, Ethiopia. Environ Challenges 5:100224

Erkossa T, Geleti D, Williams TO, Laekemariam F, Haileslassie A (2020) Restoration of grazing land to increase biomass production and improve soil properties in the Blue Nile basin: effects of infiltration trenches and Chloris Gayana reseeding. Renewable Agric Food Syst 37:S64–S72

FAO (1986) Highlands reclamation study-Ethiopia. Final report, volume 1 and 2. Italy, Food and Agriculture Organization, Rome

FAO (2020) Main report. In: Reforming China's healthcare system.https://doi.org/10.4324/978131 5184487-1

FDRE (Federal Democratic Republic of Ethiopia) (2011) Ethiopia's climate-resilient green economy green economy strategy document. 22 pp. Addis Ababa, November 2011. Retrieved 21 July 2016, from http://www.undp.org/content/dam/ethiopia/docs/

Fenta AA, Tsunekawa A, Haregeweyn N, Poesen J, Tsubo M, Borrelli P, Vanmaercke M, Broeckx J, Yasuda H, Kawai T, Kurosaki Y, Panagos P (2020) Land susceptibility to water and wind erosion risks in the East Africa region. Sci Total Environ 703:135016

Garzanti E, Andò S, Vezzoli G, Megid AAA, El Kammar A (2006) Petrology of Nile River sands (Ethiopia and Sudan): sediment budgets and erosion patterns. Earth Planet Sci Lett 252:327–341

Gashaw T, Tulu T, Argaw M, Worqlul AW (2017) Evaluation and prediction of land use/ land cover changes in the Andassa watershed, Blue Nile Basin, Ethiopia. Environ Syst Res 6:17

Gebere SB, Alamirew T, Merkel BJ, Melesse AM (2015) Performance of high-resolution satellite rainfall products over data scarce parts of Eastern Ethiopia. Remote Sensing 7(9):11639–11663

Gebrehiwot SG, Bewket W, Gärdenäs AI, Bishop K (2014) Forest cover change over four decades in the Blue Nile Basin, Ethiopia: comparison of three watersheds. Reg Environ Chang 14(1):253–266

Gelagay HS, Minale AM (2016) Soil loss estimation using GIS and Remote sensing techniques: a case of Koga watershed, Northwestern Ethiopia. Int Soil Water Conserv Res 4(2):126–136

Gessesse BB, Bewket W, Bräuning A (2015) Determinants of farmers' tree-planting investment decisions as a degraded landscape management strategy in the central highlands of Ethiopia. Solid Earth 7:639–650

Ghorbanzadeh O, Meena SR, Blaschke T, Aryal J (2019) UAV-based slope failure detection using deep-learning convolutional neural networks. Remote Sens 11:2046

Gashaw T, Worqlul AW, Dile YT, Addisu S, Bantider A, Zeleke G (2020) Evaluating potential impacts of land management practices on soil erosion in the Gilgel Abay watershed, upper Blue Nile basin. Heliyon 6(8):e04777

Girmay G, Moges A, Muluneh A (2020) Estimation of soil loss rate using the USLE model for Agewmariayam Watershed, northern Ethiopia. Agric Food Secur 9:9

Gizaw D, Leta KM (2019) Spatial and temporal distribution of sediment yield-case study Nashe, Blue Nile Basin, Ethiopia. J Water Sustain 9(2):23–34

Gizaw D, Lulseged T, Wuletawu A, Tilahun A, Anthony W (2021) Effects of land management practices and land cover types on soil loss and crop productivity in Ethiopia: a review. Int Soil Water Conserv Res 9(4):544–554

Haregeweyn N, Tsunekawa A, Nyssen J, Poesen J, Tsubo M, Tsegaye Meshesha D, Schütt B, Adgo E, Tegegne F (2015) Soil erosion and conservation in Ethiopia: a review. Prog Phys Geogr 39:750–774

Haregeweyn N, Tsunekawa A, Poesen J, Tsubo M, Meshesha DT, Fenta AA, Nyssen J, Adgo E (2017) Comprehensive assessment of soil erosion risk for better land use planning in river basins: case study of the Upper Blue Nile River. Sci Total Environ 574:95–108

Hurni H (1985) Erosion—productivity—conservation systems in Ethiopia. In: Pla Sentis I (ed) International conference on soil conservation, 4th ed. Maracay, Venezuela, pp 654–674

Hurni H (1988) Degradation and conservation of the resources in the Ethiopian highlands. Mt Res Dev 8:123–130

Hurni H (1993) Land degradation, famines and resource scenarios in Ethiopia. In: Pimentel D (ed) World soil erosion and conservation. Cambridge University Press, Cambridge, pp 27–62

Hurni H (2005) Decentralised development in remote areas of the Simen mountains, Ethiopia. IP2 working paper no. 1. NCCR north-south dialogue series. Bern, Switzerland: Swiss National Centre of Competence in Research (NCCR) North-South. Also available at: http://www.nccr-north-south.unibe.ch/publications/Infosystem/Online%20Dokumente/UploadDecentralised%20Development%20Simen%20Ethiopia_Hurni%202005%281%29.pdf. Accessed on 5 January 2010.

Hurni H, Tato K, Zeleke G (2005) The implications of changes in population, land use, and land management for surface runoff in the Upper Nile basin area of Ethiopia. Mt Res Dev 25(2):147–154

Hurni H, Solomon A, Amare B, Berhanu D, Ludi E, Portner B, Birru Y, Gete Z (2010) Land degradation and sustainable land management in the Highlands of Ethiopia. In: Hurni H, Wiesmann U (eds) With an international group of co-editors. Global change and sustainable development: a synthesis of regional experiences from research partnerships. Perspectives of the Swiss National Centre of Competence in Research (NCCR) North-South, University of Bern, vol 5. Geographica Bernensia, Bern, Switzerland, pp 187–207

Hurni K, Zeleke G, Kassie M, Tegegne B, Kassawmar T, Teferi E, Moges A, Tadesse D, Ahmed M, Degu Y (2015) Soil degradation and sustainable land management in the rainfed agricultural areas of Ethiopia: an assessment of the economic implications. Report for the economics of land degradation initiative

Kabite G, Muleta MK, Gessesse B (2022) Impacts of combined and separate land cover and climate changes on hydrologic responses of Dhidhessa River basin, Ethiopia. Int J River Basin Manage 2022:2101464.https://doi.org/10.1080/15715124.2022.2101464

Kebede B, Atsushi T, Nigussie H, Enyew A, Kindiye E, Derege T, Meshesha M, Tsubo TM, Ayele AF (2020) Determining C- and P-factors of RUSLE for different land uses and management practices across agro-ecologies: case studies from the Upper Blue Nile basin, Ethiopia. Phys Geogr 42(2):160–182

Kenneth GR, George RF, Glenn AW, Jeffrey P (1991) Revised universal soil loss equation, soil and Water Conservation Society. J Soil Water Conserv 46(1):30–33

Khosravi K, Panahi M, Golkarian A, Keesstra SD, Saco PM, Bui DT, Lee S (2020) Convolutional neural network approach for spatial prediction of flood hazard at national scale of Iran. J Hydrol 591:125552

Khosravi K, Rezaie F, Cooper JR, Kalantari Z, Abolfathi S, Hatamiafkoueieh J (2023) Soil water erosion susceptibility assessment using deep learning algorithms. J Hydrol 618(2023):129229

Kidane D, Alemu B (2015) Review of the effect of upstream land use practices on soil erosion and sedimentation in the Upper Blue Nile Basin, Ethiopia. Res J Agricul Environ Manage 4(2):55–68

Kisi O, Dailr AH, Cimen M, Shiri J (2012) Suspended sediment modeling using genetic programming and soft computing techniques. J Hydrol 450–451:48–58

Kisi O, Genc O, Dinc S, Zounemat-Kermani M (2016) Daily pan evaporation modeling using chi-squared automatic interaction detector, neural networks, classification and regression tree. Comput Electron Agric 122:112–117

Kuok KKK, Mah DYS, Chiu PC (2013) Evaluation of C and P factors in universal soil loss equation on trapping sediment: case study of Santubong river. J Water Resour Prot 5(12):1149–1154

Laflen JM, Flanagan DC (2013) The development of U. S. soil erosion prediction and modeling. Int Soil Water Conserv Res 1(2):1–11

Legese AG, Attila N, Hailu KA (2022) Soil loss estimation and severity mapping using the RUSLE model and GIS in Megech watershed, Ethiopia. Environ Challenges 8(2022):100560

Lemann T, Zeleke G, Amsler C, Giovanoli L, Suter H, Roth V (2016) Modelling the effect of soil and water conservation on discharge and sediment yield in the upper Blue Nile basin, Ethiopia. Appl Geogr 73:89–101

Liyew M, Tsunekawa A, Haregeweyn N (2019) Exploring land use/land cover changes, drivers and their implications in contrasting agro-ecological environments of Ethiopia. Land Use Policy 87(June):104052

Maalim FK, Melesse AM (2013) Modeling the impacts of subsurface drainage systems on Runoff and Sediment Yield in the Le Sueur Watershed, Minnesota. Hydrol Sci J 58(3):1–17

Maalim FK, Melesse AM, Belmont P, Gran K (2013) Modeling the impact of land use changes on runoff and sediment yield in the Le Sueur Watershed, Minnesota Using GeoWEPP. CATENA 107:35–45

Malede DA, Alamirew T, Kosgie JR, Andualem TG (2023) Analysis of land use/land cover change trends over Birr River Watershed, Abbay Basin, Ethiopia. Environ Sustain Indic 17:100222

Mekuria W, Wondie M, Amare T, Wubet A, Feyisa T, Yitaferu B (2018) Restoration of degraded landscapes for ecosystem services in North-Western Ethiopia. Heliyon 4(8)

Melesse AM, Ahmad S, McClain ME, Wang X, Lim YH (2011) Suspended sediment load prediction of river systems: an artificial neural network approach. Agric Water Manag 98(5):855–866

Melaku ND, Renschler CS, Flagler J, Bayu W, Klik A (2018) Integrated impact assessment of soil and water conservation structures on runoff and sediment yield through measurements and modeling in the Northern Ethiopian highlands. CATENA 169:140–150

Mengesha Z, Mohammed SM, Demeke S, Anwar A, Adem ML (2018) Assessment of soil erosion using RUSLE, GIS and remote sensing in NW Ethiopia. Geoderma Reg 12

Mengie B, Teshome Y, Dereje T (2019) Effects of soil and water conservation practices on soil physicochemical properties in Gumara watershed, Upper. Ecol Process 1–14

Mersha BD, Zeleke G, Gebrehiwot SG (2022) Assessing the effect of sustainable land management on improving water security in the Blue Nile Highlands: a paired catchment approach. Environ Monit Assess 194:197

Mhiret DA, Dagnew DC, Guzman CD, Alemie TC, Zegeye AD, Tebebu TY, Langendoen EJ, Zaitchik BF, Tilahun SA, Steenhuis TS (2020) A nine-year study on the benefits and risks of soil and water conservation practices in the humid highlands of Ethiopia: the Debre Mawi watershed. J Environ Manage 270:110885

Moges DM, Bhat HG (2017) Integration of geospatial technologies with RUSLE for analysis of land use/cover change impact on soil erosion: case study in Rib watershed, north-western highland Ethiopia. Environ Earth Sci 76

Moges MA, Zemale FA, Alemu ML, Ayele GK, Dagnew DC, Tilahun SA, Steenhuis TS (2016) Sediment concentration rating curves for a monsoonal climate: upper Blue Nile. SOIL 2:337–349

Mohammed H, Alamirew T, Assen M, Melesse AM (2015) Modeling of sediment yield in Maybar gauged watershed using SWAT, northeast Ethiopia. CATENA 127:191–205

Mosavi A, Sajedi-Hosseini F, Choubin B, Taromideh F, Rahi G, Dineva A (2020) Susceptibility mapping of soil water erosion using machine learning models. Water 12:1995

Msagahaa JJ, Ndomba PM, Melesse AM (2014) Modeling sediment dynamics: effect of land use, topography and land management. In: Melesse AM, Abtew W, Setegn S (eds) Nile River basin: ecohydrological challenges, climate change and hydropolitics, pp 165–192

Muche H, Temesgen M, Yimer F (2013) Soil loss prediction using USLE and MUSLE under conservation tillage integrated with 'fanya juus' in Choke Mountain, Ethiopia. Int J Agricul Sci 3(10):46–52

Naomi R (2020) Effects of afforestation and crop systems on the water balance in the highlands of Ethiopia

Nigus DM, Chris SR, Jared FW, Bayu AK (2018) Integrated impact assessment of soil and water conservation structures on runoff and sediment yield through measurements and modeling in the Northern Ethiopian highlands. CATENA 169

Nkonya E, Mirzabaev A (2016) Economics of Land Degradation and Improvement-A Global Assessment for Sustainable Development (Ephraim NkJoachim von Braun AM (ed.)). Springer, Cham Heidelberg New York Dordrecht London

Nyssen J, Poesen J, Haile M, Moeyersons J, Deckers J, Hurni H (2009) Effects of land use and land cover on sheet and rill erosion rates in the Tigray highlands, Ethiopia. Z Geomorphol 53(2):171–197

Panagos P, Borrelli P, Meusburger K, Alewell C, Lugato E, Montanarella L (2015) Estimating the soil erosion cover-management factor at the European scale. Land Use Policy 48:38–50

Panagos P, Borrelli P, Meusburger K, Yu B, Klik A, Lim KJ, Yang JE, Ni J, Miao C, Chattopadhyay N, Sadeghi SH, Hazbavi Z, Zabihi M, Larionov GA, Krasnov SF, Gorobets AV, Levi Y, Erpul G, Birkel C, Naipal V, Ballabio C (2017) Global rainfall erosivity assessment based on high-temporal resolution rainfall records. Sci Rep 7(1):1–12

Preetha P, Al-Hamdan A (2022) A union of dynamic hydrological modeling and satellite remotely-sensed data for spatiotemporal assessment of sediment yields. Remote Sensing 14(2)

Raza A, Ahrends H, Habib-Ur-Rahman M, Gaiser T (2021) Modeling approaches to assess soil erosion by water at the field scale with special emphasis on heterogeneity of soils and crops. Land 10:422

Renard K, Ferreira V (1993) RUSLE model description and database sensitivity. J Environ Qual 22:458–466

Renard KG, Foster GR, Weesies GA, McCool DK, Yoder DC (1997) Predicting soil erosion by water: a guide to conservation planning with the revised universal soil loss equation (RUSLE). In: Agricultural handbook, vol 703. U.S. Department of Agriculture, Washington, DC, p 407

Rice SP, Church MA, Rennie CD (2012) Bedload transport measurements using the acoustic Doppler current profiler (ADCP) in large, gravel-bed rivers. Geomorphology 139:444–455

Schmidt E, Zemadim B (2015) Expanding sustainable land management in Ethiopia: scenarios for improved agricultural water management in the Blue Nile. Agric Water Manag 158(2015):166–178

SCRP (Soil Conservation Research Project) (1996) Soil erosion hazard assessment for land evaluation. Research Report, SCRP, Addis Ababa

Setegn S, Dargahi B, Srinivasan R, Melesse A (2010) Modeling of sediment yield from Anjeni-gauged watershed, Ethiopia using SWAT model. J Am Water Resour Assoc 46:514–526

Setegn SG, Srinivasan R, Dargahi B, Melesse AM (2009) Spatial delineation of soil erosion prone areas: application of SWAT and MCE approaches in the Lake Tana Basin, Ethiopia. Hydrol Process 23(26):3738–3750

Sewnet A, Abebe G (2017) Land use and land cover change and implication to watershed degradation by using GIS and remote sensing in the Koga watershed, North Western Ethiopia. Hussein 2009.https://doi.org/10.1007/s12145-017-0323-5

Shiferaw B, Holden ST (1999) Soil erosion and smallholders' conservation decisions in the highlands of Ethiopia. World Dev 27:739–752

Sinshaw BG, Belete AM, Mekonen BM, Wubetu TG, Anley TL, Alamneh WD, Atinkut HB, Gelaye AA, Bilkew T, Tefera AK, Dessie AB, Fenta HM, Beyene AM, Bizuneh BB, Alem HT, Eshete DG, Atanaw SB, Tebkew MA, Birhanu MM (2021) Watershed-based soil erosion and sediment yield modeling in the Rib watershed of the Upper Blue Nile basin, Ethiopia. Energy Nexus 3:100023

SLMP (2013) Sustainable Land Management Project (SLMP) II Revised final draft document on environmental and social management framework (ESMF). Addis Ababa September, p 70

Smyth A, Dumanski J (1993) FESLM: an international framework for evaluating sustainable land management. World Soil Resources Report, p 74

Sonneveld B (2002) Land under pressure: the impact of water erosion on food production in Ethiopia. Shaker Publishing (PhD dissertation). Netherlands

Sonneveld BGJS, Keyze MA, Stroosnijder L (2011) Evaluating quantitative and qualitative models: an application for nationwide water erosion assessment in Ethiopia. Environ Model Softw 26:1161–1170

Sutcliffe JV, Parks YP (1999) The hydrology of the Nile. IAHS Special Publication 5. Oxfordshire, IAHS, Wallingford, UK

Taddese G (2001) Land degradation: a challenge to Ethiopia. Environ Manage 27(6):815–824

Tadesse G, Zavaleta E, Shennan C, Fitzsimmons M (2014) Policy and demographic factors shape deforestation patterns and socio-ecological processes in southwest Ethiopian coffee agroecosystems. Appl Geogr 54:149–159

Tan Z, Leung LR, Li H, Tesfa T (2018) Modeling sediment yield in land surface and earth system models: model comparison, development, and evaluation. J Adv Model Earth Syst 10:2192–2213

Taye G, Poesen J, Van Wesemael B, Goosse T, Teka D, Deckers J, Hallet V, Haregeweyn N, Nyssen J, Maetens W (2013) Effects of land use, slope gradient, soil, and water conservation techniques, on runoff and soil loss in a semi-arid environment. J Phys Geogr 34(3):236–259

Taye G, Poesen J, Vanmaercke M, Van Wesemael B, Tesfaye S, Nyssen J, Haregeweyn N (2017) Determining RUSLE P- and C-factors for stone bunds and trenches in rangeland and cropland, northern Ethiopia. Land Degrad Dev 29(3):812–824

Tedla S, Lemma K (1998) Environmental management in Ethiopia: have the national conservation plans worked? Environmental Forum Publications Series, Number 1. The Organization for Social Science Research in Eastern and Southern Africa (OSSREA), Addis Ababa

Tefera B, Ayele G, Atnafe Y, Jabbar MA, Dubale P (2002) Nature and causes of land degradation in the Oromiya Region: a review. Socio-economics and policy research working paper 36. ILRI (International Livestock Research Institute), Nairobi, Kenya, p 82

Terefe T, Chala D, Belay S, Bamlaku A, Moges K (2019) Land use/land cover dynamics in response to various driving forces in Didessa sub-basin, Ethiopia. GeoJournal 4

Tian P, Zhu Z, Yue Q, He Y, Zhang Z, Hao F, Guo W, Chen L, Liu M (2021) Soil erosion assessment by RUSLE with improved P factor and its validation: case study on mountainous and hilly areas of Hubei Province, China. Int Soil Water Conserv Res 9(2021):433–444

Tibebe D, Bewket W (2010) Surface runoff and soil erosion estimation using the swat model in the Keleta watershed, Ethiopia. Land Degrad Dev 22(6):551–564

Tien B, Pradhan D, Lofman B, Revhaug OI (2012) Landslide susceptibility assessment in vietnam using support vector machines, decision tree, and Naïve Bayes models. Math Probl Eng 2012:1–26

Tilahun SA, Guzman CD, Zegeye AD, Ayana ES, Collick AS, Yitaferu B, Steenhuis TM (2014) Spatial and temporal patterns of soil erosion in the semi-humid Ethiopian highlands: a case study of Debre Mawi watershed. In: Melesse A, Abtew W, Setegn S (eds) Nile River Basin. Springer

Titti G, Napoli GN, Conoscenti C, Lombardo L (2022) Cloud-based interactive susceptibility modeling of gully erosion in Google Earth Engine. Int J Appl Earth Observ Geoinform 115:103089

Tsegaye L, Bharti R (2021) Soil erosion and sediment yield assessment using RUSLE and GIS-based approach in Anjeb watershed, Northwest Ethiopia. SN Appl Sci 3(5):1–19

Tsige MG, Malcherek A, Seleshi Y (2022) Improving the modified universal soil loss equation by physical interpretation of its factors. Water 14:1450

Vu DT, Tran XL, Cao MT, Tran TC, Hoang ND (2020) Machine learning based soil erosion susceptibility prediction using social spider algorithm optimized multivariate adaptive regression spline. Measurement 164:108066

Wang X, Shang S, Yang W, Melesse AM (2008) Simulation of an agricultural watershed using an improved curve number method in SWAT. Trans ASABE 51(4):1323–1339

Wassie SB (2020) Natural resource degradation tendencies in Ethiopia: a review. Environ Syst Res 9(1):1–29

Williams JR (1975) Sediment-yield prediction with Universal Equation using runoff energy factor. In: Present and prospective technology for predicting sediment yield and sources, U.S. Department Agriculture, ARS-S-40, pp 244–252

Wischmeier WH, Smith DD (1978) Predicting rainfall erosion losses. A guide to conservation planning. The USDA Agricultural Handbook No. 537, Maryland

Wischmeier WH, Smith DD (1965) Predicting rainfall erosion losses from crop land east of the rocky mountains guide for selection of practices soil and water conservation. U.S. Department of Agriculture, Agricultural Handbook No 282

Woldegiorgis B (2020) A history and policy analyses of Forest Governance in Ethiopia and REDD+. July, 0–79

Yaekob T, Tamene L, Gebrehiwot SG, Demissie SS, Adimassu Z, Woldearegay K, Mekonnen K, Amede T, Abera W, Recha JW, Solomon D (2022) Assessing the impacts of different land uses and soil and water conservation interventions on runoff and sediment yield at different scales in the central highlands of Ethiopia. Renewable Agric Food Syst 37:S73–S87

Yalew SG, Mul ML, Van Griensven A, Teferi E, Priess J (2016) Land-use change modelling in the Upper Blue Nile basin. Environments 1–16

Yesuph AY, Dagnew AB (2019) Soil erosion mapping and severity analysis based on RUSLE model and local perception in the Beshillo Catchment of the Blue Nile Basin, Ethiopia. Environ Syst Res 8:17

Yibeltal M, Tsunekawa A, Haregeweyn N, Adgo E, Meshesha DT, Aklog D, Masunaga T, Tsubo M, Billi P, Vanmaercke M, Ebabu K, Dessie M, Sultan D, Liyew M (2019) Analysis of long-term gully dynamics in different agro-ecology settings. CATENA 179:160–174

Zeleke G, Hurni H (2001) Implications of land use and land cover dynamics for mountain resource degradation in the Northwestern Ethiopian Highlands. Mt Res Dev 21(2):184–191

Zeleke G, Kassie M, Pender J, Yesuf M (2006) Stakeholder Analysis for Sustainable Land Management (SLM) in Ethiopia: assessment of opportunities, strategic constraints, information needs, and knowledge gaps. Environmental Economics Policy Forum for Ethiopia (EEPFE), International Food Policy Research Institute (IFPRI), Addis Ababa, Ethiopia

Chapter 12
Remote Sensing-Based Agricultural Drought Monitoring in the Eastern Abbay Basin

Gebremariam Adane Getu and Worku Zewdie

Abstract Drought studies in Ethiopia showed an increasing frequency and intensity of agricultural drought specifically in Eastern Amhara [North Wello Zone (NWZ)] over time. Most of these studies are based on ground point measurements, which are limited and unevenly distributed with sparse networks, to analyze and monitor drought. However, remote sensing-based agricultural drought assessments have been inadequate to perform in Amhara region especially in NWZ for the last seasons. Therefore this study mainly aimed to monitor the spatial and temporal patterns of agricultural drought based on satellite-based remote sensing data and Geographic Information Systems (GIS). This study mainly used the Normalized Difference Vegetation Index (NDVI) and Land Surface Temperature (LST) from Moderate Resolution Imaging Spectroradiometer (MODIS) and Climate Hazard Group Infrared Precipitation with Station (CHIRPS) data on different sources. This study also used the satellite-derived products of the Vegetation Condition Index (VCI), Temperature Condition Index (TCI), Vegetation Health Index (VHI), and correlation between NDVI and LST from 2002 to 2021 in NWZ. The CHIRPS precipitation data were used to classify whether the study area was dry or wet during the particular period. TCI results revealed the stress of vegetation from moderate to extreme drought conditions in 2009 and 2015 over most parts of the study area. However, in 2006 and 2021, near-favorable or good vegetation conditions during the main cropping season. The VCI results also showed that the northeastern, eastern, southern, and parts of the central lowlands of the NWZ were dominated by low vegetation conditions. The VHI results also revealed that agricultural drought conditions occurred in 2009 and 2015 with different severities and spatial extents. Similarly, 65% of NWZ was hit by extreme drought and 10% by severe drought in 2015 and 63% by extreme drought and 15% by severe drought in 2009. The correlation analysis of LST-NDVI was

---

G. A. Getu (✉)
Ethiopian Meteorology Institute, Addis Ababa, Ethiopia
e-mail: gebremariamadane@gmail.com

W. Zewdie
Remote Sensing Department, Ethiopian Space Science and Geospatial Institute, Addis Ababa, Ethiopia

negative in 2006, 2009, and 2015. The output indicated that water limits vegetation growth by showing LST-NDVI negative relationship. This analysis confirmed remote sensing data's contribution to monitoring and assessing agricultural drought risk areas and vulnerable societies. The combined analysis of VHI is recommended rather than using single indices to monitor agricultural drought in NWZ and in other places with similar agroecology.

Keywords Agricultural drought · NDVI · LST · VHI · Drought monitoring

Introduction

Drought is a severe shortage of water caused by below-normal rainfall compared to the long-term climatology (WMO 2012) or abnormally dry weather caused by below-normal rainfall (Bayissa et al. 2018). Drought can affect agricultural activities, water resources, ecosystem services, tourism, biodiversity, livestock husbandry, and other natural resources (Mastrangelo et al. 2012). Drought affects most parts of African continents and occurring frequently with overwhelming impacts on food shortages, loss of life, migration from rural to urban areas, and reduced agricultural production as well as loss of lives of humans and livestock (Nam et al. 2018; Kogan and Guo 2016).

Drought can be commonly classified into four main categories: meteorological (deficit of precipitation compared to long-term average condition for a specified location and timescales), agricultural (deficit of water in soil for the plant growth during the growing stages), hydrological (deficiencies of surface runoff and groundwater among reservoirs, lakes, ponds, and rivers), and socioeconomic (decrease in natural resources and human economy due to the above three types of drought) (Wilhite and Glantz 1985; Mishra and Singh 2010). All four drought types are interlinked. Agricultural drought occurs in two ways: because of the shortage of precipitation that reduces the amount of water in the soil, which results in crop failure, and because of the temperature increase that causes an increase in evapotranspiration (Mishra and Singh 2010). Regularly, agricultural drought is a period when the soil moisture is insufficient to meet the water requirements of plant roots and hinders appropriate agricultural management in a specific time or phase (Hagenlocher et al. 2019). Agricultural drought reduces crop production and can be monitored by different drought-monitoring meteorological and vegetation indices (Legesse 2010).

Studies on various aspects of drought has been reported (Seka et al. 2022; Zeleke et al. 2022; Bayissa et al. 2021; Zelelew et al. 2018; Abiy et al. 2019). Several drought monitoring meteorological indices have been developed to date, including the palmer drought severity index (PDSI) (Palmer 1965), rainfall anomaly index (RAI) (Van Rooy 1965), standardized precipitation index (SPI) (McKee et al. 1993), and standardized precipitation evapotranspiration index (SPEI) (Vicente-Serrano et al. 2010). Bayissa et al. (2018) have used some meteorological indices to characterize drought in the Upper Blue Nile basin.

The meteorological parameters that are used to monitor drought can be station-based weather parameters or satellite-based blended datasets. In most developing countries, including Ethiopia, in situ-based meteorological stations are scarce, have a coarse resolution, are less continuous, have observational errors, and are unevenly distributed throughout the country (Hashim et al. 2016). Satellite-based remote sensing drought monitoring contributes to these limitations by offsetting the uneven distribution of in situ stations, incorporating imagery to a coarse resolution, and covering large areas (Peng et al. 2017). Remote sensing-based satellite imageries from different sources have the ability to monitor and assess agricultural drought. The Advanced Space Borne Thermal Emission and Reflection Radiometer (ASTER), Moderate Resolution Imaging Spectroradiometer (MODIS), Sentinel, Landsat, and Satellite Pour Observation de la Terre (SPOT Satellite) were some of the data sources to monitor agricultural drought. The combination of satellite data products for drought monitoring is appropriate and substantial to take mitigation and adaptation measures, early warning, and early action as well as for preparedness (Tran et al. 2017). In this regard, various indices were developed to monitor agricultural drought based on remotely sensed images at regional and global scales. Among these indices are the land surface temperature (LST) (Feldhake et al. 1996), temperature condition index (TCI) (Kogan 1997; Kogan et al. 2004), normalized difference vegetation index (NDVI) (Holben et al. 1980), vegetation condition index (VCI) (Kogan 1990), vegetation health index (VHI) (Kogan 2002), enhanced vegetation index (EVI) (Huete 1988), and leaf area index (LAI) (Chen and Cihlar 1996).

Determining the onset, secession, and spatial extent of drought hazards is very complex and challenging. Consequently, identifying and understanding drought-prone areas, depending on the spatial and temporal pattern of historical drought in Ethiopia, particularly in eastern Amhara, has great importance. Monitoring activity is vital to take appropriate measures of mitigation and adaptation strategies to reduce the future drought risk over the region by using integrated satellite data and GIS tools.

Satellite-based remote sensing data by GIS techniques are commonly used and are considered the most popular drought monitoring tools to deliver near real-time information through several drought indices (Chopra 2006; Gadisso 2007). These methods are commonly employed for monitoring and identifying agricultural drought and crop yield estimation (Legesse and Suryabhagavan 2014). Consequently, agricultural drought can be monitored by analyzing different remote sensing-based vegetation indices and meteorological drought monitoring indices (Alahacoon et al. 2021). Such methods have paramount importance in assessing and monitoring the impact of agricultural droughts on crops that are predominantly grown and cultivated in Ethiopia, specifically in NWZ.

Analyzing the spatial and temporal variation in agricultural drought has a significant contribution to reducing the impact of agricultural drought. The occurrence of drought can cause food insecurity and social instability in the local community. Therefore, this study assesses and monitors the dimensions of agricultural drought by using a satellite-based vegetation condition index.

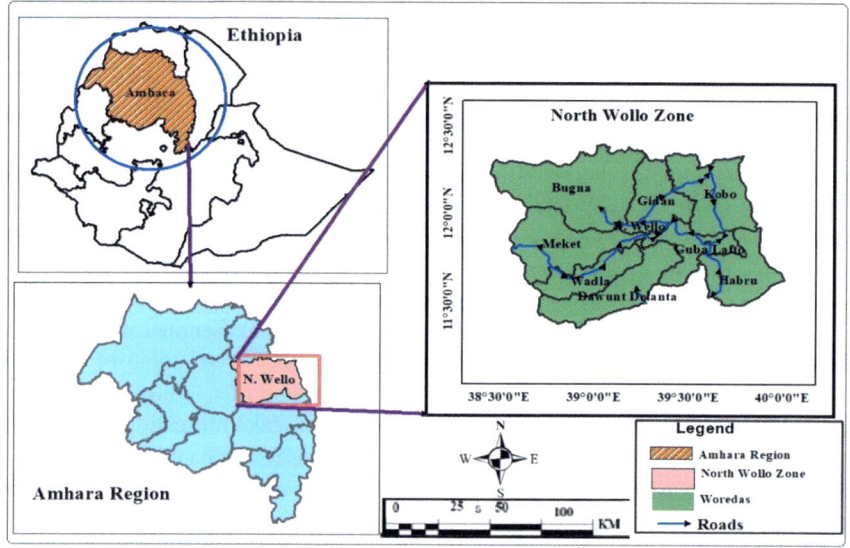

Fig. 12.1 Map of the study area

Methodology

Description of Study Area

Location

This study was performed in the North Wello Zone (NWZ) of the Amhara National Regional State (ANRS). NWZ is among the 12 zones in ANRS and is located between longitudes of 38°30′00″ E to 40°00′00″ E and latitudes of 11°30′00″ N to 12°30′00″ N (Fig. 12.1). The study area covers approximately 12,755.14 km^2, and it covers an estimated of 20%, from the total area of ANRS (Gebre et al. 2017). The NWZ is divided into 14 rural districts locally called Weredas and four town administrations. The capital city and the economic center of the NWZ is Woldia (CSA 2008).

Climate

The climate of the NWZ is depicted by diverse climatic environments that vary from semiarid and arid climate in the northern and eastern parts to cold and humid areas in the southern and central parts. It is one of the hottest in the lowlands and the coldest in the highlands, obtaining erratic rainfall. NWZ is characterized by two seasons or bimodal rainfall regimes, of kiremt (June–September) and the Belg season from (February to May) (Korecha and Barnston 2007; Lemma 1996; Degefu 1987).

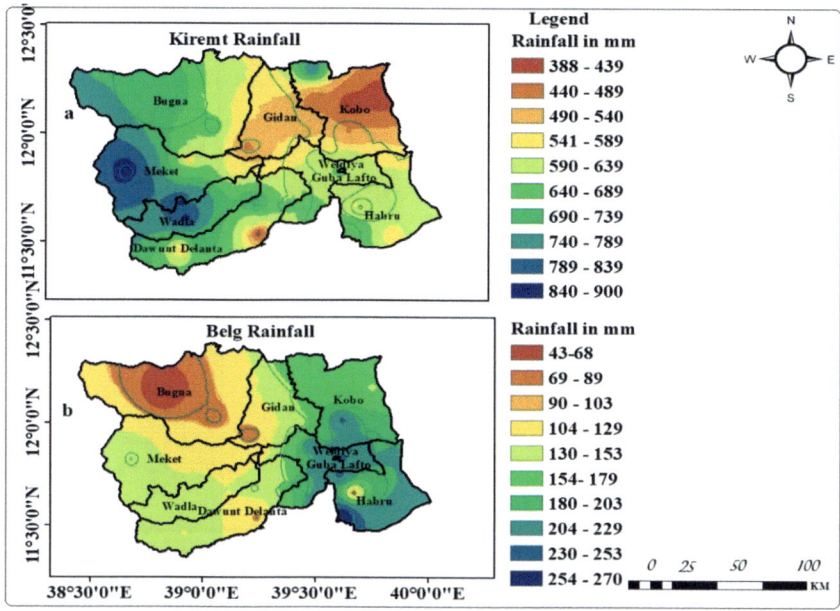

Fig. 12.2 Spatial distribution of mean seasonal rainfall in Kiremt (**a**) and Belg (**b**) for NWZ (2002–2021)

Rainfall in the Study Area

Climatologically, eastern Amhara, particularly NWZ, is characterized by a bimodal rainfall regime with short and small amounts of rain in the Belg season and long rainy seasons in the Kiremt season. These NWZ can be characterized by a minimum of 300 mm of rainfall and the highest maximum rainfall 1271.51 mm (recorded in 2010) and an average of 634 mm of seasonal rainfall in the main rainy season (Fig. 12.2a). However, in the Belg season, it received up to 270 mm of rainfall (Fig. 12.2b). The seasonal rainfall of NWZ was inconsistent and erratic in both the Kiremt and Belg seasons with spatial and temporal distribution from 2002 to 2021. Significant variability and unpredictability were observed over parts of eastern and certain central areas, especially in the short rainy season. During Kiremt season, the average rainfall in most parts of the region is 634 mm. The Belg season rainfall is short and periodic. As indicated in the seasonal rainfall distribution map (Fig. 12.2b), most parts of the region except some parts of eastern, the rest had received less amount of rainfall. The erratic nature and deficit of rainfall from the expected normal rainfall is an indicator of drought in the Belg Season.

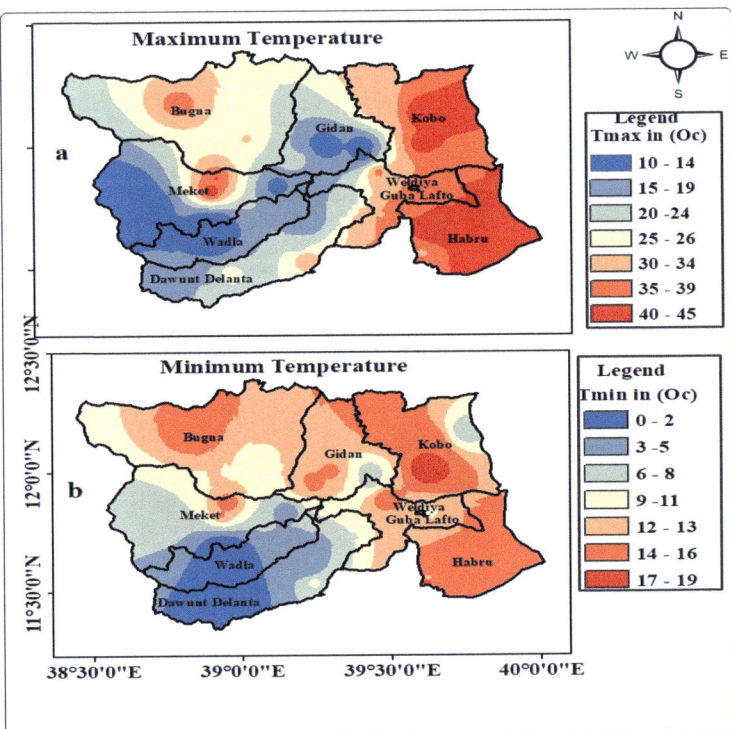

Fig. 12.3 The spatial distribution of maximum (**a**) and minimum (**b**) temperatures in the Kiremt season from 2002 to 2021 for NWZ

Temperature in the Study Area

The variation in the minimum temperature ranges from 0 to 19 °C, and the maximum temperature variation runs between 10 and 45 °C in the Kiremt season, but the variation increases in the Bega and Belg seasons over the study area (Fig. 12.3a, b). The highest temperature was recorded up to 45 °C in the lowland areas of Kobo, Habiru, and Bugna woredas, which accelerates the rate of evaporation in the soil moisture to the plant (Fig. 12.3a). In addition to frequent drought, the zone is mainly affected by the minimum temperature (frost), which affects wheat and barley crops at the flowering and ripening stages. The lowest temperature is mostly recorded in the highland areas of Meket (Gashena station) and neighboring woredas of Wadla Delanta and Dawunt woredas, with approximately less than 5 °C in the Kiremt season, whereas in the Bega season, the minimum temperature decreases up to zero or negative (Fig. 12.3b).

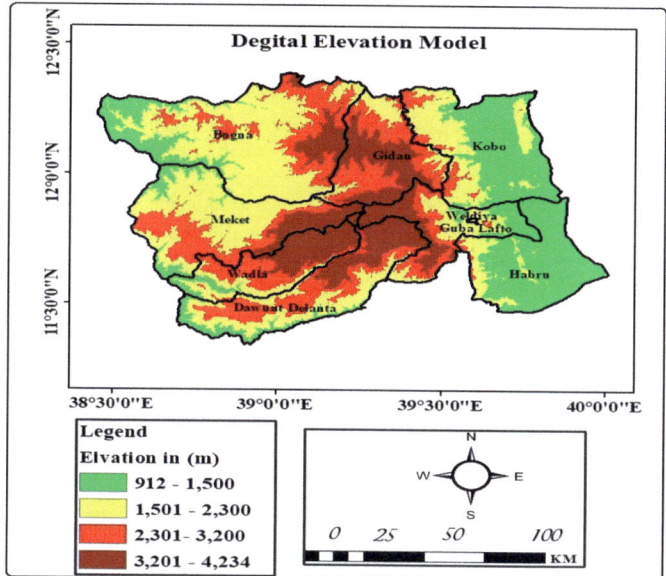

Fig. 12.4 Topographic map of the study area

Topography

Most parts of the study area are complex and have difficult topographic structures. It encompasses high mountainous and deeply incised canyons, valleys, and plateaus with steep slopes that dominate most parts of the zone, which severely limits the size of the cultivated areas (Hailu 2013). The NWZ is characterized by different topographic structures and significantly affects the annual cycle of precipitation. The altitude of the area varies from 912 to 4234 m above the mean sea level. The highest altitude is in Lasta (Mount Abuna Yosef), and the lowest altitudes are located in the Habru districts. Depending on the altitudinal variation, the study area covers different agro-ecological regions, namely, low land or Kola (912–1500) meters, mid-altitude or Woyna Dega (1500–2300) meters, highland or Dega (2300–3200) meters and Wurch (3200–4234) meters (Fig. 12.4a).

Vegetation

In highland areas, eucalyptus trees are commonly planted for construction and charcoal production. The factors that increase the population growth can cause to loss and degradation of the natural vegetation over time. In the northern highlands, the land use and land cover conditions frequently change due to the increasing in population growth and increasing agricultural practices. The highlands and the steep slope parts

Table 12.1 Land use/land cover classes and percentages (Zeleke et al. 2021)

Land use type	2010	In percentage (%)	2018	In percentage (%)
Forest	59530.24	4.870934	75,458.23	6.174208
Cropland	771521.2	63.12807	763,576.2	62.47799
Settlement	481.37	0.039387	866.61	0.070909
Grassland	362261.9	29.6413	353,955	28.96161
Wetland	1023.74	0.083765	1023.74	0.083765
Another land	27333.9	2.236538	27,272.54	2.231517

of the NWZ areas are commonly categorized as conservation areas through afforestation. Six categories of land use/land cover were identified based on IPCC definition (IPCC 2013). These classes are forest, grassland, cropland, wetland, settlement, and other classifications (Table 12.1).

Soils

Soil is a key component for plant and vegetation growth by storing water, minerals, and nutrients. Soil varies by parent materials, climate, vegetation and cultural impacts, and interactions among these factors (Legesse and Suryabhagavan 2014). Since the topography is a very rugged, steep slope and dominantly mountainous, soil erosion, deforestation, and land degradation are serious problems in the NWZ. In the higher elevation of the study area, there are various soil types, such as Cambisols, Luvisols, Vertisols, Xerosols, Leptosols, Regosols, and Nitisols. Cambisols are the dominant soil type in the study area, accounting for 46% of the total area (Zeleke et al. 2021). The Cambisol soil is mostly red to red-brown (locally called "Keyate") soils that have similar properties to Nitisols. It is dominant in the highland areas and is more fertile, with a high water holding capacity. However, the major limitation of Cambisol soil is the low organic matter and susceptibility to erosion in lowland areas (Elias and Fantaye 2000). Through repeated and continuous plowing and cultivation, the soil could deplete and lose its nutrients and change to a black color (Hodges 2010).

Agriculture

Most parts of the NWZ are unsuitable for agricultural practice due to the topographic effect (Hailu 2013). The region is categorized as a frequently drought-prone area with a food-insecure community.

The highland areas, locally called "Dega", receive their rainfall in the Belg season and are suitable for crop production, especially for cereal crops such as sorghum and barley. In NWZ, the major agricultural products are teff, sorghum, barley, maize, wheat, chickpea, and bean crops. Among these crops, teff, sorghum, maize, chickpea, and bean were cultivated in lowlands or the Kolla and Woyna Dega climatic zones of the eastern, southeastern, and western half of the areas. However, the central and southern halves of the NWZ are not suitable for agriculture due to their landscapes.

In general, rain-fed crop cultivation and livestock rearing are the main livelihoods of the society. However, agricultural activities in the zone are characterized by erratic rainfall, a high rate of evapotranspiration, and a low water holding capacity in the soil, causing crop failure that leads to an acute decline in food security, which frequently occurs in the zone. Sheep and horses in the highland and goats in the lowland areas are dominant in the study area. Cows and oxen are the most common livestock breeding in the areas.

Data

CHIRPS-Based Precipitation Data

The historical climatic precipitation data were collected from the Ethiopian Meteorology Institute (EMI) former NMA from in situ meteorological gauge station data. Rainfall stations are scarce and unevenly distributed in the country including this study area. Consequently, this study used the satellite gridded precipitation dataset obtained from (fp://fp.chg.ucsb.edu/pub/org/chg/products/CHIRPS-2.0/). The CHIRPS precipitation data products have a spatial resolution of 5 km developed by mixing the high-resolution cold cloud duration estimates of precipitation with point station rainfall (Funk et al. 2015). This dataset was evaluated over Eastern Africa, showing a higher skill with a lower bias (Dinku et al. 2018).

The CHIRPS data are appropriate for drought monitoring and early warning systems. The study area has a complex and heterogeneous topography and variation in climate characteristics. The CHIRPS-derived SPI and RAI have the ability to identify the wet and dry years to show the temporal distribution of drought in the study period of NWZ.

Enhanced/expected MODIS (eMODIS) Terra NDVI

Agricultural drought conditions were examined using Terra eMODIS NDVI (MOD13Q1) data. The data were obtained from the NASA USGS LP DAAC data center (https://lpdaac.usgs.gov/data_access/data_pool) from 2002 to 2021. The dataset has a spatial resolution of 250 m and a temporal resolution of 16 days. The Terra eMODIS NDVI data are better than Aqua data to monitor agricultural drought (Brown et al., 2015). The Aqua data are more liable to noise than the Terra data, and the variation in the internal cloud mask employed for compositing rules make Terra preferable (Brown et al. 2015). According to Rhee et al. (2010), eMODIS NDVI data is widely used vegetation indices for drought monitoring. NDVI-based data are essential for computing VCI, which is used to improve the identification, measurement, and monitoring of drought. Healthy vegetation absorbs much in the visible region and reflects strongly in the NIR. Kogan (1995a, b) stated that satellite-based NDVI data are important for analyzing and assessing VCI to monitor agricultural drought. NDVI can be computed mathematically as follows (Eq. 12.1):

$$\text{NDVI} = \frac{\text{NIR} - \text{RED}}{\text{NIR} + \text{RED}} \tag{12.1}$$

where RED is visible reflectance of the red band and NIR is the near-infrared reflectance. NDVI is an important drought indicator and is used for mapping vegetation dynamics and studying the vegetation greenness of the satellite image within a pixel. The values of NDVI vary from -1 to 1, where NDVI values ranging from approximately 0.1 or below are considered barren land, sand, snow, or rock, and NDVI values greater than 0.5 can be considered vegetated areas or dense forests (Gross 2005).

$$\text{eMODIS NDVI} = \text{Float}\left(\frac{\text{smoothed eMODIS NDVI} - 100}{100}\right) \tag{12.2}$$

The NDVI data were extracted for the Kiremt season for the years 2002–2021.

Land Surface Temperature (LST)

eMODIS LST (MOD11A2) emissivity with 8-day temporal and 1 km spatial resolution (aggregated into monthly and seasonal) data with NASA USGS LPDAAC data center (https://lpdaac.usgs.gov/data_access/data_pool). Frey et al. (2012) stated that LST developed during the daytime is preferable to nighttime (Aqua) to obtain more specific and detailed information. According to Frey et al. (2012), the eMODIS LST has a higher value of information than the AVHRR sensor. This is due to its up-to-date algorithm, periodic data acquisition, satellite azimuthal angle and zenith, and ease of interpretation of the products. This study utilized eMODIS LST (MOD11A2) data to analyze TCI and VHI, which is an improved and combined agricultural drought-monitoring index. Moreover, eMODIS LST is used to assess and monitor the effect of temperature on the vegetation canopy in the study area from 2002 to 2021.

Table 12.2 Data used for the analysis

No	Data type	Data source	Spatial resolution	Temporal resolution	Duration
1	eMODIS NDVI (MOD13Q1)	USGS earth data	250 m	16 days	2002–2021
2	eMODIS LST (MOD11A2)	USGS earth data	1 km	8 days	2002–2021
3	CHIRPS precipitation data	Satellite	5 km	Dekadal data	2002–2021
4	Major crop yield data (Maize, Teff, Wheat, Sorghum, and Barley)	CSA	Zonal bases	Annual crop data	2002–2018

Annual Crop Yield

Crop production is very sensitive and can be affected by agricultural drought (Kogan 1997; Legesse and Suryabhagavan 2014). To determine the association between crop yield and the existing agricultural drought conditions caused by climate variability and climate change, continuous and measured data are needed. Accordingly, the relationship between crop yield data and satellite-based drought indices is analyzed and correlated among the crop-growing seasons in the region. Seasonal or annual cereal crop yield per hectare data are used to quantify the impact of drought on the production of teff, wheat, maize, barley, and sorghum in NWZ. The ancillary crop data were collected from CSA at a zonal level from 2002 to 2018 (Table 12.2).

Methodology

Preprocessing of Remote Sensing Data

The eMODIS NDVI and LST with spatial resolutions of 250 m and 1 km and temporal resolutions of 16 days and 8 days, respectively, were used for this study. All pixel-based NDVI and LST satellite images were extracted in the generic binary format of hierarchical data format (HDF) and sinusoidal projection.

The LST satellite data were transformed and converted from degrees Fahrenheit (°F) to degrees centigrade (°C) with the mathematical formula. For pixel-based single LST images (LST = raw data * 0.02) − 273.15, each LST image was resampled from a 1 km spatial resolution to 250 m to match the NDVI dataset.

Rainfall Anomaly Index (RAI)

The RAI is also one of the meteorological drought monitoring indices that incorporates the magnitudes of positive and negative precipitation anomalies. The RAI uses positive and negative rainfall anomalies. First, the observed precipitation data were ordered in ascending and descending arrangements, and the first ten highest values were selected from the averaged data to form a threshold for positive anomalies. The ten lowest values from the averaged data were selected to form a threshold for negative anomalies, and the ten most extreme positive and negative anomalies had values of -3 and $+3$. The RAI is an easy way to calculate, with single precipitation data that can be analyzed on monthly, seasonal, and annual timescales (Van Rooy 1965). Hence, the RAI can be calculated as follows.

$$\text{RAI} = 3\left[\frac{N - \overline{N}}{\overline{M} - \overline{N}}\right], \text{ this is for positive anomalies} \qquad (12.3)$$

$$\text{RAI} = -3\left[\frac{N - \overline{N}}{\overline{X} - \overline{N}}\right], \text{ this is for negative anomalies} \qquad (12.4)$$

where N = the recent monthly/seasonal or yearly rainfall data, in other words, of the month/season or year generated in (mm); $\overline{N}$ = monthly/seasonal or yearly average precipitation for the long year timescale in (mm); $\overline{X}$ = the mean of the ten highest monthly/seasonal or yearly precipitation (mm); $\overline{X}$ = average of the ten lowest monthly/seasonal or yearly precipitations of the historical series (mm); and positive anomalies have above average values and negative anomalies have below average values in the timescale

Standardized Precipitation Index (SPI)

There are different drought monitoring meteorological indices to quantify and measure drought events in the world. The SPI is among the most commonly used drought monitoring indices. McKee et al. (1993) formulated the SPI to monitor and quantify water deficits as well as the severity of meteorological drought, which can be calculated at distinctive timescales, such as 1, 3, 6, 12, 24, and 48 months. The SPI on a short timescale, i.e., from 1 month and 3 months, is preferable for the monitoring of meteorological drought. SPI timescales from 3 months to 6 months are usually used for agricultural drought, whereas the SPI for the long term, i.e., from 12 months to 24 months, can be effective and efficient for detecting hydrological drought monitoring. Hence, the SPI simultaneously indicates dry conditions at one time and wet situations at another timescale. This study uses the timescales of seasonal-based SPI3 (June-September) months for agricultural drought monitoring to identify the dry and wet seasons within the time window (2002–2021).

The SPI can be computed by fitting the normal precipitation data to a Gamma probability distribution function. The SPI is widely used for a particular timescale and for a specified location, and the gamma probability distribution function is converted to a normal distribution function with a mean (μ) is equal to zero and standard deviation (σ) is equal to 1 for a given amount of rainfall. The observed cumulative precipitation difference from the climatological average is an equivalent and normally distributed function. Therefore, this study also uses long-term CHIRPS data to quantify the spatial and temporal variabilities in precipitation from 2002 to 2021. A drought event occurs at any time and space. The probability density function of the gamma distribution is defined as:

$$g(X) = \frac{1}{\beta^\alpha \gamma(\alpha)} x^{\alpha-1} e^{-\frac{x}{\beta}} \quad \text{for} \quad x > 0, \tag{12.5}$$

where $\alpha > 0$ is a shape parameter, $\beta > 0$ is a scale parameter, $x > 0$ is the amount of precipitation, and $\gamma(\alpha)$ is the gamma function (Guttman 1999; Lloyd-Hughes and Saunders 2002).

Vegetation Condition Index (VCI)

The VCI is one of the drought monitoring vegetation indices used to measure the severity, spatial extent, and duration of agricultural drought. Its value specifies how the vegetation is improving or declining in response to meteorological weather conditions (Mishra and Singh 2010). Although NDVI can be used for monitoring vegetation and stressed crops, analysis challenges often arise due to shifts in vegetation levels and environmental factors, such as climate characteristics, soil type, and vegetation condition in a given area. Therefore, VCI is an essential parameter that is vastly appropriate for calculating vegetation stress, observing the response of plant greenness, and measuring the weather component (Singh et al. 2003). To identify the weather impact on crops, the NDVI value alone is challenging. Therefore, it is important to design the VCI for monitoring and assessing drought (Kogan 1995a)

$$\text{VCI} = 100 * \left(\frac{\text{NDVI} - \text{NDVI}_{\text{Min}}}{\text{NDVI}_{\text{Max}} - \text{NDVI}_{\text{Min}}} \right) \tag{12.6}$$

where NDVI is the smoothed 8-day NDVI to monthly and seasonal bases and NDVI_{min} and NDVI_{max} are the corresponding historical minimum and maximum values of the data, respectively. The maximum value of the VCI is in the range of 100 for a favorable crop environment, and the minimum is close to 0, indicating bad crop conditions for vegetation greenness.

Temperature Condition Index (TCI)

Temperature and extreme moisture stresses can be analyzed using TCI. TCI can be computed by using long-term LST data. LST-derived satellite images of TCI are widely used and common for monitoring drought and desertification. It is important to determine the vegetation stress that can be caused by temperature variations (Nicholson and Farrar 1994). The increase in LST can affect vegetation growth, condition, and crop productivity in the region, and negative values can be calculated inversely to the VCI for agricultural drought monitoring (Kogan 1995a, b). Mathematically, the TCI can be expressed by Eq. (12.7).

$$\text{TCI} = 100 * \left(\frac{\text{LST}_{max} - \text{LST}}{\text{LST}_{max} - \text{LST}_{min}} \right) \quad (12.7)$$

where LST_{max} LST_{max} and LST_{min} are the smoothed multiyear maximum and minimum LST values at any given time period. The TCI can have similar drought categories as the VCI; the maximum value of the TCI is in the range of 100% for unfavorable vegetation conditions and thermal stress in vegetation, and the minimum is close to 10%, indicating good vegetation conditions or vegetation greenness. Temperature has mostly affected crops in the early stages, but a continuing rise in temperature leads to drought (Kogan 1995a, b).

Vegetation Health Index (VHI)

Kogan (1995a, b) and other researchers indicated that the VHI is used to identify the spatial extent, duration, severity, and magnitude of agricultural drought impacts on different geographical regions. Tsiros et al. (2004) stated that the combined effect of VCI and TCI is used for generating VHI. It is a common agricultural drought monitoring index that is used to assess and monitor agricultural drought impact and early warning by considering vegetation and temperature. It is calculated to decide crop health conditions by combining VCI and TCI. Both the TCI and VCI values should have the same scale and dimensions. The increasing VCI values indicate favorable conditions for crop performance (Parviz 2016). However, an increase in the TCI value indicates that the condition is not favorable for vegetation and crops (Kogan 1995a, b). The highest VHI value indicates the highest vegetation dynamics or greenness in the area. VHI can be calculated by the following Eq. (12.8).

$$\text{VHI} = \alpha(\text{VCI}) + (1 - \alpha)\text{TCI} \quad (12.8)$$

where the values of α are the contributing factors for both VCI and TCI. In most studies, the VHI drought index is calculated by keeping the weighted value α at 0.5 for both the TCI and VCI. The VHI value is scattered in the range of 0–100.

Correlation Analysis Between the Normalized Difference Vegetation Index (NDVI) and Land Surface Temperature (LST)

The Pearson correlation coefficient is used to measure the degree of relationship between two different variables and the association between them. Pearson's correlation coefficient value ranges between − 1 and 1. A value of − 1 indicates a strong negative relationship and implies a perfect negative relationship between the two variables. If the correlation coefficient is 0, there is no relationship between the two variables, and if the Pearson correlation coefficient value is 1, the relationship is strong and perfectly positive.

$$r = \frac{N(\sum(\text{NDVI} * \text{LST})) - \sum(\text{NDVI})\sum(\text{LST})}{\sqrt{N\left[\sum \text{NDVI}^2 - \left(\sum \text{NDVI}\right)^2\right]\left[N\sum \text{LST}^2 - \left(\sum \text{LST}\right)^2\right]}} \quad (12.9)$$

where r is the Pearson correlation coefficient, N is the number of pairs of variables in the event, ΣNDVI* LST is the sum of the products of the paired variables (NDVI and LST), ΣNDVI is the sum of the NDVI variables, ΣLST is the sum of the LST variables, ΣNDVI^2 and ΣLST^2 is the sum of the squared NDVI and LST variables in the analysis.

The Pearson's correlation coefficient (r) between NDVI and LST can reveal the time and place of restraining effect of water on vegetation growth. NDVI helps to identify healthy and stressed vegetation and to reveal the growth status of vegetation. NDVI is used to measure general vegetation health and good vegetation status (Zhang et al. 2013). It is an effective indicator of moisture conditions and good vegetation status (Ji and Peters 2003). On the other hand, LST is used as reflectance for indicating the soil moisture content, evapotranspiration of the surface, and water stress of vegetation. Therefore, this study uses the correlation between NDVI and LST to indicate when water is the limiting factor for vegetation growth.

The correlation of NDVI and LST can be used to decide the containing factor for vegetation growth based on the season and location. A negative correlation of NDVI-LST shows water as the limiting factor for vegetation growth, but if it is positive, energy is the limiting factor for vegetation growth (Sun and Kafatos 2007). Here, the relationship between NDVI and LST was implemented at a pixel level at a seasonal timescale for the study area to identify whether water is the limiting factor for vegetation growth. The Pearson correlation coefficient was calculated for each pixel in the vegetation period by using the available eMODIS NDVI and LST satellite-derived data from 2002 to 2021.

Results and Discussion

The Spatial and Temporal Distribution of SPI

Figure 12.5a, b shows the SPI distribution on a monthly and seasonal basis of meteorological data over NWZ from 2002 to 2021. The seasonal SPI can be employed to detect the drought or dry seasons and non-drought or wet seasons of events in the Kiremt and Belg seasons. As indicated in the figure, the red to light green color specifies drought-affected areas in both the Kiremt and Belg seasons. An extreme and severe drought occurred in Kobo and Gidan and in pocket areas of neighboring woredas in Kiremt (Fig. 12.5a). However, in the Belg season, most parts of the zone except parts of Habru, Guba Lafto, and parts of Kobo Woreda were dominated by extreme to moderate drought situations (Fig. 12.5b). The spatial distribution of rainfall varies from season to season, resulting in short- and long-term drought events in NWZ from 2002 to 2021. On the other hand, the greatest positive value of SPI in the spatial distribution map depicted by dark blue indicates that the region was considered non-drought and favorable for agricultural activity. In the Kiremt season, the southwestern, western, and northwestern regions (Fig. 12.5a) and in the Belg season, the tip areas of the southern and southeastern regions of the study area had favorable conditions for vegetation (Fig. 12.5b).

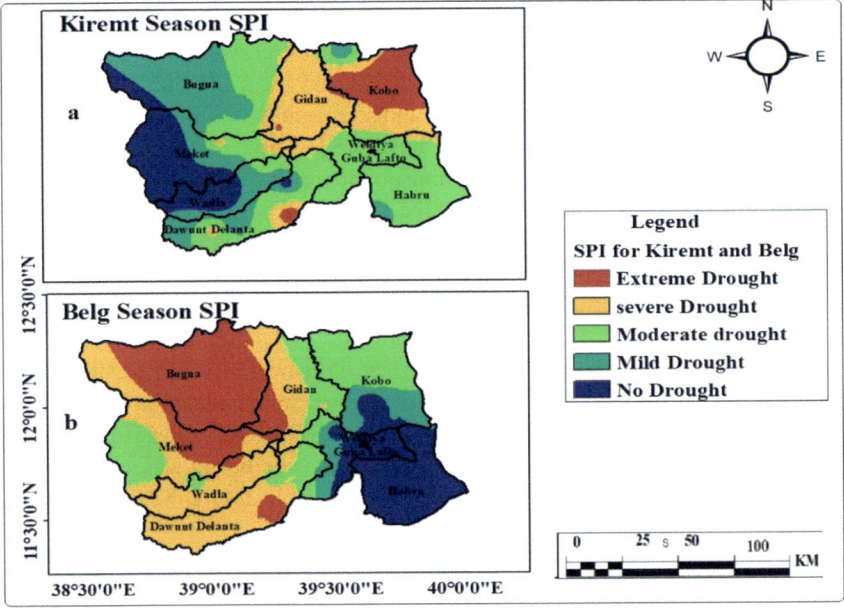

Fig. 12.5 The spatial distribution of SPI for Kiremt season (**a**) and Belg season (**b**) rainfall from 2002 to 2021 over NWZ

The Spatial Distribution and Temporal Trend of RAI

The RAI was calculated to examine the spatial distribution and temporal pattern of rainfall in the Kiremt and Belg seasons in NWZ. The results showed that the intensity of rainfall was negative (extreme to moderate drought) in most parts of Meket, Wadla, Delanta, and Kobo Woredas in the Kiremt season (Fig. 12.6a). Most parts of the Zone except pocket areas of Northern Kobo and Western and Southern Meket were observed with low rainfall in the Belg season (Fig. 12.6b). The RAI map showed that most parts of the areas were under intense and extreme to moderate drought in both the Kiremt and Belg seasons. The negative values of RAI mean there is less than median or average precipitation, especially in the low land areas of NWZ. On the other hand, the spatial distribution map showed a positive RAI, indicating that the region is considered to have non-drought conditions in most parts of Bugna and Habru, pocket areas of Kobo, Meket, and Guba Lafto Woredas in the Kiremt season (Fig. 12.6a). The tip areas of Kobo and Meket Woredas received a low amount of rainfall in the Belg season (Fig. 12.6b).

The temporal pattern of RAI indicates the temporal variability and erratic nature of rainfall in the main crop-growing season (Fig. 12.7). Hence, an increase in drought events was observed in 2002, 2004, 2009, 2011, and largely in 2015, with RAI

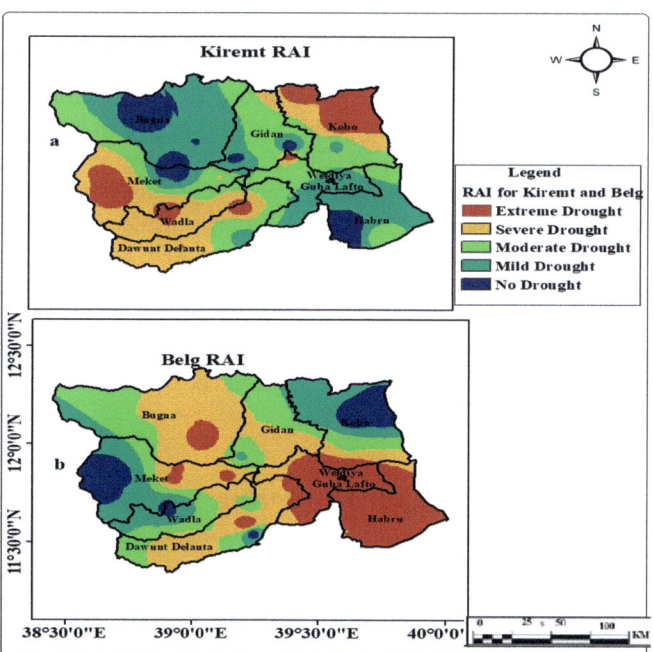

Fig. 12.6 The spatial distribution of RAI in the Kiremt season (**a**) and Belg season (**b**) from 2002 to 2021 over NWZ

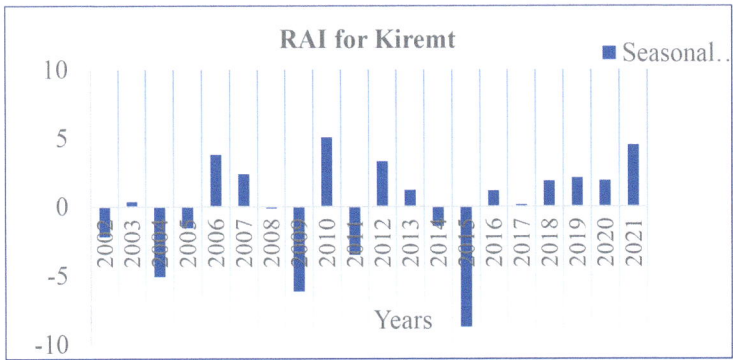

Fig. 12.7 The temporal trend of the Kiremt seasonal mean RAI from 2002 to 2021 for the study area

values of − 8.5 in the main crop-growing season. In line with this, Wassie et al. (2022) reported the same result in NWZ.

NDVI-Based Agricultural Drought Monitoring

The spatial distribution and extent of the vegetation conditions during the main crop-growing season over NWZ are shown in Fig. 12.8a–d. The NDVI results indicated that in most areas, there was low vegetation resulting from agricultural drought events in 2009 and 2015 over most parts of the zone, especially in low land areas with different severity levels (Fig. 12.8a, b). The low NDVI values in NWZ indicate low vegetation vigor, sparse vegetation cover, and poor vegetation conditions during 2009. However, in contrast to the low NDVI, improvements in intensity and vigorous vegetation conditions or near normal vegetation conditions were observed during the main rainy season in 2006 and 2021 (Fig. 12.8c, d). The result of NDVI analysis showed that the decrease in vegetation condition (NDVI) is in line with the decrease in crop production in the main crop-growing season. Eze et al. (2020) also obtained the same result in the surrounding areas of the Raya Azebo districts.

VCI-Based Agricultural Drought Monitoring

In Fig. 12.9a–d, the spatial distribution of agricultural drought is depicted and analyzed based on VCI values during the main crop-growing season over the past 20 years. The persistence of agricultural drought in the study area showed that the crop-growing season signifies poor vegetation growth with vegetation stress. Agricultural drought conditions occurred in most parts of NWZ in the main crop-growing

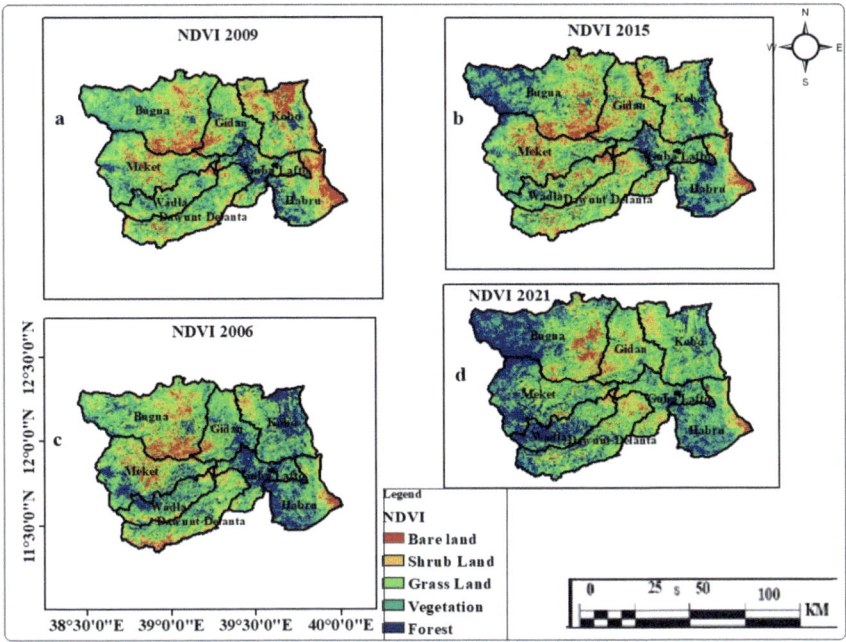

Fig. 12.8 Seasonal NDVI for the dry years 2009 (**a**) and 2015 (**b**) and the wet years 2006 (**c**) and 2021 (**d**)

periods of 2009 and 2015. The analysis results indicated that most parts of Kobo, Habiru, Dawunt, and parts of Meket and Bugna Woredas faced drought years in 2009 and 2015 (Fig. 12.9a, b). Such drought events affected the vegetation status and greenness, which resulted in a reduction in crop production. However, wet or approximately normal years in 2006 and 2021 were identified for the crop-growing season (Fig. 12.9c, d). During wet and normal conditions, the vegetation greenness and status were good; consequently, crop production increased. The analyses of the seasonal NDVI and VCI showed the same result with a small spatial variation in the study area. NDVI-derived VCI drought monitoring indicated lower vegetation conditions and resulted in a decrease in crop production in the main crop-growing season. The current finding is in line with the results of Gebrehiwot et al. (2011), which were conducted in the Amhara and Tigray regions.

TCI-Based Agricultural Drought Monitoring

The TCI is anticipated to be the relative long-term average seasonal maximum and minimum temperature envelopes to the current year LST over the study area. Figure 12.10a and b shows the existence of vegetation stress in vegetation over most

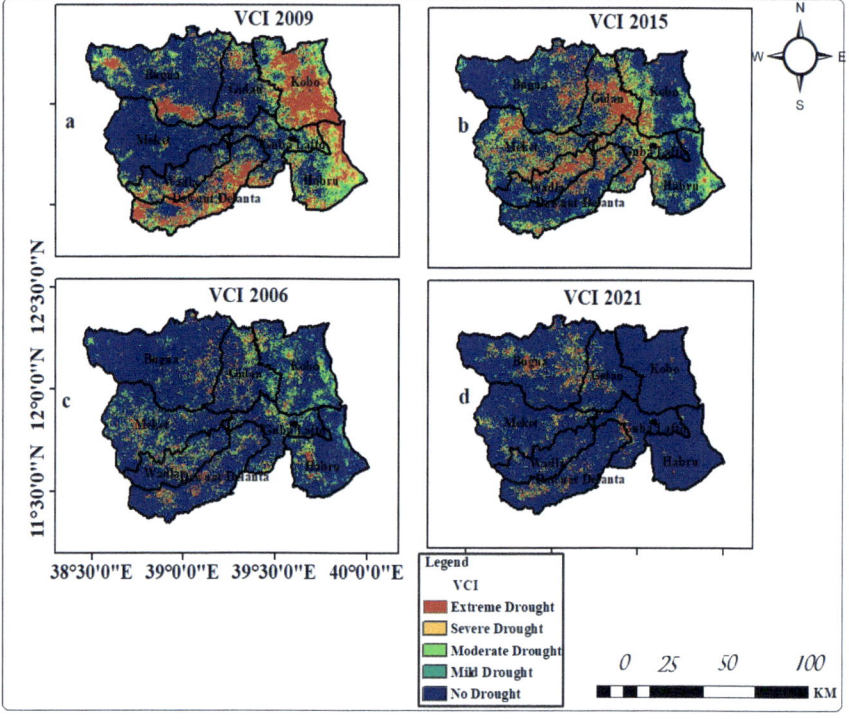

Fig. 12.9 VCI for the dry years 2009 (**a**) and 2015 (**b**) and approximately wet years 2006 (**c**) and 2021 (**d**)

parts of the study area. On the other hand, the lowest temperatures in the highland and midland areas indicate favorable thermal conditions or normal conditions for vegetation in the main crop-growing season. The result depicted in dark blue reveals that moderate to extreme drought conditions occurred in 2009 and 2015 (Fig. 12.10a, b). However, the reddish color in most parts of the study area indicates mild drought to normal or optimum conditions in the region (Fig. 12.10c, d). In the main crop-growing seasons of 2006 and 2021, the TCI results showed near-favorable vegetation conditions over NWZ. Therefore, the TCI can detect and identify both normal or optimum vegetation conditions and extreme dryness or severe drought years over the NWZ. Vegetation is under thermal stress conditions when the temperature is high and when there is a deficit in precipitation and, consequently, a decrease in crop productivity (Gaznayee and Al-Quraishi 2019).

The computed values of the TCI indicated that high temperature and high thermal stress or drought conditions were observed over most parts of the study area. TCI is used to detect and monitor agricultural drought by considering the effect of temperature on vegetation and decreasing soil moisture by increasing evapotranspiration over the area, which results in crop failure. Gebrehiwot et al. (2011) and Gidey et al. (2018a) conducted studies and found similar results in Ethiopia and Eastern Amhara.

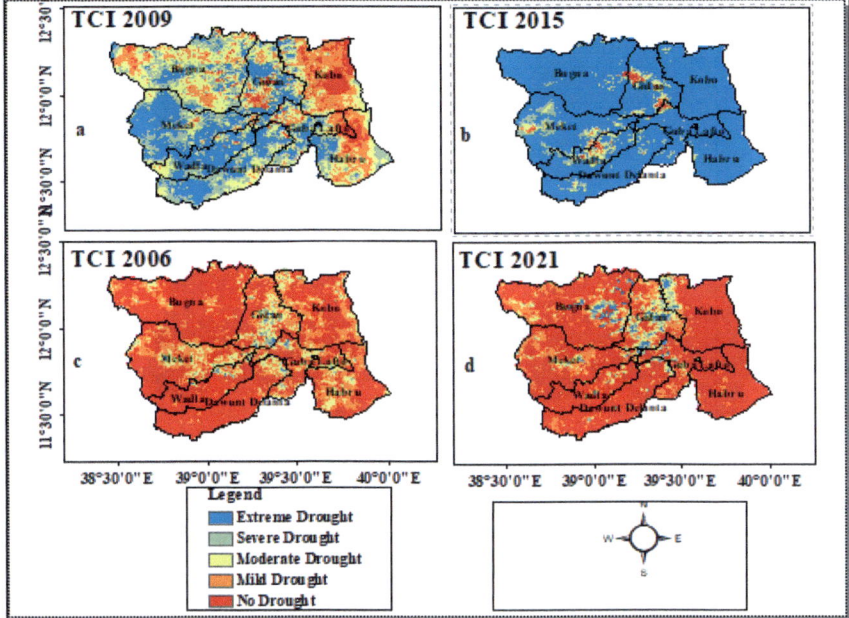

Fig. 12.10 TCI-based agricultural drought monitoring for the dry years 2009 (**a**) and 2015 (**b**) and the wet years 2006 (**c**) and 2021 (**d**)

The TCI value is directly influenced by LST, and the highest temperatures in the crop-growing season indicate poor vegetation conditions or vegetation stress, while the lowest temperatures indicate mostly favorable conditions or normal conditions for the crop-growing season (Gidey et al. 2018b).

VHI-Based Agricultural Drought Monitoring

The spatial extent, duration, affected area, and intensity of agricultural drought can be monitored using satellite-based combined effects of the TCI and VCI. Therefore, the VCI and TCI largely contributed to assessing and monitoring vegetation health and stress depending upon the LST and NDVI that are used to monitor agricultural drought. The VHI results indicated that the vegetation was under thermal stress or in poor vegetation conditions across most parts of the zone in 2002–2021 for the main crop-growing season. Figure 12.11a and b revealed that in the drought years of 2009 and 2015, the vegetation health was under extreme thermal stress and very poor vegetation greenness in almost all parts of the zone with different levels of intensity and extent. Accordingly, the results of the VHI analysis showed that very poor vegetation conditions to moderately poor vegetation conditions were observed over most parts of the zone, while certain pocket areas showed good and very good

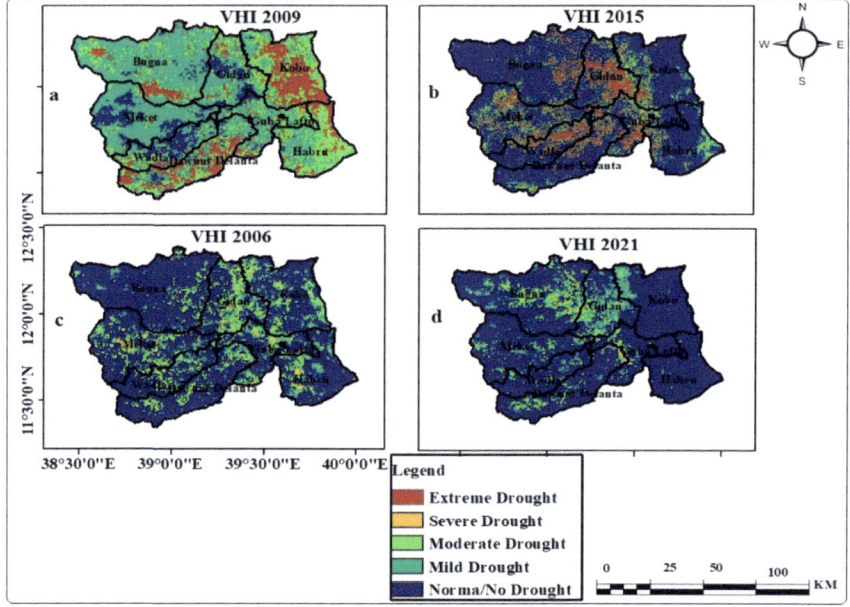

Fig. 12.11 VHI-based drought monitoring for the dry years 2009 (**a**) and 2015 (**b**) and the wet years 2006 (**c**) and 2021 (**d**)

vegetation conditions during 2009 and 2015. However, in Fig. 12.11c and d, in the main crop-growing seasons of 2006 and 2021, good to very good vegetation conditions were observed in most parts of the NWZ. The results confirmed that 65% of extreme drought and 10% of severe drought occurred in NWZ in the 2009 crop-growing season. Generally, the increase in rainfall in the cropping season raises the VHI, results in an increase in crop production, and signifies a sustainable economy for the study area. Gadisso (2007) conducted an assessment of drought in the Blue Nile Basin in which the severity, detection, and early warning of agricultural drought could be estimated based on VHI by considering the effect of temperature on vegetation response.

Relationship Between Pixel-Based Seasonal NDVI and LST

Figure 12.12a–d shows the seasonal variation in the NDVI-LST pixel-based relationship in the Kiremt season for the years 2002–2021 in NWZ. The results revealed that when NDVI increases, LST decreases and vice versa. The LST shows lower values at higher NDVI (dense vegetation) areas because it induces more evapotranspiration and cools the surface. However, low NDVI values are mostly found at high LST because the vegetation is under high water stress. Karnieli et al. (2010) also found that in areas deficient in water for vegetation growth, the relationship between NDVI and LST was negative.

During the Kiremt 2009 and 2015 seasons, the correlations between NDVI and LST were $r = -0.51$ and $r = -0.43$, respectively. The coefficient of determination (R^2) showed 18.2%, 28%, and 26% of the variations in the LST in the study area in 2009, 2015, and 2006 in the Kiremt season, respectively (Fig. 12.12a–c). Ghobadi et al. (2015) and Guha et al. (2019) evaluated the relationship between NDVI and LST and observed a strong negative relationship in the Kiremt season. As indicated in Fig. 12.12a–c, in the highest peaks of LST, lower values of NDVI exist, while the high peaks of NDVI correspond to low peaks of LST.

Figure 12.12a–c shows that the simple linear regression coefficient $a1$ indicates that when the dependent variable (i.e., LST) increases by one unit, the independent variable (i.e., NDVI) decreases by -25.86 units in 2009, -40.52 units in 2015, and -31.93 units in 2006. Hence, the correlation between NDVI and LST is a negative linear regression, and the variation in NDVI is very sensitive to change. The change in NDVI can cause a change in LST. Since the NDVI-LST correlation is negative and has an inverse relationship, this study can conclude that water is the limiting factor for vegetation growth in the study area (Yue et al. 2007). In general, the results revealed that as LST increases, NDVI values decrease and may cause vegetation stress, which will lead to agricultural drought. This situation indicates agricultural drought risk areas in the northern highlands of eastern Amhara, particularly in NWZ.

The Temporal Pattern Between SPI. RAI, NDVI, and Crop Production

In Fig. 12.13a, the seasonal NDVI anomaly was calculated from the measurement for 2001 to 2021 during the growing season. As indicated in Fig. 12.13b, the vegetation cover (seasonal NDVI) also has a strong and significant positive increase over time in the Kiremt season. The higher the deviation in NDVI, the higher the meteorological drought resulting in higher crop yield loss in 2009, 2015, and 2017 and vice versa. This result agrees with Eze et al. (2020), who conducted a similar analysis in the Raya Azebo districts.

The computed temporal pattern of the meteorological indices of RAI showed that the deficit of rainfall over the study area significantly affects the vegetation condition and results in a decrease in crop production in the main crop-growing season. The observed significant positive relationship between rainfall and NDVI in the main rainy season signifies a similar trend over the study area in 2002–2021. There was also a strong relationship between rainfall and crop production. Figure 12.13c shows that when the intensity of rainfall increases, crop production increases. Conversely, when rainfall decreases, crop production also decreases. In addition to rainfall, the satellite-derived NDVI has a significant correlation with crop production. Rainfall can affect NDVI and crop production for the NWZ in the main crop-growing season.

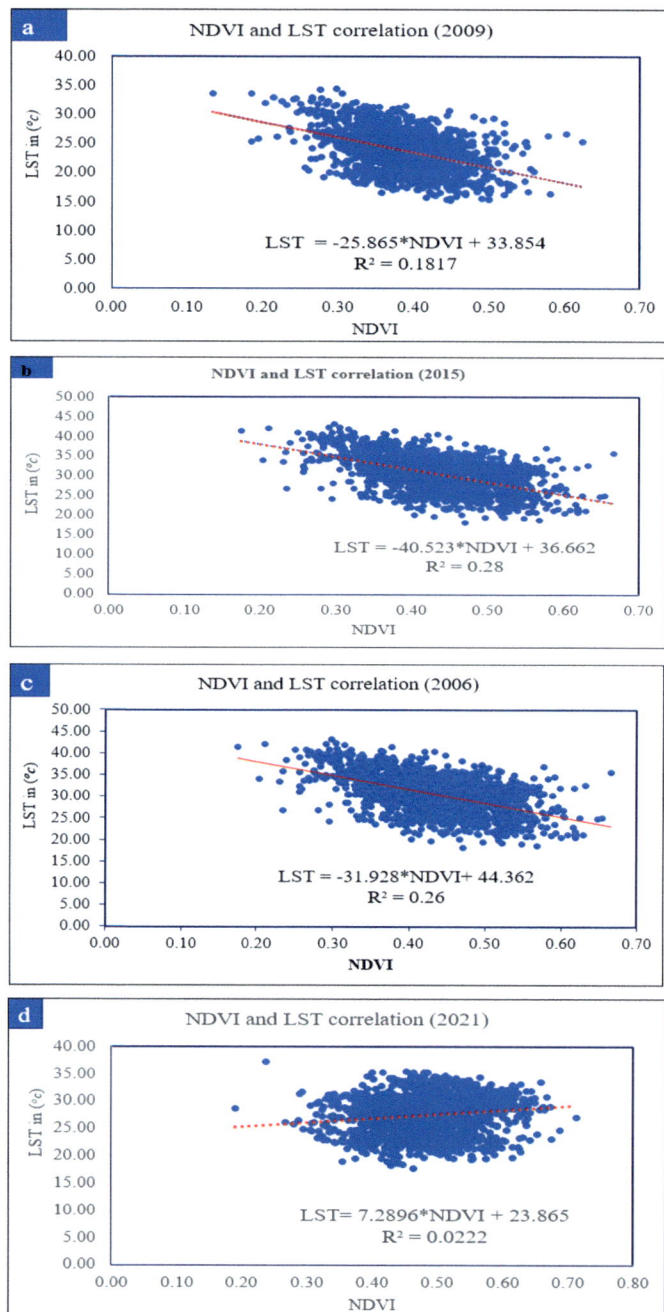

Fig. 12.12 The pixel-based correlation between seasonal mean NDVI and LST for the drought years 2009 (**a**) and 2015 (**b**) and the wet years 2006 (**c**) and 2021 (**d**) over NWZ

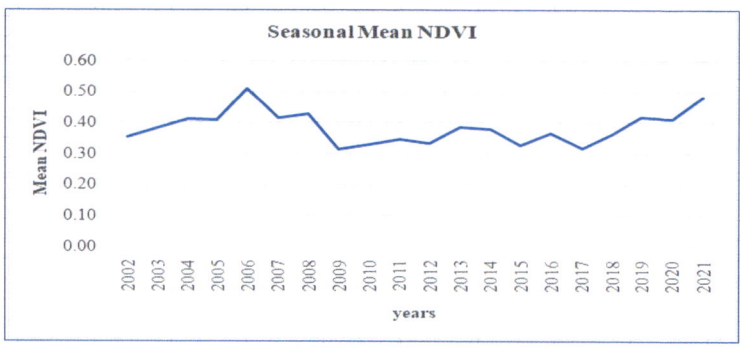

(a). The temporal trend of seasonal means NDVI in NWZ.

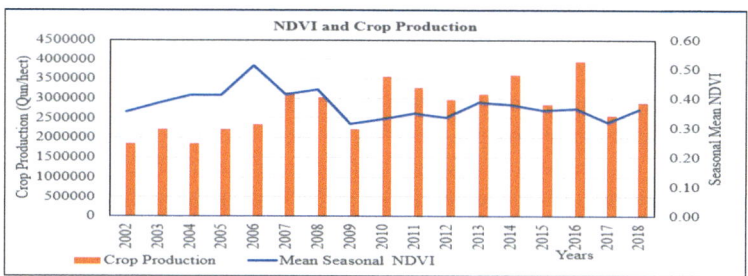

(b). Temporal variability of seasonal NDVI and crop production.

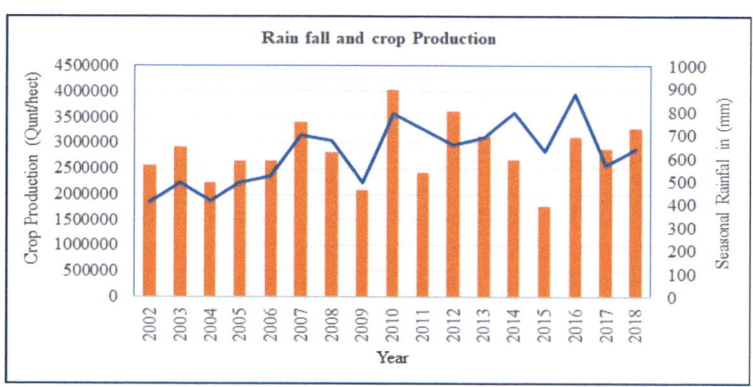

(c). Temporal variability of seasonal rainfall and crop production.

Fig. 12.13 **a** The temporal trend of seasonal means NDVI in NWZ. **b** Temporal variability of seasonal NDVI and crop production. **c** Temporal variability of seasonal rainfall and crop production

Legesse and Suryabhagavan (2014) also indicated that rainfall and satellite-derived NDVI have a significant relationship with crop production in the East Shewa Zone. The seasonal variabilities in precipitation during the main cropping season are more critical and determine the NDVI condition. Areas with low vegetation conditions were stricken to low crop production for rain feed agriculture (Segele and Lamb 2005). Indeed, previous studies indicated that in the NWZ, which is semiarid and arid lowlands, there was no year without drought with different variations in time (Kogan and Guo 2016).

Conclusion and the Way Forward

Drought is a natural hazard that affects day-to-day activities and the livelihood of vulnerable communities. The effect of drought can reduce the production of crops, the availability of water, food, and fodder, and social instabilities. In Ethiopia, drought hazards can cause the migration and displacement of people, famine, and malnutrition, resulting in loss of lives. Hence, the monitoring and assessment of the effect of agricultural drought in NWZ is vital. The consideration of drought-related climate constraints using satellite and ground-based indices such as precipitation, soil moisture, surface temperature, actual evapotranspiration, humidity, and wind is crucial for agricultural drought monitoring. However, ground-based drought monitoring indices have shortcomings due to unevenness and limited distribution of gauge stations to capture the spatial distribution and temporal variabilities. Remote sensing-based techniques can be utilized by combining ground measurements for agricultural drought monitoring and early warning systems. These techniques can provide continuous and near real-time information and cover large areas.

Remote sensing and geospatial tool-based drought monitoring indices are the most crucial indicators of agricultural drought, especially in arid and semiarid areas. This study aimed to monitor and assess agricultural drought by using the VHI. In addition to VHI, the correlation analysis between NDVI and LST can indicate when and where vegetation growth and health are limited, mainly related to the water deficit in the NWZ.

The eMODIS NDVI and LST are widely and extensively used for drought monitoring, identification, detection, and assessment of drought conditions in the main crop-growing period. TCI was calculated and produced using historical eMODIS relative maximum and minimum LST compared with the current status of the thermal infrared brightness temperature and emissivity per pixel. The main advantage of using TCI is to evaluate the impact of rising temperature on plant moisture stress. The increase in LST and decrease in rainfall in the lowland areas result in unfavorable conditions for vegetation growth due to thermal stress and high evapotranspiration. This situation mostly contributes to a soil moisture deficit in the plant root zone, and agricultural drought starts its triggering point. However, the decrease in LST and the increase in NDVI revealed good vegetation conditions and an increase in crop productivity.

In this study, meteorological drought monitoring indices such as SPI and RAI were used to detect and quantify the precipitation deficit in the crop-growing season. The computed SPI and RAI from seasonal mean rainfall data were assigned to each grid cell of the study area and reclassified based on drought severity classes. The variation in the SPI and RAI showed that in the past 20 years, drought did not have a similar trend, and in some years, it can be extreme, severe, and moderate. This is an indicator that drought can occur at any given time with different magnitudes and extents over eastern Amhara, particularly NWZ.

This study showed that there was a strong relationship between the vegetation health indices, seasonal rainfall, and crop yield within the temporal pattern of drought over the crop-growing season. The decrease in seasonal RAI and SPI over the main crop-growing season leads to a decrease in NDVI and crop production. The worst drought years were in the 2009 and 2015 crop-growing seasons. The correlation analysis between NDVI and LST also identified when and where vegetation growth was primarily limited due to water deficit. The negative correlation between LST and NDVI in the main crop-growing season of the study area indicates that water is the limiting factor for vegetation growth.

Understanding the impact of agricultural drought is crucial to developing significant strategies and plans, taking early action and early response, developing mitigation and adaptation strategies, and prioritizing drought-affected areas. Analyzing the interactions between the critical precipitation period at early seedling stages and the critical sunshine period at stem elongation, vegetation, and ripening stage for the major crop yield is crucial. Hence, it is essential to incorporate the effects of sunshine hour data on plant growth for a better output. This study also recommends using satellite-based combined drought monitoring indices rather than using only single indices to monitor and assess agricultural drought. It is also vital to create collaboration among institutions that include policymakers and governmental and nongovernmental bodies for operational services of drought monitoring, assessment, and early warning systems. In addition, prioritization and implementation of site-specific mitigation and adaptation strategies should be put in place in the affected areas.

References

Abiy AZ, Melesse AM, Abtew W (2019) Teleconnection of regional drought to ENSO, PDO, and AMO: Southern Florida and the everglades. Atmosphere 2019(10):295

Alahacoon N, Edirisinghe M, Ranagalage M (2021) Satellite based meteorological and agricultural drought monitoring for agricultural sustainability in Sri Lanka. Sustainability 13(6):3427

Bayissa Y, Maskey S, Tadesse T, Van Andel SJ, Moges S, Van Griensven A, Solomatine D (2018) Comparison of the performance of six drought indices in characterizing historical drought for the Upper Blue Nile Basin, Ethiopia. Geosciences 8(3):81

Bayissa Y, Moges S, Melesse AM, Tadesse T, Abiy AZ, Worqlul A (2021) Multi-dimensional drought assessment in Abbay/Upper Blue Nile Basin: the importance of shared management and regional coordination efforts for mitigation. Remote Sens 13(9):1835. https://doi.org/10.3390/rs13091835

Brown JF, Howard D, Wylie B, Frieze A, Ji L, Gacke C (2015) Application-ready expedited MODIS data for operational land surface monitoring of vegetation condition. Remote Sens 7(12):16226–16240

Central Statistical Agency (CSA) (2008) Ethiopian census of population and housing: population and housing census abstract of the country. Retrieved in 2007 from CSA, Addis Ababa

Chen JM, Cihlar J (1996) Retrieving leaf area index of boreal conifer forests using Landsat TM images. Remote Sens Environ 55:153–162

Degefu W (1987) Some aspects of Meteorological drought in Ethiopia. In: Michael H, Glantz (eds) Drought and hunger in Africa. Cambridge University Press London, pp 23–36

Dinku T, Funk C, Peterson P, Maidment R, Tadesse T, Gadain H, Ceccato P (2018) Validation of the CHIRPS satellite rainfall estimates over eastern Africa. Q J R Meteorol Soc 144:292–312

Elias E, Fantaye D (2000) Managing fragile soils: a case study from North Wollo, Ethiopia. Managing Africa's Soils No 13

Eze E, Girma A, Zenebe A, Kourouma JM, Zenebe G (2020) Exploring the possibilities of remote yield estimation using crop water requirements for area yield index insurance in a data-scarce dryland. Environment 183:1261

Feldhake CM, Glenn DM, Peterson DL (1996) Pasture soil surface temperature response to drought. Agrono J 88(4):652–656

Frey CM, Kuenzer C, Dech S (2012) Quantitative comparison of the operational NOAA-AVHRR LST product of DLR and the MODIS LST product V005. Int J Remote Sens 33(22):7165–7183

Funk C, Peterson P, Landsfeld M, Pedreros D, Verdin J, Shukla S, Husak G, Rowland J, Harrison L, Hoell A, Michaelsen J (2015) The climate hazards infrared precipitation with stations—a new environmental record for monitoring extremes. Sci Data 2(1):1–21

Gadisso BE (2007) Drought assessment for the Nile Basin using Meteosat second generation data with special emphasis on the upper Blue Nile Region 91. http://www.itc.nl/library/papers_2007/msc/wrem/gadisso.pdf

Gaznayee HAA, Al-Quraishi AMF (2019) Analysis of agricultural drought, rainfall, and crop yield relationships in Erbil province, the Kurdistan region of Iraq based on Landsat time-series MSAVI2. J Adv Res Dyn Control Syst 11(12):536–545

Gebre E, Berhan G, Lelago A (2017) Application of remote sensing and GIS to characterize agricultural drought conditions in North Wollo Zone, Amhara Regional State, Ethiopia. J Nat Sci Res 7(17):41–50

Gebrehiwot T, van der Veen A, Mathuis B (2011) Spatial and temporal assessment of drought in the Northern highlands of Ethiopia. Int J Appl Earth Obs Geoinf 13(3):309–321

Ghobadi Y, Pradhan B, Shafri HZM, Kabiri K (2015) Assessment of spatial relationship between land surface temperature and landuse/cover retrieval from multitemporal remote sensing data in South Karkheh Subbasin, Iran. Arab J Geosci 8:525–537

Gidey E, Dikinya O, Sebego R, Segosebe E, Zenebe A (2018a) Modelling the spatiotemporal meteorological drought characteristics using the Standardized Precipitation Index (SPI) in Raya and its environs, Northern Ethiopia. Earth Syst Environ 2:265–279

Gidey E, Dikinya O, Sebego R, Segosebe E, Zenebe A (2018b) Analysis of the long—term agricultural drought onset, cessation, duration, frequency, severity and spatial extent using Vegetation Health Index (VHI) in Raya and its environs, Northern Ethiopia. Environ Syst Res

Gross D (2005) Monitoring agricultural biomass using NDVI time series. Food and Agriculture Organization of the United Nations (FAO), Rome, Italy

Guha S, Govil H, Diwan P (2019) Analytical study of seasonal variability in land surface temperature with normalized difference vegetation index, normalized difference water index, normalized difference built-up index, and normalized multiband drought index. J Appl Remote Sens 13(2)

Guttman NB (1999) Accepting the standardized precipitation index: a calculation algorithm. JAWRA J Am Water Resour Assoc 35:311–322

Hagenlocher M, Meza I, Anderson CC, Min A, Renaud FG, Walz Y, Siebert S, Sebesvari Z (2019) Drought vulnerability and risk assessments: state of the art, persistent gaps, and research agenda. Environ Res Lett 14:083002

Hailu S (2013) The impact of disaster risk management interventions in humanitarian programme on household food security: the case of East Africa, Ethiopia, Amhara Region, North Wollo Zone

Hashim M, Reba NM, Nadzri MI, Pour AB, Mahmud MR, Yusoff AMRM, Ali MI, Jaw SW, Hossain MS (2016) Satellite-based run-off model for monitoring drought in Peninsular Malaysia. Remote Sens 8(8):1–25

Hodges SC (2010) Soil fertility basics: NC certified crop advisor training. Soil Science Extension, North Carolina State University, Raleigh, NC, pp 1–75

Holben BN, Tucker CJ, Fan CJ (1980) Spectral assessment of soybean leaf area and lear biomass. Photogramm Eng Remote Sens 46(5):651–656

Huete AR (1988) A soil-adjusted vegetation index (SAVI). Remote Sens Environ 25(3):295–309

IPCC (2013) Impacts of changes in climate and land use/land cover under IPCC RCP scenarios on streamflow in the Hoeya River Basin, Korea. Sci Total Environ 452–453(March):181–195. 10.1016/j

Ji L, Peters AJ (2003) Assessing vegetation response to drought in the northern Great Plains using vegetation and drought indices. Remote Sens Environ 87(1):85–98

Karnieli A, Agam N, Pinker RT, Anderson M, Imhoff ML, Gutman GG, Panov N, Goldberg A (2010) Use of NDVI and land surface temperature for drought assessment: merits and limitations. J Clim 23(3):618–633

Kogan FN (1990) Remote sensing of weather impacts on vegetation in nonhomogeneous areas. Int J Remote Sens 11(8):1405–1419

Kogan FN (1995a) Application of vegetation index and brightness temperature for drought detection. Adv Space Res 15:91–100

Kogan FN (1995b) Droughts of the late 1980s in the United States asderived from NOAA polar orbiting satellite data. Bull Am Meteor Soc 76:655–668

Kogan FN (1997) Global drought watches from space. Bull Am Meteor Soc 78:621–636

Kogan FN (2002) World droughts in the new millennium from AVHRR-based vegetation health indices. Eos 83(48):3–7

Kogan F, Guo W (2016) Early twenty-first-century droughts during the warmest climate. Geomat Nat Haz Risk 7(1):127–137

Kogan F, Stark R, Gitelson A, Jargalsaikhan L, Dugrajav C, Tsooj S (2004) Derivation of pasture biomass in Mongolia from AVHRR-based vegetation health indices. Int J Remote Sens 25(14):2889–2896

Korecha D, Barnston AG (2007) Predictability of June–September rainfall in Ethiopia. Mon Weather Rev 135(2):628–650

Legesse G (2010) Agricultural drought assessment using remote sensing and GIS techniques. Unpublished MSc thesis Addis Ababa University, Addis Ababa, 76pp

Legesse G, Suryabhagavan K (2014) Remote sensing and GIS based agricultural drought assessment in East Shewa Zone, Ethiopia. Trop Ecol 55:349–363

Lemma G (1996) Climate classification of Ethiopia. Meteorological research report series. National Meteorological Services Agency of Ethiopia, Addis Ababa, pp 1–76

Lloyd-Hughes B, Saunders MA (2002) A drought climatology for Europe. Int J Climatol 22(13):1571–1592

Mastrangelo AM, Mazzucotelli E, Guerra D, Vita P, Cattivelli L (2012) Improvement of drought resistance in crops: from conventional breeding to genomic selection. In: Crop stress and its management: perspectives and strategies, pp 225–259

McKee TB, Doesken NJ, Kliest J (1993) The relationship of drought frequency and duration to time scales. In: Proceedings of the 8th conference on applied climatology, Anaheim, CA, America. Meteorological Society, Boston, pp 179–184

Mishra AK, Singh VP (2010) A review of drought concepts. J Hydrol 391(1–2):202–216

Nam WH, Tadesse T, Wardlow BD, Hayes MJ, Svoboda MD, Hong EM, Pachepsky YA, Jang MW (2018) Developing the vegetation drought response index for South Korea (Vegdri-skorea) to assess the vegetation condition during drought events. Int J Remote Sens 39(5):1548–1574

Nicholson SE, Farrar TJ (1994) The influence of soil type on the relationships between NDVI, rainfall, and soil moisture in semiarid Botswana. Remote Sens Environ 50:107–120

Palmer WC (1965) Meteorological drought. Office of Climatology, Washington, D.C.

Parviz L (2016) Determination of effective indices in the drought monitoring through analysis of satellite images. J Agricul Forest 62(1):305–324

Peng J, Loew A, Merlin O, Verhoest NEC (2017) A review of spatial downscaling of satellite remotely sensed soil moisture. Rev Geophys 55(2):341–366

Rhee J, Im J, Carbone GJ (2010) Monitoring agricultural drought for arid and humid regions using multisensor remote sensing data. Remote Sens Environ 114(12):2875–2887

Rooy MP (1965) A rainfall anomaly index independent of time and space. Notos 14:43–48

Segele ZT, Lamb PJ (2005) Characterization and variability of Kiremt rainy season over Ethiopia. Meteorol Atmos Phys 89(1–4):153–180. https://doi.org/10.1007/s00703-005-0127-x

Seka AM, Zhang J, Prodhan FA, Melesse AM et al (2022) Hydrological drought impacts on water storage variations: a focus on the role of vegetation changes in the East Africa region. A systematic review. Environ Sci Pollut Res. https://doi.org/10.1007/s11356-022-23313-0

Singh RP, Roy S, Kogan F (2003) Vegetation and temperature condition indices from NOAA AVHRR data for drought monitoring over India. Int J Remote Sens 24:4393–4402

Sun D, Kafatos M (2007) Note on the NDVI-LST relationship and the use of temperature-related drought indices over North America. Geophys Res Lett 34(24):1–4

Tran HT, Campbell JB, Tran TD, Tran HT (2017) Monitoring drought vulnerability using multispectral indices observed from sequential remote sensing (Case Study:TuyPhong, BinhThuan, Vietnam). GIScience Remote Sens 54:167–184

Tsiros E, Domenikiotis C, Spiliotopoulos M, Dalezios NR (2004) Use of NOAA/AVHRR-based vegetation condition index (VCI) and temperature condition index (TCI) for drought monitoring in Thessaly, Greece. In: EWRA symposium on water resources management: risks and challenges for the 21st century, Izmir, Turkey, pp 2–4

Vicente-Serrano SM, Beguería S, López-Moreno JI (2010) A multiscale drought index sensitive to global warming: the standardized precipitation evapotranspiration index. J Clim 23(7):1696–1718

Wassie SB, Mengistu DA, Berlie AB (2022) Trends and spatiotemporal patterns of meteorological drought incidence in North Wollo, Northeastern highlands of Ethiopia. Arab J Geosci 15(12):1–21

Wilhite DA, Glantz MH (1985) Understanding the drought phenomenon: the role of definitions planning for drought. Water Int 10:111–120

WMO (2012) Standardized precipitation index user guide. World Meteorological Organization, WMO-No. 1090

Yue S, Pilon P, Cavadias G (2007) Power of the Mann-Kendall and Spearman's rho tests for detecting monotonic trends in hydrological series. J Hydrol 259:254–271

Zeleke Tessera G, Teshome M, Ayele L (2021) Land degradation neutrality assessment using geospatial techniques in North Wello Zone, Northern Ethiopia

Zeleke EB, Melesse AM, Kidanewold BB (2022) Assessment of climate and catchment control on drought propagation in the Tekeze River Basin, Ethiopia. Water 14:1564

Zelelew DG, Ayimute TA, Melesse AM (2018) Evaluating the response of in situ moisture conservation techniques in different rainfall distributions and soil-type conditions on sorghum production and soil moisture characteristics in drought-prone areas of Northern Ethiopia. Water Conserv Sci Eng 3(3):157–167. https://doi.org/10.1007/s41101-018-0045-7

Zhang Y, Peng C, Li W, Fang X, Zhang T, Zhu Q, Chen H, Zhao P (2013) Monitoring and estimating drought-induced impacts on forest. Environ Rev 115:103–115

Chapter 13
Quantifying Spatiotemporal Drought Dynamics Under Climate Change in the Abbay River Basin, Ethiopia

Getachew Tegegne and Assefa M. Melesse

Abstract Changes in drought intensity and frequency have a greater detrimental effect on hydrological ecosystem services. Consequently, this study examined how droughts in the Abbay River Basin's meteorological, agricultural, and hydrological components will be influenced by climate change. This study used the standardized precipitation index (SPI) to explore the impact of climate change on the frequency of meteorological drought (using the 3-month SPI), agricultural drought (using the 6-month SPI) and hydrological drought (using the 12-month SPI). The analysis of the SPI using five GCMs in the Abbay River Basin revealed that the effects of meteorological drought are more exacerbated in the agricultural and hydrological sectors as the duration increases. The frequency of droughts will increase in the northern regions while decreases in the southern regions of the Abbay River Basin. Overall, the results showed that the risk of drought in the study basin will rise in the middle future (2031–2060) and potentially decline in the far future (2061–2090) under both SSP2-4.5 and SSP5-8.5. The analysis of climate change impact on the intensity–duration–frequency (IDF) curve revealed that the frequency of drought occurrence will be more than 20% and 15% in 2031–2060 under SSP2-4.5 and SSP5-8.5, respectively. Hence, this study recommends that significant efforts be devoted in a timely manner to minimizing the vulnerability of droughts to climate change. Therefore, this study recommends that early and considerable measures be made to reduce the susceptibility of agricultural and hydrological systems to the extreme droughts under climate change.

Keywords Standardized precipitation index · Climate change · Droughts · Abbay River Basin

G. Tegegne (✉)
Department of Civil Engineering, Sustainable Energy Center of Excellence, Addis Ababa Science and Technology University, Addis Ababa, Ethiopia
e-mail: getachewtegegne21@gmail.com; getachew.tegegne@aastu.edu.et

A. M. Melesse
Department of Earth and Environment, Institute of Environment, Florida International University, Miami, FL, USA

© The Author(s), under exclusive license to Springer Nature Switzerland AG 2025
A. Melesse et al. (eds.), *Abbay River Basin*, Springer Geography,
https://doi.org/10.1007/978-3-031-65241-7_13

Introduction

The droughts severity and frequency of occurrence are dramatically changing as a result of global warming (Tegegne et al. 2019, 2020a, b, 2022; Tegegne and Melesse 2020, 2021). The International Disaster Database (EM-DAT) has reported that climate change and the associated extreme events occurred between 1998 and 2017 killed approximately 1.3 million people and caused approximately 4.4 billion people to need emergency assistance, while 91% of the disasters were caused by droughts, floods, and other extreme events. Thus, climate change impact assessment on drought is vital to proactively manage drought-associated risks in meteorological, agricultural, and hydrological systems. Several global climate models (GCMs) have been developed to project the future possible drought impacts on the meteorological, agricultural, and hydrological sectors. Several climate change impact studies on drought have been reported in various regions over the past ten years. Several studies on the effects of climate change on droughts have been published in various regions over the past ten years (Raje and Mujumdar 2010; Wang et al. 2011; Burke and Brown 2008; Park et al. 2015; Touma et al. 2015; Dai 2011, 2013; Masih et al. 2014; Naumann et al. 2015, 2018; Rhee and Cho 2016; Tegegne et al. 2019; 2020a, b, 2022; Tegegne and Melesse 2020). The majority of previous researches concluded a rising tendency in the drought severity and frequency in various regions, including tropical and subtropical areas (Touma et al. 2015), the Mediterranean region (Naumann et al. 2015, 2018), Africa (Dai 2013; Masih et al. 2014; Tegegne et al. 2020a; Tegegne and Melesse 2021), Central America (Naumann et al. 2018), and West and Southern Asia (Wang et al. 2017). Naumann et al. (2018) reported an increase in the frequency of droughts by 5–10 times compared to the historical frequency in the Mediterranean basin, Central America, West and Southern Asia, most of Africa, and Oceania.

Depending on the drought indices employed, there can be variations in the projections of the spatiotemporal dynamics of droughts under climate change. Based on a shortage of water balance components, drought is typically divided into three categories: meteorological, agricultural, and hydrological. For example, meteorological droughts lack precipitation, agricultural droughts lack soil moisture, and hydrological droughts lack streamflow (Tallaksen and Van Lanen 2004; Wang et al. 2011). The consequences of agricultural and hydrological droughts can further translate into operational and socioeconomic droughts. In addition, operational and socioeconomic droughts can also be associated with excess demand due to population growth or in the adequate design and management of artificial reservoirs. Hydrological drought can greatly affect several sectors, such as irrigated agriculture, ecosystems, drinking water, energy, and industry, among many others. Conversely, meteorological drought can affect rain-fed agriculture, while agricultural drought can adversely affect both rain-fed and irrigated agricultural systems. Note that it might be difficult to prevent an impact of meteorological drought owing to global warming, but proactive actions can be taken into account to mitigate and reduce hydrological drought (Barker et al. 2016).

The Abbay River is the primary source of the Nile River, which is shared by eleven African countries. The Abbay River from Ethiopia contributes approximately 60% of the total inflow into the Aswan reservoir in Egypt. The Abbay River water is essential for both the upstream and downstream riparian countries with competing water needs for hydropower, irrigation, industry, and other ecosystem services. The varying nature of the climate has a significant impact on these competing water needs. Several studies have assessed the climate change potential impact on the hydrological systems in the Abbay River Basin (Tegegne et al. 2020a; Tegegne and Melesse 2021; Beyene et al. 2010; Kim and Kaluarachchi 2009; Elshamy et al. 2009; Taye et al. 2011; Worqlul et al. 2018). According to Taye et al. (2011), there are discrepancies in the effects of climate change on the hydrological and ecosystem services in the Abbay River Basin. The primary goal of this research is to examine how climate change may affect meteorological, agricultural, and hydrological droughts in the Abbay River Basin using a set of scenarios from the most recent Coupled Model Intercomparison Project Phase 6 (CMIP6). Furthermore, several climate change impact studies have assessed agricultural and hydrological droughts by using the SPI with precipitation data only (McKee et al. 1993) since droughts often arise from a shortage of precipitation and precipitation droughts can propagate to agricultural and hydrological droughts. Thus, this study used meteorological drought indicators to quantify agricultural and hydrological droughts (Stagge et al. 2015; Wang et al. 2015).

Description of Data and Study Area

The primary tributary of the Nile River, the Abbay River Basin (see Fig. 13.1), is shared by eleven African nations and is home to more than 238 million people. The Abbay River from Ethiopia contributes approximately 60% of the flow at the Aswan reservoir in Egypt, indicating that the river is the economic backbone of the riparian nations downstream (Sudan and Egypt). The measured daily rainfall data from 21 gauge stations recorded in the 1985–2014 period were used for downscaling and bias correcting the climate data. The measured rainfall data from 21 locations were retrieved from the Ethiopian National Metrological Agency. The Abbay River Basin is located between latitudes 7°45′ to 12°45′N and 34°05′ to 39°45′E. Its natural area is approximately 176,000 km^2, and its elevation ranges from 500 to 4160 m above mean sea level. The average annual precipitation in the study basin ranges from 1200 to 1600 mm, while the mean annual potential evapotranspiration varies between 1000 and 1800 mm/year. The Abbay River Basin is dominantly used for agricultural production. Hence, it is important to investigate the climate change potential impact on hydrological ecosystem services, mainly meteorological, agricultural, and hydrological ecosystem services.

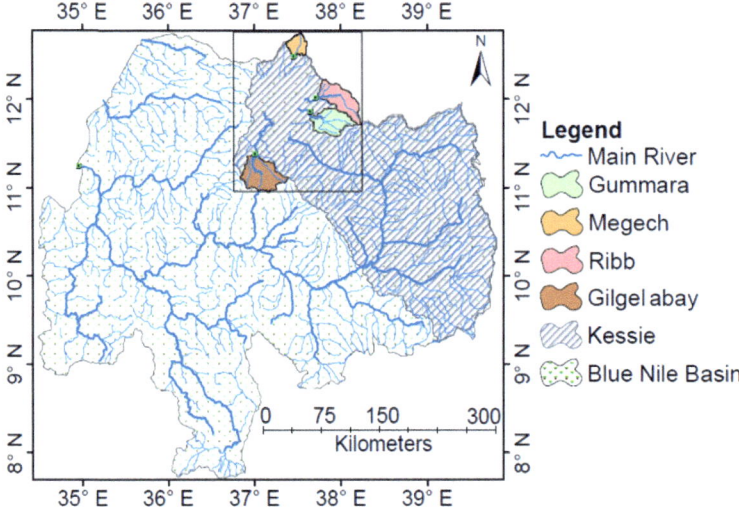

Fig. 13.1 Map shows the large-scale basin (Abbay River Basin), medium-scale basin (Kessie Basin), and small-scale catchments (Gummara, Ribb, Gilgel Abay, and Megech) in the study basin

Methods

Overview of Climate Change Impact Assessment on Droughts

The frequency of droughts using precipitation can be determined based on the commonly used threshold level methods that were developed to characterize droughts when precipitation declines from a defined threshold level. This research employed the SPI (McKee et al. 1993) to quantify the magnitude and frequency of droughts. The GCMs, from the CMIP6 with a shared socioeconomic pathways scenario, suggested by previous studies were used to project future climate and quantify the droughts impacts on the meteorological, agricultural, and hydrological sectors. The World Climate Research Programmes—CMIP6 was used to extract the climate data under two different forcing scenarios: medium (SSP2-4.5) and strong (SSP5-8.5). The precipitation data are downscaled from CMIP6 five climate models (see Table 13.1). The quantile delta mapping (QDM) bias correction technique is employed for correcting the random and systematic errors in the daily rainfall outputs from the GCMs. The drought is quantified in the past and future periods, and the future projected changes in the droughts are then assessed in relation to the past period.

Table 13.1 List of global circulation models employed in this study

S.No.	GCM	Country	Resolution [degree]
1	CMCC-ESM2	Italy	0.9 × 1.25
2	GFDL-ESM4	USA	1. × 1.25
3	MIROC6	Japan	1.4 × 1.4
4	MPI-ESM1-2-LR	Germany	1.85 × 1.85
5	MRI-ESM2-0	Japan	1.125 × 1.125

Precipitation Bias Correction

The daily precipitation data at 21 gauging stations are downscaled and bias corrected from 5 GCMs. The coarse resolution of the GCMs precipitation output had first been spatially disaggregated to a finer spatial scale by an inverse distance weighting interpolation technique, and then bias correction was applied by using an algorithm that matched the modelled and measured data cumulative distribution functions (CDFs) with the quantile delta mapping (QDM) algorithm originally developed by Cannon et al. (2015). Johnson and Sharma (2015) reported the necessity of bias correction application on the climate data for drought analysis. They concluded that the drought projections with the raw climate model simulations resulted in an incorrect representation of the region. Thus, the large distributional bias from the outputs of five GCM forcing data sets was corrected using the QDM approach. The QDM has the capability of preserving the projected changes of all quantiles of the simulated distribution. According to Cannon et al. (2015), the QDM technique is effective in replicating the bias-corrected projected trends in climate extremes as closely as the trends first anticipated by GCMs. The QDM algorithm comprises two sequential steps: first, projected model simulations are quantile-detrended and bias-corrected to reproduce the observations by quantile mapping; second, projected quantiles relative changes are superimposed on the bias-corrected quantiles (Cannon et al. 2015), as in the equation below:

$$\Delta_m(t) = \frac{F_{m,p}^{-1}\left[F_{m,p}^{(t)}(x_{m,p}(t))\right]}{F_{m,h}^{-1}\left[F_{m,p}^{(t)}(x_{m,p}(t))\right]} = \frac{x_{m,p}(t)}{F_{m,h}^{-1}\left[F_{m,p}^{(t)}(x_{m,p}(t))\right]}, \quad (13.1)$$

where $\Delta_m(t)$ is the quantile relative changes, $F_{m,p}^{(t)}$ is the CDF of the modelled data in the projection period, $x_{m,p}(t)$ is the raw downscaled data at time t in the projection period, and $F_{m,h}$ is the CDF of the modelled data in the historical period. The relative changes are then multiplied by the historical bias-corrected data to produce the bias-corrected future projections, as shown by the equation below:

$$\hat{x}_{m,p}(t) = F_{o,h}^{-1}\left[F_{m,h}(x_{m,p}(t))\right]\Delta_m(t), \quad (13.2)$$

where $\hat{x}_{m,p}(t)$ is the bias-adjusted data at time t for the projection period and $F_{o,h}$ is the observed data CDF in the historical period.

Climate Projection Using the Spatiotemporal Reliability Ensemble Averaging Algorithm

This research adopted the spatiotemporal reliability ensemble averaging (ST-REA) algorithm developed by (Tegegne et al. 2019). Giorgi and Mearns' (2002) original reliability ensemble averaging (REA) proposal is based on two factors: the GCMs capability to converge within the entire GCMs-projected quantiles range (convergence weight), and the GCMs ability to replicate the historical observed quantiles (performance weight). The distance of a GCM's projected change from the change in the multi-model ensemble average is used to determine the convergence weight, whereas the performance weight is calculated from the model bias. Therefore, the ensemble mean algorithm weight is computed by multiplying two reliability factors: the model performance in terms of bias (Bi) and the convergence (Di). Keep in mind that if the associated GCM bias (Bi) and the distance of the individual model projected change from the ensemble mean (Di) are within the natural variability range ($\mathcal{E}$), the reliability of the model's performance and convergence is deemed to be 1. Giorgi and Mearns (2003) provided the detailed procedure of the REA approach. The ST-REA approach proposed by Tegegne et al. (2019) reflects both the temporal and spatial variabilities simultaneously during the ensemble weight computation as expressed below:

$$S - TREA_i : w_i^{ST} = \left\{ \left[\frac{\varepsilon^{ST}}{\sqrt{(B_i^{ST})^2 + Var_i^{ST}}} \right]^m \left[\frac{\varepsilon^{ST}}{abs\left(D_i^{ST}\right)} \right]^n \right\}^{[1/(m \times n)]}, \quad (13.3)$$

where the superscripts and subscripts S and T represent the variables over space and time, respectively, and Var_i^* indicates the variability of the GCM$_i$. B_i^{ST}, Var_i^{ST}, D_i^{ST}, and $\mathcal{E}^{ST}$ can be expressed as follows:

$$B_i^{ST} = \frac{1}{J+T} \left[\sum_{j=1}^{J} \sum_{t=1}^{T} \left(X_{i,j,t} - O_{j,t}\right) \right] \quad (13.4)$$

$$Var_i^{ST} = \frac{1}{J+T} \left[\sum_{j=1}^{J} \sum_{t=1}^{T} \left(X_{i,j,t} - \frac{1}{J+T} \sum_{j=1}^{J} \sum_{t=1}^{T} \left(X_{i,j,t}\right)\right)^2 \right] \quad (13.5)$$

$$D_i^{ST} = \Delta X_i^{ST} - \frac{\sum_{i=1}^{N} w_i^{ST} \times \Delta X_i^{ST}}{\sum_{i=1}^{N} w_i^{ST}} \quad (13.6)$$

$$\varepsilon^{ST} = \max \begin{Bmatrix} O_{1,1} & O_{1,2} & \cdots & O_{1,T} \\ O_{2,1} & O_{2,2} & \cdots & O_{2,T} \\ \vdots & \vdots & \ddots & \vdots \\ O_{J,1} & O_{J,2} & \cdots & O_{J,T} \end{Bmatrix} - \min \begin{Bmatrix} O_{1,1} & O_{1,2} & \cdots & O_{1,T} \\ O_{2,1} & O_{2,2} & \cdots & O_{2,T} \\ \vdots & \vdots & \ddots & \vdots \\ O_{J,1} & O_{J,2} & \cdots & O_{J,T} \end{Bmatrix} \quad (13.7)$$

where X and O, respectively, indicate the projected and measured precipitation. The subscripts i, j, and t, respectively, indicate the GCM name, site, and time. $\max\{\overline{O}_1^S, \ldots, \overline{O}_T^S\}$ and $\max\{\overline{O}_1^T, \ldots, \overline{O}_J^T\}$, respectively, represent the spatially and temporally maximum values of the mean precipitation after linearly detrending the data; $\min\{\overline{O}_1^S, \ldots, \overline{O}_T^S\}$ and $\min\{\overline{O}_1^T, \ldots, \overline{O}_J^T\}$, respectively, show the minimum spatially and temporally values of the average precipitation after linearly detrending the data. Δ is the projected changes, and N is the number of models.

Drought Projection and Analysis

Droughts are the outcomes of shortages in precipitation (meteorological drought), soil water content (agricultural drought), and streamflow (hydrological drought), respectively (Wang et al. 2011). The SPI is considered to quantify the drought (McKee et al. 1993), which is a popular indicator for drought assessment based on monthly precipitation aggregates at different time steps (Wang et al. 2011; Stagge et al. 2015). The commonly used drought durations are 1, 2, 3, 6, 9, 12, and 24 months. For example, the SPI for the 9-month duration can be computed using the following steps: (1) the precipitation from month $j - 8$ to month j is aggregated, (2) the probability density function (PDF) is fitted to the aggregated precipitation, which typically follows a gamma distribution (Wilks and Eggleston 1992; Wang et al. 2011), (3) the PDF is subsequently converted into a normal PDF, and (4) finally, the SPI can be computed from the standard normal PDF (McKee et al. 1993). The detailed procedure for SPI computation is provided by McKee et al. (1993). This study used the drought classification proposed by McKee et al. (1993): extreme wet (SPI $\geq$ 2), very wet (1.5 $\leq$ SPI < 2), moderate wet (1 $\leq$ SPI < 1.5), nearly normal ($-$ 1 $\leq$ SPI < 1), moderate drought ($-$ 1.5 $\leq$ SPI < $-$ 1), severe drought ($-$ 2 $\leq$ SPI < $-$ 1.5), and extreme drought (SPI $\leq -$ 2). Note that meteorological drought is characterized by precipitation event computed from 1 to 3 months, while agricultural drought is assessed by using droughts computed from 3 to 6 months, and hydrological drought is computed from 9- to 24-month time scales. Finally, the potential impact of climate change on meteorological, agricultural, and hydrological droughts is quantified by constructing the intensity–duration–frequency (IDF) curve based on the procedures followed by Wang et al. (2011). The future projected spatial variability and the frequency of droughts are computed relative to the historical period.

Results and Discussion

Projected Changes in Precipitation and Temperature

The QDM used for bias correcting the precipitation and temperature is calibrated and validated, respectively, using the observations from the 1985–2009 and 2010–2014 periods. The analysis shows that the QDM was effective in reducing the biases from simulated precipitation and temperature outputs. The climate change impacts on precipitation and temperature in both study basins were assessed and showed the projected increases under all selected GCMs and future time periods. The future mean annual precipitation (averaged across all-weather stations and GCMs) exhibits substantial increases in 2031–2060 and 2061–2090, respectively, by 6% and 4% under SSP2-4.5 and 4% and 10% under SSP5-8.5. On the other hand, the overall minimum and maximum temperature changes averaged over all-weather locations and GCMs with respect to the baseline temperature reached 3.14 °C and 3.42 °C under SSP5-8.5 in the future period (2061–2090), respectively.

Spatiotemporal Dynamics of Droughts

In the ungauged basins, it is difficult to perform climate change impact assessment on the agriculture considering the soil water content and hydrology based on the runoff. In this direction, different climate change impact studies have used extreme weather indicators to quantify the agricultural and hydrological droughts (Wang et al. 2015; Stagge et al. 2015). Hence, the climate change potential impact on the agricultural and hydrologic systems is computed using the 6- and 12-month SPI outputs, respectively. Moreover, this study also assessed the climate change influence on the IDF of meteorological (3-month SPI), agricultural (6-month SPI), and hydrological (12-month SPI) drought events.

Temporal Variability of Drought

Figures 13.2, 13.3 and 13.4 show the temporal variations of all droughts over all the selected gauging precipitation stations. Note that stations 1 through 21 represent Addis Zemen, Ambo, Arjo, Asosa, Bahirdar, Bedele, Begi, Chagni, Dangla, Debrebrehan, Debremarkos, Debretabor, Fiche, Gimbi, Gonder, Mankush, Mota, Nefas Mewcha, Nekmte, and Pawe, respectively. The results showed that the SPI for both emission scenarios tended to have the same extreme frequencies. The results also showed that the temporal variability (gap between minimum and maximum values in the box plot) of SPI in the future time periods (2031–2060 and 2061–2090) will decline compared to the historical periods (1985–2014) under both emission scenarios. The temporal variability in the box plot shows that the SPI median value in the future periods (2031–2060 and 2061–2090) will be reduced slightly compared to the median value in the historical period under both SSP2-4.5 and SSP5-8.5, indicating the overall increase in drought frequencies in the future periods.

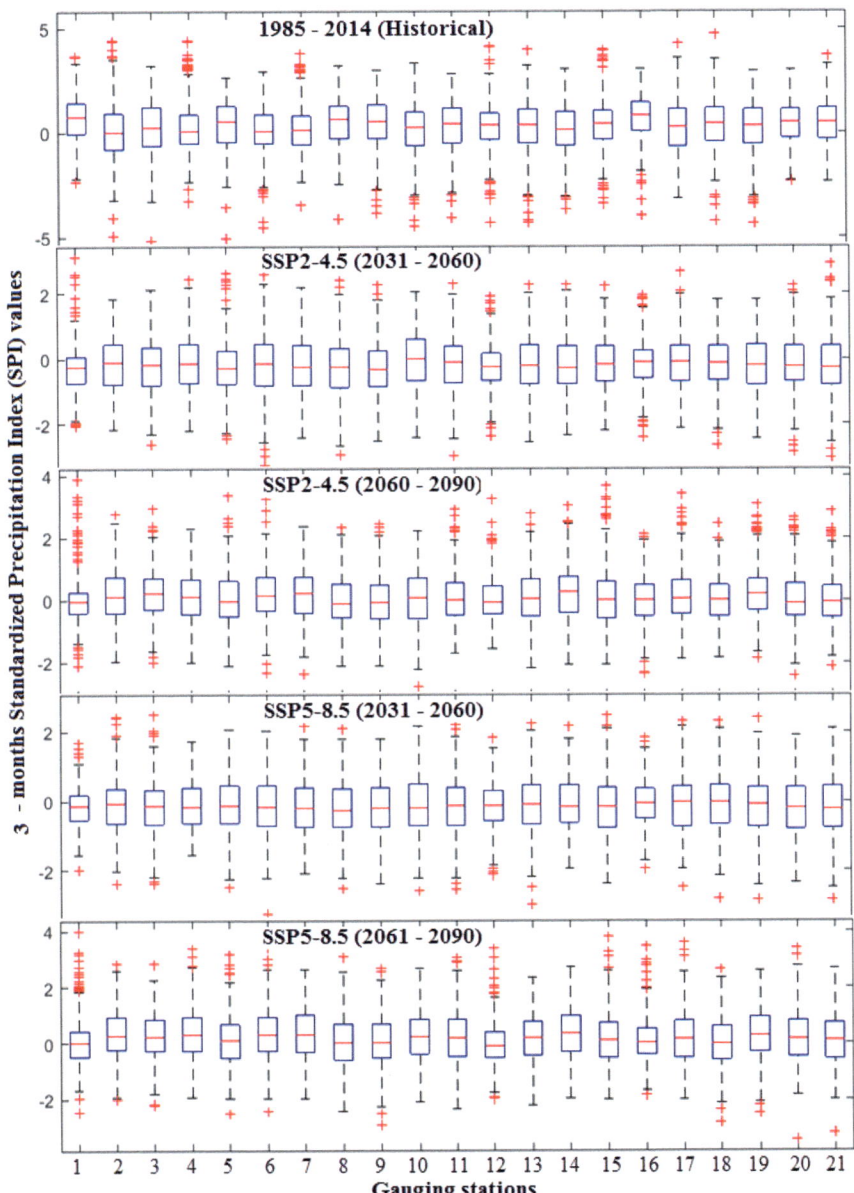

Fig. 13.2 Temporal variability in the 3-month standardized precipitation index (SPI) over each selected precipitation gauging station under both emission scenarios (SSP2-4.5 and SSP5-8.5)

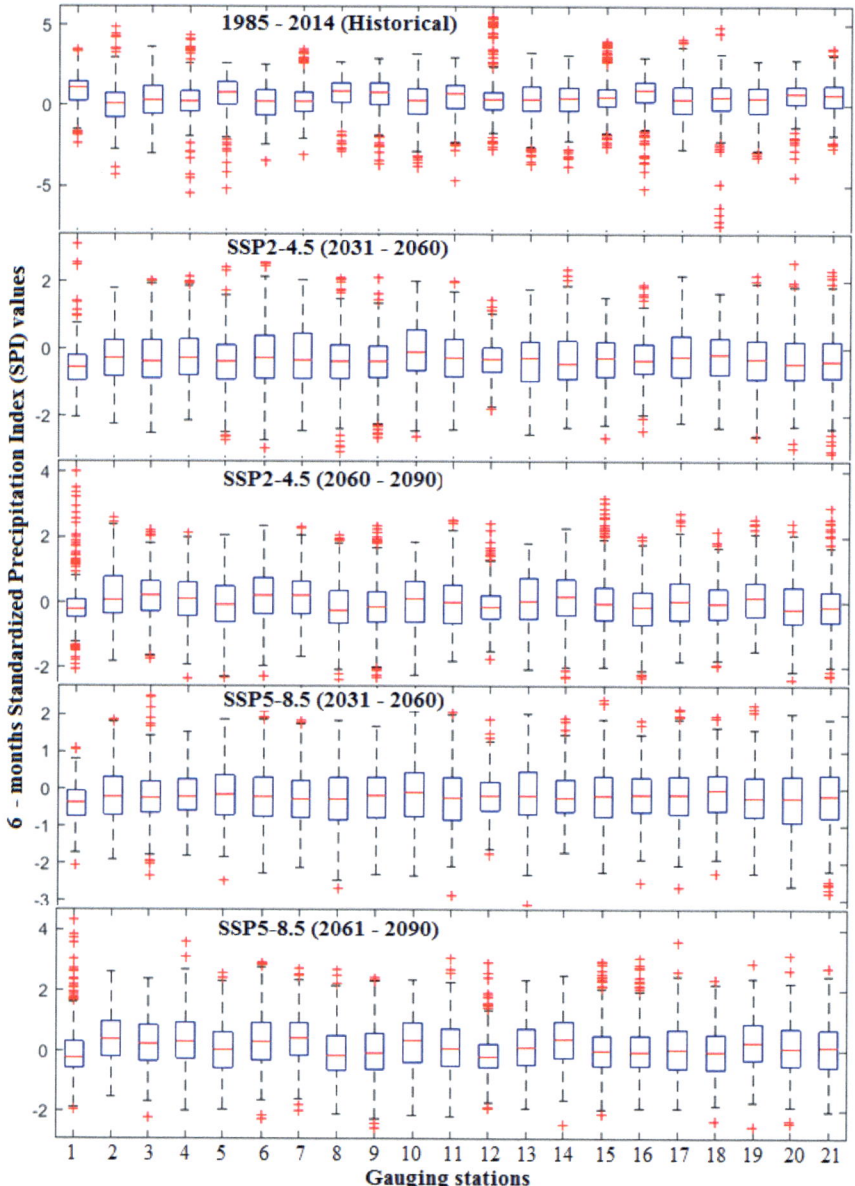

Fig. 13.3 Temporal variability in the 6-month standardized precipitation index (SPI) over each selected precipitation gauging station under both emission scenarios (SSP2-4.5 and SSP5-8.5)

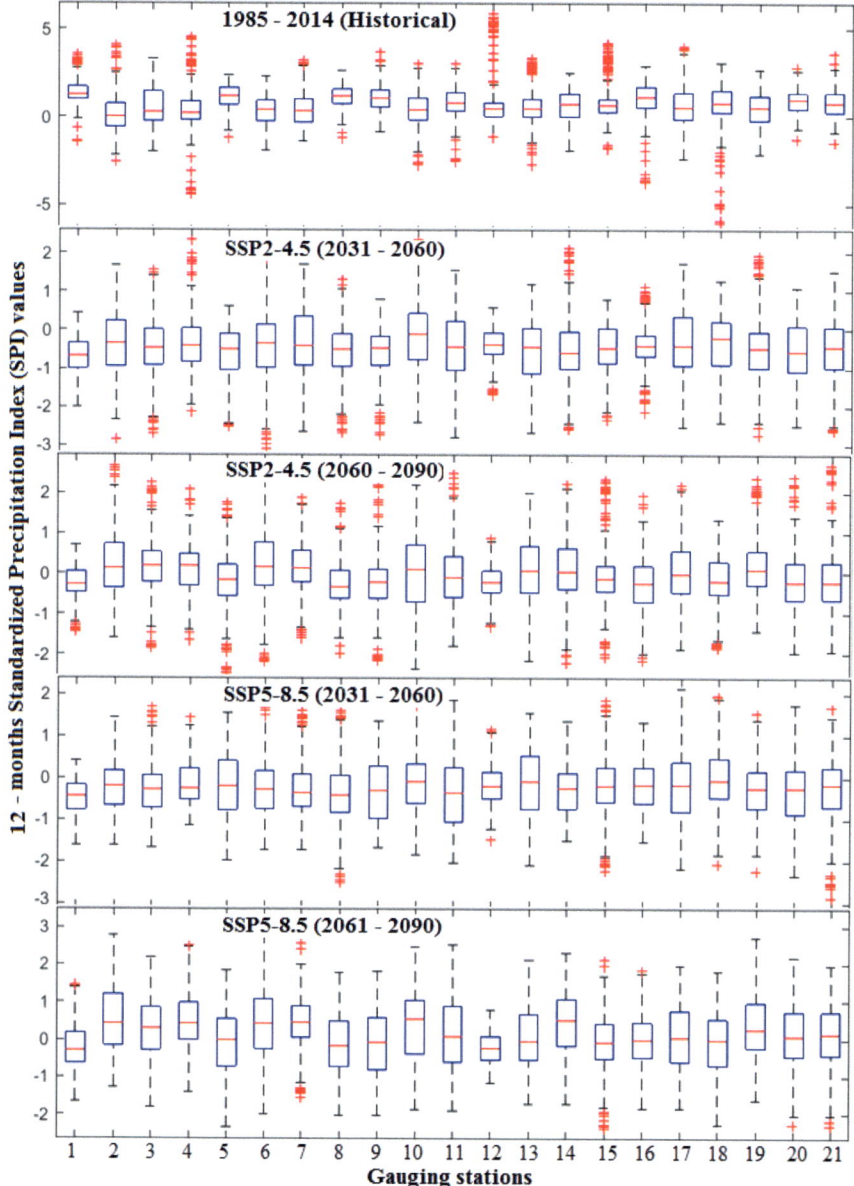

Fig. 13.4 Temporal variability in the 12-month standardized precipitation index (SPI) over each selected precipitation gauging location under both SSP2-4.5 and SSP5-8.5 scenarios

Spatial Variability of Droughts

The intensity and frequency of drought (SPI < − 1), normal drought (− 1 ≤ SPI ≤ 1), and no drought (SPI > 1) were computed in the 2031–2060 and 2061–2099 and assessed with respect to the 1985–2014 period (see Figs. 13.6, 13.7 and 13.8). The severity and frequency of SPI for 3-month (towards meteorological drought), 6-month (towards agricultural drought), and 12-month (towards hydrological drought) were computed to investigate the specific areas in the Abbay River Basin that need critical climate change impact mitigation and adaptation plans. The findings demonstrated that as drought duration increased, so did its severity. This suggests that because the agricultural and hydrological systems are severely impacted by the precipitation anomalies over comparatively longer time periods, the impact of climate change is more pronounced on them. In general, to investigate the areas with severe meteorological, agricultural, and hydrological droughts, the projected changes in the frequency of time in each no drought and drought event expressed as the percentage of months calculated in the 2031–2060 and 2061–2090 periods with respect to the 1985–2014 period were quantified.

Spatial Variability of Meteorological Drought

Meteorological drought can be characterized using the 3-month SPI values. The meteorological drought frequency was found to spatially vary between − 29.98 and 84.51% in 2031–2060, − 66.22 and 33.32% in 2061–2090 under SSP2-4.5, and − 30.82 and 268.93% in 2031–2060 and − 58.13 and 224.12% in 2061–2090 under SSP5-8.5. The maximum meteorological drought frequency was observed in the north-western areas of the Abbay River Basin (Fig. 13.5). In particular, the Lake Tana and Beles basins will face frequent droughts compared to historical observations. Conversely, the south-eastern parts of the Abbay River Basin will encounter relatively fewer projected changes of the meteorological drought frequency. The frequency of no droughts (SPI > 1) will decline in all parts of the Abbay River Basin under SSP2-4.5, with spatial variation between − 91.47 and − 34.79 in 2031–2060 and between − 84.26 and − 13.73 in 2061–2090. On the other hand, the projected frequency of no droughts will decline in the southern regions of the study area under SSP5-8.5. Moreover, the frequency of occurrence of normal meteorological droughts (− 1 ≤ SPI ≤ 1) will increase in the future under both emission scenarios.

Spatial Variability of Agricultural Drought

The frequency of agricultural droughts can be quantified using the 6-month SPI values. The spatial variation in the frequency of agricultural droughts (SPI < − 1) over the Abbay River Basin was found to vary between − 11.27 and 187.98% in 2031–2060, − 77.56 and 77.67% in 2061–2090 under SSP2-4.5, − 27.41 and 639.93% in 2031–2060 and − 74.96 and 488.19% in 2061–2090 under SSP5-8.5. Drought frequency will increase in the north-western regions of the Abbay River Basin (Tana-Beles basins), whereas it will decrease in the southern regions (Fig. 13.6). The frequency of no droughts (SPI > 1) will decline in the majority of the Abbay River Basin. Moreover, the projected frequency of normal agricultural drought occurrence

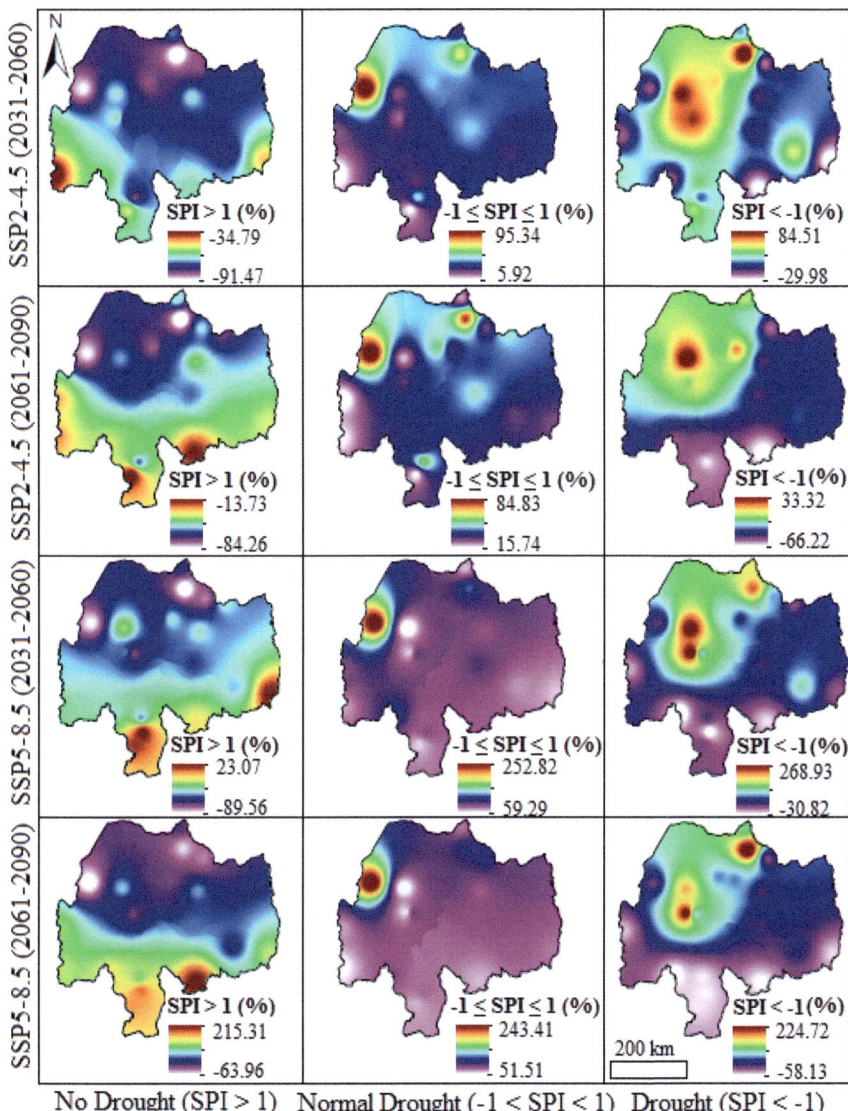

Fig. 13.5 Projected changes in the frequency of months with no meteorological drought (SPI > 1), near normal meteorological drought ($-1 \leq$ SPI ≤ 1), and meteorological drought (SPI < -1) events calculated in the 2031–2060 and 2061–2090 periods relative to the 1985–2014 period over the Abbay River Basin

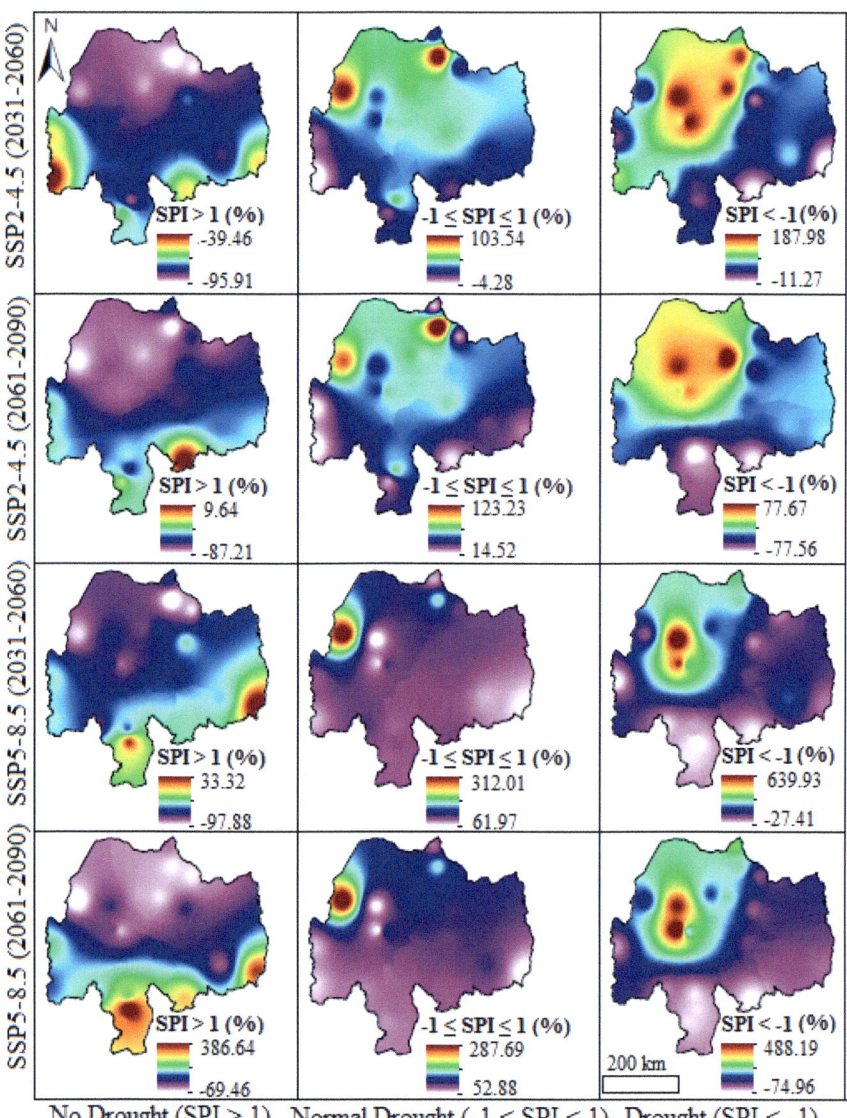

Fig. 13.6 Projected changes in the frequency of months with no agricultural drought (SPI > 1), normal agricultural drought ($-1 \leq$ SPI ≤ 1), and agricultural drought (SPI < -1) events calculated in the 2031–2060 and 2061–2090 periods relative to the 1985–2014 over the Abbay River Basin

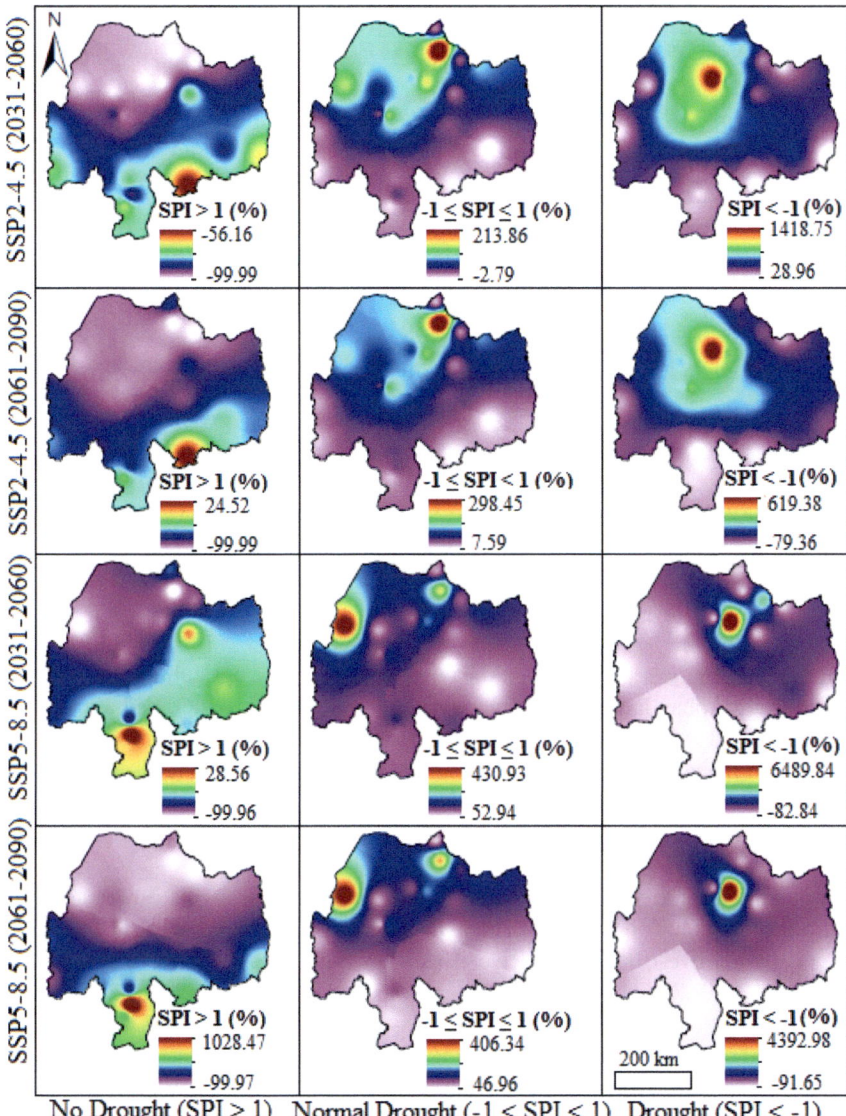

Fig. 13.7 Projected changes in the frequency of months with no hydrological drought (SPI > 1), normal hydrological drought ($-1 \leq$ SPI ≤ 1), and hydrological drought (SPI < -1) events calculated in the 2031–2060 and 2061–2090 periods relative to the 1985–2014 period over the Abbay River Basin

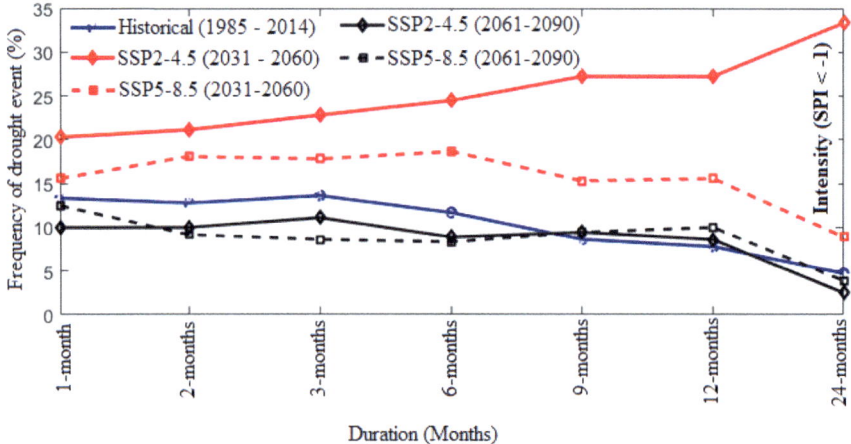

Fig. 13.8 Meteorological (SPI with 1–3 months), agricultural (SPI with 3–6 months), and hydrological (SPI with 9–24 months) drought intensity ($I < -1$)—duration—frequency curve for the Abbay River Basin during baseline (1985–2014) and projected periods of (2031–2060) and (2061–2090) under both emission scenarios. The droughts are quantified based on the precipitation averaged over the entire basin and all GCM outputs

will increase in most parts of the Abbay River Basin. The agricultural droughts in the northern Abbay River Basin are expected to rise in 2031–2060 and 2061–2090 compared to 1985–2014. The increased frequency of agricultural drought in the majority of the northern regions of the Abbay River Basin will have a severe impact on the region's rain-fed agricultural productivity.

Spatial Variability of Hydrological Drought

The potential impacts of climate change on hydrological droughts can be assessed using the 12-month SPI values. The hydrological drought (SPI < -1) frequency of occurrence over the Abbay River Basin was found to spatially vary between 28.96 and 1418.75% in 2031–2060, -79.36 and 619.38% in 2061–2090 under SSP2-4.5, -82.84 and 6489.84% in 2031–2060 and -91.65 and 4392.98% in 2061–2090 under SSP5-8.5. It is worth noting that the Addis Zemen, Bahirdar, and Dangla areas will experience the greatest increase in the frequency of hydrological droughts. Conversely, the hydrological droughts frequency of occurrence tends to decrease in most parts of the Abbay River Basin under both emission scenarios (Fig. 13.7). The frequency of no droughts (SPI > 1) showed a decrease and increase in much of the northern and southern Abbay River Basin, respectively. The projected frequency of normal hydrological drought occurrence will spatially vary between -2.79 and 213.86% in 2031–2060 and 7.59 and 298.45% in 2061–2090 under SSP2-4.5 and between 52.94 and 430.93% in 2031–2060 and between 46.96 and 406.34% in 2061–2090 under SSP5-8.5. This result revealed that the frequency of normal hydrological droughts will increase in much of the Abbay River Basin. It is apparent from these results that the hydrological droughts in (2031–2060 and 2061–2090) compared to (1985–2014) tend to decrease in most regions of the Abbay River Basin. The decrease in the frequency of hydrological drought occurrence can be beneficial to the improvement of hydropower and agricultural production by means of water harvesting.

Intensity–Duration–Frequency (IDF) of Droughts

Figure 13.7 depicts frequency of droughts projected changes (SPI < − 1) calculated in the 2031–2060 and 2061–2090 periods with respect to 1985–2014. The drought frequency of occurrence will rise in the middle future (2031–2060) under both emission scenarios. The frequency of occurrence of meteorological (1–3 months SPI), agricultural (3–6 months SPI), and hydrological (12–24 months SPI) droughts will increase by more than 20% in 2031–2060 under SSP2-4.5 (Fig. 13.8). Droughts in 2031–2060 under SSP2-4.5 showed an increase in the frequency of occurrence as the duration increased. The results also showed that the drought frequency of occurrence will decrease in 2061–2090 under both emission scenarios relative to the historical drought conditions, except for the increase in hydrological drought corresponding to the 12-month SPI (Fig. 13.8). Overall, the findings revealed that the risk of drought in the Abbay River Basin will increase in middle future (2031–2060) and potentially decline in far future (2061–2090) under both emission scenarios.

Conclusion

In this study, the potential impact of climate change on drought was assessed using the 3-month (towards meteorological drought), 6-month (towards agricultural drought) and 12-month (towards hydrological drought) SPI in the Abbay River Basin. This could improve our understanding of how the impact propagates from one system to the other. The precipitation from five GCMs was downscaled and bias corrected using the QDM. The five GCM precipitation values were then combined using the spatiotemporal reliability ensemble averaging approach. The analysis revealed that the frequency of droughts in the Abbay River Basin will potentially decrease in the south and increase in the north. Furthermore, under both SSP2-4.5 and SSP5-85, the probability of drought in the Abbay River Basin will increase in the mid-term (2031–2060) and perhaps decrease in the long-term (2061–2090). The analysis also revealed that the frequency of drought occurrence will increase by more than 20% and 15%, respectively, under SSP2-4.5 and SSP5-8.5 in the mid-term (2031–2060) period. This indicated that vulnerability to various systems might be further aggravated if the changes in the climatic extremes continue in the future. As a result, institutional reforms and management objectives must be implemented in a timely way.

Acknowledgements The Sustainable Energy Center of Excellence, Addis Ababa Science and Technology University funded this research under project code number IG 30/2022.

Conflict of Interest: The authors state that there is no conflict of interest in the publishing of this work.

References

Barker LJ, Hannaford J, Chiverton A, Svensson C (2016) From meteorological to hydrological drought using standardized indicators. Hydrol Earth Syst Sci 20:2483–2505

Beyene T, Lettenmaier DP, Kabat P (2010) Hydrologic impacts of climate change on the Nile River Basin: implications of the 2007 IPCC scenarios. Clim Change 100:433–461

Burke EJ, Brown SJ (2008) Evaluating uncertainties in the projection of future drought. J Hydrometeorol 9:292–299

Cannon AJ, Sobie SR, Murdock TQ (2015) Bias correction of GCM precipitation by quantile mapping: how well do methods preserve changes in quantiles and extremes? J Clim 28:6938–6959

Dai A (2011) Drought under global warming: a review. Wiley Interdisc Rev: Clim Change 2:45–65

Dai A (2013) Increasing drought under global warming in observations and models. Nat Clim Chang 3:52

Elshamy ME, Seierstad IA, Sorteberg A (2009) Impacts of climate change on Blue Nile flows using bias-corrected GCM scenarios. Hydrol Earth Syst Sci 13:551–565

Giorgi F, Mearns LO (2002) Calculation of average, uncertainty range, and reliability of regional climate changes from AOGCM simulations via the "reliability ensemble averaging" (REA) method. J Clim 15(10):1141–1158

Giorgi F, Mearns LO (2003) Probability of regional climate change based on the Reliability Ensemble Averaging (REA) method. Geophys Res Lett 30

Johnson F, Sharma A (2015) What are the impacts of bias correction on future drought projections? J Hydrol 525:472–485

Kim U, Kaluarachchi J (2009) Climate change impacts on water resources in the Upper Blue Nile River Basin, Ethiopia 1. JAWRA J Am Water Resour Assoc 45:1361–1378

Masih I, Maskey S, Mussá F, Trambauer P (2014) A review of droughts on the African continent: a geospatial and long-term perspective. Hydrol Earth Syst Sci 18:3635–3649

McKee TB, Doesken NJ, Kleist J (1993) The relationship of drought frequency and duration to time scales. In: Proceedings of the 8th conference on applied climatology, pp 179–183

Naumann G, Spinoni J, Vogt JV, Barbosa P (2015) Assessment of drought damages and their uncertainties in Europe. Environ Res Lett 10:124013

Naumann G, Alfieri L, Wyser K, Mentaschi L, Betts R, Carrao H, Spinoni J, Vogt J, Feyen L (2018) Global changes in drought conditions under different levels of warming. Geophys Res Lett 45:3285–3296

Park C-K, Byun H-R, Deo R, Lee B-R (2015) Drought prediction till 2100 under RCP 8.5 climate change scenarios for Korea. J Hydrol 526:221–230

Raje D, Mujumdar P (2010) Hydrologic drought prediction under climate change: uncertainty modelling with Dempster-Shafer and Bayesian approaches. Adv Water Resour 33:1176–1186

Rhee J, Cho J (2016) Future changes in drought characteristics: regional analysis for South Korea under CMIP5 projections. J Hydrometeorol 17:437–451

Stagge JH, Kohn I, Tallaksen LM, Stahl K (2015) Modelling drought impact occurrence based on meteorological drought indices in Europe. J Hydrol 530:37–50

Tallaksen LM, Van Lanen HA (2004) Hydrological drought: processes and estimation methods for streamflow and groundwater. Elsevier

Taye MT, Ntegeka V, Ogiramoi N, Willems P (2011) Assessment of climate change impact on hydrological extremes in two source regions of the Nile River Basin. Hydrol Earth Syst Sci 15:209

Tegegne G, Kim YO, Lee JK (2019) Spatiotemporal reliability ensemble averaging of multimodel simulations. Geophys Res Lett 46:12321–12330

Tegegne G, Melesse AM (2020) Multimodel ensemble projection of hydroclimatic extremes for climate change impact assessment on water resources. Water Resour Manag 1–17

Tegegne G, Melesse AMJEP (2021) Comparison of trend preserving statistical downscaling algorithms toward an improved precipitation extremes projection in the headwaters of Blue Nile River in Ethiopia 8:59–75

Tegegne G, Melesse AM, Alamirew T (2020a) Projected changes in extreme precipitation indices from CORDEX simulations over Ethiopia, East Africa. Atmos Res 247:105156

Tegegne G, Melesse AM, Worqlul AW (2020b) Development of multimodel ensemble approach for enhanced assessment of impacts of climate change on climate extremes. Sci Total Environ 704:135357

Tegegne G, Mellesse AMJT, Climatology A (2022) Multimodel ensemble projection of precipitation over South Korea using the reliability ensemble averaging, 1–10

Touma D, Ashfaq M, Nayak MA, Kao S-C, Diffenbaugh NS (2015) A multimodel and multi-index evaluation of drought characteristics in the 21st century. J Hydrol 526:196–207

Wang D, Hejazi M, Cai X, Valocchi AJ (2011) Climate change impact on meteorological, agricultural, and hydrological drought in central Illinois. Water Resour Res 47

Wang H, Rogers JC, Munroe DK (2015) Commonly used drought indices as indicators of soil moisture in China. J Hydrometeorol 16:1397–1408

Wang Z, Li J, Lai C, Zeng Z, Zhong R, Chen X, Zhou X, Wang M (2017) Does drought in China show a significant decreasing trend from 1961 to 2009? Sci Total Environ 579:314–324

Wilks DS, Eggleston KL (1992) Estimating monthly and seasonal precipitation distributions using the 30-and 90-day outlooks. J Clim 5:252–259

Worqlul AW, Dile YT, Ayana EK, Jeong J, Adem AA, Gerik T (2018) Impact of climate change on streamflow hydrology in headwater catchments of the Upper Blue Nile Basin, Ethiopia. Water 10:120

Part V
Water Use and Watershed Management

Chapter 14
Irrigation Development in the *Abbay* River Basin, Ethiopia

Sisay Demeku Derib

Abstract The *Abbay* River Basin is the main basin, with 176,000 km^2 in the Blue Nile River. It originates from four main tributaries in the highlands around Lake Tana and drains to the border of Sudan through the Grand Renaissance Dam, GERD. The *Abbay* River Basin has 13 main tributaries and runs about 900 km of river courses from Lake Tana to leave Ethiopia. This part of the book tries to address the natural resources and irrigation potential as compared to the overall country's potential in the first two parts. The third and last part tried to share a case study in the upstream Lake Tana basin. This part covers irrigation water loss and productivity in a small irrigation system under a mixed crop-livestock farming system, as well as water availability and stress at the watershed level. There is huge water loss from traditional irrigation water management. However, a very small amount of the water loss through leakage and canal overflow produced grass around the canal banks and drainage basins. This could benefit feed for cattle during the dry season, when feed supply is a big problem. Productivity of irrigation water (kg m^{-3}) was seen with an improved measuring approach that can consider both straw and grain since straw production is also equally important for the mixed farming system. Together with low land productivity (kg ha^{-1}) due to shortage and misuse of irrigation water, there are rooms to increase irrigation water productivity. The water availability and stress under watershed level have been assessed with different land uses, watershed treatments, and irrigation water abstraction.

Keywords Irrigation · *Abbay* · Blue Nile · Water productivity · Ethiopia

---

S. D. Derib (✉)
Civil Engineering Department, Construction Quality & Technology Center of Excellence, Addis Ababa Science and Technology University, Addis Ababa, Ethiopia
e-mail: sisay.demeku@aastu.edu.et

© The Author(s), under exclusive license to Springer Nature Switzerland AG 2025
A. Melesse et al. (eds.), *Abbay River Basin*, Springer Geography,
https://doi.org/10.1007/978-3-031-65241-7_14

Introduction

Very few parts of the fresh water on our globe are scarce for agriculture in terms of space and time. Irrigation is the best and most important human activity to make this scarce water for human life in most parts of the world in general and the *Abbay* Basin in particular. Three-fourths of the earth is covered by water. Our water is in the global hydrologic cycle, occurring in three phases at different geographical locations. It is not only stored under the surface, on the surface, and in the atmosphere of the earth, but also flows from storage to storage. Understanding the nature of water is very important for making a decision on irrigation and other uses. About 97.5% of this water occurs in saline oceans and lakes and 2.5% in fresh water (three-quarters of this amount is present in polar ice caps and glaciers). About 0.8% of the world's fresh water is available in rivers, lakes, groundwater, the atmosphere, soil moisture, and biological cells (Shiklomanov 1993). These figures indicate that the amount of freshwater is small as compared to the total water bodies. Oceans, seas, ice caps, glaciers, groundwater, rivers, streams, lakes, swamps, soil moisture, water vapor, and the biosphere are the natural water reservoirs of the earth. Water is also flowing along these natural storages. In addition to the small amount of freshwater on our planet, the spatiotemporal and residence times of water in storage are also challenges for the availability of water for irrigation and other uses. The sun's energy, supported by gravitational forces, changes the phase and geographic location (Marcinek 2007) of the water parts in which we are interested. The average time a water molecule spends within the reservoirs of our water within the hydrologic cycle is referred to as residence time. It is a measure of the average age of the water in a given reservoir. This time is estimated at 2–10 days for atmospheric moisture (Ruud and Tuinenburg 2017) and 3–6 years for coastal water (Liu et al. 2019). Gyamera (2014) referred to the residence times of natural reservoirs as oceans (3200 years), glaciers (20–100 years), groundwater (100–10,000 years), lakes (50–100 years), rivers (2–6 months), soil moisture (1–2 months), and the atmosphere (9 days). The residence time of water in soil reservoirs indicates how frequently the available water can renew soil moisture for the rainfed agriculture of *Abbay* Basin. Although a considerable amount of annual rainfall is in the basin, rainfed farming is challenged even to a total loss of crop production when there are weeks of dry spells during the initial growing stages. This occurrence of crop failure has been common in Ethiopia for centuries as a hydrology-induced soil moisture drought (Temam et al. 2009). The drought occurrence is getting more frequent than decades of severe droughts in the last forty years to five consecutive failed rainy seasons (OCHA 2023).

There are different categories of fresh water according to the state of water availability in the basin: blue, green, environmental, virtual, and peak waters. These partitions of rainfall are very important in making decisions on the application of irrigation. The green water is the precipitation that adds more or less temporarily to soil water storage and is consumed by ecosystems through evapotranspiration, while the blue water is the surface and groundwater stored in and flowing from rivers, lakes, aquifers, and dams that can be exploited for human use. The green water flow is from terrestrial biomass-producing systems like crops, forests, grasslands, and savannas, while the blue water flow is in rivers, through base flow from groundwater, and through wetlands. The green water flow is the sum of the nonproductive part of the actual evaporation and the productive part of the actual transpiration (Falkenmark and Rockström 2006). Another water category that is more related to blue water is the environmental water requirement (EWR). It is blue water that is planned for the maintenance of freshwater ecosystem functions and the services they provide to humans (Smakhtin et al. 2005). This proportion of blue water is very important to consider for irrigation from transboundary water like the *Abbay* River. Virtual water, also known as embedded, embodied, or indirect water, is the amount of water needed for the production of the products (Hoekstra 2003). This concept is also very important when making a decision on the selection of crops for irrigation. Peak water is the concept of discussion on water with the analogy of "peak oil" that leads to the point of maximum production or generation of blue water renewing rate in the hydrologic cycle (Gleick and Meena 2010). This concept might also be another concern for the diversion of blue water for irrigation in future.

The author of this chapter has observed the interest of salt, groundwater, and political waters in the blue water of the *Abbay* River. The articulation of Ethiopia as a "water tower" of the region and a country of less water stress with relatively higher available water per capita (https://data.apps.fao.org/aquastat/?lang=en assessed on July, 2023) leads to wrong political implications for the *Abbay* River blue water use. Egypt is considered a country only dependent on Nile water flows, forgetting the application of ample groundwater and saltwater potential, while Ethiopia's high transboundary blue water flow is considered to calculate the water stress parameters. The green water availability as productive transpiration of the high amount of annual rainfall in Ethiopia and the *Abbay* River Basin needs to be considered again for irrigation development.

Humans interfere with the natural phenomenon of water balance to satisfy crop and/or food and industrial vegetation production water needs through irrigation. Irrigation might be supplemental, deficit, or full water supply. The interference can be done by constructing dams to fill storage gaps and by providing transportation soil moisture supply to fill green water gaps. This chapter tries to discuss the irrigation potential of the *Abbay* River Basin with respect to the potential share of Ethiopia at the country level. Results from a case study on irrigation at farm and watershed levels were presented to see how green water shortage and management were happening on the ground in Abbay Basin. The presentations of the chapters can show the perspectives of the above water categories on how to exploit the water resources effectively and sustainably.

Natural Resources and Irrigation Status of Ethiopia

Natural Resources

Natural resources are one of the three resources (Simachew 2020), including human and financial/technological capital resources for the survival of human beings. The land and water resources of Ethiopia are not well developed, and Ethiopia is one of the countries with limited access to basic needs for life. This is because of a lack of financial and technological resources and the political unrest in the region (Woldesenbet et al. 2022). Minimal attention is given to the country's development policy despite the chronic problem of drought for a long time. Water harvesting in the form of micro dams and tanks has been part of the food security program since the late 1990s, but the amount of water collected had to be too small to have a significant impact (Demeke et al. 2004).

The population of the country is going to double within three decades. It makes Ethiopia the 12th and 2nd most populous country in the globe and Africa, respectively (https://data.worldbank.org/). The population was estimated to be 75.7 million in 2008 and 136.8 million in 2037 (CSA 2013). Most of the population in Ethiopia lives in highland areas of the country, with 85% being rural and dependent on subsistence and drought-affected rainfed agriculture with a low level of productivity (Awulachew et al. 2007; Haile and Kasa 2015).

Effective, efficient, and sustainable utilization of these resources is the way to get out of poverty and satisfy the basic needs of Ethiopians and the region. Ethiopia is reported as one of the countries in the world that is endowed with available and untapped natural resources. The landmass of the country is estimated to be 1.13 million km^2, with fertile soil, large but erratic annual rainfall and surface water, a favorable climate to grow diverse plant species throughout the year, wildlife, and other resources (Alemu 2017; Baye 2017). Ethiopia is also called the water tower of East and North Africa, and it has five major river basins that flow from its highlands to neighboring countries (see Fig. 14.1 and Table 14.1). The ten river basins discharge about 122–124 BMC annually (Awulachew et al. 2007), and the rest are dry basins. Except for the Awash River basins with 4.9 BMC, all the rest discharge their flows to the Mediterranean Sea, Indian Ocean, and Lake Turkana in Kenya, flowing seasonally with low diversion for irrigation and other uses. That is why considering this amount of flow without withdrawal as a water stress parameter leads to the wrong implication.

In the 1.13 million km^2 landlocked area of Ethiopia, more than 85% of the population and 70% of the livestock are hosted by 36% of the highland area with rugged topography and relatively better seasonal rainfall (Demeke et al. 2004). Another study has also suggested that 81% of the country's territory can be used for agricultural activities (Belay 2017). Approximately 66% of the land is said to be suitable for agricultural production. The other 36% of lowland areas that are suitable for irrigation still have a sparse population and are not properly utilized. According to the FAO database for 2020, approximately 34.1% of the land is agricultural, and

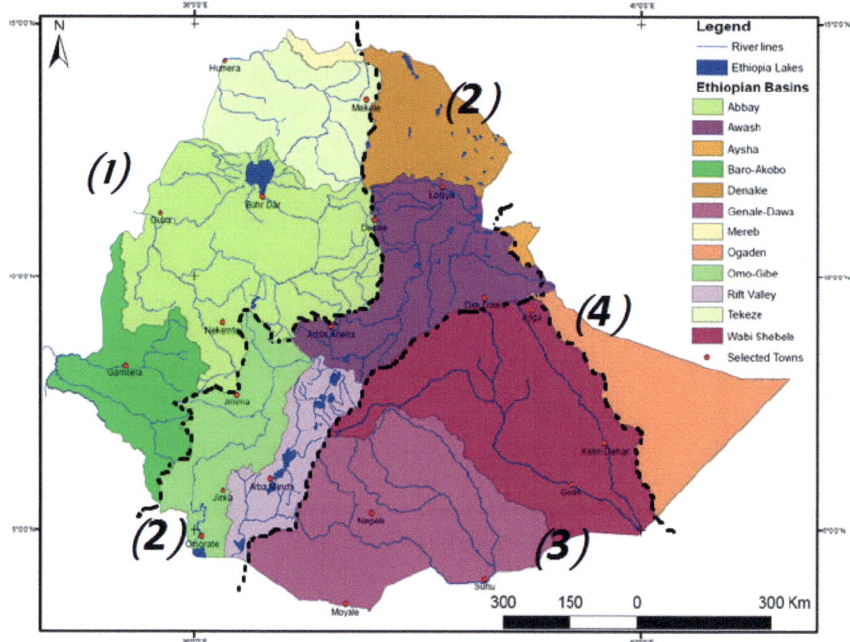

Fig. 14.1 Major drainage systems of Ethiopia with their river basins: (1) Ethiopian Blue Nile with *Abbay* River Basin, (2) Rift Valley, (3) *Shebele-Juba*, and (4) East Coast. *Source* Map modified form of Berhanu et al. (2014)

only 16.2 Mha (14%) is arable land. Most of them are managed by subsistence rainfed agriculture. The other land-use types comprised cultivated land (14.7%), woody vegetation (forest and bush land) (11.6%), and unproductive land (3.7%). The cultivated agricultural land of Ethiopia currently under cultivation is approximately 12 million ha (MoA 2011). Moreover, even if the potential and actual irrigated areas are not precisely investigated (Belay and Woldeamlak 2013), estimates of irrigable land in Ethiopia vary between 1.5 and 4.3 million ha, or on average, approximately 3.5 Mha (Sadoff 2019; Werfring et al. 2004). Some estimates show that only 15 Mha of land is under cultivation, and only approximately 4–5% is irrigated, with existing irrigation schemes covering approximately 640,000 ha (Kassa and Tesfa 2020).

The other natural resource not yet exploited efficiently is rainfall, which is the source of the discharges from the river basin. It is estimated that to be approximately 936.4 BMC per year (https://www.fao.org/aquastat/en assessed on 20/03/2023). False narratives have been articulated and communicated about the availability of this enormous amount of annual rainfall. They narrate that Ethiopia has ample rainfall, so diverting the blue river water is not important, especially from the 84 BMC historical Blue Nile flow to downstream countries. However, the availability of water, by definition, is its accessibility in space and time, both in quantity and quality, for a given purpose. This amount of annual rainfall is erratic and seasonal

Table 14.1 Relative contributions of the *Abbay* River Basin to the major drainage systems of Ethiopia

Major drainage system	River basins	Area (% of total)	Runoff (% of total)	Potential irrigable land (% total)	Gross hydroelectric potential (% total)
Nile Basin		32.4	69.51	53.70	63.55
	Abbay	17.6	43.9	22.8	50.8
	Baro-Akobo	6.5	18.9	28.5	8.9
	Setit-Tekeze/Atbara	7.8	6.1	2.3	3.9
	Mereb	0.5	0.6	0.0	0.0
Rift Valley		14.9	23.25	9.56	26.97
	Awash	9.9	3.7	3.75	2.9
	Denakil	6.5	0.7	0.00	0.0
	Omo-Gibe	6.9	14.4	1.90	23.6
	Central Lake	4.6	4.5	3.90	0.5
Shebelli-Juba		32.7	7.24	36.74	9.48
	Wabi-Shebelle	17.6	2.5	6.66	3.5
	Genale-Dawa	15.0	4.7	30.08	6.0
North East Coast		7.0	0	0	
	Ogaden	6.8	0		
	Aysha	0.2			
Total		100.0	100	100	100

Sources Data from FAO (2005) and different river master plan studies

in nature. As the topography of Ethiopia is rugged, a considerable amount of rainfall has the chance to fall in a place where it is difficult to practice agriculture by changing this rainfall into productive transpiration. Therefore, the rainfall will evaporate with no or few benefits for Ethiopians. The other important spatial factor is that the rain falls on a small part of the south-west part of the country, mostly on the Blue Nile Basin (see Fig. 14.1). That is why 30% of the land contributes more than 70% of the total runoff, and 50% and 60% of the irrigation and hydropower potential of the country, respectively (see Fig. 14.2). That is why approximately 86% of the Nile flow at Aswan High Dam is contributed by 13% of the Blue Nile Basin in Ethiopia within 3–4 months of the year.

Irrigation in Ethiopia

Irrigation can be defined as an artificial application of water to soil for the purpose of supplying the moisture essential in the plant root zone to prevent stress that may

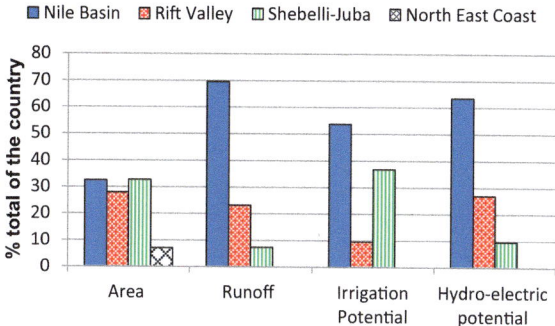

Fig. 14.2 Relative potential of river basins in Ethiopia. *Sources* Data taken from different river basin master plan studies and (FAO 2005). The Nile Basin encompasses the Ethiopian Nile Basin, which drains through the *Mereb*, *Tekezie*, *Abbay*, *Baro*, and *Akobo* Rivers to the main Nile River

cause reduced yield and/or poor quality of harvest for crops (Reddy 2010). Different authors, such as Awlachew et al. (2007), Makombe et al. (2007), Hagos et al. (2009), and Bacha et al. (2011), stressed that supplementary and spate irrigation have been practiced by smallholder farmers in Ethiopia for centuries. The bilateral agreement between the government of Ethiopia and the Dutch company jointly known as HVA-Ethiopia sugar cane plantation is the case for modern irrigation in the early 1950s (MoA 2011). Nevertheless, it is rare to have a fully equipped large-scale irrigation scheme in the country compared to the era of sugarcane irrigation at Wonji and Metehara, where irrigation development in Ethiopia is classified into three types based on the size of the command area (MoWR 2002):

- Small-scale irrigation systems (< 200 ha).
- Medium-scale irrigation systems (200–3000 ha).
- Large-scale irrigation systems (> 3000 ha).

Approximately half of the irrigation schemes in Ethiopia are categorized as small-scale irrigation (Makombe et al. 2007). MoA (2011) reported that approximately 10–12% of the total irrigable potential is currently under production using traditional and modern irrigation schemes. Enormous differences in irrigation potentials and actually irrigated lands on different studies, such as 3.7 million ha and 197,000 ha according to Awulachew et al. (2007), 3.5 million ha and 626,116 ha according to Hagos et al. (2009), respectively, and basin-wide potentials of irrigation as indicated in Table 14.2, are indications of a low level of irrigation development in the country and future trends for the survival of the existing problems.

Table 14.2 Area, surface water, irrigation, and hydroelectric power potential share of the *Abbay* River Basin

(ENB)	Area (% of total)		Runoff (% of total)		Potential irrigable land (% total)		Gross hydroelectric potential (% total)	
	1*	2*	1*	2*	1*	2*	1*	2*
Abbay	17.6	54.2	43.9	63.2	22.8	42.5	50.8	80.0
Baro-Akobo	6.5	20.1	18.9	14.2	28.5	53.1	8.9	14.0
Setit-Tekeze	7.8	24.1	6.1	8.8	2.3	4.3	3.9	6.1
Mereb	0.5	1.6	0.6	0.8	0.0	0.0	0.0	0.0

Source Modified from river master plan studies and FAO (2005)
ENB stands for Ethiopian Nile Basins, 1* is at country level and 2* is at ENB level

The *Abbay* Basin

Location and River Network

The *Abbay* Basin has been called by different names in different studies. It is called the Blue Nile (IWMI 2012), Upper Blue Nile (Jung et al. 2017), and Blue Nile in Ethiopia (USBR 1964; Yilma and Awulachew 2009). The *Abbay* River Basin is defined here in this paper as the basin area that contributes to the border of Ethiopia through the Grand Ethiopian Renaissance Dam (GERD). This means that the area draining through the Dinder and Rahad Rives that join the Blue Nile River near Khartoum is not considered. These two river basins are wrongly considered part of the *Abbay* River Basin. They have their own potential and catchment characteristics to be considered separately for better development. The same analysis has to be performed on the Angereb and Shinfa Rivers, which are connected to the Atbara River in Sudan and are different from the Ethiopian *Tekeze* River Basin. The correction of such basins on mapping for hydrometry and resource valuation has improved our understanding of sustainable development.

The *Abbay* River Basin, outlet at the boarder of Sudan, is subdivided into 13 subbasins based on the major rivers in the basin and their tributaries. The basin covers an area of 176,000 km^2 (Abebe et al. 2022; Hareghweyn et al. 2016; Yilma and Awulachew 2009; Tekleab et al. 2013). The *Abbay* River Basin contributes approximately 60% of the annual streamflow to the main Nile River (Jung et al. 2017). The basin has a significant influence on periodic drought (e.g., Tadesse et al. 2015; Taye et al. 2024), regional food security (e.g., Shukla et al. 2014; Tadesse et al. 2015; McNally et al. 2019), and land-use management (e.g., Gebrehiwot et al. 2011). The *Abbay* River Basin shares 815,581 ha of irrigation and 78,820 GWH/yr of hydropower potential (Fig. 14.3).

Data were obtained and organized from different master plan studies and the FAO AQUASTAT database (FAO 2005). The annual runoff discharged from the river basins amounts to approximately 124 BMC. The potential for irrigation and hydroelectric power has been estimated at about 36,000 km^2 and 155,000 kwh.

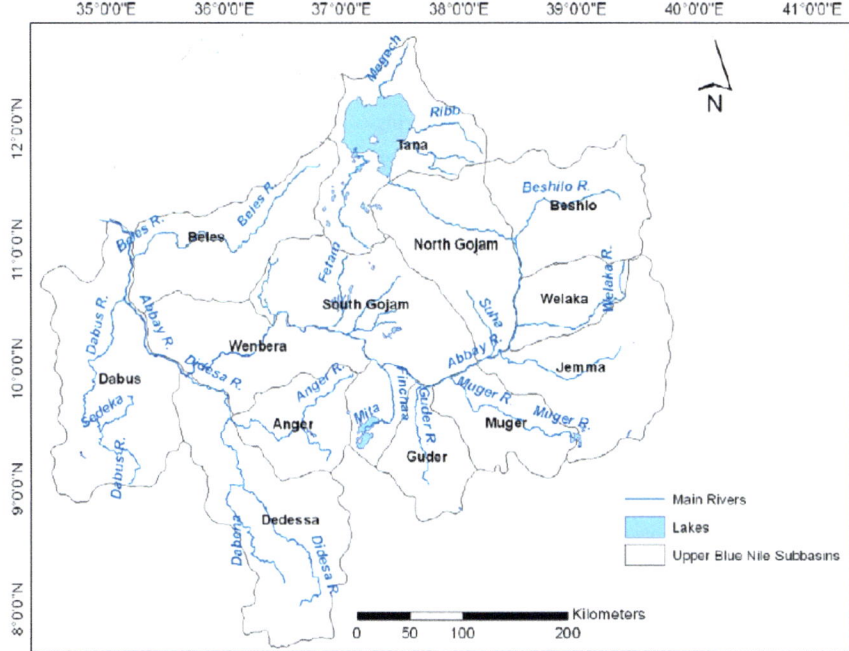

Fig. 14.3 Main subbasins of the *Abbay* River. *Source* The map modified from Yilma and Awulachew (2009)

The *Abbay* River Basin covers approximately 17.6% of the land of the country and contributes 44%, 23%, and 64% of the runoff, the irrigation, and hydropower potential of the country, respectively (see Fig. 14.2 and Table 14.1).

These figures have significant meaning for the country. Development of the hydropower from the *Abbay* River Basin can develop the agricultural and domestic water supply of the other basin. The government of supplies many diesel pumps for pumping water from rivers and groundwater. The fuel cost should increase the cost of agricultural products, increasing the existing inflation in addition to releasing CO_2 from burning the fuels. There are observed potable water supply problems for most of the towns in the country due to an energy shortage for pumping groundwater.

Irrigation Potentials in the Abbay River Basin

The *Abbay* River Basin covers an area of 199,812 km^2, 54.8 BMC per annum runoff, 8155.82 km^2 irrigation potential, and 78,820 Gwh per annum of hydroelectric potential. It covers approximately 18% and 54% of the area, 44 and 64% of the runoff, 23 and 43% of irrigation potential, and 51 and 80% of the hydroelectric power potentials of the country and Ethiopian Nile River Basin (ENB), respectively (see Table 14.3).

Table 14.3 Water stress index scenarios

No	Code	Descriptions
1	$LU_1\_G_N01$	Land use without considering *Gumara* irrigation project (GIP) in 2001
2	$LU_1\_G_N50$	Land use without considering *Gumara* irrigation project (GIP) in 2050
3	$LU_2\_G_N01$	Land use with GIP and green water availability (G_N) in 2001
4	$LU_2\_G_N50$	Land use with GIP and green water availability (G_N) in 2050
5	$LU_2\_G_NE_{WR}01$	Land use with GIP, G_N, and environmental requirement (E_{WR}) in 2001
6	$LU_2\_G_NE_{WR}50$	Land use with GIP, G_N, and environmental requirement (E_{WR}) in 2050
7	$LU_2\_G_NY_{LD}01$	Land use with GIP, G_N, and all flow yield (Y_{LD}) in 2001
8	$LU_2\_G_NY_{LD}50$	Land use with GIP, G_N, and all flow yield (Y_{LD}) in 2050

GN (green water) stands for part of the rainfall that is available for plant growth (FAO 1986) or effective rainfall on the productive land of the watershed (lower extreme case)

EWR (environmental water requirement) stands for 20% of the water yield of the river flow (YLD) for irrigation (as blue water availability) leaving the remaining 80% for environmental demand, EWR (Richter et al. 2011)

YLD (water yield) stands for diverting the whole river flow or considering all flow as available for irrigation (higher extreme case)

The total water needed was calculated based on the population in 2001 and 2050 indicated as 01 and 50, respectively

These figures, with relatively small area coverage and larger surface water, irrigation, and hydropower potentials, can show the importance of *Abbay* River for the country relative to other basins.

Pushing Factors to Exploit the Blue Water Potentials of the Abbay River

According to the USAID report (2018) in Bayissa (2021), the overall gross domestic product (GDP) of Ethiopia depends heavily on rainfed agriculture, which is sensitive to the seasonal variability of rainfall. Unfortunately, drought is a frequently recurring phenomenon in Ethiopia and has a rigorous impact on human lives and the socioeconomic sector. Frequent droughts have affected the socioeconomic sector, which largely relies on rainfed agriculture and is less resilient to drought. For example, Bayissa et al. (2021) demonstrated the drought-vulnerable parts of the basin and the historic drought events in the basin. The 1984 and 2015 droughts were some of the severe drought years in the basin that affected the socioeconomic sectors and annual water budget. This indicates that failure to rainfed agriculture is directly related to failure to GDP and failure to GDP leads to poverty, violations of human rights, degradation of natural resources, and political unrest. The *Abbay* River is the major tributary for the Nile River and contributes more than 60–69% of the total annual flow (Conway and Geogr 2000; Yilma and Awulachew 2009; Conway 2005). This fact leads to intensified blue water diversion for irrigation and hydropower, which is going to receive prior attention in the basin.

In addition to unreliable rainfed agriculture in the rainy undulating highland area with a higher population growth rate, there is an eight to nine time's denser population distribution (Derib 2015; Yilma and Awlachew 2009) in this area as compared to the lowland areas of the basin with better irrigation potential. The lowland irrigation potential area is covered by dense to open woodland with a sparse population, while the erodible, sloppy highlands are intensively cultivated with a dense population (Yilma and Awlachew 2009). This led to poor land use and water management, which became the recurring challenges that aggravate food and water resource insecurity and land degradation in the basin.

Rainfall deficits are one of the harmful effects of climatic shocks that affect agricultural productivity. The deficit guides the use of irrigation water supply to improve productivity (Birthal and Hazrana 2019). Positive rainfall deficit (rainfall surplus) values indicate that the rainfall is higher than evapotranspiration and that a sufficient amount of water is available from rainfall, indicating that irrigation supplementation is not needed. However, the rainfall deficit values indicate that evapotranspiration is higher than rainfall, indicating the need for additional irrigation water for crop production, either with full or supplemental irrigation.

More than fifteen years of monthly rainfall, temperature, wind speed, sunshine hours, and relative humidity data from 58 meteorological stations were available at the Ethiopian National Meteorological Agency (ENMA). Most of these station data are reported and documented under CLIMWAT, a climatic database of the Food and Agriculture Organization of the United Nations (FAO). From these stations, effective rainfall versus potential evapotranspiration (Eto) data for six stations (https://www.fao.org/land-water/databases-and-software/climwat-for-cropwat/en/ assessed on June 20, 2023) are manipulated and presented to show the deficiency and surplus of rainfall as one of the best pushing factors to divert the blue water of the *Abbay* Basin to green water, irrigation application. The monthly rainfall was aggregated on an annual basis for each station. The stations are selected across the elevation and annual rainfall distribution, as indicated in Fig. 14.4.

The monthly rainfall distributions clearly indicate the rainfall and rainfall-evapotranspiration deficit regimes. All the sample station data show that surplus rain falls only for three and a half months of the year, from June to mid-September. After mid-September, the rainfall amount goes to less than 0.5 mm per month, which is much less than interception losses without replenishing the green water depletion. This long-term trend implies either developing a short-variety crop that can give production within three and a half months or accepting the low productivity of rainfed agriculture, which is what is observed in the ground for farming in the *Abbay* River Basin.

The above observation is even without consideration of the variability of the rainfall within months, especially during the onset of the rainfall in June and the cessation of rainfall in mid-September. The erratic nature of rainfall has made it variable with these growing seasons, so farmers have suffered a lot to replant again and again to cope with this variability. They plant maize in June. If the rainfall is

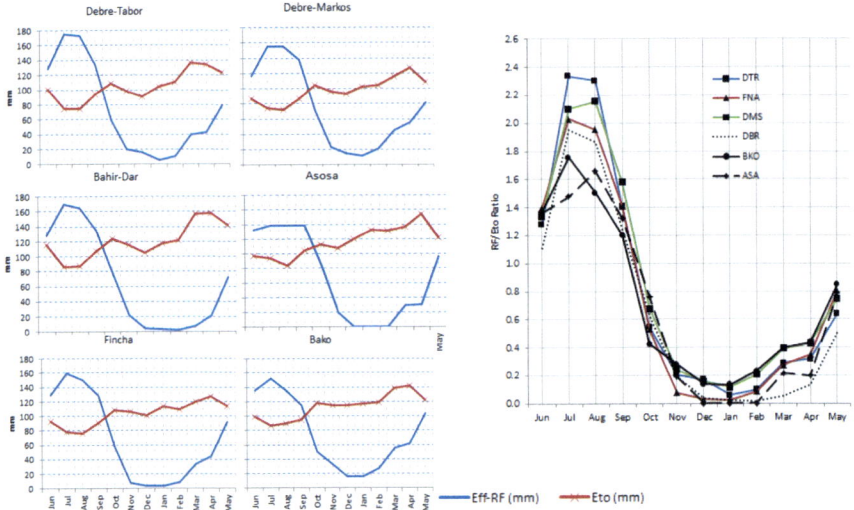

Fig. 14.4 Monthly rainfall, potential evapotranspiration, and their ratio as an index of main climatic stations in the *Abbay* River Basin. Where DTR stands for Debere-Tabor (2410), FNA stands for Fincha (2320), DMS stands for Debre-Markos (2510), DBT stands for Bahir-Dar (1770), BKO stands for Bako (1590), and ASA stands for Assosa (710) climatic stations. Numbers in brackets are for altitude (m) of the stations above sea level

below the demand for green water for germination, they have been forced to replant other short-variety crops like wheat and teff during July. If the July rainfall is too small to replenish the soil moisture needed for the germination of wheat and teff, the farmers are forced to replant legume crops like chickpeas for the third time. If failure occurs to get the minimum amount of green water within the time frame, total rainfed farming failure occurs even though the seasonal and monthly rainfall-evapotranspiration balance looks satisfactory. Such challenges of rainfed farming in the *Abbay* Basin are now frequently occurring due to the existing climatic change problem. In such a challenging rainfed situation, supplemental irrigation is an option to use the erratic rainfall effectively. More of such challenges from farm to watershed levels in the basin are presented in the next section of the chapter using a case study at the upstream watershed of the *Abbay* River Basin.

Case Study: Irrigation and Watershed Development in Lake Tana Basin of the *Abbay*

A study was performed on the potential of irrigation from farm to watershed level at the Lake Tana watershed of the *Abbay* River Basin (Derib 2015; Derib et al. 2011) (Fig. 14.5). They tried to indicate how water was applied from the diversion canal to four representative crops, starting from land preparation to production. The

14 Irrigation Development in the *Abbay* River Basin, Ethiopia

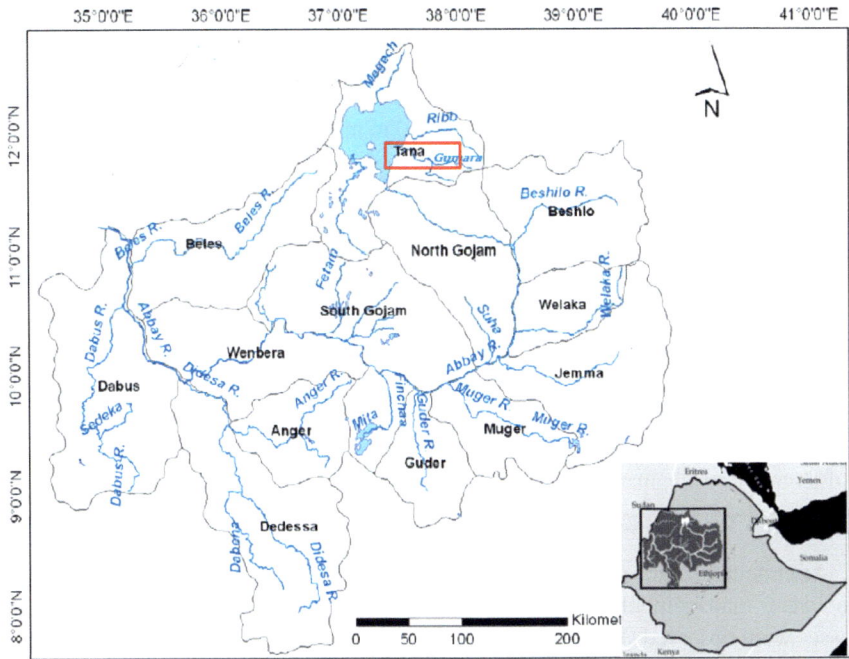

Fig. 14.5 Case study area of Gumara watershed in the *Abbay* River Basin. *Source* The map modified from Yilma and Awulachew (2009)

production and productivity of crops and water have been compared to identify the gaps for improvement. Details on the application of blue water, together with supporting watershed management, have been modeled to see different scenarios on water availability for 2050. Some of the extracts from these studies are presented on the next subtopics.

Small-Scale Irrigation Water Management

Derib et al. (2011) carried out a study on a small-scale irrigation scheme in an area of 90 ha in 2009–2010. Since there is no improvement in irrigation water management or infrastructure in the area, the lessons from this case study are still important for the ongoing irrigation infrastructure like the *Rib* and *Megech* irrigation projects in the basin. The area was found upstream of the *Abbay* Basin, called Lake Tana (Fig. 14.5).

The study was carried out to assess water abstraction and food and feed water productivity with the following objectives:

(i) To assess the irrigation water loss and water needed to produce biomass.
(ii) To assess feed and food water productivity.

(iii) To identify opportunities for improving irrigation efficiency and productivity.

The irrigation scheme, named Guanta, was constructed by the local government in 2001 with a stone masonry diversion structure and a 1555 m main canal (conveying water from the diversion) and an unlined 850 m main canal and a 1341 m secondary canal (conveying water from the main canal). Excess irrigation water was directly released into an abandoned area as drainage basins and wetlands with enclosed gully-like natural flood basins. Wetlands were changed to wetlands due to the overflow of water from the secondary canals and drainage basins. Approximately 21 ha of land upstream of the main canal supply was covered by pump irrigation in 2009.

The Water Use Association (WUA), which rules for water price, canal maintenance, and water, were not functional, leading to the canal not being maintained on time. Water allocation was done randomly, mainly through agreements among some influential and wealthy farmers. Another study performed by Deneke et al. (2011) reported a lack of transparency in scheme boundaries and land redistribution, rule enforcement mechanisms, theft, and corruption.

(a) **Canal Water Loss from the Scheme**

After detailed data collection during irrigation, as indicated in Derib et al. (2011), and water management as described in the above section, the following canal water losses were observed (Fig. 14.6).

Significant water loss was observed, with the highest lost from the main canal and the lowest from field canals (Fig. 14.6). However, the same amount of water from the main canal was distributed to many field canals at the same time. Taking the average 30 l/s flow rate of the season from the main canal to the field canal as a reference, approximately 26% of the water in the field canals was lost. This higher loss was due to the improper size and design of the field canal by the farmers.

However, grass production was seen along the main and secondary canals, with average grassland borders of 6.6 m in width for feed during the dry season. The grass production along earthen canals due to canal seepage and overflow was considered

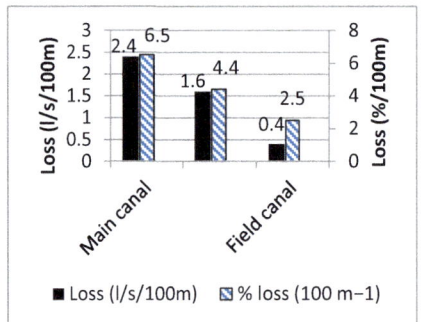
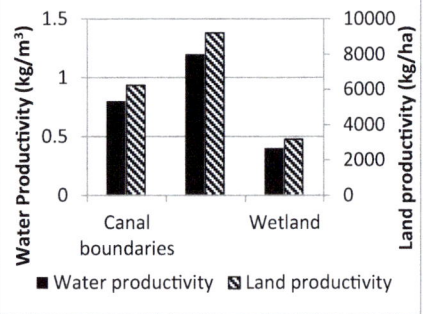

Fig. 14.6 Canal water losses (left) due to water surface evaporation and seepage from the Guanta small-scale irrigation scheme and their grass productivity (right). *Source* Data taken from (Derib et al. 2011)

(Fig. 14.6). The fodder produced along earthen canals, drainage basins, and wetlands from seepage, overflows, and drained water benefited cattle feed in the mixed farming system. Wetlands and drainage basins were not observed in the motor pump irrigation area. This is because the water has had pumping costs, and farmers were conscious of saving a drop of water to save it from overtopping and drainage. They used deficit irrigation on their farm during pumping.

Water abstracted and used by grass, the actual ET, and the water productivity of grassland varied with farm positions from 0.4 to 1.2 kg m^{-3}, which was below the productivity of rainfed wetlands. The land productivity was quite high, ranging from 3000 to 9000 kg ha^{-1}.

The drainage basin was the most productive, whereas the wetland had the lowest productivity (see Fig. 14.7). The field data collected through the whole dry season showed that only approximately 0.05% of the water lost from the canals was actually used for grass production. The other part was lost through canal storage, deep drainage, water surface evaporation, and flow back to the river system. The natural method of reducing water loss from canals for feed production to the mixed farming system is considered and recommended for any similar irrigation infrastructure. If more productive and notorious fodder species could be applied with proper design and planning along these land units in the command area, the productivity of seepage and overflow water might increase and assist livestock productivity.

(b) **Field Water Loss of the Scheme**

The above subsection has presented and discussed water losses and their footprints during transmission from the diversion sources to the farm. This subsection discusses the results of relative water supply (RWS) on the four main crops. The RWS, as the ratio of water supplied to water demanded in the irrigation season, was used as a measuring index for field water loss for selected crops as well as for pump and gravity irrigation (see Fig. 14.8). If this measuring index is around unity, the water supplied for irrigation meets just the demand, and irrigation water loss is almost null. Values exceeding unity indicate water loss; otherwise, it is deficit irrigation.

Fig. 14.7 Drainage basin (left) and wetland (right). *Source* Photo taken by the author and published by Derib et al. (2011)

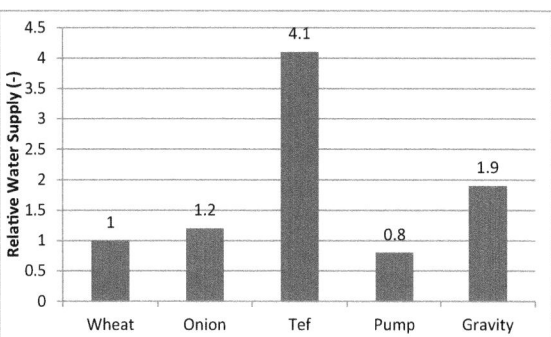

Fig. 14.8 Relative water supply for different crops and irrigation types. *Source* Graph constructed using data from Derib et al. (2011)

The results showed that seasonal irrigation water applications for wheat and onion met the demand with RWS values of 1.04 and 1.18, respectively. The figures indicate that the total amount of water applied was similar to the crop needs, regardless of whether the water distribution and irrigation interval were not properly designed. With significantly different RWS values of 4.1 and 1.90, the average amount of water applied was four and two times higher than the requirements for teff and gravity irrigation, respectively. This value ranged up to seven for certain teff plots. The relative water supply was significantly lower ($p \leq 0.05$) for motor pumps than for gravity irrigation. This indicated that farmers underirrigated their farms when they were using pumps and overirrigated their farms when they were using gravity irrigation. In the meantime, farmers prefer onion, a market crop, to less economical crops of wheat and teff.

In addition to RWS, which shows the total seasonal status, drainage losses beyond soil field capacity during each irrigation time were observed, as shown in Fig. 14.9. The pump-irrigated onion fields showed 2.5% drainage losses due to overirrigation during some irrigation. The highest field drainage losses were observed (approximately 79%) on teff. The reason was that teff was irrigated by less efficient flood irrigation.

(c) **Water and Land Productivity of the Scheme**

Irrigation water productivity was approached differently in this case study. Effective rainfall was considered, and its productivity was counted as irrigation water productivity (IWP). On the other hand, actual evapotranspired water (water demand) was considered, and its productivity was counted as ETP. The comparison of the IWP and ETP can increase our understanding of field water losses in terms of productivity concepts.

Another approach considered is valuing both grain/bulb and straw (food and feed) or both for water productivity analysis. When the total water supplied by irrigation and depleted by actual ET is used to compute grain or straw (but not both), it is termed a "conventional approach", while if both grain and straw were considered in water productivity, it is termed an "improved approach" (Haileslassie et al. 2009). The improved approach increases water productivity, which reflects the real situation

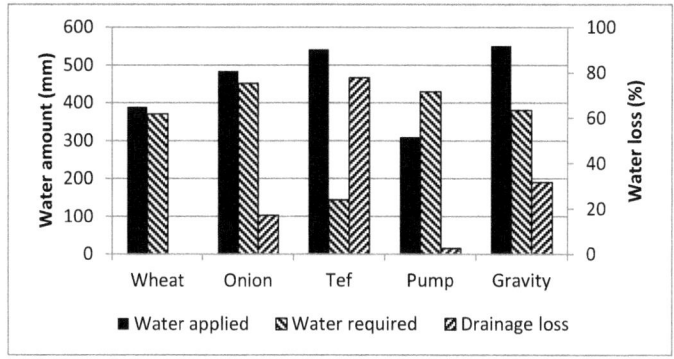

Fig. 14.9 Irrigation water application and requirement for different crops, pump, and gravity irrigation. *Source* Graph constructed from Derib et al. (2011)

on the ground, in which both straw and grain (feed and food) production using the same water is very important in the dry season of a mixed farming system.

Land productivity (as explained in kg/ha) for grain yield for teff and wheat was insignificant at 770 and 759 kg ha^{-1}, respectively, as well as straw yield at 2048 kg ha^{-1} compared to 1864 kg ha^{-1} for wheat. The onion bulb yield was 5903 kg ha^{-1} and was significantly higher and different in terms of water productivity than of the other cereals (see Fig. 14.10).

The higher IWP than EWP (see Fig. 14.10) shows the effect of irrigation water application losses as indicated by RWS parameters. Therefore, onion and teff had statistically similar EWPs, while they had statistically different IWPs considering the improved water productivity. The conventional IWP ranged from 0.18 to 1.39 kg m^{-3} while the improved IWP ranged from 0.68 to 1.78 kg m^{-3}. Generally, higher water loss and lower land and water productivity have been observed compared to other studies (Akkuzu et al. 2007; Bakry and Awad 1997; Mohammadi et al. 2019).

During the field research time throughout the whole production season, the following observations were taken that give more understanding behind the above figures of water losses and productivity. Cumulative field canal losses were highest as compared to losses from the 100-m measurement canal segment. This is because these canals were destroyed during tillage that the canal banks were not stabilized like the main and secondary canals. This led to more water logging than seepage from other canals. Some strong farmers applied night irrigation to overcome a crowd during the day. They used hand-powered torches that were too dim to see properly. Therefore, overtopping from canals during night irrigation was common, increasing water logging and higher water losses that were not measured. On the other hand, large farmlands within 8–15 m from the field canals were out of production due to water shortages. Farmers were aware of irrigation and the increasing water demand due to the irrigation area extension to the downstream part of the command area. These extensions and applications of pumps from the main canals and steam flow upstream of the diversion have made the management of the irrigation water in the

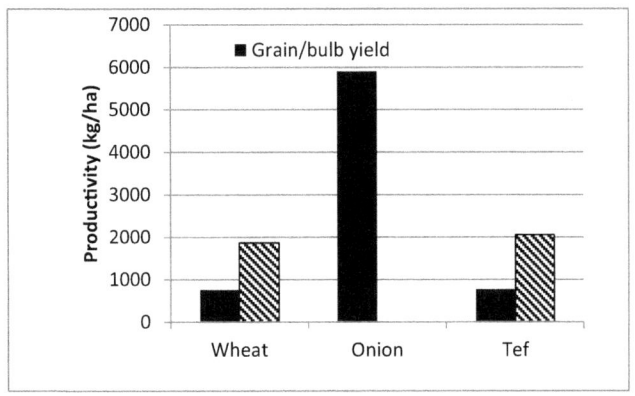
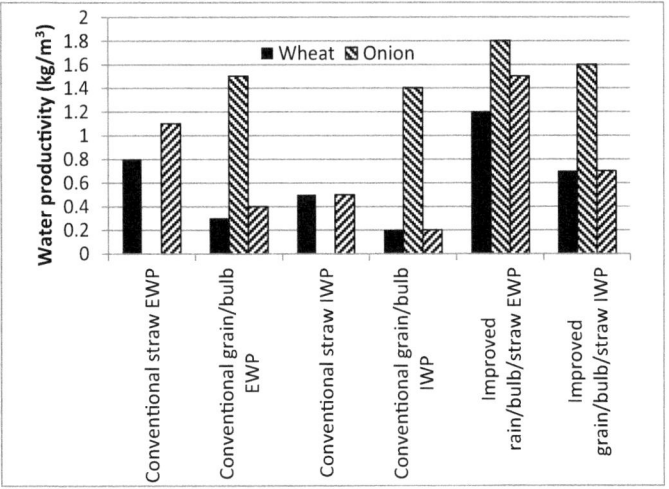

Fig. 14.10 Land (kg ha^{-1}) (top) and water (in kg m^{-3}) (bottom) productivity of the main crops in the Guanta irrigation scheme. Where EWP: evapotranspired water productivity; and IWP: irrigation water productivity. *Source* Graph constructed from Derib et al. (2011)

command area more complicated. This leads to increased irrigation intervals, which results in crop water stress and cracks in the vertic soils. In addition, due to decreased canal flow capacity in the secondary and field canals, leakage and the time needed for sufficient irrigation increased, and farmers were forced to conduct night-time irrigation, resulting in large losses due to inefficiency and an unpredicted canal flow rates during the night.

Water was a more constraining factor than land around the scheme during the irrigation season. This is because there is ample downstream plain and fertile land that is out of production six months a year (December to June). On the other hand, water from night stream flow, springs, and shallow groundwater was still not used properly. Rather than misusing water at night, it is very simple to use night storage.

A simple profile leveling survey of the Guanta River bank showed up to 20,000–45,000 m³ potential for night-time water storage, which could be used during the day. This storage can be possible if a simple barrage structure on the narrow cross-section of the stream could be innovated.

Another problem related to pump irrigation is the availability and cost of fuel for motor pumps. Almost all farmers were unable to cover the cost of motor pumps and fuel. Farmers invested approximately 50% of their produce in kind as a share crop for pump irrigation, with an increasing trend of pumping activities. However, because of higher production costs, pump irrigation typically results in lower financial water productivity than downstream gravity irrigation. This means, there is a stronger possibility of improving water productivity in downstream gravity irrigation than in upstream pump irrigation. Currently, pump irrigation is common in most streamside irrigation systems in the country. The government has a plan to shift fuel pumps to solar pumps (Otoo et al. 2018). Exploitation of the electric potential of the country is also one of the best alternative solutions to supply power for such kinds of pumped irrigation projects in future.

Irrigation and Water Availability in 2050

Modeling Approach

Derib (2015) tried to see the availability status of green water in 2050 using different water and watershed management scenarios. The study was carried out using the results of farm-level irrigation study that were shared in the above subsection, the irrigation feasibility study (MoWR 2008), and the best watershed management options. The calibrated Soil and Water Assessment Tool (SWAT) and Water Stress Indices (WSIs) were applied to the water balance and water stress status, respectively, under the scenarios developed.

SWAT Model Development

The SWAT model was calibrated to the 1520 km² Gumara watershed using 1992–2001 climate and hydrometric data. A 30-m-resolution digital elevation model (DEM) as well as land use and soil data were used for subwatershed and hydrological response unit (HRU) discretization with 328 subwatersheds and 917 HRUs. The model was fitted very well for the measured river discharge giving 0.75 Nash–Sutcliffe efficiency (NSE), 6% bias (PBIAS), and 0.3 root mean square error (RMSE) to observation standard deviation (SRS) values. The model uncertainty was handled using the uncertainty parameters of SWAT (p_factor of 0.79 and r_factor of 0.48).

Land-Use Scenario Development

Two land-use scenarios were developed using field survey data from 2008 and 2009, scanned maps from the feasibility study of Gumara Irrigation Project (GIP) (MoWR 2008), and information from the land-use policy of the country. These land-use

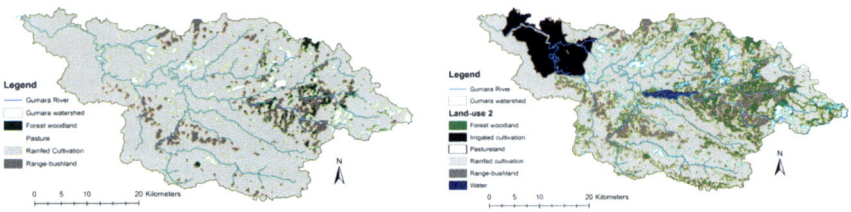

Fig. 14.11 Land-use map of the Gumara watershed without (left) and with (right) the Gumara Irrigation Project (GIP). *Source* Data taken from MoWR (2008) and developed by the author for Derib (2015)

scenarios developed were: land use up to 2008 and land use considering GIP planned by the government to be implemented in the near future.

The planned GIP in 2008 (and not yet implemented) on one of the tributaries of the Gumara River has a 3.51 km^2 inundated area and 14,000 ha of command land on the downstream side of the watershed. The irrigation project plan will change the land use: cultivated land will decrease from 87 to 78% of the watershed, bush rangeland will decrease from 7 to 5%, water bodies will increase from 0.1 to 0.8%, irrigated land coverage will increase from 0.14 to 8%, and forest will increase from 2.5 to 4.3% (Fig. 14.15). Detailed data on population, water abstraction, irrigation, and watershed development for SWAT modeling were used from the feasibility document (MoWR 2008).

Finally, two land uses and three water availability status scenarios were developed and used for the population of 2001 and 2050 to develop eight water stress scenarios. The following scenarios (see Fig. 14.11 and Table 14.3) were developed and applied to SWAT to estimate water availability impacts on the water balance and water stress.

Water Stress Indices

The most frequently used categories (Smakhtin et al. 2005) were used to identify the level of water scarcity with the following Water Stress Indices (WSIs), as the ratio of water demanded to water supplied, using long-term mean annual runoff and considering environmental flow.

1. *WSI > 1*: Overexploited (current water use is tapping into EWR)—environmentally water scarce basins.
2. *0.6 ≤ WSI < 1*: Heavily exploited (0–40% of the utilizable water is still available in a basin before EWR conflicts with other uses)—environmentally water stressed basins.
3. *0.3 ≤ WSI < 0.6*: Moderately exploited (40–70% of the utilizable water is still available in a basin before EWR conflicts with other uses).
4. *WSI < 0.3*: Slightly exploited.

14 Irrigation Development in the *Abbay* River Basin, Ethiopia

Water Balance Shift Due to Land-Use Changes

Temporal Water Balance Shift:

The annual water balance of the Gumara watershed is shown in Fig. 14.16 with and without the Gumara Irrigation Project (GIP). Approximately 95% of the annual rainfall left the watershed through river discharge or yield (YLD; 752 mm) and AET (648 mm). The remaining 5% was stored in the deep groundwater. This storage was approximately 61 mm (92 mm^3) per year. River discharge and AET accounted for 51% and 44% of the annual rainfall, respectively, under the existing land-use conditions. A shift from river discharge and groundwater storage to AET was observed due to GIP and watershed treatment methods. Watershed management and the planned irrigation project shifted an additional 99 mm (151 mm^3) of the annual yield to AET. However, 106 mm (161 mm^3) water was additionally evapotranspired due to GIP. The balance was filled by deep groundwater recharge. Therefore, groundwater storage was decreased by 4 mm (7 mm^3) when watershed treatment and GIP were implemented in the model (see Fig. 14.12).

As indicated in Fig. 14.13 time series graph, the rising limb, and the peak of the hydrograph were regulated due to GIP. Evapotranspiration increases during the dry period using GIP. Therefore, an additional 154 mm^3 of water is evapotranspired in the dry season based on 130 mm^3 YLD regulations during the wet season. The difference of 24 mm^3 in the AET is from the rainfall in the dry season. Both Figs. 14.13 and 14.14 show that the natural YLD was altered without affecting the 20% presumptive standard for environmental flow requirements.

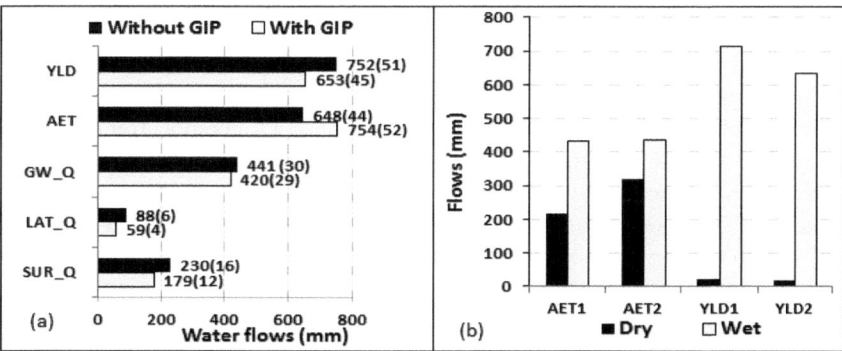

Fig. 14.12 Annual water flows without and with the Gumara irrigation project (GIP): **a** annual and **b** seasonal. *Source* Data developed by the author for a thesis series (Derib 2015) where numbers in brackets are the percent annual rainfall covered by each component (YLD is the total river discharge through the outlet of the watershed, AET is the actual evapotranspiration, GW_Q is the groundwater flow, LAT_Q is the lateral flow, and SUR_Q is the surface water flow to the channel. The numbers 1 and 2 indicate land-use scenarios without and with the Gumara irrigation project)

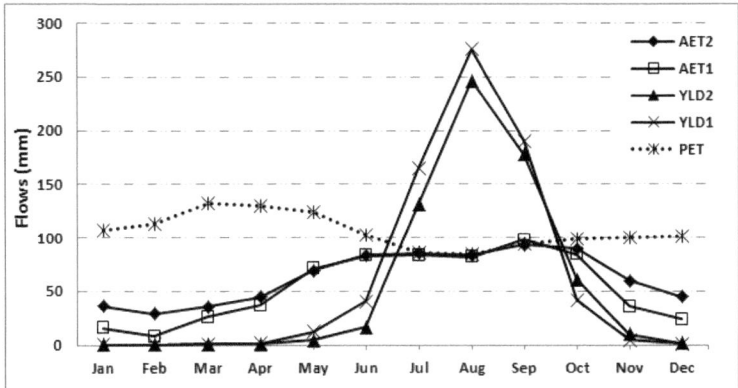

Fig. 14.13 Grand Ethiopian Renaissance Dam. Where YLD is total discharge through the outlet of the watershed, AET is actual evapotranspiration, and PET is potential evapotranspiration. The numbers 1 and 2 indicate land-use scenarios without and with the Gumara irrigation project. *Source* Data developed by the author for a thesis series (Derib 2015)

Spatial Patterns of Water Flow Shifts

Watershed treatment practices were considered together with the GIP study recommendations and according to the land-use policy of Ethiopia. The practices are contouring of land units with slopes between 15 and 30%, terracing of slopes steeper than 30%, and afforestation of hillsides steeper than 60%, which led to differences in surface and groundwater flows. The data obtained from the GIP, including dam storage, irrigation command area, and catchment land use recombination from the feasibility study, were utilized in the SWAT model. The results related to water balance and water stress are presented in Fig. 14.14.

An approximately 1.8% increase in AET in the irrigation command area, the reservoir, and around the reservoir area. The annual evaporation from the open water surface of the reservoir is approximately 1492 mm. The annual average AET increased by 73 mm (varying from 0 to 962 mm from HRU to HRU). An approximately 1.2% increase was observed in YLD in reservoir area, which decreased by 74 mm (varying from 0 to 784 mm from HRU to HRU) at the watershed level (results not shown here) with respect to land management interventions and GIP. These land management practices also decreased surface runoff by 49% on average and increased groundwater and lateral flows by 27% and 20%, respectively, at the reservoir and hillslopes around the reservoir and upstream areas. These spatial water balance shifts around the reservoir and upstream slope areas indicate the impact on the protection of the reservoir and downstream flooding, which are the challenges to human life in the area. However, there has been no progress rather than the feasibility study for 15 years until now.

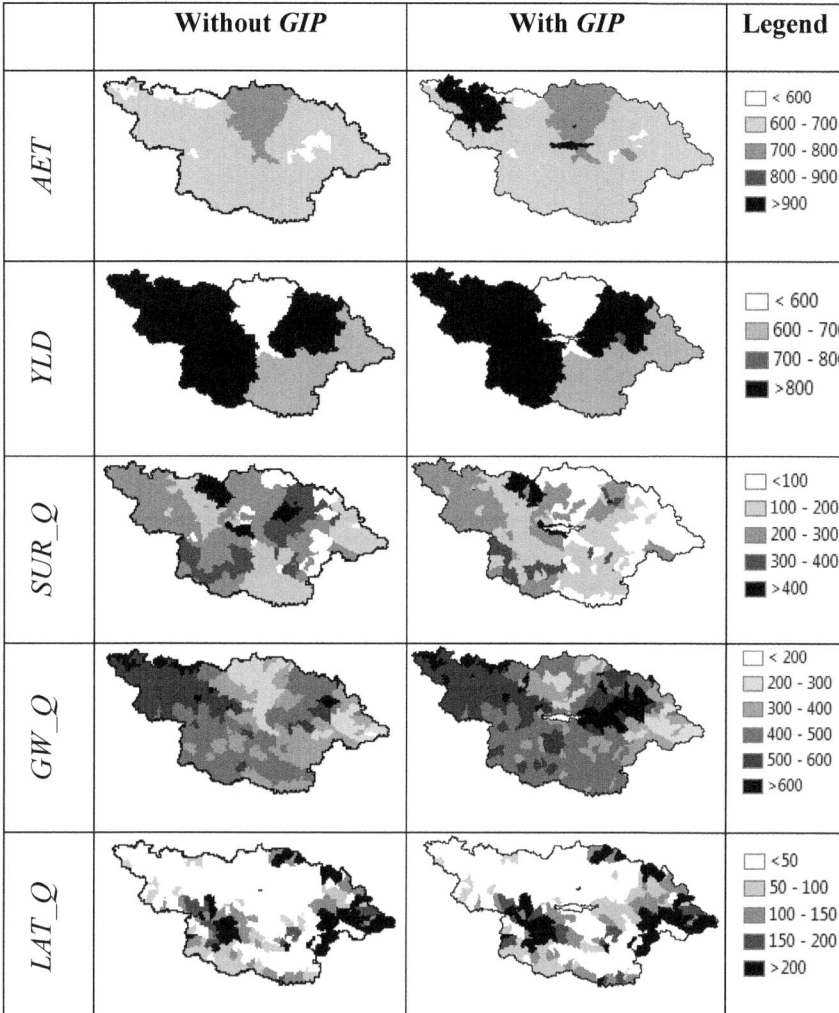

Fig. 14.14 Water balance components (mm y^{-1}) without and with the Gumara irrigation project (GIP) and watershed management interventions. Where AET is actual evapotranspiration, YLD is discharge (yield) through the outlet of the watershed, SUR_Q is surface water flow, GW_Q is groundwater flow, and LAT_Q is lateral flow through the soil layer. *Source* Data developed by the author for a thesis series (Derib 2015)

Water Availability and Scarcity

Available water was categorized into three groups in this study: green water (approximated by part of the actual evapotranspiration from the soil moisture), green water plus 20% of the river flow (YLD), and green water plus all the river flow.

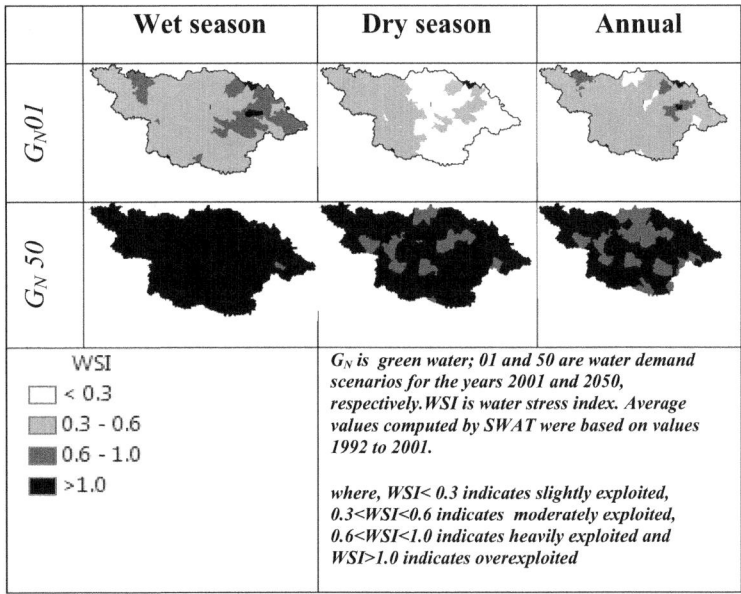

Fig. 14.15 Water Stress Index (WSI) spatial status using business as usual (without GIP) land-use data during wet and dry seasons and on 2001 and 2050. *Source* Data developed by the author for a thesis series (Derib 2015)

Figure 14.15 shows the Water Stress Indices of the existing land-use scenario without GIP using green water as available water during dry and wet seasons, as well as at an annual level in 2001 and 2050 under basic water requirement conditions. As indicated in the figure, most of the subwatersheds belong to the class with a WSI lower than 0.6 under the current rainfed agriculture during the wet season. The existing business as usual scenario for now indicates a slightly to moderately exploited water stress status range at seasonal and annual levels, i.e., 40–70% of the utilizable water is still available in a basin before environmental water requirements (EWR) conflict with other uses downstream. However, for 2050, the stress status is highly changed to overexploited in most parts of the watershed, especially during the wet season, i.e., the current green water from rainfed agriculture from June to September is not enough to meet the 2050 demands of society. Water is highly scarce (WSI > 0.6) in the upstream part of the watershed during this season. However, green water is not scarce in this area during the dry season (WSI < 0.3) in some parts. This result shows that the green water during the wet season from the existing crops, pastures, and woodlands can fulfill the basic water demand of the watershed in both the wet and dry seasons. All the subwatersheds will be under extremely water scarce conditions (WSI > 1.0) in 2050 if the current rainfed land-use activities are continued with the existing low water productivity.

The spatial distribution of Water Stress Indices based on watershed management and the planned irrigation project interventions is shown in Fig. 14.16. The water

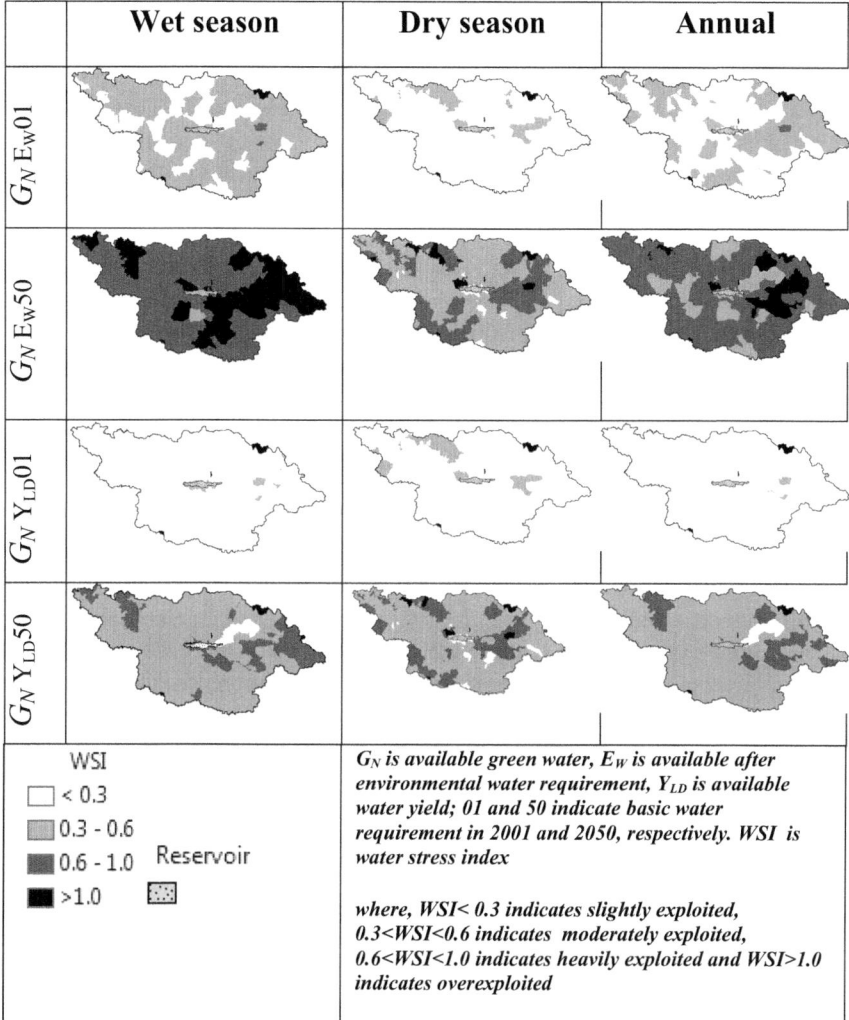

Fig. 14.16 Water Stress Indices (WSIs) based on planned irrigation projects and watershed management interventions. *Source* Data developed by the author for a thesis series (Derib 2015)

stress level is improved when blue water is withdrawn in addition to green water to fulfill the basic water needs of the population. There is less water stress on 2001 demands if we divert all stream flows and moderate stress when environmental flow is allowed to be met. 20% of the YLD in green water improved water availability and decreased the Water Stress Index from moderately exploited ($0.3 < \text{WSI} < 0.6$) to slightly exploited ($\text{WSI} < 0.3$) for some of the subwatersheds. In this case, much of the available water (40–70%) was still available for other water needs beyond the basic water requirements in 2001. However, most of the subwatersheds will still be overexploited (WSI > 1) in 2050 if only green water is used. The watershed will be environmentally scarce in 2050, and the contribution of the watershed to downstream livelihoods will be limited.

Watershed management, together with diverting the blue water considering the environmental flow, shows better improvement in meeting the demand for water in future. Improving vegetation cover together with afforestation of the steep slopes increases actual evapotranspiration and groundwater recharge. Microbasin water harvesting structures have shown good land cover and increased biomass production by minimizing discharge in northeast Ethiopia (Derib et al. 2009). Shrubland is considered the best choice for minimizing runoff and soil erosion in China compared to alfalfa pastureland (Wei et al. 2007). The authors suggest grassland and woodland for runoff and soil erosion management rather than large-scale alfalfa plantations. Around the study area, legume trees, alfalfa, Napier, and vetiver grasses were proposed and used (Gebreselassie et al. 2009). However, careful selection of crops and trees has to be done with respect to environmental benefits and water productivity optimization.

The AET during the dry season was lost through unproductive evaporation since the land is bare and there is almost no production of food and feed during this season. This is for two reasons. The first and most important reason is the small rainfall amount and duration and the resulting low soil moisture (green water), which was not enough to supply the required AET for food and feed production in the dry season. The second reason was that the farmers had no additional technology, such as irrigation infrastructure and low-water-demanding crops, in the dry season. However, the contribution of the existing small amount of available blue water from rivers, springs, and wells for domestic uses and livestock drinking was not considered in the green water analysis. Shifting 6% of the rainfall from annual YLD to productive evapotranspiration, GIP, and associated watershed management interventions made another 2% of the evaporated annual rainfall productive in the dry season in the irrigation command area. It played a role in increasing water availability for the community without compromising the environmental flow. The unproductive green water in the dry season can be shifted to productive transpiration using supplemental irrigation.

Green water from rainfall on the farm is the only available water for the existing rainfed agricultural system in the study area, which is still occurring now in 2023. Based on the experience of the author and field observations, the most productive green water was that of the wet season on farmlands. The understanding at continental and international levels is slightly far from this truth, especially by downstream countries, since Ethiopia has wrongly entitled with enough rainfall volume not to divert any water from the *Abbay* River. This case study is an indicator to lead us to a deep understanding of the system. A more time-tuned scale of water availability, such as monthly level water stress analysis, is important for better understanding and decision-making. However, this requires agricultural water demand data at a monthly level. This is only possible with a detailed study of crop water requirements.

The availability of water using annual rainfall and river flows is misleading for the *Abbay* water negotiations. Green water in the soil of productive land and environmental water requirements can be considered as the second option for calculating water availability in the *Abbay* Basin. Using the 20% rule of the presumptive standard for environmental flow protection (Richter et al. 2011), 20% of the YLD was added to the green water of the rainfed system as available water for water sharing. However, this presumptive standard is difficult to implement in the existing Nile hydropolitics. The standard can minimize approximately 10.5 km^3 of the water from Nile flow at Aswan if it is implemented on the whole Ethiopian Blue Nile watershed. The sum of green water and YLD was the other extreme option used to calculate the available water for each subwatershed.

Conclusion

In addition to physical water stress, Nile water is now in more political tension than ever. The Ethiopian highland contributes approximately 86% of the Nile flow at the Aswan High Dam, while the country uses less than 5% of its total internal renewable water (FAO AQUASTAT 2005) and 3% of the Blue Nile runoff (Mason 2004). The largest user of this flow, Egypt, is dependent on 98% of the Nile water at the moment and needs to shift to its groundwater and saline water resources from the Mediterranean and Red Sea coasts. It also needs reconsidering the political water that Egypt and Sudan agreed to use the Nile flow in 1929 and 1959, but the agreements are not binding for all the riparian countries in the Nile Basin.

Ethiopia gains 936 km^3 of annual rainfall and discharges 122 km^3 (14%) of this rainfall, where 90% of this flow is transboundary (FAO AQUASTAT database; http://www.fao.org/nr/water/aquastat/countries_regions/ Cited 12/08/2023). Per capita, 14,200 (5300) m^3 and 1800 (698) m^3 rain and river flow water, respectively, are calculated for 2001 (2050). The effective rainfall that accounts for a total 814 km^3 with 12,300 (4600) m^3 per capita in 2001 (2050) is a very large amount compared to Egypt's total water availability of 68.3 km^3 with 979 (504) m^3 per capita in 2001 (2050). However, as can be observed at the headwaters of the Blue Nile, approximately 53% of the annual rainfall is directed to river flow, and the green water is not enough to support the basic water needs in future if the existing rainfed water productivity does not improve. This indicates that there are some sources and sinks of river flow in the Blue Nile. For example, a study concerning the Lake Tana basin (15,096 km^2) showed that approximately 30% of the rainfall is discharged through the outlet (Setegn et al. 2008). Another modeling study carried out by Engida (2010) in the same basin using 8 subwatersheds (area varying from 103 to 15,120 km^2) showed a variation in discharge contribution from 24 to 60% of the annual rainfall. Green and blue water managements must be designed based on these differences. Rainfed agricultural systems are not productive enough to support future life due to the large discharge contribution, low green water productivity, and high population density in the highland rainfed area. It needs direction on the lowland downstream areas of the *Abbay* River diverting blue water, considering the environmental water requirements.

The integral understanding of global and regional water balance on different time scales calls for another way of thinking to alleviate the consumptive water scarcity and the existing hydropolitical stress. Even watershed management and blue water withdrawal can improve water availability in the area, but they will not solve the water stress in society and the environment in future. As recommended by many studies (e.g., Waterbury and Whittington (1998), Whittington (2004), Mason (2004), Arsano (2007), Martens (2011)), basin-wide integration and efficient water use in the Nile Basin can benefit local livelihoods like the *Abbay* Basin and the environment. Non-water consumptive uses such as hydropower production, fishing, and tourism can benefit local livelihoods while environmental water is not negatively affected. An extensive Blue Nile water development project in Ethiopia, the "Grand Ethiopian Dam Project (GERDP)", started in April 2011 on the Blue Nile River. It is designed to generate 6000 MW of electric power, covering 74 km^3 in a reservoir covering 1680 km^2 (EEPC 2013). The project is non-water consumptive since it is designed only for power generation. As it is located in a sparsely populated (19 persons km-2) (CSA 2011; Derib 2015) and inaccessible river valley area, it will attract human life after completion so that the green water burden of the densely populated highland and cities will be alleviated to a certain extent. Fish production, navigation, tourism, and business activities related to the stored water may be livelihood means for the community. There is also a chance to use the generated power to develop the groundwater of the Ethiopian lowlands outside the Nile Basin for irrigation and drinking water infrastructure. Ethiopia has ample potential and a diversity of non-water consumptive alternatives without appreciably harming the water share of downstream users. In addition, downstream countries also have the potential for Nubian groundwater, ocean water for desalination, and coastline non-water consumptive trade alternatives. This basin-wide integration should relieve the existing quantitative and political water stress of the Nile River.

Acknowledgements The German Federal Ministry for Economic Development Cooperation (Bundesministerium für Wirtschaftliche Zusammenarbeit-BMZ) through the Center for Development Research (ZEF), Bonn University and the International Water Management Institute (IWMI) for financial support, the IWMI and Amhara Region Agricultural Research Institute (ARARI) for providing the materials and facilitating field work, and the National Meteorological Agency and Ministry of Water Resources of Ethiopia for providing secondary data are highly acknowledged for part of the research results presented under the case study.

References

Abebe SA, Qin T, Zhang X, Li C, Yan D (2022) Estimating the water budget of the Upper Blue Nile River Basin with water and energy processes (WEP) model. Front Earth Sci 10:1–14

Akkuzu E, Unal HB, Karatafi BS (2007) Determination of water conveyance loss in the Menemen open canal irrigation network. Turk J Agric 31:11–22

Alemu MM (2017) Current trends of investment effect on land-use practices of Ethiopia. Open Access Libr J 04(01):1–14

Arsano Y (2007) Ethiopia and the Nile Dilemmas of national and regional hydropolitics. PhD Dissertation, Swiss Federal Institute of Technology

Awulachew SB, Yilma AD, Loulseged M, Loiskandl W, Ayana M, Alamirew T (2007) Water resources and irrigation development in Ethiopia. International Water Management Institute, Colombo, Sri Lanka, 78p (Working Paper 123)

Bacha D, Regasa N, Ayalneh B, Abonesh T (2011) Impact of small-scale irrigation on household poverty: empirical evidence from the Ambo district in Ethiopia. Irrig Drain 60:1–10

Bakry MF, Awad AM (1997) Practical estimation of seepage losses along earthen canals in Egypt. Water Resour Manag 11:197–206

Baye T (2017) Poverty, peasantry and agriculture in Ethiopia. Ann Agrar Sci 15:420–430

Bayissa Y, Maskey S, Tadesse T, Van Andel SJ, Moges S, Van Griensven A, Solomatine D (2018) Comparison of the performance of six drought indices in characterizing historical drought for the upper Blue Nile basin, Ethiopia. Geosciences 8(81):1–26

Bayissa Y, Moges S, Melesse A, Tadesse T, Abiy AZ, Worqlul A (2021) Multi-dimensional drought assessment in *Abbay*/Upper Blue Nile Basin: the importance of shared management and regional coordination efforts for mitigation. Remote Sens 13(9):1–20

Belay M, Woldeamlak B (2013) Traditional irrigation and water management practices in highland Ethiopia: case study in Dangila woreda. Irrig Drain 62(2013):435–448

Belay Z (2017) Land resource, uses, and ownership in Ethiopia: past, present and future. Int J Sci Res Eng Technol 2(1):17–24

Berhanu B, Seleshi Y, Melesse AM (2014) Surface water and groundwater resources of ethiopia: potentials and challenges of water resources development. In: Melesse A, Abtew W, Setegn S (eds) Nile River Basin: ecohydrological challenges, climate change and hydropolitics. Springer, New York

Birthal PS, Hazrana J (2019) Crop diversification and resilience of agriculture to climatic shocks: evidence from India. Agric Syst 173:345–354

Conway D, Geogr J (2000) The climate and hydrology of the Upper Blue Nile River. 166:49–62

Conway D (2005) From headwater tributaries to international river: observing and adapting to climate variability and change in the Nile basin. Glob Environ Chang 15:99–114

CSA (2011) Population size by sex, area and density by region, Zone and Wereda. Central Statistics Authority (CSA) Addis Ababa, Ethiopia

CSA (2013) Population projections for Ethiopia for 2007–2037. Central Statistical Agency, Addis Ababa

Demeke M, Fantu G, Ferede T (2004) Agricultural development in Ethiopia: are there alternatives to food aid? Department of Economics, Addis Ababa University. https://www.semanticscholar.org/paper/. Assessed on July 2023

Deneke T, Mapedza E, Amede T (2011) Institutional implications of governance of local common pool resources on livestock water productivity in Ethiopia. Exp Agric 46:99–111

Derib SD (2015) Balancing water availability and water demand in the Blue Nile: a case study of Gumara watershed in Ethiopia. Ecology and Development Series Bd. 95. PhD Thesis, University of Bonn, Bonn

Derib SD, Assefa T, Berhanu B, Zeleke G (2009) Impacts of microbasin water harvesting structures in improving vegetative cover in degraded hillslopes areas of northeast Ethiopia. Rangeland J 31(2):259–265

Derib SD, Descheemaeker K, Haileslassie A, Amede T (2011) Irrigation water productivity as affected by water management in a small-scale irrigation scheme in the Blue Nile Basin, Ethiopia. Expl Agric 47:39–55

EEPC (2013) Grand Ethiopian Renaissance Project progress report, Ethiopian Electric Power Corporation (EEPC). http://www.hidasse.gov.et/c/document_library/get_file?p_l_id=11731&folderId=11740&name=DLFE-202.pdf. Assessed 23 Jun 2013

Engida A (2010) Hydrological and suspended sediment modelling in the Lake Tana Basin, Ethiopia. PhD Dissertation, Université de Grenoble

Falkenmark M, Rockström J (2006) The New Blue and Green Water Paradigm: breaking new ground for water resources planning and management. J Water Resour Plan Manag 132(3):129–132

FAO (1986) Irrigation water management. Training manual No. 3, Food and Agriculture Organization (FAO) of the United Nations, Via delle Terme di Caracalla, Rome, Italy

FAO (2005) Irrigation in Africa in figures AQUASTAT Survey FOA, Rome. ISBN 92-5-105414-2

FAO AQUASTAT (2005) Irrigation in Africa in figures: Ethiopia–AQUASTAT Survey. http://www.fao.org/nr/water/aquastat/countries_regions/ETH/CP_ETH.pdf. Cited 18 Jul 2013

Gebrehiwot T, Anne V, Ben M (2011) Spatial and temporal assessment of drought in the Northern highlands of Ethiopia. Earth Obs Geoinf 13(3):309–321

Gebreselassie Y, Amdemariam T, Haile M, Yamoah C (2009) Lessons from upstream soil conservation measures to mitigate soil erosion and its impact on upstream and downstream users of the Nile River. Upstream-Downstream Project in the Blue Nile Intermediate Results Dissemination Workshop held on 5–6 February 2009, International Water Management Institute (IWMI), Nile Basin and East Africa Office, Addis Ababa

Gleick PH, Meena P (2010) Peak water limits to freshwater withdrawal and use Pacific National Academy of Sciences (PNAS). 107(25):11155–11162

Gyamera EA (2014) .Hydrological studies of the university of Cape Coast school of agriculture research station at Twifo Wamaso. Glob Res J Geogr 2(1):010–066

Hagos F, Makombe G, Namara RE, Awulachew SB (2009) Importance of irrigated agriculture to the Ethiopian economy: capturing the direct net benefits of irrigation. International Water Management Institute, Colombo, Sri Lanka, 37p (IWMI Research Report 128)

Haile GG, Kasa AK (2015) Irrigation in Ethiopia: a review. Acad J Agric Res 3(10):264–269

Haileslassie AD, Peden S, Gebreselassie T, Amede K, Descheemaeker (2009) Livestock water productivity in mixed crop–livestock farming systems of the Blue Nile basin: assessing variability and prospects for improvement. Agr Syst 102:33–40

Haregeweyn N, Tsunekawa A, Tsubo M et al (2016) Analysing the hydrologic effects of region-wide land and water development interventions: a case study of the Upper Blue Nile basin. Reg Environ Change 16(4):951–966

Hoekstra AY (2003) Virtual water: an introduction. In: Proceedings of the international expert meeting on virtual water trade, delft, value of water research report series no 12

IWMI (2012) The Nile River Basin water, agriculture, governance and livelihoods. Routledge 711 Third Avenue, International Water Management Institute (IWMI) New York, NY 10017

Jung HC, Getirana A, Policelli F, McNally A, Arsenault KR, Kumar S, Tadesse T, Peters-Lidard CD (2017) Upper Blue Nile basin water budget from a multi-model perspective, J Hydrol 535–546

Kassa M, Tesfa GA (2020) Review of irrigation practice in Ethiopia, lessons from Israel. Irri Drainage Syst Eng 9(1):2–6

Liu X, Dunne JP, Stock CA, Harrison MJ, Adcroft A, Resplandy L (2019) Simulating water residence time in the coastal ocean: a global perspective. Geophys Res Lett 46(13):910–919

Makombe G, Kelemework D, Aredo D (2007) A comparative analysis of rainfed and irrigated agricultural production in Ethiopia. Irrig Drain Syst 21:35–44

Marcinek J (2007) Hydrological cycle and water balance—a global survey. In: Lozán JL, Grassl H, Hupfer P, Menzel L, Schönwiese C-D (eds) Global change: enough water for all? Wissenschaftliche Auswertungen, Hamburg

Martens AK (2011) Impacts of global change on the Nile Basin options for hydropolitical reform in Egypt and Ethiopia. IFPRI discussion paper 01052, International Food Policy Research Institute (IFPRI)

Mason SA (2004) From conflict to cooperation in the Nile Basin: interaction between water availability, Water Management in Egypt and Sudan, and International Relations in the Eastern Nile Basin, conflict sensitive interviewing and dialogue workshop methodology. PhD Dissertation, Swiss Federal Institute of Technology

McNally A, Kristine V, Laura H, Augusto G, Jossy J, Shraddhanand S, Kristi A, Christa P, James PV (2019) Acute water-scarcity monitoring for Africa. Water 11(10):1968. https://doi.org/10.3390/w11101968

MoA (2011) Small-scale irrigation situation analysis and capacity needs assessment, (Ministry of Agriculture) Natural Resources Management Directorates, Addis Ababa, Ethiopia

Mohammadi A, Atefeh PR, Abbasi N (2019) Field measurement and analysis of water losses at the main and tertiary levels of irrigation canals: Varamin Irrigation Scheme. Iran Glob Ecol Conserv 18:1–10

MoWR (Ministry of Water Resources) (2002) Water Sector Development Program (WSDP), Addis Ababa, Ethiopia

MoWR (Ministry of Water Resources) (2008) Gumara irrigation project feasibility study report. Ministry of water resources, Addis Ababa, Ethiopia

OCHA (2023) ETHIOPIA: drought situation update #1. UN Office for the Coordination of Humanitarian Affairs (OCHA) https://reliefweb.int/report/ethiopia/ethiopia-drought-situation-update-1-10-march-2023. Assessed on July 2023

Otoo M, Lefore N, Schmitter P, Barron J, Gebregziabher G (2018) Business model scenarios and suitability: smallholder solar pump-based irrigation in Ethiopia. In: Agricultural water management—making a business case for smallholders. International Water Management Institute (IWMI), Colombo, Sri Lanka, 67p (IWMI Research Report 172). https://doi.org/10.5337/2018.207

Reddy RN (2010) Irrigation engineering. Gene-Tech Books, New Delhi, pp 110002

Richter BD, Davis MM, Apse C, Konrad C (2011) A presumptive standard for environmental flow protection. River Res Applic 28:1312–1321

Ruud J, Tuinenburg OA (2017) The residence time of water in the atmosphere revisited. Hydrol Earth Syst Sci 21:779–790

Sadoff C (2019) Managing water resources to maximize sustainable growth: a World Bank water resources assistance strategy for Ethiopia. The World Bank, NW

Setegn SG, Srinivasan R, Dargahi B (2008) Hydrological modelling in the Lake Tana Basin, Ethiopia using SWAT model. Open Hydrology J 2:49–62

Shiklomanov LA (1993) World freshwater resources. In: Gleick PH (ed) Water in crisis: a guide to world's freshwater resources, Oxford University Press, New York, pp 13–24

Shukla S, McNally A, Husak G, Funk C (2014) A seasonal agricultural drought forecast system for food-insecure regions of East Africa. Hydrol Earth Syst Sci 18:3907–3921

Simachew BW (2020) Natural resource degradation tendencies in Ethiopia: a review. Wassie Environ Syst Res 9(33):1–29

Smakhtin V, Revenga C, Döll P (2005) Taking into account environmental water requirements in global-scale water resources assessments. Comprehensive Assessment Research Report 2. IWNI, Comprehensive Assessment Secretariat, Colombo, Sri Lanka

Tadesse T, Brian DW, Jesslyn FB, Mark DS, Michael JH, Brian F, Denise G (2015) Assessing the vegetation condition impacts of the 2011 Drought across the U.S. Southern great plains using the vegetation drought response index (VegDRI), pp 53–169. https://doi.org/10.1175/JAMC-D-14-0048.1

Taye MT, Ellen D (2024) Hydrologic extremes in a changing climate: a review of extremes in East Africa. Curr Clim Change Rep 10:1–11

Tekleab S, Mohamed Y, Uhlenbrook S (2013) Hydro-climatic trends in the Abay/Upper Blue Nile basin, Ethiopia. Phys Chem Earth 61:32–42

Temam D, Uddamer V, Mohammadi G, Annette Hernandez E, Ekwaro-Osire S (2009) Long-term drought trends in Ethiopia with implications for dryland agriculture. Water 11(12):1–22

USBR (United States Bureau of Reclamation) (1964) Land and water resources of the Blue Nile Basin. Main Report, United States Department of Interior Bureau of Reclamation, Washington, DC

Waterbury J, Whittington D (1998) Playing chicken on the Nile? The implications of micro dam development in the Ethiopian highlands and Egypt's New Valley project. Nat Resour Forum 22(3):166–163

Wei W, Chen L, Fu B, Huang Z, Wu D, Gui L (2007) The effect of land uses and rainfall regimes on runoff and soil erosion in the semiarid loess hilly area, China. J Hydrol 335:247–258

Werfring A, Lemperiere P, Boelee E (2004) Typology of irrigation in Ethiopia. In: Proceedings of inception workshopon IWMI-BOKU-Siebersdorf-EARO-Arbaminch University collaborative study on the impact of irrigation development on poverty and the environment, 26–30 April 2004, Addis Ababa, Ethiopia

Whittington W (2004) Visions of Nile basin development. Water Policy 6:1–24

Woldesenbet E, Gebreluel G, Bedasso B (2022) Economic development and political violence in Ethiopia. GEG working paper 145. The Global Economic Governance Programme, University of Oxford

Yilma AD, Awulachew SB (2009) Characterization and Atlas of the Blue Nile Basin and its Sub basins. International Water Management Institute (IWMI), Colombo, Sri Lanka

Chapter 15
The Hydropolitics and Legal Dimension of Ethiopia's Right to Utilize the Abbay River

Firehiwot Sintayehu, Yusuf Ali Mohammed, and Melak Melkamu

Abstract Ethiopia, despite contributing 86% of the Abbay Waters, has not been able to access its equitable and reasonable share of the river basin because of Egypt's strategy of maintaining an unfair status quo through a claim of 'established, historical, and acquired right.' The paper intends to address the legal and political basis by which Ethiopia can ensure its equitable and reasonable share from the Abbay River basin by outlining various basin-wide treaties, international watercourse laws, and political discourse. Here, agreements concluded within the Abbay basin such as the Cooperative Framework Agreement (CFA) and Declarations of Principles (DoPs) on the GERD clearly stipulate the equitable and reasonable use principle in line with Ethiopia's interest. The challenge here is regarding the implementation of such legal frameworks, which have been inhibited by the unfulfilled criteria of not being ratified by six signatory states in the case of the CFA and the continuous reference of Egypt to 'historical rights.' Ethiopia continues to push for its interest in ensuring equitable and reasonable shares of the Abbay River by employing counter hegemony strategies of deconstructing Egypt's securitization through discourses such as 'the right to development,' cooperation, and pan-Africanism. If Ethiopia is to uphold its equitable and reasonable use on Abbay, there is a need to work on proactive diplomacy and push for the CFA to enter into force as well as investments in water development projects such as the GERD that will enhance the bargaining power of the country vis-à-vis downstream countries.

Keywords Hydropolitics · International law · Hydro hegemony · Securitization · Abbay waters · Nile River

---

F. Sintayehu (✉)
Department of Political Science and International Relations, Addis Ababa University, Addis Ababa, Ethiopia
e-mail: Firehiwot.s@gmail.com

Y. A. Mohammed
Department of Public Law, Institute of Social Science, Anadolu University, Eskisehir, Türkiye

M. Melkamu
Department of Political Science and International Relations, Wollo University, Dese, Ethiopia

© The Author(s), under exclusive license to Springer Nature Switzerland AG 2025
A. Melesse et al. (eds.), *Abbay River Basin*, Springer Geography,
https://doi.org/10.1007/978-3-031-65241-7_15

Introduction

Ethiopia is not only a hub for the African Union headquarters but also a hub of the Abbay River, popularly known by an outsider as the Nile River, contributing to the lion's share of the river basin (Arsano 2011; Lie 2010). The Abbay River is the longest river in the world. It is shared by eleven riparian countries, namely Ethiopia, Eritrea, Egypt, Sudan, South Sudan, Kenya, Uganda, Tanzania, Burundi, Rwanda, and the Democratic Republic of Congo. Tvedt (2004) states that the Nile is well-known in the literature as it has been the subject of numerous poems and books.

Hydrology of the Nile River Basin has been studied very well, and enough data and knowledge have been generated. These studies cover important hydrological and water management issues which can be used for discussion in the negotiations for Nile water management (Melesse et al. 2009, 2010, 2011, 2014; Abtew et al. 2009a, b; Abtew and Melesse 2014a, b, c; Melesse 2011). Despite the body of knowledge generated by various studies, the water allocation, management, and sharing have been unresolved.

Abbay is an important source of livelihood in the riparian countries through which it flows. The river is also the subject of heated political debate between the upstream and downstream basin countries. Ethiopia, a riparian state giving birth to the whole bulk of Abbay Water contributing to 75–90% of the total Nile River flow, has been in a difficult position in using this transboundary watercourse (Dellapenna 1997). This has been particularly because of the various political strategies that Egypt has been applying to maintain a status quo whereby it remains the main user of the waters of the Nile.

One major strategy employed by Egypt to maintain the status quo is through the act of securitizing the Nile issues using threats made by different Egyptian leaders. In June 2013, the ex-president of Egypt, Mr. Mohammed Morsi made a speech on national television that Egypt is willing to use any mechanism including force if the Nile flow is reduced "by a drop." This speech was in response to Ethiopia's commencement of the construction of the Grand Ethiopian Renaissance Dam (GERD), which has renewed the discussion over sharing the Nile waters thus far dominated by Egyptian use. Moreover, Egypt put the doctrine of 'historical right' to its new constitution adopted in 2014, which was regarded by upstream countries as an act of Egypt's unwillingness to negotiate on the Nile waters.

Despite such provocative acts by Egypt, Ethiopia has been working towards challenging the status quo of unfair water use on the Nile River through various diplomatic initiatives aiming at cooperatively managing the waters of the Nile River as well as acts of countering the 'hydrohegemony' of Egypt. This chapter deliberates upon the hydropolitical strategies employed by Ethiopia towards utilizing the Abbay Waters in an equitable and reasonable manner along with the international and regional legal frameworks that support the country's claim.

Conceptual and Theoretical Discussion

Conceptual Discussion

Most of the river basins in the world are shared by more than one sovereign state. Tiwary (2006) states that there are approximately 261 international river basins that cover 45% of the earth. Moreover, approximately 145 states have a territory within Transboundary Lake or river basins (UNEP-DHI and UNEP 2016). Shared river basins are considered sources of conflicts considering the scarcity of freshwater sources globally (Starr 1991).

There is a debate over the tendency of riparian states of international river basins to cooperate. Those who argue that river basins are the sources of conflicts put forward scarcity as a justification. Nearly three-fourths of the earth is covered with water. Nonetheless, freshwater that can be utilized by human beings is only 3%. Majority of this (2.5%) is inaccessible as it is found in the form of ice or found as groundwater (Baker et al. 2016). Agriculture is the dominant user of freshwater resources, consuming 70%. Agriculture is followed by the use of 20% of freshwater sources in the world. Domestic use is found in third place, at 8% (World Business Council for Sustainable Development 2005).

One challenge for the availability of freshwater is its uneven distribution. Approximately 60% of freshwater is found in only nine countries of the world. Distribution even within these countries is uneven (Baker et al. 2016). Similar trends are observed in Ethiopia, which is referred to as the water tower of Africa. Most of the major river basins of the country that contribute to 80–90% of the flow are found in the western and southwestern parts, where 40% of the population resides, while less than 20% of the flow of water is in the river basins located in the eastern, southeastern, central, and some parts of the northern region, which houses 60% of the population (Worku and Giweta 2018).

The dynamisms of interstate conflict and cooperation over the utilization of transboundary river basins are studied by hydropolitics. The term 'hydropolitics' was used for the first time by Waterbury (1979). The debate over defining hydropolitics has not settled, as scholars such as Elhance argue that it systematically studies conflicts and cooperation between states on the use of transboundary water resources, while others such as Turton (2002) define hydropolitics in a broader fashion as the manner in which the society allocates values in relation to water. Turton's definition includes a wide range of actors in hydropolitics ranging from the individual to international actors such as the state and non-state actors. Moreover, the issues that are included in hydropolitics are framed to be endless. Second, the horizontal dimension of range: hydropolitics covers almost endless interactions between water and issues such as gender, ecosystem, and food security. This study focuses on interstate conflict and cooperation dynamics in the Abbay basin, as states are major actors in the hydropolitics of the basin.

Hydrohegemony and Securitization Theories

Hydrohegemony is a theory that dominates the study of hydropolitics in general and the Abbay basin specifically. According to Menga (2016), the modern definition of hegemony is attributed to Vincenzo Gioberti, who defined it as 'that sort of supremacy, preeminence, superiority, not legal nor juridical in the strict sense of the word, but morally efficient, that among several congeneric, unilingual and compatriot provinces, one exercises over the others.' Gramsi also contributed significantly to the coining of the concept of hegemony. He used hegemony to explain the relations between the state and civil society. For him, hegemony represents a dominant's class success shaping perspectives and setting ideologies to achieve intellectual and moral leadership over others (Menga 2016).

The concept of hegemony has also joined hydropolitics, and many have used the hydrohegemony framework in their analysis of inter-riparian country relations. Zeitoun and Warner (2006) discuss the different tactics that may be applied by hydrohegemons to establish a stronger position and maintain the status quo. The first is coercive compliance-producing mechanisms in which hydrohegemons may use military force or coercion pressure against the relatively weaker riparian states. The second tactic is utilitarian compliance-producing mechanisms whereby hydrohegemons incentivize the act of compliance with a hydrohegemon's preferred state of affairs. The third mechanism is normative compliance-producing mechanisms. Here, the role of a hegemon would be to come up with an agreement that will maintain the status quo that is advantageous to the hydrohegemon. The last strategy is hegemonic compliance-producing mechanisms, which may include securitization, knowledge construction, sanctioned discourse, and coercive resources. Securitization promotes issues around the river basin as national security concerns, which leads to the silencing of critical voices. Securitization also leads to the construction of knowledge in a particular manner. Cascao (2009) argues that the confusion that Egyptian governments creates through knowledge construction to their own citizen international donors and to their riparian competitors or friends creates more room to manoeuvre by reducing apparent external pressure. The coercive resources of hydrohegemons are the power resources that enable them to resist forces with the objective of equalizing power discrepancies. These resources may include international support, the ability to obtain funds and human capital, among others.

The hydrohegemony framework has been applied to the analysis of many river basins, including Abbay. On the Abbay basin, authors such as Zeitoun and Warner (2006), Cascao and Zeitoun (2010), and Gebrehiwot (2020) applied hydrohegemony by acknowledging Egypt's dominance in the basin as the main user of the basin and the various acts of coercion and provision of incentives in its effort towards maintaining the status quo. The framework has a way of putting those who are considered 'nonhegemons' applying counter hegemony strategies that include recourse to morality and international law, desecuritization, economic development, alternative funding sources, negotiations, and generation of positive-sum outcomes.

Transformative mechanisms of counterhegemony aim to transform a hegemonic order through undermining the acceptance bestowed upon the hydrohegemony using various mechanisms. These may include questioning the constructed knowledge and coming up with an alternative reality which may inform emerging political agendas (Zeitoun et al. 2016). The prominent examples of counterhydrohegemon are the decision made by Ethiopia on the draft of the Cooperative Framework Agreement (CFA) in 2010 to sign it without Egypt and Sudan, as well as the country's decision to launch the construction of the GERD in 2011 (Zietoun et al. 2016).

Cascao (2009) discussed the changing power relations in the Abbay basin even before the coming of the GERD by taking into consideration factors such as Ethiopia and the equatorial states having improved economic growth and political stability. She also highlights the joining of new regional actors such as the NBI and China as a factor here.

Securitization theory is about the intent of some states to broaden the security agenda to include threats apart from military ones. The theory was designed by the Copenhagen School of Security Studies. According to Waever (1995), a prominent scholar from the Copenhagen School, the intent of states to securitize various issues is characterized by the pressure to deal with a particular issue as a result of the threat that the state is encountering. Historically, the military sector has been securitized, while in more recent times, the other realms of life have been securitized. Buzan et al. (1998) note that an issue is securitized when it is portrayed as an existential threat requiring actions that are extraordinary.

Actors of securitization want to securitize the agenda to frame a particular issue as a threat to the sovereignty of a state, which enables them to mobilize maximum effort to curb it. This is a strategy used by states to enable them to claim a 'special right' that will be defined by the state and its elites. The power holders in a state use securitization as a mechanism of gaining control. By upholding such an argument, Waever (1995) argues that securitization is a socially constructed phenomenon.

Securitization is understood as a speech act. Here, security is formed through speech and in a manner that is defined by a particular set of actors. Waever (1995) argues that it is through utterance of security that state actors move a particular issue as a security agenda and thereby claim a special right to make use of any mechanism to deter the threat.

Securitization is not always regarded as a positive move as may be perceived. The example raised by the Copenhagen School is the move to securitize issues such as the environment, where the use of water may also be raised. Buzan (1983) argues that the reason behind securitizing the environment has to do with the magnitude of threats posed as well as the urgency to obtain unprecedented responses to them. However, securitizing the environment has been criticized from various vantage points. Moss (1992) argues that security threats are to be addressed centrally by the state and that such an approach will not work to address global environmental problems. The other reasoning forwarded by Waever (1995) as the disadvantage of securitizing the environment is creating the 'us' versus 'them' spirit, which externalizes the problem and results in developing enmity between the different blocks formed.

The securitization strategy refers to constructing an issue into a national security concern (Waever 1995), and Egypt initially framed the Nile as a survival and existential matter to its citizens. Ethiopia, on the other hand, has been trying to desecuritize the Nile River, particularly through the 'right to development' discourse, in which Ethiopia claims that the utilization of Abbay Water is for its socioeconomic development with no intention of harming downstream countries (Gienanth 2020).

International and National Legal Frameworks: Ethiopia's Right to Utilize Abbay Waters

Given the international and national legal regimes and saving other riparian states' perspectives, this section is devoted to explaining the legally justified rights of Ethiopia to use the Abbay Waters equitably and reasonably.

Ethiopia's Rightful Claim to Use the Abbay Waters in the Context of Basin-Wide Treaty Regimes of the Nile River Basin

Ethiopia has been asserting its rightful claim against occupying powers since the colonial era, an era when the Nile River Legal regime started to come into existence. In this colonial period, as the international legal regimes of Africa were subsequently convoluted by the occupying powers, the influence of the United Kingdom (UK) on the Nile River Basin is undeniable. Thus, aiming to secure the entire Nile River course, the UK concluded several strategic agreements with its counter-occupying powers and a couple of riparian countries. Accordingly, the UK had brokered, the 1891, 1901, 1906, 1919, and 1925 treaties with Italy, Belgium, and France, while the 1902, 1929, and 1952 accords with Ethiopia and Egypt. To put the record straight, albeit a number of attempts, Ethiopia has been the only uncolonized state in Africa, therefore, was at liberty to assert its objection from the gate go against all those colonial treaty regimes considered unjust and undermining its right to use the Nile River equitably and reasonably (Ram 1977).

Bearing in mind the above facts, to have a clear picture of what Ethiopia is claiming rightfully and fiercely standing against, it is crucial to scrutinize several significant colonial and postimmediate colonial treaties.

To start with the Agreement between Ethiopia and UK on the Frontiers between the Sudan, Ethiopia, and Eritrea (1902), although its ultimate objective was meant to delineate about 1600 km Ethio-Sudan boundary, it parenthetically included the matter of the Nile River, thereby, contained a hitherto contentious provision, Article III (Ullendorff 1967). As the agreement was crafted both in English and Ethiopian

Language (Amharic), what the word 'arrest' entails in the English[1] and Amharic versions[2] started becoming confusing and thus became a huge point of controversy. The UK intended to prevent Ethiopia's right to utilize the waters of the Nile definitively while Ethiopia assumed the intention was to only deter a complete halt to the flow of the river (Woldetsadik 2015). These conflicting assumptions and interpretations led contracting parties to a deadlock; hence, the agreement ended up being rejected by Ethiopia (Mtua 2017; Paisley and Henshaw 2013; Woldetsadik 2015). Although this is the case, on various contemporary negotiation platforms, the utmost downstream riparian states, particularly Egypt, keep bringing up the Agreement on the Frontiers between the Sudan, Ethiopia, and Eritrea (1902) as an argument to restrain Ethiopia from undertaking any projects over the Abbay River course. This assertion does not hold water, as the 1902 accord is a done deal; thus, neither Ethiopia nor the UK was/is going to abide by it.

Aside from the above, the other concerning legal regime is the Exchange of Notes between the UK and Egypt in Regard to the Use of the Waters of the River Nile for Irrigation Purposes (1929). This regime not only laid the ground for Egypt's 'established, historical and acquired right' discourse but also entitled the entire whole Nile River flow, including Abbay Water, to the two utmost down riparian countries. It, accordingly, allocates 92.3% and 7.7% for Egypt and Sudan, respectively, leaving no single drop in share for upstream riparian states (Johnston 2009; Waterbury 1997). Furthermore, per the aforesaid treaty regime, these down riparian states held the right to conduct water development projects on the Nile river while maintaining 'the right to veto any construction projects' on the upper basin courses (Mtua 2017). "As this treaty was agreed upon in a note exchanged only between Egypt and the UK, representing Sudan and the other East African riparian colonies, unsurprisingly, Egypt asserts the 1929 Agreement remains to be enforced per the theory of Universal State Succession,"[3] while the former colonies refuse to do so invoking the 'clean state' principle,[4] and the *'Nyerere'* Doctrine of State Succession[5] (Arsano 2011).

Regarding the case of Ethiopia, it is needless to mention that when the Exchange of Notes in Regard to the Use of the Waters of the River Nile for Irrigation Purposes (1929) deal was brokered, the UK was neither an occupying power nor invited Ethiopia to the negotiating table; thus, by no means of international law regulation and argument abided Ethiopia (VCLT 1969). Moreover, saving Ethiopia's objection to the established, historical, and acquired right claim, Article 6 (1) (e) of the UN Watercourses Convention (1997) and Article 13 (2) (e) of the Berlin Rules (2004), among others, coupled with several international courts' decisions, notably the *Land*

---

[1] See Agreement on the Frontiers (1902).

[2] See Woldetsadik (2013).

[3] According to Universal State Succession, a state is bound to succeed rights and duties of its predecessor (Fiedler 1987; Mohammed 2017).

[4] According to the clean state (tabula rasa) principle, a seceding state is not obliged to observe or enforce a treaty regime of its predecessor (Janig 2018; VCST 1978, Art. 16).

[5] According to the *Nyerere* doctrine, a state is not categorically bound to a colonial treaty regime which the occupying power concluded on its behalf (Mohammed 2017; Okoye 1972).

and Maritime Boundary case (2002), *Territorial Dispute case* (1994), and *Lake Lanoux Arbitration* (1957), reveal the inexistence of independent criteria entailing hegemonical monopolization of an entire particular river flow under the auspices of an established, historical, and acquired right. Rather, it underlines its mere, otherwise, significance in considering it only as one single factor, together with other relevant components, which cumulatively pitch in to attain an equitable and reasonable utilization ultimatum (Kimenyi and Mbaku 2015; Mohammed 2017; Stebek 2007).

Although the legality and viability of the colonial legal regimes, particularly the Agreement on the Frontiers between the Sudan, Ethiopia, and Eritrea (1902) and Exchange of Notes in Regard to the Use of the Waters of the River Nile for Irrigation Purposes (1929), are so controversial, the UK itself admitted its unfairness (Okoth-Owiro 2004). The UK's attitude not only opens the door for colonial treaties to be quashed but also uncovers, confirms, and validates Ethiopia's long-standing rightful claim to use the Abbay Waters.

The above being the case, the postcolonial treaty regimes, although brokered between or/and among independent sovereign riparian states, are highly induced by the colonial mindset, thus, hinder to pave the way for contemporary negotiations (Bulto 2008). A textbook example signifying this fact can be deduced from the Agreement for the Full Utilization of the Nile Waters (1959), CFA (2010), and DoPs on the GERD (2015). Moreover, the ongoing negotiation over the first filling and operation of the GERD is also evident in this regard (Tekuya 2021).

Although the Agreement between Egypt and Sudan for the Full Utilization of the Nile Waters (1959) accord is the first postimmediate colonial treaty brokered between two independent sovereign states, it is just an extended and revised reference of the Exchange of Notes in Regard to the Use of the Waters of the River Nile for Irrigation Purposes (1929) regime, which came into life up on the pushing demand of a newly independent state, Sudan (Bulto 2008). As the naming of the Agreement for the Full Utilization of the Nile Waters (1959) speaks for itself, the two utmost downstream riparian states agreed to fully utilize the entirety of the Nile River flow between themselves. Assuming the entire Nile water to be 84 billion cubic meters (BCM), Egypt and Sudan agreed to share 55.5 BCM and 18.5 BCM, respectively, while considering the remaining 10 BCM to be lost in the form of evaporation (Agreement for the Full Utilization 1959; Bulto 2008; Mtua 2017). The upstream riparian states, including Ethiopia, were never invited or consulted during the Agreement for the Full Utilization of the Nile Waters (1959), thus, vehemently voiced their objection against the treaty regime, and asserted that the treaty only binds the contracting parties, *pacta tertiis nec nocent nec prosunt,* hence neither conferring rights nor imposing an obligation on the other noncontracting riparian states (VCLT 1969, Art. 34).

Coming to the CFA (2010) regime, unlike the previous treaty regimes, it was crafted in the presence and active participation of all basin riparian countries; thus, it was considered the first ever-inclusive legal regime the Nile River Basin ever had. Although the regime was envisioned to pursue a water allocation strategy and institute a basin-wide Nile River Basin Commission (NBC) by in-lining itself with the

UN Watercourses Convention (1997), its dream is short-lived as previous agreements concluded on the Nile water use in 1902, 1929, and 1959 remained to be a source of disagreement (Tekuya 2019). Although upstream riparian states pursue their negotiations intending to substitute the foregoing colonial treaty regimes by the CFA, the downstream riparian states assert otherwise to maintain and consider the colonial accords non-negotiable starting benchmark, accordingly, seeking to get endorsed in the colonial legacy into the CFA (Mekonnen 2010).

Furthermore, their basin-wide negotiation reached a deadlock when the nonlegal and political term named 'water security' principle was incorporated under Articles 3 (15) and 14 of the CFA (2010). In addition to the questionable legality of the 'water security' principle itself, the two downstream riparian states submitted their reservation on Article 14 (b), which was primarily drafted as *'not to significantly affect the water security of any other Nile Basin State,'* and instead insisted on replacing it with a new phrase/statement, which read as *'not to adversely affect the water security and current uses and rights of any other Nile Basin State,'* which was utterly disapproved by Ethiopia and other upstream riparian states, turning their decade-long talks back to square one (Lie 2010; Mohammed 2017). Leaving these unresolved issues to be settled by the NBC, the CFA (2010) was opened for signature on 14 May 2010, and thus far, six riparian states have signed the Agreement, among which four riparian states—Ethiopia, Rwanda, Tanzania, and Uganda—have ratified it (Lie 2010; Mohammed 2017; "NBI," n.d.). Although the CFA (2010) is not enforced yet pending two more riparian states' ratification, the intention of Ethiopia and those other ratifying riparian states indicate not only their pursuit of the right to use the Abbay Waters and the other Nile tributaries but also their commitment to abiding by basin-wide watercourse law.

Aside from the above legal regimes, feeling frustrated with the Nile Déjà vu, driven by pushing developmental demands and affirming that its action would not violate international watercourse laws, Ethiopia embarked on the construction of a huge dam, the GERD, alongside the Abbay Watercourse in April 2011 (Tekuya 2020). However, the downstream riparian states portray the commencement of the dam construction 'as [a] threat for water security,' thus becoming a subject of controversy hitherto (Abtew and Dessu 2019).

After several years of uncertainty and negotiations, the three key riparian countries—Egypt, Ethiopia, and Sudan—came up with the DoPs on the GERD (2015), mainly to set up a framework and settle issues relating to the first filling and operation of the dam. Intending to conduct comprehensive scientific studies on the first filling and operation of the GERD, these riparian states had established several expert groups, inter alia, the International Panel of Experts (IPoE), and 'BRLi Group' and 'Artelia' firms through the Technical National Committee (TNC) and National Independent Scientific Research Group (NISRG) (Tekuya 2021).

Although many negotiations were, accordingly, held on various levels, the riparian states have not been able to come to terms with, inter alia, drought mitigation strategies, and dispute resolution mechanisms, thus going back to square one (Tekuya 2021). After the Washington-led negotiations had failed to bear fruits, the AU (2020) officially seized the matter and chaired the negotiation table, but with no breakthrough yet.

The above facts and scrutinized anatomy boldly underscore a consistent and clear assertion of Ethiopia's rightful claim to use the Abbay Waters. Thus, what Ethiopia has been consistently averring was nothing but its rightful claim to use the Abbay Waters equitably and reasonably. Having these asseverations, the following section thoroughly digs into 'how concrete Ethiopia's rightful claim is' from international watercourse law perspectives.

Ethiopia's Rightful Claim of Using the Abbay Waters in the Context of International Watercourse Laws

As briefly provided in the above section, Ethiopia has been consistently asserting its rightful claim to use Abbay Waters since the inception of the colonial legal regimes. The basis of its right can be tracked down from the theoretical doctrine of a transboundary watercourse. Unlike downstream riparian countries, which strictly adhere to absolute territorial integrity[6] and prior appropriation rule,[7] Ethiopia's stand is deeply rooted in limited territorial sovereignty doctrine.[8]

Being triggered by the action of downstream riparian states, at first although Ethiopia's attitude, which was inferred from the '*aide-memoire*' and 'pronouncement,' was perceived to incline with the absolute territorial sovereignty doctrine,[9] 'it did not categorically adhere to it' (Bulto 2008; Mohammed 2022a). A careful reading of these '*aide-memoire*' and 'pronouncement'[10] reveals the very objective of Ethiopia to pursue its rightful equitable share (Bulto 2008; Mohammed 2022a). Therefore, Ethiopia undoubtedly clings to the limited territorial sovereignty doctrine and thus pursues the use of the Abbay Waters equitably and reasonably.

---

[6] Absolute territorial integrity is a legal doctrine that confers an absolute right to a downstream riparian state to—enjoy the total flow of a river course without getting interrupted—and veto any hydraulic development that goes against its interest (McIntyre 2010; Mohammed 2022b).

[7] Prior appropriation is a legal doctrine that refers to a scenario in which a particular riparian state, irrespective of its basin position, is entitled to have prior right over an amount of water flow that it uses first before other riparian state starts using so (Lazerwitz 1993).

[8] Limited territorial sovereignty/integrity entails all upstream and downstream riparian states to use a shared watercourse equitably and reasonably within their respective territory (McIntyre 2010; Mohammed 2022b).

[9] Absolute territorial integrity is a legal doctrine that confers an absolute right to an upstream riparian state to utilize a whole water resource of a transboundary river as it wishes, irrespective of its consequence (McIntyre 2010; Mohammed 2022b).

[10] At the UN Water Conference organized in Argentina in 1977, Ethiopia stated that if there is no treaty regime in a basin, every riparian state is unilaterally entitled to pursue its hydraulic development on a river course within its respective sovereign boundary (Bulto 2008; Mohammed 2022a).

The equitable and reasonable use[11] and no-harm rule,[12] coming into existence from limited territorial sovereignty/integrity doctrinal thoughts, unquestionably, serve as the guiding the contemporary legal regimes of a transboundary watercourse (Mohammed 2022a, b). Like every riparian state, but with so much weighting justification, Ethiopia claims to use Abbay Waters primarily based on the equitable and reasonable use principle while giving due consideration to the no-harm rule principle. Although the equitable and reasonable use principle entails and accentuates Ethiopia with an inherent customary right to use the Abbay Watercourse, given the controversial issues exhibited over the precedence of equitable and reasonable use vis-à-vis the no-harm rule, it is important to set a clear background as to the scope of no-harm rule application and to what extent Ethiopia's legal responsibility extends.

Attributing environmental and economic value to the no-harm rule principle, overwhelming international environmental legal regimes and scholarly writings, although showing significant support for its environmental protection, back down to do so for its economic value, mainly due to the absence of practical scenarios in 'upholding absolute prohibition of harms' (Mohammed 2022b).

Moreover, a scholar named Caflisch, while uplifting the primary nature of the equitable and reasonable use principle, downplays the absoluteness of the no-harm rule in contemporary transboundary watercourse management, arguing the lower availability of transboundary watercourses, which could satisfy the booming population and economy. Thus, engendering harm would not be a matter in question; instead, equitable allocation of shared watercourses among riparian countries would be a surpassing matter (Stebek 2007).

The above concert justification underscores the nonabsolute nature of the no-harm rule, providing the existence of several mitigating factors and compensation for greater materialized damage that occurred in violation of 'a legally protected interest' (McCaffrey 2001; Stebek 2007).

Ruling out the primacy of the no-harm rule, although already averred in the above, the contemporary status of equitable and reasonable use principle under international watercourse law is to be addressed, in black and white, as follows.

The guiding legal regimes of international transboundary watercourses, several international courts' decisions, significant commentaries, and authoritative experts, among others, have endorsed the paramountcy of equitable and reasonable use principles. A cumulative reading of Articles 5, 6, 7, 15, 16, 17, and 19 of the UN Watercourses Convention (1997), Articles IV, V, VII, X, XXIX [4] of the Helsinki Rules (1996), Articles 10.1, 12, 13, 14, 16 of the Berlin Rules (2004), among others, uncover, in line with the above thoughts and analysis, the primacy of equitable and reasonable use principle (Mohammed 2022b). In particular, the principal guidance

---

[11] Equitable and reasonable use is a principle, derived from a well-established doctrine of limited territorial sovereignty/integrity, which entitles every upper and lower riparian state to consciously use an international watercourse equitably and reasonably (Mohammed 2022b).

[12] Like equitable and reasonable use, no-harm rule is a principle originating from the limited territorial sovereignty/integrity doctrine, which dictates a riparian state to be conscious when using a transboundary river within its territory not to cause significant harm against its ecology and/or fellow riparian state interest (Mohammed 2022a, b; User's Guide Fact Sheet Series: Number 5 n.d.).

provided under Articles 7 (2) and 10 (2) of the UN Watercourses Convention (1997) amplifies the qualification of the no-harm rule within the context of the equitable and reasonable use principle, even in case significant harm occurs and/or conflict arises between/among riparian states.

In a similar tone, the International Law Commission (ILC) and International Law Association (ILA) commentary unanimously uphold the supremacy of the equitable and reasonable use principle (ILA 2004; McIntyre 2010; Stebek 2007). Moreover, several international courts' decisions, particularly the *River Oder case* (1929) and *Gabčíkovo-Nagymaros case* (1997), set a clear precedent and affirm the dominion of equitable and reasonable use principle. Moreover, the overwhelming number of authoritative scholars have largely made a consensus that the equitable utilization rule in international water should prevail (Dellapenna 2001).

Given the above facts and justifications, articulating Ethiopia's rightful claim in using the Abbay Waters deserves to be addressed here. To start with the bottom line, the supremacy of the equitable and reasonable use principle and its primary qualification in managing a contemporary transboundary watercourse is what inherently entitles Ethiopia to pursue its rightful claim to use the Abbay Waters. However, for obvious provided reasons, the two utmost downstream states, Egypt and Sudan, are in control of the whole bulk of the Nile River flow and Abbay Waters in particular. What hurts Ethiopia the most is, out of its whole 123 BCM annual water surface, including the Abbay Waters, just 'a mere [3%] remains in the country, while [of 97% leftovers], 70 [%], and [27%] flows away [toward the] Nile Basin and other frontiers,' respectively (Arsano 2007). Consequently, the fast-growing *population of Ethiopia* (2022) coupled with its pushing economic, social, and political factors leave Ethiopia no option but to use the Abbay Waters equitably and reasonably, thus, in turn, might cause harm to the downstream riparian countries, as using a single drop of water, out of a total river flow, would automatically amount to harm against downstream riparian countries, which, of course, is legally tolerable under international watercourse law.

Strategies Employed by Ethiopia to Utilize Abbay Waters

The downstream riparian states, particularly Egypt, employed strategies of discursive struggle to endure the existing status quo by producing storyline pieces of evidence to support historical right discourses, while Ethiopia engaged in emancipatory struggle, which tried to change historic doctrines by using discourses such as equitable and reasonable utilization of the Abbay River, decolonization of the Nile River, framing natural right discourses to utilize Abbay, and other strategies discussed in the following sections of the paper.

Egypt's Construction and Deconstruction Strategies

The 'dialectics of construction and deconstruction' mechanism is a counterhydrohegemonic strategy derived from the work of Gramsci, which implies the replacement of Egypt's historic bloc through Ethiopia's new bloc by ideational and bargaining power (Cascao 2009; Gienanth 2020). Deconstruction strategy is supposed to necessitate the contestations of the Egyptian historic narration of hydrohegemon on the Nile River, while construction discourses recite the creation of a new Ethiopian counterhegemonic in the hydropolitical base (Gienanth 2020).

Egypt employed hegemonic compliance strategies such as securitization, knowledge construction, and sanctioning discourses to make use of its ideational power to shape the discourse surrounding the Abbay Water in its goodwill (Gienanth 2020). This narrative has been supported by the myth that Egypt is 'the gift of Nile,' which was first coined by the Greek historian Herodotus and used as a base rock for Egyptian hydrohegemony over the Nile Basin, and its very existence still depends upon this historic description of the Nile (Arsano 2007; Gebeto 2010). One of the hydroexpert and legal counsel to the Ministry of Foreign Affairs of Egypt argued that '*Egypt always care about the Grand Ethiopian Renaissance Dam because its existence entirely depends on the Nile.*' He further argued that 104 million Egyptian populations entirely depend on a single source of water, the Nile; 98% of all of Egypt's water resources come from this world's longest river, and more than 85% of the Nile water flows from Ethiopia, as he said. Therefore, Egypt uses securitizing strategies by believing that the GERD poses both water security and national security. Egypt's water security is closely linked to the 1959 agreement; under the 1959 agreement, a drop of water loose was a threat to Egyptian water security. Egypt also linked the GERD to national security concerns, and this strategy will continue as an impediment for the future development of not only ongoing GERD constructions but also upper basin developmental projects as a whole.

Knowledge construction is also a strategy of hydrohegemony in which Egypt has created different perspectives of its own hydrosituation to international communities and international donors as well as for its domestic citizens (Cascao 2009). For instance, upper stream riparian projects, such as the Grand Ethiopia Renaissance Dam, are portrayed as triggering great environmental harm to Egypt as well as the Nile region (Gienanth 2020). The normative compliance strategies of Egypt were rooted in the colonial history of the Nile, which recounts that 'the scramble of Africa supposes the scramble of the Nile.' The European colonizers were provoked that controlling Egypt and the Suez Canal depended upon their capacity to dominate over the Nile River. Great Britain was able to have the upper hand over the Nile River from its origins to the Mediterranean Sea through it colonies with the exception of Ethiopia which remained an independent riparian of the Nile River (Tvedt 2011). In order to surpass the challenge of not controlling the Nile in Ethiopia, Great Britain employed various strategies. One of the strategies was concluding agreements such as the 1929 Nile agreement between Britain and Egypt (Arsano 2007).

These external legal agreements are highly contested and have resulted in imbalanced water-sharing arrangements for all upstream riparian states. Egypt's strategy of putting forwards historical claims has denied Ethiopia's right to use the Abbay Waters and made a historic injustice over the share of the Nile water courses (Melak 2022). Therefore, according to the informant, the GERD itself is a counterhegemonic step to deconstruct such historic injustice by enabling Ethiopia to utilize Nile River resources. From a legal perspective, the notion of 'historical right' is nonexistent in the international water law; rather, it is a simple fiction. In fact, the claim that Abbay Water is 'Ethiopia's natural right' is more logical than the obsolete Egyptian historical rights since the Nile River flows from the Ethiopian hub (EBC 2020[13]).

Egypt, as a so-called hydrohegemon, also employed a coercive pressure strategy by using economic, military, and diplomatic threats against Ethiopia to maintain its dominance over the Nile River. As a result, in the economic front, from the initial stage of the construction of the Grand Ethiopian Renaissance Dam, Egypt has lobbied bilateral donors such as the World Bank (WB) and International Monetary Fund (IMF) as well as other donor states and institutions, which has succeeded in convincingly blocking any assistance for the construction of the project. However, Ethiopia has embarked on counterstrategy by attracting foreign finance, especially from China (as a new external player in the economic development of Ethiopia), in different mega projects. The Ethiopian diaspora must also be taken into consideration because its global mobilization has been crucial in both raising awareness of Ethiopia's situation among other nations and has been financing various mega dam projects. Therefore, the diaspora has served as Ethiopia's new counterhegemonic tactic.

Egypt also employed a military strategy since the eighteenth and nineteenth centuries. However, its goal was not effective under the Blue Nile River because of successive defeat by Ethiopia (Arsano 2007). Egypt mainly employed a strategy of covert actions to restore her historic dominance from Sadat to EL-Sisi said an informant from the Institute for Strategic Affairs. For instance, the Egyptian president Abdel Fattah El-Sisi has always brought a military threat to the public media expressing that threat towards Ethiopia since the construction of the Grand Ethiopian Renaissance Dam.

Ethiopian Deconstruction Strategies

Egypt works on a realignment with Sudan for working against Ethiopia. Moreover, they employ strategies of misinformation, propaganda, and misrepresentation of facts, as Professor Brook Hailu said in a political science discussion with Addis dialogue (EBC 2021). He argued that Egypt is also exerting pressure through social

---

[13] EBC (2020). Addis dialogue GERD discussion With Dr Yacob Arsano (PHD). In *YouTube*. https://www.youtube.com/watch?v=31SIrnst5Ns

media to get more audience from the Arab world, African states, and western audiences to confuse by blaming Ethiopia to disrupt the negotiation process and to stop the GERD operations. Regarding misrepresentations of facts, for instance, Egyptians claim that Guba, the place where the GERD is under construction, belongs to Sudan's region, which is a recent addition crafted by Egyptians when the second and the third filling of the GERD is operated. Professor Brook refers to such a strategy as dangerous and odd compared to past actions, which implies that Egypt uses Sudan as an instrument. Egyptians also work to disunite the Ethiopian people by supporting any insurgent groups both domestically and externally by manipulating even minor differences as a total disunifying factor by building a wall between Ethiopians.

Ethiopia's deconstruction of the Egyptian claims of 'securitization' is mainly through the 'right to development' in which Ethiopia claims that the utilization of Abbay Water, particularly the construction of the Grand Ethiopia Renaissance Dam, is for its socioeconomic development with no intention of harming downstream countries (Gienanth 2020). Deconstructions of Egypt's colonial normative compliance also occurred through engaging in the Nile Basin Initiative (NBI), CFA, and Declarations of Principles (DoPs), which are discussed in the cooperative discourse strategies.

Construction strategies were possible, in which Ethiopia would have to create not only a historic 'bloc' but also its own hydropolitical base that can be attained via the narrations of public support both externally and domestically. The Abbay River is central to Ethiopia's socioeconomic development and eradicates poverty to achieve national pride and unity (Melak 2022). The following part of the paper further discusses the different strategies of Ethiopia to utilize Abbay Waters. Each mechanism is either the deconstruction or construction tactics to contest the hydrohegemonic control of the downstream states over the Nile water.

The Right to Development Discourse Strategies

Independent states have permanent sovereignty over the use of their natural resources, and the right to development is recognized and protected in an extensive framework of international, regional, and domestic instruments (Kamga and Nagang 2021). The right to development by utilizing a country's natural resources for economic, social, and cultural development is a cardinal principle of international law (Ibid). The United Nations (UN) General Assembly Resolution 1803 (XVII) adopted on 14 December 1962 is well-defined as permanent sovereignty over natural resources, which provides a solid base for absolute autonomy in asserting natural resource ownership rights, which should be exercised in the interest of national development and the well-being of the people of a country. The sovereignty of natural resources is also recognized in the covenants of human rights that all people have the right to freely utilize their natural resources, which no one is deprived of. The interrelationships between the sovereignty of natural resources (the right to natural resources and the right to development) are clearly confined in the UN Declaration on the Right to Development.

This implies that depriving the utilization of people's natural resources would amount to a violation of their right to development.[14] The right to development incorporates the right to self-determination and permanent sovereignty over one's own natural resources that all people would freely dispose of their natural resources conceived as a means of subsistence and are not be deprived, which are enshrined in article 1(2) of the international covenants on human rights and in article 21(1) of the African Charter.[15]

Ethiopia has legal and sovereign rights to use her water potential to escape from historic poverty narrations; the Abbay River, particularly GERD, is an imperative and existential matter for the life of Ethiopians said, Dr. Eng. Silesh Bekele, the former Ethiopian Water, Irrigation and Energy Minister of Ethiopia to United Nation Security Council about the negotiations of Grand Ethiopian Renaissance Dam.

He strongly stated that:

> …the life of Ethiopians that languish on Saharan desert attempting to cross into Europe; the migrants in the middle East that sacrifice their youth to bring a better day for a better life; the young boys and girls in migrant prisons in Africa and beyond, the barefooted migrant that you see returning to their homeland in mass deportation from the middle East deserve a dignified life. (United Nations 2021)

Thus, the Grand Ethiopia Renaissance Dam is a matter of survival for Ethiopia as over 65% of its population are without electricity, millions of youth are without a job, and the country only has surface water as he said. This impressive argument is a direct response to Egyptians' so-called securitization strategy.

Ethiopia's Pan-Africa Discourse in Diplomatic Negotiations

Following the cooperative discourse agenda, the most widely documented Ethiopian counterhegemony strategy was an international diplomatic approach in the form of reactive and active (proactive) diplomacy (Cascao 2008). Historically, since 1879, Emperor Menelik II and his wife, empress Tayitu, had international diplomacy and public opinion campaigns against European' colonial attempts to control Abbay sources (Jonas 2011). Moreover, Ethiopia's reactive diplomatic strategies went through switching sides during the Cold War against Egypt's hydroimperialist position, and emperor Haileselassie I sent letters of protest to the international community against the Egyptian New Valley Project (Cascao 2008). Since the 1990s, Ethiopia began proactive diplomatic strategies by having bilateral and regional agreements and treaties over the Abbay Waters (Cascao 2008). For instance, on 23 December 1991, Ethiopia and Sudan agreed on equitable right of the uses of the Abbay Waters without causing appreciable harm with one another and worked together for better management of water (Waterbury 2002).

---

[14] See Declaration on the Right to Development (DRTD), African Charter on Human and Peoples' Rights; UN Declaration on the Rights of Indigenous Peoples.

[15] ICCPR, note 6; ICESCR, note 6;African Charter, note 14.

Most importantly, when the GERD was inaugurated in 2011, Ethiopia shifted from the use of reactive diplomacy to active (proactive) diplomatic strategies. For instance, Ethiopia and Egypt have been negotiating for more than a decade on key technical and legal issues over GERD disputes, all of which are initiated by the good will of Ethiopia that ensures the proactive diplomatic strategies of Ethiopia. Some of the major attempts at negotiation processes are the foundations of an International Panel of Experts (IPoE) in 2012 and the establishment of the TNC in 2014.

Throughout the negotiation process, undefined national interest, lack of a common framework, and unilateral actions hinder mutual consensus between the riparian states on the negotiation process of the Nile River, particularly over GERD disputes. Therefore, mediation through the principle of preventive diplomacy is necessary based on African approaches focusing on realizing not only the principles of mutual gains but also ensuring interconnected benefits (Faibt 2019). However, the two countries (Ethiopia and Egypt) have divergent perceptions on who and how preventive diplomacy is carried out in the negotiation process over GERD disputes.

Ethiopia believes that GERD's diplomatic negotiation is a technical issue; technical problems can be solved by science; and scientists and water experts of the three countries can solve technical problems scientifically, so the issue should not go beyond the three countries since preventive diplomacy is a kind of political intervention and consists of measures to avoid escalated conflict and tensions (Melak 2022). If the three countries do not resolve the issue on their own, it is possible that the African Union and other Nile Basin riparian countries could play a role in the due course of preventive diplomacy to facilitate the negotiations.

The African Union is maturing in mediating and solving African problems on its own.[16] Ethiopia believes that non-African actors such as the United Nations, USA, or other actors cannot impose solutions to African issues, as both Egypt and Ethiopia are African states that are members of the Africa Union; hence, the union is able to mediate the common problem within African norms and identity.[17]

After the 2018 Ethiopian political transition, political debates on the GERD gained momentum, and Egypt changed the negotiation process from technical negotiation to political negotiation by the strategies of regionalization and internationalization of the issues of the Grand Ethiopian Renaissance Dam (Melak 2022). Regionalization of the issue of GERD is related to the Arabization of the Abbay Waters by linking the issue to Arab states; the Abbay Water is a matter not only for Egypt and Sudan but also for the entire Red Sea and the Arabian Gulf states. The Arab League sees the issue of GERD as a security issue, and they want to go straight into the Grand Ethiopian Renaissance Dam negotiations process (Melak 2022). Internalization refers to taking the matter to the Security Council and making it global, he added. Ethiopia's stance is that GERD is an African issue, and African problems need African solutions such

---

[16] Ethiopia's Renaissance Dam: Has diplomacy failed? Zoom discussion with Mohammed Girma, a visiting lecturer at the University of Roehampton and a political commentator on the Horn of Africa. From Inside story, Aljazeera, Jul 4, 2021. Available on: https://aljazeera.com/program/inside-story/2021/7/4/ethipia's-renaissance-dam-has-failed.

[17] Ibid.

as African brotherhood and sisterhood that share the same river. There is a need to change the political, socioeconomic, and environmental landscape of the Nile Basin, and the CFA and NBI are good examples that need to be supported by AU-led negotiations (Melak 2022). In this regard, the establishment of NBI, the ratifications of CFA, and the agreement of Declarations of Principles have supported the concept of African Solutions to African Problems to negotiate the Abbay River disputes.

References

Abtew W, Dessu SB (2019) The grand Ethiopian renaissance dam on the Blue Nile. Springer International Publishing, Cham. https://doi.org/10.1007/978-3-319-97094-3

Abtew W, Melesse AM (2014a) Nile River Basin hydrology. In: Melesse AM, Abtew W, Setegn S (eds) Nile River Basin: ecohydrological challenges, climate change and hydropolitics. pp 7–22

Abtew W, Melesse AM (2014b) Climate teleconnections and water management. In: Nile River Basin. Springer International Publishing, pp 685–705

Abtew W, Melesse AM (2014c) Transboundary Rivers and the Nile. In: Nile River Basin. Springer International Publishing, pp 565–579

Abtew W, Melesse A, Desalegn T (2009a) Spatial, inter and intra-annual variability of the Blue Nile River Basin Rainfall. Hydrol Process 23(21):3075–3082

Abtew W, Melesse A, Desalegn T (2009b) El Niño Southern Oscillation link to the Blue Nile River Basin hydrology. Hydrol Process 23(26):3653–3660

Agreement between the United Arab Republic and the Republic of Sudan for the full utilization of the Nile Waters, Signed at Cairo on 8 November 1959 (1959)

Agreement on Declaration of Principles between the Arab Republic of Egypt, the Federal Democratic Republic of Ethiopia and the Republic of the Sudan on the Grand Ethiopian Renaissance Dam Project (Signed at Cairo on 23 March 2015)

Agreement on the Nile River Basin Cooperative Framework (opened for signature in May 2010)

Arsano Y (2007) Ethiopia and the Nile: Dilemmas of national and regional hydropolitics. Center for Security Studies, Swiss Federal Inst. of Technology, Zurich

Arsano Y (2011) Negotiations for a Nile-cooperative framework agreement. Institute for Security Studies, Paper 222. Retrieved from https://www.files.ethz.ch/isn/136717/PAPER222.pdf

AU (2020, June 26) Communiqué of the Extraordinary African Union (AU) Bureau of the Assembly of Heads of State and Government video-teleconference Meeting on the Grand Ethiopian Renaissance Dam (GERD)|African Union. Retrieved June 13, 2023, from https://au.int/en/pressreleases/20200626/hosg-communique-meetinng-grand-ethiopian-renaissance-dam-gerd

Baker BH, Omer A, Aldridge CA (2016) Water: availability and use. Mississippi State University Extension Service. Retrieved from https://www.researchgate.net/publication/324226678_Water_Availability_and_us

Berlin Rules on Water resources, adopted International Law Association (2004)

Bulto TS (2008) Between ambivalence and necessity in the Nile Basin: occlusions on the path towards a basin-wide treaty. Mizan Law Rev 2(2):201–228. https://doi.org/10.4314/mlr.v2i2.56149

Buzan BG, Waever O, de Wilde JH (1998) Security: a new framework for analysis. London/Boulder, CO, Lynne Rienner

Buzan B (1983) People, states and fear: the national security problem in international relations. Brighton, Harvester Press/Chapel Hill, North Carolina University Press

Cascao AE (2008) Ethiopia–challenges to Egyptian hegemony in the Nile basin. Water Policy 10(S2):13–28

Cascao AE (2009) Changing power relations in the Nile river basin: unilateralism vs. cooperation? Water Altern 2(2):245–268

Cascao AE, Zeitoun M (2010) Power, hegemony and critical hydropolitics. In: Earle A, Jägerskog A, Ojendal J (eds) Transboundary water management earthscan

Chebud YA, Melesse AM (2009a) Numerical modeling of the groundwater flow system of the Gumera Sub-Basin in Lake Tana Basin, Ethiopia. Hydrol Process 23(26):3694–3704

Chebud YA, Melesse AM (2009b) Modeling lake stage and water balance of Lake Tana, Ethiopia. Hydrol Process 23(25):3534–3544

Chebud Y, Melesse AM (2013) Stage level, volume, and time-frequency change information content of Lake Tana using stochastic approaches. Hydrol Process 27(10):1475–1483. https://doi.org/10.1002/hyp.9291

Convention on the Law of the Non-navigational Uses of International Watercourses (Adopted on 21 May 1997 by the UN General Assembly, entered into force on 17 August)

Dellapenna J (1997) The Nile as a legal and political structure. Retrieved from https://www.researchgate.net/publication/288014044_%27The_Nile_as_a_legal_and_political_structure%27

Dellapenna J (2001) The customary international law of transboundary fresh waters. Int J Glob Environ 1(3–4):264–305. https://doi.org/10.1504/IJGENVI.2001.000981

Dessu SB, Melesse AM (2012) Modeling the rainfall-runoff process of the Mara River Basin using SWAT. Hydrol Process 26(26):4038–4049

Dessu SB, Melesse AM (2013) Impact and uncertainties of climate change on the hydrology of the Mara River Basin. Hydrol Process 27(20):2973–2986

Dessu SB, Melesse AM, Bhat M, McClain M (2014) Assessment of water resources availability and demand in the Mara River Basin. CATENA 115:104–114

EBC (2020) Addis dialogue GERD discussion With Dr Yacob Arsano (PHD). In YouTube. https://www.youtube.com/watch?v=31SIrnst5Ns. Last visited on 7/5/2021

EBC (2021) Addis dialogue: professor Brook Hailu Beshah on GERD, Sudan, Egypt and upcoming election. In YouTube. https://www.youtube.com/watch?v=OiUfD61zHh0. Last visited on 17/7/2021

Ethiopia Population 2022 (Demographics, Maps, Graphs). (2022, October 23). Retrieved October 23, 2022, from World Population Review website: https://worldpopulationreview.com/countries/ethiopia-population

Exchange of Notes between Her Majesty's Government in the United Kingdom and the Egyptian Government in Regard to the Use of the Waters of the River Nile for Irrigation Purposes on 7 May 1929

Faibt R (2019) How mediation based on African approaches to conflict resolution can transform the conflict over the Nile. Conflict Trends 2019:29–37

Fiedler W (1987) State succession. In: Bernhardt R (ed) Encyclopedia of public international law: states-responsibility of states-international law and municipal law, vol 10. Elsevier Science, Amsterdam, pp 446–456. Retrieved from https://books-library.net/files/books-library.online-02262152Dz8X7.pdf

Gabčíkovo-Nagymaros Project (Hungary vs. Slovakia), No. 92 (International Court of Justice September 25, 1997)

Gebeto PJ (2010) No more thirst: the citizens of the Nile. Nairobi, Author

Gebrehiwot K (2020) Hydro-hegemony, an antiquated notion, in the contemporary Nile river basin: the rise of water utilization in upstream riparian countries. Heliyon 6(9):e04877

Gienanth E (2020) An analysis of Ethiopian and Egyptian discourses surrounding the Grand Ethiopian Renaissance Dam [Unpublished master's thesis]. University of Amsterdam, Netherlands

Helsinki Rules on the Uses of Waters of International Rivers, adopted by International Law Association, (1996).

ILA (2004) Berlin conference on water resources law. International Law Association. Retrieved from International Law Association website: https://www.internationalwaterlaw.org/documents/intldocs/ILA/ILA_Berlin_Rules-2004.pdf

Janig P (2018) 1978—The 1978 Vienna convention, the clean slate doctrine and the decolonization of sources. Austrian Rev Int Eur Law 23. Retrieved from https://papers.ssrn.com/abstract=3677220

Johnston E (2009) Factors influencing a basin-wide agreement governing the Nile River (Master Dissertation, Simon Fraser University). Simon Fraser University. Retrieved from https://summit.sfu.ca/item/9400

Jonas R (2011) The battle of Adwa: African victory in the age of empire. Cambridge, Massachusetts, London, Harvard University Press

Kamga SD, Ngang CC (2021) The natural resource and right to development dilemma. In: Carol Chi N, Serges DK (eds) Natural resource sovereignty and the right to development in Africa, pp 1–15

Kimenyi MS, Mbaku JM (2015) International water law and the Nile River Basin. In: The search for a new legal regime. Governing the Nile River Basin, pp 60–71. Brookings Institution Press. JSTOR. Retrieved from http://www.jstor.org/stable/https://doi.org/10.7864/j.ctt130h973.9

Lake Lanoux Arbitration (France v. Spain). R.I.A.A. 281 (Arbitral Tribunal November 16, 1957)

Land and maritime boundary between Cameroon and Nigeria (Cameroon v. Nigeria: Equatorial Guinea intervening), No. 194 (International Court of Justice October 10, 2002)

Lazerwitz DJ (1993) The flow of international water law: the international law commission's law of the non-navigational uses of international watercourses. Indiana J Glob Legal Stud 1(1):247–271

Lie JHS (2010) Supporting the Nile Basin initiative: a political analysis "Beyond the River." Norwegian Institute of International Affairs. Retrieved from https://www.academia.edu/2243972/Supporting_the_Nile_Basin_Initiative_A_Political_Analysis_Beyond_the_River

McCaffrey SC (2001) The law of international watercourses: non-navigational uses. Oxford University Press

McIntyre O (2010) International water law: concepts, evolution and development. In: Transboundary water management. Routledge

Mekonnen DZ (2010) The Nile Basin cooperative framework agreement negotiations and the adoption of a 'water security' paradigm: flight into obscurity or a logical cul-de-sac? Eur J Int Law 21(2):421–440. https://doi.org/10.1093/ejil/chq027

Melesse AM (2011) Nile River Basin: hydrology, climate and water use. Springer Science & Business Media

Melesse AM, Loukas AG, Senay G, Yitayew M (2009) Climate change, land-cover dynamics and ecohydrology of the Nile River Basin. Hydrol Process 23(26):3651–3652

Melesse AM, Abtew W, Desalegne T, Wang X (2010) Low and high flow analysis and wavelet application for characterization of the Blue Nile River system. Hydrol Process 24(3):241–252

Melesse AM, Abtew W, Setegn S, Dessalegne T (2011) Hydrological variability and climate of the Upper Blue Nile River Basin. In: Melesse A (ed) Nile River Basin: hydrology, climate and water use. Springer Science Publisher Chapter 1, pp 3–37. https://doi.org/10.1007/978-94-007-0689-7_1

Melesse A, Abtew W, Setegn SG (2014) Nile River Basin: ecohydrological challenges, climate change and hydropolitics. Springer Science & Business Media

Menga F (2016) Reconceptualizing hegemony: the circle of hydro-hegemony. Water Policy 18(2):401–418

Mohammed YA (2017) The "water security" principle under Nile Basin CFA: the "water security" principle and its ramification. LAMBERT Academic Publishing, Germany. Retrieved from https://www.amazon.com/Water-Security-Principle-Under-Basin/dp/3330352965

Mohammed YA (2022a) The endless controversies of The Nile River Basin in the context of international transboundary watercourse doctrines. Soc Sci Univ Ankara Law J 4(2):895–930. https://doi.org/10.47136/asbuhfd.1050465

Mohammed YA (2022b) The guiding legal regime and institutional arrangement of transboundary watercourse: a review. Int J Water Manage Diplomacy 1(4):75–90

Moss R (1992) Environmental security? The illogic of centralized state responses to environmental threats. In: Geopolitical perspectives on environmental security. The Studies and Research Center on Environmental Policies-GEPRE, Université Laval, Quebec, Cahier pp 92–05

Mtua G (2017) Bilateral treaties on the Nile River and their impacts on international relations (Master Dissertation, Tumaini University Makumira). Tumaini University Makumira. Retrieved from https://www.academia.edu/37288651/BILATERAL_TREATIES_ON_THE_NILE_RIVER_AND_THEIR_IMPACTS_ON_INTERNATIONAL_RELATIONS

NBI (n.d.) Retrieved 5 Oct 2022, from https://nilebasin.org/

Okoth-Owiro A (2004) The Nile treaty: state succession and international treaty commitments, a case study of the Nile Water treaties. Konrad Adenauer Foundation

Okoye FC (1972) International law and the New African States. Sweet & Maxwell Ltd, London. Retrieved from https://www.amazon.co.uk/International-Law-African-States-Africa/dp/042116140X

Paisley RK, Henshaw TW (2013) Transboundary governance of the Nile River Basin: past, present and future. Environ Dev 7:59–71. https://doi.org/10.1016/j.envdev.2013.05.003

Ram KV (1977) The survival of Ethiopian independence. J Hist Soc Niger 8(4):131–141

Schäfer PJ (2013) Securitization and discourse. In: Human and water security in Israel and Jordan. SpringerBriefs in Environment, Security, Development and Peace, vol 3. Springer, Berlin, Heidelberg. https://doi.org/10.1007/978-3-642-29299-6_4

Setegn SG, Srinivasan R, Dargahi B, Melesse AM (2009a) Spatial delineation of soil erosion prone areas: application of SWAT and MCE approaches in the Lake Tana Basin, Ethiopia. Hydrol Process 23(26):3738–3750

Setegn SG, Srinivasan R, Melesse AM, Dargahi B (2009b) SWAT model application and prediction uncertainty analysis in the Lake Tana Basin, Ethiopia. Hydrol Process 24(3):357–367

Setegn SG, Bijan Dargahi B, Srinivasan R, Melesse AM (2010) Modelling of sediment yield from Anjeni Gauged watershed, Ethiopia using SWAT. JAWRA 46(3):514–526

Starr J (1991) Water wars. Foreign Policy 82

Stebek EN (2007) Eastern Nile at crossroads: preservation and utilization concerns in focus. Mizan Law Rev 1(1):33–59. https://doi.org/10.4314/mlr.v1i1.55613

Tekuya M (2019) Governing the Nile Under climatic uncertainty: the need for a climate-proof basin-wide treaty. Nat Resour J 59(2):321

Tekuya M (2020, March 30) Ethiopia does not need Egypt's permission to start filling GERD. Retrieved 23 Oct 2022, from Ethiopia Insight website: https://www.ethiopia-insight.com/2020/03/30/ethiopia-does-not-need-egypts-permission-to-start-filling-gerd/

Tekuya M (2021) Sink or swim: alternatives for unlocking the grand Ethiopian Renaissance Dam Dispute. Columbia J Transnational Law 59(1):65–116

Territorial Dispute (Libyan Arab Jamahiriya/Chad), No. 83 (International Court of Justice February 3, 1994)

Territorial Jurisdiction of the International Commission of the River Case. (United Kingdom, Czechoslovakia, Denmark, France, Germany, Sweden V. Poland), No. 23 (The Permanent Court of International Justice August 20, 1929)

The World Business Council for Sustainable Development (WBCSD) (2005) Facts and trends: water. Retrieved from https://docs.wbcsd.org/2005/08/WaterFactsAndTrends.pdf

Tiwary R (2006) Conflicts over international waters. Econ Polit Wkly 41(17)

Treaties between the United Kingdom, between United Kingdom, Italy and Ethiopia Relative to the Frontiers between the Soudan, Ethiopia, and Eritrea (signed at Addis Ababa on 15 May 1902), (1902)

Turton A (2002) Hydropolitics: the concept and its limitations. In Turton A, Henwood R (eds) Hydropolitics in the developing world: a Southern African perspective. Pretoria, The African Water Issues Research Unit (AWIRU)

Tvedt T (2004) The river Nile in the age of the British. Political ecology and the quest for economic power. London, I.B. Tauris

Tvedt T (2011) Hydrology and empire: the Nile, water imperialism and the partition of Africa. J Imp Commonw Hist 39(2):173–194
Ullendorff E (1967) The Anglo-Ethiopian treaty of 1902. Bull Sch Orient Afr Stud Univ Lond 30(3):641–654
UNEP-DHI, UNEP (2016) Transboundary river Basins: status and trends. United Nations Environment Programme (UNEP), Nairobi
United Nations (2021) Ethiopia on the grand Ethiopian renaissance dam (GERD) - media stakeout (8 July 2021). In YouTube. https://www.youtube.com/watch?v=QqfPMFVZ4bk. Last visited on 20/8/2021
User's Guide Fact Sheet Series, Number 5. No Significant Harm Rule (n.d.) Retrieved 5 Oct 2022, from UN Watercourses Convention website: https://www.unwatercoursesconvention.org/documents/UNWC-Fact-Sheet-5-No-Significant-Harm-Rule.pdf
Vienna convention on the law of treaties (adopted at Vienna on 23 May 1969 by the UN General Assembly, entered into force on 27 January 1980) (1969)
Vienna convention on succession of states in respect of treaties (adopted at Vienna on 23 August 1978 by the UN General Assembly, entered into force on 6 November 1996) (1978)
Waever O (1995) Securitization and desecuritization. In Lipschutz RD (ed) On security. New York, Columbia University Press, pp 46–87
Waterbury J (1997) Between unilateralism and comprehensive accords: modest steps toward cooperation in International River Basins. Int J Water Resour Dev 13(3):279–290. https://doi.org/10.1080/07900629749692
Waterbury J (1979) Hydropolitics of the Nile Valley. Syracuse, N.Y, Syracuse University Press
Woldetsadik TK (2013) International watercourses law in the Nile River Basin: three states at a crossroads. Routledge, London. Retrieved from https://www.routledge.com/International-Watercourses-Law-in-the-Nile-River-Basin-Three-States-at/Woldetsadik/p/book/9781138573116
Woldetsadik TK (2015) Anglo-Ethiopian treaty on the Nile and the Tana Dam concessions: a script in legal history of Ethiopia's diplomatic confront (1900–1956). Mizan Law Rev 8(2):271–298. https://doi.org/10.4314/mlr.v8i2.1
Worku Y, Giweta M (2018). Can we imagine pollution free rivers around Addis Ababa city, Ethiopia? What were the wrong-doings? What action should be taken to correct them? J Pollut Effects Control 06(03). https://doi.org/10.4172/2375-4397.1000228
Yitayew M, Melesse AM (2011) Critical water resources management Issues in Nile River Basin. In: Melesse A (ed) Nile River Basin: hydrology, climate and water use. Springer Science Publisher, Chapter 20, pp 401–416. https://doi.org/10.1007/978-94-007-0689-7_20
Zeitoun M, Warner J (2006) Hydro-hegemony – a framework for analysis of trans-boundary water conflicts. Water Policy 8(5):435–460
Zeitoun M, Cascão A, Warner J, Mirumachi N, Matthews N, Farnum R, Menga F (2016) Transboundary water interaction III: contesting hegemonic arrangements. Int Environ Agreem 271–294

Chapter 16
Land Management and Productivity in the Abbay Basin

Mengistie Mersha

Abstract The rural community's livelihood of the Abbay basin heavily relies on agriculture, which in turn depends on land resources. The purpose of this article is, therefore, to assess the current condition of land resources in the Abbay basin. Therefore, this study aimed at characterizing the land resource; the type, nature, and magnitude of land degradation; impacts of land degradation; land management practices; determinants of applying land management practices and productivity of the land in terms of certain crop yields focusing on the Bechet Watershed of the Abbay basin. In addition to their effect on the productivity of the land in situ, the implementation of land management practices such as terracing and tree planting are believed to have significant impacts on the future fate of the Great Ethiopian Rennaissance Dam. This paper, therefore, escalates the efforts of land management practices so far implemented in the basin and recommends how such practices will be strengthened.

Keywords Land resource · Land degradation · Land management · Land productivity · Abbay basin · Ethiopia

Introduction

Natural resources are intertwined in every aspect of our lives. Barnett and Morse (2013) reported that from the air we breathe to the water we drink, we depend on Earth's natural resources to survive. We use some of these resources in their natural state for food, shelter, and clothing; others become products that improve our standard of living. Land resources, among others, are a vital resource on which all human kinds rely for living. Using natural resources wisely can help humans live more economically and more in harmony with Earth, not just for current consumption but for years to come.

M. Mersha (✉)
Yom Postgraduate College, Addis Ababa, Ethiopia
e-mail: mengiste123@yahoo.com

As clearly stated by Ogendo (2006), land is a major asset in modern as well as traditional societies. It is especially a critical factor for millions of extremely poor people who live in rural areas and depend on agriculture for their livelihood. Land is not only an economic resource but also a central issue in the creation of individual and collective identity and the formation of social, cultural, and religious life. It is an enormous political resource as well, defining power relations between and among individuals, families, and communities under established systems of governance.

For most rural residents in developing countries, land continues to remain the primary means of generating livelihoods. It is the most important asset that rural smallholders can use as sources of wealth, and for their relatives or coming generations. As a result, the way property rights to land are defined has been seen as an important determinant of sustainable land management, economic growth, and poverty alleviation. This is because access to land and its effective use have great importance for poverty reduction, economic growth, and investment (Deininger 2003).

Rural land, particularly, has a large economic value in Ethiopia, where the economy is mainly based on agriculture, which is the main source of livelihood for more than 85% of the total population (Zerga 2016). The issue of land has, therefore, become a main debatable political issue in Ethiopia (Davies 2008). The major aim of this chapter is to provide insight into the current status of land degradation, management practices, and productivity in the Abbay basin of Ethiopia.

Land Degradation

Land degradation is defined as the long-term loss of ecosystem function and productivity caused by disturbances from which the land cannot recover unaided (Bai et al. 2008). Land degradation is widespread and severe, mainly in the highlands of Ethiopia 1500 m above sea level, which represents approximately 40% of the total land area but is home to 90% of the total population and 70% of the livestock. In this regard, Jansen et al (2007) carried out a study on soil degradation and sustainable land management in the rain-fed agricultural areas of Ethiopia and found that the present annual net erosion across the study areas measured by the Unit Stream Power Erosion Deposit (USPED) Model is -940 million tons or -18 tons/ha. The study also focused on croplands of the Abbay basin and estimated its annual net erosion to be -380 million tons (-20.2 tons/ha) using the same model. According to this study, the share of croplands lying on slopes steeper than 8% totals 77%. As slopes steeper than 8% are liable to erosion, this study recommends that conservation structures be built on an additional 59% of the cropland for all sloping cropland to be conserved (18% of the cropland is already conserved).

Another study by Alemu et al. (2007) reported the severity of land degradation in the highlands of the Abbay basin, which estimated the net soil loss from croplands to be approximately 100 tons/ha, leading to a net cropland soil loss of approximately 1100 million tons per annum from the highlands. Generally, this study levels that approximately half of the Ethiopian highlands are "significantly eroded" and over one-fourth is "seriously eroded". It also concluded that over 2 million hectares of farmlands in the Ethiopian highlands have reached the "point of no return" in the sense that they are unlikely to sustain crop production in the future. In addition, the study indicated large areas of the Abbay basin in the administrative regions of Wollo, Gondar, and Gojjam to be "hot spots" of land degradation.

The population continues to increase rapidly in the highlands of the Abbay basin and exerts a burden on the supplies of agricultural land, particularly arable land for cultivation and pasture (Gashaw et al. 2014). There are signs of diminishing farm size and fragmentation due, among others, to high population pressure and limited livelihood options other than rain-fed agriculture. Additionally, overgrazing, deforestation, and high population density have led to massive soil degradation, leading to low productivity (Temesgen et al. 2014). This situation created an excess demand for farmland, which is manifested in the form of pushing cultivation onto marginal lands (i.e., steep slopes, low rainfall zones) and increasing land rentals. However, landlessness is still becoming a growing crucial problem in the rural community.

Types and Drivers of Land Degradation in the Abbay Basin

The physical, chemical, and biological degradation of soil, biodiversity, and agrodiversity degradation, water resource degradation, deforestation, land use and land cover changes, and climate variability and changes are the common formas of land degradation in Ethiopia (Gessesse 2013). Erosion by water is a severe agent of soil degradation in Ethiopia. Soil erosion is the most significant environmental challenge for ensuring food security for the increasing population and for sustainable development (Wagayehu 2003). Regarding soil erosion, a considerable volume of information has been produced since the mid-1980s (Barbier 2000; Eyasu 2003). However, reliable and consistent data on the extent and rate of soil loss (tonnes/ha/year) is lacking. The rate and extent of soil erosion rate from arable land reported by different studies were not consistent. As such, the estimated current rates of soil erosion in Ethiopia vary between 42 and 300 tons/ha/year (Gessesse 2013).

The wide range of estimates of soil erosion rate could be due to the complex patterns of spatial and temporal variations of the soil erosion factors and the inherent conceptual and methodological challenges in the estimation. There is considerable variability in erosion rates over time and place depending on the agroecological zone and soil type. Soil erosion occurs at varying rates and with varying degrees in different parts of the country. Deforestation, forest burning, and expansion of cultivated lands to marginal lands have also contributed to the widespread problem of land degradation in the country. Approximately 70% of Ethiopia's highland population and an area of over 40 million hectares are affected by land degradation (Melaku 2013), indicating the scale and extent of the problem confronting the country.

Natural factors coupled with the effects of a long history of settlement, prevailing farming methods, and increasing population pressure, which forces people to cultivate even steeper slopes, have exacerbated the devastating land and resource degradation in the Abbay basin (Berhanu and Fayera 2005; Askale 2005). The increasing population at an alarming rate, leads the people to expand their farms to hillsides and ecologically fragile areas, and forces the people to use crop residues and dung for fuel rather than using them as sources of organic fertilizer to improve soils, which results in the reduction of land management activities such as following, planting trees and investing in conservation structures (Lakew et al. 2000).

The extent of deforestation is severe and has a long history in the Abbay basin attributed to the changing of natural landscape due to the subsistence farming and high population pressure in the basin for millennia. As such, the major causes of deforestation were the substantial change in land use and land cover, which included conversion of forest to grassland and cropland. Particularly, increasing cropland was observed during the second half of the twentieth century, mainly at the expense of grassland and forestland (Hans et al. 2010). Approximately 20 thousand hectares of forest were cleared annually in the Abbay basin for the purpose of fuel wood, logging, and construction (ILRI 2000; Lakew et al. 2000). As such, deforestation in the Abbay basin has been dwindling from day to day due to population growth, overgrazing pressure, and lack of strong forest management policy. In Ethiopia, still there is no forest resource management policy, strategy, and proclamation to control deforestation and illegal forest product movement and encroachments (Mulatie et al. 2015). In addition, the increased demand land for pasture, shelter, food crops, urbanization, and the eventual conversion of natural forests to croplands are the other contributing factors to the severe deforestation in the Abbay basin.

The basin has suffered from recurrent and severe droughts and pest invasions since long time ago. Studies showed that there has been no single year since 1950 where there was no drought, particularly in the eastern part of the basin (USAID 2000). This frequent drought reduced the growth of vegetation, aggravating the runoff, removal of fertile topsoil, and reduced agricultural production and finally resulting in land degradation.

The livelihood of the community living in the Abbay Basin depends on livestock production as alternative sources of income, and thus it is the major component of the economy in the Abbay Basin.

The basin is home to approximately 35% of the total country's livestock population. In addition, based on an agricultural sample survey of 2012/13, high population density of livestock was found in this basin (Samson and Frehiwot 2014). Amhara region, which comprises large proportion of the Abbay basin, stands first in the number of goats and second in the number of cattle, sheep, asses, horses, and poultry (CSA 2008). As such, livestock in the region provide approximately 16.4 million tonnes of manure annually, which is equivalent to 114 thousand tonnes of nitrogen. However, these manure have been mainly used for fuel rather than manure.

Free grazing or uncontrolled grazing and browsing too much livestock for a long period on land (overgrazing) are the dominant grazing systems in the region and are the major reasons for deforestation and land degradation. Overgrazing contributes to soil degradation by increasing soil compaction and demanding frequent tillage to prepare fields for crops, which further aggravates soil erosion. Particularly, overgrazing in the hillsides, fragile lands, and on and cultivated land aggravates soil compaction, low moisture retention, and high runoff, which results in the formation of gullies, excessive vegetation removal, and crop yield reduction (Lakew et al. 2000). In addition, it harms conservation efforts, as trampling animals often damages physical conservation structures such as stone terraces and soil bunds. Biological conservation practices such as grass strips and tree plantations are also being destroyed or trampled, reducing the chance for establishment and regeneration. It is more destructive during the rainy season when other sources of feed (e.g., stubble grazing and crop residues) are in short supply (Samuel et al. 2002), during the period of high runoff production, and when the soil is very sensitive to washing by floods.

As described above, the livelihood of the population living in the Abbay basin heavily relies on mixed agricultural practices such as crop production and livestock rearing. The poor agricultural system in basin coupled with the large human population size, aggravates land degradation in the basin. Moreover, livestock rearing depends on free grazing, which causes the degradation of forests, soil, and water as well as the distraction of land management measures.

Furthermore, land degradation in the Abbay basin has been aggravated by inadequate property rights (Askale 2005). Studies showed that the absence of secure property rights of land and natural resources accelerates land degradation (Berhanu and Fayera 2005). According to Assefa (2010), inadequate land rights, sociocultural factors, backward methods of agricultural practices, and ever-increasing population pressure, exacerbated the devastating land resource degradation in the region. Moreover, Samuel et al. (2002) reported that livelihood and land management, and increasing pressure on grazing land are greatly influenced by land redistribution.

According to the constitution of Ethiopia, all natural resources and urban and rural land are the owner of the state and its people. The constitution allows utilization of the land and natural resources by individuals and firms. To improve land tenure security and promote better land management and attract more investment, a land registration and title certification program should be implemented. The land certification was supported by the Swedish Government (Sida) as part of a rural development program in the Amhara region (Berhanu and Fayera 2005). However, even though the people of the region have been certified for their land, they are still not assured about the ownership of the land and they suspect prospects of future land redistribution, which undermine their investment in land improvements. This tenure insecurity emanates mainly from the frequent land redistribution, which has been ongoing since 1974 in an effort to balance land holdings and quality across households (Ehui et al. 2003). Therefore, to solve such confusion and build trust, more work needs to be done to raise awareness among the beneficiary communities and landholders.

Poverty is one of the underlying driving factors for land degradation in Abbay Basin for several reasons. People exert more pressure on land when access to alternative sources of livelihood is lacking. In addition, when people lack alternative fuel sources, they are forced to deforestation and burning of dung and crop residues. Electricity and kerosene are expensive and, in most cases, and even not available in most parts of the rural areas in the basin (Lakew et al. 2000). The other causes of land degradation include lack of strong local institutions and organizations, low perceptions and attitudes of the local community about the problem, and other agricultural extension issues.

Increasing population pressure and poor land management are the main drivers of land degradation in Ethiopia (Berry 2003; Genanew and Alemu 2012). The increased population in the Abbay basin, resulted in declining arable lands, which in turn led to the expansion of agricultural land to marginal areas. According to Berry (2003), the patterns of land ownership and government control, low levels of investment in agriculture and animal husbandry, poor rural infrastructure and markets, and low levels of technology are the underlying causes of land degradation in Ethiopia. Weak policy and low capacity to implement government interventions also contribute to land degradation (The Global Mechanism 2007; Wagayehu 2003).

The less-than-desired and largely unsustainable effect of a series of physical conservation structures such as terraces, bunds, and tree planting is explained by a lack of policy action or framework that is essential to address or minimize the effect of the externalities of benefits or costs associated with participation or lack of participation in such programs by farmers. This is due to the negligence of policy and institutional factors in several conservation programs financed through food or cash aid projects (Eyasu 2003; Berry 2003), which is deepened by lack of evidence-based and action-oriented research (Von Braun et al. 2013).

Another key driver of the problem is the limited capacity and/or obligation to solve the problem correctly and on time. Unreliable, incomplete, or inadequate involvements strengthen the problem while lowering the capacity of farmers and the local authorities to fundamentally deal with the problem. Most interventions focused on addressing the symptoms of the problem such as reducing the human cost of the problem, and distress sales of assets than providing long-lasting solutions. However, by focusing only on short-term solutions, inaction or the postponement of real actions (i.e., actions by beneficiaries and authorities to address the root cause of the problem) are encouraged.

A study on the proximate and underlying drivers of land degradation in Eastern Africa including Ethiopia, was carried out by Kirui and Mirzabaev (2014). Table 16.1 shows the summary for Ethiopia. The major proximate drivers of land degradation in Ethiopia include topography, unsustainable agricultural practices, and land cover change (forests, woodlands, and shrubland conversion to new agricultural land uses). On the other hand, the underlying drivers of land degradation in the country include weak policy and regulatory environment and institutions, poverty, demographic growth, low empowerment of local communities, infrastructural development, and unclear user rights, particularly land tenure.

Table 16.1 Proximate and underlying drivers of land degradation in Ethiopia

Proximate drivers	Underlying drivers	References
Topography, unsustainable agriculture, fuel wood consumption, conversion of forests, woodlands, and shrub-lands to new agricultural land (deforestation)	Weak regulatory environment and institutions, demographic growth, unclear user rights, low empowerment of local communities, poverty, infrastructural development, population density	Pender et al. (2001), Jagger and Pender (2003), Holdenet al. (2004), Rudel et al. (2009), Bai et al. (2008), Belay et al. (2014), Tesfa and Mekuria (2014)

Source Kirui and Mirzabaev (2014)

Impacts of Land Degradation

Land degradation in the Abbay Basin is a severe environmental problem that negatively affects the socioeconomic of the community and ecosystem services decline such as declines in both surface and groundwater water availability; declines in natural biomass and vegetative ground cover; soil degradation (chemical, physical and biological) and environmental impacts. As the economy of the region heavily depends on agriculture and its products, land degradation (i.e., soil loss, change in water quantity and quality, vegetation loss) greatly influences the livelihood of the community. It is estimated that the annual rate of soil loss in the region due to water erosion is approximately 119 million tons, which accounts for 70% of the total soil loss in the country as a whole (IFSP 2004). Approximately 29% of the total area of the region experiences high erosion rates (51.2 t/ha per year); 31% experiences moderate erosion rates (16.50 t/ha per year); 10% experiences very high erosion rates (>200 t/ha per year); and the remaining 30% experiences low erosion rates (<16 t/ha per year) (Lakew et al. 2000). The situation is becoming disastrous as marginal lands are being converted to cultivated land including very steep slopes (Tesfahun and Osman 2003) while grazing land is becoming scarce, and the remaining are under extreme grazing pressure (IFSP 2004), resulting in low and declining agricultural productivity and continuing food insecurity and rural poverty (Assemu and Shigdaf 2014).

Poverty drives the populations to over utilize the remaining natural resources, triggering a vicious cycle and accelerating land degradation. The declining soil fertility and food inscarcity resulted in the migration of the rural population to the urban areas. Therefore, land degradation is a major challenge to biodiversity, ecological sustainability, and ecosystem stability. Land degradation declines the ecosystem services of the land like nutrient cycling, the global carbon cycle, and the hydrological cycle. Even though the Abbay basin has a wide biodiversity of flora and fauna, the quality and quantity of biodiversity resources are declining due to deforestation, soil erosion, water resources degradation, loss of soil fertility, increased demand for arable land, and livestock pressure (Mulatie et al. 2015).

The current forest resources of the Amhara region are estimated to be 5.91% of the total area of the region, but the coverage is declining due to increasing population growth and overgrazing pressure. Land degradation, where vegetative cover is removed and the soil surface is exposed to the impact of raindrops greatly modifies hydrological processes. On the other hand, land degradation greatly deteriorates the quantity and quality of water resources. As a result, many perennial springs and streams in the region have become drying and seasonal (Lakew et al. 2000). Water bodies are prone to rapid sedimentation, thus reducing the storage capacity of irrigation and drinking water and hindering the generation of hydroelectric power (IFSP 2004).

Land Management Practices

Agricultural sector is the pillar of the Ethiopian economy and as such, special consideration was given by the government to lead the economic transformation of the country (Atsbaha and Tessema 2012). However, land degradation in general and soil erosion in particular persist as the major challenges that unpleasantly affect the agricultural performance of the country and thus call for improved land management practices (Bewket 2003). As such, the issue of land degradation got attention long time ago, and the first step against it emerged during the reign of Menelik II.

Aware of the prevailing deforestation and land degradation by the warlords accompanying the king and by the growing population, emperor Menelik II introduced the eucalyptus tree into Ethiopia in response to a critical wood shortage within the rapidly expanding capital city, Addis Ababa (Amare 2010; Yitebitu 2010). Then, the mandate of environmental protection was given to the then ministry of agriculture, which took the responsibility to manage all forest resources. However, studies showed that natural resource management has received special attention and institutional form only since the 1970s, which is after the drought incident in the northern part of Ethiopia (Amsalu 2006; Zeleke et al. 2006; Haregeweyn et al. 2015). In the 1970s, international organizations supported the implementation of land conservation programs to reduce soil degradation, enhance agricultural productivity, reduce food insecurity, and reduce poverty (Gashaw et al. 2014). However, special emphasis was given to the construction of physical structures such as terraces and stone bunds, which were not effective by itself (Alemu and Kidane 2014).

Sustainable land management requires the continuous effort of all stakeholders sustainably for its effectiveness. As such, land management has been one of the central agendas of the Ethiopian government for the last few years. Consequently, there was an effort to actively engage the rural and the urban community for implementing land management practices in the country in general and in the basin in particular. The Amhara region is one of Ethiopia where land management practices have been extensively implemented. This is due to the fact that the majority of its land mass drains to the Abbay basin, which requires land management, and the region as a whole is under high population pressure requiring proper management. As such, the federal government is also promoting land management efforts in the Amhara region to reduce the sedimentation and siltation effects on the Grand Ethiopian Renaissance Dam.

The land management practices that have been implemented in Ethiopia include the indigenous and introduced with different degrees of acceptability, areal coverage, and benefits (Zeleke et al. 2006). The indigenous knowledge includes the complex practices and judgments made by local people based on experience passed from one generation to the next with few changes, modifications, and assimilating new ideas (Oudwater and Martin 2003). The widely used and preferred indigenous land management practices include terracing, agroforestry (woodlot), grass strips, zero grazing, trash lines, minimum tillage, contour ploughing, animal manure, fallowing, and biological or agronomic methods such as crop rotation, cereal-legume intercropping, mulching, and residues of crop production (Ayalew et al. 2009).

The physical structure of land management technologies with standard length, width, and height are recommended in Ethiopia (Blata 2010), which includes soil/stone bunds, bench terraces, inorganic fertilizer, check dams, waterways, cut-off drains, area closure and closed gullies, hillside terraces, fanya juu, and organic fertilizer.

Due to aggravated land and soil degradation, land management has become a way of survival in recent decades, although it has been used by Ethiopian farmers since ancient times. According to Haregeweyn et al. (2015), for enhancing agricultural development and rural livelihood and minimize land degradation, various conservation strategies have been introduced since the 1960s. The 1970s and 1980s, the period when large-scale conservation projects were underway, were remarkable in the history of conservation in the country. After the incidence of a devastating drought and famine in the northern and eastern parts of Ethiopia in the 1970s and 1980s, the intervention of the public in soil conservation practices was started. Hence, the government of Ethiopia, in collaboration with several development partners like the World Food Program, World Bank, and African Development Bank, made huge investments in promoting SLM practices over the past decades (Bewket 2007b).

Over decades, several soil and water conservation activities have been designed and implemented in highly degraded areas and food-insecured areas mainly through food-for-work programs (Asrat et al. 2004; Moges and Taye 2017). The physical soil conservation structures such as bunds, terraces, and check dams in cultivated fields, as well as planting trees in hillside areas were given more attention over the years (Bewket 2007a). Similarly, Danano (2010) reported that the main SLM practices implemented in the Ethiopian highlands include physical measures like bunds, terraces, check dams, and in situ soil moisture conservation structures, as well as biological measures, such as planting trees and area enclosures for natural regeneration. Since 2010, most SLM activities have been realized through community mass mobilization at watershed levels (Wolancho 2015; Adego et al. 2018). Mass mobilization is a strategy pursued to mobilize all land users in a community (men, women, and youth) to collectively address soil erosion and the declining productivity of farmlands (Danano 2010; Wolancho 2015).

Although the achievements were remarkable in quantitative terms, the impacts of land management on reducing land degradation are below expectations, 2006and land degradation continued to be serious challenges in the basin (Admassie 2000, cited in Aklilu). The ineffectiveness of the land conservation measures could be due to several reasons. One of the reasons is that the physical structures were often poorly adopted, hardly maintained and sometimes even removed by farmers (Kassie et al. 2010). According to Aklilu (2006), the failure of land conservation measures was due to the introduction of measures that did not consider local conservation and farming practices and in many cases did not fit in with traditional methods, and since interventions normally include activities such as reforestation and terrace construction, they are generally characterized by high initial costs that poor farmers cannot afford and by benefits that only become apparent in the long run. Moreover, the extensive and uniform application of similar soil and water conservation (SWC) measures disregarded local agro-climatic and socioeconomic variations.

Research over the past decades has revealed that the adoption and implementation of sustainable land management practices by Ethiopian farmers are constrained by personal, socioeconomic, biophysical, and institutional factors (Asrat et al. 2004; Shiferaw et al. 2009; Adimassu et al. 2012; Abebe and Sewnet 2014; Teshome et al. 2016). Personal factors include age, education, perceived effects of erosion, and perception of technological attributes (Tadesse and Belay 2004; Abebe and Sewnet 2014). The socioeconomic factors that affect adoption of conservation measures include labor, farmland size, social capital, access to information, and sources of income (Amsalu and de Graaff 2007; Abebe and Sewnet 2014; Miheretu and Yimer 2017).

Moreover, slope, soil type, soil fertility, soil depth, topography and rainfall major biophysical factors seriously affect adoption of conservation measures (Tadesse and Belay 2004; Amsalu and de Graaff 2007; Shiferaw et al. 2009; Miheretu and Yimer 2017). Likewise, a top-down approach, lack of consideration of farmers' priority needs, poor extension services, and farming systems, and a high emphasis on the promotion of physical structures are some of the institutional factors constraining SLM (Bewket 2007b; Shiferaw et al. 2009; Kassie et al. 2010; Weldemariam et al. 2013).

Determinants of Land Management Practices

The rural community in the Abbay basin has long time experiences of implementing different land management practices. The rugged topography of the environment is the major factor pushing the rural community to implement land management practices, which enables them to maximize the productivity of their farmland. However, effectiveness and the status of land management practice in the Abbay basin are far below the needy. Different scholars have identified various determinant factors that impede the rural community from applying land management practices. For the sake of convenience, this paper categorized the major determinant factors into three classes as follows.

Economic Factors

The major requirement of implementing soil conservation practices includes necessary labor, capital, and technological inputs, and if they notice a fast economic benefit (Shiferaw 2011). Individuals with little current income and an inability to obtain capital for conservation investments may not be keen or able to sacrifice income to maximize expected net returns over a long period. Similarly, individuals in undefined economic circumstances will be motivated to use short planning horizons because they are unable to predict future costs and prices (Tesfaye et al. 2014). As poor farmers generally own less land, they are more often involved in off-farm activities such as petty trade. This can decline their interest in investing in soil conservation practices. According to Hagos and Holden (2006), small farm holdings and land fragmentation may challenge farmers' interest in undertaking some kind of land improvement. For example, farmers may find the cost of transportation of manure or other organic materials to distant and small plots not worth the significant effort needed. In addition, an investment that can be easily damaged by free-ranging livestock or subject to theft (such as trees) is less likely to be made far from the household where it is different to protect them.

Policy and Institutional Factors

A suitable policy environment is a precondition for being able to implement a natural resource management process that satisfies the objectives specified by the profits of interest. Government policies are not translated into action unless there is the political will to make them work. Therefore, the situation in many countries today is that plans are made for the conservation of natural resources, but they have little practical effect. Generally, many developing countries have a large portion of land that was previously reserved. As authoritarian management has declined, population pressure and land hunger have increased. Therefore, the chance of escaping punishment for illegal encroachment on reserved land. The constraint of land was often to preserve the income or power of the ruling elites; there are also many examples where the land was deliberately withheld from the settlement because it was ecologically unsuitable.

Sociocultural Factors

Land was free and enough during the past couple of decades when population of the country was small. After population increases in population just means to bring more land into use. Obtaining this new land is not a simple task, and it resulted in the expansion of farming activities to erosion-prone marginal areas, serious deforestation, and a decrease in fallow period and continuous cultivation (Habtamu 2006; Hussen 2006). Many people in developing countries can barely eke out a living from their land through hard work such as hard tillage. They know that traditional management has kept them and their predecessors alive and that they have nothing to spare for gambling on a new method. It is difficult for them to change their techniques even for immediate benefits such as higher yields and less soil loss. It is still more difficult for them to adopt a practice that requires an investment, especially if the benefits are delayed or disturbed over several years.

The establishment of conservation practices under such conditions requires a reliable guarantee that those people will not starve to death if the new practice fails. Short-term tenancy prevents the adoption of many desirable practices. Theoretically, landowners should be willing to invest in sound long-term practices, but many owners are too far removed from the land to realize what practices are needed. Short-term tenancy makes it easy for both tenants and owners to overlook problems when those problems reach a critical stage. Another constraint to the sustainability of conservation practices is the social significance of the cattle population. The role played by livestock in adding pressure on the land resource base varies greatly from one country to another. In areas where cattle are symbols of status and evidence of wealth and have religious significance, the focus is given to quantity rather than quality. Associated with this are low standards of livestock management and low levels of production. The total effect of these combined factors is unnecessary stress on the ecological system.

Sustainable Land Management Practice

Currently, sustainable land management is emerging as a new concept. This is because the ever-increasing population is posing pressure on land resources and demanding high food production. Sustainable land management is simply about people looking after land - for the present and the future. The main objective of sustainable land management is thus to integrate people's coexistence with nature over the long-term so that the provisioning, regulating, cultural, and supporting services of ecosystems are ensured (Liniger et al. 2011).

It is indicated in the previous discussion that land degradation is persistent and continues at a high rate and extent. For such a sustaining problem, it is essential to take sustainable measures. Therefore, the concept of sustainable land management seems to emanate from this very nature of the problem itself. According to Ogbazghi et al. (2011), the overall intent of the concept of "sustainable land management" is to bring sustainable use of land resources through the reduction (stagnation if possible) of land degradation. In a wider sense, sustainable land management is expected to bring triple-win solutions: increased land productivity, improved livelihoods, and improved ecosystems being environmentally friendly (Liniger et al. 2011).

Generally, the concept of "sustainable land management" is defined in different ways by different sources. For instance, Herweg et al. (1998) defined it as "the use of land resources such as soils, water, animals and plants for the production of goods—to meet changing human needs—while assuring the long-term productive potential of these resources and the maintenance of their environmental functions". On the other hand, sustainable land management has been defined by Liniger et al. (2011) as 'the adoption of land use systems that, through appropriate management practices, enables land users to maximize the economic and social benefits from the land while maintaining or enhancing the ecological support functions of the land resources". Similarly, Mitiku et al. (2006) stated sustainable land management as the use of land to meet changing human needs (agriculture, forestry, conservation) while ensuring long-term socioeconomic and ecological functions of the land. Even if these definitions seem different, the central themes included in each definition are more or less similar.

Land Management and Productivity: Case in the Bechet Watershed

The productivity of agricultural land in the Abbay basin is found to be a function of various interconnected factors. These factors are directly or indirectly associated with land ownership modalities, land management practices, and other economic factors, such as land holding size, land fertility, land fragmentation, access to assets (livestock, farming tools, labor force), and access to services (extension and credit).

The land holding system of the rural community is currently improved through the recent land registration and certification program, which slightly improved land tenure security and thereby land productivity. The land certificate has motivated landholders to engage in various sustainable land and soil management practices. Consequently, output (quintal/hectare) from the farming land has shown a significant increment after the land certification and land management practices. The other determining factor for farming land productivity is holding size. The majority of the farmlands are smaller in size and highly fragmented, which are real threats to land productivity.

Prior studies regarding the role of land management practices in land productivity have mixed results. In their study in the Tigray region, Pender and Gebremedhin (2006) found higher crop yields on farmlands treated with land management practices compared with plots without such practices. Similarly, Kassie et al. (2007) obtained a positive association between land management practices and crop output. On the other hand, a study by Kassie et al. (2009) found an inverse association between land management practices and crop yield in a rain-fed part of Ethiopia. In a study by Mengistie (2023), a paired-samples t-test was run to see the mean difference in yield amount for different crops before and after the implementation of land management practices in the Abbay Basin. As indicated in Table 16.2, the mean output difference after land management practices is found to be the highest for teff and sorghum, which are 1.98 and 1.83 quintals, respectively. The result of an independent t-test also indicated that the mean output differences in teff and sorghum before and after land management practices were statistically significant at $t\ (353) = 26.717$, $p = 0.000$, and $t\ (164) = 15.588$, $p = 0.000$, respectively.

Generally, the mean yield difference for pea, maize, and wheat before and after land management practice was higher than 1 quintal, while for barley, bean, vetch, chickpea, and Niger seed, it was lower than 1 quintal. Land management intervention resulted in the lowest mean output difference for Niger seed, which was only 0.64 quintal and statistically significant at $t\ (92) = 12.649$, $P = 0.000$. Even though the amount of mean output difference before and after land management practices may vary, it is evident that there is an improvement in mean output difference for all crops considered (Table 16.2).

Agricultural land productivity is the result of various factors. Among these, the availability of important assets such as livestock, farming tools, and the farming labor force is crucial for its success. In these regards, the study indicated that the farming community has limitations in possessing the required amount of livestock, farming tools, and farming labor, although there is a disparity between male-headed and female-headed households. Comparatively speaking, it is possible to state that male-headed households are by far in a better position regarding access to such essential production assets as livestock, farming tools, and farming labor. Hence, the study indicated that the differences in farm output between male-headed and female-headed households are partly associated with differences in asset possession.

The productivity of farming land is also a function of accessing extension and credit services. The study assessed that almost all the farming community has access to both extension and credit services. However, the study identified that there are some knowledge and skill gaps among extension workers, which led few farmers not to accept some extension packages. On the other hand, the study showed that the farming community has access to informal financial institutions and traditional money lenders, which are, however, mentioned by the rural community as highly exploitative for the unfairly high interest they collect. Formal financial institutions (banks) are closed for farmers to access credit but open only for saving. In summary, the availability of extension and credit services with all their limitations is helpful in increasing farmland productivity.

A study by Mengistie (2023) in the Abbay basin focused on teff and sorghum output from farmlands because these are the most dominant cereal crops in the study area. Hence, several environmental, institutional, and socioeconomic factors are found to have crucial roles in determining the output of teff and sorghum crops. The study revealed that the productivity of teff and sorghum from a given farmland is highly influenced by factors such as sex, age, family size, education level, sharecropping, land rent, fertility level of the land, slope of the land, number of oxen owned, amount of chemical fertilizer used, participation in off/nonfarm activities, access to credit service and possession of farming tools.

Table 16.2 Difference in output for different crops before and after land management practices

Paired samples test

Output difference for major crops before and after LMP (Qt)	Mean Area cultivated (Ha)	Mean output after LMP (Qt.)	Mean output before LMP (Qt.)	Paired Differences			t	df	Sig. (2-tailed)
				Mean Output diff. (Qt)	Std. Dev	Std. Error Mean			
The output of barely after—before LMP	0.2067	3.833	3.060	0.773	0.3153	0.068	11.247	20	0.000
The output of wheat after—before LMP	0.2484	4.588	3.560	1.028	0.4922	0.045	22.589	116	0.000
The output of teff after—before LMP	0.6801	9.801	7.823	1.978	1.3930	0.074	26.717	353	0.000
The output of maize after—before LMP	0.7973	4.808	3.603	1.205	0.9036	0.048	25.058	352	0.000
The output of sorghum after—before LMP	0.2211	8.336	6.509	1.827	1.5057	0.117	15.588	164	0.000
The output of bean after—before LMP	0.2524	3.612	2.727	0.885	0.5360	0.058	15.037	82	0.000
The output of pea after—before LMP	0.2329	3.933	2.878	1.055	0.6144	0.092	11.524	44	0.000
The output of vetch after—before LMP	0.2189	3.938	2.942	0.996	0.6413	0.060	16.577	113	0.000
The output of chickpeas after—before LMP	0.2489	4.215	3.295	0.920	0.6060	0.082	11.258	54	0.000
The output of Niger seed after—before LMP	0.2067	2.508	1.863	0.645	0.4911	0.051	12.649	92	0.000

Source Mengistie (2023)

Concluding Remarks

Land resources play a pivotal role in developing countries such as Ethiopia, where the majority of the population is engaged in agriculture. The Abbay basin in Ethiopia is characterized by rugged terrain resulting from basic geomorphic processes. Hence, the land resources in this basin are experiencing land degradation. Aware of the impacts of the land degradation process, the rural community is engaged in various land management practices that are appreciable.

Land management practices in the Abbay basin have multiple positive impacts. First, the productivity of agricultural lands will be enhanced if they are treated with land management practices. Second, the application of land management practices reduces the rate and magnitude of soil erosion, which in turn significantly safeguards the Great Ethiopian Renaissance Dam from siltation. Hence, it is difficult for the government of Ethiopia and the rural population in the Abbay basin to strengthen the land management practices in the basin.

References

Abebe ZD, Sewnet MA (2014) Adoption of soil conservation practices in north Achefer district, northwest Ethiopia. Chinese J Popul Resour Environ 12(3):261–268

Adego T, Simane B, Woldie GA (2018) Sustainability, institutional arrangement and challenges of community-based climate smart practices in northwest Ethiopia. Agriculture and Food Security

Admassie Y (2000) Twenty years to nowhere. Property rights, land management, and conservation in Ethiopia. Red Sea Press: Asmara

Adimassu Z, Kessler A, Hengsdijk H (2012) Exploring determinants of farmers' investments in land management in the Central Rift Valley of Ethiopia. Appl Geogr 35:191–198

Aklilu A (2006) Caring for the land best practices in soil and water conservation in Beressa watershed, highlands of Ethiopia. Tropical Resource Management Papers, No 76

Alemu B, Kidane D (2014) The implication of integrated watershed management for rehabilitation of degraded lands: case study of Ethiopian Highlands. J Agric Biodivers Res 3(6):78–90

Alemu GT, Berhanie Ayele Z, Abelieneh Berhanu A (2007) Effects of land fragmentation on productivity in Northwestern Ethiopia. Adv Agric

Amare G (2010) Eucalyptus farming in Ethiopia: the case of Eucalyptus farm and village woodlots in Amhara Region

Amsalu A, de Graaff J (2007) Determinants of adoption and continued use of stone terraces for soil and water conservation in an Ethiopian highland watershed. Ecol Econ 61:294–302

Amsalu A (2006) Caring for the land: best practices in soil and water conservation in beressa watershed, highlands of Ethiopia, Thesis, Wageningen UR

Askale T (2005) Land registration and women's land rights in Amhara Region, Ethiopia. Securing land rights in Africa. Research Report 4. IICR. Addis Ababa, Ethiopia

Asrat P, Belay K, Hamito D (2004) Determinants of farmers' willingness to pay for soil conservation practices in the southeastern highlands of Ethiopia. Land Degrad Dev 15:423–438

Assefa B (2010) The effect of rural land certification in securing land rights: a case of Amhara Region, Ethiopia. MSc Thesis submitted to the International Institute of Geo-Information Science and Earth Observation. ITC, Netherlands

Assemu T, Shigdaf M (2014) The effect of land degradation on farm size dynamics and crop-livestock farming system in Ethiopia: a review. J Soil Sci 4:1–5

Atsbaha G, Tessema B (2012) A review of Ethiopian agriculture: roles, policy and small-scale farming systems
Ayalew A, Deininger K, Holden S, Zevenbergen J (2009) Rural land certification in Ethiopia: process, initial impact, and implications for other African countries. World Dev 36(10):1786–1812
Bai ZG, Dent DL, Olsson L, Schaepman ME (2008) Global assessment of land degradation and improvement. 1. Identification by remote sensing. International Soil Reference and Information Centre (ISRIC), Wageningen, The Netherlands
Barbier EB (2000) The economic linkages between rural poverty and land degradation: some evidence from Africa. Agric Ecosyst Environ 82(1):355–370
Barnett HJ, Morse C (2013) Scarcity and growth: the economics of natural resource availability. Routledge
Belay KT, Van Rompaey A, Poesen J, Van Bruyssel S, Deckers J, Amare K (2014) Spatial analysis of land cover changes in eastern Tigray (Ethiopia) from 1965 to 2007: are there signs of a forest transition? Land Degradation and Development. https://doi.org/10.1002/ldr.2275
Berhanu A, Fayera A (2005) Research report 3 land registration in Amhara Region, Ethiopia. Securing Land Rights in Africa. Addis Ababa, Ethiopia
Berry L (2003) Land degradation in Ethiopia: its extent and impact. A study commissioned by the GM with WB support
Bewket W (2003) Towards integrated water shade management in highland Ethiopia the chemoga watershed case study. Tropical Resource Management. Paper, No 44/2003
Bewket W (2007a) Soil and water conservation intervention with conventional technologies in north-western highlands of Ethiopia: acceptance and adoption by farmers. Land Use Policy
Bewket W (2007b) Soil and water conservation intervention with conventional technologies in north-western highlands of Ethiopia: acceptance and adoption by farmers. Land Use Policy 24(2):404–416. https://doi.org/10.1016/j.landusepol.2006.05.004
Blata ST (2010) Land degradation and farmers' perception: the case of Limo Woreda, Hadya Zone Of SNNPR, Ethiopia (doctoral dissertation, school of graduate studies environmental science program thesis submitted to school of graduate studies, AAU)
CSA (2008) Summary and Statistical Report of the 2007 population and housing census results. United Nations Population Fund (UNFPA). Addis Ababa, Ethiopia
Danano D (2010) Sustainable land management technologies and approaches in Ethiopia. Sustainable land management project, natural resources management sector, ministry of agriculture and rural development of the Federal Democratic Republic of Ethiopia. Addis Ababa
Davies SJ (2008) The political economy of land tenure in Ethiopia. Doctoral dissertation, University of St Andrews
Deininger K (2003) Land policies for growth and poverty reduction. A World bank Policy Research Report
Ehui SK, Ahmed MM, Berhanu G, Benin SE, Nin Pratt A, Lapar MaL (2003) 10 Years of livestock policy analysis. Policies for improving productivity, competitiveness and sustainable livelihoods of smallholder livestock producers. ILRI (International Livestock Research Institute), Nairobi, Kenya
Eyasu E (2003) National assessment on environmental roles of agriculture in Ethiopia. Unpublished Research Report Submitted to EEA, Addis Ababa
Gashaw T, Bantider A, Silassie HG (2014) Land degradation in Ethiopia: causes, impacts and rehabilitation techniques. Environ Earth Sci 4(9):98–104
Genanew BW, Alemu M (2012) Investments in land conservation in the Ethiopian highlands: a household plot-level analysis of the roles of poverty, tenure security, and market incentives. Int J Econ Financ 4(6). https://doi.org/10.5539/ijef.v4n6p32
Gessesse B (2013) Characterization and modelling of landscape transformation for optimizing agricultural land use in the ethiopian highlands: a case study of modjo watershed. PhD Thesis, Addis Ababa University, Ethiopia and Friedrich-Alexander University of Erlangen-Nürnberg, 91054 Erlangen, Germany

Habtamu E (2006) Adoption of physical soil and water conservation structures in Anna Water Shade, Hadiya Zone Ethiopia, A Thesis Submitted to School of Graduate Studies, Institute of Regional and Local Development Studies, AAU

Hagos F, Holden S (2006) Tenure security, resource poverty, public programs, and household plot-level conservation investments in the highlands of northern Ethiopia. Agric Econ 34(2):183–196

Hans H, Solomon A, Amare B, Berhanu D, Eva L, Brigitte P, Birru Y, Gete Z (2010) Global change and sustainable development: land degradation and sustainable land management in the highlands of Ethiopia. https://doi.org/10.13140/2.1.3976.5449

Herweg K, Steiner K, Slaats J (1998) Sustainable land management—Guidelines for impact monitoring. Workbook and Toolkit. Bern

Haregeweyn N, Tsunekawa A, Nyssen J, Poesen J, Tsubo M, Tsegaye M, Tegegne F (2015) Soil erosion and conservation in Ethiopia: a review. Prog Phys Geogr 39(6):750–774

Holden S, Shiferaw B, Pender J (2004) Nonfarm income, household welfare, and sustainable land management in a less-favoured area in the Ethiopian highlands. Food Policy 29:369–392

Hussen H (2006) Land use change and challenges of land degradation in Adaba Area, Bale Zone. A thesis submitted to school of graduate studies, Institute of Regional and Local Development Studies, AAU

IFSP (Integrated Food Security Program) (2004) Status report on the use of Vetiver Grass for soil and water conservation by GTZ IFSP South Gonder, Ethiopia. Integrated Food Security Program South Gonder. Bureau of Agriculture, Amhara Region, Bahir Dar

ILRI (International Livestock Research Institute) (2000) Policy for sustainable land management in the highlands of Ethiopia. Addis Ababa, Ethiopia

Jagger P, Pender J (2003) The role of trees for sustainable management of less-favored lands: the case of eucalyptus in Ethiopia. Forest Policy Econ 3(1):83–95

Jansen H, Hengsdijk H, Legesse D, Ayenew T, Hellegers P, Spliethoff P (2007) Land and water resources assessment in the Ethiopian Central Rift Valley. Alterra Rep 1587:81

Kassie M, Pender J, Yesuf M, Kohlin G, Bluffstone R, Mulugeta E (2007) Impact of soil conservation on crop production in the Northern Ethiopian Highlands. International Food Policy Research Institute, Washington, DC

Kassie M, Holden S, Köhlin G, Bluffstone R (2009) Economics of soil conservation adoption in high-rainfall areas of the Ethiopian Highlands. Addis Ababa, Ethiopia: Environment for Development

Kassie M, Zikhali P, Pender J, Köhlin G (2010) The economics of sustainable land management practices in the Ethiopian highlands. J Agric Econ 61:605–627

Kirui OK, Mirzabaev A (2014) Economics of land degradation in Eastern Africa (No. 128). ZEF Working Paper Series. Center for Development Research (ZEF), University of Bonn, Germany

Lakew D, Menale K, Benin S, Pender J (2000) Land degradation and strategies for sustainable development in the Ethiopian highlands: Amhara Region. Socio-Economics and policy research working Paper 32. ILRI (International Livestock Research Institute), Nairobi, Kenya

Liniger H, Mekdaschi R, Hauert C, Gurtner M (2011) Sustainable land management in practice: guidelines and best practices for Sub-Saharan Africa, FAO, Rome, Italy

Melaku T (2013) Sustainable land management program in Ethiopia: linking local REDD+ projects to national REDD+ strategies and initiatives. Power point presentation made by National Program Coordinator of SLMP. April 29–May 1, 2013, Hawassa, Ethiopia

Mengistie M (2023) Linkages among ownership modality, sustainable management and productivity of agricultural land in Dejen woreda, North-West Ethiopia. PhD dissertation (unpublished)

Miheretu and Yimer (2017) Determinants of farmers' adoption of land management practices in Gelana subwatershed of Northern highlands of Ethiopia. Ecol Process 6:19

Mitiku H, Herweg KG, Stillhardt B (2006) Sustainable land management: a new approach to soil and water conservation in Ethiopia

Moges DM, Taye AA (2017) Determinants of farmers' perception to invest in soil and water conservation technologies in the North-Western Highlands of Ethiopia. Int Soil Water Conserv Res 5:56–61

Mulatie M, Tsegaye S, Mulu G, Bayleyegn A, Assefa M (2015) GIS and remote sensing-based forest resource assessment, quantification, and mapping in Amhara Region, Ethiopia. Landscape Dynamics, Soils and Hydrological Processes in Varied Climates, Switzerland. https://doi.org/10.1007/978-3-319-18787-7_2

Ogbazghi W, Stillhardt B, Herweg KG (2011) Sustainable land management—a textbook with a focus on Eritrea, Geographica Bernensia and Hamelmalo Agricultural College, Bern and Keren

Ogendo O (2006) Workshop on land tenure security for poverty reduction in Eastern and Southern Africa. Organized by IFAD/UNOPS/Ministry of Lands, Housing and Urban Development, Government of Uganda. Kampala, Keynote Address, 27–29 June

Oudwater N, Martin A (2003) Methods and issues in exploring local knowledge of soils. Geoderma 111(3):387–401

Pender J, Gebremedhin B, Benin S, Ehui S (2001) Strategies for sustainable development in the Ethiopian highlands. Am J Agr Econ 83(5):1231–1240

Pender J, Gebremedhin B (2006) Land management, crop production, and household income in the highlands of Tigray, Northern Ethiopia: an econometric analysis

Rudel TK, Schneider L, Uriarte M, Turner BL, DeFries R, Lawrence D, Grau R (2009) Agricultural intensification and changes in cultivated areas, 1970–2005. Proc Nat Acad Sci 106(49):20675–20680

Samson L, Frehiwot M (2014) Spatial analysis of cattle and shoat population in Ethiopia: growth trend, distribution and market access. Springerplus 2014(3):310. https://doi.org/10.1186/2193-1801-3-310

Samuel B, Pender J, Simeon E (2002) Policies for sustainable land management in the east african highlands. In: EPTD Workshop Summary Paper NO. 13. Summary of Papers and Proceedings of the Conference held at the United Nations Economic Commission for Africa, Addis Ababa, Ethiopia

Shiferaw A (2011) Estimating soil loss rates for soil conservation planning in the Borena Woreda of South Wollo Highlands, Ethiopia. J Sustain Dev Africa 13(3):87–106

Shiferaw BA, Okello J, Reddy RV (2009) Adoption and adaptation of natural resource management innovations in smallholder agriculture: reflections on key lessons and best practices. Environ Dev Sustain 11:601–619

Tadesse M, Belay K (2004) Factors influencing adoption of soil conservation measures in Southern Ethiopia: the case of Gununo area. J Agric Rural Dev Trop Subtrop 105:49–62

Temesgen G, Amare B, Silassie HG (2014) Land degradation in Ethiopia: causes, impacts and rehabilitation techniques. J Environ Earth Sci 4(9):98–104

Tesfa A, Mekuriaw S (2014) The effect of land degradation on farm size dynamics and crop livestock farming system in Ethiopia: a review. Open J Soil Science

Tesfahun F, Osman A (2003) Challenge and prospects of food security in Ethiopia. In: Proceedings of the food security conference. UNNCC, Addis Ababa, Ethiopia, 13–15 August

Tesfaye A, Negatu W, Brouwer R, Van der Zaag P (2014) Understanding soil conservation decision of farmers in the Gedeb watershed, Ethiopia. Land Degradation Development 25(1):71–79

Teshome A, de Graaff J, Kassie M (2016) Household-Level determinants of soil and water conservation adoption phases: evidence from North-Western Ethiopian highlands. Environ Manage 57:620–636

The Global Mechanism (2007) Increasing finance for sustainable land management. The Global Mechanism of the UNCCD-Via Paolo di Dono 44-00142 Rome, Italy. Available online at www.global-mechanism.org

USAID (2000) Amhara national regional state food security research assessment report. Collaborative Research Work, Addis Ababa, Ethiopia

Von Braun J, Gerber N, Mirzabaev A, Nkonya EM (2013) The economics of land degradation (No. 147910). University of Bonn, Center for Development Research (ZEF), Bonn, Germany

Wagayehu B (2003) Economics of soil and water conservation: theory and empirical application to subsistence farming in the eastern Ethiopian highlands. In: Doctoral thesis, Swedish University of Agricultural Sciences, Uppsala, Sweden

Weldemariam D, Kebede M, Taddesse M, Gebre T (2013) Farmers' perceptions and participation on Mechanical soil and water conservation techniques in Kembata Tembaro Zone: the Case ofKachabirra Woreda, Ethiopia. Int J Adv Struct Geotech Eng 2(04):2319–5347

Wolancho KW (2015) Evaluating watershed management activities of campaign work in Southern nations, nationalities and peoples' regional state of Ethiopia. Environ Syst Res

Yitebitu M (2010) Eucalyptus trees and the environment: a new perspective in times of climate change

Zeleke G, Kassie M, Pender J, Yesuf M (2006) Stakeholder analysis for sustainable land management (SLM) in Ethiopia: assessment of opportunities, strategic constraints, information needs, and knowledge gaps. In Environmental Economics Policy Forum for Ethiopia (EEPFE), Addis Ababa

Zerga B (2016) Land resource, uses, and ownership in ethiopia: past, present and future. Int J Sci Res Eng Technol 2(1):2395–2566

Chapter 17
Lesson Learned from Over 50 years of Watershed Management in the Abbay Basin, Ethiopia

Ermias Teferi, Tibebu Kassawmar, Woldeamlak Bewket, and Gete Zeleke

Abstract The chapter discusses the efforts made by the Government of Ethiopia and development partners to promote watershed management (WM) interventions that incorporated various technical, institutional, and policy improvements in Ethiopia in general and the Abbay River Basin in particular. Over the past five decades, Ethiopia has shifted toward more participatory and livelihood-focused approaches, as well as adopting an integrated WM paradigm. These changes demonstrate a willingness to learn from previous experiences and the advantages of adapting approaches based on lessons learned along the way. While these conservation interventions have yielded ecological benefits, there have been notable shortcomings in large-scale implementation. The impact of WM interventions, particularly terraces, on surface runoff, soil loss, and grain yield is analyzed. Although terraces have shown success in reducing soil loss, erosion rates on treated plots are still above tolerable limits, indicating the need for further improvement or integration with other technologies. While terraces are effective at reducing runoff at the plot scale, catchment-scale results vary due to multiple contributing factors. Additionally, the review explores the impact of WM measures on crop yield, highlighting the success of various practices such as conservation tillage in achieving uniform soil moisture distribution and reducing waterlogging effects. Integration of different watershed management technologies like conservation tillage and terracing is recommended to achieve the desired outcomes. Overall, the review highlights the need for further research, catchment-scale experiments, and scenario analysis to better understand long-term dynamics and optimize WM practices.

E. Teferi (✉)
School of Development Studies, Addis Ababa University, Addis Ababa, Ethiopia
e-mail: ermias.teferi@aau.edu.et

E. Teferi · T. Kassawmar · W. Bewket · G. Zeleke
Water and Land Resource Center, Addis Ababa, Ethiopia

W. Bewket
Department of Geography and Environmental Studies, Addis Ababa University, Addis Ababa, Ethiopia

T. Kassawmar
School of Earth Sciences, Addis Ababa University, Addis Ababa, Ethiopia

Keywords Watershed management · Abbay · Surface runoff · Conservation tillage · Soil loss · Grain yield

Introduction

Land degradation in Ethiopia poses a significant concern as it adversely affects food production, water supply, energy supply, and overall ecosystem services. Rural development and poverty reduction in Ethiopia are considerably hindered by the extensive land degradation present in the country (Sonneveld and Keyzer 2003; Hurni et al. 2015). According to Teferi et al. (2023), most part of the country's land surface (83%) falls in the critical or fragile land degradation classes. Of the total degraded area, about 22% of the country's land surface is exposed to high and very high land degradation levels. For the Abbay basin, studies of erosion rates have produced widely varying estimates, with more recent research suggesting an annual sediment load of estimates of erosion rates vary considerably across different studies, with recent work indicating a sediment load of 30.5 tons/ha/yr (WLRC 2019).). This will definitely affect the lifespan of the Grand Ethiopian Renaissance Dam (GERD).

Land degradation affects the livelihood security of millions of rural Ethiopians by reducing soil fertility and agricultural yields (Sonneveld and Keyzer 2003; Hurni et al. 2015). Studies of erosion rates have produced widely varying estimates, with more recent research suggesting an annual sediment load of estimates of erosion rates vary considerably across different studies, with recent work indicating a sediment load of 30.5 tons/ha/yr (WLRC 2019). In the last 50 years, the Government of Ethiopia and various development partners have implemented a growing number of initiatives aimed at tackling land degradation issues in the country. These initiatives have included the establishment of new programs/initiatives, institutions, strategies, and guidelines. Nevertheless, numerous challenges still require attention. Therefore, it becomes imperative to thoroughly document the valuable insights gained from previous WM interventions.

A watershed refers to an area that channels rainfall and surface water to a shared outlet. This makes it a logical hydrological unit for integrated management of land and water resources. The watershed-based land resource management approach is a coordinating framework for managing land and water resources through the efforts of different stakeholders with the goal of tackling the most pressing issues within watersheds, taking into account both groundwater and surface water resources. The watersheds provide a hydrological boundary to focus efforts on the highest priorities holistically. The water flowing in a watershed interconnects upstream and downstream areas, and provides livelihood support to farmers holding unequal use rights, making people, the ecosystem, and animals integral parts of watersheds (Wani et al. 2009).

In Ethiopia, large-scale programs of restoration of degraded lands were started in the mid-1970s and have continued to date with varying scales and foci. The major

WM-related efforts in Ethiopia include the Food-for-Work (FFW) program (1973–2002), Managing Environmental Resources to Enable Transitions to Sustainable Land Use project (MERET 2003–2011), the Productive Safety Net Program (PSNP 2005–present), Community Mobilization through free-labor days (1998–present), the Sustainable Land Management Project (SLMP 2008–2018), Resilient Landscapes and Livelihoods Project (RLLP 2019–2024), and the Climate Action through Landscape Management project (CALM 2019–2024). The important components of the WM interventions implemented in the Abbay basin involved: (a) construction of ex-situ water harvesting structures such as farm ponds for supplementary irrigation, (b) construction of in-situ water harvestings structures such as terraces, bunds, check dams, and cut-off drains, (c) implementation of agronomic measures such as conservation tillage and mulching, and (d) afforestation and revegetation of fragile and hillside areas.

Although these efforts have resulted in many ecological benefits, the initiatives had some serious shortcomings. During the 1980s and in the early 1990s, the focus of almost all interventions was on reducing soil erosion through physical soil and water conservation measures, and the participatory approach was given less emphasis. The absence of short-term benefits in implementing WM practices by poor farmers has been noted as the major limitation of past efforts (Adimassu et al. 2017). The latest WM programs have a broader goal of tackling challenges like food insecurity and poverty, beyond just managing water resources. Improving the livelihoods of local communities is highlighted by realizing the fact that in the absence of these, sustainable management of land and water resources would be elusive. In light of these considerations, watershed programs have expanded beyond just soil and water conservation. They now encompass activities like improving productivity through homestead development, horticulture, livestock rearing, and gender equality initiatives.

This chapter discusses the lessons learned from over 50 years of WM activities in Ethiopia and identifies gaps in addressing the problem of land degradation. It consists of seven sections. This section introduces the issue. The next section explains about the approach and settings. The third section discusses the WM programs and initiatives in Ethiopia. The fourth section explores WM in the Abbay basin. The fifth section describes the hydrological and agronomic impacts of WM. The sixth section deals with the downstream impacts and implications of WM interventions. The final section summarizes the lessons learned.

Approaches and Setting

Approaches

In order to gain a thorough understanding of watershed management, a desk review procedure was followed. The review focuses on peer-reviewed publications in journals and other relevant technical and programmatic publications ("grey" literature) on watershed management. During the initial stage of review of the literature, three inclusion criteria were applied: (1) Search criteria indicated in the search string in Table 17.1; (2) the publication was written in English between 1970 and 2022; and (3) the publication focuses on the Abbay basin.

When collecting the data, these criteria were used to create search terms to find publications related to watershed management and impact indicators (Table 17.1). Both academic literature and relevant technical and programmatic publications (grey literature) were targeted. The search string summarized in Table 17.1 was used to search peer-reviewed literature (i.e., scientific journals) from the Google Scholar database. The search was limited to the period from January 1970 to 2022. The abstract, title, and keywords of each publication were used as retrieval/search units. In cases where the three areas did not contain enough information to determine if the publication should be included, the full text of the publication was examined. To capture ("grey" literature), we searched, among others, the International Water Management Institute Library Catalogue, the FDRE Ministry of Agriculture website, World Bank e-library (https://elibrary.worldbank.org/), and United Nations (http://www.un-ilibrary.org) repositories. A total of 450 grey literature were obtained from Google Scholar, which is widely known as a good source of grey literature (Giustini and Boulos 2013). After screening, 47 publications were retained for the review based on relevance from watershed management perspective.

Table 17.1 Search string used in Scopus included for the reviewal

Theme	Search string
General issues on watershed management	"Watershed management" OR "soil and water conservation" OR "Sustainable Land Management" OR "landscape restoration" AND "Ethiopia"
Watershed management impact indicators	"Conservation" AND "agriculture" OR "tillage" AND "blue" AND "Nile" OR Abbay
	"Soil and water conservation" OR "terrace" OR "bund" AND "blue" AND "Nile" OR "Abbay"
	"Rainwater" OR "water" AND "harvesting" AND "blue" AND "Nile" OR "Abbay"

Study Area

The Abbay Basin is a crucial river system in Ethiopia, responsible for over 50% of the country's total annual surface runoff while only covering 17.5% of its land area. It provides water resources for various sectors such as agriculture, hydropower, and domestic use, making accurate measurement and management of river flow vital for their sustainability and effectiveness. It is located in the center-western part of Ethiopia, between latitude 7°45′ and 12°46′ N, and longitude 34°06′ and 40°00′ E (Fig. 17.1). The basin, also known as the Upper Blue Nile, covers an area of 199,800 km^2 within Ethiopia including Dinder and Rehad sub-basins. Elevation ranges from 475 m a.s.l. at the Sudanese border to 4257 m a.s.l. at the summit of Mount Guna on the northern basin boundary. The river, which originates at Lake Tana, accounts for approximately 55% of the annual renewable surface water resources of Ethiopia and accounts for 60% of the Nile flows reaching Egypt.

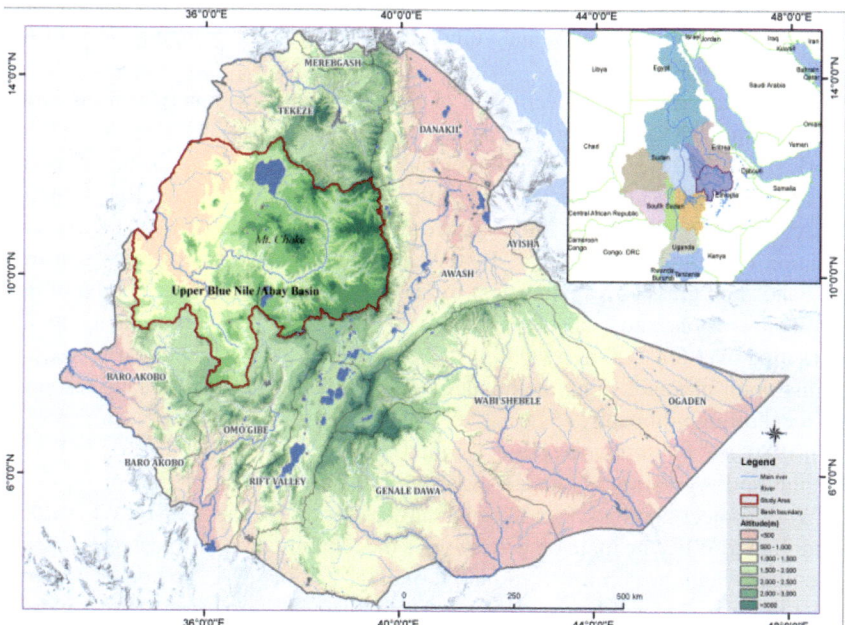

Fig. 17.1 Location map of the Abbay basin

National Watershed Management Initiatives

In Ethiopia, large-scale programs of restoration of degraded lands were started in the mid-1970s and have continued to date with varying scales and foci (Table 17.2). The important components have been the rehabilitation of degraded lands and management of rainwater, and involved: (a) construction of ex-situ water harvesting structures such as farm ponds for supplementary irrigation, (b) construction of in-situ water harvestings structures such as terraces, bunds, check dams, and cut-off drains, (c) implementation of agronomic measures such as conservation tillage and mulching, and (d) afforestation and revegetation of fragile and hillside areas.

Following the severe drought that persisted for three consecutive years between 1972 and 1974/75, a Food-for-Work (FFW) project was initiated with support from the United Nations World Food Program (WFP) in 1974 in the northern part of Ethiopia (mainly Tigray and Wollo areas). The program, which was started in the form of relief assistance, gradually shifted to supporting development activities to address to address land degradation, which was identified as an underlying cause of the problem of food shortages and vulnerability to recurrent droughts (World Bank 1985; WFP 1989).

In a similar effort, the Government of Ethiopia, with assistance from the World Bank, formulated another project named "Drought Areas Rehabilitation Project" (World Bank 1974). The project had nine components ranging from infrastructure development to the establishment of the *Sirinka* Pilot Catchment Rehabilitation Project (SPCRP), which was intended to establish a long-term strategy for the rehabilitation of drought-prone areas. By 1984, the project had managed to construct soil bunds and grass strips equivalent to 180 km that protected over 600 ha of land from soil erosion, and it also designed new conservation tillage implements and distributed over 600,000 tree seedlings (World Bank 1985). The project served as a turning point in establishing the linkage between the degradation of land resources and local drought impacts.

In 1980, a WFP-supported project, called 'Project 2488 (Rehabilitation of Forest, Grazing and Agricultural Lands)' was launched as the culmination of its Food-for-Work project that had been running earlier, during the late 1970s (WFP 1989). According to WFP, the main objectives were the rehabilitation of forest, grazing, and agricultural lands which involved land terracing, tree planting, and other improvements to farmer-owned lands. Activities were designed to increase crop yields by reducing land degradation, and thereby improve food security. Over its 20-year lifespan, Project 2488 laid the foundations for the following project, the Managing Environmental Resources to Enable Transitions to Sustainable Land Use (MERET) project (Nedessa and Wickrema 2010). This also involved changes in focus and approaches: (i) Food-for-Work for large infrastructure development and forestry (1981–1993); (ii) introduction of participatory approaches for activities identification (1994–2000); and (iii) focus on livelihoods (from 2000/2001).

After its formal launch in 2003, the MERET project supported more than 50 activities in three broad areas: (i) physical and biological measures of soil and water

Table 17.2 Watershed management initiatives since the mid-1970s in Ethiopia

Project name and years of operation	Activities related to green water management	Project site	Approach	Source of information
WFP food-for-work projects (1974–1980)	Reforestation, soil and water conservation	Drought-prone areas	Emergency operations that responded to food crises	(WFP 1989)
Drought areas rehabilitation project (1974–1984)	Construction of diversion canals, improved moldboard plow, establishment of soil bunds and grass strips, contour farming, and tree planting on degraded hills	Parts of Tigre and Wollo Provinces which were affected by drought	• Principally responding to the needs of drought-affected people, developing a long-term strategy for the development of the highlands • Top-down approach	(World Bank 1974; World Bank 1985)
Project 2488-rehabilitation of forest, grazing, and agricultural lands (1980–2002)	Afforestation, on-farm and hill-side terracing, area closure and gully control	Ethiopian highlands on 117 watersheds, 3.5 million ha	• 1980–1993: (i) top-down government managed watershed approach; (ii) focus on large watershed areas; (iii) very limited level of long-term planning with activities defined based on available food aid; (iv) total lack of ownership by the community of assets created • 1994–2000: (i) a more community-friendly smaller scale of planning, (ii) introduction of local-level participatory planning Approach (LLPPA) with a focus on smaller watershed activities • From 2000: participatory monitoring and evaluation practices, integrating agricultural packages with income-generating activities	(Nedessa and Wickrema 2010)

(continued)

Table 17.2 (continued)

Project name and years of operation	Activities related to green water management	Project site	Approach	Source of information
Peasant agriculture development program (PADEP): Bure-Silala soil conservation and watershed management (1988–1997)	Construction of physical works such as gully control structures, water ponds, soil bunds, area closure, afforestation in appropriate areas, and improved farming practices	North-west Ethiopia (Bure-Silala)	• Top-down approach in planning and implementation • Watershed-based (10,000 ha Bure watershed in Gojam, a 2500 ha sub-watershed–Silala) • Conservation-based development approach, recommended by the Ethiopian Highland Reclamation Study, which sees conservation measures as an important step in the attempt to increase agricultural productivity among small-scale farmers	(World Bank 1988)
Managing environmental resources to enable transition (MERET) (2003–2006)	• Soil and water conservation measures, livelihood improvement activities, and capacity development • Production and dissemination of the 'Community-Based Participatory Watershed Development' (CBPWD) guidelines	Ethiopia's chronically food insecure 72 woredas: Tigray (17), Amhara (23), Oromia (16), SNNP (12), Somali (3), Dire dawa (1)	• Community-driven and refined LLPPA • Focus on smaller watersheds (500–600 ha) • Systematic targeting using vulnerability analysis and mapping • Shift from technical focus to capacity building and income generation	(Nedessa and Wickrema 2010; MoARD 2005a, b)

(continued)

Table 17.2 (continued)

Project name and years of operation	Activities related to green water management	Project site	Approach	Source of information
Managing environmental resources to enable transitions through partnerships and land users solidarity (MERET-PLUS) (2007–2011)	• More emphasis on community capacity building, homestead production, and income generation • Soil and water conservation measures, soil fertility management, agroforestry and forestry, income generation, homestead gardens and crop diversification, rainwater harvesting, and small-scale irrigation	Highly degraded and food-insecure areas: 65 *woredas* in the regions of Tigray, Amhara, Oromia, SNNPR, Diredawa, and Somali	Participatory and community-based watershed development	(WFP 2009)
Community-based integrated natural resources management in Lake Tana watershed (2009–2018)	• Helping communities prepare and implement 650 watershed management plans • Establishing a database of existing land-use patterns and natural resources • Rehabilitating severely degraded lands, supporting soil and water conservation measures	Amhara region, 21 Woredas, with a total area of 1.5 million hectares	Participatory and community-based watershed development	(FDRE 2019)
Productive safety net program (PSNP) (2004-to date)	• Public works focus on soil and water conservation measures, development of water infrastructure	Food-insecure households in drought-prone areas across the country	Integrated community-based watershed development	(MoARD 2014)

(continued)

Table 17.2 (continued)

Project name and years of operation	Activities related to green water management	Project site	Approach	Source of information
SLMP-I (2008–2013)	• Farmland and homestead development • Communal land and gully rehabilitation • Community infrastructure development such as water harvesting systems	High potential areas in 45 selected watersheds	Participatory and community-based watershed development	(World Bank 2008)
SLMP-II (2013–2018)	• Sustainable natural resource management in public and communal lands • Homestead and farmland development, livelihoods improvements, and climate smart agriculture	135 watersheds in six regions, i.e., Oromia, Amhara, Tigray, SNNP, Gambela, and Benishangul Gumuz	Participatory and community-based watershed development	(World Bank 2013)
Resilient landscapes and livelihoods project (RLLP) (2019–2024)	• Biophysical watershed restoration with a set of associated activities supporting sustainable livelihoods in restored landscapes	Ethiopian highlands in 152 major watersheds	Participatory and community-based watershed development	(World Bank 2018)
Climate action through landscape management (CALM) program (2019–2024)	• Adoption of sustainable land management practices and to expand access to secure land tenure in non-rangeland rural areas	Ethiopian highlands in 500 major watersheds	Participatory watershed management and rural land administration	(World Bank 2019)

conservation, (ii) livelihoods, and (iii) capacity building. A significant achievement of this program was developing and distributing the "Community-Based Participatory Watershed Development" guidelines in 2005 (MoARD 2005a, b). These guidelines have become the authoritative manual and training resource for watershed management practices across Ethiopia. As a project, MERET has made substantial achievements in improving livelihood and food security opportunities for drought-stricken areas and it also created capacity for other land management projects. However, MERET was able to reach only about 4% of the areas requiring soil and water conservation in the country (Nedessa and Wickrema 2010).

MERET Plus ('MERET through Partnerships and Land Users Solidarity') was implemented in highly degraded and food-insecure areas: 65 *woredas* in the regions of Amhara, Dire Dawa, Oromia, SNNPR, Dire Dawa, Somali, and Tigray (WFP 2009). The project locations were selected through vulnerability analysis and mapping (VAM), evaluating agroecology and farming systems, and field evidence, in coordination with relevant government agencies at all levels. MERET Plus sought to tackle land degradation and introduce improved land management practices and skills in highly eroded and food-insecure regions. It also aimed to diversify income opportunities while ensuring the sustainability of natural resources (WFP 2009; Nedessa and Wickrema 2010). This aim is similar to the previous MERET project, but MERET Plus focuses on building effective partnerships for SLM and community-led asset creation targeted at the resource-poor. (WFP 2009). Its package also includes soil and water conservation measures, agroforestry and forestry, soil fertility management, agroforestry and forestry, income generation, homestead gardens development, and crop diversification, rainwater harvesting and small-scale irrigation, crop diversification, and income generation.

In 2008, the Government of Ethiopia developed a multi-year (2009–2024) Strategic Investment Framework for Sustainable Land Management (ESIF-SLM) to guide the prioritization, planning, and implementation, by both public and private sectors, of current and future investments in SLM (MoARD 2010). Since 2008, ESIF-SLM has directed efforts to tackle land degradation, decrease vulnerability to climate shocks, provide land tenure security, and address knowledge and institutional capacity limitations at local, regional, and national levels. The first SLM Project (SLMP-I 2008–2013) was designed for the implementation of the ESIF-SLM framework; and its second phase, SLM Project-II (2013–2018), has also been already completed. ESIF-SLM is currently in phase 3, which is running from 2019 to 2024, and it includes the Resilient Landscapes and Livelihoods Project (RLLP) and the Climate Action through Landscape Management (CALM) project. The RLLP implements core investments in restoring watersheds along with associated activities to support sustainable livelihoods in rehabilitated landscapes across 152 major watersheds in the Ethiopian highlands. The CALM program strives to boost the adoption of sustainable land management practices and expand access to secure land tenure in non-rangeland rural zones (World Bank 2019).

Selected Watershed Management Initiatives in the Abbay Basin

In addition to those programs and projects described in Sect. 17.3, there are several significant projects in the Abbay basin with explicit watershed management components. These include the Koga Irrigation and Watershed Management Project (ADB 2001); the Tana Beles Integrated Water Resources Development Project (World Bank 2008); the Community-Based Integrated Natural Resource Management Project; and the Eastern Nile Watershed Management Project. The principal technical document guiding the design and implementation of the interventions is the Community-Based Participatory Watershed Development Guidelines prepared under the auspices of the Ministry of Agriculture (MoARD 2005a, b).

According to Zeleke et al. (2023), the coverage of existing terraced landscape in the Abbay basin is about 2.8 Mha, and this is out of more than 10.3 Mha of land that is said to require terracing (Table 17.3 and Fig. 17.2a). Nationwide, the GoE's target to achieve landscape restoration is 22 million ha by 2030 as it is mentioned in GTP II. Although the scale of SWC activities in the Abbay basin has increased significantly since the 1973/74 drought, only a fraction of the land in need of terracing has been treated so far. The current coverage of terraces is the cumulative result of many initiatives over the years under development-partner-financed projects and mainstream government programs, which are discussed in Sect. 17.3. The existing physical conservation structures are found mainly in Gojjam, Wello and Debre Birhan areas (Fig. 17.2a).

Constructing physical structures alone is not sufficient for effective watershed management. While building structures can help control soil erosion and runoff along hillslopes by reducing slope length and preventing concentrated runoff, it is just one component of a comprehensive watershed management approach. To achieve maximum effectiveness, physical soil and water conservation (SWC) measures should be complemented with biological and agronomic measures. Figure 17.2 shows areas potentially suitable for area closure, vegetation on SWC structure, and agronomic measures. The methodology followed to derive these maps has been described in detail in Zeleke et al. (2023)

Biological measures, such as vegetation on terraces (Fig. 17.2b) and area closure (Fig. 17.2c), play crucial roles in land management. These measures can enhance

Table 17.3 Extent of suitable area for different types of WM measures in the Abbay basin

Type of WM measures	Area (ha)
Terraced landscape	2,778,002
Area that needs terracing	7,529,300
Area closure	4,176,628
Vegetation on SWC	10,305,431
Agronomic measures	7,446,667

Source Authors' analysis based on data from Zeleke et al. (2023)

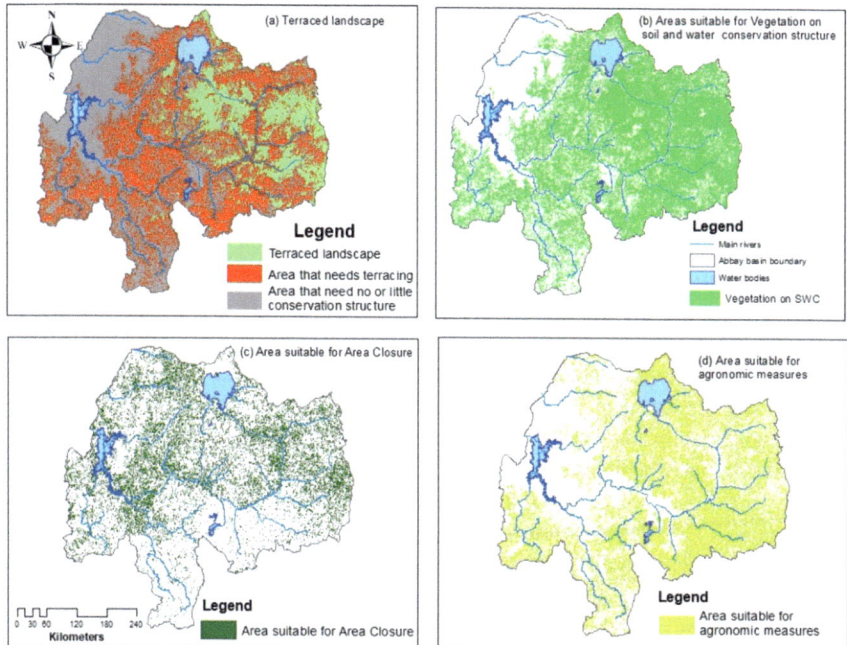

Fig. 17.2 Suitable areas for watershed management interventions in the Abbay basin: **a** The coverage of terraced landscape and areas that needs terracing, **b** vegetation on soil and water conservation structure, **c** area closure, and **d** agronomic measures. *Source* Author's analysis based on data from Zeleke et al. (2023)

the benefits provided by physical structures through providing immediate benefits. Additionally, agronomic measures (Fig. 17.2d), including integrated soil fertility and soil moisture management techniques, are essential for watershed management. The integration of these measures with physical structures was explored, recognizing the growing evidence supporting the superior benefits of combining biological and agronomic measures with physical SWC structures.

Hydrological and Agronomic Impacts of WM in the Abbay Basin

Impacts of Conservation Tillage on Soil Moisture

Spatial Variability of Soil Moisture: Asmamaw et al. (2012) reported a significant difference in soil moisture content at the lower side (0.323 ± 0.003 m^3 m^{-3}) as compared to the upper side (0.305 ± 0.003 m^3 m^{-3}) of *fanya juu* bunds under conventional tillage. However, no significant difference in soil moisture content between the upper side ($0.275 \pm 0.003 \text{m}^3 \text{ m}^{-3}$) and the lower side ($0.278 \pm 0.002 \text{m}^3 \text{ m}^{-3}$) of the bund was observed under conservation tillage (Table 17.4). This indicates that conservation farming could cause the soil water to be stored uniformly on the upper and lower sides of the bunds. Implementation of terraces alone may not affect the desired outcome of improved green water management. For example, crops gown behind *fanya juu* terraces usually appear yellow with stunted growth under conventional tillage. Besides, waterlogging is a common problem behind *fanya juu* (Temesgen et al. 2012). However, if plow pans are disrupted through conservation farming, better soil aeration could be achieved through improved drainage.

Soil Moisture as a Function of Soil Depth: According to Temesgen et al (2012), the impact of conservation tillage is visible in the lower layer of the soil profile. Soil moisture in traditional tillage was significantly higher than that of conservation tillage in the upper layer of the soil profile (0–15 cm). However, in conservation tillage, significantly higher soil moisture was observed at the lower layer (at 15–30 cm) as compared to the traditional tillage. This is mainly explained by the increase in soil penetration resistance as depth increases (i.e., 1 megapascals (MPa) at the soil surface versus 3 MPa at 15 cm depth). A soil penetration resistance experiment carried out in Enerata watershed in Gozamn district (in the Abbay basin) shows that a rise in penetration resistance starts at 10 cm (i.e., the average depth of operation of the Maresha plow) and the resistance peaks at 20 cm depth (Temesgen et al. 2012). Another experiment conducted by Tebebu et al. (2017) revealed that soil penetration resistance value increased with depth in Anjeni and Debre Mewi watersheds for cultivated land. A difference of 1 MPa was observed between the top soil layer (0–15 cm) and bottom soil layer (15–30 cm). A soil penetration resistance value of 2 MPa indicates the presence of hardpan, where roots cannot penetrate, and soil water movement is restricted.

Table 17.4 Impacts of conservation tillage on surface runoff, soil moisture, soil loss, infiltration, grain yield, and biomass expressed as percent deviations (%) from the conventional tillage

Type of conservation tillage	Crop type	Surface runoff	Soil moisture	Infiltration	Soil loss	Grain Yield	Biomass	Source
Berken tillage	Maize	−53%	NA	102%	−53	5%	48%	(Muche 2020)
Winged subsoiler tillage with *fanya juu*	Wheat	−48%	−3% at 0–15 cm depth 9% at 15–30 cm depth	46%	−37%	35%	40%	(Asmamaw et al. 2012; Temesgen et al. 2012)
	Tef	−15%	NA	NA	−9%	28%	14%	(Temesgen et al. 2012)
Deep tillage with hoe	Maize	−58%	NA	88%	−42%	10%		(Hussein et al. 2019)

NA—indicate if no data

Impacts of Conservation Tillage on Surface Runoff and Soil Loss

Compared with traditional tillage, plots treated with conservation tillage showed better moisture retention, low surface runoff, and soil loss. The role of conservation tillage in reducing runoff and soil loss has been demonstrated by different authors (Table 17.4). Berken tillage and deep tillage reduce surface runoff by 53% and 58%, respectively, as compared to conventional tillage on maize plots by allowing more infiltration through disrupting plow pan (Hussein et al. 2019). Surface runoff reductions were 15% on tef plots and 48% on wheat plots (Temesgen et al. 2012). This may indicate that conservation tillage treatment could be effective more on wheat fields in terms of reducing surface runoff and soil loss. Higher negative deviation values are associated with higher effectiveness of treatments in reducing surface runoff and soil loss.

Conservation tillage showed potential benefits to improve infiltration. For example, Berken tillage allowed a 102% increase in infiltration on maize plots and winged subsoiler induced a 45% increase in infiltration on wheat plots (Table 17.4). Significant differences in cumulative infiltration in the soils were reported between winged subsoiler (16.92 ± 0.17 cm) and traditional tillage (11.6 ± 0.11 cm) treated plots in Enerata watershed (Asmamaw et al. 2012).

Impacts of Conservation Tillage on Biomass and Grain Yield

Table 17.4 presents the effects of conservation tillage on grain and biomass yields of wheat, tef, and maize in the Abbay basin. In Robit watershed, Berken tillage significantly increased biomass yield by 48%, but slightly increased grain yield by 5% compared to conventional tillage (Muche et al. 2017) and deep tillage with hoe yielded a 10% increase in grain yield compared to traditional tillage (Hussien et al. 2019). An experiment conducted at Enerata watershed by Temesgen et al. (2012) showed that the average biomass and grain yields from the experimental plots treated with winged subsoiler were higher than those from traditional tillage although the differences were not statistically significant at $\alpha = 0.05$. This could be due to high variability in soil fertility, as the replications occurred across different farmers' fields.

Impacts of Terraces/Bunds on Surface Runoff, Soil Loss, and Crop Yield

Several studies in different parts of the Abbay basin showed statistically significant reductions in soil loss for a majority of SWC treatments when compared with sites without conservation measures (Table 17.5). The reduction in soil loss ranged from

25% at Enerata (*fanya juu*) to 86% at Dibatie (soil bund with grass strip) (Temesgen et al. 2012; Herweg and Ludi 1999). Comparisons of *fanya juu* and soil bund in their effectiveness of reducing soil loss show mixed results. For example, at Andit Tid, a reduction of soil loss by 41% and 63% was observed on soil bund and *fanya juu*, respectively. At Anjeni, a 68% reduction of soil loss on *fanya juu* and a 66% reduction of soil loss on soil bunds were observed. This indicates that there are no significant soil loss differences between *fanya juu* and soil bund. Also, Herweg and Ludi (1999) noted no significant soil loss differences between most SWC treatments, and hence, there is no 'best' measure as such. Another important point could be absolute soil erosion rates on treated plots might still be above a given tolerance level, and there is a need for further development of SWC technologies (Herweg and Ludi 1999).

Runoff was considerably reduced at all sites reviewed here; thus, the goal of moisture conservation was met (Table 17.5). The reduction in surface runoff ranged from 2% at Andit Tid (*fanya juu*) to 50% at Anjeni (*fanya juu*). The reduction of runoff due to higher infiltration rates resulted from SWC measures on treated plots. However, at the catchment level, there were mixed results (Hurni et al. 2005). In Anjeni, Minchit catchment, the rainfall–runoff coefficient did not substantially decrease during the period 1984–2000. The catchment was treated with intensive soil and water conservation measures since 1986, the runoff coefficient, however, showed an insignificant trend toward less runoff, from about 47% at the beginning of conservation in 1984 to about 42% in 2000 (Hurni et al. 2005). Therefore, implementation of soil and water conservation may not necessarily lead to a significant decrease in total annual catchment runoff rates over time. This suggests that additional catchment-scale experiments and scenario analyses are required to understand catchment-scale dynamics over long years.

Significant variability is reported in the impact of soil and water conservation measures on crop yield. As opposed to the expected benefit of conservation measures on grain yield, an experiment at Andit Tid showed a reduction in grain yield (Herweg and Ludi 1999). The highest increase in grain yield (28%) has been reported at Enerata on a plot treated with *fanya juu* (Temesgen et al. 2012). Herweg and Ludi (1999) noted that at Anjeni grain yields from on-farm experimental plots treated with conservation measures rarely increased during the first three to five years of SWC works.

Downstream Impacts of WM of the Abbay Basin

The hydrologic effect of WM interventions is generally to delay surface runoff and increase infiltration. At the catchment scale, this is expected to result in reductions in flood peaks and surface runoff volumes and an increase in dry season flows. In the Abbay basin, soil and water conservation substantially decreased surface runoff rates on cultivated plots that utilized conservation practices versus those without conservation practices. At the catchment level, however, there were mixed results because

Table 17.5 Average relative impact (%) of SWC (in-situ water harvesting) measures on soil loss, runoff, crop yield, and biomass compared with local cultivation practices

Location	SWC structure	Runoff	Soil loss	Grain yield	Biomass	Source
Andit Tid on 24% slope (1987–1991)	*Fanya juu*	−2	−63	−50	−45	(Herweg and Ludi 1999)
	Soil bund	−5	−41	−12	−11	
	Grass strip	−33	−73	−39	−37	
Anjeni on 28% slope (1986–1990, 1992)	Fanya juu	−33	−68	+4	−5	
	Soil bund	−32	−66	−13	−13	
	Grass strip	−41	−72	0	+8	
Anjeni on 12% slope (1986–1990, 1992)	Fanya juu	−50	−81	+14	+5	
	Soil bund	−40	−63	−6	−12	
	Grass strip	−19	−57	+14	+11	
Anjeni at catchment level	Fanya juu	−11	?	?	?	(Hurni et al. 2005)
Enerata on 9–11% slope 2011	Fanya juu with conservation tillage	?	−25	+32	+28	(Temesgen et al. 2012)
Guder on 15% slope	Fanya juu	−32	−72	–	–	(Ebabu et al. 2019; Sultan et al. 2018)
	Soil bund	−27	−67	–	–	
	Soil bund with grass strip	−29	−77	–	–	
Aba Gerima On 15% slope	Fanya juu	−29	−61	–	–	
	Soil bund	−20	−60	–	–	
	Soil bund with grass strip	−22	−66	–	–	
Dibatie on 15% slope	Fanya juu	−35	−63	–	–	
	Soil bund	−29	−68	–	–	
	Soil bund with grass strip	−43	−86	–	–	
Debre Mewi on 10% slope	Soil bund	−36	−57	–	–	(Amare et el. 2014)
	Soil bund with local grass	−17	−26	–	–	

apparently, many factors contribute to the increase or decrease in river discharge, which includes base flows.

At an experimental plot (6 m × 30 m) level, Hurni et al. (2005) reported a long-term (1984–2000) average reduction of surface runoff by 39% at Anjeni (Minchet catchment) due to soil and water conservation measures on cultivated land compared with non-conserved cultivated land (Table 17.6). However, at the catchment scale, the runoff coefficient showed an insignificant decreasing trend, from about 47% at the beginning of conservation in 1984 to about 42% in 2000. This suggests that the conservation work in Anjeni contributed to enhanced dry season flow in this catchment. In another sub-humid catchment in *Maybar* (located on the ridge of Abbay basin and Awash basin), for an observation period of eight years (1982–1989), the runoff-reduction effect was more pronounced. The runoff coefficient changed from 32% in 1982 (1431 mm rainfall) to 15% in 1989 (1406 mm rainfall) after intensified soil and water conservation had been carried out in 1983 (SCRP 2000a). In contrast, in a humid Hulet Wenz catchment at Andit Tid research station, the runoff coefficient increased from 30% in 1983 (1547 mm rainfall) to 43% in 1992 (1472 mm rainfall) after the implementation of conservation structures.

No study has explored why mixed results are obtained at the catchment scale, such as in the SCRP research stations. The mixed findings are perhaps related to the geologic formation of the catchments. A catchment may respond fast or slowly after WM interventions depending on its geologic formation and acquirer characteristics.

The geology of a watershed is recognized as one of the key factors in determining base flow index (BFI) calculations. (Longobardi and Villani 2008; Bloomfield et al. 2009). Likewise, research by Mwakalila et al. (2002) showed base flow index has a robust association with climate and geology. Abebe and Foerch (2006) demonstrated a link between the climatic, morphologic, and geologic characteristics of a watershed and its base flow index in Ethiopia's Wabi Shebele river basin. They exhibited a strong relationship between base flow index and geology. Watersheds with high rainfall or low evapotranspiration on granite or basalt substrates tend to have greater base flow. Nyssen et al. (2010) reported a rise in the groundwater table after conservation measures (stone bunds and check dams) in the May ZegZeg catchment (Tekeze river basin, northern Ethiopia) where the geology is Antalo limestone layers overlain by Amba Aradam sandstone. However, it is very difficult to come up with a conclusive statement by taking only a three-month data after the catchment management. Akale et al. (2019) computed the groundwater flow index (GWFI = annual subsurface flow/total flow) for the period 2010–2015 based on hydrological model results of Tikur-Wuha watershed (in the Abbay Basin) during (2010–2011)

Table 17.6 Impact of GWM intervention on runoff coefficient based on data from SCRP research stations in the Abbay basin

Research station	Impact of GWM Intervention	Geology	Source
Andit Tid (Area: 477.3 ha Climate: Humid)	The runoff coefficient increased from 30% in 1983 to 43% in 1992	Volcanic rocks: hyolites, trachites, tuffs, and basalts	(SCRP 2000b)
Anjeni (Area: 113.4 ha Climate: sub-humid)	The runoff coefficient reduced from 47% at the beginning of conservation in 1984 to about 42% in 2000	Tertiary olivine basalt and tuff	(Hurni et al 2005)
Maybar (Area: 112.8 ha Climate: sub-humid)	The runoff coefficient reduced from 32% in 1982 to about 15% in 1989	Volcanic Trapp series with alkali-olivine basalts	(SCRP 2000a)

and (2012–2015) WM interventions. The average GWFI was less (an average of 62%) before implementation than after (an average 64%). The catchment appeared to respond fast because the GWFI showed a slight increase in the early phase of the implementation (the year 2012), and showed a decrease in the year 2015.

It is uncertain whether the surface runoff captured by in-situ watershed management practices can be transformed into baseflow, thereby increasing streamflow during dry seasons. Thus, upgrading rainfed agriculture through investments in in-situ and ex-situ WM systems may result in water trade-offs with downstream users and ecosystems (Calder 1999). However, the effects on downstream streamflow from small water storage systems are very minimal (e.g., Schreider et al. 2002; Sreedevi et al. 2006). When in-situ watershed management is implemented, water is utilized near the original source (rainfall) and, as a result, less water is lost as runoff and soil loss is reduced. Hence, upstream capture and use of rainwater saves water that might otherwise be lost by evaporation along the way downstream without any beneficial use. The effect of large-scale adoption of both in-situ and ex-situ WM practices on blue water resources downstream is not known and it is still a subject of discussion.

A "win–win" WM option for both downstream water users and upstream rainfed farmers would be to focus more on practices related to evaporation management or vapor shift (i.e., turn E into T and a higher T/ET ratio), which can increase crop yields and improve water productivity without affecting downstream water users. The interventions involved in vapor shift are dry planting, mulching, conservation tillage, agroforestry, intercropping, and vegetative bunds.

Lessons Learned, Summary, and Research Needs

Lessons Learned

After the famines in the 1970s and 1980s, the Ethiopian government, with support from donors and NGOs, undertook ambitious land conservation initiatives that incorporated several watershed management interventions. In the past three decades, Ethiopia has increasingly adopted far more participatory (farmer-led) approaches, a livelihood focus, and an integrated watershed management paradigm than in the past. These changes demonstrate the readiness for learning from experience and the benefits of changing approaches based on lessons learned along the process. Although those conservation interventions have resulted in many ecological benefits, the large-scale efforts had some serious shortcomings. Those shortcomings include the following.

- The past interventions have not achieved a transition from reversing land degradation to a goal of increasing and sustaining land and water productivity.
- Despite the importance of the resource, factors of implementing WM particularly land tenure and governance has not been given adequate policy attention.

- There have of course been a few measures in the form of sustainable land management and those have yielded encouraging results in increasing crop yield and water productivity. There is thus, huge potential to increase both yield and water productivity.
- Given the significance of the Abbay basin, investments in WM are scarce. Much more needs to be done to address the areas that have not been treated with any WM measures.
- Farmers are reluctant to keep the implemented SWC measures sustainably because most technologies are not adapted to the local situation and were donor-driven. Interventions such as homestead development and agroforestry have shown promising results in providing immediate benefits.
- Implementing the SWC measures requires labor, and farmers are reluctant to implement labor-intensive measures without getting immediate benefits. Farmers' participation in the selection of sites and technologies for interventions is still inadequate.
- Past interventions and research have focused on single-practice interventions as opposed to integrated multiple interventions. For example, the integration of terracing with conservation tillage might bring about a larger impact on crop yields.
- Finally, although past interventions were mostly watershed-based, they have not considered upstream–downstream linkages.

Summary

In calling for improved watershed management, four broad categories of methods have been reviewed: in-situ water harvesting, ex-situ water harvesting, vapor shift (evaporation management), and crop management. The WM practices range from soil amendments, conservation tillage practices, soil and water conservation practices, use of mulches and crop residue, to runoff harvesting techniques. The most widely implemented WM practice in the Abbay basin is found to be terracing. The coverage of existing terraced landscape is about 2.8 million ha (27%), and this is out of at least around 10.3 m ha of land that requires terracing. Various nationwide WM initiatives supported by multiple development partners since the mid-1970s have contributed to the current coverage of terraces. The major WM-related efforts in Ethiopia include the Food-for-Work (FFW) program (1973–2002), the MERET project (2003–2011), the PSNP program (2005–present), community mobilization through free-labor days (1998–present), the SLM program and its SLM projects (2008–2018), the RLLP, (2019–2024), and the CALM program (2019–2024). These initiatives indicate that there are fertile grounds for improved implementation of WM and it can be integrated with various existing projects/programs.

This review has shown the impact of these interventions, particularly terraces/bunds on surface runoff, soil loss, and grain yield. Although terraces/bunds are more successful in reducing soil loss for the majority of SWC treatments when compared with sites that did not receive conservation measures, absolute soil erosion rates on treated plots remained above tolerable limits. This suggests the need for integration of terraces with other technologies or further development of the terracing practice in terms of design. Terraces/bunds are found to be effective in reducing runoff considerably at the plot scale. However, at the catchment scale, mixed results have been reported probably because many factors contribute to the increase or decrease in river discharge, which includes base flows. This suggests additional catchment-scale experiments and scenario analysis are required to understand catchment-scale dynamics over long years. The review revealed significant variability, and hence inconclusiveness, also in results regarding the impact of soil and water conservation measures on crop yield.

This review has also shown that conservation tillage practices are successful in causing uniform spatial soil moisture distribution in the upper and lower sides of bunds, thereby reducing the waterlogging effect of *fanya juu*. Thus, the integration of different WM technologies such as conservation tillage and terracing will result in the desired outcome of uniform soil moisture distribution in farms. Depth-wise, conservation tillage reduces soil penetration resistance and increases infiltration by breaking down the hardpan. Conservation tillage integrated with terrace has been effective in reducing surface runoff and soil loss and increasing infiltration and grain yield. It is also known to be a "win–win" WM option for both downstream water users and upstream rainfed farmers, which can increase crop yields and improve water productivity without affecting downstream water users. Other interventions with similar effects to conservation tillage are dry planting, mulching, agroforestry, intercropping, and vegetative bunds.

Further Research Needs

Water has been left out in different watershed management programs of the country, and this has led to weak water management investments in rainfed agriculture areas. Increasing the productivity of the whole landscape requires a shift in paradigm from a narrow focus on erosion control to a broader blue-green water management approach in a watershed. Alterations in land use in upstream areas can impact downstream water flows, potentially causing undesirable trade-offs between water for food production in upper watersheds (green water use) and available blue water downstream. Upstream areas, often rainfed areas, are seen primarily as blue water-generating zones. Water resource planning spanning from individual farms to the basin level is needed, incorporating the direct and indirect social and ecological services produced by blue and green water flows. This necessitates research and development efforts. This review suggests the following as important research areas.

- While numerous studies exist on individual soil and water conservation measures, research on integrated watershed management impacts is limited, especially in the

Abbay basin. Thus, studies on the hydrological impacts of WM interventions and their implications for livelihoods are needed for improved WM.
- Soil and water conservation measures reduced surface runoff rates considerably on cultivated plots that utilized WSM practices versus those without WSM practices. At the catchment level, however, mixed results are reported. One can therefore assume that a thorough implementation of soil and water conservation may lead to a decrease in total annual catchment runoff rates over time (reduced green water storage). However, the interventions may have a positive effect on agricultural production, carbon storage, and soil biodiversity. Therefore, a new way of looking into the impacts of interventions on green water, blue water, and crop yield (food security) is critically important.
- No or little attention has been given to WM interventions other than those practices which reduce soil loss. For example, it is widely known that conservation agriculture increases grain yield by promoting root development and infiltrating more rainfall deeper into the soil profile, especially in soils with compacted, low permeability subsoil layers. However, little is known about the level of impact on the hydrology and agronomic effects of conservation tillage.
- While rainwater is the major contributor to livelihoods in the Abbay basin, little attention has been given to its management in the programs, policies, and strategies. A future SLM program in the Abbay basin should aim at applying an integrated approach to rainwater management that recognizes the vital function of both green and blue hydrological flows in sustaining direct and indirect ecological services and benefits for rural communities.

Acknowledgements This work was supported by the Water Security and Sustainable Development Hub funded by the UK Research and Innovation's Global Challenges Research Fund (GCRF) [grant number: ES/S008179/1].

References

Abebe A, Foerch G (2006) Catchment characteristics as predictors of base flow index (BFI) in Wabi Shebele river basin, east Africa. In: Proceedings of conference on international agricultural research for development. Citeseer, pp 1–8

ADB (2001) Koga irrigation and watershed management project: appraisal report

Adimassu Z, Langan S, Johnston R, Mekuria W, Amede T (2017) Impacts of soil and water conservation practices on crop yield, run-off, soil loss and nutrient loss in Ethiopia: review and synthesis. Environ Manage 59:87–101

Akale AT, Dagnew DC, Moges MA, Tilahun SA, Steenhuis TS (2019) The effect of landscape interventions on groundwater flow and surface runoff in a watershed in the upper reaches of the Blue Nile. Water 11:2188. https://doi.org/10.3390/w11102188

Amare T, Zegeye AD, Yitaferu B, Steenhuis TS, Hurni H, Zeleke G (2014) Combined effect of soil bund with biological soil and water conservation measures in the northwestern Ethiopian highlands. Ecohydrol Hydrobiol 14:192–199. https://doi.org/10.1016/j.ecohyd.2014.07.002

Asmamaw DK, Leye MT, Mohammed AA (2012) Effect of winged subsoiler and traditional tillage integrated with Fanya Juu on selected soil physico-chemical and soil water properties in the northwestern highlands of Ethiopia. East African J Sci 6:105–116

Bloomfield JP, Allen DJ, Griffiths KJ (2009) Examining geological controls on baseflow index (BFI) using regression analysis: an illustration from the Thames Basin, UK. J Hydrol 373:164–176

Calder IR (1999) The blue revolution: land use and integrated water resources management. Earthscan

Ebabu K, Tsunekawa A, Haregeweyn N, Adgo E, Meshesha DT, Aklog D, Masunaga T, Tsubo M, Sultan D, Fenta AA, Yibeltal M (2019) Effects of land use and sustainable land management practices on runoff and soil loss in the Upper Blue Nile basin, Ethiopia. Sci Total Environ 648:1462–1475. https://doi.org/10.1016/j.scitotenv.2018.08.273

FDRE (2019) Community-based integrated natural resources management project (CBINReMP). GEF Terminal Evaluation Report

Giustini D, Boulos MNK (2013) Google scholar is not enough to be used alone for systematic reviews. Online J Public Health Inform 5:214

Herweg K, Ludi E (1999) The performance of selected soil and water conservation measures—case studies from Ethiopia and Eritrea. CATENA 36:99–114. https://doi.org/10.1016/S0341-8162(99)00004-1

Hurni H, Tato K, Zeleke G (2005) The implications of changes in population, land use, and land management for surface runoff in the Upper Nile Basin Area of Ethiopia. Mt Res Dev 25:147–154. https://doi.org/10.1659/0276-4741(2005)025[0147:TIOCIP]2.0.CO;2

Hurni K, Zeleke G, Kassie M, Tegegne B, Kassawmar T, Teferi E, Moges A, Tadesse D, Ahmed M, Degu Y, Kebebew Z, Hodel E, Amdihun A, Mekuriaw A, Debele B, Deichert G, Hurni H (2015) The economics of land degradation. Ethiopia case study. Soil degradation and sustainable land management in the rainfed agricultural areas of Ethiopia: an assessment of the economic implications. Rep Econ L Degrad Initiat

Hussein MA, Muche H, Schmitter P, Nakawuka P, Tilahun SA, Langan S, Barron J, Steenhuis TS (2019) Deep tillage improves degraded soils in the (sub) humid Ethiopian highlands. Land 8:159. https://doi.org/10.3390/land8110159

Longobardi A, Villani P (2008) Baseflow index regionalization analysis in a mediterranean area and data scarcity context: role of the catchment permeability index. J Hydrol 355:63–75

MoARD (2005a) Community based participatory watershed development : a guideline Part 1

MoARD (2005b) Community based participatory watershed development : a guideline annex Part 2

MoARD (2010) Ethiopian strategic investment framework for sustainable land management, Addis Ababa

MoARD (2014) Productive safety net programme phase IV. Programme implementation manual

Muche H, Abdela M, Schmitter P, Nakawuka P, Tilahun SA, Steenhuis T, Langan S (2017) Application of deep tillage and Berken Maresha for hardpan sites to improve infiltration and crop productivity. ICAST conference, Bahrdar University

Muche H (2020) Application of deep tillage and Berken maresha to break hardpan, improve infiltration and crop productivity, Robit watershed, Ethiopia (Doctoral dissertation)

Mwakalila S, Feyen J, Wyseure G (2002) The influence of physical catchment properties on baseflow in semi-arid environments. J Arid Environ 52:245–258

Nedessa BB, Wickrema S (2010) Disaster risk reduction: experience from the MERET project in Ethiopia. Revolut Food Aid Food Assist 139–156

Nyssen J, Clymans W, Descheemaeker K, Poesen J, Vandecasteele I, Vanmaercke M, Zenebe A, Van Camp M, Haile M, Haregeweyn N (2010) Impact of soil and water conservation measures on catchment hydrological response—a case in north Ethiopia. Hydrol Process 24:1880–1895

Schreider SY, Jakeman AJ, Letcher RA, Nathan RJ, Neal BP, Beavis SG (2002) Detecting changes in streamflow response to changes in non-climatic catchment conditions: farm dam development in the Murray-Darling basin, Australia. J Hydrol 262:84–98

SCRP (2000a) Area of Maybar, Wello, Ethiopia: long-term monitoring of the agricultural environment 1981–1994. Soil Conservation Research Programme (SCRP), University of Berne Switzerland

SCRP (2000b) Area of Andit Tid, Shewa, Ethiopia: long-term monitoring of the agricultural environment, 1982–1994, Soil Conservation Research Programme (SCRP)

Sonneveld BGJS, Keyzer MA (2003) Land under pressure: soil conservation concerns and opportunities for Ethiopia. L Degrad Dev 14:5–23. https://doi.org/10.1002/ldr.503

Sreedevi TK, Wani SP, Sudi R, Patel MS, Jayes T, Singh SN, Shah T (2006) On-site and off-site impact of watershed development: a case study of Rajasamadhiyala, Gujarat, India. J SAT Agric Res 2:1–44

Sultan D, Tsunekawa A, Haregeweyn N, Adgo E, Tsubo M, Meshesha DT, Masunaga T, Aklog D, Fenta AA, Ebabu K (2018) Efficiency of soil and water conservation practices in different agroecological environments in the Upper Blue Nile Basin of Ethiopia. J Arid Land 10:249–263. https://doi.org/10.1007/s40333-018-0097-8

Tebebu TY, Bayabil HK, Stoof CR, Giri SK, Gessess AA, Tilahun SA, Steenhuis TS (2017) Characterization of degraded soils in the humid Ethiopian highlands. Land Degrad Dev 28(7):1891–1901. https://doi.org/10.1002/ldr.2687

Teferi E, Bantider A, Zeleke G, Bewket W (2023). Land degradation in Ethiopia: an assessment using a composite land degradation index method. Working Paper No. 4. Water and Land Resource Centre, Addis Ababa University: Addis Ababa

Temesgen M, Uhlenbrook S, Simane B, van der Zaag P, Mohamed Y, Wenninger J, Savenije HHG (2012) Impacts of conservation tillage on the hydrological and agronomic performance of *Fanya juus* in the upper Blue Nile (Abbay) river basin. Hydrol Earth Syst Sci Discuss 9:1085–1114. https://doi.org/10.5194/hessd-9-1085-2012

Wani SP, Sreedevi TK, Rockström J, Ramakrishna YS (2009) Rainfed agriculture–past trends and future prospects. Rainfed Agric Unlocking potential 1–35

WFP (1989) Mid-Term evaluation by a WFP/FAO/ILO/UN mission of project Ethiopia 2488 Exp. II rehabilitation of forest, grazing, and agricultural lands, vol 1. Draft report. Addis Ababa, Ethiopia

WFP (2009) Mid-term evaluation of the ethiopia country programme 10430.0 (2007–2011): final report, world food programme, office of evaluation

WLRC (2019) Modelling hydrologic and soil erosion processes in Abbay Basin, Ethiopia. Water and Land Resource Centre, Addis Ababa University, Addis Ababa

World Bank (1974) Ethiopia—drought area rehabilitation project (English). The World Bank, Washington, DC

World Bank (1985) Ethiopia—drought area rehabilitation project. The World Bank, Washington, DC

World Bank (1988) Ethiopia—peasant agricultural development project (English). The World Bank, Washington, DC

World Bank (2008) Sustainable land management project (Ethiopia). The World Bank, Washington, DC

World Bank (2013) Ethiopia—second phase of the sustainable land management project (English). The World Bank, Washington, DC

World Bank (2018) Ethiopia resilient landscapes and livelihoods project project information. The World Bank, Washington, DC. PIDISDSA21985

World Bank (2019) Ethiopia climate action through landscape management (CALM). The World Bank, Washington, DC

Zeleke G, Teferi E, Bantinder A, Kassawmar T, Bewket W (2023) Mapping current coverage of physical soil and water conservation structures and estimating areas requiring different sustainable land management measures in Ethiopia. In: Working Paper No. 5. Water and Land Resource Centre, Addis Ababa University: Addis Ababa

Part VI
Recent Advances in Remote sensing for Basin Management

Chapter 18
Enhancing Earth Observation Application to Derive Hydrological Modelling Parameters in the Abbay Basin

Berhan Gessesse, Gebeyehu Abebe, Wubetu Anley, and Worku Zewdie

Abstract The Abbay Basin is one of the hydrologically data-scarce basins in Ethiopia. However, hydrological information and model simulation outputs require sufficient and near real-time datasets to evidence-based decision about the management of water resources. Earth observation (EO) applications is advancing in large basin-level hydrological planning and development by offering data and information for spatial-based best decision support services and systems. Unfortunately, the application of EO for water resource management and development remains a major limitation for many developing countries. Subsequently, Ethiopia has not been fully utilizing this opportunity and many impediments remain a critical challenge for EO data application for the water resource management and development sector. Therefore, this chapter aimed to explore the status of EO product applications for water monitoring and management efforts in the Abbay Basin. The application of EO has increased from time to time and EO products could contribute greatly to efficient and effective use of hydrological parameters and information retrieval, and facilitate the improved decision-making process in the water sector for long-term development in the basin. Specifically, satellite datasets such as remote sensing; meteorological satellites, and global navigation satellite systems (GNSS) show a decisive role in obtaining the spatiotemporal dynamics information and quantitative measurements of the different hydrological parameters such as the extents of surface water bodies; suspended minerals; bathymetric mapping; land and water surface temperature; precipitation; aerosols; clouds and water vapour as well as dam safety

---

B. Gessesse (✉)
Remote Sensing Department, Ethiopian Space Science and Geospatial Institute, Addis Ababa, Ethiopia
e-mail: berhanavu@gmail.com

Department of Geography and Environmental Studies, Kotobe University of Education, Addis Ababa, Ethiopia

G. Abebe · W. Anley · W. Zewdie
Remote Sensing Department, Space Science and Geospatial Institute, Addis Ababa, Ethiopia

© The Author(s), under exclusive license to Springer Nature Switzerland AG 2025
A. Melesse et al. (eds.), *Abbay River Basin*, Springer Geography,
https://doi.org/10.1007/978-3-031-65241-7_18

monitoring in the Abbay Basin. After introducing the status of EO product application in water-related issues, the paper presented the status of EO applications in hydrological parameter retrieval as well as input for hydrological simulation models and possible remedial suggestions essential to building capacity for EO application for water resource monitoring and development in the Abbay Basin.

Keywords Abbay basin · Hydrological parameters · Surface water extent · Hydrological models · Ethiopia

Introduction

Earth contains plenty of water and it looks like a 'blue planet' as water reflects much of the blue portion of the electromagnetic spectrum. Water is the only well-known resource that can naturally exist in the form of gas, liquid, and solid state. Strahler (2013) reported that the land mass portion of the Earth makes up approximately 29%, while the other 71% consists of water bodies, and the amount of water on the Earth is approximately 1.39 billion Km^3. Large saline oceans contain around 97% of the water on Earth. Freshwater streams, rivers, lakes, and reservoirs make up only ~0.02% of the water on Earth. Icecaps (2.2%), subsurface aquifers (0.6%), and glaciers (0.8%) contain the remaining water and water vapour in the Earth's atmosphere (0.001%). More specifically, all lakes, rivers, and swampy areas combined only account for a small fraction (<1%) of the Earth's total freshwater reserve that can be used for vital resources for terrestrial ecosystem services and functions (Shiklomanov 1993; Jensen 2014; Steffen et al. 2015; Schelwald-van and Reijerkerk 2009; Albert et al. 2021).

The quantity of freshwater on the planet Earth is limited, and it exhibits spatiotemporal variability at a global scale, driven mainly by fluctuations in precipitation, water runoff, and evapotranspiration (Steffen et al. 2015; Albert et al. 2021). This shows that freshwater supply is not evenly distributed on Earth and only a limited number of countries have enough freshwater resources to meet the demands of their people (Gleick 1996; Schelwald-van and Reijerkerk 2009; UNESCO 2009; Sipes 2010). All these components of water are vital for domestic supply, livestock, and aquaculture practices mainly for irrigation, industrial, mining, and hydroelectric power generation (Shiklomanov 1993; Schelwald-van and Reijerkerk 2009; Sipes 2010).

Yang et al. (2023) and Papa et al. (2023) reported that the lion's share of freshwater resources has been used for irrigation agriculture and power generation in different parts of the world. Like other parts of landmasses on the planet, Africa is especially endowed with river basins and several inland lakes. Accordingly, water is becoming an important strategic commodity in Sub-Saharan Africa (SSA) and an essential resource for the growth of the region's economy, and structural integrations as well as in the management of natural ecosystem functions and services. The water resource fact sheet of the continent revealed that it is endowed with plentiful freshwater resources including large inland lakes and river systems. However, the distribution of this plentiful renewable resource across the continent is not equal (Newby 2010) and millions of Africans still suffer from water shortages throughout the continent African Ministers' Council on Water (AMCOW 2012).

The geographical location advantage of Ethiopia in the tropics ensures the availability of varied natural resources including water resources in the country (Gessesse 2013). The dome-shaped nature of the Ethiopian topography, with high rising mountains and a high tableland features of plateaus at the central part and descends in all directions to the surrounding lowlands that circumscribe the plateaus is a peculiar characteristic of the Ethiopian landscape for the formation of many drainage basins and water resource systems. The majority of the rivers have erratic flow patterns, and the average annual precipitation, surface runoff, and evaporation are all extremely mild over the whole drainage system of the nation. The major water systems in Ethiopia include natural lakes, reservoirs (dams), springs, many swamps, marshes, floodplains, and relatively good groundwater potential. Thus, Ethiopia's water bodies account for approximately 70,000 km^2 area while wetlands cover approximately 1.14% of the nation's total land area (Ayenew 2009).

MOWR et al. (2004) also claimed that Ethiopia receives between 1.3 and 1.6 trillion m^3 of rainwater, of which 112 billion m^3 and 2.6 billion m^3 of surface water and groundwater, respectively, are potentially usable. To understand the spatial distribution of freshwater coverage, the four basins (i.e., Abbay, Tekeze, Barao-Akobo, and Omo-Gibe Basins) provide 83% of the potential surface water. These sub-basins are found in the northwestern, western, and southwestern parts of the country, which represent roughly 40% of the total area coverage. Ethiopia's annual renewable freshwater resources amount to some 124.6 billion m^3/yr. Only 3% of this amount of water is left in the country and 97% of it is lost every year as runoff to the lowlands of neighbouring countries and as evaporation to the atmosphere without providing enough advantages for the Ethiopian people (MOWR et al. 2004).

Regrettably, research findings concerning the monitoring of the surface and groundwater, water resource quantity and quality, and hydrological modelling in the Abbay Basin in particular and Ethiopia at large have not been precisely investigated due to a lack of sufficient and high spatial and temporal resolution hydrometrological datasets. More specifically, accurate, precise, and timely discharge and recharge estimation datasets are not available in the Abbay basin to enable the determination of hydraulic properties of the surface and the groundwater potential of the basin. Water quality studies and information are insufficient in the Basin, and most of the existing studies have been localized to small geographic areas. Even, the 'Rapid Assessment of Water Quality Analysis' (Tadesse et al. 2010) of a sample-based national pilot survey conducted in Ethiopia covered large geographic areas, the study lacks a detailed investigation of the quantity and quality of water information status at the national level including the Abbay Basin. Accordingly, well-organized and complete datasets and information regarding surface and groundwater availability that are necessary for rational planning in the water sector are found to be inadequate, inaccurate, and often not available in the country.

In recognition of the scientific gaps stated, this chapter aimed to highlight the role of EO datasets in closing gaps in data scarcity for water monitoring and management in the Abbay Basin. In line with this, Jensen (2007) reported that there are several types of satellite products and these datasets are alternatively useful for constructing accurate hydrological models. This is because EO products have incredible potential to characterize the quantity and quality of water. In this regard, Van Dijk and Renzullo (2011), Klemas and Pieterse (2015), and Yang et al. (2023) argued that satellite inputs such as remote sensing; Gravity Recovery And Climate Experiment Satellite missions (GRACES); meteorological satellites, and Global Navigation Satellite Systemes (GNSS) products and services contribute a lot in obtaining quantitative measurements and spatial information regarding different hydrologic variables such as surface extents of water bodies; water constituents (organic and inorganic suspended minerals); bathymetric mapping using passive and active optical as well as microwave remote sensing sensors; water surface temperature measurement; precipitation measurement; aerosol and cloud monitoring; and water vapour and snow monitoring. Therefore, this chapter highlights the contribution as well as the applicability of EO-based driven datasets to characterize, monitor, and map water resources across the Abbay Basin.

Methods of Literature Selection

A systematic search of the literature on the application of EO to derive hydrological parameters and inputs for hydrological model simulations in the Abbay Basin was conducted on Web of Sciences, Scopus, Google Scholar, and Ethiopia's Ministry of Water and Energy website. These databases consist of peer-reviewed published scientific papers and technical reports. The title, abstract, and keyword were used for searching using the following search string: 'Earth observation (EO)', 'hydrological modelling', 'hydrological parameters', 'catchment characterization', 'land surface evaporation', 'soil moisture monitoring', 'surface water extent monitoring; 'Grand Ethiopian Renaissance Dam', 'soil erosion', 'water quality analysis', 'groundwater characterization' 'hydrological modelling' and 'Upper Blue Nile', 'Abbay basin', and 'Abbay basin'. As a result, over 148 articles were identified. However, 97 articles were selected and used for this review chapter after applying all exclusion criteria and removing duplicates.

Abbay Basin

Ethiopia occupies a region of around 1.103 million square kilometres in the Horn of Africa. Topographically, Ethiopia is characterized by a huge central plateau in the centre that is surrounded by lowland plains. These highlands are divided into several parts by the wide and deep valleys of the major river drainage systems. The country is the home of 12 major drainage basins, 9 major rivers, 12 large lakes, and several manmade reservoirs. Most of the rivers are international; however, the Awash and Omo-Ghibe rivers are national rivers and they terminate in the saline lakes of Lake Abe and Lake Rudolf, respectively. In general, the western drainage basin is the largest of all drainage systems in Ethiopia, drains 40% of the total area of the country, and carries 60% of the annual water flow.

The Abbay (Blue Nile) Basin is part of the western drainage System of Ethiopia, and a substantial portion of the Ethiopian Highlands (175,000 km^2) is covered by the Abbay Basin, which ranges in altitude from less than 500 m above sea level along the border with Sudan to more than 4200 m above sea level in the basin's centre and eastern escarpment. The Abbay (Blue Nile) River, which rises around Lake Tana and has its outlet at the border to Sudan, flows approximately 1450 kms, and joins the White Nile River in Khartoum, Sudan, to form the Nile River. In addition, several rivers such as Muger, Temecha, Didesa, Dabus, Beles, and many others are major rivers that drain into the Abbay Basin.

About 60% of the Nile's yearly flow is produced by the Abbay basin (Conway 2000). According to Dile et al. (2018), the climate of the basin changes greatly both regionally and across time. According to Dile et al. (2018), the climate of the basin varies greatly both spatially and across time. From June through September, the lengthy rainy season, the Abbay Basin has its most precipitation. The basin also receives a substantial amount of precipitation during the brief rainy season, which lasts from March to May. On average, the basin has 1200 mm of annual rainfall in the southwest and 1600 mm in the northeast (Kim and Kaluarachchi 2009).

Figure 18.1 depicts the basin's elevation map. The basin is characterized by elevation variations that range between 500 and 4200 m above sea level. The current rapidly expanding population provides enormous difficulties to the already strained ecological systems. Furthermore, as a result of rising temperatures and altered precipitation patterns brought on by climate change, the area would probably see recurring problems with access to food and water as well as complicated water politics, which will likely result in conflict and humanitarian disasters (Burrows and Kinney 2016; Asseng et al. 2018). Subsequently, it is vital to comprehend the connections between the hydrological components and characterize their geographical and temporal variabilities given the potential future issues the region may face.

EO for Monitoring Hydrological Parameters

Precipitation

In basins with limited data, like the Abbay basin, recent studies have shown the possible use of global satellite products for computing precipitation values (Lakew et al. 2017; Worqlul et al. 2015, 2017, 2018; Ayehu et al. 2018; Taye et al. 2020; Mohammed et al. 2022). In comparison to a poorly populated network of rain gauge

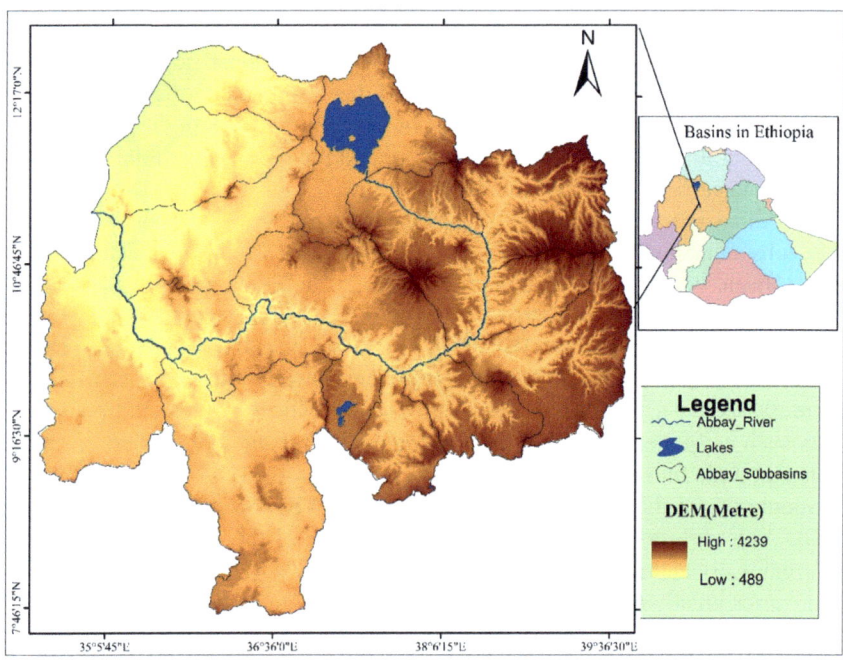

Fig. 18.1 Location map of the Abbay Basin

stations in the basin, Worqlul et al. (2015) analysed high-resolution satellite rainfall data as input to hydrological models. The outcomes demonstrated that stream flow across the Abbay Basin could be reproduced using both precipitation predictions from Climate Forecast Systems Reanalysis (CFSR) and observed information from rain gauge stations. To assess the regional and temporal patterns of meteorological dryness for the Upper Blue Nile Basin, Bayissa et al. (2017) suggested using satellite-derived rainfall products. The same authors noted that nine meteorological stations having Climate Hazards Group InfraRed Precipitation with Station Data (CHIRPS) had a higher correlation coefficient ($r > 0.84$) on a monthly period. Most of the stations in CHIRPS at dekadal, monthly, and seasonal times showed outstanding bias scores (almost one) and mean error. Contrarily, African Rainfall Climatology and Time Series (TARCAT) and Tropical Applications of Meteorology using SATellite Data and Ground-Based Observations (TAMSAT) outperformed CHIRPS while Precipitation Estimation from Remotely Sensed Information using Artificial Neural Networks (PERSSIAN) performed poorly across the board. As a result, the CHIRPS rainfall product was chosen and employed in this study to evaluate the geographical and temporal variability of meteorological drought. The authors argue that it may be viable to employ PERSIANN as a different information source for creating grid-based drought monitoring tools for the basin that might aid in creating early warning systems.

In a different study, Lakew et al. (2017) assessed the potential of satellite and reanalysis precipitation products for applications involving water resources in the Abbay Basin, Ethiopia. The study's findings revealed that the precipitation products consistently outperform the simulation modes of the products when it comes to runoff estimation. In the Abbay Basin, Worqlul et al. (2017) investigated the potential use of Climate Forecast System Reanalysis (CFSR), Tropical Rainfall Measuring Mission-Multi-satellite Precipitation Analysis (TRMM-TMPA), and ground-based rainfall datasets as input for hydrological models. The results revealed that both the gauged and the CFSR reanalysis data were capable of reproducing the streamflow when calibrated separately to the discharge data. The hydrological model Hydrologiska Byrns Vattenbalansavdelning (HBV) uses the corrected MPEG dataset as input to simulate the discharge of the Gilgel Abbay and Gumara watersheds in Ethiopia's Upper Blue Nile basin. The findings demonstrated that the measured rainfall variability was caught by the MPEG satellite rainfall to the tune of 81% and 78%, with a bias that consistently underestimated measured rainfall by 60%. After using a linear bias correction, the study demonstrated the possible application of MPEG SRE in water budget studies (Worqlul et al. 2018).

In a different research, Ayehu et al. (2018) compared the Abbay Basin rain gauge data to satellite-based rainfall predictions of CHIRPS rainfall products to assess their viability. The African Rainfall Climatology (ARC-2) and Tropical Applications of Meteorology utilizing SATellite and Ground-Based Observations (TAMSAT-3) products have been used as a standard and compared with CHIRPS. The outcomes showed the CHIRPS product's potential for a range of practical applications, including investigations of the variability and patterns of rainfall in the study region. For the Abbay Basin's Lake Tana basin, Fenta et al. (2018) assessed the precision of three satellite products. The findings demonstrate that rainfall occurrences were understated by the three satellite products, whereas TAMSAT fairly accurately represented rainfall occurrence in both regions. The results also showed that TAMSAT and CHIRPS performed reasonably well in estimating the amount of rainfall at daily, dekadal, and monthly time scales (high efficiency, low random errors, and bias 10%), whereas ARC did not (high random errors, low efficiency, and bias >20%) at any time scale.

On the other hand, Lakew et al. (2020) compared and evaluated five satellite-based products for estimating global precipitation, including the Climate Prediction Centre Morphing Technique (CMORPH), Tropical Rainfall Measuring Mission (TRMM), Multi-Satellite Precipitation Analysis 3B42 version 7 (TMPA), ERA-Interim (ERAI), Global Precipitation Climatology Centre (GPCC), and Multi-Source Weighted Ensemble Precipitation. The potential use of high-resolution satellite rainfall products for hydrological simulation as well as both the Tropical Rainfall Measuring Mission Multi-satellite Precipitation Analysis (TMPA-3B42v7) and CHIRPS satellite rainfall products were assessed for daily stream flow simulation (Belayneh et al. 2020). The findings of the analysis confirmed that, during the calibration periods, the performance results of the HEC-HMS model showed R^2 of 0.78 and an ENS of 0.69 for CHIRPS-2 and R^2 of 0.79 and an ENS of 0.76 for the TMPA-3B42v7 satellite rainfall products.

In the Gilgel Abbay watershed, Abbay Basin, Andualem et al. (2020a, b) assessed the potential of IMERG on the GPM of rainfall generation by all three IMERG runs using gauge-based gridded data. According to the findings, IMERG early rainfall outperforms the other two satellite products on yearly, monthly, and daily timescales. By using bias correction, the authors suggested that satellite IMERG early GPM be used in data-limited regions for a variety of applications related to the development of water resources.

In an other research, Taye et al. (2020) sought to assess the possible use of satellite rainfall estimates from 1981 to 2018 (CHIRPS-v2 and MSWEP-v2) for monthly meteorological drought assessments across the Abbay Basin. The authors concluded that a better early warning system and drought-resistant seed technology were needed to improve farming practices in the basin.

Abebe et al. (2020) also made an effort to assess the performance of high-resolution precipitation datasets over the Abbay Basin for the year's 2007 to 2016 using a spatial resolution of $0.1°$ and a daily temporal resolution datasets. Due to their high resolution and assimilation of physiographic data in their data-generating processes, ENACTS (calibrated with the majority of Ethiopia's quality gauges) and CHIRPS-2 demonstrated and outperformed well. Both the continuous and categorical indices utilized indicated that IMERG6 and MSWEP2.2 had the next best performance. Due to issues that may be related to the SM2RAIN algorithm's incorrect interpretation of soil moisture signals, SM2RAIN-ASCAT1.1 exhibits the least proficiency everywhere.

Using CHIRPS satellite rainfall estimates, Mohammed et al. (2022) estimated the spatiotemporal variability and trend of rainfall in the Beshilo sub-basin of the Abbay Basin (UBNB) from 1981 to 2019. The standardized anomaly index (SAI) revealed the presence of moderate rainfall variability between the years with negative and positive anomalies in 53.84% and 46.15% of the years analysed, respectively. Analysis of the yearly and Kiremt rainfall patterns revealed a rising tendency and declining trends in Belg and Bega rainfall.

Catchment Characterization

Mainly caused by human, land use/land cover changes are studied extensively to understand their impact in water and energy fluxes. Watersheds with less vegetation covers tend to increase surface runoff and reduce baseflow recharge.

Catchment characterization is an important parameter for monitoring the surface and subsurface hydrology of a river basin. In their study, using EO products and local perceptions, Yesuph and Dagnew (2019) assessed the spatiotemporal land use and land cover (LULC) dynamics in the Gedalas watershed, Abbay Basin. They found that the watershed had undergone significant LULC changes, which are likely to continue in the future as a result of various socioeconomic activities and natural factors.

Gashaw et al. (2017) looked into the Andassa watershed's LULC dynamics analysis from 1985 to 2015 and forecasted the LULC scenarios for 2030 and 2045. The CA–Markov model was used to predict LULC and the hybrid classification approach was used to extract thematic data from satellite pictures. According to the findings, cultivated land increased from 62.7% in 1985 to 73.1% in 2000 and 76.8% in 2015. Between 1985 and 2015, the built-up area also saw a minor growth. The percentages of forest, shrubland, and grassland decreased from 3.5 to 1.9%, 26.2 to 15.3%, and 7.6 to 4.9%, respectively, between the years 1985 and 2015. The expansion of cultivated land and built-up area, and the declining trends of forest, shrubland, and grassland will continue in the 2030 and 2045 periods. In the same watershed, Gashaw et al. (2019) assessed the hydrological effects of LULC changes between 1985 and 2015 and projected the LULC change influence on the hydrological situation in 2045.

Tassew et al. (2019) conducted a different investigation to assess the use of the HEC-HMS model for flow simulation in the Gilgel Abbay watershed of the Abbay Basin. The model is acceptable for hydrological simulations in the Gilgel Abbay Catchment, as demonstrated by the comparison of the observed and simulated hydrographs, the model's performance (NSE = 0.884), and their correlation (R^2 = 0.925). Guzha et al. (2018) also used a mix of field observations, field interviews, and remotely sensed data to examine the dynamics and driving forces of land cover change in the Muger sub-basin, Ethiopia, during the years 1986 to 2020. According to the study's findings, the forest cover from 1986 has decreased by 49.90%. From 1.15% in 1986 to 0.57% in 2020, bare land decreased. The percentages of cultivated land and shrubland rose from 68.86% to 70.44% and from 14.39% to 20.27%, respectively, at the same time.

In a different investigation, Galata (2020) assessed LULC changes and their origins from 1987 to 2017, in the Hangar watershed of Abbay Basin, Ethiopia. The findings showed that the research regions cultivated land and built-up area had increased. However, between 1987 and 2017, there was a decline in the amount of woodland, rangeland, grassland, and water bodies. In their research, Dibaba et al. (2020) used the ensemble mean of four regional climate models (RCMs) from the CORDEX-Africa datasets to estimate future LULC and future climatic scenarios. The study's findings showed that under both typical concentration paths (RCP4.5 and RCP8.5), the ensemble mean of the four RCMs predicted a decrease in future precipitation and a temperature rise.

Using TOPMODEL, Gumindoga et al. (2014) assessed the hydrological consequences of changing land cover types and the runoff contributions from each kind of land cover in Ethiopia's Upper Gilgel Abbay basin. The results showed that the land cover types of agriculture, forest, and grassland, which predominate in the catchment, varied in terms of the highest peak flow as well as the annual stream flow volume. The results also showed that satellite images provide substitute land surface data for land surface parameterization and rainfall-runoff modelling. In the Koga watershed in Abbay Basin in northwestern Ethiopia, Menale and Zeleke (2018) also assessed LULC change and its implications for watershed degradation using GIS and remote sensing. The results showed that the cultivated and settlement areas increased by 7054.6 ha while the grassland and bushland areas decreased by 4846.5 ha and

3376 ha, respectively. The wetland also decreased, going from 580.2 hectares to 68.3 ha.

In general, considering the LULC and catchment characterization studies using multi-temporal satellite imagery, the Abbay basin underwent significant changes in LULC over the past three decades as cultivated and built-up areas grew at the expense of vegetation cover (grassland, shrubs, and forest cover). Fast deforestation and vegetation degradation led to fast erosion, which had an impact on the pace of sedimentation in the basin and led to structures like the Koga dam in the Tana sub-basin. A notable example of how LULC changes brought on by human activity have a significant influence on a basin's surface and subsurface hydrology is clearly observed in the Abbay basin.

Land Surface Evaporation

Evapotranspiration at the land surface is a crucial biophysical parameter used to measure the hydrological cycle and the surface energy balance of the planet. For the purpose of forecasting stream flow in the Lake Tana Sub-basin of Abbay Basin, Setegn et al. employed the efficacy of the soil water assessment tool (SWAT) model using EO data to compute some paramters. According to the study's findings, base flow is a significant contributor to the overall discharge within the Tana Sub-basin, contributing more than surface runoff, while evapotranspiration accounts for more than 60% of basin losses. In accordance, Allam et al. (2016) used a combination of satellite data of rainfall, terrestrial water storage, and river-flow gauge readings to estimate real evapotranspiration over the upper Blue Nile Basin. The technique, according to the authors, may be used in basins with less observational data but comparable future water shortage and climate change issues.

The hydrological cycle's components—discharge, precipitation, evapotranspiration, and storage—were estimated using a methodology described by Abera et al. (2017); and the findings revealed that the basin experiences 1360 ± 230 mm of precipitation annually on average. The annual water budget is divided into three categories: evapotranspiration makes up 56% of it, runoff makes up 33%, and storage makes up between 10% and +17%. SWAT was used by Takele et al. (2021) to simulate the hydrological responses of the Abbay Basin. The sequential uncertainty fitting (SUFI-2) approach in SWAT-CUP was used to calibrate and evaluate the model. During the calibration and validation periods, the findings revealed a significant correlation between the observed and simulated stream flows.

The findings also showed that 49.5% of precipitation was lost by evapotranspiration, whereas 22.43% of precipitation contributed to stream flow as a surface flow. Takele et al. (2022) provided an integrated climatic and hydrological model in a different research to assess the effects of climate change on the Abbay's water resources. The findings showed that the anticipated temperature has a statistically significant upward trend. The yearly evapotranspiration rose by around 10.4% as a result of the rapid temperature rise. Due to this and the trend towards lessening rainfall, stream flow, surface runoff, and water yield might all decrease by up to 54%, 31%, and 31%, respectively. In general, this section has discussed how remote sensing may be used to estimate land surface evaporation, one of the hydrological cycle's components. Although multi-spectral observations have also been used to recover land surface evaporation, previous research on the topic has focused on the use of thermal infrared data and the land surface heat balance.

Soil Moisture Monitoring

Satellite technologies have been providing soil moisture data globally for the last 37 years (Wagner 2007). The development of global precipitation products has offered new possibilities for monitoring soil moisture at continental, regional, national, and basin levels. For instance, in their study, Gumindoga et al. (2020) used the Surface Energy Balance System (SEBS) and EO data and the TOPographic driven MODEL (TOPMODEL) to predict soil moisture in the Mbire area in Zimbabwe. The integration of geostatistical techniques, remote sensing data, and hydrologic models, according to the authors, is promising for planning and managing soil moisture and water resources in data- and water-scarce contexts.

In their study, Ayehu et al. (2020) recently used artificial neural networks (ANN) to analyse Sentinel-1 SAR and Landsat satellite data for a residual soil moisture prediction model development for wheat crop agricultural areas of the Abbay Basin. The residual soil moisture was calculated using a semi-empirical backscattering model of soil and plant, empirical correlations generated from Sentinel-1 SAR data, and soil roughness characteristics. Additionally, the study's results supported the notion that using Sentinel-1 SAR and Landsat sensor products as input datasets for the ANN model significantly enhanced soil moisture content prediction. For many environmental phenomena, including hydrological, meteorological, and agricultural uses, soil moisture is a crucial variable. Estimating soil moisture at the basin level, like in the Abbay Basin, has been one of the most fascinating research agendas in the field of remote sensing.

Surface Water Extent Monitoring and Mapping

For the water, energy, and biogeochemical cycles as well as the preservation of the Earth's biosphere, the size and variety of surface water bodies are of utmost significance. A significant portion of the land surface in many parts of the world is covered by water bodies. These water bodies exhibit strong seasonal and inter-annual variability, which is important for the carbon cycle (Hastie et al. 2021; Hubau et al. 2020), biogeochemical processes (Borges et al. 2015), and water-related disasters like the risk of flooding.

It has historically been difficult to precisely assess surface water extent and distribution across different spatial and temporal dimensions, especially when using in-situ based ground monitoring networks. For land surface water variability monitoring and mapping, in-situ measurements are used to characterize the movement (height, extent, discharge) and quality of water in river channels, lakes, and wetlands. However, in-situ based measure gauge networks are rare and irregularly dispersed throughout the Abbay Basin, or even within any hydrological basin, particularly inaccessible areas with difficult access or security concerns.

Ground-based measured hydrological data availability has considerably decreased recently, notably in Africa, and in-situ gauge networks are often expensive to maintain, especially for developing countries (Tramblay et al. 2021). Additionally, even if data are there, access to them may be constrained by governmental organizations (Chawla et al. 2020), and they are frequently unavailable to the scientific community due to political circumstances or conditions involving transboundary water sharing (Papa et al. 2010). Finally, yet importantly, in-situ data are not able to monitor all water properties such as significant flood occurrences, wetland-river connection, or the variability of several small lakes/ponds in the same area (Alsdorf et al. 2007).

According to Kebede et al. (2006), estimations of rainfall, runoff, outflow, and evaporation are used to calculate the yearly water budget for Lake Tana in the Abbay Basin. Based on the modelling of lake level change (1960–1992) by a monthly time step, the yearly water budget of Lake Tana was provided. The findings demonstrated that the level of Lake Tana, in contrast to the life-threatening lakes in the Ethiopian Rift Valley or the other great lakes of tropical Africa, is less susceptible to variations in rainfall and changes in catchment features in its present hydrologic state. In their study, Dessie et al. (2015), presented a daily water balance analysis for Lake Tana. Considering scenario-based model simulation, the effect of the floodplain on this lake's water balance was assessed. The Gilgel Abbay River is responsible for around 60% of the lake's inflow, according to the findings. The findings also demonstrated the significance of floodplains, their impact on the lake's water balance, and the necessity of taking into account these impacts when designing irrigation systems.

In a different research, Abera et al. (2017) recommended using the JGrass-NewAge system and other EO products to better estimate the hydrological cycle's components, such as precipitation, evapotranspiration, discharge, and storage. To estimate the water balance of Lake Tana and its surrounding catchments, Duan et al. (2018) looked at the feasibility of using openly accessible satellite data in addition to conventional meteorological data. The study's findings demonstrated that measurements and projected annual runoff for two gauged sub-basins varied by no more than 4%. The disparity between the expected and measured annual outflows from Lake Tana was about 12%. The viability of utilizing the CHIRPS-v2 dataset for surface water monitoring over Tana Lake was recently assessed by Alemu et al. (2020). According to the findings, there was an unexplained loss of 0.6 km^3 per year, or 20 cm of water across the Lake Tana region annually. For a better understanding of the unexplained water losses and a more accurate estimation of the volume of subsurface flow leaving the lake, the authors concluded that combined hydrological and geology research are necessary.

Surface Water Extent mapping Use Case: The Grand Ethiopian Renaissance Dam (GERD) landscape: EO products can contribute greatly to the development of a nation by establishing scientific knowledge, building technological competencies, stimulating innovations, and offering data and information for monitoring the spatial distribution and extent of water surface for the best decision support services. Jensen (2007) argued that EO solutions such as satellites imagery and GNSS play a crucial role in obtaining the spatiotemporal dynamics and quantitative information of different hydrologic variables such as suspended mineral monitoring, water constituent quantification, bathymetric mapping, water surface temperature measurement, precipitation, aerosol and cloud characterization, water vapour measurement, dam safety monitoring, and surface extent of water body mapping. Unfortunately, Ethiopia has not been able to fully benefit from the application and contribution of EO datasets, and numerous obstacles continue to be major challenges for the management and development of surface water.

In this use case, the EO data application contributes to the historical development of the GERD aerial extents from its first filling phase to the predicted final filling phase. The objective of this use-case analysis was to highlight the contribution of EO data application for water resource mapping and monitoring mainly to delineate the 570, 590, 610, and 640 m elevation-based spatial extents of the GERD and generate the spatiotemporal extents of the GERD reservoir for water resource monitoring and management efforts. The GERD landscape is found in the western part of Ethiopia, from 90 49′ 30″ to 110 20′ 15″ north latitude and from 340 56′ 35″ to 350 47′ 00″ east longitude (Fig. 18.2).

Satellite remote sensing products obtained from spaceborne optical imaging were used to map and monitor the extent and changing aspects of surface water in the GERD landscape. In this use case, the extraction of the GERD reservoir landscape and the aerial extent of the reservoir were achieved using Senteneil-2 and ALOS PALSAR12.5 DEM data. Senteneil-2 satellite remote sensing data were used for surface water layer extraction in the GERD environment. However, we calculated the water level in the GERD Reservoir using the ALOS World 3D version 3.2 provided by the Japan Aerospace Exploration Agency (JAXA) at 570, 590, and 610 m for the first, second, and third filling phases since July 2021, respectively. The final filling of the surface water extent has been modelled using ALOS World 3D version 3 DEM of the 640 m contour line (Table 18.1).

The statistical information presented in the above table shows the surface water extent of the GERD during the first, second, and third phases of water filling and the predicted surface water extent of the final filling of the reservoir using satellite images and DEM datasets. The results of this analysis derived from the ALOS global DEM confirmed that the base surface elevation is 500 m a.s.l. at the bottom of the GERD height site. To estimate the area changes, the Senteniel-2 and ALOS DEM datasets were utilized to detect the aerial extent in the area of the GERD reservoir during the filling process (Table 18.1). The spatial extent of the GERD reservoir since the first filling phase is represented in Table 18.1 and Fig. 18.2. Furthermore, the analysis showed that the area covered by water was approximately 214.37, 614.86, and 1056.73 km^2 during the 1st, 2nd and 3rd filling phases of the dam, respectively.

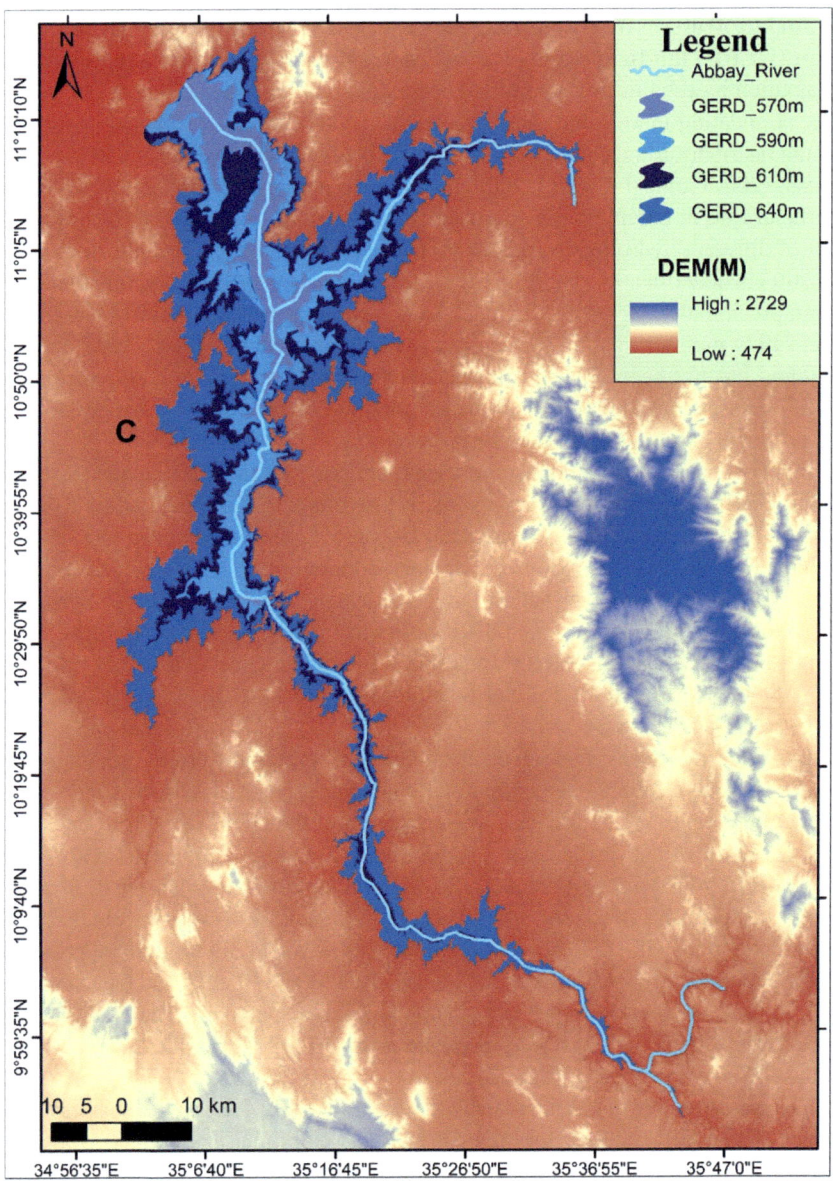

Fig. 18.2 Location map of the GERD and its surface water extent based on first filling phase (570 m), second filling phase (590 m), third filling phase (610 m) and the predicted aerial extent of the GERD after the final filling process (640 m)

Table 18.1 Surface water extent of the Grand Renaissance Ethiopian Dam

	GERD reservoir filling phase	Area (km^2)
Before June 2020	Before first filling phase	*
End of August 2020	During first filling phase	214.37
End of August 2021	During second filling phase	614.86
End of August 2022	During third filling phase	1056.73
*	After the final filling phase	1891.19

In addition, the aerial surface water extent of the dam will reach approximately 1891.19 km^2 after the final filling of the dam.

Therefore, this section of the chapter highlights the contribution of medium resolution satellite products (i.e., both Sentinel-1 and 2 ALOS World 3D version 3.2 DEM datasets) for the mapping of water extent of the GERD reservoir with reasonable accuracy. However, in the future, it will be advisable to use high-resolution satellite imagery and 3D data complemented with ground survey inputs to generate precise and accurate surface water extent, the volume of water, and the dynamics of water in the dam.

Soil Erosion

Various studies used laboratory, field scales and modelling approaches to understand soil erosion and sediment loads to water bodies in different regions (Aga et al. 2018, 2019; Defersha and Melesse 2012; Defersha et al. 2010, 2012; Maalim and Melesse 2013; Maalim et al. 2013). However, using a newly developed methodological framework, Haregeweyn et al. (2017) examined the variability of gross soil loss and sediment yield rates in the Abbay Basin under current and anticipated future conditions. The findings showed that the basin generates an average soil loss rate of 27.5 t ha^{-1} yr^{-1} and a gross soil loss of ca. 473 Mt yr^{-1}, of which, at least 10% comes from gully erosion and 26.7% leaves Ethiopia. The effects of LULC and climatic changes on soil erosion in the Muga watershed, Abbay Basin, were assessed by Belay and Mengistu (2021) utilizing an integrated approach of the CA-Markov chain, climate, and soil erosion models. According to the authors, due to LULC change, the rate of soil erosion in the Muga Watershed increased over time, going from 19.7 tons per hectare per year (t ha^{-1} year^{-1}) in 2017 to 20.7 t ha^{-1} year^{-1} in 2033 owing to LULC change.

Additionally, the rainfall erosivity factor may rise by the mid-2050s, which might result in a greater rate of soil erosion. Therefore, due to the enhanced erosive capacity of the upcoming severe rainfall, the soil loss rate in the Muga watershed is anticipated to increase to 22.0 t ha^{-1} year^{-1} and 22.8 (t ha^{-1} yr^{-1}), respectively, under the RCP4.5 and RCP8.5 scenarios. The average annual soil loss rate rose by 13.2% and 15.7% under RCP4.5 and RCP8.5, respectively, when the combined impacts of LULC and climate change are taken into account, which is much larger than the effects of LULC and climate change alone.

Gashaw et al. (2020) examined how the rate of soil erosion changed in the Gilgel Abbay watershed of the upper Blue Nile basin in Ethiopia under various land management practices. Three scenarios—baseline, intense cultivation, and extensive cultivation—were used to assess the effects of land management practices on soil erosion. At the baseline scenario, the mean annual soil erosion was estimated at ~32.8 t ha^{-1} yr^{-1}, which is equivalent to a loss of approximately 13.66 M t yr^{-1} from the entire watershed.

Elnashar et al. (2021) also suggested a Revised Universal Soil Loss Equation framework that was applied in the Google Earth Engine cloud platform (RUSLE-GEE) for 90 m spatial resolution soil erosion evaluation; the outcomes showed that the mean soil loss rates were 39.73, 57.98, and 6.40. t ha^{-1} yr^{-1} for the entire Blue Nile, Upper Blue Nile, and Lower Blue Nile Basins, respectively. In their study, Erkossa et al. (2015) calculated the on-site cost of soil erosion using the productivity change technique in the Dapo, Meja, and Mizewa watersheds in the upper Abbay Basin, Ethiopia. The findings showed a large decline in production and a considerable loss in farmer income. Ebabu et al. (2019) researched runoff and soil loss from various land use types and assessed the efficacy of various SLM practices by monitoring runoff and sediment from 42 runoff plots in various agroecologies in the Abbay Basin, Ethiopia. Based on the land use type, agroecology, and SLM techniques, the results showed that runoff and soil loss differed substantially.

Similarly, Getu et al. (2022) estimated the rate of soil erosion in the Megech watershed using the Revised Universal Soil Loss Equation (RUSLE) model linked with ArcGIS for efficient planning and decision-making procedures. The average annual soil loss and erosion hotspots in the watershed were calculated using the six RUSLE model input parameters: erosivity, erodibility, slope length and steepness, cover management, and erosion control practices. The finding revealed that there is a total soil loss of 1,399,210 t yr^{-1} from the watershed with a mean annual soil loss of 32.84 t ha^{-1} yr^{-1}. The findings also indicated that more than 82% of the watershed was in the high-risk category, which highlights the urgency of taking rapid land management measures. Endalamaw et al. (2021) also used the RUSLE model linked with multi-criteria assessment (MCE) to estimate the yearly soil loss rate and identify high erosion-risk locations in the Gilegel Beles watershed. To calculate soil loss, LULC, soil maps, DEM (30 m), and 25 years of rainfall data from 9 rain gauge meteorological stations were used and the results revealed that the annual soil loss of the watershed ranges from 0 to 511.2 t ha^{-1} year^{-1} with an average of 28.68 t ha^{-1} year^{-1}. As a result, it is highly important to extract soil erosion model elements such as rainfall erosivity, soil erodibility, slope length, slope steepness; land use and land cover maps, cover management, and erosion control variables from weather monitoring satellite and EO satellite data.

Water Quality Analysis

The physical, chemical, and/or biological characteristics of water are often referred to as the quality of the water. The outflow of nonpoint chemicals from all human and natural activities affects the quality of the water. All human and natural activities contribute nonpoint substances to discharge by water that alters its quality. These nonpoint substances include suspended sediments, coloured dissolved organic matter (CDOM), chlorophyll-a, and contaminants. Additionally, according to Muhammad et al. (2019), EO products may calculate the following variables: electrical conductivity (EC), ammonia nitrogen (NH3-N mg/L), chemical oxygen demand (COD), biochemical oxygen demand (BOD mg/L), and sea surface salinity.

In their work, Essayas et al. (2014) highlighted the potential of Moderate Resolution Imaging Spectroradiometer (MODIS) images for calculating silt concentration in Lake Tana close to the Gumara River. Their findings demonstrated that EO products, such as MODIS images, are potentially useful and affordable instruments for monitoring suspended sediment concentration and obtaining a history of concentration for assessing the impact of best management practices. They discovered that there was a considerable rise in concentration in the lake during the devastating drought of 2002 and 2003. Thus, it is important to balance the ease of access and processing, sensitivity to changes in watercolour, and temporal resolution of the remotely sensed data when using EO pictures to estimate sediment concentration.

Over the course of the one-year research period using EO data, a substantial shift in the concentration of Chl-a and total suspended material (TSM) was seen in Lake Tana. The greatest value was noted near the conclusion of the rainy season from August to October 2020, despite the fact that the Chl-a concentration varied by season (Tadesse et al. 2022). The smallest Chl- concentration occurred in February 2020 (4.45 mg m^{-3}) during the dry season. The northeastern and eastern regions of the lake, where the mouths of the three main and several smaller tributary gutters are located, received the highest level of attention during that time. During the stormy season, attention grew sharply throughout the whole east reinforcement and the southwestern section, reaching levels exceeding 40 mg m^{-3} by the end of August. Throughout the experimental period, TSM attention fluctuated in both location and time, exhibiting a spatial pattern similar to Chl-a. However, May had the lowest mean values recorded (30.1 gm^{-3}), and the highest during the late wet season, in September (62.5 gm^{-3}).

The northern and eastern parts of the lake often have advanced TSM levels (Tadesse et al. 2022). The research by Mamaru et al. (2017) demonstrated that the average turbidity of lake water rose between 1999 and 2013, as determined by Landsat ETM+ for the month of December. The projected average sediment input into the lake was 38 t ha^{-1} yr^{-1}, which was 46% higher than the national average of 26 t ha^{-1} yr^{-1}. The estimated yearly sediment influx had the same pattern as that which was seen for the same period in the three clusters. According to Dersseh et al. (2019) investigation using remote sensing satellite images, total nitrogen (TN) concentrations in Lake Tana were highest in August during the rainy season and lowest in December and March during the dry season (2.7 mg L^{-1} in August, 2.6 mg L^{-1} in December, and 1.9 mg L^{-1} in March). These points lead us to the conclusion that multi-mission EO satellite systems offer a wide range of uses for observing water quality.

Groundwater Characterization

Due to its consistent availability, acceptable natural quality, and ease of direct diversion to the underprivileged population more affordably and swiftly, groundwater is a key source of water supply. According to estimates by MacDonald et al. (2012), the groundwater potential of the African continent is 100 times higher than its freshwater potential. Similarly, Mengistu et al. (2019), Ethiopia has an estimated 40 billion cubic metres of accessible groundwater, and groundwater meets more than 90% of Ethiopia's home and industrial water demands. On the other hand, Alemu et al. (2020), morphometric factors such as the compactness coefficient, elongation ratio, circularity ratio, and length of overland flow directly correlate with the presence of groundwater. Both urban and rural residents of the Abbay Basin rely on groundwater as a supply of household water. Due to population expansion, the community in the Muga watershed has a significant issue with inadequate public water supply. The practical use of groundwater resources will have a substantial effect on the improvement of the standard of living in the community. It significantly improves the sustainable management of groundwater resources in the studied region to generate a groundwater potential map. So, this work shows how remote sensing, GIS, and drainage morphometry are useful for determining groundwater potential.

The spatial distribution of land subsidence, groundwater withdrawal, and compressible layer thickness in the Abbay Basin may be determined using EO, GIS, and statistical methods. Timketa and Gemechis (2022) estimate that the groundwater potentials for 33.6% of the Upper Abbay Basin and Guder River (2315 km^2) are good and very good, respectively. 100% (6883.5 km^2), 22.9% (1578.8 km^2), 20.2% (1388.7 km^2), and 23.3% (1601.0 km^2) of the Guder sub-basin's total designated area come under the categories of 'good', 'moderate,' and 'poor,' respectively. Groundwater potential zones are typically high to very high in the western, northeastern, and southern portions of the Upper Blue Nile Basin of the Guder River Basin, and low to very low in the central, northeastern, and southern regions. Zones with low to extremely poor groundwater potential were in the watershed's centre perimeter mountainous regions.

Similar to this, remote sensing, GIS, and statistical methods were used to define the extremely high potential zone of the groundwater prospective zone in the Muga watershed, which is situated in the southeast corner of Mount Choke at the headwater of the Abbay basin (Alemu et al. 2020). The extremely high potential zone spans 52.7 km^2 in total. They discovered that a good prospective zone encompasses 52.7 km^2 (37.6%). On the other hand, 387.9 km^2 or 55% of the study watershed is made up of poor and extremely poor zones. In the western and eastern parts of the Muga watershed, inadequate groundwater potential is caused by a lack of rainfall and geological features. Approximately 27.5% (3224 km^2) of groundwater potential, 30.6% (3592 km^2) of moderate potential, 26.9% (3151 km^2) of high potential, and 15% (1765 km^2) of very high potential were found in the basin's entire area according to this research. The southern and eastern edges of Lake Tana have a lot of potential for groundwater. In summary, the northern, eastern, and southern escarpments were where the basin's promising locations were discovered and the Abbay Basin groundwater potential zone may be defined with the aid of applications of EO datasets, GIS, and statistical methods.

EO Data for Hydrological Modelling

The development of distributed hydrological models offers an effective tool for managing water resources that are changing in contexts. The most prevalent physically based water balance and sediment transport model, distributed hydrological models, requires many high-resolution input data. Currently, the data requirements for distributed hydrological simulation models can be satisfied thanks to the ongoing advancement of EO data accessibility. EO data can acquire grid-based ground observations periodically, which improves the spatiotemporal resolutions of data as compared to the conventional hydrological data measuring approach. In this regard, studying terrestrial hydrological conditions is facilitated by the development of land surface and hydrological models (van Dijk et al. 2014). Khaki et al. (2017) have made the case that hydrological models are crucial for long-term simulations of diverse terrestrial hydrological components. Additionally, they are crucial for foretelling changes in water storage and hydrological processes at different geographical scales.

Within the context of the hydrological cycle, hydrological processes vary both spatially and temporally. These variabilities in hydrological components have conventionally been bundled together with processes that vary in time during the modelling of hydrological processes (Velásquez et al. 2023). However, Dubayah et al. claim that several variables, including uncertainty in model forcings, model parameters, beginning and boundary conditions, and simplicity of the representation of processes, might affect how well land surface models work. Hydrological modelling was done using station-based observation records to address this. To enhance model estimates, uncertainty in model forcings, model parameters, and simplicity of the representation of processes, more datasets are linked with models.

Alfieri et al. (2022) made the case that EO data might offer a lot of helpful solutions for monitoring water resources in this regard. However, as EO technology has developed, it is now possible to obtain geographically distributed datasets for hydrological modelling on surface characteristics, including vegetation, soil parameters, land use, geological features, and hydrological data like precipitation or soil moisture, which are factors in surface-atmosphere exchanges and can be measured and assessed from space with various types of instruments (Wu et al. 2017; Beck et al. 2017; Dembélé et al. 2020a, b). These observations have been used in operational surface and ground hydrology, water resources engineering, and surface and ground hydrology forecasting utilizing spatial-based hydrological modelling. Modern industrialized nations have focused their hydrologic data collection on streamflow, precipitation, and basic surface meteorological data, which are sufficient for the design and forecasting needs of water resource engineers. Specifically, the design of water supply and flood protection works requires long-term records for river flows, and the forecasting of floods requires (spatially) accurate precipitation measurements.

For instance, Alfieri et al. (2022), Wu et al. (2017) and Dembélé et al. (2020b), argued that EO data needs for operational hydrology mainly the change in soil moisture over a specified time interval; considering the basic water balance equation, parameters like evapotranspiration, which is the sum of evaporation from bare soil, could be computed from EO datasets. Time series of runoff and perhaps precipitation, along with a climatological estimate of monthly evapotranspiration, are crucial components of water supply and/or flood control systems that are generated from EO information. For simulating land-atmospheric hydrologic interactions, it is also necessary for EO to provide datasets for both model parameters, such as surface air temperature, humidity, precipitation, and radiation.

Although EO datasets and methods cannot be used to directly quantify runoff, Dubayah et al. identified two main applications for EO in hydrologic and runoff modelling: i) supplying input data, such as soil moisture or delineated land use classes, that are used to define runoff coefficients and ii) determining watershed geometry, drainage network, and other map-type information for distributed hydrologic models and for empirical flood peak, annual runoff, or low flow equations. The Abbay Basin can benefit from the use of optical and microwave EO datasets to derive the necessary hydrological forcing and parameterization variables, such as precipitation (both liquid and solid), soil moisture, evapotranspiration, air temperature, humidity, incoming solar radiation from the atmosphere, and down welling longwave (thermal) radiation from the atmosphere that can be used for the surface runoff, surface water, and energy balance formulations and inputs for.

Concluding Remarks and Future Research Direction

Large-scale geographical and temporal variability may be seen in the hydrology of the Horn of Africa in general, and Ethiopia in particular. To plan, develop, manage, and use water resources sustainably, this chapter focuses on the contribution of EO datasets and products for extracting hydrological metrics and inputs for spatially distributed hydrological models. Currently, EO datasets have become a much more user-oriented discipline than it was in the past. However, enhancing land surface models while studying Earth and hydrological dynamics is made possible by EO-based derived solutions. This is crucial to raising the calibre of model predictions, which are impacted by several variables, including inaccurate input data, erratic model forcings, and unclear parameter values. In recent years, a wealth of datasets from multi-session weather and EO satellites have offered the chance to enhance both model estimates and parameter estimation procedures.

Weather satellites as well as optical and microwave EO satellite datasets are the major sources of information for hydrological parameters and water and energy balance models in data-scarce environments to derive geophysical data parameters such as precipitation, temperature, evapotranspiration, storage, discharge, and soil moisture. Soil moisture, surface temperature, runoff, and latent, sensible, and ground heat fluxes are included in the output from the water and energy balance models. For monitoring water resources, land use, land cover, river networks, rainfall, evapotranspiration, and several atmospheric variables that may be measured from space with various types of equipment, EO products can progressively offer highly helpful observations.

Specifically, the findings of the review confirmed that most of the hydrological models developed in the Abbay Basin were used to evaluate monthly and seasonal water balance, surface energy flex, surface runoff simulation, sediment production analysis, hydrological modelling, flood risk simulations and simulate the impact of climate change on the rainfall-runoff processes in the basin. This finding affirmed that the enormous potential of EO data applications for hydrological modelling is slowly coming to fruition in Ethiopia in general and the Abbay Basin in particular. For example, EO-based hydrological parameters such as digital elevation models, weather, soil, LULC, and streamflow data were highly utilized by researchers as the most critical input data for the development of different hydrological simulation models in the Abbay Basin. The reasons for this are as follows: (i) there is an enormous effort made by researchers to use the EO dataset in the basin; (ii) hydrological model simulation method development and model validation have steadily incorporated attempts to combine ground data with various EO-based spatial and temporal resolutions.; and (iii) Increases in hydrologic model structure and a reevaluation of how researchers in the Basin should use such data are both results of the use of dispersed model forcing fields produced by EO data.

In summary, EO datasets offer unique opportunities for Ethiopian researchers to challenge the limitations of hydrological datasets in the country, conceptualize water and energy balance, and develop state-of-the-art multi-mission environmental and hydrological simulation models to understand hydrological processes. Nonetheless, in the future, substantial consideration should be given to employing EO datasets, products, and advanced algorithms that are essential for determining the parameters of hydrological processes including (i) forcing variables such as precipitation, incoming solar radiation, and thermal radiation are highly needed to drive water and energy balance as well as hydrological models; (ii) parameterizing the evaporation, transpiration, and sensible heat variables in model simulations, EO-based meteorology sensors that measure surface air temperature, surface humidity, and surface wind should be integrated with EO datasets., and (iii) additional parameters for the processes, such as infiltration, bare soil evaporation, surface runoff, and base flow recession, that have an impact on the water and energy balance must be obtained from EO records and properly integrated with hydrological simulation models.

References

Abebe SA, Qin T, Yan D, Gelaw EB, Workneh HT, Kun W, Dong B (2020) Spatial and temporal evaluation of the latest high-resolution precipitation products over the Upper Blue Nile River Basin, Ethiopia. Water 12:3072. https://doi.org/10.3390/w12113072

Abera W, Formetta G, Brocca L, Rigon R (2017) Modelling the water budget of the Upper Blue Nile basin using the JGrass-NewAge model system and satellite data. Hydrol Earth Syst Sci 21:3145–3165. https://doi.org/10.5194/hess-21-3145-2017

African Ministers' Council on Water (AMCOW) (2012) Status report on the application of integrated approaches to water resources management in Africa

Aga AO, Chane B, Melesse AM (2018) Soil erosion modelling and risk assessment in data scarce rift valley lake regions, Ethiopia. Water 10:1684. https://doi.org/10.3390/w10111684

Aga AO, AM Melesse, Chane B (2019) Estimating the sediment flux and budget for a data limited rift valley lake in Ethiopia. Hydrology 6(1). https://doi.org/10.3390/hydrology6010001

Albert JS, Destouni G, Duke-Sylvester SM et al (2021) Scientists' warning to humanity on the freshwater biodiversity crisis. Ambio 50:85–94. https://doi.org/10.1007/s13280-020-01318-8

Alemu ML, Worqlul AW, Zimale FA, Tilahun SA, Steenhuis TS (2020) Water balance for a tropical lake in the volcanic highlands: lake tana, Ethiopia. Water 12:2737. https://doi.org/10.3390/w12102737

Alfieri L, Avanzi F, Delogu F, Gabellani S, Bruno G, Campo L, Libertino A, Massari C, Tarpanelli A, Rains D, Miralles DG, Quast R, Vreugdenhil M, Wu H, Brocca L (2022) High-resolution satellite products improve hydrological modelling in northern Italy. Hydrol Earth Syst Sci 26:3921–3939. https://doi.org/10.5194/hess-26-3921-2022

Allam MM, Jain Figueroa A, McLaughlin DB, Eltahir EA (2016) Estimation of evaporation over the upper b lue n ile basin by combining observations from satellites and river flow gauges. Water Resour Res 52(2):644–659

Alsdorf DE, Rodríguez E, Lettenmaier DP (2007) Measuring surface water from space. Rev Geophys 45:RG2002. https://doi.org/10.1029/2006RG000197

Andualem TG, Malede DA, Ejigu MT (2020a) Performance evaluation of integrated multisatellite retrieval for global precipitation measurement products over Gilgel Abay watershed, Upper Blue Nile Basin, Ethiopia. Model Earth Syst

Andualem TG, Malede DA, Ejigu MT (2020b) Performance evaluation of integrated multi-satellite retrieval for global precipitation measurement products over Gilgel Abay watershed, Upper Blue Nile Basin, Ethiopia. Model Earth Syst Environ 6:1853–1861. https://doi.org/10.1007/s40808-020-00795-w

Asseng S, Kheir AMS, Kassie BT, Hoogenboom G, Andelaal AIN, Haman DZ, Ruane AC (2018) Can Egypt become self-sufficient in wheat? Environ Res Lett 13(9):094012. https://doi.org/10.1088/1748-9326/aada50

Ayehu GT, Tadesse T, Gessesse B, Dinku T (2018) Validation of new satellite rainfall products over the Upper Blue Nile Basin, Ethiopia. Atmos Meas Tech 11:1921–1936. https://doi.org/10.5194/amt-11-1921-2018

Ayehu G, Tsegaye T, Gessesse B, Yibeltal Y, Melesse AM (2020) Combined use of Sentinel-1 SAR and landsat sensors products for residual soil moisture retrieval over agricultural fields in the upper blue Nile Basin, Ethiopia. Sensors 20(11). https://doi.org/10.3390/s20113282

Ayenew T (2009) Natural lakes of Ethiopia. Addis Ababa University Press, Addis Ababa

Bayissa Y, Tadesse T, Demisse G, Shiferaw A (2017) Evaluation of satellite-based rainfall estimates and application to monitor meteorological drought for the Upper Blue Nile Basin, Ethiopia. Remote Sensing 9(7):669

Beck HE, Vergopolan N, Pan M, Levizzani V, van Dijk AIJM, Weedon GP, Brocca L, Pappenberger F, Huffman GJ, Wood EF (2017) Global-scale evaluation of 22 precipitation datasets using gauge observations and hydrological modeling. Hydrol Earth Syst Sci 21:6201–6217. https://doi.org/10.5194/hess-21-6201-2017

Belayneh A, Sintayehu G, Gedam K, Muluken T (2020) Evaluation of satellite precipitation products using HEC-HMS model. Model Earth Syst Environ 6:2015–2032

Belay T, Mengistu DA (2021) Impacts of land use/land cover and climate changes on soil erosion in Muga watershed, Upper Blue Nile basin (Abay), Ethiopia. Ecol Process 10(1):1–23

Borges AV, Darchambeau F, Teodoru, CR, Bouillon S (2015) Globally significant greenhouse-gas emissions from African inland waters. Nat Geosci 8(8):20–2015

Burrows K, Kinney PL (2016) Exploring the climate change, migration and conflict nexus. Int J Environ Res Public Health 13(4):443. https://doi.org/10.3390/ijerph13040443

Chawla I, Karthikeyan L, Mishra AKA (2020) Review of remote sensing applications for water security: quantity, quality, and extremes. J Hydrol 585:124826. https://doi.org/10.1016/j.jhydrol.2020.124826

Conway D (2000) The climate and hydrology of the Upper Blue Nile River. Geogr J 166(1):49–62. https://doi.org/10.1111/j.1475-4959.2000.tb00006.x

Defersha MB, Melesse AM (2012) Effect of rainfall intensity, slope and antecedent moisture content on sediment concentration and sediment enrichment ratio. CATENA 90:47–52

Defersha MB, Quraishi S, Melesse AM (2010) Interrill erosion, runoff and sediment size distribution as affected by slope steepness and antecedent moisture content. Hydrol Earth Syst Sci Discuss 7:6447–6489

Defersha MB, Quraishi S, Melesse AM (2011) The effect of slope steepness and antecedent moisture content on interrill erosion, runoff and sediment size distribution in the highlands of Ethiopia. Hydrol Earth Syst Sci 15:2367–2375. https://doi.org/10.5194/hess-15-2367-2011

Defersha MB, Melesse AM, McClain M (2012) Watershed scale application of WEPP and EROSION 3D models for assessment of potential sediment source areas and runoff flux in the Mara River Basin, Kenya. CATENA 95:63–72

Dembélé M, Ceperley N, Zwart SJ, Salvadore E, Mariethoz G, Schaefli B (2020a) Potential of satellite and reanalysis evaporation datasets for hydrological modelling under various model calibration strategies. Adv Water Res 143:103667. https://doi.org/10.1016/j.advwatres.2020.103667

Dembélé M, Schaefli B, van de Giesen N, Mariéthoz G (2020b) Suitability of 17 gridded rainfall and temperature datasets for large-scale hydrological modelling in West Africa. Hydrol Earth Syst Sci 24:5379–5406. https://doi.org/10.5194/hess-24-5379-2020

Dersseh MG, Kibret AA, Tilahun SA, Worqlul AW, Moges MA, Dagnew DC, Abebe WB, Melesse AM (2019). Potential of water hyacinth infestation on Lake Tana, Ethiopia: a prediction using a GIS-based multi-criteria technique

Dessie M, Verhoest NE, Pauwels VR, Adgo E, Deckers J, Poesen J, Nyssen J (2015) Water balance of a lake with floodplain buffering: Lake Tana, Blue Nile Basin, Ethiopia. J Hydrol 522:174–186

Dibaba WT, Demissie TA, Miegel K (2020) Watershed hydrological response to combined land use/land cover and climate change in highland Ethiopia: finchaa catchment. Water 12(6):1801

Dile YT, Tekleab S, Ayana EK, Gebrehiwot SG, Worqlul AW, Bayabil KH, Yimam YT, Tilahun SA, Daggupati P, Karlberg L, Srinivasan R (2018) Advances in water resources research in the Upper Blue Nile basin and the way forward: a review. J Hydrol 560:407–423. https://doi.org/10.1016/j.jhydrol.2018.03.042

Duan Z, Gao H, Ke C (2018) Estimation of lake outflow from the poorly gauged Lake Tana (Ethiopia) using satellite remote sensing data. Remote Sensing 10(7):1060

Ebabu K, Tsunekawa A, Haregeweyn N, Adgo E, Meshesha DT, Aklog D, Yibeltal M (2019) Effects of land use and sustainable land management practices on runoff and soil loss in the Upper Blue Nile basin, Ethiopia. Sci Total Environ 648:1462–1475

Elnashar A, Zeng H, Wu B, Fenta AA, Nabil M, Duerler R (2021) Soil erosion assessment in the Blue Nile Basin driven by a novel RUSLE-GEE framework. Sci Total Environ 793:148466

Endalamaw NT, Moges MA, Kebede YS, Alehegn BM, Sinshaw BG (2021) Potential soil loss estimation for conservation planning, upper Blue Nile Basin, Ethiopia. Environ Chall 5:100224

Erkossa T, Wudneh A, Desalegn B, Taye G (2015) Linking soil erosion to on-site financial cost: lessons from watersheds in the Blue Nile basin. Solid Earth 6:765–774. https://doi.org/10.5194/se-6-765-2015-108

Essayas E, Philpot W, Melesse A, Steenhuis T (2014) Bathymetry, Lake Area and volume mapping: a remote-sensing perspective. In: Melesse A, Abtew W, Setegn S (eds) Nile River Basin. Springer, Cham. https://doi.org/10.1007/978-3-319-02720-3_14

Fenta AA, Yasuda H, Shimizu K, Ibaraki Y, Haregeweyn N, Kawai T, Ebabu K et al (2018) Evaluation of satellite rainfall estimates over the Lake Tana basin at the source region of the Blue Nile River. Atmos Res 212:43–53

Galata AW (2020) Analysis of land use/land covers changes and their causes using Landsat data in hangar watershed, Abay basin, Ethipioa. J Sedim Environ 5(4):415–423

Gashaw T, Tulu T, Argaw M, Worqlul AW (2017) Evaluation and prediction of land use/land cover changes in the Andassa watershed, Blue Nile Basin, Ethipioa. Environ Syst Res 6(1):1–15

Gashaw T, Tulu T, Argaw M, Worqlul AW (2019) Modelling the impacts of land use–land cover changes on soil erosion and sediment yield in the Andassa watershed, upper Blue Nile basin, Ethiopia. Environ Earth Sci 78:1–22

Gashaw T, Worqlul AW, Dile YT, Addisu S, Bantider A, Zeleke G (2020) Evaluating potential impacts of land management practices on soil erosion in the Gilgel Abay watershed, upper Blue Nile basin. Heliyon 6(8):e04777

Gessesse B (2013) Characterization & modelling of landscape transformation for optimizing agricultural land use in the ethiopian highlands: a case study of modjo watershed. PhD Thesis, Addis Ababa University, Ethiopia and Friedrich-Alexander University of Erlangen-Nürnberg, 91054 Erlangen, Germany

Getu LA, Nagy A, Addis HK (2022) Soil loss estimation and severity mapping using the RUSLE model and GIS in Megech watershed, Ethiopia. Environ Chall 8:100560

Gleick PH (1996) Water resources. In: Schneider SH (eds) Encyclopedia of climate and weather, vol 2. Oxford University Press, New York, pp 817–823

Gumindoga W, Murwira A, Rwasoka DT, Jahure FB, Chikwiramakomo L (2020) The spatio-temporal soil moisture variation along the major tributaries of Zambezi River in the Mbire District, Zimbabwe. J Hydrol Reg Stud 32. https://doi.org/10.1016/j.ejrh.2020.100753

Gumindoga W, Rientjes THM, Reggiani P, Makurira H, Haile AT (2020) Hydrologic evaluation of bias corrected CMORPH rainfall estimates at the headwater catchment of the Zambezi River. Phys Chem Earth Parts a/b/c 115:21–32. https://doi.org/10.1016/j.pce.2019.11.004

Guzha AC, Rufino MC, Okoth S, Jacobs S, Nóbrega RLB (2018) Impacts of land use and land cover change on surface runoff, discharge and low flows: evidence from East Africa. J Hydrol: Reg Stud 15:49–67

Haregeweyn N, Tsunekawa A, Poesen J, Tsubo M, Meshesha DT, Fenta AA, Adgo E et al (2017) Comprehensive assessment of soil erosion risk for better land use planning in river basins: a case study of the Upper Blue Nile River. Sci Total Environ 574:95

Hastie A, Lauerwald R, Ciais P, Papa F, Regnier P (2021) Historical and future contributions of inland waters to the Congo basin carbon balance. Earth Syst Dyn 12:37–62. https://doi.org/10.5194/esd-12-37-2021

Hubau W, Lewis SL, Phillips OL, Affum-Baffoe K, Beeckman H, Cuní-Sanchez A, Daniels AK, Ewango CE, Fauset S, Mukinzi JM, Sheil D (2020) Asynchronous carbon sink saturation in African and Amazonian tropical forests. Nature 579(7797):80–87. https://doi.org/10.1038/s41586-020-2035-0

Jensen JR (2007) Remote sensing of the environment: an earth resource perspective (2ed). NJ, Prentice Hall, Upper Saddle River

Jensen JR (2014) Remote sensing of the environment: an Earth resource, vol 56. Wiley, pp 56–78

Kebede S, Travi Y, Alemayehu T, Marc VJJOH (2006) Water balance of Lake Tana and its sensitivity to fluctuations in rainfall, Blue Nile basin, Ethiopia. J Hydrol 316(1–4):233–247

Khaki M et al (2017) Assessing sequential data assimilation techniques for integrating GRACE data into a hydrological model. Adv Water Resour 107:301–316. https://doi.org/10.1016/j.advwatres.2017.07.001

Kim U, Kaluarachchi JJ (2009) Climate change impacts on water resources in the Upper Blue Nile River Basin, Ethiopia1. JAWRA J Am Water Resour Assoc 45:1361–1378. https://doi.org/10.1111/j.1752-1688.2009.00369.x

Klemas V, Pieterse A (2015) Using remote sensing to map and monitor water resources in arid and semiarid regions. Springer International Publishing Switzerland. In: Younos T, Parece TE (eds) Advances in watershed science and assessment, the handbook of environmental chemistry vol 33, pp 33–60. https://doi.org/10.1007/978-3-319-14212-82

Lakew HB, Moges SA, Asfaw DH (2017) Hydrological evaluation of satellite and reanalysis precipitation products in the Upper Blue Nile Basin: a case study of Gilgel Abbay. Hydrology 4(3):39

Lakew HB, Moges SA, Asfaw DH (2020) Hydrological performance evaluation of multiple satellite precipitation products in the upper Blue Nile basin, Ethiopia. J Hydrol: Reg Stud 27:100664

Maalim FK, Melesse AM (2013) Modeling the impacts of subsurface drainage systems on Runoff and Sediment Yield in the Le Sueur Watershed, Minnesota. Hydrol Sci J 58(3):1–17

Maalim FK, Melesse AM, Belmont P, Gran K (2013) Modeling the impact of land use changes on runoff and sediment yield in the Le Sueur Watershed, Minnesota using GeoWEPP. CATENA 107:35–45

MacDonald AM, Bonsor HC, Dochartaigh, BÉO, Taylor RG (2012) Quantitative maps of groundwater resources in Africa. Environ Res Lett 7:024009

Mamaru AM, Petra S, Seifu T, Tammo SS (2017) Water quality assessment by measuring and using landsat 7 ETM+ images for the current and previous trend perspective: Lake Tana Ethiopia. J Water Res Prot 9(12):1564–1585. https://doi.org/10.4236/jwarp.2017.912099

Menale AS, Zeleke GA (2018) Land use and land cover change and implication to watershed degradation by using GIS and remote sensing in the Koga watershed, North Western Ethiopia. Earth Sci Inf 11(1). https://doi.org/10.1007/s12145-017-0323-5

Mengistu HA, Demlie MB, Abiye TA (2019) Review: groundwater resource potential and status of groundwater resource development in Ethiopia. Hydrogeol J 27:1051–1065. https://doi.org/10.1007/s10040-019-01928-x

Ministry of Water Resources (MoWR), United Nations Educational, Scientific, and Cultural Organization (UNESCO) and Generations Integrated Rural Development Consultants (GIRDC) (2004) United Nations Educational, Scientific, and Cultural Organization World Water Assessment Program: National Water Development Report for Ethiopia. Addis Ababa, Ethiopia

Mohammed JA, Gashaw T, Tefera GW, Dile YT, Worqlul AW, Addisu S (2022) Changes in observed rainfall and temperature extremes in the upper Blue Nile basin of Ethiopia. Weather Clim Extremes 37:100468. https://doi.org/10.1016/j.wace.2022.100468

Newby J (2010) The facts of water in Africa. Retrieved on 10 October 2015, from www.panda.org/livingwaters

Papa F, Durand F, Rossow WB, Rahman A, Bala SK (2010) Satellite altimeter-derived monthly discharge of the Ganga-Brahmaputra River and its seasonal to interannual variations from 1993 to 2008. J Geophys Res 115(C12). https://doi.org/10.1029/2009JC006075

Papa F, Crétaux JF, Grippa M et al (2023) Water resources in Africa under global change: monitoring surface waters from space. Surv Geophys 44:43–93. https://doi.org/10.1007/s10712-022-09700-9

Schelwald-van LS, Reijerkerk L (2009) Water: a way of life. Taylor & Francis Group, London, UK

Shiklomanov I (1993) World freshwater resources. In: Gleick PH (ed) Water in crisis: a guide to the world's fresh water resources. Oxford University Press, New York

Sipes JL (2010) Sustainable solutions for water resources: policies, planning, design, and implementation. John Wiley & Sons Inc., Hoboken, New Jersey

Steffen W, Richardson K, Rockström J, Cornell SE, Fetzer I, Bennett EM, Biggs R, De Carpenter SR et al (2015). Planetary boundaries: guiding human development on a changing planet. Science 347:1259855. https://doi.org/10.1126/science.1259855

Strahler A (2013) Introducing physical geography, 6th edn. John Wiley & Sons, Inc., Boston, Massachusetts, USA

Tadesse D, Desta A, Geyid A, Girma W, Fisseha S, Schmoll O (2010) Rapid assessment of drinking-water quality in the Federal Republic of Ethiopia: country report of the pilot project implementation in 2004–2005. World Health Organization, Geneva, United Nations Children's.Addis Ababa,Ethiopia

Tadesse M, Sara H, Sokratis P, Isabel C (2022) Water quality and water hyacinth monitoring with the Sentinel-2A/B Satellites in Lake Tana (Ethiopia). Remote Sens 14(19):4921. https://doi.org/10.3390/rs14194921

Takele GS, Gebre GS, Gebremariam AG, Engida AN (2021) Hydrological modeling in the Upper Blue Nile basin using soil and water analysis tool (SWAT). Model Earth Syst Environ 1–16

Takele GS, Gebrie GS, Gebremariam AG, Engida AN (2022) Future climate change and impacts on water resources in the Upper Blue Nile basin. J Water Clim Change 13(2):908–925

Tassew BG, Belete MA, Miegel K (2019) Application of HEC-HMS model for flow simulation in the Lake Tana basin: the case of Gilgel Abay catchment, upper Blue Nile basin, Ethiopia. Hydrology 6(1):21

Taye M, Sahlu D, Zaitchik BF, Neka M (2020) Evaluation of satellite rainfall estimates for meteorological drought analysis over the upper Blue Nile basin, Ethiopia. Geosciences 10(9):352

Timketa A, Dand Gemechis AD (2022) Assessment of groundwater potential zones of upper blue nile river basin using multi-influencing factors under gis and rs environment: a case study on guder watersheds, Abay Basin, Oromia Region, Ethiopia. Geofluids. https://doi.org/10.1155/2022/1172039

Tramblay Y, Rouché N, Paturel JE, Mahé G, Boyer J-F, Amoussou E, Bodian A, Dacosta H, Dakhlaoui H, Dezetter A, Hughes D, Hanich L, Peugeot C, Tshimanga R, Lachassagne P (2021) ADHI: the African database of hydrometric indices (1950–2018). Earth Syst Sci Data 13:1547–1560. https://doi.org/10.5194/essd-13-1547-2021

United Nations Educational, Scientific and Cultural Organization (UNESCO) (2009) Water in a changing world. The United Nations World Water Development Report 3. UNESCO and Earthscan Publishing, London

Van Dijk AIJM, Renzullo LJ (2011) Water resource monitoring systems and the role of satellite observations. Hydrol Earth Syst Sci 15(39–55):201

van Dijk AIJM, Renzullo LJ, Wada Y, Tregoning PA (2014) Global water cycle reanalysis (2003–2012) merging satellite gravimetry and altimetry observations with a hydrological multi-model ensemble. Hydrol Earth Syst Sci 18:2955–2973. https://doi.org/10.5194/hess-18-2955-2014

Velásquez N, Vélez JI, Álvarez-Villa OD, Salamanca SP (2023) Comprehensive analysis of hydrological processes in a programmable environment: the watershed modeling framework. Hydrology 10:76. https://doi.org/10.3390/hydrology10040076

Worqlul AW, Collick AS, Tilahun SA, Langan S, Rientjes THM, Steenhuis TS (2015) Comparing TRMM 3B42, CFSR and ground-based rainfall estimates as input for hydrological models, in data scarce regions: the Upper Blue Nile Basin, Ethiopia. Hydrol Earth Syst Sci Discuss 12(2):2081–2112

Worqlul AW, Yen H, Collick AS, Tilahun SA, Langan S, Steenhuis TS (2017) Evaluation of CFSR, TMPA 3B42 and ground-based rainfall data as input for hydrological models, in data-scarce regions: The upper Blue Nile Basin, Ethiopia. CATENA 152:242–251

Worqlul AW, Ayana EK, Maathuis BH, MacAlister C, Philpot WD, Leyton JMO, Steenhuis TS (2018) Performance of bias corrected MPEG rainfall estimate for rainfall-runoff simulation in the upper Blue Nile Basin, Ethiopia. J Hydrol 556:1182–1191

Wu H, Adler RF, Tian Y, Gu G, Huffman GJ (2017) Evaluation of quantitative precipitation estimations through hydrological modeling in IFloodS River Basins. J Hydrometeorol 18:529–553. 10.1175/ JHM-D-15-0149.1

Yang P, Wu L, Cheng M, Fan J, Li S, Wang H, Qian L (2023) Review on drip irrigation: impact on crop yield, quality, and water productivity in China. Water 15:1733. https://doi.org/10.3390/w15091733

Yesuph AY, Dagnew AB (2019) Land use/cover spatiotemporal dynamics, driving forces and implications at the Beshillo catchment of the Blue Nile Basin, North Eastern Highlands of Ethiopia. Environ Syst Res 8(1):1–30

Chapter 19
Linking Earth Observation to Crop Area Mapping and Yield Estimation in the Abay Basin

Gebeyehu Abebe and Berhan Gessesse

Abstract Crop monitoring, mapping, and yield estimation has continued to be the focus of agricultural remote sensing studies. In this regard, multi-source imagery from Earth Observation satellites, often combined with ancillary data from ground observation, can provide repetitive and synoptic views of key crop parameters including crop nutrient and growth status. Different methods using earth observation data have been developed for this purpose. Approaches vary from purely empirical such as optical/microwave based and the synergistic methods between optical and microwave data to more complex ones, including assimilating Earth Observation (EO) data with crop growth model (CGMs) through continuously infusing EO data into the model system. The present review aims to offer a comprehensive and systematic review considering the different approaches of crop area mapping, monitoring, and yield estimation methods and its possible application from the Abay basin perspective, which differ in terms of the assumptions they entail and complexity as well as their requirement for field observation and other ancillary data. In this respect, the overview of the remote sensing methods is timely as such method is being applied in the operational crop mapping, retrieval of crop growth status, and crop monitoring by a series of satellite platforms such as Landsat 8 OLI and Sentinels 1 and 2. In general, EO data have the potential to improve the estimation accuracy of crop area, soil properties, canopy state variables, and crop yield based on quantitative remote sensing such as the assimilation of remote sensing variables and crop simulation models in the data-scarce smallholder agriculture such as the Abay basin, Ethiopia. Finally, considering the repetitive and synoptic views of key crop parameters of the EO satellites and the various methods such as empirical regressions between historical yield and in-season variables derived from remotely sensed data, assimilation of

---

G. Abebe (✉)
Department of Natural Resources Management, Debre Berhan University, Debre Berhan, Ethiopia
e-mail: gebeyehuabebe2010@gmail.com

B. Gessesse
Remote Sensing Department, Ethiopian Space Science and Geospatial Institute, Addis Ababa, Ethiopia

Departments of Geography and Environmental Studies, Kotebe Metropolitan University, Addis Ababa, Ethiopia

remote sensing and dynamic crop growth models (CGMs), or on integration between empirical regression model with CGMs, we propose new methods or future directions for crop area mapping monitoring and yield estimation improvement in the Abay basin using agricultural remote sensing approaches.

Keywords CGM · Crop monitoring · Microwave and optical sensors · Yield estimation · Abay basin

Introduction

Crop monitoring during the growing season, crop mapping, and estimating crop yields are important for the assessment of seasonal production (Doraiswamy et al. 2003). In this context, the rapid advancements of earth observation (EO) satellites have provided the opportunity to improve regional crop monitoring and yield forecasting (Jiang et al. 2014) and several studies derived useful information about agricultural fields from EO data. Compared with ground-based methods, such as survey sampling and visual observation, EO is highly suitable for mapping and monitoring crop conditions over large areas (Xin et al. 2013). EO applications have been studied extensively during the past decades, providing important information about the growth and development of crops. Recently EO data have been extensively applied to a wide range of applications such as crop yield estimation and forecasting (Ferencz et al. 2004).

Currently, EO sensors provide us data acquired from wide spectral range, various orbits and in different spatial and temporal resolutions (Atzberger 2013). Several EO-based methods were developed for crop monitoring, mapping, and yield estimation in different parts of the world (Wojtowicz et al. 2016; Becker-Reshef et al. 2010; Svotwa et al. 2014). Hence, the use of EO has proved to be very important in mapping and monitoring the growth of crops (Prasad et al. 2006). In this respect, both optical and microwave EO datasets have been used to monitor, map, and estimate crop yield, which are discussed in the following sub-topics.

The rest of the chapter is organized as follows: Sect. 19.2 highlights the empirical methods of crop area estimation and mapping using EO products. Section 19.3 outlines the empirical EO-based methods used in crop growth monitoring are presented, while the empirical EO-based methods for crop yield estimation is presented in Sect. 19.4. Section 19.5 provides highlights into the integration of EO with other methods such as meteorological parameters and the crop growth model (CGM). Finally, the concluding remarks of the chapter are presented in Sect. 19.6. Four operational applications have been selected, which are described as follows: (1) crop area estimation and mapping; (2) crop biophysical parameters retrieval; (3) crop calendar and phenology analysis, and (4) crop yield estimation. Many other remarkable applications exist, such as crop variety identification, crop disease detection, detection of nutrient deficiencies, crop damage and loss assessment, vegetation vigor and drought monitoring, irrigation management and precision agriculture, however, were not included in this review chapter to keep the paper more concise.

Crop Area Estimation and Mapping

Optical EO Data for Crop Area Estimation and Mapping

Optical data were extensively used for crop area estimation and mapping (Hao et al. 2015; Skakun et al. 2017; Mosleh et al. 2015). Hao et al. (2015) presented a methodology for multi-temporal MODIS data to study the effect of the time series with different lengths on crop mapping. Eight time series ranging from one month to eight months were tested, and they used the Random Forest (RF) to calculate the variable importance for all features including multi-spectral data, NDVI, NDWI, and phonological metrics.

Zhou et al. (2015) using multi-temporal Chinese HJ-1 CCD imaging data proposed the object-oriented method (OOM) and data mining (DM) for mapping sugarcane over large areas and found an overall accuracy of 93.6% and Kappa coefficient of 0.85. Rimal et al. (2018) studied cropping patterns of Nepal using multi-temporal MODIS NDVI data for the year 2016. Skakun et al. (2017) examined the possible use of NDVI derived from time-series MODIS images and growing degree days (GDD) data for winter crop mapping. Their validation using the Crop Data Layer (CDL) of the US Department of Agriculture (USDA) for Kansas, and ground measurements for Ukraine showed that accuracies of 90% could be obtained in mapping winter crops before harvesting date.

On the other hand, Mulyono (2016) presented a methodology for the detection of sugarcane plantations using Support Vector Machines (SVM) and phenology profiles using EVI value from Landsat 8 OLI data. Verma et al. (2017) also investigated the feasibility of using Indian EO Satellite (IRS-P6) LISS IV imaging data to discriminate the sugarcane and found that the Decision Tree method provided the best performance. Sonobe et al. (2018) conducted crop classifications using Sentinel-2 MSI, and the results showed that the SVM attained a better overall accuracy of to 92.0% compared to 89.3% achieved by the RF algorithm.

Recent studies have used the fusion of higher temporal frequency MODIS data with higher spatial resolution Landsat data for crop mapping and areal estimation. For instance, Li et al. (2014) used Landsat TM data (30 m) with MODIS NDVI (250 m) product time-series VI data for crop identification and area estimation in Hungary. The finding of this study demonstrated that the spectral data from high-spatial resolution images and time-series NDVI data played complementary roles in crop mapping and identification. In another study, Heupel et al. (2018) proposed a crop classification algorithm using crop phenological development and the reflectance characteristics based on time-series EO data of Landsat-7 and -8, Sentinel-2A, and RapidEye.

Useya and Chen (2018) compared the performance of and pixel and decision levels image fusion using ensemble classification and Landsat 8, Landsat 7, and Sentinel-2 data, and the results showed that decision-level method achieved 85.4% better accuracy compared to 82.5% attained by pixel-level image fusion. In their study Piedelobo et al. (2019), presented crop classification approach by a fusion of time series and Landsat-8 OLI and Sentinel- 2 MSI data; the results showed that more accurate results of classification were obtained by Ensemble Bagged Trees (EBT) algorithm. In another study, Eggen et al. (2016), used multi-temporal Landsat imageries to classify LULC in the Blue Nile/Abay basin, Ethiopia. The authors suggested the use of time-series Landsat data to detect LULC patterns on a complex and fragmented smallholder agricultural landscape.

SAR EO Data for Crop Area Estimation and Mapping

Traditionally, EO for crop mapping has depended upon optical imageries; however, the recent use of SAR data provides new possibilities for crop mapping, in areas where persistent cloud cover is a challenge for acquiring quality optical images (Mosleh et al. 2015). Access to SAR datasets such as ALOS PALSAR, RADARSAT-2, TerraSAR-X, RISAT-1, Cosmo-Skymed, and Sentinel-1 could play a significant role in crop mapping (Shao et al. 2001), since, SAR sensors have the capability of acquiring images under cloud cover, snow, and rain conditions (Mosleh et al. 2015). Currently, SAR sensors that provide spatial resolution similar to that of high-resolution VIS/IR satellites are appropriate for local crop area estimation, mapping, and precision agriculture.

Hence, recent studies have recognized SAR as a valuable tool for crop mapping. For instance, Bargiel (2017) using Sentinel-1 images, proposed a new classification approach that incorporates phenological development of crops. The results showed that multi-temporal Sentinel-1 images provided for improvised classification accuracies of canola, grasslands, sugar beets, maize, and potatoes using Random Forest, Maximum Likelihood, and PSP classification methods. Jiang et al. (2019) developed a methodology for mapping sugarcane using the Sentinel 1A SAR imaging data; the results showed an accuracy of Kappa coefficient above 0.9 using multi-temporal data.

In another study, Useya and Chen (2019) used multi-temporal Sentinel-1 data to detect cropping patterns on small-scale farmlands in Zimbabwe. Results demonstrated the operational application of Sentinel-1 SAR imaging data to map cropping patterns on small-scale farmlands. Recently, Dey et al. (2020) proposed the potential use of multi-temporal RADARSAT-2 data using random forest (RF) and extreme gradient boosting (XGB) classifiers for crop mapping in Canada and India. The results showed that the overall accuracy (OA) of the multi-temporal RADARSAT-2 dataset was better than the day-wise data.

Similarly, Gao et al. (2019) used multi-temporal sentinel-1 data for crop type classification in the Gansu province, China, and the results showed that the proposed method achieved an OA above 90%. Arias et al. (2020) evaluated the possible use of multi-temporal Sentinel-1 data for crop classification and the results showed OA higher than 70% with VH, VV, and VH/VV bands as the input. However, there is no conclusive evidence that a single SAR or optical dataset is optimal for crop area estimation and mapping applications. Integrating information from microwave and optical data provides promising ways to take full advantage of the available and upcoming EO satellites (McNairn and Brisco 2004). Studies suggested an improvement in crop area estimation and mapping, with the integration of optical and SAR imaging data compared with either optical data or single SAR data alone.

Optical and SAR Data Fusion for Crop Area Estimation and Mapping

Several studies used optical imaging data for crop area estimation and mapping. However, optical radiation cannot penetrate clouds, and this remains a major drawback for optical sensors (Van Tricht et al. 2018). Clouds and cloud shadows, thus, affect the quality of optical imageries and lead to missing data in optical time series. Alternative approaches, particularly strategies that exploit the fusion of optical and SAR data should be explored to overcome the impact of cloud and cloud shadow (Joshi et al. 2016a). The C-band, for instance, can penetrate clouds and has wavelengths well above the size of cloud particles (Van Tricht et al. 2018). Besides, optical, and microwave bands vary in their sensitivity to different crop characteristics, and data from these sensors could provide complementary data (McNairn and Brisco 2004; Joshi et al. 2016a).

As a result, in addition to the independent use of optical and microwave data, many studies have been conducted upon fusion of both data together for crop mapping (Forkuor et al. 2014; Van Tricht et al. 2018; Talema and Hailu 2020). For instance, Forkuor et al. (2014) attempted to map crops and crop groups in northwestern Benin, West Africa, using the integration of multi-temporal RapidEye and TerraSAR-X data. The results showed that integration of RapidEye and TerraSAR-X data improved classification accuracy by 10–15% compared to the use of RapidEye only. Blaes et al. (2005) evaluated the possible application of ERS and Radar sat images to map crop types using and photograph interpretation and per-parcel classification approaches. Van Tricht et al. (2018) developed an optimized RF classifier algorithm to produce a crop map for Belgium based on the combined use of Sentinel-1 and Sentinel-2 data. They found that integration of radar and optical imagery outperformed a classification based on single-sensor inputs.

In another study, Talema and Hailu (2020) presented a methodology for fusion of Sentinel-1A data and Sentinel-2A VI for mapping crop fields in the Fogera district, Northwestern Ethiopia. The result showed that fusion of VIs from Sentinel-2A imaging data with Sentinel-1 SAR multi-temporal backscatter improved the accuracy of crop classification by 0.08. Using multi-temporal Landsat 7/8 and Sentinel

1/2 images, Wang et al. (2020) proposed a pixel- and phenology-based classification for mapping sugarcane in the Guangxi province, China. The results showed the potential of the proposed method in mapping croplands and cropping intensity in the fragmented crop fields and complex landscapes.

Hence, the combined use of SAR and optical sensors has led to intensive research activities toward the application of EO technologies. Synergistic use of optical and SAR data provides valuable information for applications such as crop type mapping, crop condition, crop monitoring, and crop yield (Joshi et al. 2016a; Forkuor et al. 2014).

EO-Data for Crop Growth Monitoring

Optical EO Data for Crop Growth Monitoring

(i) Estimation of Crop Growth Parameters

The progress in remote sensing techniques has provided alternative approaches to monitor crop growth parameter during its growth stages. Previous studies have shown the potential use of optical EO for estimating crop biophysical parameters such as Leaf Area Index (LAI), Nitrogen uptake, total chlorophyll (Chl) content, biomass, and so on; suggesting that the retrieved crop biophysical parameters can be used to monitor status and productivity of different crops (Fu et al. 2014, Liu et al. 2012, Jin et al. 2013, Xiao et al. 2011, Prabhakara et al. 2015). For example, Xiao et al. (2011) used MODIS time-series data (MOD09A1) for real-time estimation of LAI; the results showed that the estimation method is able to efficiently produce a relatively smooth LAI data. Liu et al. (2012) used Landsat data to estimate green LAI of three crop types (corn, soybean and wheat) at a regional scale; the results demonstrated that the estimated LAI agreed well with the observed LAI.

In another study, Kross et al. (2015) using EO data estimated LAI and biomass of corn and soybean; the analysis showed that continuous crop LAI monitoring can be achieved with a combined use of RapidEye, Landsat, and SPOT imaging data. Zhou et al. (2016) using HJ-CCD VIs and the RF algorithm estimated wheat biomass; the results showed that the RF achieved more accurate estimates of wheat biomass compared with the SVR and ANN algorithms. Boschetti et al. (2017) used MODIS data and a rule-based algorithm for automatic deriving of essential temporal information on the rice crop. The results showed that PhenoRice algorithm is robust for extracting temporal descriptions of rice crop which is suitable for regional crop monitoring on a seasonal basis.

The development of hyperspectral EO/imaging spectroscopy, which provides contiguous narrow spectral measurements of reflected light, opens a new opportunity for monitoring crop parameters (Yue et al. 2017). For instance, Jin et al. (2013) estimated wheat LAI, N uptake, and total Chl content using in-situ spectral data; the results showed that the spectral indices could be used to accurately estimate wheat growth parameters. Prabhakara et al. (2015) used CROPSCAN sensor data to estimate the percent ground cover and biomass of winter cover crops; the results showed that taking into account senescence, index saturation, and frost burn on leaves can improve the estimation accuracy of percent ground cover and biomass for winter crops.

Fu et al. (2014) evaluated hyperspectral indices combined with multivariate techniques and band depth parameters; the results showed that Partial Least Square Regression (PLSR) analysis with the optimal VIs and band depth parameters significantly improved winter wheat biomass estimation. Tilly et al. (2015) used plant height and in-situ hyperspectral data to estimate barley biomass; the results showed the potential of hyperspectral data for estimations of barley biomass. Elarab et al. (2015) proposed the possible use of imagery from AggieAir platform with RVM algorithm to estimate Chl concentration for operational applications in precision agriculture.

Yue et al. (2017) proposed snapshot hyperspectral sensor mounted on UAV to estimate above-ground biomass (AGB); the results showed that crop height from the UAV sensor can improve AGB estimation. Cheng et al. (2017) compared dry matter indices (DMIs) with chlorophyll indices (CIs) to estimate the biomass of individual components such as leaves, stems, and result of the analysis showed that DMIs performed better as compared to CIs.

Recently, the development of light detection and ranging (LIDAR) technology has offered new possibilities for monitoring crop growth parameters. For instance, Eitel et al. (2016) used an ATLS to estimate crop biomass; the results showed the feasibility of ATLS technology for crop biomass monitoring and mapping. De Souza et al. (2017) proposed the use of crop surface models (CSMs) extracted from UAV point cloud data for estimating sugarcane height. The results of the analysis showed that the method is use full for estimating the average height of an entire field crop at once instead of ground measurements.

(ii) Crop Calendar and Phenology Analysis

Crop phenology is a key parameter for crop monitoring, yield estimation, and crop simulation (Misra et al. 2020). In this context, monitoring crop phenology can be used an indicator of crop productivity and plays an important role in crop management such as irrigation and fertilizer applications, (Sakamoto et al. 2013). Satellite observation data have a vital role in characterizing spatiotemporal patterns of vegetation phenology at local, regional, and global scales. However, repeated observations using medium to coarse spatial resolution satellite sensors, such as AVHRR, MODIS, and Landsat are required to monitor crop phenology (Misra et al. 2020).

In their study, Sakamoto et al. (2005), estimated the planting, heading, and harvesting stage of rice crop by detecting inflection points. In another study, Sakamoto et al. (2010) developed a method for detecting maize and soybean phenology using time-series MODIS data and a two-step filtering approach. The results showed that agronomic stages of corn and soybeans could be estimated by establishing the relationship between ground-based, agronomic stages of these crops and multi-temporal VI.

In their study, Meng et al. (2009) developed a model for estimating the phenological stage of winter wheat using MERIS data, and the results showed that the error for the heading date and flowering date of winter wheat was less than 3 days. You et al. (2013) demonstrated the use of NOAA AVHRR imagery data and threshold method in detecting the start and end of the growing season for 43 agricultural zones in China. The results show that the developed method had better accuracy compared with other methods. In their study, Pan et al. (2015a, b) developed a method for crop phenology using multi-temporal NDVI from HJ-1 A/B image and their analysis showed that the crop season start/end derived from HJ-1 A/B NDVI data was comparable with ground agro-metrological observation.

Xu et al. (2017) evaluated the feasibility of optimizing the settings of BISE, thresholds, and smoothing algorithms using the calibration dataset of ground-based, crop growth stage observations for estimating phenological stages of different crops. In their study, Huang et al. (2019) derived the start of the season (SOS) and the end of the season (EOS) of rice, wheat, and maize crops using the optimal thresholds method, and the results showed that the modified dynamic threshold method is more applicable to extract crop SOS/EOS.

Onojeghuo et al. (2018) developed a data fusion model to downscale the MODIS data to extract a 30 m multi-temporal NDVI products throughout the rice-growing season. The authors concluded that the downscaled NDVI data were able to characterize the development of paddy rice across wide area over multiple growing seasons. In addition to multispectral-based VIs, several other EO derived variables including solar-induced chlorophyll fluorescence (SIF) (Jeong et al. 2017), vegetation optical depth (VOD) retrieved from microwave sensors (Jones et al. 2011), and SAR polarimetric parameters (McNairn et al. 2018) were also successfully used for monitoring crop phenology.

SAR Data for Crop Growth Monitoring

(i) Estimation of Crop Growth Parameters

Recent studies have reported the potential use of microwave EO for estimating crop parameters such as LAI, AGB, crop height, and water content (Hosseini et al. 2015; Kumar et al. 2018; Pichierri et al. 2018; Liao et al. 2018). For instance, Hosseini et al. (2015) estimated corn and soybeans LAI using RADARSAT-2 (C-band) and Uninhabited Aerial Vehicle Synthetic Aperture Radar (UAVSAR) (L-band); the result demonstrated that a high correlation coefficients between ground measured and estimated values. Choudhury and Chakraborty (2006) evaluated the capability of RADARSAT time-series data to monitor rice crop; the results demonstrated the feasibility of RADARSAT data for rice monitoring.

Kumar et al. (2018) estimated winter wheat crop parameters using Sentinel-1A data and RFR, SVR, ANNR, and LR algorithms; the results showed the potential use of Sentinel-1A data combined with the RFR model for estimating winter wheat parameters. Pichierri et al. (2018) used multi-temporal Pol-InSAR data to estimate crop biomass, water content (VWC), and canopy structure; the results showed the sensitivity of the Pol-InSAR parameters to biomass and canopy structure.

Liao et al. (2018) demonstrated the feasibility of RADARSAT-2 polarimetric data to estimate crop height and fractional vegetation cover (FVC) for corn and wheat. The results showed that the corn height and FVC are strongly correlated with SAR parameters at the early growing stage. Mandal et al. (2018) evaluated the feasibility of water cloud model and multi-output support vector regression (MSVR) using Sentinel-1 data to retrieve crop parameters such as PAI and wet biomass. The results showed that the proposed methodology has a high potential for simultaneous PAI and biomass retrieval.

In another study, Abdikan et al. (2018) attempted to evaluate the performance of Sentinel-1 backscatter image data for crop monitoring; the results showed that the Sentinel-1 data has a high potential for multi-temporal analyses for crop monitoring and mapping. Nasirzadehdizaji et al. (2019) used multi-temporal Sentinel-1 data to evaluate the sensitivity of SAR parameters to canopy coverage (CC) and height and of different crops; analysis of the results showed the potential of Sentinel-1 parameters for estimation of CC and height of the maize. Prudente et al. (2019) also attempted to retrieve the parameters of wheat and soybean crops using Sentinel-1 SAR interferometric and polarimetric and parameters.

(ii) Crop Calendar and Phenology Analysis

The operational application of multi-temporal optical data for retrieving crop calendar and phenological development has long been recognized. Recently launched EO satellites such as Sentinel-1 SAR sensor allow researchers to monitor crop phenology over large areas with high accuracy. Alemu and Henebry (2017) used a blended dataset from the passive microwave radiometers AMSR-E (Advanced Microwave Scanning Radiometer on EOS) and AMSR2 (AMSR), rainfall data from TRMM (Tropical Rainfall Measuring Mission), and ETa (actual evapotranspiration) data estimated from the simple surface energy balance model (SSEB) to characterize cropland dynamics in Eastern Africa. The results showed that the time series of the land surface variables displayed unimodal seasonality at study sites in Ethiopia and South Sudan, in contrast to bimodality at sites in Tanzania.

In another study, Kumar et al. (2017) evaluated the sensitivity of m-χ scattering powers for retrieving phenological stages of corn and wheat; the results demonstrated that the m and χ features along with the scattering powers are affected by the uncertainties in the transmission of a perfectly circular polarized wave. McNairn et al. (2018) evaluated the possible use of TerraSAR-X dual-polarization and RADARSAT-2 quad polarization combined with a dynamic filtering framework to estimate crop phenological stages. These results showed that the developed method could be used for monitoring the phonological development of canola over large areas. Wang et al. (2019a, b) proposed the possible use of random forest algorithm and RADARSAT-2 polarimetric parameters to monitor crop phenology; the results showed that the retrieved phenology from multiple polarimetric parameters using the random forest agrees well with the ground observations. Using the simulated Compact-Pol Synthetic Aperture Radar (CP-SAR) data from RADARSAT-2 analyzed the scattering powers from the iS-Ω decomposition for various crops at different growth stages (Kumar et al. 2020).

Schlund and Erasmi (2020) estimated the phenological stages of wheat fields, in Germany using Sentinel 1 SAR backscatter coefficient and interferometric coherence. The results showed the possible application of Sentinel 1 VH/VV ratio for monitoring the phenological stages of wheat fields. Phung et al. (2020) used multi-temporal Sentinel-1 data to estimate planting date, age, and harvest date of rice in the Mekong Delta, Vietnam. The results showed the capability of using sentinel 1 data to retrieve rice phenological information including its age, planting, and harvest dates. Yang et al. (2021) using Sentinel-1A imaging data attempted to obtain the spatiotemporal distribution of rice phenology, and the results demonstrated that the VH backscatter has better performance for accurate estimations of paddy rice phonological stages.

Optical and SAR Data Fusion for Crop Growth Monitoring

(i) Estimation of Crop Growth Parameters

Optical and radar systems are complementarily, yet there is a difference in their imaging and information contents (Forkuor et al. 2014). Previous studies have shown that the combined optical and SAR data performed better than those derived from either the SAR optical or optical data only (Koppe et al. 2012; Jin et al. 2015; Molijn et al. 2019; Alebele et al. 2020). Koppe et al. (2012) in their study on winter wheat biomass estimation reported improvements in parameter estimation based on the fusion of hyperspectral data and microwave backscatter.

In another study, Gao et al. (2013) estimated LAI, height, and biomass of maize based on the fusion of VIs from HJ-1 and RADARSAT-2 data; the results demonstrated that the method based on the fusion of optical and SAR data has provided better accuracy than using only the optical or microwave observation. Jin et al. (2015) evaluated the feasibility of using the combined Huanjing (HJ) and RADARSAT-2 data to estimate LAI and biomass in winter wheat. The results showed that the combined multi-temporal optical and radar parameters had a good relationship with winter wheat LAI and biomass. Torbick et al. (2017) used multi-temporal Sentinel-1A Interferometric Wide (IW) images to map rice extent, crop calendar, inundation, and cropping intensity in Myanmar; the results demonstrated the operational application of multi-temporal sentinel-1A data to monitor rice production over a large region.

Pratap et al. (2019) used time-series Sentinel 1-A SAR and Sentinel-2 optical data in combination with a modified water cloud model (MWCM) to estimate the LAI of wheat crops in the Varanasi district, India. The results showed that the estimated values of LAI by MWCM were found to show a better correlation than those values of LAI obtained by WCM at VV and VH polarization. Molijn et al. (2019) attempted to evaluate the dynamic behavior of signals from five SAR sensors and optical sensors with growing sugarcane; the results demonstrated that the highest agreements were found between the C-band SAR sensors and between the optical sensors.

Alebele et al. (2020) proposed a methodology to estimate above-ground biomass (AGB) of rice in the Jiangsu Province, China using multi-target Gaussian process regressor stacking (MGPRS) and the combined imaging data from Sentinel-1A and 2A imageries. The results showed that the combined indices performed better than those derived from either the Sentinel 2A or Sentinel 1A. Luo et al. (2020) attempted to estimate LAI and biomass of maize by combining spectral and texture features of optical and SAR data; the results demonstrated the capability of combining spectral and texture features for improving the estimation accuracy of LAI and biomass in maize. More recently Abebe et al. (2022a) developed a methodology to estimate LAI and above-ground biomass (AGB) of sugarcane in the Awash basin, Ethiopia, using Gaussian Process Regression (GPR) and optical and SAR data fusion (Landsat 8 and Sentinel-1A). The results showed that combined data from optical and SAR imaging based on the GPR could improve LAI and biomass estimation.

(ii) Crop Calendar and Phenology Analysis

Recently, Zhi et al. (2017) proposed a methodology for estimation of crop phenology based on the fusion of SAR and multispectral data. The overall phenological estimation accuracy (OPEA) based on the fusion of SAR and multispectral data (86.59%); better than that based on either the SAR signatures or the optical VIs only. More recently, Mercier et al. (2020) attempted to predict wheat and rapeseed phenological stages by the synergistic use of Sentinel-1 SAR and Sentinel-2 optical data. The results showed that combining Sentinel-1 and Sentinel-2 data allowed better identification of the beginning and end of tillering for wheat and the beginning and end of ripening for rapeseed.

EO Data for Crop Yield Estimation

Several EO-based methods were developed for estimating crop yield in different parts of the world. Several studies have been conducted on the application of empirical methods to crop yield estimation from remotely sensed and in-situ data.

Optical Data for Crop Yield Estimation

Several studies have under taken to explore the usefulness of optical EO sensors for estimating crop yield (Li et al. 2014; Sibley et al. 2014; Ferencz et al. 2004). These optical sensors have the potential of obtaining multi-spectral data over croplands that can be used for deriving multi-temporal VIs (Mosleh et al. 2015). In this regard, VIs is widely used for monitoring various crop characteristics, primarily due to their simplicity in application and ease of data processing (Nguy-Robertson et al. 2014). The principal assumption in use of remotely sensed data for crop yield estimation is that the spectral data is strongly related with canopy parameters, which are related to the final yield at the critical stage of the growth. So far, several vegetation indices have been developed from optical satellite images, which can be used to estimate crop yield.

For instance, Bégué et al. (2010) used ground observations and SPOT4 and SPOT5 time-series data to study the spatiotemporal variability of sugarcane fields to select appropriate spectral indicators for yield forecast. The results showed that maximum NDVI and integrated NDVI methods gave comparable RMSE at the field-scale, i.e., 13.2 t/ha and 15 t/ha, respectively. Duveiller et al. (2013) conducted research for sugarcane yield estimation and monitoring using the SPOT fAPAR time-series dataset, and they found that sugarcane yield can be estimated with a RMSE of 1.5 t/ha. Kouadio et al. (2014) proposed a spring wheat yield forecasting scheme at the eco-district scale using an integrated crop yield forecasting approach, and the results showed that forecasting error could be reduced using MODIS VIs.

Jurecka et al. (2016) conducted a study to estimate crop yield at the field level using VIs obtained from Landsat OLI, Sentinel 2, drone, and field green seeker; the results showed that satellite data can obtain better results than the drone data. In another study, Fernandes et al. (2017) used time series of NDVI and neural networks algorithm to estimate sugarcane yield production in Brazil (t/ha). Zhou et al. (2017) attempted to predict rice grain yield using VIs extracted from the multispectral and digital images from UAV sensors; the results showed that both multispectral and digital images for rice growth and grain yield estimation.

In their study, Pinheiro Lisboa et al. (2018) developed a sugarcane yield estimation model using NDVI and the concentration of leaf-tissue nutrients. The results showed that the models could be useful for monitoring spatiotemporal crop yield changes induced by straw removal. Skakun et al. (2019) evaluated the possible use of multi-source optical data for crop yield estimation; the results showed that the combined use of Landsat 8 and Sentinel-2A/B data performed than a single sensor (2016–2018). In another study, Abebe et al. (2022b) proposed a support vector regression (SVR) based on a combined Landsat 8 and sentinel 2A data; the results showed that the empirical prediction error could be significantly reduced by making use of the combined Landsat 8 and Sentinel 2A data based on the SVR algorithm.

The application of optical imagery for crop yield estimation was reported in earlier studies. However, the availability of recent multi-temporal satellite imageries such as Landsat 8 OLI and Sentinels-2A provides new insights into the potential of improved VIs derived from those new optical imageries for yield estimation studies. Even though it has been argued that RS may not be suitable for operational applications in developing countries due to the very small farm sizes and complex agricultural systems, this difficulty is likely to persist shortly because of RS's inability to estimate yield in a complex small holder agriculture systems. The increased availability of high-spatial-resolution multispectral satellite and airborne data (including UAV acquired data) at reasonable costs makes this technique a viable and interesting alternative for crop yield forecasting in small holding agricultural system such as the Abay basin (upper Blue Nile).

SAR Data for Crop Yield Estimation

Several studies have been conducted since the 1990's for crop yield estimation using SAR imagery and the findings revealed encouraging results. In their study, Shao et al. (2001) evaluated the performance of time-series RADARSAT-1 imaging data at different growth stages for rice yield estimation in Guangdong Province, China; the results showed an accuracy of 91% compared with the actual yield. In their study, Li et al. (2003) developed a model using multi-temporal RADARSAT-1 images acquired during different growth stages for rice yield estimation in Guangdong Province, China, and the results demonstrated that Radarsat ScanSAR data has a high potential to estimate rice yield in a large area. Chen and McNairn (2006) developed a neural network-based algorithm for estimating rice yield using the relationship between rice growth and radar backscatter from RADARSAT-1 data, and found promising accuracy (94%) compared with the actual yield.

In their study, Zhang et al. (2017) attempted to estimate rice production from Radarsat-2 data using rice canopy scattering model with genetic algorithm; the proposed methods demonstrated the capability of Radarsat-2 imaging data for retrieving rice parameters and operational yield estimation. Clauss et al. (2018) evaluated the performance of Sentinel-1 SAR time-series data to estimate rice production in the Mekong Delta, Vietnam. A Random Forest algorithm was developed between backscatter coefficients and in-situ grain yield data collected over 357 rice farms. The analysis of the results showed that the potential of Sentinel-1 data to estimate rice grain yield at the local scale. In another study, Wang et al. (2019a, b) attempted to estimate rice grain yield using a Sentinel-1A-based SAR simple difference (SSD) index, and the results showed the potential of Sentinel-1A-based SAR index to estimate rice grain yield.

Optical and Microwave Data Fusion for Crop Yield Estimation

Recently, several studies have been conducted upon synergistic use of optical and SAR datasets for estimating crop yield. For instance, Heinzel (2006) demonstrated that integrating multi-source optical and SAR data significantly improves the accuracy of the yield prediction. In another study, Fieuzal et al. (2017) used multi-temporal optical and SAR imaging data and neural networks algorithms to estimate corn yield. The results showed that R^2 of 0.69 and RMSE of 7.0 q/ha are during the development of the central stem by the combined use of both optical and SAR signals. In their study, Mateo-Sanchis et al. (2019) proposed the combination of multi-sensor (optical and microwave) EO data for crop yield estimation; the results showed potential of using of EVI and VOD at the same time, with machine learning algorithms.

Integration of EO Data with Other Methods

Many crop yield estimation techniques have been proposed that make use of EO and other methods. According to Mosleh et al. (2015), the synergy between EO and other methods could broadly be classified into two categories: (i) meteorological parameters; and (ii) crop growth models (CGM). These categories are briefly discussed below in the following sub-topics.

Integration of EO Data with Meteorological Parameters

This method is based on the integration of meteorological parameters with EO methods for estimating crop yield/production. For instance, in a study of crop yield estimation, Prasad et al. (2007) used 10-day composite of NDVI data derived from NOAA AVHRR images integrated with meteorological parameters (i.e., surface temperature and rainfall) and soil moisture to predict crop yield over India; the prediction model demonstrated a high correlation coefficient that achieved an accuracy of greater than 90%. Sarma et al. (2008) evaluated the significance of meteorological variables (annual rainfall, southern oscillation index, sea surface temperature, and growing degree day) combined with NDVI derived from AVHRR imagery to develop a rice yield prediction model for Andra Pradesh, India. The results showed that the estimated rice yield was consistent with the actual yield. Huang et al. (2014) conducted a study in Yunnan province of China, and they proposed a crop yield estimation method using meteorological factors and NDVI data in small regions where crop type is unknown exactly.

In their study, Saeed et al. (2017) used weather and MODIS NDVI data to forecast wheat yield in Punjab province, India; the results showed that the sunshine hours in combination with NDVI performed best for estimating wheat. Recently, Ma et al. (2018) developed a Deep Learning (DL) algorithm, the Stacked Sparse Autoencoder (SSAE), for rice production estimation using climatic and MODIS data. Various scenarios were developed for the selection of optimal features in terms of the length of the crop season and aggregation periods. Their study reported that with the combination of the best scenario and the set of best parameters, the SSAE model showed an encouraging prediction accuracy (i.e., RMSE% of 6.89%), compared with the ANN model (i.e., RMSE% of 8.03%). More recently, Prasad et al. (2021) used VIs derived from multi-sensor EO satellites and a random forest (RF) algorithm to estimate cotton yield in the state of Maharashtra, India. The results demonstrated the possible use of RF to integrate and process a large number of inputs from different sources and avoid fitting the model.

Integration of EO Data with Crop Growth Model (CGM)

In predicting crop yield effectively, integration of EO data with crop growth model (CGM) has a great advantage when compared with traditional methods. Due to the synoptic and continuous view of EO, empirical relationships between crop yield and remotely sensed data, particularly VIs were often used in earlier studies (Yang et al. 2004; Ma et al. 2011). Previous studies demonstrated that the integration of remotely sensed data and CGM have become progressively recognized as a potential tool for crop yield forecasting (Huang et al. 2016; Jiang et al. 2014; Jin et al. 2015). This approach depends on the retrieval of crop parameters such as LAI and AGB from remotely sensed data; the different techniques to combine a CGM with remotely sensed data were first described by Ma in 1988. According to Delécolle et al. (1992), four methods of data assimilation have been identified:

(i) the direct use of a driving variable estimated from EO in the model;
(ii) the updating of a state variable of the model derived from EO;
(iii) the re-initialization of the model, i.e., the adjustment of an initial condition;
(iv) the re-calibration of the model i.e., the adjustment of model parameters.

However, Ma et al. (2011) in their study suggested mainly two strategies of integration: The first strategy, reinitializes crop model parameters by minimizing the differences between crop state variables simulated by the crop model and derived from EO; the second strategy, coupling a radiation transfer model to a crop model. Ma et al. (2011) coupled the WOFOST model with SAIL-PROSPECT model through LAI to simulate SAVI. The results showed that the spatial distribution of simulated weight of storage organ at the potential production level was more consistent to official yields.

Previous studies reported that the synergistic use of EO data with CGM could be used to improve the accuracy of crop yield estimation at local and regional scale. Curnel et al. (2011) assimilated wheat LAI derived from remotely sensed data into the WOFOST model using a re-calibration-based technique; the results demonstrated that EO data can be used to improve crop yield estimations. In another study, Ma et al. (2013) assimilated HJ-1 CCD NDVI data into WOFOST with the EnKF algorithm to estimate winter wheat; the results showed that winter wheat yield estimation was significantly improved.

In their study on sugarcane yield forecasting in smallholders farming conditions, Morel et al. (2014) compared the empirical relationship with a growing season-integrated NDVI, the Kumar–Monteith efficiency model, and a forced-coupling method with the sugarcane model MOSICAS and satellite-derived fAPAR. However, their results showed that the method based on the empirical integrated NDVI gave the most accurate estimation of crop yields. The HJ-1A/B satellite data was assimilated into the CERES-Wheat model using the ensemble-based 4DVar algorithm; and improved estimates of winter wheat yield in field plots were reported (Jiang et al. 2014). Huang et al. (2015) assimilated LAI from landsat TM and MODIS data into the WOFOST model to predict winter wheat at a regional scale using a four dimensional variational (4DVar) algorithm. Their results showed that assimilating the time-series LAI improved the estimation of wheat yield at both field and regional scales.

In another study, Huang et al. (2016) integrated multi-temporal LAI data with a 30 m resolution into the WOFOST model with a Kalman Filter (KF) algorithm and found improved accuracy of winter wheat yield estimation compared with the conventional approaches. Jin et al. (2016) proposed the synergistic use of winter wheat biomass derived from EO data with the AquaCrop model to improve biomass and yield estimations; the results showed that the estimated yield was in good agreement with the measured yield.

Mokhtari et al. (2018) assimilated the LAI derived from VIs and satellite surface incoming solar radiation into the SWAP model to improve crop yield estimation. Setiyono et al. (2018) integrated MODIS and SAR data into ORYZA model to estimate crop yield, and found that the predicted yield agreed well with the official yield data in the Red River Delta of Vietnam. Novelli et al. (2019) integrated the Environmental Policy Integrated Climate (EPIC) model with the LAI derived from the Sentinel-2 imagery using a re-calibration data assimilation approach to estimate winter wheat yield; the results showed that the assimilation of LAI with the EPIC model provided an improvement in yield estimation. Recently, Abebe et al. (2022c) proposed the assimilation of LAI with the WOrld FOod STudies (WOFOST) model to estimate sugarcane yield using an Ensemble Kalman Filter (EnKF) algorithm; the results showed that the accuracy of sugarcane yield estimated by the WOFOST model was significantly improved after LAI assimilation retrieved from the fusion of Landsat 8 Sentinel 1A data.

Furthermore, the integration of CGM and EO data enable crop monitoring at national and regional scale and also can improve the accuracy of phenology date estimates compared to other approaches such as phenology matching (Zeng et al. 2016). For instance, Vintrou et al. (2014) demonstrated that the combined use of CGM and EO parameters provide a better estimate of crop phenology in the data-scarce West African countries. Zeng et al. (2016) incorporated CGM and phenology match methods to detect the phenological stages of two crops (corn and soybeans). The study showed that the simulated vegetation growth rate based on a temperature and photo-period response function was more appropriate to growth dynamics of both crop types than the calendar (day of year, DOY) and minimized the influence of inter-annual climatic fluctuations that can negatively affect VI-based crop phenology methods.

Summary

EO technology is crucially developing and wide-ranging trends are visible in EO for agriculture these days. In this context, the main conclusions and future outlooks of this review manuscript can be summarized as follows:

- This review covered a wide array of topics in crop monitoring, mapping, and yield estimation using EO imagery. In this regard, low- to medium-resolution optical EO data and related techniques have been widely used for crop monitoring, mapping, and yield estimation but might not be suitable in an area with consistent cloud cover. Complex cropping systems with fragmented agricultural landscape has also hindered crop monitoring, mapping, and yield estimation efforts at local and regional. But, the increased availability of high-resolution optical and radar imagery either at reasonable cost or free of charge such as RapidEye and Spot 6, landsat 8 OLI, Sentinel 1, and Sentinel 2 makes this approach a possible alternative for crop monitoring, mapping, and yield estimation in the small holder cereal crops such as the Abay basin, Ethiopia. In fact, optical and SAR data provide complementary information related to different crop characteristics. However, the main purpose of this review is not a complete overview or a supplement to the literature review in previous publications but to explore the possible future direction of crop monitoring, mapping, and yield estimation using EO data from the perspectives of cereal crops in the Abay basin, Ethiopia.
- The combined use of EO data and meteorological parameters is also an interesting alternative in crop monitoring, and yield estimation. Meteorological variables such as annual rainfall, temperature, evapotranspiration, etc., combined with vegetation indices such as NDVI product derived from satellite imagery to develop the statistical spectro agro-metrological model in predicting the crop yield. Some of these models of them are relatively easy to use and parameterize and can be applicable to some developing countries like Ethiopia. Another trend associated the development of crop monitoring and yield estimation is the synergistic use of crop parameters derived from remotely sensed data with CGM. Combining EO data with CGMs will improve the reliability and accuracy of crop monitoring and yield estimation, but model calibration might need extensive data input.
- From this review, it is understood that the use of remote sensing techniques might improve the spatiotemporal monitoring and analysis of small holder cereal crops in the Abay Basin. Importantly, this study synthesized the empirical studies carried out by other researchers in the field of remote sensing for crop mapping, monitoring, and yield estimation using different models and methods. The models, approaches, and methods in this review will be used to develop a remote sensing based approach for crop mapping, monitoring, and yield estimation in the Abay Basin, Ethiopia.
- The literature review has also helped the researcher to identify the research objectives and research questions. In the process, it has also provided an understanding of the various variables, and methodology as well as the importance of EO and multi-approach strategies in the area of crop mapping, monitoring, and yield estimation. This review will also be helpful to improve the knowledge and understanding of policy- and decision-makers about the benefits of agricultural remote sensing for crop mapping, monitoring, and yield estimation with notable application prospect, which in turn helps to design the future Ethiopian Strategic agricultural development framework.

Acknowledgements The authors would like to thank the anonymous reviewers who provided valuable comments on an earlier draft of this manuscript.

References

Abebe G, Tadesse T, Gessesse B (2022a) Estimating leaf area index and biomass of sugarcane based on Gaussian process regression using Landsat 8 and Sentinel 1A observations. Int J Image Data Fusion 1–31

Abebe G, Tadesse T, Gessesse B (2022b) Combined use of Landsat 8 and Sentinel 2A imagery for improved sugarcane yield estimation in Wonji-Shoa, Ethiopia. J Indian Soc Remote Sensing 50(1):143–157

Abebe G, Tadesse T, Gessesse B (2022c) Assimilation of leaf Area Index from multisource earth observation data into the WOFOST model for sugarcane yield estimation. Int J Remote Sens 43(2):698–720

Abdikan SAYGIN, Sekertekin A, Ustunern M, Sanli FB, Nasirzadehdizaji R (2018) Backscatter analysis using multi-temporal Sentinel-1 SAR data for crop growth of maize in Konya Basin, Turkey. Int Arch Photogramm Remote Sens Spat Inf Sci 42:9–13

Alebele Y, Zhang X, Wang W, Yang G, Yao X (2020) Estimation of canopy biomass components in paddy rice from combined estimation of canopy biomass components in paddy rice from combined optical and SAR data using multi-target Gaussian Regressor Stacking. Remote Sensing 12. https://doi.org/10.3390/rs12162564

Alemu WG, Henebry GM (2017) Land surface phenology and seasonality using cool earthlight in croplands of eastern Africa and the linkages to crop production. Remote Sensing 9(9):914. https://doi.org/10.3390/rs9090914

Arias M, Campo-Bescós MÁ, Álvarez-Mozos J (2020) Crop classification based on temporal signatures of Sentinel-1 observations over Navarre province, Spain. Remote Sensing 12(2):278

Atzberger C (2013) Advances in remote sensing of agriculture: context description, existing operational monitoring systems, and major information needs. Remote Sensing 5:949–981

Bargiel D (2017) A new method for crop classification combining time series of radar images and crop phenology information. Remote Sens Environ 198:369–383. https://doi.org/10.1016/j.rse.2017.06.022

Becker-Reshef I, Vermote E, Lindeman M, Justice C (2010) A generalized regression-based model for forecasting winter wheat yields in Kansas and Ukraine using MODIS data. Remote Sens Environ 114:1312–1323

Bégué A, Lebourgeois V, Bappel E, Todoroff P, Pellegrino A et al (2010) Spatio-temporal variability of sugarcane fields and recommendations for yield forecast using NDVI. Int J Remote Sens 31(20):5391–5407. https://doi.org/10.1080/01431160903349057

Blaes X, Vanhalle L, Defourny P (2005) Efficiency of crop identification based on optical and SAR image time series. Remote Sens Environ 96(3–4):352–365. https://doi.org/10.1016/j.rse.2005.03.010

Boschetti M, Busetto L, Manfron G, Laborte A, Asilo S, Pazhanivelan S, Nelson A (2017) PhenoRice: a method for automatic extraction of spatio-temporal information on rice crops using satellite data time series. Remote Sens Environ 194:347–365. https://doi.org/10.1016/j.rse.2017.03.029

Chen C, Mcnairn H (2006) A neural network integrated approach for rice crop monitoring. Int J Remote Sens 27(7):1367–1393

Cheng T, Song R, Li D, Zhou K, Zheng H, Yao X, Tian Y, Cao W, Zhu Y (2017) Spectroscopic estimation of biomass in canopy components of paddy rice using dry matter and chlorophyll indices. Remote Sensing 9(4). https://doi.org/10.3390/rs9040319

Choudhury I, Chakraborty M (2006) SAR signature investigation of rice crop using RADARSAT data. Int J Remote Sens 27(3):519–534

Clauss K, Ottinger M, Leinenkugel P, Kuenzer C (2018) Estimating rice production in the Mekong Delta, Vietnam, utilizing time series of Sentinel-1 SAR data. Int J Appl Earth Obs Geoinf 73:574–585

Curnel Y, de WitA JW, Duveiller G, Defourny P (2011) Potential performances of remotely sensed LAI assimilation in WOFOST model based on an OSS Experiment. Agric for Meteorol 151:1843–1855

De Souza CHW, Lamparelli RAC, Rocha JV, Magalhães PSG (2017) Height estimation of sugarcane using an unmanned aerial system (UAS) based on structure from motion (SfM) point clouds. Int J Remote Sens 38(8–10):2218–2230. https://doi.org/10.1080/01431161.2017.1285082

Delécolle R, Maas S, Guérif M, Baret F (1992) Remote sensing and crop production models: present trends. ISPRS J Photogramm Remote Sens 47(3):145–161. https://doi.org/10.1016/0924-271 6(92)90030-D

Dey S, Mandal D, Dingle L, Banerjee B, Kumar V, Mcnairn H, Bhattacharya A, Rao YS (2020) In-season crop classification using elements of the Kennaugh matrix derived from polarimetric RADARSAT-2 SAR data. Int J Appl Earth Obs Geoinf 88:102059. https://doi.org/10.1016/j.jag.2020.102059

Doraiswamy PC, Moulin S, Cook PW, Stern A (2003) Crop yield assessment from remote sensing. Photogramm Eng Remote Sens 69(6):665–674. https://doi.org/10.14358/PERS.69.6.665

Duveiller G, López-Lozano R, Baruth B (2013) Enhanced processing of 1-km spatial resolution fAPAR time series for sugarcane yield forecasting and monitoring. Remote Sensing 5(3):1091–1116. https://doi.org/10.3390/rs5031091

Eggen M, Ozdogan M, Zaitchik BF, Simane B (2016) Land cover classification in complex and fragmented agricultural landscapes of the Ethiopian highlands. Remote Sensing 8(12):1020

Eitel JUH, Magney TS, Vierling LA, Greaves HE, Zheng G (2016) An automated method to quantify crop height and calibrate satellite-derived biomass using hypertemporal lidar. Remote Sens Environ 187:414–422

Elarab M, Ticlavilca AM, Torres-Rua AF, Maslova I, McKee M (2015) Estimating chlorophyll with thermal and broadband multispectral high-resolution imagery from an unmanned aerial system using relevance vector machines for precision agriculture. Int J Appl Earth Obs Geoinf 43:32–42. https://doi.org/10.1016/j.jag.2015.03.017

Ferencz C, Bognar P, Lichtenberger J, Hamar D, Tarcsai G, Timár G, Ferencz-Árkos I et al (2004) Crop yield estimation by satellite remote sensing. Int J Remote Sens 25(20):4113–4149. https://doi.org/10.1080/01431160410001698870

Fernandes JL, Ebecken NFF, Esquerdo JCDM (2017) Sugarcane yield prediction in Brazil using NDVI time series and neural networks ensemble. Int J Remote Sens 38(16):4631–4644. https://doi.org/10.1080/01431161.2017.1325531

Fieuzal R, Sicre CM, Baup F (2017) Estimation of corn yield using multi-temporal optical and radar satellite data and artificial neural networks. Int J Appl Earth Obs Geoinf 57:14–23. https://doi.org/10.1016/j.jag.2016.12.011

Forkuor G, Conrad C, Thiel M, Ullmann T, Zoungrana E (2014) Integration of optical and synthetic aperture radar imagery for improving crop mapping in Northwestern Benin, West Africa. Remote Sensing 6:6472–6499. https://doi.org/10.3390/rs6076472

Fu Y, Yang G, Wang J, Song X, Feng H (2014) Winter wheat biomass estimation based on spectral indices, band depth analysis and partial least squares regression using hyperspectral measurements. Comput Electron Agric 100:51–59. https://doi.org/10.1016/j.compag.2013.10.010

Gao S, Niu Z, Huang N, Hou XH (2013) Estimating the leaf area index, height and biomass of maize using HJ-1 and RADARSAT-2. Int J Appl Earth Obs Geoinf 24:1–8

Gao H, Wang C, Wang G, Li Q, Zhu J (2019) A new crop classification method based on the time-varying feature curves of time series dual-polarization Sentinel-1 data sets. IEEE Geosci Remote Sens Lett 17(7):1183–1187. https://doi.org/10.1109/LGRS.2019.2943372

Hao P, Zhan Y, Wang L, Niu Z, Shakir M (2015) Feature selection of time series MODIS data for early crop classification using random forest: a case study in Kansas, USA. Remote Sensing 7(5):5347–5369. https://doi.org/10.3390/rs70505347

Heinzel V (2006) Synergetic use of optical and ERS-2 data for crop yield retrieval. Cent Remote Sens Land Surf 28:30

Heupel K, Spengler D, Itzerott S (2018) A progressive crop-type classification using multitemporal remote sensing data and phenological information. PFG–J Photogramm Remote Sens Geoinf Sci 86:53–69

Hosseini M, Mcnairn H, Merzouki A, Pacheco A (2015) Estimation of Leaf Area Index (LAI) in corn and soybeans using multi-polarization C- and L-band radar data. Remote Sens Environ 170:77–89. https://doi.org/10.1016/j.rse.2015.09.002

Huang J, Dai Q, Wang H, Han D (2014) Empirical regression model using NDVI, meteorological factors for estimation of wheat yield in Yunnan, China. CUNY Academic Works. http://academicworks.cuny.edu/cc_conf_hic/5

Huang J, Tian L, Liang S, Ma H, Becker-Reshef I, Huang Y, Su W, Zhang X, Zhu D, Wu W (2015) Improving winter wheat yield estimation by assimilation of the leaf area index from Landsat TM and MODIS data into the WOFOST model. Agric for Meteorol. https://doi.org/10.1016/j.agrformet.2015.02.001

Huang J, Sedano F, Huang Y, Ma H, Li X, Liang S, Tian L, Zhang X, Fan J, Wu W (2016) Assimilating a synthetic Kalman filter leaf area index series into the WOFOST model to improve regional winter wheat yield estimation. Agric for Meteorol 216:188–202

Huang X, Liu J, Zhu W, Atzberger C, Liu Q (2019) The optimal threshold and vegetation index time series for retrieving crop phenology based on a modified dynamic threshold method. Remote Sensing 11. https://doi.org/10.3390/rs11232725

Jeong S, Schimel D, Frankenberg C, Drewry DT, Fisher JB, Verma M, Berry JA, Jiang H, Li D, Jing W, Xu J, Huang J, Yang J, Chen S (2017) Early season mapping of sugarcane by applying machine learning algorithms to Sentinel-1A/2 time series data: a case study in Zhanjiang City, China. Remote Sensing 11(7):861. https://doi.org/10.3390/rs11070861

Jiang H, Li D, Jing W, Xu J, Huang J, Yang J, Chen S (2019) Early season mapping of sugarcane by applying machine learning algorithms to Sentinel-1A/2 time series data: a case study in Zhanjiang City, China. Remote Sens 11(7):861

Jiang Z, Chen Z, Jin C, Liu J, Ren J, Li Z, Sun L, Li H (2014) Application of crop model data assimilation with a particle filter for estimating regional winter wheat yields. J Sel Topics Appl Earth Obs Remote Sens 7

Jin X, Yang G, Xu X, Yang H, Feng H et al (2015) Combined multi-temporal optical and radar parameters for estimating LAI and biomass in winter wheat using HJ and RADARSAR-2 data. Remote Sensing 7:13251–13272. https://doi.org/10.3390/rs71013251

Jin X, Kumar L, Li Z, Xu X, Yang G, Wang J (2016) Estimation of winter wheat biomass and yield by combining the aquacrop model and field hyperspectral data. Remote Sensing 8(12). https://doi.org/10.3390/rs8120972

Jin X-L, Diao W-Y, Xiao C-H, Wang F-Y, Chen B, Wang K-R, Li S-K (2013) Estimation of wheat agronomic parameters using new spectral indices. PLOS One 8(8):e72736. https://doi.org/10.1371/journal.pone.0072736

Jones MO, Jones LA, Kimball JS, McDonald KC (2011) Satellite passive microwave remote sensing for monitoring global land surface phenology. Remote Sens Environ 115:1102–1114

Joshi N, Baumann M, Ehammer A et al (2016) A review of the application of optical and radar remote sensing data fusion to land use mapping and monitoring. Remote Sensing 8(70). https://doi.org/10.3390/rs8010070

Jurecka F, Hlavinka P, Lukas V, Trnka M, Zalud Z (2016) Crop yield estimation in the field level using vegetation indices. In: Proceedings of international PhD students conference, (MENDELNET 2016), pp 90–95

Koppe W, Gnyp ML, SiMon D, Li F, Miao Y, Chen X, Jia L, Bareth G (2012) Multi-Temporal hyperspectral and radar remote sensing for estimating winter wheat biomass in the North China Plain. PFG 3, 0281. https://doi.org/10.1127/1432-8364/2012/0117

Kouadio L, Newlands NK, Davidson A, Zhang Y, Chipanshi A (2014) Assessing the performance of MODIS NDVI and EVI for seasonal crop yield forecasting at the ecodistrict scale. Remote Sensing 6(10):10193–10214. https://doi.org/10.3390/rs61010193

Kross A, Mcnairn H, Lapen D, Sunohara M, Champagne C (2015) Assessment of RapidEye vegetation indices for estimation of leaf area index and biomass in corn and soybean crops. Int J Appl Earth Obs Geoinf 34:235–248. https://doi.org/10.1016/j.jag.2014.08.002

Kumar V, McNairn H, Bhattacharya A, Rao YS (2017) Temporal response of scattering from crops for transmitted ellipticity variation in simulated compact-pol SAR data. IEEE J Sel Topics Appl Earth Obs Remote Sens 10:5163–5174

Kumar P, Prasad R, Gupta DK, Mishra VN, Vishwakarma AK, Yadav VP, Avtar R et al (2018) Estimation of winter wheat crop growth parameters using time series Sentinel-1A SAR data. Geocarto Int 33(9):942–956

Kumar V, Mandal D, Bhattacharya A, Rao YS (2020) Crop characterization using an improved scattering power decomposition technique for compact polarimetric SAR data. Int J Appl Earth Obs Geoinf 88:102052

Li Y, Liao Q, Li X, Liao S, Chi G, Peng S (2003) Towards an operational system for regional-scale rice yield estimation using a time-series of RADARSAT ScanSAR images. Int J Remote Sens 24:4207–4220

Li Q, Cao X, Jia K, Zhan M, Dong Q (2014) Crop type identification by integration of high-spatial resolution multispectral data with features extracted from coarse resolution time-series vegetation index data. Int J Remote Sens 35(16):6076–6088. https://doi.org/10.1080/01431161.2014.943325

Liao C, Wang J, Shang J, Huang X, Liu J, Huffman T (2018) Sensitivity study of Radarsat-2 polarimetric SAR to crop height and fractional vegetation cover of corn and wheat. Int J Remote Sens 39:1475–1490

Liu J, Pattey E, Jégo JRG (2012) Assessment of vegetation indices for regional crop green LAI estimation from Landsat images over multiple growing seasons. Remote Sens Environ 123:347–358

Luo P, Liao J, Shen G (2020) Combining spectral and texture features for estimating leaf area index and biomass of maize using Sentinel-1/2 and Landsat-8 Data. IEEE Access 8:53614–53626. https://doi.org/10.1109/ACCESS.2020.2981492

Ma G, Huanga J, Wub W, Fan J, Zoub J, Wud S (2011) Assimilation of MODIS-LAI into the WOFOST model for forecasting regional winter wheat yield. Math Comput Model. https://doi.org/10.1016/jmcm.2011.10.038

Ma Q, Wang J, Shang J, Want P (2013) Assessment of multi-temporal RADARSAT-2 polarimetric SAR data for crop classification in an urban/rural fringe area. In: Proceedings of the second international conference of agro-geoinformatics held 12–16 August 2013, Fairfax, USA, pp 314–319

Ma J, Nguyen C, Lee K, Heo J (2018) Regional-scale rice-yield estimation using stacked auto-encoder with climatic and MODIS data: a case study of South Korea. Int J Remote Sens 00(00):1–21

Mandal D, Kumar V, Bhattacharya A, Rao YS, McNairn H (2018) Crop biophysical parameters estimation with a multi-target inversion scheme using the Sentinel-1 SAR data. In: IGARSS 2018–2018 IEEE international geoscience and remote sensing symposium. IEEE, pp 6611–6614. https://doi.org/10.1109/IGARSS.2018.8518700

Mateo-Sanchis A, Piles M, Muñoz-Marí J, Adsuara JE, Pérez-Suay A, Camps-Valls G (2019) Synergistic integration of optical and microwave satellite data for crop yield estimation. Remote Sens Environ 234:111460

McNairn H, Brisco B (2004) The application of C-band polarimetric SAR for agriculture: a review. Can J Remote Sensing 30(3):525–542

McNairn H, Jiao X, Pacheco A, Sinha A, Tan W, Li Y (2018) Estimating canola phenology using synthetic aperture radar. Remote Sens Environ 219:196–205

Meng J, Wu B, Li Q, Du X, Jia K (2009) Monitoring crop phenology with MERIS data—a case study of winter wheat in monitoring crop phenology with MERIS data—a case study of winter wheat in North China Plain. In: Progress in electromagnetics research symposium, Beijing, China, March 23–27, 2009, February 2015

Mercier A, Betbeder J, Baudry J, Le Roux V, Spicher F, Lacoux J, Hubert-Moy L et al (2020) Evaluation of Sentinel-1 & 2 time series for predicting wheat and rapeseed phenological stages. ISPRS J Photo Remote Sens 163:231–256

Misra G, Cawkwell F, Wingler A (2020) Status of phenological research using Sentinel-2 data: a review. Remote Sensing 12(17):2760

Mokhtari A, Noory H, Vazifedoust M (2018) Improving crop yield estimation by assimilating LAI and inputting satellite-based surface incoming solar radiation into SWAP model. Agri Meteorol 250:159–170

Molijn RA, Iannini L, Vieira Rocha J, Hanssen RF (2019) Sugarcane productivity mapping through C-band and L-band SAR and optical satellite imagery. Remote Sensing 11(9):1109

Morel J, Todoroff P, Bégué A, Bury A, Martiné JF, Petit M (2014) Toward a Satellite-based system of sugarcane yield estimation and forecasting in smallholder farming conditions: a case study on Reunion Island. Remote Sensing 6:6620–6635. https://doi.org/10.3390/rs6076620

Mosleh MK, Hassan QK, Chowdhury EH (2015) Application of remote sensors in mapping rice area and forecasting its production: a review. Sensors 15:769–791. https://doi.org/10.3390/s150100769

Mulyono S (2016, November) Identifying sugarcane plantation using LANDSAT-8 images with support vector machines. IOP Conf Ser: Earth Environ Sci 47(1):012008

Nasirzadehdizaji R, Sanli FB, Abdikan S (2019) Sensitivity analysis of multi-temporal Sentinel-1 SAR parameters to crop height and canopy coverage. Appl Sci. https://doi.org/10.3390/app9040655

Nguy-Robertson AL, Peng Y, Gitelson AA, Arkebauer TJ, Pimstein A, Herrmann I, Karnieli A, Rundquist DC, Bonfil DJ (2014) Estimating green LAI in four crops: potential of determining optimal spectral bands for a universal algorithm. Agric Meteorol 192:140–148

Novelli F, Spiegel H, Sandén T, Vuolo F (2019) Assimilation of Sentinel-2 leaf area index data into a physically-based crop growth model for agronomy, p 9

Onojeghuo AO, Blackburn GA, Wang Q, Atkinson PM, Kindred D, Miao Y, Onojeghuo AO, Blackburn GA, Wang Q, Peter M (2018) Rice crop phenology mapping at high spatial and temporal resolution using downscaled MODIS. Giss Remote Sens 1–19. https://doi.org/10.1080/15481603.2018.1423725

Pan Z, Huang J, Zhou Q, Wang L, Cheng Y (2015) Mapping crop phenology using NDVI time-series derived from HJ-1 A/B data. Int J Appl Earth Obs Geoinf 34:188–197

Pan Z, Huang J, Zhou Q, Wang L, Cheng Y, Zhang H, Blackburn GA, Yan J, Liu J (2015) Mapping crop phenology using NDVI time-series derived from HJ-1 A/B data. Int J Appl Earth Obs Geoinf 34:188–197

Phung HP, Nguyen LD, Nguyen-Huy T, Le-Toan T, Apan AA (2020) Monitoring rice growth status in the Mekong Delta, Vietnam using multitemporal Sentinel-1 data. J Appl Remote Sens 14(1):014518. https://doi.org/10.1117/1.JRS.14.014518

Pichierri M, Hajnsek I, Zwieback S, Rabus B (2018) On the potential of Polarimetric SAR Interferometry to characterize the biomass, moisture and structure of agricultural crops at L-, C-and X-Bands. Remote Sens Environ 204:596–616

Piedelobo L, Hernández-López D, Ballesteros R, Chakhar A, Del Pozo S, González-Aguilera, D, Moreno MA (2019) Scalable pixel-based crop classification combining Sentinel-2 and Landsat-8 data time series: case study of the Duero river basin. Agric Syst 171:36–50

Pinheiro Lisboa I, Melo Damian J, Roberto Cherubin M, Silva Barros PP, Ricardo Fiorio P, Cerri CC, Eduardo Pellegrino Cerri C (2018) Prediction of sugarcane yield based on NDVI and concentration of leaf-tissue nutrients in fields managed with straw removal. Agronomy 8(9):196. https://doi.org/10.3390/agronomy8090196

Prabhakara K, Dean Hively W, McCarty GW (2015) Evaluating the relationship between biomass, percent groundcover and remote sensing indices across six winter cover crop fields in Maryland, United States. Int J Appl Earth Obs Geoinf 39:88–102. https://doi.org/10.1016/j.jag.2015.03.002

Prasad AK, Singh RP, Tare V, Kafatos M (2007) Use of vegetation index and meteorological parameters for the prediction of crop yield in India. Int J Remote Sens 28:5207–5235

Prasad AM, Iverson LR, Liaw A (2006) Newer classification and regression tree techniques: bagging and random forests for ecological prediction. Ecosyst 9:181–199

Prasad NR, Patel NR, Danodia A (2021) Crop yield prediction in cotton for regional level using random forest approach. Spat Inf Res 29(2):195–206

Pratap YV, Rajendra P, Ruchi B (2019) Leaf area index estimation of wheat crop using modified water cloud model from the time-series SAR and optical satellite data. Geocarto Int 1–12. https://doi.org/10.1080/10106049.2019.1624984

Prudente VHR, Oldoni LV, Vieira DC, Cattani CEV, Sanches ID (2019) Relationship between SAR/SENTINEL-1 polarimetric and interferometric data with biophysical parameters of agricultural crops. In: The international archives of the photogrammetry, remote sensing and spatial information sciences, vol XLII-3/W6, 2019 ISPRS-GEOGLAM-ISRS Joint Int. Workshop on "Earth Observations for Agricultural Monitoring", 18–20 February 2019, New Delhi, India REL, XLII(February), pp 18–20

Rimal B, Zhang L, Rijal S (2018) Crop cycles and crop land classification in Nepal using MODIS NDVI. Remote Sens Earth Syst Sci 1(1):14–28

Saeed U, Dempewolf J, Becker-reshef I, Khan A, Wajid SA (2017) Forecasting wheat yield from weather data and MODIS NDVI using Random Forests for Punjab province, Pakistan. Int J Remote Sens 38(17):4831–4854

Sakamoto T, Yokozawa M, Toritani H, Shibayama M, Ishitsuka N, Ohno H (2005) A crop phenology detection method using time-series MODIS data. Remote Sens Environ 96:366–374

Sakamoto T, Wardlow BD, Gitelson AA, Verma SB, Suyker AE, Arkebauer TJ (2010) A Two-Step Filtering approach for detecting maize and soybean phenology with time-series MODIS data. Remote Sens Environ 114:2146–2159

Sakamoto T, Gitelson AA, Arkebauer TJ (2013) MODIS-based corn grain yield estimation model incorporating crop phenology information. Remote Sens Environ 131:215–231

Sarma AALN, Lakshmi Kumar TV, Koteswararao K (2008) Development of an agro-climatic model for the estimation of rice yield. J Ind Geophys Union 12:89–96

Schlund M, Erasmi S (2020) Sentinel-1 time series data for monitoring the phenology of winter wheat. Remote Sens Environ 246

Setiyono TD, Quicho ED, Gatti L, Campos-Taberner M, Busetto L, Collivignarelli F, García-Haro FJ, Boschetti M, Khan NI, Holecz F (2018) Spatial rice yield estimation based on MODIS and Sentinel-1 SAR Data and ORYZA crop growth model. Remote Sensing 10:1–20. https://doi.org/10.3390/rs10020293

Shao Y, Fan X, Liu H, Xiao J, Ross S, Brisco B, Brown R, Staples G (2001) Rice monitoring and production estimation using multi-temporal RADARSAT. Remote Sens Environ 76:310–325

Sibley AM, Grassini P, Thomas NE, Cassman KG, Lobell DB (2014) Testing remote sensing approaches for assessing yield variability among maize fields. Agron J 106(1):24–32

Skakun S, Franch B, Vermote E, Roger JC, Becker-Reshef I, Justice C, Kussul N (2017) Early season large-area winter crop mapping using MODIS NDVI data, growing degree days information and a Gaussian mixture model. Remote Sens Environ 195:244–258. https://doi.org/10.1016/j.rse.2017.04.026

Skakun S, Vermote E, Franch B, Roger JC, Kussul N, Ju J, Masek J (2019) Winter wheat yield assessment from Landsat 8 and Sentinel-2 data: incorporating surface reflectance, through phenological fitting, into regression yield models. Remote Sensing 11(15):1768

Sonobe R, Yamaya Y, Tani H, Wang X, Kobayashi N, Mochizuki KI (2018) Crop classification from Sentinel-2-derived vegetation indices using ensemble learning. J Appl Remote Sens 12(2):026019. https://doi.org/10.1117/1.JRS.12.026019

Svotwa E, Masuka AJ, Maasdorp B, Murwira A, Masocha M (2014) Estimating tobacco crop area and yield in Zimbabwe using operational remote sensing and statistical techniques. Int J Agri Res Rev 2(5):084–091

Talema T, Hailu BT (2020) Mapping rice crop using sentinels (1 SAR and 2 MSI) images in tropical area: a case study in Fogera wereda, Ethiopia. Remote Sens Appl Soc Environ 18:100290

Tilly N, Aasen H, Bareth G (2015) Fusion of plant height and vegetation indices for the estimation of barley biomass, pp 11449–11480. https://doi.org/10.3390/rs70911449

Torbick N, Chowdhury D, Salas W, Qi J (2017) Monitoring rice agriculture across Myanmar using time series Sentinel-1 assisted by Landsat-8 and PALSAR-2. Remote Sensing 9(2):119. https://doi.org/10.3390/rs9020119

Van Tricht K, Gobin A, Gilliams S, Piccard I (2018) Synergistic use of radar Sentinel-1 and optical Sentinel-2 imagery for crop mapping: a case study for Belgium. Remote Sensing 10(10):1642. https://doi.org/10.3390/rs10101642

Useya J, Chen S (2018) Comparative performance evaluation of pixel-level and decision-level data fusion of Landsat 8 OLI, Landsat 7 ETM+ and Sentinel-2 MSI for crop ensemble classification. IEEE J Select Top Appl Earth Observ Remote Sens 1-11. https://doi.org/10.1109/JSTARS.2018.2870650

Useya J, Chen S (2019) Exploring the potential of mapping cropping patterns on smallholder scale croplands using sentinel-1 SAR data. Chin Geogra Sci 29(4):626–639

Verma AK, Garg PK, Prasad KH (2017) Sugarcane crop identification from LISS IV data using ISODATA, MLC, and indices based decision tree approach. Arab J Geosci 10(1):16. https://doi.org/10.1007/s12517-016-2815-x

Vintrou E, Bégué A, Baron C, Saad A, Lo Seen D, Traoré S (2014) A comparative study on satellite- and model-based crop phenology in West Africa. Remote Sensing 6(2):1367–1389. https://doi.org/10.3390/rs6021367

Wang H, Magagi R, Goïta K, Trudel M, McNairn H, Powers J (2019a) Crop phenology retrieval via polarimetric SAR decomposition and Random Forest algorithm. Remote Sens Environ 231:111234

Wang J, Dai Q, Shang J, Jin X, Sun Q, Zhou G (2019b) Field-scale rice yield estimation using Sentinel-1A Synthetic Aperture Radar (SAR) data in coastal saline region of Jiangsu Province, China. Remote Sensing 11(19):2274

Wang J, Xiao X, Liu L, Wu X, Qin Y, Steiner JL (2020) Mapping sugarcane plantation dynamics in Guangxi, China, by time series Sentinel-1, Sentinel-2 and Landsat images. Remote Sens Environ 247:111951. https://doi.org/10.1016/j.rse.2020.111951

Wojtowicz M, Wojtowicz A, Piekarczyk J (2016) Application of remote sensing methods in agriculture. Commun Biometry Crop Sci 11:31–50

Xiao Z, Liang S, Wang J, Jiang B, Li X (2011) Real-time retrieval of Leaf Area Index from MODIS time series data. Remote Sens Environ 115(1):97–106. https://doi.org/10.1016/j.rse.2010.08.009

Xin Q, Gong P, Yu C, Yu L, Broich M, Suyker AE, Myneni RB (2013) A production efficiency model-based method for satellite estimates of corn and soybean yields in the Midwestern US. Remote Sens 5:5926–5943. https://doi.org/10.3390/rs511592

Xu X, Conrad C, Doktor D (2017) Optimising phenological metrics extraction for different crop types in Germany using the moderate resolution imaging spectrometer (MODIS). Remote Sensing 9:254

Yang P, Tan GX, Zha Y, Shibasaki R (2004) Integrating remotely sensed data with an ecosystem model to estimate crop yield in North China. In: The international archives of the photogrammetry, remote sensing and spatial information sciences, Istanbul, Turkey, vol XXXV

Yang H, Pan B, Li N, Wang W, Zhang J, Zhang X (2021) A systematic method for spatio-temporal phenology estimation of paddy rice using time series Sentinel-1 images. Remote Sens Environ 259:112394

You X, Meng J, Zhang M, Dong T (2013) Remote sensing based detection of crop phenology for agricultural zones in China using a new threshold method. Remote Sensing 5(7):3190–3211. https://doi.org/10.3390/rs5073190

Yue J, Yang G, Li C, Li Z, Wang Y, Feng H, Xu B (2017) Estimation of winter wheat above-ground biomass using unmanned aerial vehicle-based snapshot hyperspectral sensor and crop height improved models. Remote Sensing 9(7):708. https://doi.org/10.3390/rs9070708

Zeng L, Wardlow BD, Wang R, Shan J, Tadesse T, Hayes MJ, Li D (2016) A hybrid approach for detecting corn and soybean phenology with time-series MODIS data. Remote Sens Environ 181:237–250

Zhang Y, Yang B, Liu X, Wang C (2017) Estimation of rice grain yield from dual-polarization Radarsat-2 SAR data by integrating a rice canopy scattering model and a genetic algorithm. Int J Appl Earth Obs Geoinf 57:75–85. https://doi.org/10.1016/j.jag.2016.12.014

Zhi Y, Yun S, Kun L, Qingbo L, Long L, Brisco B, Yang Z, Shao Y, Li K, Liu Q, Liu L, Brisco B (2017) An improved scheme for rice phenology estimation based on time-series multispectral HJ-1A/B and polarimetric RADARSAT-2 data. Remote Sens Environ 195:184–201. https://doi.org/10.1016/j.rse.2017.04.016

Zhou Z, Huang J, Wang J, Zhang K, Kuang Z, Zhong S, Song X (2015) Object-oriented classification of sugarcane using time-series middle-resolution remote sensing data based on AdaBoost. PLoS ONE 10(11):e0142069

Zhou X, Zhu X, Dong Z, Guo W (2016) Estimation of biomass in wheat using random forest regression algorithm and remote sensing data. CJ 4(3):212–219. https://doi.org/10.1016/j.cj.2016.01.008

Zhou X, Zheng HB, Xu XQ, He JY, Ge XK, Yao X, Cheng T, Zhu Y, Cao WX, Tian YC (2017) Predicting grain yield in rice using multi-temporal vegetation indices from UAV-based multi-spectral and digital imagery. ISPRS J Photo Remote Sens 130:246–255. https://doi.org/10.1016/j.isprsjprs.2017.05.003

Chapter 20
Recent Development and Innovative Tools for Climate and Hydrological Data Collection and Analysis in the Abbay Basin

Yonas Getaneh, Wuletawu Abera, Getachew Tesfaye Ayehu, Degefie Tibebe, and Lulseged Tamene

Abstract The Abbay Basin is the largest river basin in Ethiopia with the highest water volume discharge of 54 BCM/year. It is a crucial source of the Nile River, which supports the livelihoods of 250 million people in Sudan and Egypt. Due to its importance, the scientific community has given significant attention to the basin. However, there is still a lack of appropriate hydroclimate data in terms of quality and quantity and a scarcity of research facilities, which has hindered the development of scientific knowledge in the region. To address this issue, the objective of the current study is to highlight innovations in climatological and hydrological data sources and analysis tools that can improve the precision and accuracy of research in the Abbay basin. Our literature review identified emerging satellite-based hydroclimate data sources that could serve as potential alternatives to address the problem of data inadequacy. Although some data sources for temperature and surface reflectance have higher accuracy and precision levels than those commonly used in the basin, there are limitations in available secondary data sources for perception and related hydrological parameters, including their spatial resolution and accuracy. We also explored the potential of emerging cloud-based computation platforms (such as Google Earth Engine) and script-based data analysis tools (such as R-based packages) to analyse

---

Y. Getaneh (✉) · W. Abera · G. T. Ayehu · D. Tibebe · L. Tamene
Alliance of Bioversity International and CIAT, Addis Ababa, Ethiopia
e-mail: y.getaneh@cgiar.org

W. Abera
e-mail: Wuletawu.Abera@cgiar.org

G. T. Ayehu
e-mail: Getachew.Tesfaye@cgiar.org

D. Tibebe
e-mail: D.Tibebe@cgiar.org

L. Tamene
e-mail: LT.Desta@cgiar.org

large and high-resolution datasets. These tools have the ability to automate the iteration of functions on large spatial extent datasets and represent distinct time intervals (hourly, daily, monthly or yearly) with separate layers. Additionally, we highlighted machine learning (ML)-based algorithms as promising tools for solving research problems in the basin, especially when long-term hydroclimatological data are not available. ML-based algorithms can provide greater accuracy in parameter estimation without requiring in-depth knowledge of the basin's physical processes and can use static basin characteristic data for hydrological parameter estimation. However, it is important to ensure that these tools are used under the correct conditions and with appropriate calibration to the context of the Abbay basin. Overall, these innovations can contribute to a better understanding of hydroclimatological processes in the Abbay basin and support sustainable water management practices.

Keywords Ethiopia · Abbay basin · Blue Nile basin · Google Earth Engine · R packages · Climatology · Hydrology · Machine learning

Introduction

The Abbay basin, which is also known as the Upper Blue Nile Basin, is situated in the northwestern region of Ethiopia and encompasses an area between 7° 40′ N and 12° 51′ N and 34° 25′ E and 39° 49′ E, as shown in Fig. 20.1. It is the largest basin in terms of area coverage, spanning 199,812 square kilometres, and it boasts the highest annual runoff of 54 billion cubic meters per year compared to the other river basins in Ethiopia (Conway et al. 2004).

Due to its significant contributions to Ethiopia's annual discharge and its large population (accounting for approximately 28% of the population), the basin holds immense strategic value for the country. Additionally, the river system is of utmost importance to the downstream countries of Sudan and Egypt, as it contributes to approximately 62% of the Nile River's total flow (Conway 2000a). The basin's climate is closely tied to altitude and proximity to the equatorial monsoon system, resulting in a variety of humid to semiarid climate zones. The annual rainfall in the basin can reach 2200 mm, with mean rainfall varying between 1200 and 1800 mm (Fig. 20.2) and gradually increasing from northeast to southwest (Kim et al. 2008). The primary rainy season, known as "Kiremt" or summer rainfall, occurs from June to September, followed by a dry season from October to January and a short rainy season known as "Belg" or spring rainfall from February to May. Approximately 70% of the annual precipitation in the basin falls during the Kiremt season (Kim et al. 2008). However, rainfall fluctuations within and between years are significant in the basin (Conway 2000b; Taye and Willems 2012), making hydrological processes complex and highly variable. The average annual temperature in the basin is 18.5 °C, with mean minimum and maximum daily temperatures of 11.4 °C and 25.5 °C, respectively. The basin's topography consists of two distinct features: the highlands (above 1500 m a.s.l), which include rugged mountainous areas in the central and eastern parts of the

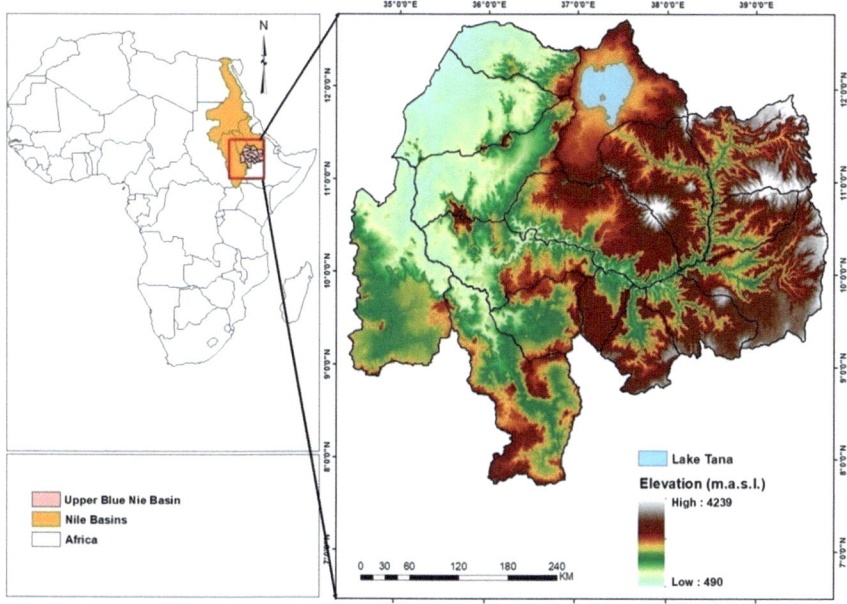

Fig. 20.1 Location of the upper Blue Nile basin (UBNB) in Africa and its elevation. The Northeastern regions have higher elevations, while the Northwestern regions have lower elevations (imagery source: SRTM Global elevation data)

basin (accounting for 60% of the total area), and the lowlands in the western parts (Yilma and Awulachew 2009).

The Abay Basin has been a subject of national and international strategic interest, drawing the attention of the scientific community throughout the twentieth century. Initial scientific accounts of the basin's hydroclimate date back to the 1930s (Cheesman 1935), with early efforts focusing on characterizing the basin's environment, including its climate, topography, river networks, and socioeconomic activities (Nilsson 1940). These early studies were primarily qualitative reports based on personal observation. However, research based on empirical data gained momentum after the 1950s, as state initiatives sought to utilize the hydrological resources of the basin. Since then, significant advancements have been made in climate and water resources research in the basin, with an increase in variable measurements, a widening of conceptual scope, and advancements in methodological complexities (Getaneh et al. 2022). While the earliest studies focused only on estimating surface runoff and river volumes using rainfall data (Hurst et al. 1933), contemporary research has progressed to estimating overall basin biophysical and hydrological parameters, such as soil, climate elements (Abera et al. 2017; Tigabu et al. 2021), water balance characteristics (Abera et al. 2017) topography (Bogale 2021), land cover (Abate

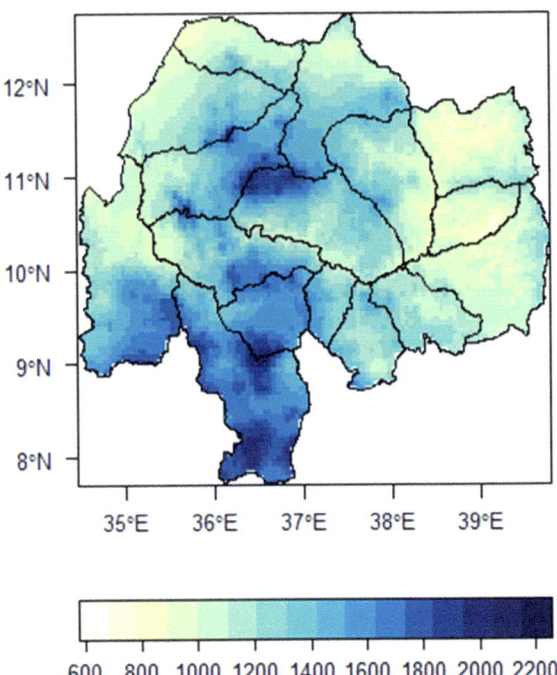

Fig. 20.2 The spatial distribution of mean annual rainfall (mm) over the Upper Blue Nile basin (UBNB) in Ethiopia (1981–2018) using CHIRPS satellite rainfall estimates (Ayehu et al. 2021; Bayissa et al. 2021; Tegegne et al. 2021)

et al. 2017), land management (Dile et al. 2016), and other offsite factors that influence basin biophysical characteristics. Research methods have also evolved, from deterministic and lumped models to advanced physically based models (Dile et al. 2018).

Although there have been some advances in hydroclimate research in the Abay basin, several limitations still exist. One significant challenge is the lack of long-term hydroclimatic data (Dile et al. 2018). The earliest climate records date back to the late nineteenth century, with the first precipitation records for Addis Ababa in 1896. However, systematic weather recording only began in the late 1950s and 1960s, following the establishment of the Meteorological Department (Conway et al. 2004). On the other hand, the earliest hydrological records date back to the 1920s, with the outflows of Lake Tana being studied by Egyptian scholars (Sutcliffe and Parks 1999). The United States government also established river gauging networks in the Blue Nile Basin between 1959 and 1964 (USBR 1964). However, it was only after the Ethiopian Water Resources Department was established in 1956 that a concerted program of river flow data collection began (Conway et al. 2004).

Despite ongoing efforts to increase the number of gauging stations and improve data quality, the available hydroclimate observation networks in the Abbay basin still fall short of the optimal data quality required for many research tasks (Taye et al. 2022; Yenehun et al. 2021). According to the World Meteorological Organization (WMO) (2018), punctual values measured by single weather stations are representative of a radius of 10–30 km with respect to their location for the analysis of mesoscale climate phenomena. However, in the Abbay basin, only 66% of the basin's area is located within 30 km of the nearest weather station. Moreover, agrometeorological and environmental applications require an observation density of less than 1 km horizontal distance, which is not available for all parts of the Abbay basin (Fig. 20.3). In addition to representativeness, the data collected from existing hydrometric station networks cannot be directly used for their intended purposes because they only measure water height and have not been converted to discharge since 1997 for most stations. Furthermore, there are major data quality issues with the flow data itself (Taye et al. 2022).

In response to the lack of appropriate hydroclimatological data in the Abbay basin, remote sensing products and other secondary geospatial data have emerged as potential alternatives to determine various hydrological and climatological parameters. There have been previous attempts to integrate weather station data with satellite-driven parameters for climate variable estimation (Ayehu et al. 2021; Bayissa et al. 2021; Tegegne et al. 2021). However, the research process is still in its trial stage, and there is no solid foundation regarding the bias and accuracy of different satellite-driven products in the context of the Abbay basin (Abera et al. 2016; Tibebe et al. 2022; Worqlul et al. 2017).

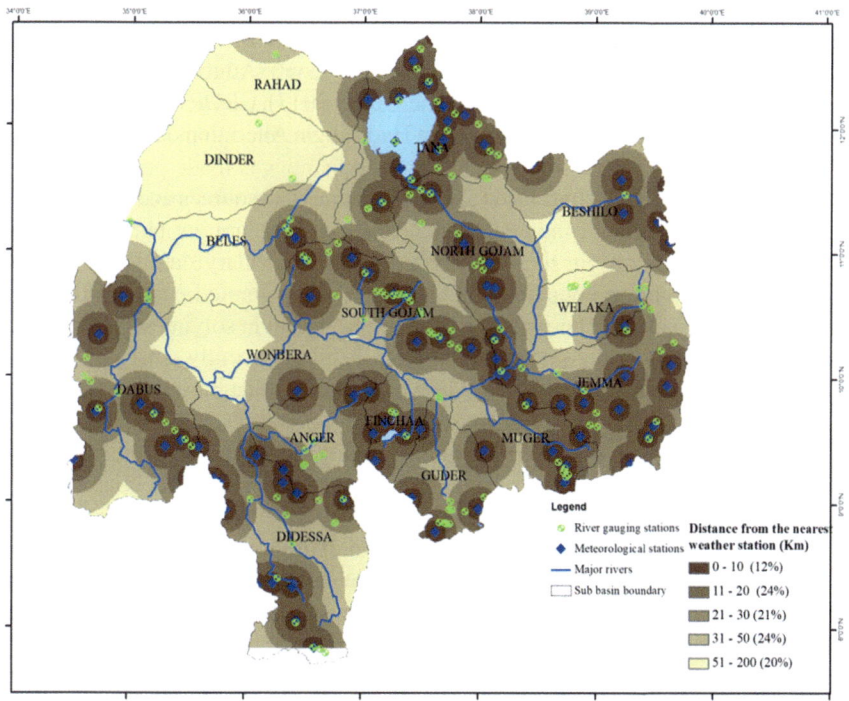

Fig. 20.3 Location of weather stations and hydrological observation networks in the Abbay basin (NB, background values represent Euclidean distance of locations from the nearby meteorological stations)

In addition to hydroclimate data sources, the use of medium-resolution remote sensing products has become a common research trend in the basin (Ayehu et al. 2020; Belete et al. 2020; Duan et al. 2018). The availability of time series satellite products, specifically Landsat time series images, has enabled the emergence of new insights and increased the accuracy of parameter estimations for many biophysical (Ali et al. 2020; Lemma et al. 2019), climatological, and hydrological variables (Abebe et al. 2021; Mulatu et al. 2022).

The other major progress in hydroclimatological research trends in the basin is the introduction of complex modelling systems, which require intensive data sources and computational resources. In the 2000s, climate and hydrological modelling in the Abbay basin was based on the use of simple statistical and conceptual models due to the limitation of data availability and computational requirements (Sutcliffe and Parks 1999). The number of input parameters was also limited to rainfall, temperature, and evapotranspiration (Conway et al. 2004). With the increasing availability of high-resolution spatial data and advances in the computational power of hardware facilities, advanced and complex hydrological models have been introduced in basin hydrological research. Some of the models previously employed in the hydrological and climatological study of the Abbay basin include the Soil and Water Assessment Tool (SWAT) (Ayele et al. 2017; Gashaw et al. 2021); Variable Infiltration Capacity (VIC) (Tariku et al. 2021), Hydrologiska Byråns Vattenbalansavdelning (HVB), Water Evaluation and Planning System (WEAP) (Adgolign et al. 2016; Ali and Zahran 2023), NewAGE JGRASS (Abera et al. 2017), CROPWAT (Bekele et al. 2019; Erkossa et al. 2014), and Indicators of Hydrologic Alterations (IHA) (Adgolign et al. 2016; Malede et al. 2022).

Although efforts have continued to use innovative data sources and data analysis tools, many of the state-of-the-art research methods have not yet been utilized by the scientific community in the basin. When using secondary data sources, there is a tendency to rely on popular satellite products with limited consideration of a wide variety of available alternatives that might be useful in solving scientific problems in the basin. In addition to data sources, there are recently emerging innovative modelling tools and alternative hydrological models that have not been utilized thus far in Abbay basin climatological and hydrological research. Many of these tools are script-based analytical tools, which enable automated computation of functions through iteration that would not be possible through traditional graphical user interface tools (Yang et al. 2022).

Therefore, the main goal of this chapter is to explore new hydroclimatological data sources and analysis tools that could enhance research in the Abbay basin. This information will assist the scientific community in selecting appropriate techniques and methods for hydroclimatological studies in the region. The study conducted a review of scientific articles, secondary data sources, and data analysis tools focusing on those relevant to the Abbay basin that have not been well utilized in previous studies. An innovative data source or analysis tool is considered if it meets one of three criteria. First, it can facilitate cost-effective research and save time compared to the methods previously used in hydroclimate research in the Abbay basin. Second, it can improve the quality of data sources by providing more precise estimates of parameters than those used in previous studies. Finally, it can fill knowledge gaps (Dile et al. 2018; Getaneh et al. 2022; Tibebe et al. 2022) or provide new scientific insights about the basin. Our study contributes to hydroclimatological research in the Abbay basin by identifying underutilized data sources and analysis tools.

Innovations in Climatological and Hydrological Data

The hydrological and climatological research trends in the Abbay basin have been based on hydroclimate data obtained from observation networks that are sparse and lack temporal continuity. This limitation has resulted in suboptimal data quality for the intended purposes (Taye et al. 2022; Tibebe et al. 2022; Yenehun et al. 2021). Therefore, it is imperative to explore alternative data sources that can enhance the precision and accuracy of scientific studies in the basin. One such alternative is secondary climate and hydrological data obtained from satellite observations. The purpose of this section is to present an overview of the latest scientific breakthroughs in hydroclimate data sources that have the potential to improve the accuracy and reliability of scientific research in the Abbay basin. Recent advancements in hydroclimate data have led to the emergence of high spatial resolution global and regional climate data sources (Dinku et al. 2022b). In contrast to the climate data sources used in previous studies in the Abbay basin, which had a maximum resolution of 5 km (Ayehu et al. 2021; Tegegne et al. 2021; Worqlul et al. 2017), these new sources provide higher spatial resolution.

Climate Data Sources

This subsection aims to highlight climate products that could serve as an alternative data source for hydrological and climatological applications in the Abbay basin.

ENACT (Enhanced National Climate Service) Climate Data: ENACTS was developed by combining data from—ground weather stations, satellite weather measurements, digital elevation models and climate reanalysis products to create high-quality, high-resolution, integrated datasets. These data are disseminated through an online platform, i.e., "Maprooms", which allows to retrieve and display of ENACTS information. ENACTS datasets are developed through collaboration of the International Research Institute for Climate and Society and Meteorological offices in Africa (Dinku et al. 2022a). The spatial resolution of the ENACTS rainfall and temperature products is approximately $0.0375°$ (about 4 km) and $0.05°$, respectively (Maidment et al. 2020). The ENACTS approach specifically engages the National Meteorology Agencies of participating countries and uses considerably more stations than any other blended product (Dinku et al. 2022a). ENACT data for Ethiopia exist from 1981 to the present at daily and dekadal temporal resolutions.

ENACT climate data are one of the optional data sources frequently used in the Abbay basin by different studies due to their relatively higher accuracy level compared with other data sources (Abebe et al. 2020).

CHIRPS (Climate Hazards Group InfraRed Precipitation with Station Data): It is a semiglobal rainfall data intended for observing global environmental changes and monitoring drought (Funk et al. 2015). The product provides daily, pentadal, dekadal, and monthly data at a 0.05° spatial resolution available at the Climate Hazards Group (ftp://ftp.chg.ucsb.edu/pub/org/chg/products) and the International Research Institute climate data library (http://iridl.ldeo.columbia.edu/). CHIRPS is developed by blending a 0.05° monthly precipitation data from Climate Hazards Group InfraRed Precipitation (CHIRP), and weather station data. In its second version, CHIRPS comes up with improved daily rainfall time series (1981–present) data set with a spatial resolution of 0.05° covering a span of 50° S to 50° N and all longitudes (Funk et al. 2015). A more detailed explanation of the technical properties of CHIRPS and its application in Africa (e.g., Ethiopia) is available in the studies of Funk et al. (2015). In addition to monitoring droughts, CHIRPS is also used for managing water resources, tracking climate change (Ceccherini et al. 2015; Deblauwe et al. 2016) and other worldwide environmental applications (Zambrano-Bigiarini et al. 2017).

WorldClim Version 2.1 Climate Data for 1970–2000: This contains spatially interpolated monthly climate data for global land areas at 1 km^2 spatial resolution. It consists of monthly temperature, precipitation, solar radiation, vapour pressure and wind speed data ranging from 1970 to 2000 (Fick and Hijmans 2017). The database was developed by combining weather station and satellite-derived covariables using the thin-plate smoothing spline algorithm. It was claimed that the algorithm was developed to improve estimates for areas with low station density. The accuracy of the products was reported to vary depending on the variables, ranging from a global correlation coefficient (between estimated and observed values) of 0.99 for temperature (the highest accuracy estimate) to 0.86 for precipitation and 0.76 for wind speed (Fick and Hijmans 2017). Despite an increase in the spatial resolution, the accuracy level of the products is not different from previously well-known climate data sources used in the Abbay basin.

The Satellite Application Facility on Land Surface Analysis (SAF-LSA): It is the product of the European Organization for the Exploitation of Meteorological Satellites (EUMETSAT), which distributes a variety of datasets with a resolution of 1–2 km^2 (Trigo et al. 2011). Land surface temperature (LST) is one of many SAF products with potential usability in the study of the Abbay basin. It is based on the radiative skin temperature over land. Because it is involved in the processes of energy and water exchange with the atmosphere, LST may play a significant role in the physics of the land surface. It is entirely a satellite-driven product and may be independent of the effect of ground observation data quality inconsistencies. The accuracy of the products was evaluated using ground observation data from East Africa, and the product performance varies depending on weather conditions, with the highest and lowest performances varying for clear sky and cloudy conditions, respectively (Dowling et al. 2022). It might be a good alternative for surface temperature products in the Abbay basin, especially if calibrated with the available weather observation data. However, despite their high spatial resolution, the temporal scope of many products is not long enough for long-term climate change applications.

TerraClimate: This is a dataset of monthly climate and water balance for global terrestrial surfaces from 1958 to 2019. A description of the product is available in studies by Abatzoglou et al. (2018). It was created by fusing high spatial resolution monthly data from the WorldClim dataset with lower spatial resolution but high temporal resolution data from CRU Ts4.0 and the Japanese 55-year Reanalysis using a climatically assisted interpolation technique (Ebita et al. 2011). All data have a monthly temporal resolution and a ~4-km spatial resolution. It contains primary and derived climate variables. The primary variables include maximum temperature, minimum temperature, vapour pressure, precipitation accumulation, downwards surface shortwave radiation, and wind speed. The derived variables are reference evapotranspiration (ASCE Penman–Monteith), runoff, actual evapotranspiration, climate water deficit, soil moisture, snow water equivalent, Palmer drought severity index, and vapour pressure deficit. One strong side of the product is the provision of ready-to-use and long-term water balance products that are not available in any other datasets. Its reliance on the WorldClim dataset makes it inherit all the problems of the WorldClim dataset, which has limited validity in data-sparse regions due to its reliance on ground observation data (Fick and Hijmans 2017). The water balance model is also criticized for its oversimplicity, which fails to account for heterogeneity in catchment biophysical characteristics and their physiological response to changing environmental conditions (Abatzoglou et al. 2018).

The Climate Prediction Center (CPC) Morphing Technique (CMORPH) Satellite Precipitation Estimate: They are reprocessed and bias-corrected perception data on an 8 km × 8 km grid over the globe (60° S–60° N) and in a 30-min temporal resolution starting from January 1998 to the present. They are purely satellite-driven products developed through a combination of three different satellite products, such as passive microwave (PMW) data of instantaneous precipitation rates, satellite-based snow/ice maps and cloud motion data from full-resolution global surface/cloud-top temperature data. The bias correction is made using weather station data from the Global Precipitation Climatology Center (GPCC) (Xie et al. 2017). One strong side of the products is their low reliance on ground observation data, which makes them free from biases due to poor observation network data. However, the spatial resolution is still lower than many of the preceding products. Bias correction results further demonstrate that the accuracy of CMORPH products varies across regions, changes over time, and has a nonlinear relationship with the intensity of the target precipitation (Dhungana et al. 2023; Ma et al. 2023; Xie et al. 2017). This signals that the product must be used with proper handling of these problems, and further bias correction based on local observation data is a prerequisite before directly applying it to a specific place.

Soil Moisture Products

In addition to climate products, the other advancement in satellite-based products is the launch of microwave satellites, the increased potentials of optical/thermal IR remote sensing, and the release of long-term soil moisture products with better spatial and temporal resolutions, which provide unique opportunities to assess the hydrology/soil moisture of the Abbay basin. Estimates of soil moisture are often retrieved from both optical/thermal IR and microwave satellite sensors (Mohanty et al. 2017; Peng and Loew 2017). However, promising progress has been made in microwave systems. Microwave regions of the electromagnetic spectrum show a unique sensitivity to the soil moisture content and are also weather-independent. Microwave remote sensing can be classified as active (e.g., synthetic aperture radar (SAR)) and passive microwave based on their source of electromagnetic energy.

The SMAP Mission (Soil Moisture Product at the Surface and Soil Root Zone): The Soil Moisture Active Passive (SMAP) satellite mission, which carries an L-band radiometer, was launched on 31 January 2015. It began to provide radiometer soil moisture (SM) products on 31 March 2015 (Entekhabi et al. 2010). The SMAP L3 is one of the soil moisture products from the SMAP mission, which is provided as the coarse product with a resolution of 36 km × 36 km (SMAPL3SMP) and the enhanced product with a spatial resolution of 9 km × 9 km (SMAPL3SMP_E) with a daily gridded global composite (Chew and Small 2020). The SMAPL3SMP is the radiometer soil moisture estimates composite of the SMAPL2SMP soil moisture product that is produced by a single channel algorithm (O'Neill et al. 2018), while the SMAPL3SMP_E was developed based on the enhanced SMAPL2SMP_E product (Chen et al. 2016; O'Neill et al. 2018). The SMAP Level-4 (L4_SM) soil moisture data product is created by combining SMAP L-band brightness temperature data into the NASA Catchment land surface model (Reichle et al. 2022; Reichle 2016). The SMAP_L4 product has been available since March 2015 and provides a 9-km resolution global estimate of surface (0–5 cm) and root zone (0–100 cm) soil moisture on a 3-h temporal interval. Since it combines information from both SMAP observations and land surface models, it is expected to be a better product than satellite estimates and model simulations alone. Very recently, in March 2023, NASA also released the SMAP-derived product containing global daily 1-km resolution surface soil moisture. The dataset is now available at the NASA National Snow and Ice Data Center Distributed Active Archive Center (NSIDC DAAC).

ESA CCI Soil Moisture Product: The European Space Agency (ESA) Climate Change Initiative (CCI) soil moisture product is provided under the framework of the European Space Agency's Water Cycle Multi-Mission Observation Strategy (WACMOS) project. The ESA CCI dataset is a merged multisatellite microwave soil moisture dataset that combines observations from the Scanning Multichannel Microwave Radiometer (SMMR), the Special Sensor Microwave Imager (SSM/I), the Tropical Rainfall Measuring Mission Microwave Imager (TMI), the Advanced Microwave Scanning Radiometer for the Earth Observing System (AMSR-E), the

Advanced Microwave Scanning Radiometer-2 (AMSR2), and the WindSat and active products including the Advanced Scatterometer (ASCAT) observations and scatterometer observations onboard the European Remote Sensing (ERS) satellite (Liu et al. 2011; Wagner et al. 2012). The ESA CCI soil moisture dataset contains daily surface soil moisture information in volumetric units at a 0.025° spatial resolution and is available from 1978 to 2018 (Liu et al. 2011).

Catchment Biophysical Data Sources

The other advances in remote sensing-based products are the increasing availability of high-resolution satellite products that could help to estimate different catchment biophysical variables with greater precision and accuracy. The following section presents an overview of these products and their potential role in the Abbay basin research applications.

Sentinel-1 Synthetic Aperture Radar (SAR) Dataset

The Sentinel-1 mission consists of a constellation of two polar-orbiting satellites that operate day and night using C-band synthetic aperture radar imaging to collect data in all weather conditions. (Torres et al. 2012). Sentinel-1A, the first satellite, was launched on April 3, 2014, and Sentinel-1B, the second satellite, was launched on April 25, 2016. The two satellites are offset by 180 degrees to allow repeat passes every six days and have a 12-day repeat cycle on each one (Geudtner et al. 2014). The products have recently emerged as the most popular land and water body mapping data sources due to their high spatial (up to 10 m horizontal distance) and temporal resolution, which makes them the highest quality freely available SAR products. Concerning its application in Abbay basin hydrological research, it might be the best alternative for biophysical and water body monitoring during the rainy season, when high cloud cover hinders the use of optical images. In addition, as a result of the RADAR's unique response to water, the product is most preferred for water body mapping using intensity thresholding methods. An example of its application in Abbay basin research is a study by Ayehu et al. (2020), who used the product to estimate residual soil moisture levels. Other specific applications include flood modelling (Elkhrachy et al. 2021), soil moisture estimation (Schönbrodt-Stitt et al. 2021), water body mapping (Shen et al. 2022), and river water delineation (Obida et al. 2019).

Phased Array L-Band Synthetic Aperture Radar (PALSAR)

The phased array L-band synthetic aperture radar (PALSAR) aboard the Advanced Land Observing Satellite (ALOS) is a geometrically and radiometrically terrain-corrected SAR product provided by the Alaska Satellite Facility Distributed Active Archive Data Center (ASF DAAC) (Jaxa 2007). The devices offer enhanced backscatter estimations for SAR sensors, which can be utilized as input for applications including land cover classification, delineating wet snow-covered areas, and monitoring deforestation. They are distributed at two resolutions, with pixel sizes of 12.5 m and 30 m. Inside the products is the Radiometric Terrain-Corrected GeoTIFF file, with pixel values representing gamma nought power in 32-bit floating point format and a DEM used for RTC processing in GeoTIFF format (Rosenqvist et al. 2007). However, the lack of temporal continuity restricts their use for time series and change detection analysis.

The Gravity Recovery and Climate Experiment Follow-On (GRACE-FO)

The Gravity Recovery and Climate Experiment (GRACE) is a satellite mission intended to quantify terrestrial water storage (TWS) variations at regional and global scales through monitored changes in the mass on the Earth's surface (Tapley et al. 2004). The mission began in April 2002 and ran until October 2017. GRACE Follow-On (GRACE-FO) project, which launched in May 2018, succeeded in the original mission and continued the record by extending the monthly time series data. (Kornfeld et al. 2019). The most critical contributions of GRACE-FO measurements are surface water, canopy water, soil water, groundwater, and snow water. With a typical spatial resolution of 300 km and an error level of 2 cm (in terms of equivalent water thickness), the GRACE and GRACE-FO data have been widely used in the field of hydrology to study catchments with a relative size of $\approx$63,000 km^2 and more (Abera et al. 2017). The role of GRACE-FO data in data-scarce regions similar to the Abbay basin is very significant. It might help to build new hydrological insights related to the volume of surface water dynamics in the basin, which would not be feasible to know using conventional ground-based experiments. However, the coarse spatial resolution of the data limits its application to smaller catchments (Table 20.1).

Innovation and Automated Tools for Climatology and Hydrological Research

The other major achievements of recent decades (in addition to the growth of diverse data archives) are the development of time- and cost-effective computational resources. This is due to an improvement in the computational power of hardware and the adoption of programming languages. Increasingly, hydrologists and climatologists are taking advantage of the advancements to derive process insights from large and complex datasets (Astagneau et al. 2021). The ability to automate the processing of a large dataset has facilitated a transition from in-depth trials in a single location towards large-sample studies (Slater et al. 2019). Many of these functionalities are not, however, properly utilized in Abbay basin research. The research approach in the basin is still based on conventional analysis tools, while data encoding, organization, and analysis are performed manually with the use of semiautomated tools (Dile et al. 2018; Yenehun et al. 2021). In addition to high operational costs, the conventional research approach is less advantageous in terms of research output reliability. One of the reasons for the limited application of cutting-edge technologies might

Table 20.1 List of innovative remote sensing products and secondary data sources relevant to the context of the Abbay basin

Data	Description	Source	Potential role in Abbay basin
ENACT	Blended station-satellite rainfall and temperature data at 4 km² resolution	https://iri.columbia.edu/resources/enacts/	Fills climate data gap
CHIRPS	A semiglobal rainfall product with a resolution of 0.005°, intended for drought monitoring and global environmental changes (Funk et al. 2015)	https://www.chc.ucsb.edu/data/chirps	Fills climate data gap
WorldClim	A spatially interpolated monthly climate data for global land areas at 1 km² spatial resolution (Fick and Hijmans 2017)	https://www.worldclim.org/data/worldclim21.html	Increase the precision and accuracy of climatological and hydrological estimations
SAF-LSA surface reflectance products	Satellite products that characterize the continental surfaces, such as radiation products, vegetation, evapotranspiration and wildfires (Trigo et al. 2011)	https://landsaf.ipma.pt/en/	Increase the precision and accuracy of climatological and hydrological parameter estimations
TerraClimate	Monthly climate and climatic water balance for global terrestrial surfaces (Abatzoglou et al. 2018)	https://www.climatologylab.org/terraclimate.html	Increase the precision and accuracy of climatological and hydrological parameter estimations
CMORPH perception estimates	Perception data on an 8 km × 8 km grid resolution (Xie et al. 2017)	https://www.ncei.noaa.gov/products/climate-data-records/precipitation-cmorph	Improve the precision and accuracy of climatological and hydrological parameter estimations
African Rainfall Climatology Version 2.0	Daily rainfall data over Africa at a 0.1° resolution (Novella and Thiaw 2013)	https://gmes.icpac.net/data-center/arc2-rfe	Fills data gap on climate variables

(continued)

Table 20.1 (continued)

Data	Description	Source	Potential role in Abbay basin
The SMAP soil moisture product	Soil moisture product at the surface and soil root zone derived from microwave-based satellite sensors (at a resolution of 9 to 36 km^2 (Entekhabi et al. 2010)	https://podaac.jpl.nasa.gov/SMAP	Fills data gap on soil moisture characteristics
ESA CCI soil moisture product	A merged multisatellite microwave soil moisture dataset (Gruber et al. 2019)	https://esa-soilmoisture-cci.org/	Fills data gap on soil moisture characteristics
Sentinel-1 SAR dataset	High-resolution SAR products at 1 6-day time interval (Torres et al. 2012)	https://asf.alaska.edu/data-sets/sar-data-sets/sentinel-1/	Fills data gap on topographic changes, soil moisture, biophysical attributes during high cloud seasons
PALSAR radar data	Geometrically and radiometrically terrain-corrected SAR product (Jaxa 2007)	https://asf.alaska.edu/data-sets/sar-data-sets/alos-palsar/	Improve the precision and accuracy of catchment topographic parameter estimations
GRACE-FO surface water volume anomaly data	Surface water volume measurements, such as surface water, canopy water, soil water, groundwater, and snow water (Tapley et al. 2004)	https://podaac.jpl.nasa.gov/GRACE?tab=mission-objectives&sections=about%2Bdata	May help to provide new scientific insight about surface and near surface water volume dynamics

be the absence of ready-to-use information about the utility of recent innovations in the basin. It might be challenging for researchers to select the relevant cutting-edge achievements from the rapidly updating, complex and huge information network. Below we briefly presented some relevant innovations and automation tools that have an important role in transforming research in the basin.

Cloud-Computing-Based Applications

The Climate Data Store (CDS) Toolbox: The CDS Toolbox is a part of the Copernicus Climate Data Store, providing facilities for users to access CDS data and build web-based applications (www.copernicus.eu). Its aim was to track expected the impact of climate variability on business sectors including energy, water management, transport and tourism. The tool links raw data to online computing power through a programming interface. It facilitates the creation of applications in Python and runs them on CDS computers through an online personal workspace. It helps to download graphs, maps and data and share online creations with other users. For its application in the Abbay basin, it might ease data search and query processes, as it facilitates browsing data from multiple sources. The variables and functionalities are also highly relevant not only to the scientific community but also to decision-makers and planners, as they help to produce customized water resource monitoring applications. The platform is, however, a programming interface, making it challenging to use with little or no coding experience.

Climate Engine: It is a web application to process and visualize satellite images and weather data. Its specific focus was to improve predictive capabilities of extreme climate events, such as drought, wildfire, and crop failure risk. It uses Google's Earth Engine for on-demand processing of satellite and climate data via a web browser. For research activities, it can be useful to retrieve climate parameters, conduct, time series analysis and share the derived results with web URL links (www.climateengine.com/). However, its usability is only limited to a few predetermined tools with limited application for complex and custom-based analysis.

Google Earth Engine

Google Earth Engine (GEE) was introduced as one of the first geospatial cloud computing platforms by Google in 2010. Since then, GEE has established itself as one of the top platforms for mining and preparing large-scale geographic data. GEE supports more geospatial data types than other cloud platforms and offers free services to all users (Mutanga and Kumar 2019). This makes it the most frequently used platform to process data in a variety of earth system science disciplines (Pérez-Cutillas et al. 2023) including; agriculture (Jaafar and Mourad 2021), water (Condeça et al. 2022), land cover/land use (Lee et al. 2018), disasters (Ghaffarian et al. 2020), climate change (Kazemi Garajeh et al. 2023), soil (Luo et al. 2022), wetland (Gxokwe et al. 2022), vegetation (Campos-Taberner et al. 2018), urbanization and other fields. The platform was already introduced in the Abbay basin research, although its functionalities were not fully exploited. Some examples of previous application of the platform in the Abbay basin research include soil erosion risk mapping (Elnashar et al. 2021), land suitability analysis (Yalew et al. 2016) and desertification mapping (Elnashar et al. 2022) (Table 20.2).

Table 20.2 Google Earth Engine-based case studies with potential relevance to climatological and hydrological applications in the Abbay basin

Field of study	Specific focus	Code availability	Reference
Fluvial geomorphology	Detecting river channel change	Available	Boothroyd et al. (2020)
Hydrology	Time series water balance computation	Available	Gemitzi and Kofidou (2022)
Landscape geomorphology	Susceptibility modelling of gully erosion	Available	Titti et al. (2022)
Hydrology	Floodwater depth estimation tool	Available	Raju et al. (2022)
Climatology	Monitoring CO and O_3 concentration	Available	Ikram et al. (2022)

Script-Based Data Analysis Tools: R Packages in Hydrology and Climatology

Other recent developments in hydroclimatological computations with notable potential importance for Abbay basin research activities include the emergence of an open-source programming language. Among the many software and statistical analysis tools, R is one of the most powerful and popular tools for statistical computing. R was created by Ross Ihaka and Robert Gentleman in the 1990s for statistical computing and it is now backed by a quickly expanding online community (Astagneau et al. 2021).

One of the principal advantages of R is its ease of use. This is from detailed documentation, extensive online resources, object-oriented programming, functional programming and the free availability of the source code under the open-source license. R is also compatible with all popular operating systems (such as Microsoft Windows, macOS, and Linux), making it perfect for both business and institutional use. The recent developments in R have benefited hydrological science through improved open science and numerical literacy. Over the past decade or so, R has developed into one of the essential tools for scientific computation in hydrology (Slater et al. 2019). Table 20.3 presents the list of selected packages that might have a significant role in facilitating hydrological and climatological studies of the Abbay basin. The packages were selected in consideration of major research gaps in the basin. The selected packages are those that might have one or more of the following roles: facilitation of data access, improved accuracy of parameter estimation and a special role in filling the structural knowledge gap.

In the hydro climatological study of the Abbay basin, one of the prime challenges while using secondary data sources is to search individual data sources from a vast network of data providers. There are packages designed to ease these tasks. Some of them are intended to browse climate data such as NASApower (Sparks 2018), Rnoaa (Chamberlain 2023), RNCEP (Kemp et al. 2012) and worldmet (Carslaw 2023). The

Table 20.3 R packages potentially useful to facilitate hydrological and climatological research in the Abbay basin

Package category	Package	Description	Specific role in Abbay basin
Hydrology and climatology data download tools	MODISTools	Makes MODIS time series satellite products easily downloadable (Tuck et al. 2014)	Facilitate biophysical data access
	MODIStsp	Enables the download, preprocessing and analysis of data from MODIS Land products (Busetto and Ranghetti 2016)	Facilitate biophysical data access
	isoWater	Provides functions to query and obtain stable H and O isotope data from the freely available water isotope data (http://waterisotopes.org)	Facilitate data access for palaeohydrological study
	smapr:	Enables programmatic access to search, retrieve, and extract NASA Soil Moisture Active Passive (SMAP) data (Joseph et al. 2023)	Facilitate hydro meteorological data access
	nasapower	Interface to browse daily meteorological data from NASA's prediction of worldwide energy resource (POWER) (Sparks 2018)	Facilitate hydro meteorological data access
	Rnoaa	Interface to NOAA weather data (Chamberlain 2023)	Facilitate hydro meteorological data access
	RNCEP:	Contains functionalities to access, organize, and display weather data from NCEP/NCAR reanalysis and NCEP/DOE reanalysis II datasets (Kemp et al. 2012)	Facilitate hydro meteorological data access

(continued)

Table 20.3 (continued)

Package category	Package	Description	Specific role in Abbay basin
	worldmet:	Provide functionalities to retrieve data from more than 30,000 surface meteorological sites throughout the world managed by the National Oceanic and Atmospheric Administration (NOAA) Integrated Surface Database (ISD) (Carslaw 2023)	Facilitate hydro meteorological data access
Hydroclimatological analysis and modelling packages	waterquality:	A function to convert satellite-based reflectance imagery into multiple water quality algorithms designed for the detection of algal blooms or the following pigment proxies: chlorophyll-a, blue–green algae (phycocyanin), and turbidity (Johansen et al. 2019)	Facilitate surface water quality monitoring with fewer observation data
	Evapotranspiration	Functions to estimate potential (PET) and actual evapotranspiration (AET) using 21 different formulations including Penman, Penman–Monteith FAO 56, Priestley-Taylor and Morton models (Guo et al. 2016)	Facilitate estimation of basin water balance parameters while studying the hydrology of poorly gauged catchments
	EGRET	Exploration and graphics for RivEr trends (EGRET) analysis tool for long-term changes in water quality and streamflow (Hirsch and De Cicco 2015)	Useful to fill knowledge gap about the presence of statistically significant water quality changes
	RSAlgaeR	Builds empirical remote sensing models of water quality variables and analyses long-term trends (Cahhansen 2019)	Facilitate surface water quality monitoring with fewer observation data

(continued)

Table 20.3 (continued)

Package category	Package	Description	Specific role in Abbay basin
	DWBmodelUN	A hydrologic modelling tool that uses the Budyko framework and the dynamic water balance model with dynamical dimension search algorithm to calibrate the model and analyse the outputs from interactive graphics (Zamora et al. 2020)	Facilitate the study of water availability and water stress in basins
	WRSS	A tool for simulation and analysis of large-scale water resources systems. (e.g., reservoirs, aquifers, etc.) based on standard operating policy (SOP) (Arabzadeh et al. 2021)	Facilitate the study of water resource utilization system in the basin
	transfR	A geomorphology-based hydrological modelling for transferring streamflow measurements from gauged to ungauged catchments (de Lavenne et al. 2023)	Improve accuracy of ungagged catchment hydrological estimations

others are useful for browsing biophysical and hydrological data, including MODIS-Tools (Tuck et al. 2014), MODIStsp (Busetto and Ranghetti 2016), smapr (Joseph et al. 2023) and isoWater (Bowen et al. 2010). With the exception of the last one, the first three products are important to drive medium-resolution surface parameters required for hydrological modelling, including surface temperature, vegetation, soil moisture and a variety of catchment biophysical indices. The function of IsoWater is to retrieve water isotope data, which are important for palaohydrological study (Bowen et al. 2010).

The hydrological analysis and modelling packages serve as a platform for data preparation, analysis and hydrological modelling. In Table 20.3, we have presented the list of packages with potential relevance in the Abbay basin. Evapotranspiration is one of these packages that contains a function to estimate evapotranspiration parameters in the absence of actual evapotranspiration measurements (Guo et al. 2016). In this regard, the tool could play a special role in filling data gaps in Abbay basin hydrological research. This is because evapotranspiration is rarely measured in the available weather stations of the Abbay basin, as many stations in the basin were not equipped with the measuring instruments.

EGRET is an important tool to verify the presence of statistically significant water quality changes across space and time (Hirsch and De Cicco 2015). Water quality change studies in the Abbay basin rarely consider the temporal dimension (Dersseh et al. 2022), and few of the available change studies were not conducted using appropriate statistical methods. A similar tool for time series water quality analysis is the RSAlgaeR package (for surface water), which helps to monitor long-term water quality changes using satellite products without the necessity of intensive ground observation data (Cahhansen 2019). In addition to its ability to fill data gaps, the tool is potentially important to bridge the knowledge gap regarding the magnitude and dimension of water quality changes in the basin, which remained unanswered due to the absence of historical water quality data.

DWBmodelUN is a tool that helps to estimate water balance parameters using the Budyko-based framework (Zamora et al. 2020). It is important for the Abbay basin research activities to estimate time series water balance fluxes together with water availability in the basin.

Another relevant tool for hydrological study of the Abbay basin is the WRSS (Water Resources System Simulator), which is important for modelling and simulation of water resources systems based on the Standard Operation Policy (SOP). The package offers ways to construct energy and water supply models alter their constituent parts, produce scenarios, and display and publish the results (Arabzadeh et al. 2021). There are numerous water resource utilization systems in the Abbay basin, including hydroelectric dams and large-scale irrigation projects, but little is known about the optimum water resource utilization levels and their side effects on the basin environment, which can be addressed using the WRSS package. Stream gauging stations in the Abbay basin are very few, and estimation of water balance components for such ungauged catchments remains a primary scientific challenge in the hydrological endeavours of the basin (Dessie et al. 2015; Duan et al. 2018; Wale et al. 2009). The transfR is an R package that is designed to ease the challenge of transferring streamflow measurements from gauged to ungauged catchments based on geomorphological principles. It works based on runoff-runoff principles. This is through combining streamflow series from available monitoring stations and estimating the streamflow elsewhere in the nearby rivers without necessitating the implementation of a full rainfall-runoff model (de Lavenne et al. 2023).

Machine Learning Applications in Hydroclimatological Research

Machine learning (ML) refers to a collection of techniques and algorithms that allow computers to enhance their performance automatically through experience. Unlike physical sciences, which rely on knowledge-driven reasoning, ML is founded on data-driven reasoning. Since the groundbreaking studies of the 1950s, the field of

ML has undergone dramatic evolution (Xu and Liang 2021), becoming an interdisciplinary area that intersects with computer science, statistics, applied mathematics, and optimization (Mosaffa et al. 2022).

In the context of the Abbay basin, data-driven ML algorithms can help address scientific issues that have remained unanswered due to the limitations of physical and simple data-driven models. Given the complex biophysical nature of the basin, theoretical knowledge about many of its hydroclimate processes is limited. As a result, it is difficult to address scientific issues using conventional knowledge-driven process-based models. However, data-driven ML models can predict many hydrological parameters without precise descriptions of the underlying physical processes (Lin et al. 2006). Moreover, many catchments in the Abbay basin are inadequately gauged, making it difficult to estimate hydroclimate parameters using conventional models that require actual measurement data. ML-based models can help address this problem by developing models that can predict the hydroclimate parameters of ungauged catchments using data from gauged catchments (Kratzert et al. 2019b). In addition, ML-based models have been found to be useful in estimating catchment flow volumes using catchment biophysical data as predictor variables in the absence of hydroclimate data across all parts of the basin (Ditthakit et al. 2021). The following sections provide an explanation of the nature of ML-based models and a synthesis of selected studies that can provide insights into the potential role of ML-based models in Abbay basin hydroclimate research.

Key Elements of Machine Learning

The objective of machine learning is to improve the performance of machines without explicit programming. This is achieved through several steps. The first step is data gathering and preparation. This is followed by a selection of a training model and algorithm. Then, the model is evaluated to make predictions and tune the hyperparameters. In the context of hydrology and climatology, ML needs accurate and reliable data relevant to the variables to be quantified or predicted. The common data required for many of the ML-based hydrological modelling includes precipitation, temperature, streamflow, and groundwater levels. To fully reflect the hydrologic system's variability, the data needs to be collected over a long period of time and at a high frequency. The temporal dimension and spatial detail of the data determine the accuracy and reliability of ML-based parameter estimation (Ghobadi and Kang 2023).

Another requirement of ML-based modelling is featuring engineering that transforms input datasets into sets of figures essential for a specific task. In hydrology, this may involve transforming raw data into features such as monthly averages, seasonal trends, and lagged variables (Kuhn and Johnson 2019).

Once the data requirements are fulfilled and feature engineering is created, the next step of ML-based modelling is to select and set up appropriate machine learning algorithms. Hydrologic modelling can make use of a variety of machine learning algorithms, such as decision trees, random forests, support vector machines, and neural networks. The decision on algorithm selection will depend on the objective of the analysis and the nature of the data input. ML algorithms are divided into three general categories: supervised, unsupervised, and Reinforcement learning (RL). In a supervised algorithm, both the input and output values are determined beforehand, and a labelled dataset is used to train the algorithms, which is then employed to classify or predict the output. In the case of unsupervised learning, algorithms are trained with unlabelled datasets. These algorithms identify occult patterns or data clusters without the assistance of a human. On the other hand, RL deals with how intelligent an agent must be to respond in order to reap the greatest rewards from its environment (Xu and Liang 2021). The major ML algorithms commonly used in hydrological and climatological applications are summarized in Table 20.4.

The final step of ML-based modelling is accuracy assessment and model evaluation. This is done to ensure that the model accurately represents patterns in the input data. There are many methods of model evaluation, the common approach is to use metrics such as coefficient of determination, root mean square error, and mean absolute error (Ghobadi and Kang 2023).

Table 20.4 Comparison of supervised, unsupervised, and reinforcement learning algorithms (from Ghobadi and Kang 2023)

Learning types	Type of data	Training	Used for	Algorithms
Supervised learning	Labelled data	Trained using labelled data (extra supervision)	Nowcasting, forecasting, classification in binary and multiple classes	Linear regression, logistic regression, RF, SVM, KNN, RNN, DNN, etc.
Unsupervised learning	Unlabelled data	Trained using unlabelled data without any guidance (no supervision)	Clustering	K—means, C—means, agglomerative hierarchical clustering, DBSCAN, Gaussian mixture models, optics, etc.
Reinforcement learning	Without predefined data	Works based on the interaction between agent and environment (no supervision)	Decision-making	Q—learning, SARSA, DQN, double DQN, dueling DQN, etc.

A Synthesis of ML-Based Case Studies Relevant to the Abbay Basin

Machine learning (ML) has numerous potential applications in hydrology, including flood prediction, streamflow forecasting, water quality monitoring, and ground water level estimation. These applications can be particularly relevant in the Abbay basin.

Flood prediction is one important application of ML in hydrology. ML models can be trained on historical data of water levels, rainfall, and other relevant variables to predict the probability and severity of floods in a specific location. Studies have shown that ML-based flood prediction algorithms can be more accurate than physical-based models. For instance, Kratzert et al. (2019a) used time-specific meteorological data and static catchment attributes as predictor variables to estimate rainfall-runoff relationships with high accuracy.

Streamflow forecasting is another important application of ML in hydrology. ML models can be trained on real-time data of rainfall, temperature, and other parameters to predict the streamflow in a particular river or stream. These predictions can be used to estimate catchment flow volumes for ungauged catchments and make temporal predictions based on historical observation data. Studies have shown that ML algorithms can effectively predict hydrological time series with nonlinear characteristics and estimate streamflow over data-poor regions.

In water quality studies, ML-based models were employed to detect the existence of contaminants and analysis water clarities. For example, Zou et al. (2020) used meteorology and water quality data to develop a water quality predictor method based on a multitime scale bidirectional LSTM network. Bui et al. (2020) tested a variety of machine learning and hybrid data-mining algorithms to predict river water quality levels using long-term water quality data and found that the model has good predictive output.

Finally, ML-based models can be used for ground water level estimation. Accurate ground water level estimations remain unease due to the nonlinear correlation of ground water level with explanatory variables, as well as their multiscale characteristics that change over time. ML algorithms, however, make it possible to effectively predict the temporal trend of groundwater levels. These models employ predictor variables such as streamflow rates, groundwater pumping rates, agricultural irrigation, and estimate groundwater levels with reasonable level of accuracy. For instance, Chen et al. (2020) found that ML-based models outperformed conventional numerical models in ground water level estimation. Sahoo et al. (2017) developed an automated hybrid artificial neural network (HANN) model that predicted ground water level dynamics using precipitation, temperature, streamflow, and climate indices as input parameters and showed superior performance compared to conventional linear and nonlinear multivariate regression models.

Summary

In this chapter, we explore how innovations in climatological and hydrological data sources, as well as data analysis tools, can improve the accuracy of hydroclimatological research in the Abbay basin. A significant research challenge in this area is the lack of appropriate data sources due to a sparse hydroclimatological observation network. However, the increasing amount of secondary data from satellite observations could help address this issue, as they offer higher spatial and temporal resolutions. In particular, we reviewed temperature and surface reflectance products and found that they could be directly observed through satellite observation systems, with few ground observation data. However, for precipitation and related water volume parameters, alternative data sources are limited, and their accuracy and bias correction should be studied before their direct application in hydroclimatological studies in the basin. Establishing satellite data validation facilities within the basin could be a feasible solution to fill the data gap, rather than covering the entire basin area with appropriate weather observation networks, which may not be achievable in the near future.

We also reviewed recent advances in cloud-based computing platforms and object-oriented data analysis tools that could be relevant for hydroclimatological research activities in the Abbay basin. Among the cloud computing platforms, we highlighted Google Earth Engine-based tools, which offer a free provision of high computational resources, enabling automated iteration of functions on high spatial resolution and large spatial extent datasets. We also reviewed R software and relevant packages, which have potential significance in facilitating research activities in the basin. However, the application level of these tools in basin research is still at an early stage, primarily due to a lack of necessary coding skills by researchers.

Finally, we explored ML-based algorithms and their potential role in Abbay basin hydroclimatological research. Our review shows that these algorithms are highly relevant to the context of the basin and can be used to address research problems that remain unanswered. They offer the possibility of making more accurate estimates of hydrological and climatological variables than conventional models, as well as the ability to estimate parameters of ungauged catchments without the need for a full theoretical understanding of basin environmental characteristics. However, their validity depends on the correct use and careful consideration of the drawbacks. For instance, many ML algorithms require large amounts of training data, which may limit their application in areas with sparse data. Overall, promoting the application of ML approaches in Abbay basin hydroclimatological research could help advance scientific understanding in the area.

Acknowledgements We acknowledge the support of the CGIAR Nexus Gains initiative and the CGIAR Sustainable Intensification of Mixed Farming System (SI-MFS) initiative.

References

Abate M, Nyssen J, Moges MM, Enku T, Zimale FA, Tilahun SA, Adgo E, Steenhuis TS (2017) Long-term landscape changes in the Lake Tana basin as evidenced by delta development and floodplain aggradation in Ethiopia. Land Degrad Dev 28:1820–1830. https://doi.org/10.1002/ldr.2648

Abatzoglou JT, Dobrowski SZ, Parks SA, Hegewisch KC (2018) TerraClimate, a high-resolution global dataset of monthly climate and climatic water balance from 1958–2015. Sci Data 5:170191. https://doi.org/10.1038/sdata.2017.191

Abebe SA, Qin T, Yan D, Gelaw EB, Workneh HT, Kun W, Liu S, Dong B (2020) Spatial and temporal evaluation of the latest high-resolution precipitation products over the Upper Blue Nile River basin, Ethiopia. Water (Basel) 12:3072. https://doi.org/10.3390/w12113072

Abebe WB, Tilahun SA, Moges MM, Wondie A, Derssch MG, Assefa WW, Mhiret DA, Adem AA, Zimale FA, Abera W, Steenhuis TS, McClain ME (2021) Ecological status as the basis for the holistic environmental flow assessment of a tropical highland river in Ethiopia. Water (Basel) 13:1913. https://doi.org/10.3390/w13141913

Abera W, Brocca L, Rigon R (2016) Comparative evaluation of different satellite rainfall estimation products and bias correction in the Upper Blue Nile (UBN) basin. Atmos Res 178–179:471–483. https://doi.org/10.1016/j.atmosres.2016.04.017

Abera W, Formetta G, Brocca L, Rigon R (2017) Modeling the water budget of the Upper Blue Nile basin using the JGrass-NewAge model system and satellite data. Hydrol Earth Syst Sci 21:3145–3165. https://doi.org/10.5194/hess-21-3145-2017

Adgolign TB, Rao GVRS, Abbulu Y (2016) WEAP modeling of surface water resources allocation in Didessa sub-basin, West Ethiopia. Sustain Water Resour Manage 2:55–70. https://doi.org/10.1007/s40899-015-0041-4

Ali DA, Deininger K, Monchuk D (2020) Using satellite imagery to assess impacts of soil and water conservation measures: evidence from Ethiopia's Tana-Beles watershed. Ecol Econ 169:106512. https://doi.org/10.1016/j.ecolecon.2019.106512

Ali RHAE-M, Zahran SAE-S (2023) Evaluation of NASA land information system in prediection stream runoff: case study of Atbara and Blue Nile sub-basins. Model Earth Syst Environ. https://doi.org/10.1007/s40808-022-01663-5

Arabzadeh R, Aberi P, Hesarkazzazi S, Hajibabaei M, Rauch W, Nikmehr S, Sitzenfrei R (2021) WRSS: an object-oriented R package for large-scale water resources operation. Water (Basel) 13:3037. https://doi.org/10.3390/w13213037

Astagneau PC, Thirel G, Delaigue O, Guillaume JHA, Parajka J, Brauer CC, Viglione A, Buytaert W, Beven KJ (2021) Technical note: hydrology modelling R packages—a unified analysis of models and practicalities from a user perspective. Hydrol Earth Syst Sci 25:3937–3973. https://doi.org/10.5194/hess-25-3937-2021

Ayehu G, Tadesse T, Gessesse B, Yigrem Y, Melesse AM (2020) Combined use of Sentinel-1 SAR and landsat sensors products for residual soil moisture retrieval over agricultural fields in the Upper Blue Nile basin, Ethiopia. Sensors 20. https://doi.org/10.3390/s20113282

Ayehu GT, Tadesse T, Gessesse B (2021) Spatial and temporal trends and variability of rainfall using long-term satellite product over the Upper Blue Nile basin in Ethiopia. Remote Sens Earth Syst Sci 4:199–215. https://doi.org/10.1007/s41976-021-00060-3

Ayele G, Teshale E, Yu B, Rutherfurd I, Jeong J (2017) Streamflow and sediment yield prediction for watershed prioritization in the upper blue nile river basin, Ethiopia. Water (Basel) 9:782. https://doi.org/10.3390/w9100782

Bayissa Y, Moges S, Melesse A, Tadesse T, Abiy AZ, Worqlul A (2021) Multi-dimensional drought assessment in Abbay/Upper Blue Nile basin: the importance of shared management and regional coordination efforts for mitigation. Remote Sens (Basel) 13:1835. https://doi.org/10.3390/rs13091835

Bekele AA, Pingale SM, Hatiye SD, Tilahun AK (2019) Impact of climate change on surface water availability and crop water demand for the sub-watershed of Abbay Basin, Ethiopia. Sustain Water Resour Manage 5:1859–1875. https://doi.org/10.1007/s40899-019-00339-w

Belete M, Deng J, Abubakar GA, Teshome M, Wang K, Woldetsadik M, Zhu E, Comber A, Gudo A (2020) Partitioning the impacts of land use/land cover change and climate variability on water supply over the source region of the Blue Nile basin. Land Degrad Dev 31:2168–2184. https://doi.org/10.1002/ldr.3589

Bogale A (2021) Morphometric analysis of a drainage basin using geographical information system in Gilgel Abay watershed, Lake Tana basin, upper Blue Nile basin, Ethiopia. Appl Water Sci 11:122. https://doi.org/10.1007/s13201-021-01447-9

Boothroyd RJ, Williams RD, Hoey TB, Barrett B, Prasojo OA (2020) Applications of Google Earth Engine in fluvial geomorphology for detecting river channel change. Wires Water. https://doi.org/10.1002/wat2.1496

Bowen JBW-GJ, Dawson TE, Tu KP (2010) Understanding movement, pattern, and process on Earth through isotope mapping

Bui DT, Khosravi K, Tiefenbacher J, Nguyen H, Kazakis N (2020) Improving prediction of water quality indices using novel hybrid machine-learning algorithms. Sci Total Environ 721:137612. https://doi.org/10.1016/j.scitotenv.2020.137612

Busetto L, Ranghetti L (2016) MODIStsp: an R package for automatic preprocessing of MODIS land products time series. Comput Geosci 97:40–48. https://doi.org/10.1016/j.cageo.2016.08.020

Cahhansen (2019) cahhansen/RSAlgae: pre-release v2. Zenodo. https://doi.org/10.5281/zenodo.2538202

Campos-Taberner M, Moreno-Martínez Á, García-Haro F, Camps-Valls G, Robinson N, Kattge J, Running S (2018) Global estimation of biophysical variables from Google Earth Engine platform. Remote Sens (Basel) 10:1167. https://doi.org/10.3390/rs10081167

Carslaw D (2023) worldmet: import surface meteorological data from NOAA integrated surface database (ISD)

Ceccherini G, Ameztoy I, Hernández C, Moreno C (2015) High-resolution precipitation datasets in South America and West Africa based on satellite-derived rainfall, enhanced vegetation index and digital elevation model. Remote Sens (Basel) 7:6454–6488. https://doi.org/10.3390/rs70506454

Chamberlain S (2023) rnoaa: "NOAA" weather data from R

Cheesman RE (1935) Lake Tana and its islands. Geogr J 85:489. https://doi.org/10.2307/1785868

Chen X, Su Y, Liao J, Shang J, Dong T, Wang C, Liu W, Zhou G, Liu L (2016) Detecting significant decreasing trends of land surface soil moisture in eastern China during the past three decades (1979–2010). J Geophys Res Atmos 121:5177–5192

Chen C, He W, Zhou H, Xue Y, Zhu M (2020) A comparative study among machine learning and numerical models for simulating groundwater dynamics in the Heihe River basin, northwestern China. Sci Rep 10:3904. https://doi.org/10.1038/s41598-020-60698-9

Chew C, Small E (2020) UCAR-CU CYGNSS level 3 soil moisture Version 1.0. NASA physical oceanography DAAC. https://doi.org/10.5067/cygnu-l3sm1

Condeça J, Nascimento J, Barreiras N (2022) Monitoring the storage volume of water reservoirs using Google Earth Engine. Water Resour Res 58. https://doi.org/10.1029/2021WR030026

Conway D (2000a) Some aspects of climate variability in the north east Ethiopian highlands—Wollo and Tigray. SEJS 23. https://doi.org/10.4314/sinet.v23i2.18163

Conway D (2000b) The climate and hydrology of the upper Blue Nile River. Geogr J 166:49–62. https://doi.org/10.1111/j.1475-4959.2000.tb00006.x

Conway D, Mould C, Bewket W (2004) Over one century of rainfall and temperature observations in Addis Ababa, Ethiopia. Int J Climatol J Roy Meteorol Soc 24:77–91

Deblauwe V, Droissart V, Bose R, Sonké B, Blach-Overgaard A, Svenning JC, Wieringa JJ, Ramesh BR, Stévart T, Couvreur TLP (2016) Remotely sensed temperature and precipitation data improve species distribution modelling in the tropics. Glob Ecol Biogeogr 25:443–454. https://doi.org/10.1111/geb.12426

Dersseh MG, Steenhuis TS, Kibret AA, Eneyew BM, Kebedew MG, Zimale FA, Worqlul AW, Moges MA, Abebe WB, Mhiret DA, Melesse AM, Tilahun SA (2022) Water quality characteristics of a water hyacinth infested tropical highland lake: Lake Tana, Ethiopia. Front Water 4. https://doi.org/10.3389/frwa.2022.774710

Dessie M, Verhoest NEC, Pauwels VRN, Adgo E, Deckers J, Poesen J, Nyssen J (2015) Water balance of a lake with floodplain buffering: Lake Tana, Blue Nile basin, Ethiopia. J Hydrol (Amst) 522:174–186. https://doi.org/10.1016/j.jhydrol.2014.12.049

de Lavenne A, Loree T, Squividant H, Cudennec C (2023) The transfR toolbox for transferring observed streamflow series to ungauged basins based on their hydrogeomorphology. Environ Model Softw 159:105562

Dhungana S, Shrestha S, Van TP, Kc S, Das Gupta A, Nguyen TPL (2023) Evaluation of gridded precipitation products in the selected sub-basins of Lower Mekong River basin. Theoret Appl Climatol 151:293–310. https://doi.org/10.1007/s00704-022-04268-1

Dile YT, Karlberg L, Daggupati P, Srinivasan R, Wiberg D, Rockström J (2016) Assessing the implications of water harvesting intensification on upstream-downstream ecosystem services: a case study in the Lake Tana basin. Sci Total Environ 542:22–35. https://doi.org/10.1016/j.scitotenv.2015.10.065

Dile YT, Tekleab S, Ayana EK, Gebrehiwot SG, Worqlul AW, Bayabil HK, Yimam YT, Tilahun SA, Daggupati P, Karlberg L, Srinivasan R (2018) Advances in water resources research in the Upper Blue Nile basin and the way forward: a review. J Hydrol (Amst) 560:407–423. https://doi.org/10.1016/j.jhydrol.2018.03.042

Dinku T, Faniriantsoa R, Cousin R, Khomyakov I, Vadillo A, Hansen JW, Grossi A (2022a) ENACTS: advancing climate services across Africa. Front Clim 3. https://doi.org/10.3389/fclim.2021.787683

Dinku T, Faniriantsoa R, Islam S, Nsengiyumva G, Grossi A (2022b) The climate data tool: enhancing climate services across Africa. Front Clim 3. https://doi.org/10.3389/fclim.2021.787519

Ditthakit P, Pinthong S, Salaeh N, Binnui F, Khwanchum L, Pham QB (2021) Using machine learning methods for supporting GR2M model in runoff estimation in an ungauged basin. Sci Rep 11:19955. https://doi.org/10.1038/s41598-021-99164-5

Dowling TPF, Langsdale MF, Ermida SL, Wooster MJ, Merbold L, Leitner S, Trigo IF, Gluecks I, Main B, O'Shea F, Hook S, Rivera G, De Jong MC, Nguyen H, Hyll K (2022) A new East African satellite data validation station: performance of the LSA-SAF all-weather land surface temperature product over a savannah biome. ISPRS J Photogramm Remote Sens 187:240–258. https://doi.org/10.1016/j.isprsjprs.2022.03.003

Duan Z, Gao H, Ke C (2018) Estimation of Lake outflow from the poorly gauged Lake Tana (Ethiopia) using satellite remote sensing data. Remote Sens (Basel) 10:1060. https://doi.org/10.3390/rs10071060

Ebita A, Kobayashi S, Ota Y, Moriya M, Kumabe R, Onogi K, Harada Y, Yasui S, Miyaoka K, Takahashi K (2011) The Japanese 55-year reanalysis "JRA-55": an interim report. Sola 7:149–152

Elkhrachy I, Pham QB, Costache R, Mohajane M, Rahman KU, Shahabi H, Linh NTT, Anh DT (2021) Sentinel-1 remote sensing data and hydrologic engineering centres river analysis system two-dimensional integration for flash flood detection and modelling in New Cairo City, Egypt. J Flood Risk Manage. https://doi.org/10.1111/jfr3.12692

Elnashar A, Zeng H, Wu B, Fenta AA, Nabil M, Duerler R (2021) Soil erosion assessment in the Blue Nile basin driven by a novel RUSLE-GEE framework. Sci Total Environ 793:148466. https://doi.org/10.1016/j.scitotenv.2021.148466

Elnashar A, Zeng H, Wu B, Gebremicael TG, Marie K (2022) Assessment of environmentally sensitive areas to desertification in the Blue Nile basin driven by the MEDALUS-GEE framework. Sci Total Environ 815:152925. https://doi.org/10.1016/j.scitotenv.2022.152925

Entekhabi D, Njoku EG, O'Neill PE, Kellogg KH, Crow WT, Edelstein WN, Entin JK, Goodman SD, Jackson TJ, Johnson J, Kimball J, Piepmeier JR, Koster RD, Martin N, McDonald KC,

Moghaddam M, Moran S, Reichle R, Shi JC, Spencer MW, Thurman SW, Tsang L, Van Zyl J (2010) The soil moisture active passive (SMAP) mission. Proc IEEE 98:704–716. https://doi.org/10.1109/JPROC.2010.2043918

Erkossa T, Haileslassie A, MacAlister C (2014) Enhancing farming system water productivity through alternative land use and water management in vertisol areas of Ethiopian Blue Nile basin (Abay). Agric Water Manage 132:120–128. https://doi.org/10.1016/j.agwat.2013.10.007

Fick SE, Hijmans RJ (2017) WorldClim 2: new 1-km spatial resolution climate surfaces for global land areas. Int J Climatol 37:4302–4315. https://doi.org/10.1002/joc.5086

Funk C, Peterson P, Landsfeld M, Pedreros D, Verdin J, Shukla S, Husak G, Rowland J, Harrison L, Hoell A, Michaelsen J (2015) The climate hazards infrared precipitation with stations–a new environmental record for monitoring extremes. Sci Data 2:150066. https://doi.org/10.1038/sdata.2015.66

Gashaw T, Dile YT, Worqlul AW, Bantider A, Zeleke G, Bewket W, Alamirew T (2021) Evaluating the effectiveness of best management practices on soil erosion reduction using the SWAT model: for the case of Gumara watershed, Abbay (Upper Blue Nile) basin. Environ Manage 68:240–261. https://doi.org/10.1007/s00267-021-01492-9

Gemitzi A, Kofidou M (2022) A Google Earth Engine tool to assess water budget and its individual components. Glob NEST J. https://doi.org/10.30955/gnj.004269

Getaneh Y, Abera W, Abegaz A, Tamene L (2022) A systematic review of studies on freshwater lakes of Ethiopia. J Hydrol Reg Stud

Geudtner D, Torres R, Snoeij P, Davidson M, Rommen B (2014) Sentinel-1 system capabilities and applications. In: 2014 IEEE geoscience and remote sensing symposium. Presented at the IGARSS 2014—2014 IEEE international geoscience and remote sensing symposium. IEEE, pp 1457–1460. https://doi.org/10.1109/IGARSS.2014.6946711

Ghaffarian S, Rezaie Farhadabad A, Kerle N (2020) Post-disaster recovery monitoring with Google Earth Engine. Appl Sci 10:4574. https://doi.org/10.3390/app10134574

Ghobadi F, Kang D (2023) Application of machine learning in water resources management: a systematic literature review. Water (Basel) 15:620. https://doi.org/10.3390/w15040620

Gruber A, Scanlon T, van der Schalie R, Wagner W, Dorigo W (2019) Evolution of the ESA CCI soil moisture climate data records and their underlying merging methodology. Earth Syst Sci Data Discuss 1–37. https://doi.org/10.5194/essd-2019-21

Guo D, Westra S, Maier HR (2016) An R package for modelling actual, potential and reference evapotranspiration. Environ Model Softw 78:216–224. https://doi.org/10.1016/j.envsoft.2015.12.019

Gxokwe S, Dube T, Mazvimavi D (2022) Leveraging Google Earth Engine platform to characterize and map small seasonal wetlands in the semi-arid environments of South Africa. Sci Total Environ 803:150139. https://doi.org/10.1016/j.scitotenv.2021.150139

Hirsch RM, De Cicco LA (2015) User guide to exploration and graphics for RivEr trends (EGRET) and dataRetrieval: R packages for hydrologic data. In: Techniques and methods. U.S. Geological Survey, Reston, VA

Hurst HE, Phillips P, Black RP, Simaika YM (1933) Ten-day mean and monthly mean discharges of the Nile and its tributaries. Government Press

Ikram NM, Afifah L, Arthatia BS, Wicaksono SJ, Maharani M, Ediyanto, Ihsanudin T, Apriyanti D (2022) Monitoring CO and O_3 concentration that caused climate change periodically using Google Earth Engine (study case: Java Island). IOP Conf Ser Earth Environ Sci 1047:012021. https://doi.org/10.1088/1755-1315/1047/1/012021

Jaafar H, Mourad R (2021) GYMEE: a global field-scale crop yield and ET mapper in Google Earth Engine based on landsat, weather, and soil data. Remote Sens (Basel) 13:773. https://doi.org/10.3390/rs13040773

Jaxa (2007) L1.0_PALSAR. NASA Alaska satellite facility DAAC. https://doi.org/10.5067/j4jvcfddpew1

Johansen R, Reif M, Emery E, Nowosad J, Beck R, Xu M, Liu H (2019) Waterquality: an open-source R package for the detection and quantification of cyanobacterial harmful algal blooms and water quality. Engineer Research and Development Center (U.S.). https://doi.org/10.21079/11681/35053

Joseph M, Oakley M, Schira Z (2023) smapr: acquisition and processing of NASA soil moisture active-passive (SMAP) data

Kazemi Garajeh M, Salmani B, Zare Naghadehi S, Valipoori Goodarzi H, Khasraei A (2023) An integrated approach of remote sensing and geospatial analysis for modeling and predicting the impacts of climate change on food security. Sci Rep 13:1057. https://doi.org/10.1038/s41598-023-28244-5

Kemp MU, Emiel van Loon E, Shamoun-Baranes J, Bouten W (2012) RNCEP: global weather and climate data at your fingertips. Methods Ecol Evol 3:65–70. https://doi.org/10.1111/j.2041-210X.2011.00138.x

Kim U, Kaluarachchi JJ, Smakhtin VU (2008) Generation of monthly precipitation under climate change for the upper Blue Nile river basin, Ethiopia. J Am Water Resour Assoc 44:1231–1247. https://doi.org/10.1111/j.1752-1688.2008.00220.x

Kornfeld RP, Arnold BW, Gross MA, Dahya NT, Klipstein WM, Gath PF, Bettadpur S (2019) GRACE-FO: the gravity recovery and climate experiment follow-on mission. J Spacecr Rockets 56:931–951. https://doi.org/10.2514/1.A34326

Kratzert F, Klotz D, Herrnegger M, Sampson AK, Hochreiter S, Nearing GS (2019a) Toward improved predictions in ungauged basins: exploiting the power of machine learning. Water Resour Res 55:11344–11354. https://doi.org/10.1029/2019WR026065

Kratzert F, Klotz D, Shalev G, Klambauer G, Hochreiter S, Nearing G (2019b) Towards learning universal, regional, and local hydrological behaviors via machine learning applied to large-sample datasets. Hydrol Earth Syst Sci 23:5089–5110. https://doi.org/10.5194/hess-23-5089-2019

Kuhn M, Johnson K (2019) Feature engineering and selection: a practical approach for predictive models, Chapman & Hall/CRC Data Science Series, 1st edn. Chapman and Hall/CRC

Lee J, Cardille J, Coe M (2018) BULC-U: sharpening resolution and improving accuracy of land-use/land-cover classifications in Google Earth Engine. Remote Sens (Basel) 10:1455. https://doi.org/10.3390/rs10091455

Lemma H, Frankl A, Griensven A, Poesen J, Adgo E, Nyssen J (2019) Identifying erosion hotspots in Lake Tana basin from a multisite soil and water assessment tool validation: opportunity for land managers. Land Degrad Dev 30:1449–1467. https://doi.org/10.1002/ldr.3332

Lin J-Y, Cheng C-T, Chau K-W (2006) Using support vector machines for long-term discharge prediction. Hydrol Sci J 51:599–612. https://doi.org/10.1623/hysj.51.4.599

Liu YY, Parinussa RM, Dorigo WA, De Jeu RAM, Wagner W, van Dijk AIJM, McCabe MF, Evans JP (2011) Developing an improved soil moisture dataset by blending passive and active microwave satellite-based retrievals. Hydrol Earth Syst Sci 15:425–436. https://doi.org/10.5194/hess-15-425-2011

Luo C, Zhang X, Meng X, Zhu H, Ni C, Chen M, Liu H (2022) Regional mapping of soil organic matter content using multitemporal synthetic Landsat 8 images in Google Earth Engine. CATENA 209:105842. https://doi.org/10.1016/j.catena.2021.105842

Maidment R, Black E, Greatrex H, Young M (2020) TAMSAT. In: Levizzani V, Kidd C, Kirschbaum DB, Kummerow CD, Nakamura K, Turk FJ (eds) Satellite precipitation measurement: volume 1, advances in global change research. Springer International Publishing, Cham, pp 393–408. https://doi.org/10.1007/978-3-030-24568-9_22

Malede DA, Alamirew T, Andualem TG (2022) Integrated and individual impacts of land use land cover and climate changes on hydrological flows over Birr River watershed, Abbay basin, Ethiopia. Water (Basel) 15:166. https://doi.org/10.3390/w15010166

Ma Q, Li Z, Lei H, Chen Z, Liu J, Wang S, Su T, Feng G (2023) Interannual variability of extreme precipitation during the boreal summer over Northwest China. Remote Sens (Basel) 15:785. https://doi.org/10.3390/rs15030785

Mohanty BP, Cosh MH, Lakshmi V, Montzka C (2017) Soil moisture remote sensing: state-of-the-science. Vadose Zone J 16:0. https://doi.org/10.2136/vzj2016.10.0105

Mosaffa H, Sadeghi M, Mallakpour I, Naghdyzadegan Jahromi M, Pourghasemi HR (2022) Application of machine learning algorithms in hydrology. In: Computers in earth and environmental sciences. Elsevier, pp 585–591. https://doi.org/10.1016/B978-0-323-89861-4.00027-0

Mulatu CA, Crosato A, Langendoen EJ, Moges MM, McClain ME (2022) Alteration of the Fogera plain flood regime due to Ribb Dam construction, Upper Blue Nile basin, Ethiopia. J Appl Water Eng Res 10:175–196. https://doi.org/10.1080/23249676.2021.1961618

Mutanga O, Kumar L (2019) Google earth engine applications. Remote Sens (Basel) 11:591. https://doi.org/10.3390/rs11050591

Nilsson E (1940) Ancient changes of climate in British East Africa and Abyssinia. Geogr Ann 22:1–79. https://doi.org/10.1080/20014422.1940.11880682

Novella NS, Thiaw WM (2013) African rainfall climatology version 2 for famine early warning systems. J Appl Meteor Climatol 52:588–606. https://doi.org/10.1175/JAMC-D-11-0238.1

O'Neill P, Bindlish R, Chan S, Njoku E, Jackson T (2018) Algorithm theoretical basis document. Level 2 & 3 soil moisture (passive) data products

Obida CB, Blackburn GA, Whyatt JD, Semple KT (2019) River network delineation from Sentinel-1 SAR data. Int J Appl Earth Obs Geoinf 83:101910. https://doi.org/10.1016/j.jag.2019.101910

Peng J, Loew A (2017) Recent advances in soil moisture estimation from remote sensing. Water (Basel) 9:530. https://doi.org/10.3390/w9070530

Pérez-Cutillas P, Pérez-Navarro A, Conesa-García C, Zema DA, Amado-Álvarez JP (2023) What is going on within google earth engine? A systematic review and meta-analysis. Remote Sens Appl Soc Environ 29:100907. https://doi.org/10.1016/j.rsase.2022.100907

Raju RT, Thampi SG, Sathish Kumar D (2022) Flood mapping using Sentinel-1 SAR data. In: Dikshit AK, Narasimhan B, Kumar B, Patel AK (eds). Springer Nature Singapore, Singapore, pp 577–590

Reichle R (2016) SMAP L4 9 km EASE-grid surface and root zone soil moisture geophysical data, version 2: 3-hourly analysis [WWW Document]. https://nsidc.org/data/spl4smgp/versions/3. Accessed 23 Mar 2023

Reichle RH, Ardizzone JV, Kim G-K, Lucchesi RA, Smith EB, Weiss BH (2022) Soil moisture active passive (SMAP) mission level 4 surface and root zone soil moisture (L4_SM) product specification document

Rosenqvist A, Shimada M, Ito N, Watanabe M (2007) ALOS PALSAR: a pathfinder mission for global-scale monitoring of the environment. IEEE Trans Geosci Remote Sens 45:3307–3316. https://doi.org/10.1109/TGRS.2007.901027

Sahoo S, Russo TA, Elliott J, Foster I (2017) Machine learning algorithms for modeling groundwater level changes in agricultural regions of the U.S. Water Resour Res 53:3878–3895. https://doi.org/10.1002/2016WR019933

Schönbrodt-Stitt S, Ahmadian N, Kurtenbach M, Conrad C, Romano N, Bogena HR, Vereecken H, Nasta P (2021) Statistical exploration of SENTINEL-1 data, terrain parameters, and in-situ data for estimating the near-surface soil moisture in a mediterranean agroecosystem. Front Water 3. https://doi.org/10.3389/frwa.2021.655837

Shen G, Fu W, Guo H, Liao J (2022) Water body mapping using long time series Sentinel-1 SAR data in Poyang Lake. Water (Basel) 14:1902. https://doi.org/10.3390/w14121902

Slater LJ, Thirel G, Harrigan S, Delaigue O, Hurley A, Khouakhi A, Prosdocimi I, Vitolo C, Smith K (2019) Using R in hydrology: a review of recent developments and future directions. Hydrol Earth Syst Sci 23:2939–2963. https://doi.org/10.5194/hess-23-2939-2019

Sparks A (2018) nasapower: a NASA POWER global meteorology, surface solar energy and climatology data client for R. JOSS 3:1035. https://doi.org/10.21105/joss.01035

Sutcliffe JV, Parks YP (1999). The hydrology of the Nile, vol 5. IAHS Special Publication No.5

Tapley BD, Bettadpur S, Watkins M, Reigber C (2004) The gravity recovery and climate experiment: mission overview and early results. Geophys Res Lett 31:n/a–n/a. https://doi.org/10.1029/2004GL019920

Tariku TB, Gan TY, Li J, Qin X (2021) Impact of climate change on hydrology and hydrologic extremes of Upper Blue Nile River basin. J Water Resour Plann Manage 147. https://doi.org/10.1061/(ASCE)WR.1943-5452.0001321

Taye MT, Willems P (2012) Temporal variability of hydroclimatic extremes in the Blue Nile basin. Water Resour Res 48

Taye MT, Haile AT, Genet A, Geremew Y, Wassie S, Abebe B, Alemayehu B (2022) Data quality deterioration in the Lake Tana sub-basin, Ethiopia: scoping study to provide streamflow and water withdrawal data. International Water Management Institute, Colombo, Sri Lanka. https://doi.org/10.5337/2022.208

Tegegne EB, Ma Y, Chen X, Ma W, Wang B, Ding Z, Zhu Z (2021) Estimation of the distribution of the total net radiative flux from satellite and automatic weather station data in the Upper Blue Nile basin, Ethiopia. Theoret Appl Climatol 143:587–602. https://doi.org/10.1007/s00704-020-03397-9

Tibebe D, Teferi E, Bewket W, Zeleke G (2022) Climate induced water security risks on agriculture in the Abbay River basin: a review. Front Water 4

Tigabu TB, Wagner PD, Hörmann G, Kiesel J, Fohrer N (2021) Climate change impacts on the water and groundwater resources of the Lake Tana basin, Ethiopia. J Water Clim Change 12:1544–1563. https://doi.org/10.2166/wcc.2020.126

Titti G, Napoli GN, Conoscenti C, Lombardo L (2022) Cloud-based interactive susceptibility modeling of gully erosion in Google Earth Engine. Int J Appl Earth Obs Geoinf 115:103089. https://doi.org/10.1016/j.jag.2022.103089

Torres R, Snoeij P, Geudtner D, Bibby D, Davidson M, Attema E, Potin P, Rommen B, Floury N, Brown M, Traver IN, Deghaye P, Duesmann B, Rosich B, Miranda N, Bruno C, L'Abbate M, Croci R, Pietropaolo A, Huchler M, Rostan F (2012) GMES Sentinel-1 mission. Remote Sens Environ 120:9–24. https://doi.org/10.1016/j.rse.2011.05.028

Trigo IF, Dacamara CC, Viterbo P, Roujean J-L, Olesen F, Barroso C, Camacho-de-Coca F, Carrer D, Freitas SC, García-Haro J, Geiger B, Gellens-Meulenberghs F, Ghilain N, Meliá J, Pessanha L, Siljamo N, Arboleda A (2011) The satellite application facility for land surface analysis. Int J Remote Sens 32:2725–2744. https://doi.org/10.1080/01431161003743199

Tuck SL, Phillips HRP, Hintzen RE, Scharlemann JPW, Purvis A, Hudson LN (2014) MODIS tools–downloading and processing MODIS remotely sensed data in R. Ecol Evol 4:4658–4668

USBR (1964) Land and water resources of the Blue Nile basin, Ethiopia. The Bureau

Wagner W, Dorigo W, De Jeu R, Fernandez D, Benveniste J, Haas E, Ertl M (2012) Fusion of active and passive microwave observations to create an essential climate variable data record on soil moisture. ISPRS Ann Photogrammetry Remote Sens Spat Inform Sci (ISPRS Ann) 7:315–321

Wale A, Rientjes THM, Gieske ASM, Getachew HA (2009) Ungauged catchment contributions to Lake Tana's water balance. Hydrol Process. https://doi.org/10.1002/hyp.7284

WMO (2018) Guide to instruments and methods of observation. World Meteorological Organization WMO. https://library.wmo.int/index.php

Worqlul AW, Yen H, Collick AS, Tilahun SA, Langan S, Steenhuis TS (2017) Evaluation of CFSR, TMPA 3B42 and ground-based rainfall data as input for hydrological models, in data-scarce regions: the upper Blue Nile basin, Ethiopia. CATENA 152:242–251. https://doi.org/10.1016/j.catena.2017.01.019

Xie P, Joyce R, Wu S, Yoo S-H, Yarosh Y, Sun F, Lin R (2017) Reprocessed, bias-corrected CMORPH global high-resolution precipitation estimates from 1998. J Hydrometeor 18:1617–1641. https://doi.org/10.1175/JHM-D-16-0168.1

Xu T, Liang F (2021) Machine learning for hydrologic sciences: an introductory overview. WIREs Water 8. https://doi.org/10.1002/wat2.1533

Yalew SG, van Griensven A, van der Zaag P (2016) AgriSuit: a web-based GIS-MCDA framework for agricultural land suitability assessment. Comput Electron Agric 128:1–8. https://doi.org/10.1016/j.compag.2016.08.008

Yang L, Driscol J, Sarigai S, Wu Q, Chen H, Lippitt CD (2022) Google Earth Engine and artificial intelligence (AI): a comprehensive review. Remote Sens (Basel) 14:3253. https://doi.org/10.3390/rs14143253

Yenehun A, Dessie M, Azeze M, Nigate F, Belay AS, Nyssen J, Adgo E, Van Griensven A, Van Camp M, Walraevens K (2021) Water resources studies in headwaters of the Blue Nile basin: a review with emphasis on lake water balance and hydrogeological characterization. Water 13:1469

Yilma AD, Awulachew SB (2009) Characterization and Atlas of the Blue Nile basin and its sub basins. International Water Management Institute

Zambrano-Bigiarini M, Nauditt A, Birkel C, Verbist K, Ribbe L (2017) Temporal and spatial evaluation of satellite-based rainfall estimates across the complex topographical and climatic gradients of Chile. Hydrol Earth Syst Sci 21:1295–1320

Zamora D, Duque N, Vega C, Arboleda P, García C (2020) DWBmodelUN: hydrological model dynamic water balance—R package. Zenodo. https://doi.org/10.5281/zenodo.3813037

Zou Q, Xiong Q, Li Q, Yi H, Yu Y, Wu C (2020) A water quality prediction method based on the multi-time scale bidirectional long short-term memory network. Environ Sci Pollut Res Int 27:16853–16864. https://doi.org/10.1007/s11356-020-08087-7

Chapter 21
Root-Zone Soil Moisture Prediction in Rainfed Systems Using Satellite-Derived Product: The Case of Abbay River Basin in Ethiopia

Getachew Tesfaye Ayehu, Tsegaye Tadesse, Berhan Gessesse, Wuletawu Abera, Degefie Tibebe, Yonas Getaneh, and Lulseged Tamene

Abstract Obtaining satellite-derived root-zone soil moisture (RZSM) data from surface observations is crucial for advancing the use of remote sensing technology and its contributions to hydrology and agriculture. Following the advancement of microwave remote sensing, the extent of soil moisture can be estimated from satellite data for large basins. However, satellite data generally provide soil moisture estimates at the first few centimetres of the soil layer. In this study, a nonlinear statistical technique is proposed to extend SMAP surface soil moisture estimates to the

---

G. T. Ayehu (✉) · W. Abera · D. Tibebe · Y. Getaneh · L. Tamene
International Center for Tropical Agriculture (CIAT), Addis Ababa, Ethiopia
e-mail: Getachew.Tesfaye@cgiar.org

W. Abera
e-mail: Wuletawu.Abera@cgiar.org

D. Tibebe
e-mail: D.Tibebe@cgiar.org

Y. Getaneh
e-mail: y.getaneh@cgiar.org

L. Tamene
e-mail: LT.Desta@cgiar.org

T. Tadesse
National Drought Mitigation Center, University of Nebraska-Lincoln, Lincoln, NE, USA
e-mail: ttadesse2@unl.edu

B. Gessesse
Remote Sensing Department, Ethiopian Space Science and Geospatial Institute, Addis Ababa, Ethiopia

W. Abera
International Center for Tropical Agriculture (CIAT), Accra, Ghana

L. Tamene
International Center for Tropical Agriculture (CIAT), Nairobi, Kenya

© The Author(s), under exclusive license to Springer Nature Switzerland AG 2025
A. Melesse et al. (eds.), *Abbay River Basin*, Springer Geography,
https://doi.org/10.1007/978-3-031-65241-7_21

soil root zone (0–100 cm) over the Abbay River basin in Ethiopia. The approach is developed by coupling the polynomial regression model and the cumulative density function (CDF) matching method. When validated using field-observed soil moisture, the prediction model was found to be reliable with a coefficient of correlation (r) ranging from 0.84 to 0.99 and an ubRMSE ranging from 0.002 to 0.039 m^3/m^3. The results suggested that systematic differences between surface and root-zone soil moisture can be adjusted statistically by employing CDF-based observation operators, which can also produce a reasonably accurate prediction of RZSM using only near-surface data. Furthermore, the comparison made between the CDF matching-based prediction model and SMAP L4 RZSM reveals the greater performance of the proposed model to predict RZSM in the study basin. CDF matching could be used as an alternative way to extend the surface soil moisture generated from remote sensing to the subsurface in areas with similar agro-climatic and soil characteristics to our study location. Thus, remote sensing technology could still provide opportunities and alternatives to estimate RZSM at different scales and with improved spatial and temporal resolutions.

Keywords Residual moisture · Root-zone · CDF · SMAP · AWS

Introduction

Ethiopia's agriculture is described by meagre production (Bekele et al. 2012) and the sector is dominated by smallholder farmers (Bekabil 2014). Unfortunately, the majority of these farmers cannot support themselves on a single harvest made during the primary crop growing season (FAO 2014; CSA 2001). Given Ethiopia's poor irrigated agriculture practices (less than 5%; Awulachew et al. 2005; World Bank 2006), planting crops in the off-season utilizing remanent soil moisture may be an alternative practical way to boost food and feed production.

The Abbay River basin in Ethiopia receives annual mean rainfall more than 2000 mm, of which, over 75% of the rainfall occurring during the *Kiremt* (June to September) growing season (Conway 2000a). As a result, once the main season cropping is harvested in the basin, some carryover moisture, also known as residual soil moisture, is still present in the soil and may be useful for extra short or medium cycle cropping in the off-season. However, for optimal crop planning and effective use of agricultural water in the off-season, crop production necessitates measuring the level of residual soil moisture available in the soil. Thus, there is an urgent demand for multitemporal soil moisture monitoring throughout the off-season.

Microwave remote sensing is appeared as an alternative and feasible technique to estimate surface soil moisture on a broad scale with improved representation in the spatial and time domains (Petropoulos et al. 2015). However, soil moisture measurements using remote sensing satellites generally provide soil moisture estimates at the first few centimetres of the soil layer. In contrast, crop production requires soil moisture information for the entire soil profile (e.g. 0–100 cm depths), which is an essential limiting factor for crop productivity (Tobin et al. 2017), rather than shallow observations from satellite instruments. Therefore, it is necessary to devise an approach to associate surface and root-zone soil moisture (RZSM) values to benefit and fulfil the demands of the agricultural sector. The direct association of surface and RZSM is challenged by the significant variability of soil moisture at the surface related to meteorological forcing, the dynamics of the subsurface moisture in connection to soil water redistribution, and the heterogeneous nature of the various activities in the soil profile (Dumedah et al. 2015). Nevertheless, satellite soil moisture estimates could be extended to provide an accurate means for RZSM prediction using different retrieval algorithms (Mahmood and Hubbard 2007) given the coupling of surface and RZSM through diffusion processes (Kornelsen and Coulibaly 2014).

Several studies have been conducted to estimate RZSM from field observations (Gao et al. 2017), satellite retrievals (Ford et al. 2014), analytical methods (Tobin et al. 2017; Wagner et al. 1999), and data assimilation techniques (Lia et al. 2010; Sabater et al. 2007). Among these methods of RZSM estimation, data assimilation (DA) has shown promise in inferring the full soil moisture profile by integrating near-surface observations (e.g. satellite observations) into a physically based hydrological model (Lia et al. 2010; Sabater et al. 2007). However, the application of the DA approach may be constrained by its computational demands, the required model inputs, and uncertainties associated with the physical descriptions of the hydrological processes (Albergel et al. 2008; Clark et al. 2008). The analytical method needs fewer input parameters and appears to be a computationally more efficient approach to estimate RZSM than the DA method (Manfreda et al. 2014; Wagner et al. 1999). For instance, the Wagner et al. (1999) exponential filter assumes that water flows between the surface and subsurface layers are proportional to the difference in soil moisture between these two layers. The most well-known analytical technique is the exponential filter method, which only needs surface soil moisture observations and one input parameter, the characteristic time length (T). For many locales, this method has been used to accurately forecast the soil moisture in the root zone based on surface data (Ford et al. 2014; Brocca et al. 2011).

Statistical techniques are even more straightforward ways of calculating RZSM because they are fully data-driven and do not consider water flux processes. For example, Kornelsen and Coulibaly (2014) proposed an ensemble of artificial neural network methods to estimate RZSM from surface measurements, while Shi et al. (2014) and Hu and Si (2014) used a linear regression model and time stability analysis to infer subsurface soil moisture from surface measurements, respectively. However, most of the statistical models employed in earlier research described surface soil moisture at a depth far beyond the penetration capability of remote sensing satellites, and their application to predicting RZSM from remote sensing retrievals is very limited. As a result, there is a need to develop a feasible and alternative statistical approach, to benefit from its simplicity, to infer RZSM (0–100 cm) using remote sensing retrievals of surface moisture. In this paper, the cumulative distribution function (CDF) matching method and the polynomial regression model are coupled to predict RZSM using SMAP surface soil moisture estimates.

According to Reichle and Koster (2014), the CDF matching method is an improved nonlinear technique that has been extensively used by various researchers (e.g. Brocca et al. 2011; Parrens et al. 2014) to eliminate systematic discrepancies in soil moisture datasets obtained from various sources (e.g. in situ measurements and remote sensing retrievals). For instance, after rescaling with the CDF technique, Liu et al. (2011) successfully linked soil moisture retrievals from various satellite missions. Similar applications of the CDF matching method were made by Han et al. (2012) and Drusch et al. (2005) for spatial upscaling and spatial transfer of point soil moisture in time and space, respectively. Since the soil moisture found in various soil layers may have come from various spatial domains or sources, the CDF matching method is probably valid in this investigation. Inferring the root-zone soil moisture based on this supposition might be accomplished by modifying the CDF of the surface soil moisture and creating the observation operators from the surface and RZSM datasets. In fact, Gao et al. (2017) recently scaled the depth of soil moisture from surface soil moisture recorded in the field using the CDF matching observation operator. However, the accuracy of the CDF matching method for deriving RZSM estimates from satellite (e.g. SMAP) surface soil moisture retrievals has not been assessed. This study evaluates the performance of the observation operator created using the CDF-polynomial model for predicting RZSM using surface soil moisture estimated from SMAP. The benefits of this method are associated with its entirely data-driven, computationally efficient, and simple nature, providing an alternative approach to the complex techniques noted previously.

The objective of this study was to (i) investigate the feasibility of combining the CDF matching method and polynomial fit model as a simple and computationally efficient approach to predict RZSM in the off-season using SMAP surface soil moisture estimates and (ii) compare the performance of the proposed RZSM prediction model to that of the SMAP L4 RZSM product.

Site Description

The Abbay River basin, which is located in the north-western part of Ethiopia, contributes the lion's share (~ 60%) of the Nile River's annual flow (Conway 2005). According to Conway (2000b), the drainage area of the basin is roughly 176,000 km^2. The basin is recognized for its diverse topography, which ranges from 423,239 m above sea level in the basin's northeast to 490 m above sea level in the basin's western portion close to the Ethiopian-Sudan border (Fig. 21.1). The climate of the study site is mainly characterized by the altitude and its closeness to the equatorial monsoonal systems. Accordingly, the study area can be classified into three distinct seasons. These are locally known as Kiremt (main rainy season) which occurs from June to September, Belg (a short rainy season) which may occur around February to May, and Bega (a dry season that runs from October to January (Conway 2000b).

The basin receives the majority of its annual precipitation (more than 70%) during the *Kiremt* season (Kim et al. 2008). The Abbay River basin receives 2200 mm or

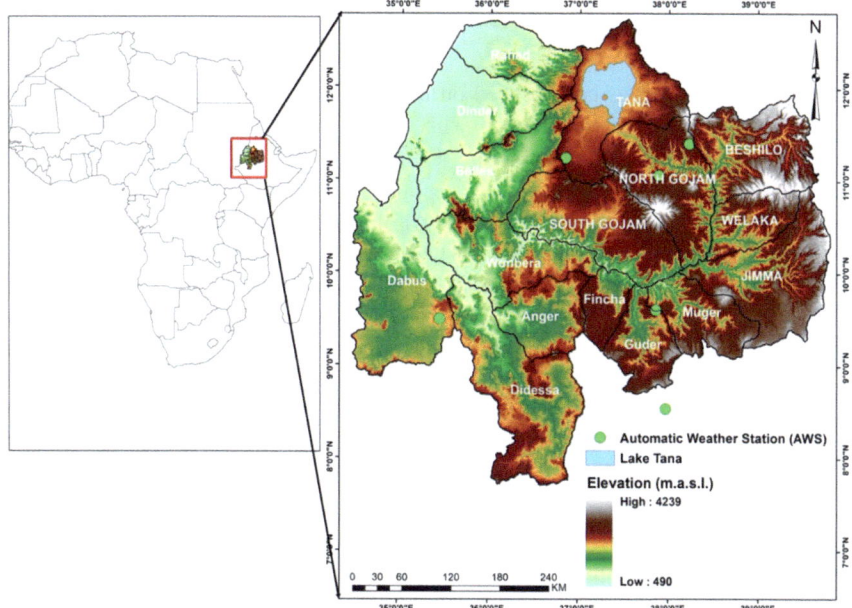

Fig. 21.1 Abbay River basin and its location in Africa are depicted in a digital elevation model (DEM). (Imagery source: SRTM Global elevation data). The Abbay River basin drains to the NW

more of rain annually, which is a sufficient quantity. According to Kim et al. (2008), the annual mean rainfall ranges between 1200 and 1800 mm, increasing from the northeast to the southwest. However, the basin is noted for having significant intra- and interannual variations in rainfall (Conway 2000b; Taye and Willems 2012). The hydrological processes in the basin are therefore very intricate and extremely varied over both geography and time. The study site's average yearly temperature is likewise close to 20.4 °C. In addition, a large portion of the population in the basin relies on rain-fed agriculture, which is subsistent and is dominated by smallholder farmers.

Dataset

Though the objective of this study is to predict residual soil moisture for informed crop production in the off-season, the data acquired over all seasons from November 2015 to June 2018 were used to increase the length of the observed datasets for calibration and validation of the prediction model. Soil moisture datasets obtained from both SMAP missions and automatic weather stations (AWS) were used in this study.

SMAP Soil Moisture Products

The SMAP mission was commenced on 31 January 2015 (Entekhabi et al. 2010), and the SMAP satellite is equipped with an L-band radiometer (which provides brightness temperature at a resolution of 36 km) to monitor the Earth's surface at sun-synchronous times (Entekhabi et al. 2013). The SMAP is one of the recent L-band satellite missions, that started delivering radiometer SM products as of 31 March 2015 (Entekhabi et al. 2014). The SMAP mission is particularly sensitive to detecting soil moisture conditions from the top few centimetres of the soil column and offers important global and regional soil moisture data (Petropoulos et al. 2015). One of the surface soil moisture products produced from the SMAP mission is the SMAP Level 3 (L3_SM), which is offered in two resolutions, the coarse (36 km) (L3_SM_P) and the enhanced (9 km) (L3_SM_P_E) products with a daily gridded global composite. To forecast root-zone soil moisture (RZSM) in the study basin, surface soil moisture values from the L3_SM_P_E (Version 2) product were used as a source. To assess the effectiveness of the proposed RZSM prediction model, the RZSM from the SMAP Levele-4 (L4_SM) product was also employed. The L4_SM is produced through incorporating SMAP L-band brightness temperature observations into the NASA catchment land surface model (Reichle et al. 2016). The L4_SM product has been accessible since 31 March 2015. It offers 3-h, 9-km global estimates of surface and RZSM. In this study, the L4_SM root-zone soil moisture (Version 4) was employed. The NSIDC website (https://nsidc.org/data/smap/smap-data.html) was used to download the L3_SM_P_E (Version 2) and L4_SM (Version 4) soil moisture products.

Indeed, soil moisture products derived from the SMAP radiometer mission have shown great skill in capturing soil moisture change at the surface and subsurface soil layers (Das and Dunbar 2019, Chan et al. 2016). According to several studies that tested SMAP soil moisture estimations against in situ soil moisture data across several sites (Table 21.1), SMAP-based surface soil moisture values are in good agreement with soil moisture values measured in the field.

Measured Data Using Automatic Weather Stations (AWS)

The National Meteorological Agency (NMA) of Ethiopia deployed AWS, from which in situ soil moisture measurements were retrieved. Along with other meteorological data, these weather stations collect data on soil moisture at various depths. In order to calibrate and validate the prediction model, six AWS stations located in and around the UBN basin were used (Fig. 21.1 and Table 21.2).

A set of measured data covering the period from November 2015 to June 2018 was used for the analysis. At each location, measurements of soil moisture were made every 15 min at depths of 20, 50 and 100 cm. The daily soil moisture average of the AWS measurements was calculated to match the daily grid values of the SMAP

21 Root-Zone Soil Moisture Prediction in Rainfed Systems Using …

Table 21.1 Partial list of the SMAP soil moisture (SM) product's statistical results at various sites

Product	Description	Grid resolution (km)	Location	UbRMSE [m^3/m^3]	References
L3_SM_P	Soil moisture (radiometer)	36	NW China	0.023–0.024	Ma et al. (2017)
L3_SM_P	Soil moisture (radiometer)	36	Tibetan Plateau	0.058–0.059	Li et al. (2018)
L3_SM_P_E	Soil moisture (radiometer, enhanced)	9	Tibetan Plateau	0.055–0.059	Li et al. (2018)
L4_SM	Soil moisture (surface and root zone)	9	SMAP core validation site	Surface (0.038) Root zone (0.030)	Reichle et al. (2017)
L4_SM	Soil moisture (surface and root zone)	9	Little Washita Watershed, USA	Surface (0.027) Root zone (0.032)	Bi et al. (2016)
L3_SM_P	Soil moisture (radiometer)	36	Little Washita Watershed, USA	0.027–0.044	Cui et al. (2018)
L3_SM_P_E	Soil moisture (radiometer, enhanced)	9	REMEDHUS network, Spain	0.039–0.040	Cui et al. (2018)
L3_SM_P	Soil moisture (radiometer)	36	Australia, India, Kenya, and USA	0.027–0.0599 Kenya (0.0453)	Montzka et al. (2017)
L2_SM_P	Soil moisture (radiometer)	36	SMAP Core validation site	An average of 0.037	Colliander et al. (2017)

Table 21.2 Summary of AWS-measured soil moisture at various soil depths for the period of November 2015 to June 2018

Station name (UBN basin)	Geographic location		N	Elevation (m.a.s.l)	Land cover type	Depth (cm)
	Long	Lat				
Dangila	36.85	11.25	183	2116	Agricultural land	20, 50, 100
Kachis	37.86	9.61	252	2557	Agricultural land	20, 50, 100
Motta	37.89	11.07	251	2417	Agricultural land	20, 50, 100
Nedjo	35.45	9.50	314	1800	Agricultural land	20, 50, 100
Simada	38.23	11.41	259	2584	Agricultural land	20, 50, 100
Woliso	37.97	8.54	312	2058	Agricultural land	20, 50, 100

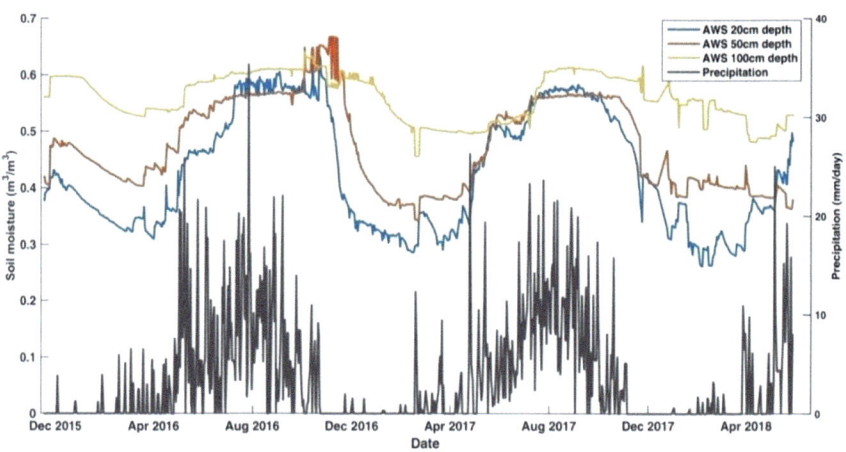

Fig. 21.2 Time series of daily average accumulated precipitation over the six observation stations (located in the UBN basin) and volumetric soil moisture recorded by the AWS at depths of 20 cm, 50 cm, and 100 cm during the period of November 2015 to June 2018

product. Furthermore, the temporal characteristics of AWS-measured data were studied through comparison with the Climate Hazards Group Infrared Precipitations with Stations (CHIRPS) rainfall product (Fig. 21.2), which has shown its greatest agreement with ground observed rainfall over the Abbay River basin (Ayehu et al. 2018). Within this domain, AWS measurements have shown a consistent temporal distribution with the CHIRPS precipitation events at 20, 50 and 100 cm soil depths (Fig. 21.2). The highest soil moisture values were observed during high rainfall events, whereas the lowest soil moisture corresponded to the dry periods. In addition, Fig. 21.2 depicts that the amount of soil water during the wet season is not considerably different across the different soil depths, and even over the month of July (the period where the basin receives the maximum rainfall), the surface soil could have more moisture than that of the subsurface. This is because during the wet season, the basin receives a high amount of rain that exceeds soil water above the surface due to saturation. Indeed, as the dry season is coming, the soil becomes dry, the amount of moisture on the surface gradually decreases due to evaporation, and more soil moisture is observed in the subsurface.

Methods

Surface soil moisture was obtained from the L3_SM_P_E soil moisture product, and the corresponding soil moisture values at different depths (20, 50, and 100 cm) were collected from the AWS of the Abbay River basin. Since AWS soil moisture values are provided as point measurements, the grid values of L3_SM_P_E surface soil moisture estimates containing the selected stations were extracted. Note that RZSM in this study refers to a depth-weighted mean of soil moisture in the 0 to 100 cm depth, which is calculated using AWS-measured soil moisture values at a depth of 20 (layer 1), 50 (layer 2), and 100 (layer 3) cm Eq. 21.1 (Gao et al. 2017):

$$\theta_p = \frac{2\theta_1 L_1 + (\theta_1 + \theta_2)L_2 + (\theta_2 + \theta_3)L_3}{2(L_1 + L_2 + L_3)} \quad (21.1)$$

where θ_p refers to the root-zone volumetric soil moisture content (m³/m³); θ_i ($i = 1$, 2 and 3) refers to the volumetric soil moisture at the ith soil layer (m³/m³); and L_i ($i = 1, 2,$ and 3) refers to the soil depth of the ith soil layers (cm).

CDF Matching Method

CDF matching is a procedure of rescaling the CDF of one dataset (e.g. SMAP-derived surface soil moisture) to match the CDF of another dataset (e.g. AWS-measured RZSM). Figure 21.3 illustrates the general principle of CDF matching, where the surface soil moisture (obtained from SMAP) cumulative probability was scaled to match that of the AWS-observed root-zone soil moisture.

$$\mathrm{cdf}_s = \mathrm{cdf}_r \quad (21.2)$$

where cdf_s and cdf_r are the CDFs of surface and root-zone soil moisture, respectively.

In principle, the two datasets (SMAP-derived surface soil moisture and AWS-observed RZSM) must be ranked first. Then, the difference in soil moisture between

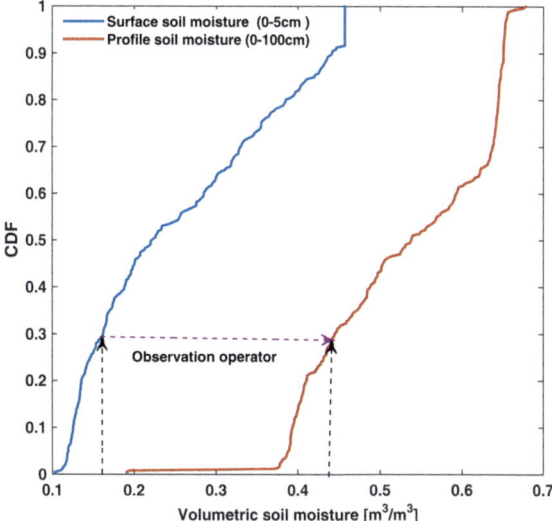

Fig. 21.3 The notion of cumulative distribution function (CDF) matching is used to convert surface soil moisture (blue) from satellite data into profile soil moisture (red) using observation operators. The arrow shows the conversion of surface soil moisture to profile soil moisture

the subsequent values of each ranked dataset needs to be calculated. Through this process, an observation operator is computed as a polynomial fit to the ranked surface soil moisture values and the corresponding differences, which modifies the systematic difference between the datasets. Subsequently, the observation operator is used to predict the RZSM (0–100 cm) content. A summary of the technical procedures to develop an observation operator is presented as follows.

Let the values θ_s (satellite-estimated surface soil moisture, i.e. SMAP_L3_P_E) and θ_{pA} (AWS-measured RZSM) be given as multitemporal observations with N points (Table 21.3). That is, θ_{s_i} and θ_{pA_i} are known, $i = 1, 2, \ldots N$. Then, CDF matching is developed:

(a) First, the datasets of θ_s and θ_{pA} were ranked.
(b) Then, their differences were computed $\Delta = \theta_{pA} - \theta_s$;
(c) Then, Δ is plotted against the SMAP L3_P_E surface soil moisture estimates (θ_s), and a fifth-order polynomial fit is used to quantify the dependence of Δ on θ_s.
(d) Fitting was performed, and polynomial prediction models were obtained.
(e) Then, the predicted difference of Δ denoted as $\check{\Delta}$ was computed and
(f) Predicted RZSM ($\widetilde{\theta_{pA}}$) could then be estimated using the observation operators to rescale SMAP_L3_P_E surface moisture estimates as $\widetilde{\theta_{pA}} = \theta_s + \check{\Delta}$.

A fifth-order polynomial was applied using the following equation (Eq.

$$\check{\Delta} = m_0 + m_1 \cdot \theta s + m_2 \cdot \theta s^2 + m_3 \cdot \theta s^3 + m_4 \cdot \theta s^4 + m_5 \cdot \theta s^5 \quad (21.3)$$

where $\check{\Delta}$ is the predicted difference of Δ and m_i ($i = 0$–5) are parameters.

It is important to note that the CDFs presented in this study were derived from multitemporal soil moisture datasets assuming that all the values in these temporal

Table 21.3 Statistical summary of the performance of predicted RZSM in both the calibration and validation phases

Stations name	Bias (m³/m³)		RMSE (m³/m³)		ubRMSE (m³/m³)		r	
	Predicted	SMAP L4	Predicted	SMAP L4	Predicted	SMAP L4	Predicted	SMAP L4
Dangila	– 2.20E−3	– 0.199	0.018	0.207	0.017	0.057	0.96	0.73
Kachis	– 1.35E−3	– 0.242	0.017	0.244	0.017	0.029	0.96	0.85
Motta	– 9.14E−4	– 0.178	0.016	0.186	0.016	0.059	0.98	0.85
Nedjo	0.011	– 0.075	0.041	0.091	0.039	0.052	0.84	0.77
Simada	– 4.10E−4	– 0.250	0.002	0.252	0.002	0.032	0.99	0.85
Woliso	4.59E−4	– 0.169	0.011	0.194	0.011	0.096	0.99	0.79

observations were equally probable (Pachepsky and Hill 2016). A pre-analysis conducted from all six stations revealed that the fifth-order polynomial fit is an optimum polynomial order in our study. As a result, the fifth-order polynomial fit was maintained and employed to develop observation operators based on the SMAP L3_SM_P_E and AWS-measured datasets. A total of 1571 data pairs were obtained for the analysis after removing the outliers and missed values. Consequently, 60% of them were used to develop the observation operator, and the remaining 40% were used for validation (Table 21.3). An observation operator was developed for each of the six stations. In addition, a single observation operator that fit all stations was also generated to ease the transferability of the model to other areas of the study site.

Statistical Tools

The polynomial models served as the observation operators to avoid the systematic difference between θ_s and θ_{pA}. The validity of this procedure was independently tested by comparing the predicted and AWS-measured RZSM. The 'R' statistical package was used in this study. The root mean square error (RMSE), the mean bias, the unbiased RMSE (unRMSE), and the coefficient of correlation (r) were the error metrics used for quantitative evaluation (Eq. 21.6). Additionally, these statistical tools are commonly used for validation of SMAP soil moisture products, which are defined as follows:

$$\text{RMSE} = \sqrt{E[(\theta_{est} - \theta_{ref})^2]} \qquad (21.4)$$

$$\text{bias} = E[\theta_{est}] - E[\theta_{ref}] \qquad (21.5)$$

$$\text{ubRMSE} = \sqrt{E[((\theta_{est} - E[\theta_{est}]) - (\theta_{ref} - E[\theta_{ref}]))^2]} \qquad (21.6)$$

$$r = \frac{E[(\theta_{est} - E[\theta_{est}])(\theta_{ref} - E[\theta_{ref}])]}{\sigma_{est} \cdot \sigma_{ref}} \qquad (21.7)$$

where θ_{est} and θ_{ref} represent the predicted and AWS-observed soil moisture values, respectively; $E[\cdot]$ is the expectation value operator; and σ_{est} and σ_{ref} are the standard deviations of the predicted and AWS-observed soil moisture, respectively.

Results

CDF-Based Profile Soil Moisture Prediction

A nonlinear observation operator, used to quantify the dependence of Δ on θ_s, was defined using a fifth-order polynomial fit. The predicted RZSM ($\widetilde{\theta}_{pA}$) was estimated using the observation validation operators by rescaling SMAP L3_SM_P_E surface moisture estimates as $\widetilde{\theta}_{pA} = \theta s + \check{\Delta}$. To gain more insight into the model performance, the cumulative distribution frequency (CDF's) of the surface (SMAP L3_SM_P_E), AWS measured, and predicted RZSM for each station as well as for an observation operator that fits for all the station are shown in Figs. 21.4 and 21.5, respectively. Figure 21.6 presents the temporal distribution of surface soil moisture (SMAP L3_SM_P_E), AWS-measured RZSM, and predicted RZSM at the six selected stations. Then, the performance of the observation operators in terms of bias, RMSE, ubRMSE, and r in the validation period is given in Table 21.3 and Fig. 21.7.

Figures 21.4 and 21.5 present the CDF of the surface, AWS-measured RZSM, and predicted RZSM for each station and a combination of all the stations for the validation phase. Generally, both the CDFs in Figs. 21.4 and 21.5 indicated that

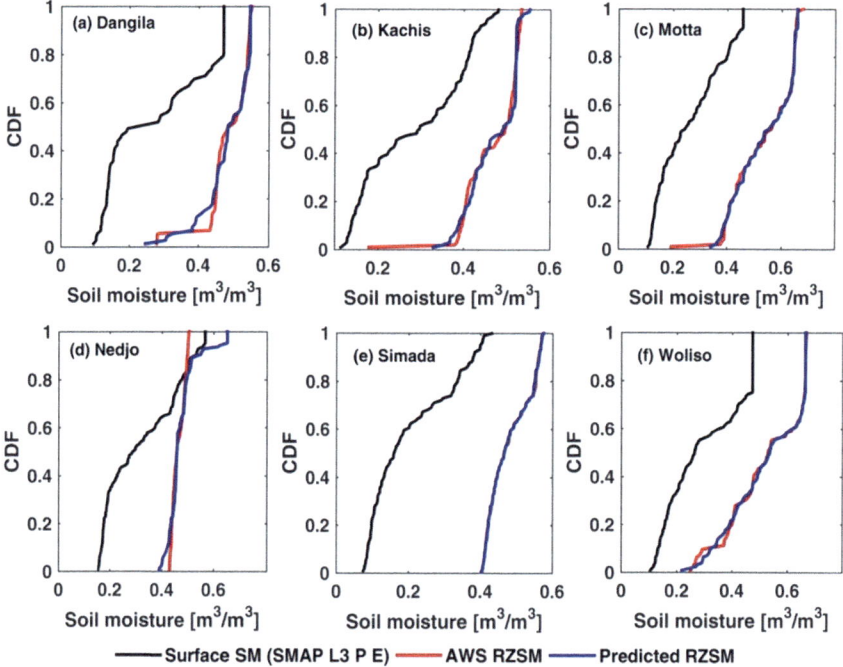

Fig. 21.4 The CDF for the validation period computed from SMAP L3_P_E surface soil moisture, AWS-measured RSM and predicted RZSM

Fig. 21.5 The CDF for the validation period computed from the SMAP L3_P_E surface, predicted and AWS-measured RZSM using a combination of all stations

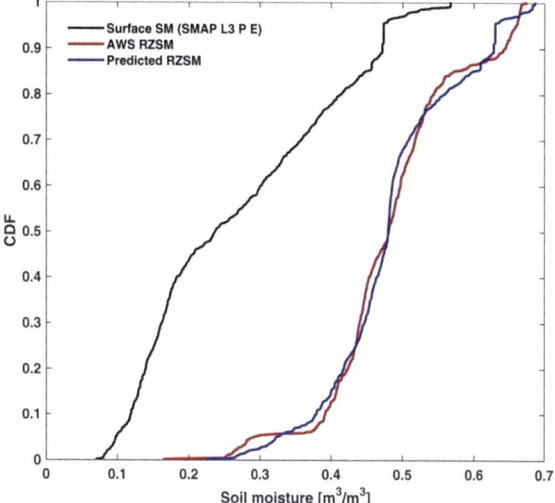

the predicted RZSM produced very close values to the AWS measurements over all stations at almost all frequencies. However, over some frequencies, e.g. the "Dangila" and "Nedjo" stations, the predicted RZSM occurred below or above the AWS-measured CDFs. In particular, the predicted RZSM at the "Nedjo" station (Fig. 21.4d) failed to capture the maximum RZSM observed by AWS.

Figure 21.6 presents the temporal distributions of surface and root-zone (both AWS observed and predicted) soil moisture over the selected stations of the study basin. In general, the predicted RZSM captured the temporal dynamics of soil moisture measured at each station. However, the predicted soil moisture is lower than that of the AWS observed during the dry season over the "Dangila" (Fig. 21.6a), "Motta" (Fig. 21.6c), and "Simada" (Fig. 21.6e) stations but higher than that of the AWS observed at the "Woliso" (Fig. 21.6f) station. The predicted RZSM over the "Nedjo" (Fig. 21.6d) station is slightly different and consistently higher than that of the AWS-observed RZSM, particularly during the wet rainfall season.

Generally, the statistical metrics of the predicted results indicated that the observation operator reliably reproduced the AWS-measured RZSM for both individual stations and a combination of stations during the validation phase (Table 21.3 and Fig. 21.7). All the correlations in Table 21.3 and Fig. 21.7 are significant at $p < 0.01$. The predicted RZSM showed strong agreement with root-zone soil moisture measured in the field and resulted in high r-values (ranging from 0.73 to 0.99), low bias, and low RMSE values. It is apparent from Table 21.3 that the predicted RZSM has a small ubRMSE (0.002–0.039 m^3/m^3) at all stations, attaining the ubRMSE of 0.04 m^3/m^3 or less required by the SMAP product. The prediction results confirm the effectiveness of the model over the study area. Indeed, it seems that the prediction accuracy of the model is dependent on the geophysical characteristics of the selected stations. A comparatively lower $r = 0.84$ and larger RMSE $= 0.041$ m^3/m^3

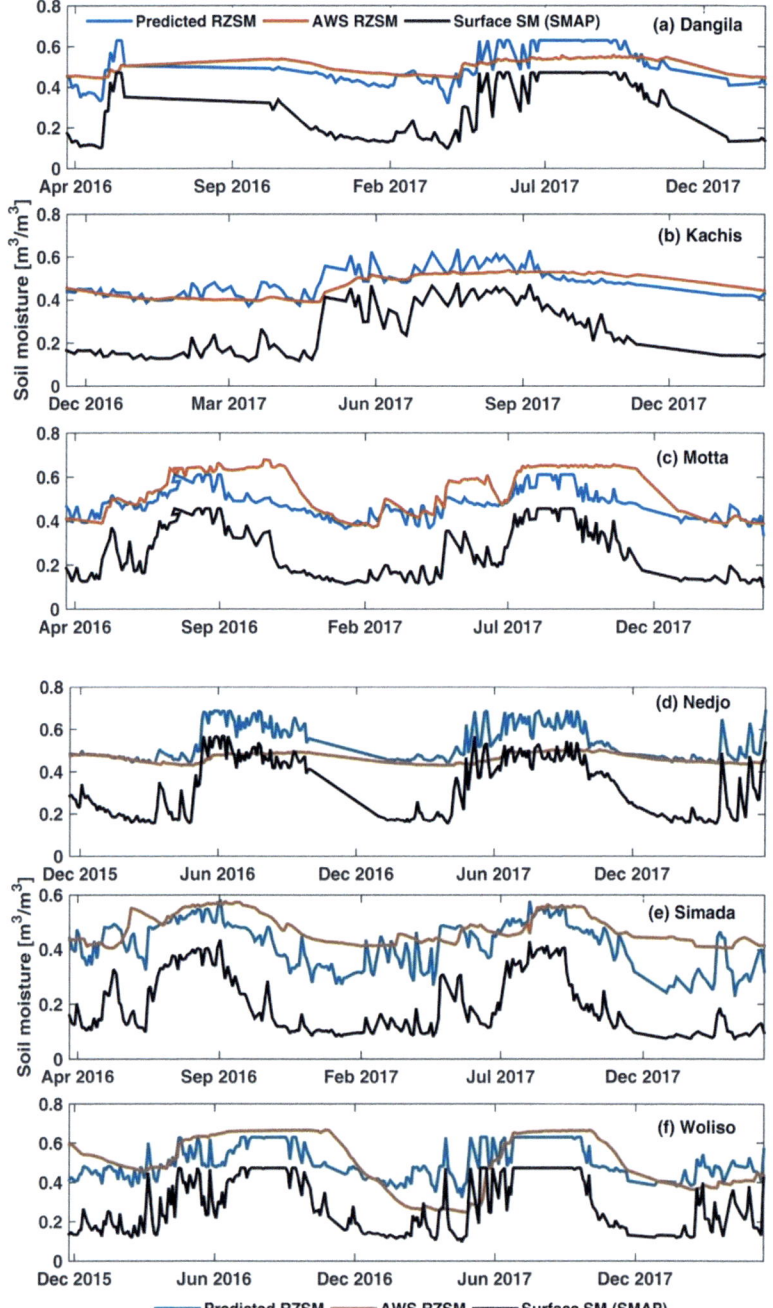

Fig. 21.6 The temporal distributions of daily surface soil moisture (SMAP L3 P E) predicted and AWS-observed RZSM at the **a** Dangial, **b** Kachis, **c** Motta, **d** Nedjo, **e** Simada, and **f** Wolsio AWS datasets acquired from November 2015 to June 2018

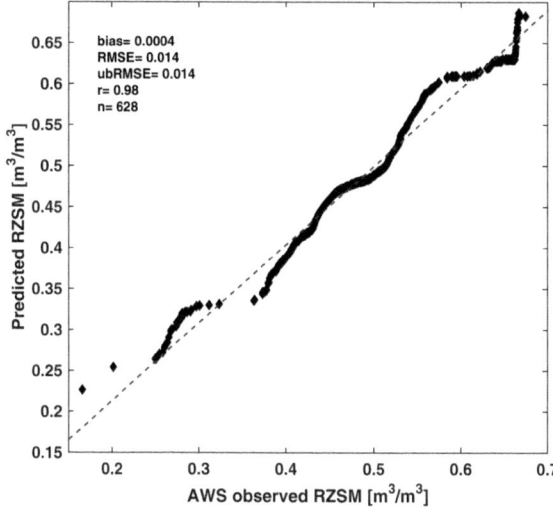

Fig. 21.7 The relation between the predicted and AWS-observed root-zone soil moisture based on the prediction model developed using the combination of all stations

and ubRMSE = 0.039 m³/m³ were observed at the relatively low elevation station of "Nedjo" during the validation period (Table 21.3). This has been further confirmed in Fig. 21.6d, in which the predicted RZSM showed a considerable deviation from the actual AWS-measured data at the "Nedjo" station. In general, better prediction accuracy was observed at the relatively higher elevation stations of "Simada" and "Kachis" (Table 21.3 and Fig. 21.6b, e).

Intercomparison Between the CDF-Based Prediction Model and SMAPL4 RZSM

An intercomparison between RZSM estimates generated using CDF matching and the SMAP L4 product was made to comparatively evaluate the performance of the prediction model using the statistical metrics of r, bias, RMSE, and ubRMSE (Table 21.3). In addition, the spatiotemporal relations between these two products were further analysed, as presented in Figs. 21.8, 21.9, 21.21.10, 21.2.11 and 21.12 below. In general, the SMAP L4 RZSM product has revealed a relatively good statistical performance compared to the reference datasets. However, compared to RZSM predicted using the CDF matching method, the SMAP L4 RZSM product has shown weak statistical performance (Table 21.3). Over all the stations, the SMAP L4 product shows a high bias (ranging from -0.075 to -0.250 m^3/m^3) when compared to the very low bias values observed by CDF-based predictions. The presence of high bias values has led SMAP L4 to a large RMSE over most weather stations in comparison with the CDF-based prediction model, with the highest RMSE $= 0.252$ m^3/m^3. In fact, the skills of the SMAP L4 root-zone soil moisture estimates are considerably improved after removing the bias and result in ubRMSE values smaller than or close to 0.04 m^3/m^3 (the SMAP mission target accuracy) over most of the AWS.

A polynomial fit developed using all the stations was used to map the spatial distribution of the RZSM in the study basin. Thus, maps of the RZSM (both for the predicted and SMAP L4) over the UBN basin during the dry months of September, October, November, and December of 2016 (using the monthly average values) are displayed in Fig. 21.8. The mean values of RZSM for these months are provided in Fig. 21.9.

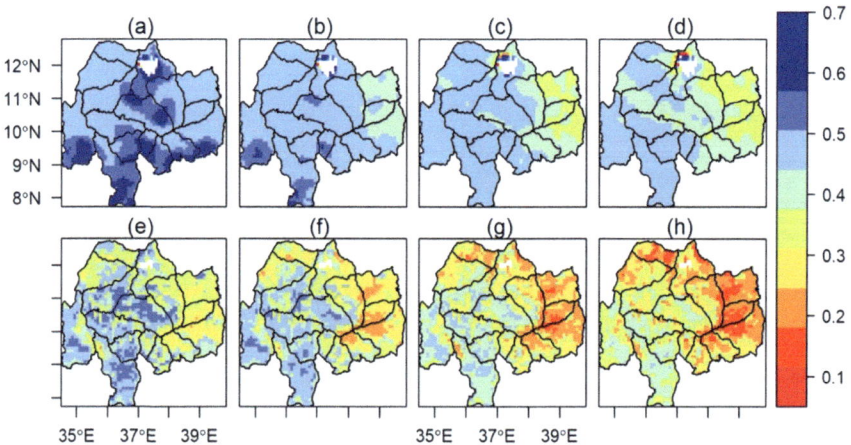

Fig. 21.8 The spatiotemporal patterns of the predicted (**a–d**) and SMAP L4 (**e–h**) RZSM (m^3/m^3) for the monthly average of September to December 2016 over the UBN basin

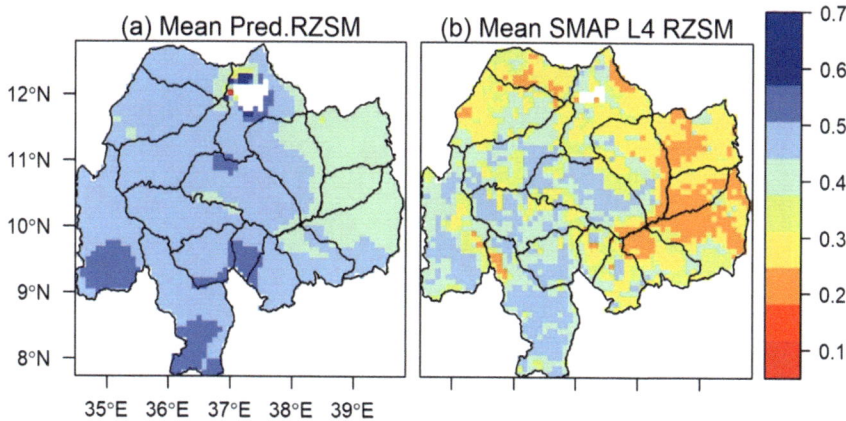

Fig. 21.9 The spatiotemporal patterns of **a** predicted RZSM and **b** SMAP L4 RZSM (m^3/m^3) for dry months (mean of September to December 2016) averaged over the UBN basin

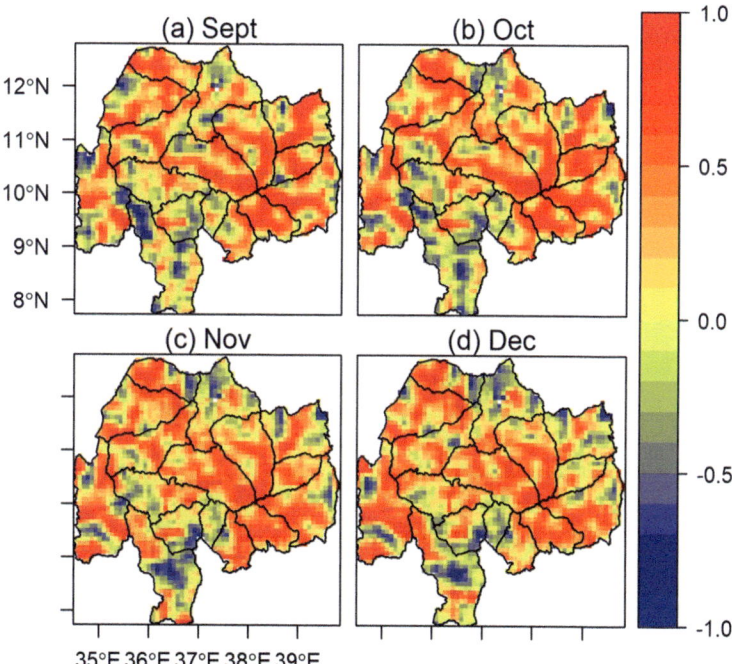

Fig. 21.10 The spatial correlation (r ranging from -1 to 1) maps between the predicted and SMAP L4 RZSM for the months of **a** September, **b** October, **c** November, and **d** December 2016 averaged over the UBN basin

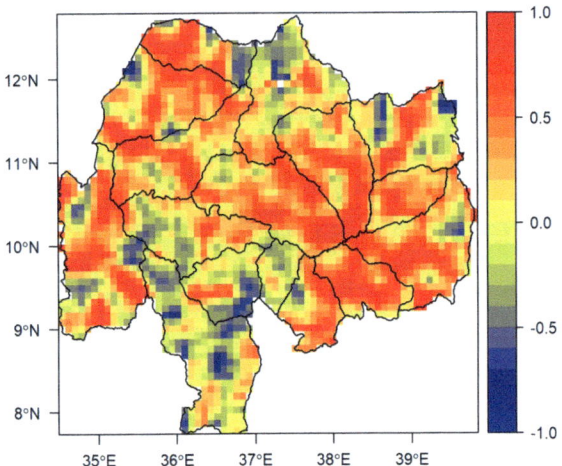

Fig. 21.11 The spatial correlation (r ranging from -1 to 1) maps between the predicted and SMAP L4 RZSM for the dry months average (September to December 2016) over the UBN basin

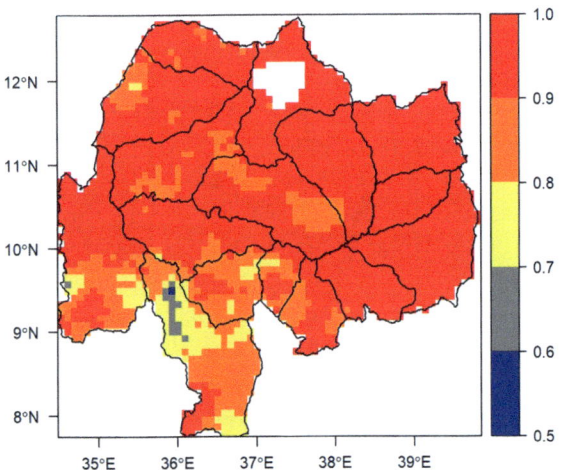

Fig. 21.12 The temporal correlation (r ranging from -1 to 1) maps between the predicted and SMAP L4 RZSM over the UBN basin

Both the predicted and SMAP L4 RZSM showed great similarity in terms of the spatial and temporal trends in the basin (Figs. 21.9 and 21.10). They indicated a decreasing trend from west to the east region (or from low- to high-elevation areas) of the basin. In addition, from Fig. 21.9, it can be observed that there is a decreasing trend of RZSM from September (when the soil has a relatively large amount of moisture) to December (known with a relatively low amount of soil moisture). Thus, the spatial and temporal distributions of RZSM seem to follow the geomorphological and meteorological conditions of the Abbay River basin in Ethiopia.

According to the spatial analysis carried out over the basin, the SMAP L4 RZSM, as opposed to the predicted RZSM, showed the most pronounced spatial variability between the wet and dry zones. In general, the SMAP L4 RZSM product generally showed drier values than the predicted RZSM for both the monthly (Fig. 21.8) and mean (Fig. 21.9) datasets, confirming the underestimation (dry bias) found in comparison with the observed RZSM values (Table 21.3). Additionally, the SMAP L4 RZSM maps (both in Figs. 21.8 and 21.9) indicated dry values in the eastern region, which is consistent with the climatic patterns of the UBN basin. This effect was particularly observed in the "Beshilo", "Welaka", and "Jimma" subbasins of the UBN basin (Fig. 21.1).

The spatial correlation maps between the predicted and SMAP L4 RZSM for monthly and dry season (September to December 2016) averages are shown in Figs. 21.10 and 21.11, respectively, while their temporal correlation map is presented in Fig. 21.12. From the spatial correlation maps, the predicted and SMAP L4 RZSM are well correlated ($r > 0.5$) over most parts of the basin. However, low correlation values are still observed over some areas, and even negative correlation values between the two products are shown in some other locations, particularly in the southern tips of the Abbay River basin. On the other hand, the temporal correlation map in Fig. 21.12 displays a higher correlation ($r > 0.8$) between the predicted and SMAP L4-estimated RZSM over most parts of the basin. A correlation value greater than 0.5 was also observed over the remaining part of the basin.

Discussion

Determining root-zone soil moisture (RZSM) information from near-surface observations is very valuable and has great importance for further augmenting the use and benefits of remote sensing technology in the field of agriculture. In this study, we presented a simple and feasible statistical technique that couples the CDF matching method and polynomial regression model (CDF-polynomial) to predict residual moisture in the root zone by extending SMAP (L3_SM_P_E) surface soil moisture estimates to support off-season farming in the Abbay River basin. In addition, an intercomparison between the CDF-based RZSM prediction model and SMAP L4 RZSM product was made to comparatively evaluate the performance of the proposed prediction model.

The prediction model was independently tested against RZSM measured at the AWS. The statistical analysis indicated that the proposed CDF-polynomial model was effective at modelling residual moisture in the root zone and resulted in great prediction accuracy in our study area, with high r, less bias and smaller RMSE and ubRMSE values. Although a slight significant difference is observed among stations due to variation in elevation, soil property, land use land cover, and length of observation periods, the proposed prediction model has effectively reproduced AWS-measured RZSM over all selected stations (Table 21.3 and Fig. 21.6). Similar statistical performance of the CDF matching method has been reported by Gao et al. (2017) using field-observed surface and profile soil moisture. These results suggest that coupling the CDF matching method and polynomial regression model could be a good alternative to the data assimilation technique to predict RZSM, with the advantage of simplicity and low computational amenities. Indeed, a relatively lower statistical performance of the prediction model has been observed at "Nedjo" station, with low elevation values in comparison to the other stations. This could be attributed to the relatively high surface evapotranspiration in lower elevation areas (Chen et al. 2016), which could possibly lead to lower coupling of surface and root-zone soil moisture. The accuracy of predicting RZSM from surface observations highly depends on the initial surface soil moisture conditions (Li et al. 2018).

On the other hand, the performance of the proposed method has been well demonstrated by the CDF plots in Figs. 21.4 and 21.5. In general, the predicted RZSM produced very close values to the AWS measurements at almost all frequencies. In this connection, the predicted RZSM has maintained the AWS-measured soil moisture patterns and shape of the cumulative density function (Figs. 21.4 and 21.5). The developed observation operator is able to reasonably follow the dynamics of AWS-measured soil moisture and remove the systematic differences between SMAP-estimated surface soil moisture and AWS-measured RZSM data. In fact, over some frequencies, e.g., at the "Dangila" and "Nedjo" stations, the predicted RZSM occurred below or above the AWS-measured CDFs. The deviation shown at some frequencies could be due to the relatively poor coupling of surface and RZSM.

Furthermore, the performance of the CDF matching RZSM prediction model was compared to that of RZSM extracted from SMAP L4 over the selected AWS stations. The intercomparison with SMAP L4 RZSM products revealed the reliability of the CDF matching method in deriving the RZSM in the study area. Although the SMAP L4 RZSM product generally showed good agreement with the reference datasets, its error estimates are larger in comparison to the proposed prediction model. The SMAP L4 product could not fully reproduce the soil moisture values observed by AWS and resulted in large dry bias values, which led to underestimations of root-zone soil moisture over all selected stations. Negative bias values for SMAP L4 root-zone soil moisture have also been reported by Bi et al. (2016). The underestimation of the SMAP L4 product may be explained by several aspects, including (i) uncertainties related to the space-borne system and the retrieval algorithm, (ii) the ancillary data used for the generation of the soil moisture dataset, and (iii) the vertical and horizontal scale mismatch between the AWS point measurement and the gridded SMAP L4 (Montzka et al. 2017).

In general, the error estimates in our model are very low in comparison with the error values reported by previous RZSM studies using exponential filter techniques (Ford et al. 2014; Tobin et al. 2017) and data assimilation methods (Dumedah et al. 2015). This could be attributed to the great skill of the observation operators in adjusting the CDF of surface and RZSM, as well as the accuracy of the SMAP satellite product. However, it should be noted that completely data-driven methods incorporate the interrelationship of training data with limited simplifying assumptions about the core processes. Thus, the CDF-polynomial statistical model was derived solely on the relations from the training dataset, and the validity of the model is highly dependent on the condition for which the model is produced. Otherwise, a physical model with data assimilation is a potential approach with more flexibility when conditions are changing (Lia et al. 2010; Sabater et al. 2007). The presented method in this study cannot replace data assimilation or analytical models for RZSM predictions. The intention of this study is to provide the CDF method as a simple and alternative approach to predict RZSM for locations where the soil moisture content information at deeper soil layers is not available or not regularly measured.

Conclusion

This study has demonstrated an improved RZSM prediction technique by extending surface soil moisture derived from SMAP satellite observations. The prediction was undertaken using the CDF matching method and the polynomial regression model. The findings of this study have highlighted the potential of the proposed model to provide better RZSM predictions in our area of interest. Furthermore, the CDF matching techniques described in this study could be used in other regions with similar Agro-climate and soil conditions to our study area. Remote sensing technology could still offer a great opportunity to estimate RZSM at different scales with improved spatial and temporal resolutions.

However, the CDF method has its own limitations, and the robustness of the observation operator could be affected by outliers in the soil moisture data series and the length of the observed datasets. Accurate estimation of RZSM based on the CDF method requires a long record of datasets, and when the training datasets are not sufficient, the matching effect between surface and RZSM might be unstable. Outliers need to be carefully monitored and removed from the dataset, and enough data needs to be secured before the building of observation operators. The model was also tested in limited locations, and future analysis needs to apply a similar model with wider geographic extents and various environmental and climate conditions. The findings in this study further pointed out questions for future studies, which include the sensitivity and performance of the CDF matching method (i) for the spatial and temporal scale of the surface soil moisture and (ii) for the penetration depth of the satellite in which surface soil moisture is derived.

Acknowledgements The authors would like to thank the National Meteorological Agency (NMA) of Ethiopia for providing soil moisture datasets observed over AWS found in the Abbay River basin. We are also grateful to NASA for providing SMAP soil moisture products. We acknowledge the funding from the CGIAR Sustainable Intensification of Mixed Farming System (SI-MFS) initiative and the International Development Association (IDA) of the World Bank to the Accelerating Impact of CGIAR Climate Research for Africa (AICCRA) project.

References

Albergel C, Rüdiger C, Pellarin T, Calvet JC, Fritz N, Froissard F, Suquia D, Petitpa A, Piguet B, Martin E (2008) From near-surface to root-zone soil moisture using an exponential filter: an assessment of the method based on in situ observations and model simulations. Hydrol Earth Syst Sci 12:1323–1337

Awulachew SB, Merry J, Kamara AB, van Koppen B, Penning de Vries F, Boelee E, Makombe G (2005) Experiences and opportunities for promoting small-scale/micro irrigation and rainwater harvesting for food security in Ethiopia. International Water Management Institute (IWMI), Colombo, Sri Lanka, p 91

Ayehu G, Tadesse T, Gessesse B, Dinku T (2018) Validation of new satellite rainfall products over the Upper Blue Nile, Ethiopia. Atmos Meas Tech 11:1921–1936

Bekabil UT (2014) Review of challenges and perspectives of agricultural production and productivity in Ethiopia. J Nat Sci Res 4:70–77

Bekele Y, Nata T, Bheemalingswara K (2012) Preliminary study on the impact of water quality and irrigation practices on soil salinity and crop production, Gergera Watershed, Atsbi-Wonberta, Tigray, Northern Ethiopia. MEJS 4(1):29–46

Bi H, Zeng J, Zheng W, Fan X (2016) Validation of SMAP soil moisture analysis product using in situ measurements over the little Washita watershed. In: IEEE international geoscience and remote sensing symposium, pp 3086–3089

Brocca L, Hasenauer S, Lacava T, Melone F, Moramarco T, Wagner W, Dorigo W, Matgen P, Martínez-Fernández J, Llorens P, Latron J, Martin C, Bittelli M (2011) Soil moisture estimation through ASCAT and AMSR-E sensors: an intercomparison and validation study across Europe. Remote Sens Environ 115:3390–3408

Central Statistical Agency (CSA) (2001) Report on the year 2000 Welfare Monitoring Survey; Central Statistical Authority: Addis Ababa, Ethiopia

Chen F, Crow WT, Colliander A, Cosh MH, Jackson TJ, Bindlish R, Reichle RH, Chan SK, Bosch DD, Starks PJ (2016) Application of triple collocation in ground-based validation of Soil Moisture Active/Passive (SMAP) level 2 data products. IEEE J Sel Top Appl Earth Obs Remote Sens 99:1–14

Clark P, Rupp D, Woods R, Zheng X, Ibbitta R, Slater A, Schmidta J, Uddstroma M (2008) Hydrological data assimilation with the ensemble Kalman filter: use of streamflow observations to update states in a distributed hydrological model. Adv Water Resour 31:1309–1324

Colliander A et al (2017) Validation of SMAP surface soil moisture products with core validation sites. Remote Sens Environ 191:215–231

Conway D (2005) From headwater tributaries to international river: observing and adapting to climate variability and change in the Nile basin. Glob Environ Change 15:99–114

Conway D (2000a) Some aspects of climate variability in the northeast Ethiopian highlands-Wollo and Tigray. Ethiop J Sci 23:139–161

Conway D (2000b) The climate and hydrology of the Upper Blue Nile River. Geogr J 166:49–62

Cui C, Xu J, Zeng J, Chen K, Bai X, Lu H, Chen Q, Zhao T (2018) Soil moisture mapping from satellites: an intercomparison of SMAP, SMOS, FY3B, AMSR2, and ESA CCI over two dense network regions at different spatial scales. Remote Sens 10(33):1–19

Drusch M, Wood F, Gao H (2005) Observation operators for the direct assimilation of TRMM microwave imager retrieved soil moisture. Geophys Res Lett 32:L15403

Dumedah G, Walker J, Merlin O (2015) Root-zone soil moisture estimation from assimilation of downscaled soil moisture and ocean salinity data. Adv Water Resour 84:14–22

Entekhabi D, Njoku EG, O'Neill PE, Kellogg KH, Crow WT, Edelstein WN, Entin JK, Goodman SD, Jackson TJ, Johnson J (2010) The soil moisture active passive (SMAP) mission. Proc IEEE 98:704–716

Entekhabi D, Yueh S, O'Neill PE, Kellogg KH (2013) SMAP handbook soil moisture active passive mapping soil moisture and freeze/thaw from space. National Aeronautics and Space Administration, Pasadena, CA, USA

Entekhabi D, Yueh S, O'Neill PE, Kellogg KH, Allen A, Bindlish R, Brown M, Chan S, Colliander A, Crow WT (2014) SMAP handbook—soil moisture active passive: mapping soil moisture and freeze/thaw from space. Jet Propulsion Laboratory, Pasadena, CA, USA

Food and Agricultural Organization (FAO) (2014) Ethiopia country programming framework. Office of the FAO Representative to Ethiopia, Addis Ababa, Ethiopia

Ford T, Harris E, Quiring S (2014) Estimating root zone soil moisture using near-surface observations from SMOS. Hydrol Earth Syst Sci 18:139–154

Gao X, Zhao X, Brocca L, Huo G, Lv T, Wu P (2017) Depth scaling of soil moisture content from surface to profile: multistation testing of observation operators. Hydrol Earth Syst Sci Discuss 1–25

Han E, Heathman GC, Merwade V, Cosh MH (2012) Application of observation operators for field scale soil moisture averages and variances in agricultural landscapes. J Hydrol 444–445:34–50

Hu W, Si C (2014) Can soil water measurements at a certain depth be used to estimate mean soil water content of a soil profile at a point or at a hill slope scale. J Hydrol 516:67–75

Kim U, Kaluarachchi J, Smakhtin V (2008) Generation of monthly precipitation under climate change for the upper Blue Nile River Basin, Ethiopia. J Am Water Resour 44:1231–1247

Kornelsen C, Coulibaly P (2014) Root-zone soil moisture estimation using data-driven methods. Water Resour Res 50:2946–2962

Li C, Lu H, Yang K, Han M, Wright J, Chen Y, Yu L, Xu S, Huang X, Gong W (2018) The evaluation of SMAP enhanced soil moisture products using high-resolution model simulations and in situ observations on the Tibetan plateau. Remote Sens 535(10):1–16

Lia F, Crow W, Kustas W (2010) Towards the estimation root-zone soil moisture via the simultaneous assimilation of thermal and microwave soil moisture retrievals. Adv Water Resour 33:201–214

Liu Y, Parinussa M, Dorigo A, De Jeu M, Wagner W, van Dijk M, McCabe F, Evans P (2011) Developing an improved soil moisture dataset by blending passive and active microwave satellite-based retrievals. Hydrol Earth Syst Sci 15:425–436

Ma C, Li X, Wei L, Wang W (2017) Multiscale validation of SMAP soil moisture products over cold and arid regions in northwestern China using distributed ground observation data. Remote Sens 9(327):1–14

Mahmood R, Hubbard G (2007) Relationship between soil moisture of near surface and multiple depths of the root zone under heterogeneous land uses and varying hydroclimatic conditions. Hydrol Process 21:3449–3462

Manfreda S, Brocca L, Moramarco T, Melone F, Sheffield J (2014) A physically based approach for the estimation of root-zone soil moisture from surface measurements. Hydrol Earth Syst Sci 18:1199–1212

Montzka C, Bogena H, Zreda M, Monerris A, Morrison R, Muddu S, Vereecken H (2017) Validation of spaceborne and modelled surface soil moisture products with Cosmic-Ray neutron probes. Remote Sens 9(103):1–30

Pachepsky Y, Hill L (2016) Scale and scaling in soils. Geoderma 287(1):4–30

Parrens M, Mahfouf F, Barbu L, Calvet C (2014) Assimilation of surface soil moisture into a multilayer soil model: design and evaluation at local scale. Hydrol Earth Syst Sci 18:673–689

Petropoulos GP, Ireland G, Petropoulos GP, Ireland G, Barrett B (2015) Surface soil moisture retrievals from remote sensing: current status, products and future trends. Phys Chem Earth Parts A/B/C 83–84:36–56

Reichle H, Koster D (2014) Bias reduction in short records of satellite soil moisture. Geophys Res Lett 31(19)

Reichle R, De Lannoy G, Koster R, Crow W, Kimball J (2016) SMAP L4 9 km EASE-grid surface and root zone soil moisture geophysical data, version 2. NASA National Snow and Ice Data Center

Reichle R, De Lannoy G, Liu Q, Koster R, Kimball J, Crow W, Ardizzone J, Chakraborty P, Collins D, Conaty A, Girotto M, Jones L, Kolassa J, Lievens H, Lucchesi R, Smith E (2017) Global assessment of the SMAP level-4 surface and root zone soil moisture product using assimilation diagnostics. J Hydrometeorol 18(12):3217–3237

Sabater J, Jarlan L, Calvet J, Bouyssel F (2007) From near-surface to root-zone soil moisture using different assimilation techniques. J Hydrometeorol 8:194–206

Shi G, Wu T, Zhao X, Li C, Wang W, Zhang Q (2014) Statistical analyses and controls of root-zone soil moisture in a large gully of the Loess Plateau. Environ Earth Sci 71:4801–4809

Taye M, Willems P (2012) Temporal variability of hydroclimatic extremes in the Blue Nile basin. Water Resour Res 48:1–13

Tobin K, Torres R, Crow W, Bennett M (2017) Multidecadal analysis of root-zone soil moisture applying the exponential filter across CONUS. Hydrol Earth Syst Sci 21:4403–4417

Wagner W, Lemoine G, Rott H (1999) A method for estimating soil moisture from ERS scatterometer and soil data. Remote Sens Environ 70:191–207

World Bank (2006) Managing water resources to maximize sustainable growth. A World Bank water resources assistance strategy for Ethiopia. The World Bank Agriculture and Rural Development Department. Report No. 36000-ET. Washington, DC, USA

Chapter 22
Downscaling ESA CCI Soil Moisture Using Sentinel-1 SAR Data: A Case Study in the Abbay River Basin in Ethiopia

Getachew Tesfaye Ayehu, Wuletawu Abera, Degefie Tibebe, Yonas Getaneh, and Lulseged Tamene

Abstract In this paper, the coarse-scale (~28 km) European Space Agency (ESA) Climate Change Initiative (CCI) soil moisture product was downscaled to a high-resolution dataset (i.e., 1 km, 250 m, and 100 m) using high-resolution Sentinel-1 SAR data and the SMAP baseline downscaling approach. A comparison was made between the automatic weather station (AWS) observed and downscaled soil moisture results in ubRMSE (correlation coefficient (r)) of 0.090 m^3/m^3 (0.51), 0.100 m^3/m^3 (0.57), and 0.090 m^3/m^3 (0.61) at 1 km, 250 m, and 100 m resolutions, respectively. The comparison with field-observed dataset yields ubRMSE (r) values of 0.022 m^3/m^3 (0.57) and 0.032 m^3/m^3 (0.22) for soil moisture estimates downscaled to 250 m and 100 m resolution, respectively. The downscaled soil moisture estimates are in good agreement with the in situ measurement and AWS observed soil moisture. Thus, the result obtained in this study is quite promising and indicates the potential of the Sentinel-1 SAR data and the downscaling algorithm to disaggregate the coarse-resolution soil moisture to finer scales in the Abbay River Basin in Ethiopia.

Keywords ESA · CCI · Soil moisture · Downscale · Sentinel-1 · SAR

G. T. Ayehu (✉) · W. Abera · D. Tibebe · Y. Getaneh · L. Tamene
International Center for Tropical Agriculture (CIAT), Addis Ababa, Ethiopia
e-mail: Getachew.Tesfaye@cgiar.org

W. Abera
e-mail: Wuletawu.Abera@cgiar.org

D. Tibebe
e-mail: D.Tibebe@cgiar.org

Y. Getaneh
e-mail: y.getaneh@cgiar.org

L. Tamene
e-mail: LT.Desta@cgiar.org

© The Author(s), under exclusive license to Springer Nature Switzerland AG 2025
A. Melesse et al. (eds.), *Abbay River Basin*, Springer Geography,
https://doi.org/10.1007/978-3-031-65241-7_22

Introduction

Soil moisture is an indispensable parameter for understanding the various processes in hydrology, agriculture, meteorology, and the Earth's climate (Munoz-Sabater et al. 2016). It is a deciding factor for crop growth and development (Dobriyal et al. 2012). Several application areas, such as agricultural production, drought monitoring, flood mapping, irrigation scheduling, and climate change, have always demanded soil moisture at different scales for skillful modeling and forecasting (AghaKouchak et al. 2015; Robinson et al. 2008; Anderson et al. 2007). Obtaining dependable soil moisture estimates with reliable accuracy over a range of spatial and temporal scales has gained wider interest in addressing the needs and demands of these application areas (Dorigo et al. 2017). However, it is very difficult to obtain such reliable soil moisture data over large areas using conventional and sparsely distributed in situ soil moisture monitoring networks.

Microwave remote sensing-based soil moisture retrieval is the most preferred and extensively used technique to monitor near-surface soil moisture at regional and global scales. Subsequent to the successful launch of various microwave satellites, such as the Soil Moisture Active Passive (SMAP) (Knipper et al. 2017), passive microwave-based Advanced Microwave Scanning Radiometer-Earth Observing System (AMSR-E) (Kawanishi et al. 2003), Environmental Satellite (ENVISAT)-Advanced SAR (ASAR) (Pathe et al. 2009), Meteorological Operational Satellite Program (METOP)-Advanced SCATterometer (ASCAT) (Albergel et al. 2009), Advanced Microwave Scanning Radiometer 2 (AMSR-2) (Li et al. 2004), and Soil Moisture Ocean Salinity (SMOS) (Kerr et al. 2002), numerous microwave-based global soil moisture datasets, including the European Space Agency Climate Change Initiative (ESA CCI) (Liu et al. 2011; Dorigo et al. 2015), have been obtainable. However, microwave-based methods can only help to retrieve surface soil moisture generally at the coarse scale of the satellite footprints at approximately several tens of kilometers. These products are unable to detect the variation in soil moisture at smaller spatial scales and cannot satisfy the demands of soil moisture dataset at high spatial resolution required for applications such as crop management, water resource management, and drought monitoring practiced at a local scale (Sun et al. 2019; Wang et al. 2014). Therefore, spatial downscaling of these soil moisture products to a kilometer or even tens of meters is needed for many regional and local hydrological and agricultural applications.

In this connection, several soil moisture downscaling techniques have been introduced and successfully used to spatially disaggregate the coarse soil moisture product to a higher spatial resolution (e.g., Das et al. 2014; Liu et al. 2021; Knipper et al. 2017; Lia et al. 2018; Kovačević et al. 2020; Warner et al. 2021; Srivastava et al. 2013). Generally, soil moisture downscaling methods can be classified into two major classes. These are the downscaling factors and downscaling functions (Sun and Cui 2020). The first methods provide the spatial variability of high-resolution soil moisture within the coarse-resolution soil moisture, while the latter methods focus on determining the method of adding the spatial variability to the coarse-resolution soil moisture (Chauhan et al. 2003). Currently, various methods of these two major classes, including the empirical polynomial fitting method (Chauhan et al. 2003; Piles et al. 2011; Choi and Hur 2012), the semi-physical evaporation-based method (Merlin et al. 2012; Malbéteau et al. 2016), and the smoothing filter-based intensity modulation (SFIM) downscaling method (Liu 2000; Peng et al. 2015) are intensively used. In most of these algorithms, high-resolution data (i.e., from optical/thermal remote sensing) is employed as downscaling factors to spatially disaggregate the coarse-scale soil moisture datasets to a finer scale (1 km or below). This is because surface soil moisture has been reported to have a significant correlation with other land surface parameters, such as the vegetation index and surface temperatures, which can be derived from optical/thermal remote sensing (Piles et al. 2016; Knipper et al. 2017). Because of the nature of optical sensors, however, these downscaling methods are always applied under clear-sky conditions (Djamai et al. 2016; Li et al. 2018). Cloud coverage that extremely affects optical/thermal remote sensing and the land surface parameters derived limits its applications to use in areas or seasons where there is dense and frequent cloud coverage.

A combination of microwave-based soil moisture datasets and high-resolution radar data has appeared as an alternative downscaling approach (Das et al. 2014; He et al. 2018). The important advantages of microwave remote sensing are its high sensitivity to soil moisture variations (Ulaby et al. 1986) and its natural capability to penetrate clouds and haze and acquire remotely sensed datasets under different weather conditions. A typical example of such a method is the NASA baseline downscaling algorithm (Das et al. 2014, 2011) proposed to disaggregate the SMAP coarse-resolution (36 km) brightness temperature/soil moisture to a medium resolution (9 km) using the relatively high-resolution (1 km) radar backscatter obtained from the SMAP mission. Using this algorithm, Das et al. (2014) have managed to downscale the low-resolution (36 km) product to a finer scale (9 km) with RMSE = 0.033 cm^3/cm^3. Later, He et al. (2018), following the collapse of SMAP's L-band radar in 2015, used these algorithms to downscale the 36 km SMAP brightness temperature/soil moisture using Sentinel-1 SAR data to finer scales of 9 km, 3 km, and 1 km with RMSEs of 0.056 cm^3/cm^3, 0.072 cm^3/cm^3, and 0.092 cm^3/cm^3, respectively.

In this study, the SMAP baseline soil moisture downscaling algorithm was adopted and used to disaggregate the ESA CCI soil moisture (~28 km) to a finer scale of 1 km, 250 m, and 100 m using the higher resolution Sentinel-1 SAR data. In this context, the overall aim of this study was (i) to generate high-resolution soil moisture data by disaggregating the coarse-scale ESA CCI soil moisture datasets using Sentinel-1 SAR data and the SMAP baseline downscaling algorithm and (ii) to test the performance of the downscaling algorithm for operational mapping of soil moisture at a higher scale that could be used for applications at a local scale.

Descriptions of the Study Area

This study is conducted over selected sites in the Abbay River Basin. The Abbay River Basin is situated in the north-western part of Ethiopia with an area between 7°40′ N and 12° 51′ N and 34° 25′ E and 39° 49′ E, as shown in Fig. 22.1. The basin is distinguished by a rugged topography and the altitude varies from 4239 to 490 m a.s.l.

The climate of the study site is mainly characterized by the altitude and its closeness to the equatorial monsoonal systems. Accordingly, the study area can be classified into three distinct seasons. These are locally known as *Kiremt* (main rainy season) that occurs from June to September, *Belg* (a short rainy season) that may occur around February to May, and *Bega* (a dry season that runs from October to January (Cheung et al. 2008; Conway, 2000). According to Conway (2000), the annual mean rainfall of the basin varies between 1200 and 1800 mm. Most of the population (about 85%) of the basin lives in rural areas (CSA 2007) and their livelihoods are mainly dependent on rain-fed agriculture. For this study, two sites named *Site one* (latitude between 10° 30′ and 12° 30′N and longitude between 36°30′ and

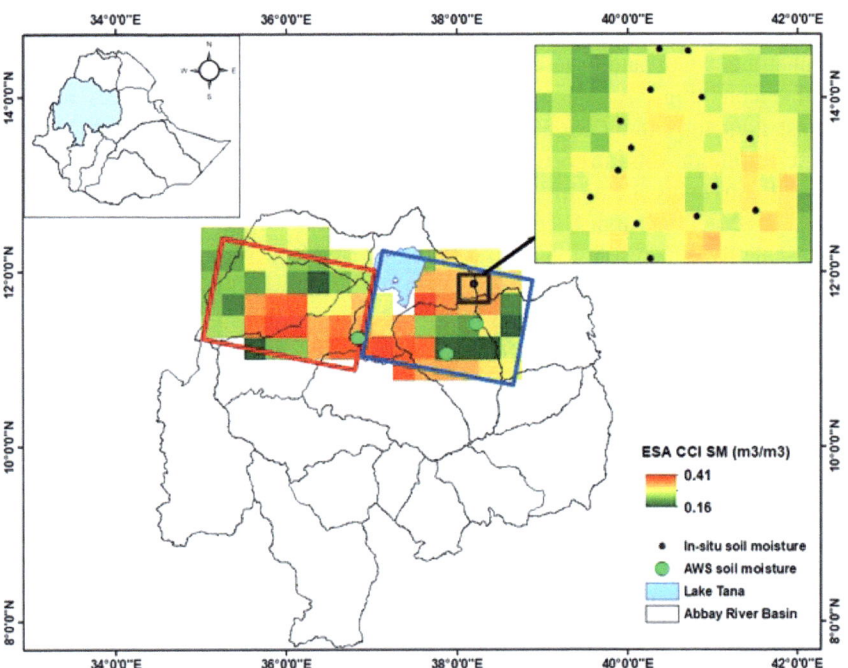

Fig. 22.1 Overview of the study site using the ESA CCI soil moisture grid cells as a background map shows selected sites. The blue (Site 1) and red (Site 2) boxes indicate the soil moisture downscaling test sites used in the study. The green bold points show the automatic weather stations (AWS), and the tiny blue points are locations for in situ measured soil moisture used for validation

39°00′E, shown in the blue box in Fig. 22.1) and *Site two* (latitude between 10° 30′ and 12° 30′N and longitude between 35° 00′ and 37°00′E, shown in the red box in Fig. 22.1) were selected as test sites to downscale the ESA CCI soil moisture and validate the performance of the downscaling algorithm.

Data

Field-Measured Soil Moisture

Surface soil moisture that was first measured for plot-based residual soil moisture monitoring (Ayehu et al. 2020) was used for validation of the downscaled soil moisture. Points that were spatially and temporally fitted with the downscaled soil moisture datasets were used for validation (Fig. 22.1). Indeed, the spatial and temporal distributions of the plot-based dataset were very limited, and it was not enough to validate the downscaled product. Thus, to further validate the performance of the downscaling algorithm and complement the absence of enough validation datasets, soil moisture observed at three automatic weather stations (AWS) of the National Meteorological Institute, named "*Motta*", "*Dangal*", and "*Simada*", found in the study area were used (Fig. 22.1). However, the AWS observes soil moisture at depths of 0–20 cm, which is beyond the penetration capability of microwave sensors, which are usually more sensitive to the top five centimeters of soil (Qin et al. 2009). This could perhaps inflate the root-mean-square error (RMSE) of the downscaling algorithms due to the relatively high amount of moisture generally found in the subsurface of the soil. However, the AWS observations are adopted in this study to see its correlation with the disaggregated soil moisture with the general assumption that the change in surface soil moisture is highly correlated with that of the near subsurface soil moisture, given the coupling of surface and subsurface soil moisture through diffusion processes (Kornelsen and Coulibaly 2014).

ECA CCI Soil Moisture

ESA CCI is a global soil moisture product produced by the European Space Agency using datasets acquired from both active and passive microwave remote sensing systems (Wagner et al. 2012; Liu et al. 2011). The ESA CCI soil moisture dataset is currently available either as an active/passive alone or combined product. It is provided as a daily product with a spatial resolution of 0.25° from 1978 to 2020 (v06.1) (Preimesberger et al. 2021). Due to the variation in individual data sources, adopted methods of combining data from different missions, and soil moisture retrieval algorithms, it has been reported that the accuracy of ESA soil moisture product obtained from either of the missions and their merged dataset are different

(Qiu et al. 2016). However, Liu et al. (2011) reported that the accuracy of the merged product is higher for either the passive or active alone, after the validation analyses, they made over the core validation sites distributed over the globe. In this study, therefore, the merged ESA CCI soil moisture data from October 2016 to March 2017 were used to produce a high-resolution surface soil moisture dataset and test the performance of the downscaling algorithm originally proposed for coarse-scale SMAP products.

Sentinel-1 SAR Data

The dual-polarized Sentinel-SAR data provided by ESA was used as a downscaling feature to disaggregate the low-resolution ESA CCI soil moisture dataset to a finer scale. The Sentinel-1 SAR data was downloaded from the Copernicus Open Access Hub, and it was acquired on a C-band SAR instrument with a frequency of 5.405 GHz. The SAR data from the Sentinel-1 satellite mission is provided with four different modes, including the Interferometric Wide-Swath (IWS) mode, which is the main operational mode on land surface. The SAR data from IWS mode is collected in 250-km swaths and it detects backscattering signals at incidence angles ranging from 29.1° to 46°. The dataset was originally generated with a ground resolution of 5 m (in the range direction) and 20 m (in the azimuth direction), and provided with a pixel size of 10 m. In the present study, Sentinel-SAR data acquired in the IWS mode from the period of October 22, 2016 to March 10, 2017 were used. Table 22.1 provides the characteristics of Sentinel-1 SAR data collected in this study.

The collected SAR data has passed through several pre-processing steps including the radiometric correction, speckle filtering, and geometric correction steps using

Table 22.1 Characteristics of Sentinel-1 SAR data used in the study area

Date of acquisitions	Polarization	Orbit	Product type
October 22, 2016	VV + VH	Descending	GRD
November 15, 2016	VV + VH	Descending	GRD
December 09, 2016	VV + VH	Descending	GRD
December 16, 2016	VV + VH	Descending	GRD
January 09, 2017	VV + VH	Descending	GRD
February 02, 2017	VV + VH	Descending	GRD
February 14, 2017	VV + VH	Descending	GRD
February 26, 2017	VV + VH	Descending	GRD
March 10, 2017	VV + VH	Descending	GRD

NB: *VH* indicates vertical transmit and horizontal receive polarization; *VV* indicates vertical transmit and vertical receive polarization; and *GRD* represents ground range, multi-look, and detected product type

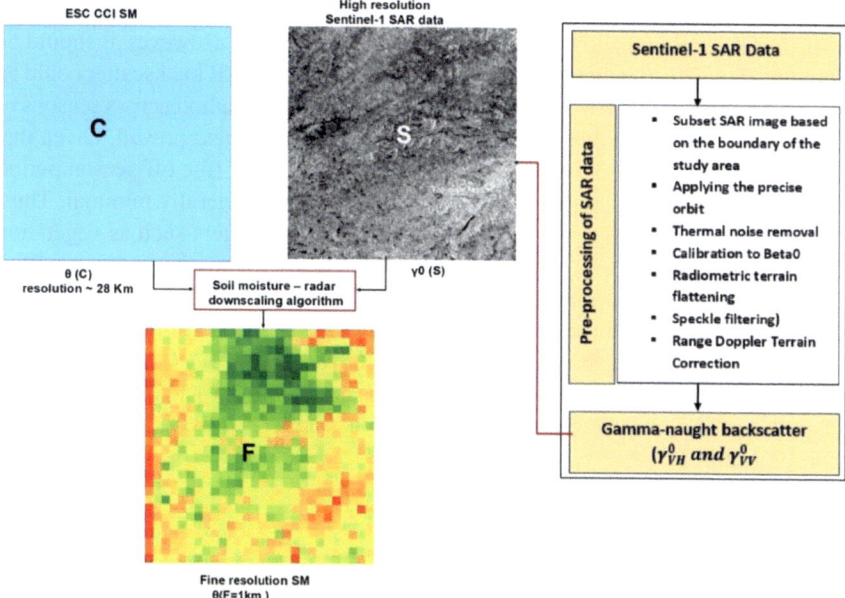

Fig. 22.2 General schematic flow to generate high-resolution volumetric surface soil moisture combining the coarse-resolution ESA CCI and the high-resolution Snetinel-1 SAR data. The ESA CCI soil moisture single grid cell with a coarse resolution of ~28 km (represented as **C**), the Sentinel-SAR data (represented as **S**) found within a grid cell of **C**, and the combined product (downscaled soil moisture) grid configuration of the finer scale (**F**) using a 1 km scale as an example

Sentinel Application Platforms (SNAP) open-source software (Fig. 22.2). The radiometric toolbox in SNAP was used to calibrate raw SAR data and convert SAR pixel values to the actual backscattering coefficient values of the scene. The effect of local incidence angle variation on the radar backscattering signal was reduced through applying a radiometric terrain flattening tool in SNAP. To downscale ECA CCI soil moisture, the Sentinel-1 SAR data were finally aggregated to a 100 m grid from their original resolution. A total of 9 temporal pairs of Sentinel-1 SAR and ESA CCI soil moisture datasets from the same day were collected over the two selected test sites.

Downscaling Algorithm

In this study, a downscaling algorithm developed by Das et al. (2011) was adopted. They have used this algorithm to downscale the SMAP soil moisture dataset using L-band SAR data. A summary of the algorithm is provided below.

When compared with surface soil moisture, which is variable, the state of land surface factors like vegetation and soil roughness changes very little over short periods of time. In this situation, there will be a considerable positive association between an increase (or reduction) in radar backscatter coefficients and a change in the surface soil moisture values or soil dielectric constant. However, it should be highlighted that the link between surface soil moisture and SAR backscatter could be impacted and be a source of inaccuracy in the downscaling method across seasons or places where fast variations in vegetation and surface roughness prevail. Given that this study was conducted from October 2016 to March 2017 (the off-season period of the study area), the practice of agricultural activities is generally minimal. Thus, we made assumptions that the effect of land surface parameters such as vegetation and soil roughness on SAR backscattering is low.

Given all these uncertainties, within the study area over a short period of time, the algorithm starts with the assumption that there are some functional relationships between volumetric soil moisture θ estimates (derived from ESA CCI soil moisture product in our case) and SAR co-polarized (Sentinel-1-1 VV) γ^0. Based on this, a hypothesis of a linear functional relationship is established over the same spatial scale (Das et al. 2014):

$$\theta = \alpha + \beta \cdot \gamma^0. \tag{22.1}$$

In Fig. 22.2, C represents the coarse-resolution (~28 km) ESA CCI, S is the high-resolution (resampled to 100 m) Sentinel-1 SAR γ^0, and F is the scale of the disaggregated (downscaled) volumetric soil moisture (i.e., 1 km, 250 m, and 100 m). The Sentinel-1 SAR data were aggregated at the footprint of C to obtain the radar backscatter values at the coarse scale. Equation (22.1) above at the scale of ESA CCI soil moisture C (~28 km) can be calculated as:

$$\theta(C) = \alpha(C) + \beta(C) \cdot \gamma^0_{vv}, (C) \tag{22.2}$$

where $\theta(C)$ is the volumetric soil moisture at the coarse resolution and $\gamma^0_{vv}(C)$ is the corresponding radar backscatter at the coarse resolution. The intercept $\alpha(C)$ and slope $\beta(C)$ parameters of the linear regression are estimated from the pairs of $\theta(C)$ and $\gamma^0_{vv}(C)$ time-series datasets over the same Earth grid.

In this study, it was imagined that the variation in local vegetation and soil roughness throughout the course of the study period has less effect on $\beta(C)$; therefore, the time-series pairs of $\theta(C)$ and $\gamma^0_{vv}(C)$ obtained from the study period (by removing grids with water bodies and permanent vegetation coverage) were combined to derive the $\alpha(C)$ and $\beta(C)$ values. To develop the ESA CCI volumetric surface soil moisture and radar-based algorithms, Eq. (22.1) above can also be theoretically evaluated at the scale of F (i.e., 1 km, 250 m, and 100 m).

$$\theta(F_i) = \alpha(F_i) + \beta(F_i) \cdot \gamma^0_{vv}(F_i), \tag{22.3}$$

where $\gamma^0_{vv}(F_i)$ is estimated from high-resolution Sentinel-1 SAR data. (i.e., average γ^0_{vv} at the scale of F). Here, θ is the unknown volumetric soil moisture value at scale F_i. $\theta(F_i)$ is the volumetric soil moisture value of a particular pixel "i" of the disaggregated scale F, and $\gamma^0_{vv}(F_i)$ is the corresponding radar backscatter value of pixel "i". Volumetric soil moisture at this scale is not available, and ESA CCI soil moisture is provided at the scale of C. Indeed, this is our target parameter to estimate, and it is called the downscaled (disaggregated) volumetric soil moisture (at resolutions of 1 km, 250 m, and 100 m). Thus, the first step in developing the downscaling algorithm is to subtract Eq. (22.2) from Eq. (22.3) (Das et al. 2014):

$$\theta(F_i) - \theta(C) = \{\alpha(F_i) - \alpha(C)\}$$
$$+ \{[\beta(F_i) \cdot \gamma_{vv}^0(F_i)] - [\beta(C) \cdot \gamma_{vv}^0(C)]\} \quad (22.4)$$

Since $\theta(F_i)$ is not available, we cannot directly estimate $\alpha(F_i)$ and $\beta(F_i)$ following the procedure at scale C in Eq. (22.2). To incorporate the effect of the variations in the slope and intercept parameters at the scale of F_i with respect to the coarse scale C, rewriting Eq. (22.4) algebraically as follows (Das et al. 2014):

$$\theta(F_i) = \theta(C) + \{\beta(C) \cdot [\gamma_{vv}^0(F_i) - \gamma_{vv}^0(C)]\}$$
$$+ \{[\alpha(F_i) - \alpha(C)] + [\beta(F_i) - \beta(C)] \cdot \gamma_{vv}^0(F_i)\} \quad (22.5)$$

The leftward of Eq. (22.5) $\theta(F_i)$ is the target variable, i.e., the disaggregated volumetric surface soil moisture at 1 km, 250 m, and 100 m F_i, $\theta(C)$ represents the ESA CCI volumetric surface soil moisture at ~28 km or scale C, the second term in the right hand of the Equation, $\{\beta(C) \cdot [\gamma_{vv}^0(F_i) - \gamma_{vv}^0(C)]\}$, calculated using regression parameter $\beta(C)$ from Eq. (22.2), and $[\gamma_{vv}^0(F_i) - \gamma_{vv}^0(C)]$ is also estimated using SAR backscatter values averaged to scales of F_i and C. The third term in the right hand of Eq. (22.5), $\{[\alpha(F_i) - \alpha(C)] + [\beta(F_i) - \beta(C)] \cdot \gamma_{vv}^0(F_i)\}$, describes for the heterogeneity of the parameters α and β within the coarse grid C, it is denoted in units of volumetric surface soil moisture and indicates the sub grid scale heterogeneity effects. High-resolution cross-polarization (VH) backscatter measurements, which are primarily sensitive to vegetation and surface characteristics, are also provided by the Sentinel-1 missions. The VH backscatter at scale F_i is a deviation from its coarse-scale aggregate $[\gamma_{vh}^0(C) - \gamma_{vh}^0(F_i)]$ and is an indicator of the sub grid heterogeneity in vegetation and soil roughness. This heterogeneity indicator $[\gamma_{vh}^0(C) - \gamma_{vh}^0(F_i)]$ can be further transferred to variation in VV backscatter by multiplying a sensitivity parameter Γ, which is defined as $\Gamma = \left[\frac{\delta \gamma_{vv}^0(F_i)}{\delta \gamma_{vh}^0(F_i)}\right] C$ for each particular grid cell C. The term $\Gamma \cdot [\gamma_{vh}^0(C) - \gamma_{vh}^0(F_i)]$ is the projection of the variation due to the heterogeneity in parameters α and β in the SAR VV polarization space. It can be converted to volumetric surface soil moisture units for use in Eq. (22.5) through multiplication by $\beta(C)$, i.e., $\beta(C) \cdot \Gamma \cdot [\gamma_{vh}^0(C) - \gamma_{vh}^0(F_i)]$. Therefore, the third term on the right of Eq. (22.5) can be approximated as (Das et al. 2014):

$$\{[\alpha(F_i) - \alpha(C)] + [\beta(F_i) - \beta(C)] \cdot \gamma_{vv}^0(F_i)\}$$
$$\approx \beta(C) \cdot \Gamma \cdot [\gamma_{vh}^0(C) - \gamma_{vh}^0(F_i)]. \quad (22.6)$$

The SAR-ESA CCI surface soil moisture downscaling algorithm is completed by substituting Eq. (22.6) for the third RHS term in Eq. (22.5) and can be written more compactly as (Das et al. 2014):

$$\theta(F_i) = \theta(C) + \beta(C) \cdot \left\{ \left[\gamma_{vv}^0(F_i) - \gamma_{vv}^0(C) \right] \right.$$
$$\left. + \Gamma \cdot \left[\gamma_{vh}^0(C) - \gamma_{vh}^0(F_i) \right] \right\}. \tag{22.7}$$

It is quite expected that the downscaled soil moisture estimates $\theta(F_i)$ are likely to have more noise than $\theta(C)$ due to the inherent errors in $\theta(C)$, $\gamma_{vv}^0(F_i)$, and $\gamma_{vh}^0(F_i)$ and the degree of uncertainty related to the estimation of the $\beta(C)$ parameter.

Statistical measures such the root-mean-square error (RMSE), bias, unbiased root-mean-square error (ubRMSE), and correlation coefficient (R) were used to assess how well the downscaling technique performed. The downscaled soil moisture data were compared against the in situ measured, AWS observed, and original ESA CCI soil moisture.

Results and Discussion

Multi-temporal soil moisture at 1 km, 250 m, and 100 m scales was generated by applying the downscaling algorithm to the coarse-resolution ESA CCI soil moisture. In this section, the main findings of the study and its associated discussion are presented as follows. We first provide the result of the estimation of the slope parameter. Then, a statistical comparison between the downscaled and reference soil moisture is presented and discussed. Finally, an overview of downscaled soil maps over the selected grid cells and study sites is provided.

Estimation of the Slope β Parameter

To estimate the $\alpha(C)$ and $\beta(C)$ parameters, the Sentinel-SAR data were first spatially aggregated to the scale of C (~28 km) and correlated with the ESA CCI dataset. Figure 22.3 shows the linear relationship between $\theta(C)$ and $\sigma_{vv}^0(C)$ of the study area, and a correlation (r) of 0.66 was found. A similar correlation analysis between SMAP soil moisture and Sentinel-1 SAR data at the scale of 36 km resulted in an r-value ranging from 0.49 to 89 (He et al. 2018). In their study, He et al. (2018) categorized the data pairs into different short window periods and generated a separate correlation coefficient value and slope parameter for each target window period to reduce the error introduced due to the change in the vegetation and soil roughness parameters. However, relatively low correlation values are still obtained over some target window periods (He et al. 2018), which could be associated with the low dynamic range in soil moisture over selected data window periods (Das et al. 2014). The correlation

Fig. 22.3 Scatter plot shows the correlation between ESA CCI soil moisture and the SAR backscatter coefficient at the scale of C

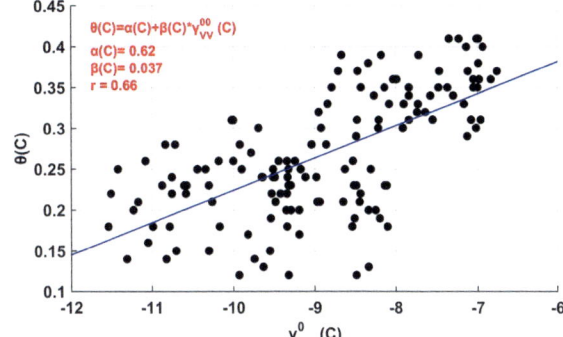

analysis made in this study assumed that the effect of land surface parameters on the correlation between ESA CCI soil moisture and Sentinel-1 SAR data is negligible. Thus, the r-value of 0.66 obtained in our study is optimal and comparatively good, which could indicate the potential of Sentinel-1 SAR data to downscale the ESA CCI soil moisture data.

The β estimated in this study is 0.037, which is within the range of the previous similar studies reported in the literature. For example, He et al. (2018) estimated β ranging from 0.031 to 0.093 over the correlation between SMAP soil moisture and Sentinel-SAR data. Velde et al. (2015) reported β ranging between 0.029 and 0.033 in a correlation between microwave-based soil moisture and L-band radar data. These results may indicate the reliability of the β estimated in this study even though previous studies demonstrated that the value of β is soil and vegetation condition dependent within the study period (Das et al. 2014; Wu et al. 2017).

Comparison Between ESA CCI and Downscaled Soil Moisture

The statistical relationship between the original ESA CCI and the downscaled soil moisture were assessed as presented in Fig. 22.4.

To compare the two estimates (i.e., the original and the downscaled soil moisture) first, the downscaled soil moisture (i.e., at 1 km, 250 m, and 100 m) was spatially aggregated to the resolution of the ESA soil moisture. The scatter plots in Fig. 22.4 show the relationship between ESA and the downscaled soil moisture at 1 km (a & d), 250 m (b & e), and 100 m (c & f), taking the data from November 11, 2016, and February 02, 2017, as an example. The statistical analysis showed great agreement between the downscaled and original ESA CCI soil moisture (Fig. 22.4). RMSE values ranging from 0.013 to 0.022 m³/m³ and r from 0.83 to 0.94 were obtained. The analysis indicates that the downscaled products perform nearly equally to the original ESA CCI soil moisture retrievals. In a similar study, Liu et al. (2021) compared the original ESA CCI with the downscaled soil moisture disaggregated using optical

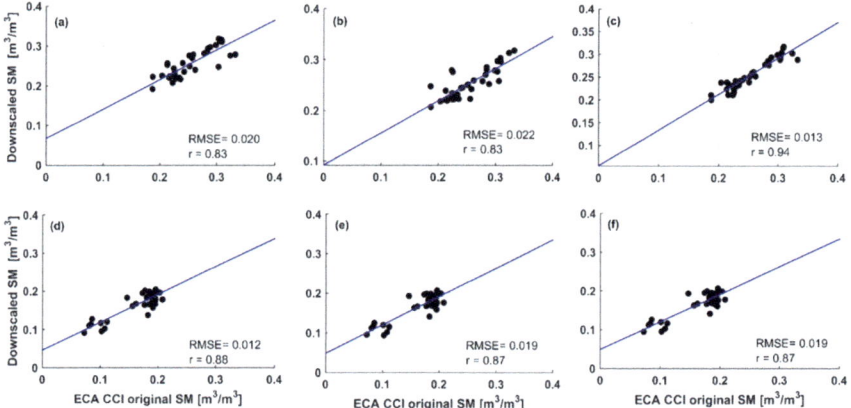

Fig. 22.4 Scatter plots between the original ESA CCI and the downscaled soil moisture (m³/m³) at 1 km (**a** & **d**), 250 m (**b** & **e**), and 100 m (**c** and **f**) on November 15, 2016, and February 02, 2017, respectively. The RMSE and r indicate the root-mean-square error and the correlation coefficients of these two soil moisture estimates

satellite-derived products and different linear and nonlinear downscaling methods and reported r-values ranging from 0.26 to 0.70 and RMSEs from 0.037 to 0.042 m³/m³ for the different methods. This may suggest the great performance and reliability of the downscaling algorithm used in this study in retaining the spatial distributions of soil moisture estimated for the coarse-scale ESA CCI product. The greater performance of the downscaling algorithm in this regard could be explained by the higher resolution and sensitivity of the SAR data used, which helped the algorithm properly disaggregate the coarse grid cells of ESA CCI.

Validation with the Observed Dataset

The soil moisture retrieval from ESA CCI and the downscaling algorithm were evaluated using both field-measured and AWS-observed soil moisture datasets (Fig. 22.1). The validation has been made to soil moisture estimates at the four scales (i.e., at the scale of ECA CCI ~28 km, at 1 km, 250 m, and 100 m). In the validation work, only grid cells with field-measured and AWS-observed soil moisture data were used. However, as shown in Table 22.3, we did not perform the required validation at the scale of ESA CCI and the 1 km downscaled soil moisture using field-measured soil moisture due to the lack of enough data pairs to perform the statistical analysis. Thus, an attempt has been made to perform a validation analysis at all four scales using the soil moisture observed at the selected AWS. From the data pairs, the bias, coefficient of correlation (r), root-mean-square error (RMSE), and unbiased root-mean-square error (ubRMSE) were estimated and are presented in Tables 22.2 and 22.3.

Table 22.2 Statistical validation of soil moisture estimates using AWS observed datasets

Scale	RMSE (m³/m³)	Bias (m³/m³)	ubRMSE (m³/m³)	r	N
28 km (ECA SM)	0.14	−0.08	0.11	0.64	39
1 km	0.21	−0.18	0.09	0.51	12
250 m	0.20	−0.18	0.10	0.57	12
100 m	0.20	−0.18	0.09	0.61	12

Table 22.3 Statistical validation of the downscaled soil moisture estimates using the field-measured datasets

Scale	RMSE (m³/m³)	Bias (m³/m³)	ubRMSE (m³/m³)	r	N
~28 km (ESA SM)	–	–	–	–	
1 km	–	–	–	–	
250 m	0.069	0.066	0.022	0.57	11
100 m	0.069	0.062	0.032	0.22	11

Comparison with AWS-observed datasets: Both the original ESA CCI soil moisture and downscaled results are generally in good agreement with the AWS-observed soil moisture at scales of ~28 km, 1 km, 250 m, and 100 m. The microwave satellite product at ~28 km was well-correlated with the AWS-observed soil moisture, with an *r*-value of 0.66. This value of *r* is comparable to the findings of Warner et al. (2021), who compared the original ESA CCI soil moisture with ground-observed soil moisture datasets and reported a correlation value of 0.69 in the USA. However, in this study, we could not obtain the RMSE of 0.031 m³/m³ reported by Warner et al. (2021). In our study, a dry bias of 0.08 m³/m³, RMSE of 0.14 m³/m³, and ubRMSE of 0.11 m³/m³ were observed, which are higher than the errors reported by Warner et al. (2021).

Indeed, this is not thought to have been triggered by an error from downscaling algorithm. This is probably because there is a significant discrepancy between the soil depth used for AWS observations (0–20 cm) and the soil layer that is sensitive to the microwave backscatter signal (the top 5 cm). Measuring reference soil moisture data at deeper layers than the penetration depth of radar could lead to underestimation of microwave-based soil moisture retrieval and of the downscaled soil moisture estimates (Dente et al. 2012). This could in turn contribute to the large RMSE compared to previous similar studies. Although the soil moisture accuracy seems reduced as the spatial resolution becomes finer, for example, *r* of 0.66 for ESA CCI to 0.51 at the 1 km scale, RMSE of 0.14 m³/m³ for ESA CCI to 0.21 m³/m³ for the 1 km scale, a better ubRMSE of 0.09, 0.10, and 0.09 m³/m³ were obtained for the downscaled soil moisture estimates at 1 km, 250 m, and 100 m, respectively. Instead, the downscaled soil moisture at 100 m has relatively better accuracy than that of the other disaggregated soil moisture estimates (i.e., 1 km and 250 m scales) over the AWS observed soil moisture (Table 22.2).

Given that the reference soil moisture used in this validation was measured at 0–20 cm soil depths, the correlation coefficient and ubRMSE values obtained are encouraging and could show the potential of the downscaling algorithm to disaggregate the coarse-resolution ESA CCI soil moisture estimate to a finer scale. Thus, a detailed validation of the downscaling algorithm using field-measured soil moisture (i.e., over the top 5 cm depths of soil) is necessary to further understand its performance and applicability. Accordingly, an attempt has been made to validate the accuracy of the downscaled soil moisture at the scales of 250 m and 100 m with the available (although small) in situ soil moisture measured at 0–5 cm depths of soil. The results and discussion are provided in the next section.

Comparison with in situ observed datasets: The statistical analysis between the field-measured data and the downscaled soil moisture at 250 m showed an encouraging result with an RMSE of 0.069 m^3/m^3, a bias of 0.066 m^3/m^3, an ubRMSE of 0.022 m^3/m^3, and an r of 0.57 (Table 22.3). These results are comparable to the findings of similar studies (e.g., He et al. 2018; Liu et al. 2021; Knipper et al. 2017; Li et al. 2018; Kovačević et al. 2020; Warner et al. 2021). For example, using a similar downscaling algorithm and Sentinel-1 SAR data, He et al. (2018) disaggregated the SMAP 36 km soil moisture product to a fine scale of 1 km with an RMSE of 0.092 cm^3/cm^3, and r of 0.66. Liu et al. (2021) and Kovačević et al. (2020) downscaled the original ESA CCI soil moisture to a 1 km scale with RMSEs of 0.023 m^3/m^3 and 0.0518 m^3/m^3 and r of 0.66, respectively, using the downscaling factors derived from optical satellites and different linear and nonlinear downscaling methods.

Although the downscaled soil moisture at the 100 m scale showed a relatively poor correlation with the field-measured data ($r = 0.22$) compared to the 250 m resolution, in terms of RMSE (0.069 m^3/m^3), bias (0.062 m^3/m^3), and ubRMSE (0.032 m^3/m^3), the 100 m scale soil moisture still showed a comparable result (Table 22.3). The ubRMSE values obtained at both 100 m (0.032 m^3/m^3) and 250 m (0.022 m^3/m^3) resolutions meet the target accuracy of ubRMSE 0.040 m^3/m^3 set for the SMAP baseline downscaling algorithm implemented in this study. The wet bias values obtained at the downscaled soil moisture (for both 250 m and 100 m scales) may indicate a slight overestimation of the disaggregated product to the soil moisture values measured in the field.

Generally, from results in Table 22.3, it seems that the finer the spatial resolution of the downscaled product the lower its accuracy in estimating field-measured soil moisture values. Although not consistent, this characteristic has also been reflected in Table 22.2. According to Wu et al. (2015) and He et al. (2018), the accuracy of the downscaled soil moisture product decreased as the spatial resolution became finer. This is probably due to the increased heterogeneity and high noise levels in the radar data at fine scales as well as the impact of surface characteristics like soil roughness on Sentinel-1 SAR backscattering. Thus, it is logical to assume that while downscaling soil moisture to increasingly finer spatial resolution, the retrieval accuracy would decline. With more validation datasets, the statistical trend shown in Table 22.3 may give some clue to the possibility of achieving better downscaling accuracy with the 1 km scale soil moisture estimates.

Soil Moisture Maps

The soil moisture maps for both the original ESA CCI and the downscaled soil moisture were developed to demonstrate the spatial and temporal variability of soil moisture at different scales over the study area (Figs. 22.5, 22.6, 22.7 and 22.8). The first two figures (i.e., Figs. 22.5 and 22.6) show the soil moisture maps produced over a specific grid cell to pinpoint individual pixels and closely compare the original and downscaled soil moisture. Figures 22.7 and 22.8 illustrate soil moisture maps of the original and downscaled soil moisture for the two sites of interest over the study periods. In this study, the typical range of volumetric soil moisture values was set from 0.02 to 0.60 m^3/m^3, as used by He et al. (2018). The soil moisture maps in Figs. 22.5 and 22.6 show that the algorithm has effectively captured the wet and dry soil moisture conditions of the study area shown over the selected grid cells of the coarse ESA soil moisture and helped to reproduce it at a higher spatial scale of 1 km, 250 m, and 100 m. The downscaled maps retained the spatial pattern of the coarse-resolution ESA CCI and captured the details better.

A similar spatial pattern of soil moisture was observed over soil moisture maps produced over the two selected sites (Figs. 22.7 and 22.8). The downscaled soil moisture reproduced the spatial and temporal variability of soil moisture observed over the coarse resolution with significantly improved spatial details within the ESA CCI footprints (Figs. 22.7 and 22.8). As seen from the same figures (Figs. 22.7 and 22.8), the soil moisture maps have generally shown a decreasing trend from October 2016 to March 2017 and generally follow the meteorological conditions of the study area. These months are indeed categorized as the off-season periods of the study area

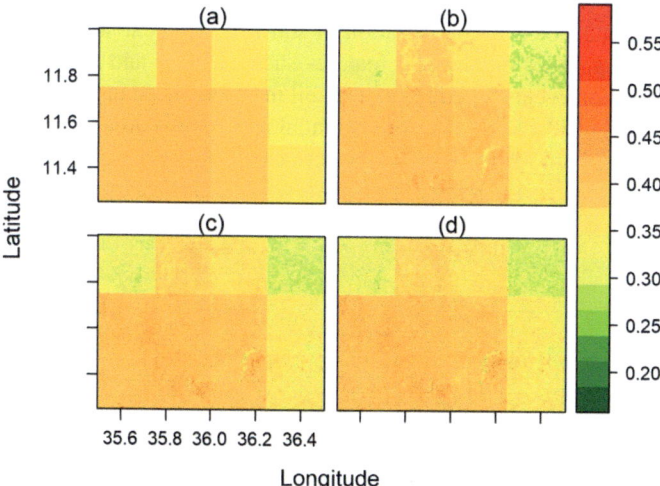

Fig. 22.5 Example of soil moisture maps over selected grid cells on October 22, 2016. **a** ESA CCI soil moisture at a grid cell of ~28 km, **b** downscaled soil moisture at 1 km resolution, **c** downscaled soil moisture at 250 m resolution, and **d** downscaled soil moisture at 100 m resolution

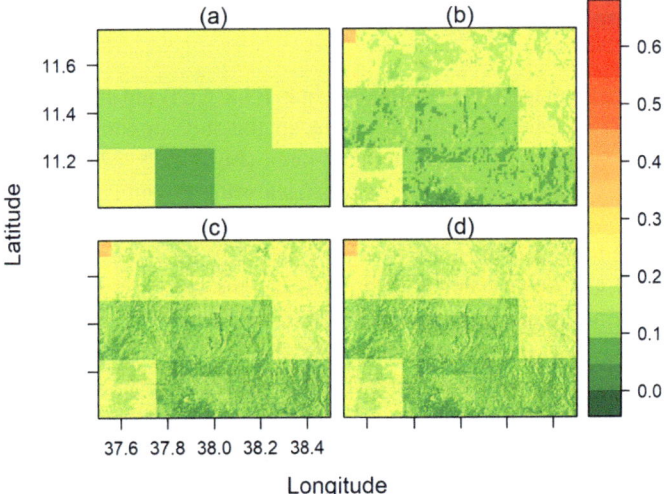

Fig. 22.6 Example of soil moisture maps over selected grid cells on January 09. **a** ESA CCI soil moisture at a grid cell of ~28 km, **b** downscaled soil moisture at 1 km resolution, **c** downscaled soil moisture at 250 m resolution, and **d** downscaled soil moisture at 100 m resolution

when there are no considerable rainfall events, and the amount of soil moisture will consistently decrease until the next wet season starts around mid-May.

A short rainy period (called "*Belg*") occurs from approximately February to May (Cheung et al. 2008; Conway, 2000) in the study area; however, it is obvious that this decreasing trend may not be true for some periods, such as February 26 and March 03, 2017 (Fig. 22.7). The spatial variation in soil moisture over the study area could also be attributed to the variation in soil texture, local surface temperature, irrigation events, other biophysical features such as slope and tillage pattern, and local precipitation events. In addition, the soil moisture maps in Figs. 22.7 and 22.8 also show good residual soil moisture potential in the study area following the main rainfall seasons. Particularly in the early periods of October and November, a good amount of residual soil moisture that could support additional cropping has been observed.

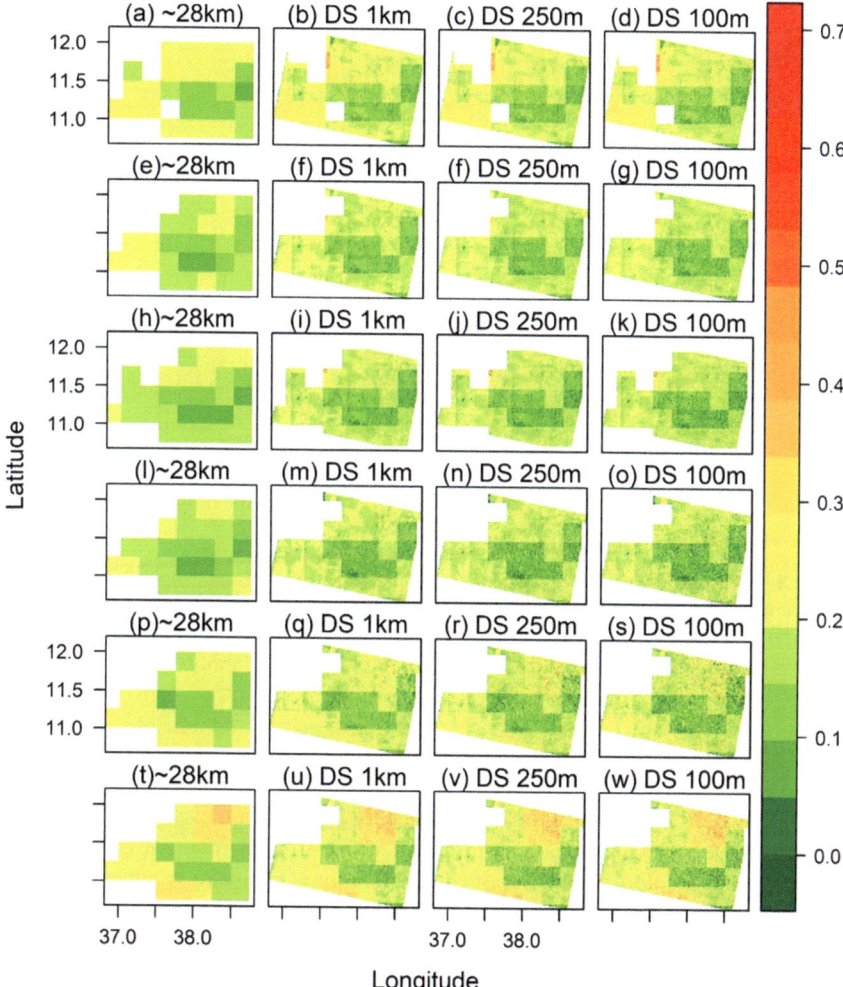

Fig. 22.7 Illustrates the ESA original and downscaled (DS) soil moisture maps of Site 1 over the study period. Figures (**a**) to (**d**) present the soil moisture maps for the date of December 16, 2016; figures (**e**) to (**g**) for the date of January 09, 2017; figures (**h**) to (**k**) for the date of February 02, 2017; figures (**i**) to (**o**) for the date February 14, 2017; figures (**p**) to (**s**) for the date February 26, 2017; and figures (**t**) to (**w**) for March 10, 2017 at the ESA CCI ('28 km), 1 km, 250 m, and 100 m resolution, respectively. The white regions (or pixels) in the soil moisture map could be because those areas are outside the coverage of SAR data, have missing values, and mask out cells with waterbodies and dense vegetation from ESA soil moisture. Some of those white pixels could also be because of the failure of the downscaling algorithms to properly estimate the soil moisture values

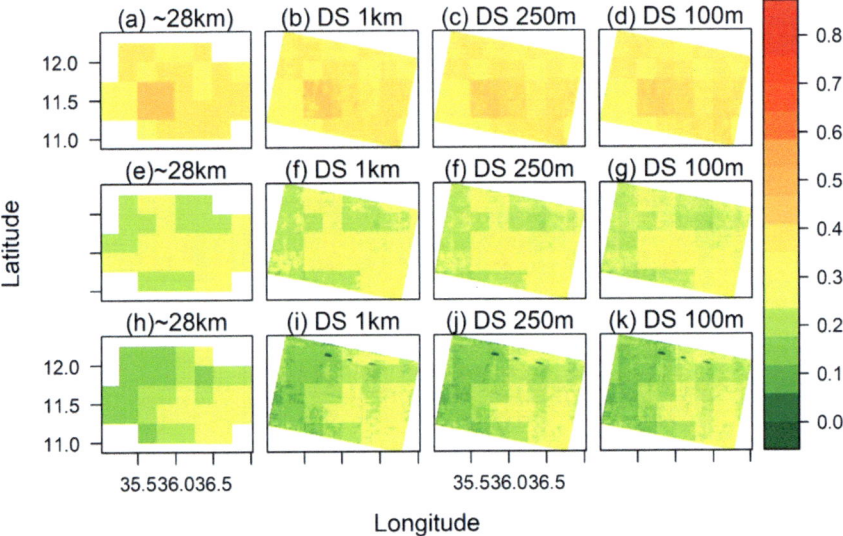

Fig. 22.8 Illustrated the ESA original and downscaled (DS) soil moisture maps of Site 2 over the study period. Figures (**a**) to (**d**) present the soil moisture maps for the date of October 22, 2016; figures (**e**) to (**g**) for the date of November 15, 2016; and figures (**h**) to (**k**) for the date of December 09, 2016, at ESA CCI ('28 km), 1 km, 250 m, and 100 m resolution, respectively

Conclusion

Although a good number of global soil moisture products are available, their application and use at local scales are very limited due to their inherent coarse spatial resolution. As a solution to most of these global soil moisture products, a downscaling algorithm that disaggregates the coarse resolution to a finer scale has been introduced for regions where there are relatively dense soil moisture observation networks. However, in other regions, such as the African continent, the lack of large and demonstrative in situ soil moisture observation networks across different agroclimatic zones of the continent still hampers the assessment and validation of most of the promising remote sensing-based soil moisture products and cannot benefit from the downscaling of these coarse-scale products for different operational applications at local scales.

This study has been initiated with the aim of testing the promising downscaling algorithm and the freely available high-resolution Sentinel-1 SAR data to disaggregate the coarse-resolution ('28 km) ESA CCI soil moisture datasets. The coarse-resolution ESA CCI soil moisture has been downscaled to a higher resolution soil moisture estimate (i.e., 1 km, 250 m, and 100 m) using the Sentinel-1 SAR data and the SMAP baseline downscaling approach. The downscaled product was validated using the available field-observed soil moisture datasets; thus, this study could be considered a showcase for sparse soil moisture network regions.

A comparison was made between the AWS observed and downscaled soil moisture results in the ubRMSE (r) of 0.090 m^3/m^3 (0.51), 0.100 m^3/m^3 (0.57), and 0.090 m^3/m^3 (0.61) at 1 km, 250 m, and 100 m resolutions, respectively. The comparison with plot-based in situ measurements reveals ubRMSE (r) values of 0.022 m^3/m^3 (0.57) and 0.032 m^3/m^3 (0.22) for the downscaled soil moisture at 250 m and 100 m, respectively. Although the validation has been made using a small number of observed datasets, the downscaled soil moisture estimates are in good agreement with the in situ measured and AWS observed soil moisture. Thus, the result obtained in this study is quite promising and indicates the potential of the Sentinel-1 SAR data and the downscaling algorithm to disaggregate the coarse-resolution soil moisture to finer scales. However, the methodology still needs further study and validation work with many measured datasets and over a longer temporal period.

Acknowledgements The authors are grateful to the National Meteorological Institute (NMI) of Ethiopia for granting soil moisture datasets observed over AWS found in the Abbay River Basin. The SAR data and ESA CCI soil moisture product are also quite important; therefore, we are grateful to the ESA for supplying them. We acknowledge the funding from the CGIAR Sustainable Intensification of Mixed Farming System (SI-MFS) initiative and the International Development Association (IDA) of the World Bank to the Accelerating Impact of CGIAR Climate Research for Africa (AICCRA) project.

References

AghaKouchak A, Farahmand AM, Melton FS, Teixeira JP, Anderson MC, Wardlow BD, Hain CR (2015) Remote sensing of drought: Progress, challenges, and opportunities. Rev Geophys 53:452–480

Albergel C, Rüdiger C, Carrer D, Calvet JC, Fritz N, Naeimi V, Bartalis Z, Hasenauer S (2009) An evaluation of ASCAT surface soil moisture products within situ observations in Southwestern France. Hydrol Earth Syst Sci 13:115–124

Anderson M, Norman JM, Mecikalski JR, Otkin JA, Kustas WP (2007) A climatological study of evapotranspiration and moisture stress across the continental United States based on thermal remote sensing: 2. Surface moisture climatology. J Geophys Res Atmos, 112, D11112

Ayehu G, Tadesse T, Gessesse B, Yigrem Y, Melesse AM (2020) Combined use of sentinel-1 sar and landsat sensors products for residual soil moisture retrieval over agricultural fields in the upper blue nile basin, ethiopia. Sensors 20(11):1–25. https://doi.org/10.3390/s20113282

Chauhan NS, Miller S, Ardanuy P (2003) Spaceborne soil moisture estimation at high resolution: a microwave-optical/IR synergistic approach. Int J Remote Sens 24(22):4599–4622

Cheung WH, Senay GB, Singh A (2008) Trends and spatial distribution of annual and seasonal rainfall in Ethiopia. Int J Climatol 28:1723–1734

Choi M, Hur Y (2012) A microwave-optical/infrared disaggregation for improving spatial representation of soil moisture using AMSR-E and MODIS products. Remote Sens Environ 124:259–269

Conway D (2000) Some aspects of climate variability in the northeast Ethiopian highlands-Wollo and Tigray. Ethiopia J Sci 23:139–161

CSA (Central Statistical Agency) (2007) The 2007 population and housing census of Ethiopia: statistical report at country level in, Edited by Commission PC

Das NN, Entekhabi D, Njoku EG (2011) An algorithm for merging SMAP radiometer and radar data for high-resolution soil moisture retrieval. IEEE Trans Geosci Remote Sens 49(5):1504–1512

Das NN, Entekhabi D, Njoku EG, Shi JJC, Johnson JT, Colliander A (2014) Tests of the SMAP combined radar and radiometer algorithm using airborne field campaign observations and simulated data. IEEE Trans Geosci Remote Sens 52(4):2018–2028

Dente L, Su Z, Wen J (2012) Validation of SMOS soil moisture products over the Maqu and Twente Regions. Sensors 12:9965–9986

Djamai N, Magagi R, Goïta K, Merlin O, Kerr Y, Roy A (2016) A combination of DISPATCH downscaling algorithm with CLASS land surface scheme for soil moisture estimation at fine scale during cloudy days. Remote Sens Environ 184:1–14

Dobriyal P, Qureshi A, Badola R, Hussain SA (2012) A review of the methods available for estimating soil moisture and its implications for water resource management. J Hydrol 458:110–117

Dorigo WA, Wagner W, Hohensinn R, Hahn S, Paulik C, Xaver A, Gruber A, Drusch M, Mecklenburg S, Oevelen PV (2015) The international soil moisture network: a data hosting facility for global in situ soil moisture measurements. Hydrol Earth Syst Sci 15:1675–1698

Dorigo W et al (2017) ESA CCI soil moisture for improved earth system understanding state-of-the art and future directions. Remote Sens Environ 203:185–215

He L, Hong Y, Wu X, Ye N, Walker JP, Chen X (2018) Investigation of SMAP active—passive downscaling algorithms using combined Sentinel-1 SAR and SMAP radiometer data. IEEE T Geosci Remote 56:4906–4918. https://doi.org/10.1109/TGRS.2018.2842153

Kawanishi T, Sezai T, Ito Y, Imaoka K, Takeshima T, Ishido Y, Shibata A, Miura M, Inahata H, Spencer RW (2003) The advanced microwave scanning radiometer for the earth observing system (AMSR-E), NASDA's contribution to the EOS for global energy and water cycle studies. Geosci Remote Sens IEEE Trans 41:184–194

Kerr YH, Waldteufel P, Wigneron JP, Martinuzzi JM, Font J, Berger M (2002) Soil moisture retrieval from space: the Soil Moisture and Ocean Salinity (SMOS) mission. IEEE Trans Geosci Remote Sens 39:1729–1735

Knipper KR, Hogue TS, Franz KJ, Scott RL (2017) Downscaling SMAP and SMOS soil moisture with moderate resolution imaging spectroradiometer visible and infrared products over southern Arizona. J Appl Remote Sens 11:026021

Kornelsen C, Coulibaly P (2014) Root-zone soil moisture estimation using data-driven methods. Water Resour. Res 50:2946–2962

Kovačević J, Cvijetinović Ž, Stančić N, Brodić N, Mihajlović D (2020) New downscaling approach using ESA CCI SM products for obtaining high resolution surface soil moisture. Remote Sens 12(7):1119. https://doi.org/10.3390/rs12071119

Li L, Njoku EG, Im E, Chang PS, Germain KS (2004) A preliminary survey of radio-frequency interference over the US in Aqua AMSR-E data. IEEE Trans Geosci Remote Sens 42:380–390

Li J, Wanga S, Gunnb G, Joossec P, Russell HAJ (2018) A model for downscaling SMOS soil moisture using Sentinel-1 SAR data. Int J Appl Obs Geoinformation 72:109–121

Liu JG (2000) Smoothing filter-based intensity modulation: a spectral preserve image fusion technique for improving spatial details. Int J Remote Sens 21(18):3461–3472

Liu YY, Parinussa RM, Dorigo WA, De Jeu RAM, Wagner W, van Dijk AIJM, McCabe MF, Evans JP (2011) Developing an improved soil moisture dataset by blending passive and active microwave satellite-based retrievals. Hydrol Earth Syst Sci 15:425–436

Liu Y, Zhu KL, Lai X, Wang J (2021) Downscaling of ESA CCI soil moisture in Taihu Lake Basin: are wetness conditions and nonlinearity important? Water & Climate Change 12(5):1564–1579

Malbéteau Y, Merlin O, Molero B, Rüdiger C, Bacon S (2016) Dis- PATCh as a tool to evaluate coarse-scale remotely sensed soil moisture using localized in situ measurements: application to SMOS and AMSR-E data in Southeastern Australia. Int J Appl Earth Observ Geoinf 45:221–234

Merlin O, Rudiger C, Bitar AA, Richaume P, Walker JP, Kerr YH (2012) Disaggregation of SMOS soil moisture in Southeastern Australia. IEEE Trans Geosci Remote Sens 50(5):1556–1571

Muñoz-Sabater J, Al Bitar A, Brocca L (2016) Chapter 18: soil moisture retrievals based on active and passive microwave data: state-of-the-art and operational applications. In: Srivastava KP, Petropoulos PG, Yann H, Kerr HY (eds) Satellite soil moisture retrieval, Elsevier

Pathe C, Wagner W, Sabel D, Doubkova M, Basara JB (2009) Using ENVISAT ASAR global mode data for surface soil moisture retrieval over Oklahoma, USA. IEEE Trans Geosci Remote Sens 47:468–480

Peng J, Niesel J, Loew A (2015) Evaluation of soil moisture downscaling using a simple thermal-based proxy—the REMEDHUS network (Spain) example. Hydrol Earth Syst Sci 19(12):4765–4782

Piles M et al (2011) Downscaling SMOS-derived soil moisture using MODIS visible/infrared data. IEEE Trans Geosci Remote Sens 49(9):3156–3166

Piles M, Petropoulos GP, Sanchez N, Gonzalez-Zamora A, Ireland G (2016) Towards improved spatiotemporal resolution soil moisture retrievals from the synergy of SMOS and MSG SEVIRI spaceborne observations. Remote Sens Environ 180:403–417

Preimesberger W, Scanlon T, Su CH, Gruber A, Dorigo W (2021) Homogenization of structural breaks in the global ESA CCI soil moisture multisatellite climate data record. IEEE Trans Geosci Remote Sens 59(4):2845–2862

Qin J, Liang S, Yang K, Kaihotsu I, Liu R, Koike T (2009) Simultaneous estimation of both soil moisture and model parameters using particle filtering method through the assimilation of microwave signals. J. Geophys. Res. 114:1–13

Qiu J, Gao Q, Wang S, Su Z (2016) Comparison of temporal trends from multiple soil moisture datasets andprecipitation: the implication of irrigation on regional soil moisture trend. Int J Appl Earth Observ Geoinf 48:17–27

Robinson DA, Campbell CS, Hopmans JW, Hornbuckle BK, Jones SB, Knight R, Ogden F, Selker J, Wendroth O (2008) Soil moisture measurement for ecological and hydrological watershed-scale observatories: a review. Vadose Zone J 7:358–389

Srivastava PK, Han D, Ramirez MR et al (2013) Machine learning techniques for downscaling SMOS satellite soil moisture using MODIS land surface temperature for hydrological application. Water Resour Manage 27:3127–3144

Sun H, Cui Y (2020) Evaluating downscaling factors of microwave satellite soil moisture based on machine learning method. Remote Sens 12:133

Sun H, Cai C, Liu H, Yang B (2019) Microwave and meteorological fusion: a method of spatial downscaling of remotely sensed soil moisture. IEEE J Sel Top Appl Earth Obs Remote Sens 12:1107–1119

Ulaby FT, Moore RK, Fung AK (1986) Microwave remote sensing: active and passive, vol 2. Radar remote sensing and surface Scattering and emission theory. Ch. 12, Artech House Publishers, Norwood, pp 962–966

Velde RVD, Salama MS, Eweys OA, Wen J, Wang Q (2015) Soil moisture mapping using combined active/passive microwave observations over the east of The Netherlands. IEEE J Sel Topics Appl Earth Observ Remote Sens 8(9):4355–4372

Wagner W, Dorigo W, de Jeu R, Fernandez D, Benveniste J, Haas E, Ertl M (2012) Fusion of active and passive microwave observations to create an essential climate variable data record on soil moisture. ISPRS Ann Photogramm Remote Sens Spat Inf Sci 7:315–321

Wang S, McKenney DW, Shang J, Li J (2014) A national-scale assessment of long-term water budget closures for Canada's watersheds. J Geophys Res Atmos 119:8712–8725

Warner DL, Guevarab M, Callahana J, Vargas R (2021) Downscaling satellite soil moisture for landscape applications: a case study in Delaware, USA. J Hydrol: Reg Stud 38:100946

Wu X, Walker JP, Rüdiger C, Panciera R (2015) Effect of landcover type on the SMAP active/passive soil moisture downscaling algorithm performance. IEEE Geosci Remote Sens 12(4):846–850

Wu X, Walker JP, Das NN, Panciera R, Rüdiger C (2017) Evaluation of the SMAP brightness temperature downscaling algorithm using active–passive microwave observations. Remote Sens Environ 155:210–221

Part VII
Future Research and Development Direction in the Basin

Chapter 23
Interactive Web Tool for Mapping Soil Organic Carbon Dynamics

Worku Zewdie, Degefie Tibebe, Abraraw Assefa, Assefa Abegaz, Berhan Gessesse, Ashenafi Ali, Wuletawu Abera, Lulseged Tamene, and Amsalu Tilaye

Abstract Soil is a fundamental constituent for plant growth and plays a vital role in carbon sequestration. However, poor land use management practices affect soil health. The conversion of forestland to other land uses, erosion, flooding, and other anthropogenic activities resulted in losses of carbon sequestrated in the above-ground trees and in the soils. Mapping and continuous monitoring of soil organic carbon improve our understanding to know the status of soils for remedial actions. Conventional soil carbon monitoring approaches involved long-term agricultural experiments or a snapshut field-based soil sample collection and laboratory analysis. Such approaches are time-consuming and laborious, costly, take too long to get relevant results within a short time frame; has limited spatial and temporal coverage, and also, it is difficult to extrapolate site-specific results for larger spatial coverage and prediction. Remote sensing data and tools contribute a lot to filling the gaps observed in conventional soil monitoring approaches. Several environmental covariates can be used from satellite imageries to model and map the spatio-temporal dynamics of soil properties including soil organic carbon dynamics. This chapter discusses the tools for web-based interactive mapping and monitoring of modeled soil organic carbon dynamics (for fifty years; 2022–2072) of cultivated and grazing lands in the landscape of Abbay Basin. The users can register on the web to obtain both datasets based on their area of interest from the Kebele to microwatershed levels over the years. This interactive web map plays a significant role in easily availing carbon content information for researchers and policymakers for making better decisions.

W. Zewdie (✉) · B. Gessesse
Remote Sensing Department, Ethiopian Space Science and Geospatial Institute, Addis Ababa, Ethiopia
e-mail: wzewdie24@gmail.com

D. Tibebe · W. Abera · L. Tamene · A. Tilaye
International Center for Tropica Agriculure (CIAT), Addis Ababa Office, Addis Ababa, Ethiopia

A. Assefa
Artificial Intelligence Institute, Addis Ababa, Ethiopia

A. Abegaz · A. Ali
Geography and Environmental Studies, Addis Ababa University, Addis Ababa, Ethiopia

© The Author(s), under exclusive license to Springer Nature Switzerland AG 2025
A. Melesse et al. (eds.), *Abbay River Basin*, Springer Geography,
https://doi.org/10.1007/978-3-031-65241-7_23

Keywords Soil carbon · Carbon sequestration · Remote sensing · Abbay Basin · Interactive web map

Introduction

Abbay Basin is among the Ethiopian basins with significant land use dynamics resulting in land degradation and losses of soil carbon (SC). A wide range of sustainability indicators for agricultural production systems has been used among which soil organic carbon (SOC) is paramount (Abegaz et al. 2022; Abegaz and van Keulen 2009). In line with, soils are known as being crucial in the exchange of carbon between the earth's surface and the atmosphere (Tao et al. 2016) and can characterize both carbon sinks and sources through forest growth and land degradation, respectively (Dixon et al. 1994). Population growth, diminishing land holdings and subsistence agriculture, and climate variability enforce competition with the remnant natural resources (Erkossa et al. 2009). Several attempts have been made to curb land degradation and maintenance of soil fertility through various measures that include afforestation and the construction of different types of physical structures for soil and water conservation purposes. However, the area is still exposed to serious agricultural land degradation, low use of technological inputs, and loss of vegetation cover (Erkossa et al. 2009).

Soil is the fundamental constituent of the terrestrial ecosystem that supplies the essential water and nutrients required for plant growth (Gavrilescu 2021). It also supports food production, sustains water quality, and plays a role as carbon sinks (Lal 2004). The land use change has significantly altered the constituent and structure of soils. Flooding and overexploitation of the topsoil surface remove the essential soil nutrients required for plant growth. Erosion and overgrazing also significantly impact the availability of SOC (Abegaz et al. 2020). Soil carbon management contributes to soil fertility, enhances soil health, and supports agricultural productivity. Moreover, soil carbon management supports carbon sequestration from the atmosphere to mitigate the greenhouse effect (Lal 2016).

Climate, soil type and management practices potentially impact the SOC dynamics (Coonan et al. 2019). The fertility and productivity of the soil are significantly influenced by the dynamics of the SC. Accordingly, the quantity of SOC significantly influences the growth and productivity of crops, because soil fertility is strongly linked to SOC content, by positively influencing soil physical, chemical, and biological properties (Abegaz and van Keulen 2009).

Nevertheless, there is a significant spatial variation in soil organic carbon among the land use types resulting from the effect of different agricultural management practices (Abegaz et al. 2020).Therefore, information on long-term SOC dynamics in relation to different management options is required to assess and evaluate their sustainability, which can support decision-making (Abegaz and van Keulen 2009).

Mapping and monitoring of the condition of soils using the conventional approach provide accurate information, but covering larger areas is costly and time-consuming

(Viscarra Rossel et al. 2016). Moreover, conventional SOC monitoring approaches involved long-term agricultural experiments or a snapshut field-based soil sample collection and laboratory analysis. Such approaches take too long to get relevant results within the time frame of present priorities; has limited spatial and temporal coverage, and also, it is difficult to extrapolate site-specific results for larger area prediction (Abegaz and van Keulen 2009). Besides, certain soil properties vary across space and time and are more difficult to incorporate into mapping and monitoring tools and also a challenge to map bigger areas employing conventional soil mapping methods. The use of earth observation data and their derived products improves modeling, mapping, and monitoring of long-term dynamics of SOC in relation to different land management options by providing continuous data on a regional to a global scale within a short time frame.

Earth observation data provide information about the soil properties by detecting the soil reflectance or emitted energy without direct contact with the soil surface that can be measured. The information can be obtained through ground spectral sensing and space spectral sensing instruments. These can be achieved by examining the spectral curve measured from the orbital sensors, the spectral curve of soil samples at the laboratory or measuring spectral reflectance in the field.

Several studies conducted on monitoring soil nutrient status (for example Abebe et al. 2020; Abegaz et al. 2020) and developing digital soil maps, but they have limitations in visualizing their outputs in digital interactive web-mapping and monitoring system to communicate to end-users. Hence, this chapter aims to deliver an interactive tool mapping and monitoring of modeled long-term dynamics of SOC as influenced by different farm management practices, aiming at supporting adoption of appropriate management practices that eventually might lead to sustainable agricultural land management (Abegaz et al. 2020; Abegaz and van Keulen 2009). The work is restricted to cultivate and grazing lands of the basin. The work is significant in providing scientific evidences in a format that decision-makers and professionals can utilize for making informed decisions related to land management.

Determination of Soil Organic Carbon Stocks

In most parts of the Abbay Basin, the crop residues are usually collected for fuelwood and animal feed. Amanuel et al.(2018) collected soil samples from four land use types (natural forest, plantation forest, other woodland, and grassland and cultivated land) to assess the variations in SOC stock among the land use types using a laboratory soil analysis approach. Accordingly, they obtained a higher overall soil organic carbon stock under natural and mixed forest land use systems. The spatial and temporal dynamics of SOC in the four land use types (cropland, grassland, shrubland, and forestland) in the Upper Blue Nile were also assessed using field soil sampling (Abegaz et al. 2016). The SOC showed a significant variation at the depth of 0–20 cm and also among the land use types. Abebe et al.(2020) also evaluated the variation of SOC and total nitrogen (TN) across the land use types (bushland, cropland, grazing land, and plantation) and topographic position collecting soil samples. The investigation obtained a significant variation in SOC and TN stocks among the land use types and topographic positions. A higher stock is obtained in bushland compared to other land use types with lower stocks at croplands. However, soil sampling and laboratory analysis to predict carbon stock are costly, time-consuming, also have limited coverage to specific areas, and are not possible to predict SOC stock in areas that are inaccessible to humans. Hence, satellite data play a significant role in monitoring carbon stocks and prediction of SOC.

The remote sensing (RS) approach lessens the concern associated with time, cost, and environmental pollution caused during soil sample collection and chemical soil laboratory analysis. The RS techniques are considered rapid, cost-effective, and non-destructive for the estimation of soil types and mapping (Vaudour et al. 2016). However, the space-based and aerial-based sensing of soil constituents has a lower signal-to-noise ratio due to the influence of atmospheric constituents, geometric distortion and spectral mixture registered from multiple signals from adjacent targets (Gholizadeh et al. 2021). However, the space-based instruments cover larger areas to measure the spectral reflectance of soil surfaces. Several open-access multitemporal and multispectral satellite sensors provide data at visible to thermal ranges to measure the characteristics of soil reflectance. The information obtained from soil reflectance is a valuable source for deriving qualitative and quantitative information on soil properties (Vaudour et al. 2016).

The SOC content varies along the depth of the soil profile. In most cases, it is dominantly found at the topsoil surfaces and it exhibits lower reflectance compared to the lower soil layers. The soils with higher organic carbon have a darker color and produce lower reflectance while soils with lower organic carbon have higher reflectance (Marques et al. 2020). However, other soil properties also influence the amount of soil reflectance due to their texture and the available soil moisture (Stenberg et al. 2010). Nonetheless, the assessment of SOC content using the RS approach has certain inadequacies owing to vegetation cover, soil moisture, roughness, and instrument configuration (Nocita et al. 2013). Hence, the correlation of spectral reflectance with the soil properties using partial least square regression considering linear relationship may not always hold true (Peng et al. 2014), and machine learning algorithms are currently implemented for better correlation analysis (Padarian et al. 2020).

Satellite data products provide an estimation of SOC in a more consistent and with minimized costs compared to the field measurements. Several covariates developed from remote sensing imageries together with geostatistical methods are more accurate than only using ordinary kriging in mapping SOC stocks (Mondal et al. 2017). In line with this, Sentinel 2 imageries were tested to map and monitor the SOC and soil texture using 10 bands and 18 spectral indices (Gholizadeh et al. 2018). The result indicated that the Sentinel 2 image is adequate to map SOC and clay in agricultural sites. The airborne hyperspectral image is also used to predict SOC content in agricultural topsoils (Hong et al. 2020). Venter et al. (2021) also used optical satellite imagery in combination with a random forest regression model to map soil organic carbon stocks over South Africa. The potential of combined optical and radar imagery to map SOC stock also tested and produced a better SOC prediction compared to using a separate prediction of the two imageries (Wang et al. 2020).

Monitoring Soil Organic Carbon Stock Changes

The change in land use is among the prominent causes for variations in the soil physical, chemical and biological contents triggering high-spatial heterogeneity. The Abbay Basin is known for its high population density and fragmented agricultural plot occupation with a significant decrease in soil fertility making agricultural activity a challenge in certain parts of the basin (Erkossa et al. 2009). The loss in fertility and diminishing size of plots for cultivation pushes people to encroach into the natural vegetation and marginal lands for farming (Malede et al. 2023). Moreover, the crop residues in most parts of the basin are collected as a source of household fuel or used as feed for their animals and the amount of above-ground litterfall and below-ground roots and biota that can potentially decompose to organic matter is very minimal (Erkossa et al. 2014). Hence, SOC content that can be added to the soil is often reduced due to these activities resulting in a significant decrease in carbon stocks. However, farmers are using additional inputs to maintain the fertility of their agricultural plots. In cultivated lands, the SOC can also be varied due to certain activities performed on the land that includes tillage, crop rotation, residue management, fallow periods, and fertilization (Venter et al. 2021).

SOC monitoring involves studying the carbon content over time delivering information regarding the soil health in line with disturbances from natural and anthropogenic effects. Soil health is a priority for higher and sustainable agricultural production and then for food security. Thus, monitoring soil organic matter in relation to the land management options that showed enhanced SOC sequestration over time is vital to secure food production and achieving sustainable development goals (SDG). Most of the SDGs are linked to soil health that incorporates SDG 1 (End Poverty), 2 (Zero Hunger), 3 (Good Health and Well-being), 6 (Clean Water and Sanitation), 7 (Affordable and Clean Energy), 9 (Industry Innovation and Infrastructure), 11 (Sustainable Cities and Communities), 12 (Responsible Consumption and Production), 13 (Climate Action), and 15 (Life on Land). Hence, continuous monitoring of soil using spectral reflectance contributes to mapping temporal and spatial changes in soils.

A significant change in total organic carbon stocks may require some years (Venter et al. 2021). Several studies showed that the change in carbon content may need more than ten years to discover substantial variation (Bellamy et al. 2005; Smith et al. 2020). The increase in soil organic carbon is a gradual process and also requires a continuous addition of inputs to sustain the changes. The measurement of carbon stocks enables us to discern the variation over time.

Earth observation data are major components to monitoring the changes in SOC stocks. The dominant factor in the changes of the carbon stock is the alteration of the land use systems particularly the deprivation of the organic matter of soils. The conventional soil maps are static depending on old datasets and may not help to monitor the dynamic changes in organic carbon content. The satellite image-derived covariates and machine learning algorithms are one source of dynamic data and methods that change temporally to monitor the changes in soil properties through time.

Machine learning algorithms are capable of monitoring the changes in the spatial distribution of SOC utilizing the soil samples collected from the field as training samples and the spectral reflectance of satellite imagery. Among the machine learning workflow, random forest (RF) is the most commonly used algorithm to distinguish the association between soil properties and covariates for carbon content mapping (Khaledian and Miller 2020; Padarian et al. 2020). This algorithm is dominantly used among machine learning algorithms due to its minimal preprocessing requirement and handling of linear and nonlinear relationships (Hengl et al. 2015; Khaledian and Miller 2020). The enhancement in algorithm development of RF and availability of open-access satellite data improves the monitoring of changes in SOC content over time.

Satellite Remote Sensing for Soil Organic Carbon Modeling

Since soil sample collection and laboratory analysis are costly, time-consuming, and laborious satellite data, and derived products are recently being used to generate SOC stock based on fewer soil samples (Vaudour et al. 2022). Satellite data cover larger areas for soil mapping with a distinct spatial resolution that fits the plot level. Several types of satellite data sources valuable for determining soil organic carbon content include multispectral (Lin et al. 2020; Vaudour et al. 2019), hyperspectral (Guo et al. 2021; Laamrani et al. 2019; Nanni et al. 2021), and radar data (Bousbih et al. 2019; Zhang et al. 2021). Soils have variations in spectral features at different wavelengths and measurements of soil properties at distinct spectral bands allow discriminating soil types (Ayala Izurieta et al. 2022; Henderson et al. 1989; Rasel et al. 2017). The difference in reflectance among the various soil types is valuable indicators for identifying the soil properties and hence for soil mapping (Huete and Escadafal 1991).

Soil organic carbon prediction using earth observation data can be categorized into at least the following four categories (Bousbih et al. 2019; Vaudour et al. 2019; Vaudour et al. 2022; Wang et al. 2018): (i) SOC prediction using only optical satellite imageries; (ii) a combination of optical and radar imageries and their derived products (e.g., soil moisture maps); (iii) combining spectral information from earth observation satellites with soil spectral libraries; and (iv) a combination of spectral reflectance of satellite imageries with derived satellite products like digital elevation models. These approaches use developed models and make validations using field-collected samples for adoption for wider area application in combination with other soil attribute covariates.

SOC can be predicted using point data samples in combination with pixel reflectance both from multispectral and hyperspectral imageries utilizing various machine learning algorithms (Emadi et al. 2020; Yang et al. 2021). Most studies limit their prediction to small areas and also consider optimal soil conditions mainly dry and bare soils (Möller et al. 2022). The fulfillment of bare soil condition is a challenge as it demands data from satellite images for mapping SOC which is difficult to obtain due to the long-term vegetation cover in most land cover types. This activity is more complicated owing to the temporal and permanent vegetation cover. The possibility of obtaining exposed soil on a few satellites imagery is very rare either the soil may be covered by seeds or covered by cloud. This limitation could be lessened by image compositing of all available satellite imageries to estimate all parameters of the soil.

The traditional way of soil carbon monitoring is accomplished through point soil sample collection based on the areas to be covered and the cost incurred for data collection. The sampling expense affects soil sampling density and spatial representativeness, particularly in developing countries, and the collected data may differ in sampling method and size of collected samples. Hence, explanatory variables are commonly used to transform the point sample into a spatial dataset. The availability of satellite data permits covering wider areas to characterize the spatial and temporal patterns of soil carbon content (Tziolas et al. 2021). Soil reflectance from spectral bands of multitemporal satellite imageries is a crucial component for predicting soil organic carbon content. However, crop residue, soil moisture, and roughness affect the precise estimation of SOC content (Dvorakova et al. 2021).

For forecasting and monitoring SOC content, topographic parameters are another explanatory variable generated from satellite data (Lamichhane et al. 2019). Various DEM-based terrain parameters were derived from satellite imageries that include Shuttle Radar Topography Mission (SRTM) at different resolutions and Advanced Land Observing Satellite (ALOS) Phased Arrayed L-band Synthetic Aperture Radar (PALSAR) images (Hengl et al. 2015; Wang et al. 2020) that can be used in soil properties mapping. The terrain attributes significantly influence SOC prediction, particularly with variations in the scale used for soil characterizations.

Soil moisture is another parameter that affects the amount of reflected and emitted energy from the soil. The water content in soil particularly in near-infrared regions of the spectrum is centered at 1.45 and 1.9 μm (Reginato et al. 1977). At these two distinct regions, there is a huge absorption of electromagnetic energy resulting in a decreased reflected energy making the soil darker. The SWIR portion of the spectrum is known for its sensitivity to surface moisture content.

The spectral reflectance of soils from different spectral bands is another source of soil information for SOC content modeling. The amount of reflectance registered varies for soil types across the spectral wavelength which is an important indicator for mapping soil properties. The visible near-infrared (VNIR) to the shortwave near-infrared (SWIR) portion of the spectral wavelength is vital sources of information for estimating SOC content. SOC shows distinct absorption characteristics in VNIR and SWIR (Nocita et al. 2015). Several factors and constituents of the soil affect the soil spectral signature, but iron oxides, organic content and water are the prominent factors that hugely influence the soil reflectance properties (Aviv 2002).

Modeling and predicting SOC content can be achieved by using linear statistical methods (Summers et al. 2011), geostatistical methods (Garc, 2005), machine learning (ML) methods (Khanal et al. 2018), and hybrid methods (Mirzaee et al. 2016). The models used have their influence on addressing SOC mapping and monitoring in agricultural fields. The linear statistical models are widely applicable to carbon modeling due to their easiness of analysis. However, in recent years there is a shift toward using ML models due to their handling of nonlinear relationships between soil carbon and environmental covariates (Huang et al. 2022).

Use Case of Interactive Digital Soil Mapping

Background

In Ethiopia, agriculture is the backbone of the GDP (about 45%) and also one of the major workforces (over 80%) involves in this sector. However, for the agricultural sector to succeed in fulfilling the needs of food demands of society and improvement of the needs of the manufacturing industry, a digital agriculture platform that supports decisions is paramount. Digital technologies are anticipated to play a vital role in the enhancement of agricultural productivity and the enablement of trade in agricultural products (Digital Ethiopia, 2025). The limitations in the agriculture sector are the lack of sound evidence and arrangements associated with the soil condition that is critical to the application of inputs that improve the efficiency of the agriculture sector. In particular, organic matter is one of the main elements of the soil content that is vital for the development of crops. Improving agricultural productivity is a key issue designed by the digital transformation to be addressed by the 2025 goals. Hence, providing digital carbon stock and projection of future sequestration potential is essential for attaining the planned objectives to increase crop yield to succeed in the digital transformation strategy.

Soil is the main terrestrial pool of carbon, storing more than twice the total of carbon than the atmosphere as decomposed plant litter and residue (Percival et al. 2000). Any variations in soil carbon storage can change the global carbon cycle and affect climate change (Conant et al. 2001). Proper land management practices have the potential to offset about a third of the annual global carbon emissions (Lal 2004). Grasslands encompass roughly 40% of the earth's land area and significantly influence the global carbon cycle (Wang and Fang 2009). Grazing intensity implicitly interrupts the growth of below- and above-ground biomass of grazing lands (Scurlock and Hall 1998) with heavy grazing critically inducing pressure on the production of above-ground biomass. Thus, accelerating soil carbon sequestration via enhanced grazing regimes in these regions is a critical measure for offsetting greenhouse gas emissions to mitigate current climate change. Recommended soil management practices and improved rangeland management enhance carbon sequestration to reduce emissions from the agriculture sector (Lal et al. 2007).

Agriculture is one of the four pillars of the development of a green economy in Ethiopia focusing on improving crop and livestock production to boost food security and farmer income while reducing greenhouse gas emissions (FDRE 2011). In Ethiopia, livestock production and cultivation of crops are major contributors to emissions from the agricultural sector which needs a balance in maintaining the carbon management of the sector. The climate-resilient green economy (CRGE) strategy of the country also prioritized limiting soil-based emissions as a mechanism to minimize carbon loss from the agriculture sector. Properly managed soils have the potential to sequester billions of tons of carbon in the soil each year (Paustian et al. 2019). The soil management options for greenhouse gas emission reduction are nutrient and crop management. These practices comprise agronomic best soil management to rise carbon storage, optimal nutrient management to advance nitrogen use efficiency, effective tillage and residue management practices, terracing and other water collecting techniques, and agroforestry practices to avert soil erosion and degradation (Erkossa et al. 2018).

The web-enabled mapping of modeled SOC dynamics and its monitoring allow users to identify the current carbon stock of user-based selection years and also projected carbon sequestration amount using different soil management options. This enhances our understanding of carbon dynamics in croplands and grazing fields for efficient carbon management and greenhouse gas emission reduction options. This web-based platform informs users of the status and trends of soil organic carbon. The platform uses the product of point observation data and other environmental covariates for modeling carbon sequestrations under different scenarios. In addition, the system can accommodate additional datasets when available at any time to better explain the status and changes in the carbon content of the soil.

Material and Methods

Study Area

Abbay commonly known as the Upper Blue Nile Basin (UBNB) is located in the northwestern part of Ethiopia between 7°40′N and 12°51′N, and 34°25′E and 39°49′E, with a total land area of 199,812 km^2(Fig. 23.1). Rugged topography in the center and eastern part and flat lowlands in the western part are the major terrain characteristics of the basin. About 60% of the basin's area is highland with elevations of $\geq$ 1500 m asl. Abbay contributes about 45% of the country's surface water resource, accommodates 25% of the population, and accounts for 20% of the landmass of the country (Erkossa et al. 2009). Cultivation and grazing lands are the largest land use and land cover units in the basin. Accordingly, this study selected this basin as a priority area to assess the status of soil carbon stock and its potential to sequester

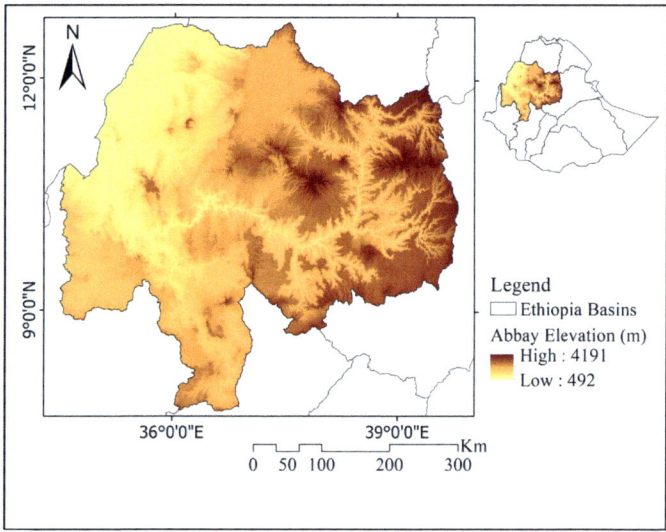

Fig. 23.1 Location map of Abbay Basin

carbon along with improved land use management. Quantification of the SOC stock and modeling of attainable SOC sequestration emphasize on croplands and grazing lands. The arable land in the basin is usually used for the cultivation of annual crops. Grazing lands comprise diverse land uses which is always devoted to livestock production with a major herbaceous species, containing intensively managed permanent pastures and hay land, widely managed grasslands and rangelands, savannahs, and scrublands.

Data and Methods

The interactive web-based system is one of the most effective ways of availing information related to the sustainable management of SOC in a very user-friendly manner for better decision-making. Accordingly, a system is developed that has several components including a geospatial database that integrates all spatial and non-spatial data related to soil organic carbon, an interactive soil organic map with improved land management options, and a highly intuitive web-based secure application. The data used to produce SOC Web-enabled interactive map includes basin-level SOC, spatially explicit basin-level SOC stock map and map of attainable SOC sequestration after 20 and 50 years of improved land management. In addition, different improved land management data and their impact on the sequestration potential of soils are integrated into the system.

The development of the system followed the Agile software approach adopting quick sustainable development, integration of SOC, SOC stock map, and attainable SOC in 20 and 50 years of improved land management. Currently, in this system, only two land use types (cropland and grassland) are considered. The output map can be visualized based on users' requests. The entire component used in the proposed system is open-source technologies. Three-tier layers are proposed for the system that includes the user interface, the web GIS application, and the backend GIS database. The database technology at the backend is organized with PostgreSQL. The Web application employed PHP(Django) as the server-side scripting language and jQuery (OpenLayers) as the front-side one. Further, MapSource is used as a framework for client-side web applications. Apache and Apache Tomcat web servers are utilized to run the web application and GIS server, respectively (Fig. 23.2).

Datasets

The datasets used in this web-enabled interactive SOC stock sequestration map include harmonized functional basin-level SOC database, basin-wide SOC stock map at 250 m resolution, and basin-wide interactive modeled attainable SOC sequestration maps after 20 and 50 years. The SOC database is a product of 12,733 soil points that are harmonized and standardized targeting soil properties including SOC, clay, bulk density, coarse fragments, and moisture content. The soil depth considered for the SOC mapping is 0–20 cm. The soil point samples were used for producing SOC stock map. The point data and main environmental variables (climate, geology, land use land cover, elevation and terrain morphology, vegetation indices) were used for modeling and mapping the SOC stock map. The SOC stock, climate data, land management options (scenarios), and soil physical parameters are being used in R-based RothC for SOC sequestration potential modeling and prediction. The simulation scenario utilized three input levels (low, medium, and high) beginning in 2022 till 2050.

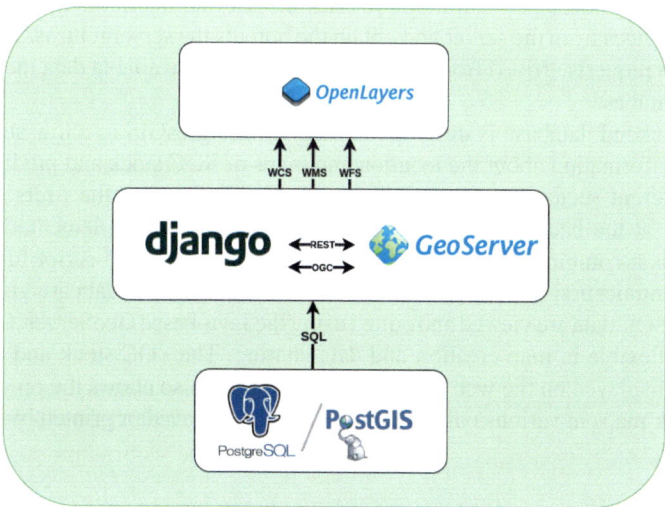

Fig. 23.2 Schematic presentation of web-enabled soil organic carbon management

In addition, the modeling activity used polygons for delineating the study area. The input datasets are the Ethiopian boundary, watershed boundary, Abbay Basin boundary, land use, and agroecology of the basin. The basin border incorporates the lower administrative boundary up to the district and Kebele level. The user can obtain the current and projected carbon stock map for selected land use types (cultivated and grassland) based on the given requirements. The interactive map selection allows users to select for any of the boundaries (the whole Abbay Basin, microwatersheds of the basin, by district, Kebele, or by agroecology/agroclimate).

To access the SOC stock map and yearly attainable SOC map, users must first register and obtain a username and password. The username/password combination is used when users intend to download maps and other produced graphs. The data downloading is possible in excel, pdf, shapefile and raster format depending on the needs of users. The user registration to this website indicates agreeing to the use condition of repository data release. The website has a good landing page to drive customers to the site to get products and services depending on their interests. The published folder on the website has an index page for the site navigation to work properly. The users enter the web search engine through the web query to satisfy their data/information needs. The website also has several functions including visualization, tooltips, identification and attribute display. It also has basic map services (zoom in, zoom out, pan), printing and standard map creation. The users can also make tables and graphs depending on their data requests and the function available on the site. There is also an analysis function that compares SOC maps produced at different years and also has the functionality for a time series analysis.

The technology implemented in the current SOC mapping follows a client–server approach. It functions on the architecture of a computer network where the user requests and receives services from the database of the centralized server. The web browser of the developed soil map provides the user an interface to allow them to request services from the server and obtain the outputs the server returns. The servers respond to requests arrived from users depending on the available data incorporated into the database.

The backend database is developed using the postgresGIS to allow storage and query of information about the location and maps of SOC stock and predicted SOC using different scenarios. All activities and data requests of the users are being processed at the backend. The backend script used in the implementation of the database is a combination of Python and Django. The backend script functions to ease communication with the server in which information and data are given by the user. The SOC data are viewed and edited using the Java-based GeoServer. GeoServer is highly flexible in map creation and data sharing. The SOC stock and attainable SOC are displayed on the web using the GeoServer. It also allows the production of SOC stock maps in various output formats that can be saved or printed by the users.

Result

Web Interface

The web user interface enables the user to communicate with content prepared for visualizing SOC content developed for the Abbay Basin running on a remote server through a web browser. The content that details the SOC maps can be downloaded from the web server and the user can interact with this content in a web browser that behaves as a client. The SOC content map is produced based on the SOC point data basin-wide SOC stock map at 250 m resolution and basin-wide interactive modeled attainable SOC sequestration maps.

The developed web interface shows and visualizes system usage on the main page. It gives detail about the content that can be seen on the main page. In addition, it also gives the service that can be acquired from the webpage depending on the need of the data/information user. The interactive map interface enables users to produce custom soil maps for their particular needs (Fig. 23.3).

Fig. 23.3 Web interface of SOC for Abbay Basin

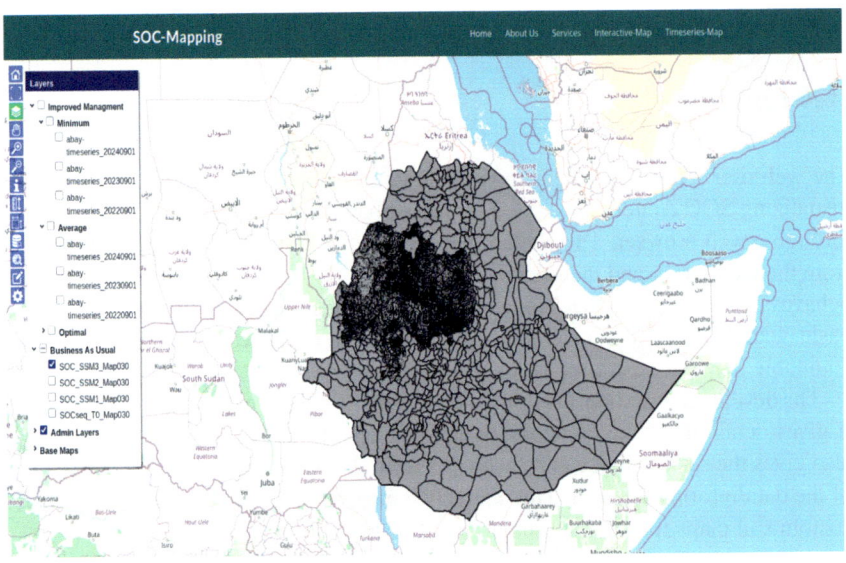

Fig. 23.4 Soil organic carbon stocks of Abbay Basin

Soil Organic Carbon Stock

Soil maps show the location and status of SOC on the ground. The service allows users to request and visualize SOC stock maps on the web by year and place. This function allows the users to view the size of the SOC content in specific spatial distribution. The users can ask for information about the content of SOC from Kebele administration to the bigger watershed level. It also allows downloading the requested information based on the user's area of interest. The system is capable of categorizing SOC maps to specify the SOC amount in the basin. The availability of a digital interactive soil map provides precise and accurate SOC information for the provision of sustainable management of cultivated land and grasslands (Fig. 23.4).

Time Series Soil Organic Carbon of Abbay Basin

This section of the interactive web viewer shows historic and future SOC stock from 2022 to 2050 in time-enabled time series maps. The possibility of obtaining long-term SOC content helps to evaluate the change in SOC dynamics in the basin over the years. Accordingly, the current web service provides and visualizes time series SOC maps on the web with percentages using graphs. The mapping is also vital in identifying and predicting the management options performed on the two selected land use types to observe anticipated changes. The maps make it possible to estimate how specific management options enhance SOC stock at the plot level. The current

23 Interactive Web Tool for Mapping Soil Organic Carbon Dynamics   591

map also displays the variation in predicted SOC content over the years incorporating the business as usual and improved management options adopted in the basin land use management systems (Figs. 23.5 and 23.6).

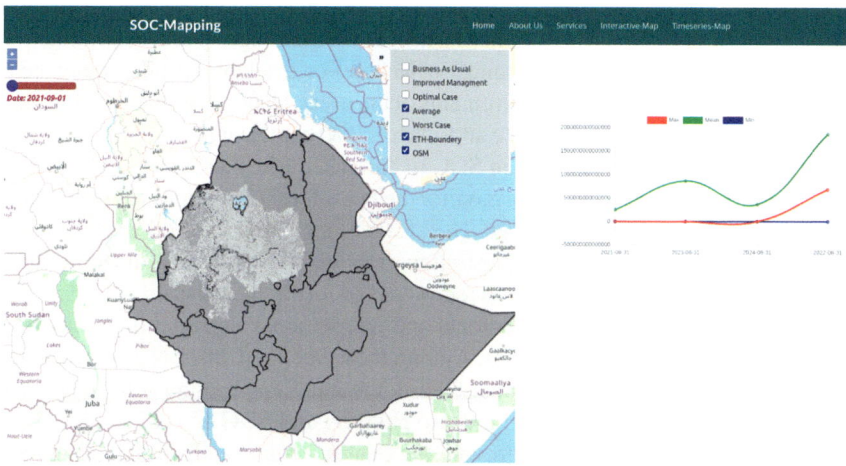

Fig. 23.5 Time series viewer of SOC stock of Abbay Basin

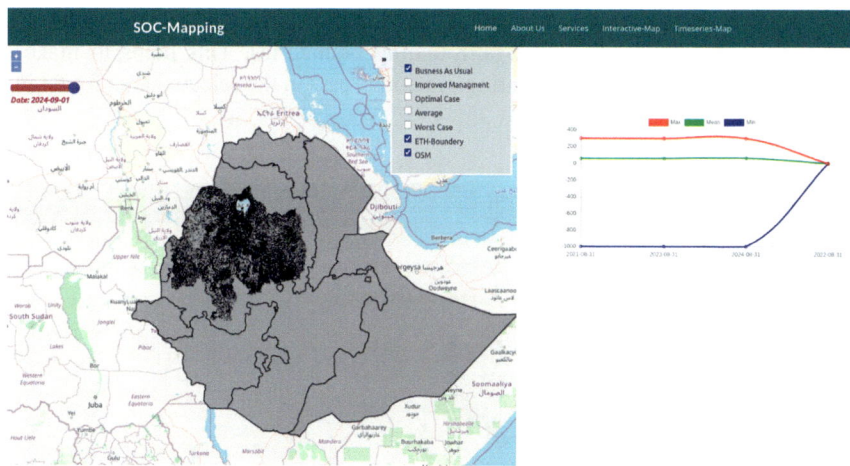

Fig. 23.6 Map of management options and attainable SOC in Abbay Basin

Selection of SOC Maps

The selection and visualization of SOC information for a certain area by drawing a rectangle or polygon and measuring distance using a line is also possible within the current web services. The SOC dynamics, in space and time, are vital to monitor the changes in the SOC content. This can be achieved through spatial discrimination using the lower administrative boundary (Kebele) or available smaller watersheds depending on the demand and requirements of the users. Hence, the developed system allows users to measure the distance and area of SOC content maps. This is helpful to identify specific areas of changes in SOC in cultivated lands and grasslands. This is an indicator for regions to properly monitor the lands for interventions for sustainable utilization of the available resources. The analysis and precise detection of the areas will help to implement carbon sequestration measures. The map also illustrates how much and where additional SOC is sequestered or lost in the basin.

Ethiopia has adopted several regional and international agreements for environmental monitoring and management. Among these, are the United Nations Sustainable Development Goals (SDGs) that focus on maintaining healthy soils to achieve the intended goals. Of these, SDG 2 (Zero Hunger) and SDG 15 (Life on land) are specifically achieved through improving the quality of land and soil to end hunger and through halting and reversing land degradation. Hence, maintaining the SOC content of soil guarantees continuous production to contribute to food security and exterminate hunger in the basin (Fig. 23.7).

The area analysis summarized information per requested Kebele, district, and microwatershed compared to the surrounding boundaries. This helps to make a SOC stock change analysis comparing previous years in the selected area for taking appropriate measures.

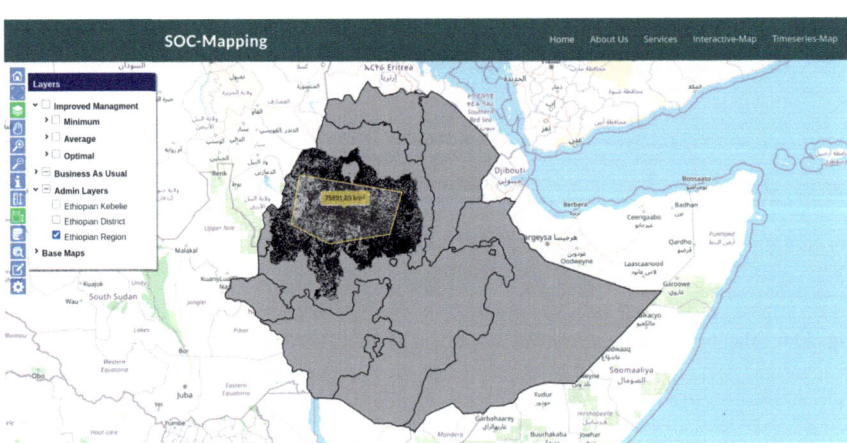

Fig. 23.7 Web-enabled carbon stock monitoring across the area of interest

Conclusion and the Way Forward

Soil organic carbon is among the nutrients that are vital for crop growth and enhancement of production and productivity. Consequently, identifying options for proper soil carbon management improves crop production and also reduces carbon emissions from the soil. This paper mapped the current carbon stock and projected carbon sequestration potential of the Abbay Basin using three scenarios. The interactive web map visualizes the current stock and projected sequestered carbon and compares the trend over the years. The interactive web-enabled soil organic carbon potential of the region is based on three scenarios, namely minimum, optimum, and best scenarios that affect the sequestration of carbon in the soil. The interactive web map is built using data from legacy soil and modeled soil potential of the basin. The site can be updated based on scenarios and additional input covariate data that can be used for building the model. It is also possible to incorporate other legacy soil data to extend to other parts of the country for better soil carbon management.

The site needs improvement with updated data and information for better prediction of carbon sequestrations with the incorporation of more relevant covariates.

The current interactive web-based carbon sequestration potential and the stock map are vital to knowing site-specific information for soil fertility management and enhancing crop production and productivity. Accordingly, it is vital to work further to develop a mobile app to reach farmers and development agents for site-specific soil management options and policy recommendations. The current work focused only on the Abbay Basin. It is advisable to develop carbon stocks and projected sequestration potential for the remaining parts of the country to support the digital transformation and reduction of greenhouse gas emissions from the agriculture sector.

Acknowledgements The authors acknowledge the financial support given by the former Ethiopian Space Science and Technology Institute (ESSTI) currently reorganized as Space Science and Geospatial Institute (SSGI) under the grant number ESSTI-RS_EP02-0621 to partially fund this work.

References

Abebe G, Tsunekawa A, Haregeweyn N, Takeshi T (2020) Effects of land use and topographic position on soil organic carbon and total nitrogen stocks in different agro-ecosystems of the Upper Blue Nile Basin

Abegaz A, van Keulen H (2009) Modelling soil nutrient dynamics under alternative farm management practices in the Northern Highlands of Ethiopia. Soil Tillage Res 103:203–215

Abegaz A, Winowiecki LA, Vågen T, Langan S, Smith JU (2016) Agriculture, ecosystems and environment spatial and temporal dynamics of soil organic carbon in landscapes of the upper Blue Nile Basin of the Ethiopian Highlands. Agric Ecosyst Environ 218:190–208. https://doi.org/10.1016/j.agee.2015.11.019

Abegaz A, Tamene L, Abera W, Yaekob T, Hailu H, Nyawira SS, Da M, Sommer R (2020) Soil organic carbon dynamics along chrono-sequence land-use systems in the highlands of

Ethiopia. Agric Ecosyst Environ 300(November 2019):106997. https://doi.org/10.1016/j.agee. 2020.106997

Abegaz A, Ali A, Tamene L, Abera W, Smith JU (2022) Modeling long-term attainable soil organic carbon sequestration across the highlands of Ethiopia. Environ Dev Sustain 24:5131–5162

Amanuel W, Yimer F, Karltun E (2018) Soil organic carbon variation in relation to land use changes: the case of Birr watershed, upper Blue Nile River Basin, Ethiopia. J Ecol Environ 42(1):1–11. https://doi.org/10.1186/s41610-018-0076-1

Aviv R (2002) Quantitative remote sensing of soil properties, 75

Ayala Izurieta JE, Jara Santillán CA, Márquez CO, García VJ, Rivera-Caicedo JP, Van Wittenberghe S, Delegido J, Verrelst J (2022) Improving the remote estimation of soil organic carbon in complex ecosystems with Sentinel-2 and GIS using Gaussian processes regression. Plant Soil 479(1–2):159–183. https://doi.org/10.1007/s11104-022-05506-1

Bellamy PH, Loveland PJ, Bradley RI, Lark RM, Kirk GJD (2005) Carbon losses from all soils across England and Wales 1978–2003. Nature 437(7056):245–248. https://doi.org/10.1038/nature04038

Bousbih S, Zribi M, Pelletier C, Gorrab A (2019) Soil texture estimation using radar and optical data

Conant RT, Paustian K, Elliott ET (2001) Grassland management and conversion into grassland: effects on soil carbon. In: Conant RT, Paustian K, Edward T (eds). Elliott. Ecol Appl 11(2):343–355. Wiley Stable URL: http://www.jstor.org/stable/3060893 REFERENCES Linked references are ava

Coonan EC, Richardson AE, Kirkby CA, Kirkegaard JA, Amidy MR, Strong CL (2019) Soil fertility and nutrients mediate soil carbon dynamics following residue incorporation. Nutr Cycl Agroecosyst. https://doi.org/10.1007/s10705-019-10037-w

Dixon RK, Brown S, Houghton RA, Solomon AM, Trexler MC, Wisniewski J (1994) Carbon pools and flux of global forest ecosystems. Science 263(5144):185–190. https://doi.org/10.1126/science.263.5144.185

Dvorakova K, Heiden U, Van Wesemael B (2021) Sentinel-2 exposed soil composite for soil organic carbon prediction. Remote Sens 13(9):1–20. https://doi.org/10.3390/rs13091791

Emadi M, Taghizadeh-mehrjardi R, Cherati A, Danesh M (2020) Predicting and mapping of soil organic carbon using machine learning algorithms in Northern Iran. Remote Sens

Erkossa T, Haileslassie A, MacAlister C (2014) Enhancing farming system water productivity through alternative land use and water management in vertisol areas of Ethiopian Blue Nile Basin (Abay). Agric Water Manag 132:120–128. https://doi.org/10.1016/j.agwat.2013.10.007

Erkossa T, Bekele S, Hagos F (2009) Characterization and productivity assessment of the farming systems in the upper part of the Nile basin. Ethiop J Nat Resour 11(2):149–167

Erkossa T, Williams TO, Laekemariam F (2018) International soil and water conservation research integrated soil, water and agronomic management effects on crop productivity and selected soil properties in Western Ethiopia. Int Soil Water Conserv Res 6(4):305–316. https://doi.org/10.1016/j.iswcr.2018.06.001

FDRE (2011) Ethiopia's climate resilient green economy. Climate Resilience Strategy

Garc L (2005) Using Geostatistical and Remote Sensing Approaches for Mapping Soil Properties 23:279–289. https://doi.org/10.1016/j.eja.2004.12.003

Gavrilescu M (2021) Water, soil, and plants interactions in a threatened environment

Gholizadeh A, Žižala D, Saberioon M, Borůvka L (2018) Soil organic carbon and texture retrieving and mapping using proximal, airborne and Sentinel-2 spectral imaging. Remote Sens Environ 218(September):89–103. https://doi.org/10.1016/j.rse.2018.09.015

Gholizadeh A, Zalidis G, Wesemael B Van (2021) Earth observation data-driven cropland soil monitoring : a review

Guo L, Sun X, Fu P, Shi T, Dang L, Chen Y, Linderman M, Zhang G, Zhang Y, Jiang Q, Zhang H, Zeng C (2021) Mapping soil organic carbon stock by hyperspectral and time-series multispectral remote sensing images in low-relief agricultural areas. Geoderma 398(March):115118. https://doi.org/10.1016/j.geoderma.2021.115118

Henderson TL, Szilagyi A, Baumgardner MF, Chen CT, Landgrebe DA (1989) Spectral band selection for classification of soil organic matter content. Soil Sci Soc Am J 53(6):1778–1784. https://doi.org/10.2136/sssaj1989.03615995005300060028x

Hengl T, Heuvelink GBM, Kempen B, Leenaars JGB, Tamene L, Tondoh JE (2015) Mapping soil properties of Africa at 250 m resolution: random forests significantly improve current predictions 1–26. https://doi.org/10.1371/journal.pone.0125814

Hong Y, Guo L, Chen S, Linderman M, Mouazen AM, Yu L, Chen Y, Liu Y, Liu Y, Cheng H, Liu Y (2020) Exploring the potential of airborne hyperspectral image for estimating topsoil organic carbon: effects of fractional-order derivative and optimal band combination algorithm. Geoderma 365(May 2019):114228. https://doi.org/10.1016/j.geoderma.2020.114228

Huang H, Yang L, Zhang L, Pu Y, Yang C, Wu Q, Cai Y, Shen F (2022) A review on digital mapping of soil carbon in cropland: progress, challenge, and prospect OPEN ACCESS A review on digital mapping of soil carbon in cropland: progress, challenge, and prospect

Huete AR, Escadafal R (1991) Assessment of biophysical soil properties through spectral decomposition techniques. Remote Sens Environ 35(2–3):149–159. https://doi.org/10.1016/0034-4257(91)90008-T

Khaledian Y, Miller BA (2020) Selecting appropriate machine learning methods for digital soil mapping. Appl Math Model 81:401–418. https://doi.org/10.1016/j.apm.2019.12.016

Khanal S, Fulton J, Klopfenstein A, Douridas N, Shearer S (2018) Integration of high resolution remotely sensed data and machine learning techniques for spatial prediction of soil properties and corn yield. Comput Electron Agric 153(July):213–225. https://doi.org/10.1016/j.compag.2018.07.016

Laamrani A, Berg AA, Voroney P, Feilhauer H, Blackburn L, March M, Dao PD, He Y, Martin RC (2019) Ensemble identification of spectral bands related to soil organic carbon levels over an agricultural field in Southern Ontario, Canada. Remote Sens. https://doi.org/10.3390/rs11111298

Lal R (2004) Soil carbon sequestration to mitigate climate change. Geoderma 123(1–2):1–22. https://doi.org/10.1016/j.geoderma.2004.01.032

Lal R (2016) Soil carbon sequestration impacts on global climate change and food security 304(5677):1623–1627

Lal R, Follett RF, Stewart BA, Kimble JM (2007) Soil carbon sequestration to mitigate climate change and advance food security. Soil Sci 172(12):943–956. https://doi.org/10.1097/ss.0b013e31815cc498

Lamichhane S, Kumar L, Wilson B (2019) Digital soil mapping algorithms and covariates for soil organic carbon mapping and their implications: a review. Geoderma 352(May):395–413. https://doi.org/10.1016/j.geoderma.2019.05.031

Lin C, Zhu A, Wang Z, Wang X, Ma, R (2020) The refined spatiotemporal representation of soil organic matter based on remote images fusion of Sentinel-2 and Sentinel-3. Int J Appl Earth Obs Geoinform 89(November 2019):102094. https://doi.org/10.1016/j.jag.2020.102094

Malede DA, Alamirew T, Kosgie JR, Andualem TG (2023) Analysis of land use/land cover change trends over Birr River Watershed, Abbay Basin, Ethiopia. Environ Sustain Ind 17(September 2022):100222. https://doi.org/10.1016/j.indic.2022.100222

Marques MJ, Mar A, Carral P, Esparza I, Sastre B (2020) Estimating soil organic carbon in agricultural gypsiferous soils by diffuse reflectance spectroscopy 1–17

Mirzaee S, Ghorbani-dashtaki S, Mohammadi J, Asadi H, Asadzadeh F (2016) Spatial variability of soil organic matter using remote sensing data. CATENA 145:118–127. https://doi.org/10.1016/j.catena.2016.05.023

Möller M, Zepp S, Wiesmeier M, Gerighausen H (2022) Scale-specific prediction of topsoil organic carbon contents using terrain attributes and SCMaP soil reflectance composites

Mondal A, Khare D, Kundu S, Mondal S, Mukherjee S, Mukhopadhyay A (2017) Spatial soil organic carbon (SOC) prediction by regression kriging using remote sensing data. Egypt J Remote Sens Space Sci 20(1):61–70. https://doi.org/10.1016/j.ejrs.2016.06.004

Nanni MR, Melo Demattê JA, Rodrigues M, Abrantes Dos Santos GLA, Reis AS, De Oliveira KM, Cezar E, Furlanetto RH, Teixeira Crusiol LG, Sun L (2021) Mapping particle size and soil organic matter in tropical soil based on hyperspectral imaging and non-imaging sensors. Remote Sens 13(9):1–19. https://doi.org/10.3390/rs13091782

Nocita M, Stevens A, Noon C, Wesemael BV (2013) Prediction of soil organic carbon for different levels of soil moisture using Vis-NIR spectroscopy. Geoderma 199:37–42. https://doi.org/10.1016/j.geoderma.2012.07.020

Nocita M, Stevens A, van Wesemael B, Aitkenhead M, Bachmann M, Barthès B, Dor EB, Brown DJ, Clairotte M, Csorba A, Dardenne P, Demattê JAM, Genot V, Guerrero C, Knadel M, Montanarella L, Noon C, Ramirez-Lopez L, Robertson J et al (2015) Soil spectroscopy: an alternative to wet chemistry for soil monitoring. Adv Agron 132:139–159. https://doi.org/10.1016/bs.agron.2015.02.002

Padarian J, Minasny B, McBratney AB (2020) Machine learning and soil sciences: a review aided by machine learning tools. Soil 6(1):35–52. https://doi.org/10.5194/soil-6-35-2020

Paustian K, Larson E, Kent J, Marx E, Swan A (2019) Soil C sequestration as a biological negative emission strategy. Front Clim 1(October):1–11. https://doi.org/10.3389/fclim.2019.00008

Peng X, Shi T, Song A, Chen Y, Gao W (2014) Estimating soil organic carbon using VIS/NIR spectroscopy with SVMR and SPA methods, 2699–2717. https://doi.org/10.3390/rs6042699

Percival HJ, Parfitt RL, Scott NA (2000) Factors controlling soil carbon levels in New Zealand grasslands is clay content important? Soil Sci Soc Am J 64(5):1623–1630. https://doi.org/10.2136/sssaj2000.6451623x

Rasel SMM, Groen TA, Hussin YA, Diti IJ (2017) Proxies for soil organic carbon derived from remote sensing. Int J Appl Earth Obs Geoinf 59:157–166. https://doi.org/10.1016/j.jag.2017.03.004

Reginato RJ, Vedder JF, Idso SB, Jackson RD, Blanchard MB, Goettelman R (1977) An evaluation of total solar reflectance and spectral band rationing techniques for estimating soil water content 82(15):2101–2104

Scurlock JMO, Hall DO (1998) The global carbon sink: a grassland perspective. Glob Change Biol 4(2):229–233. https://doi.org/10.1046/j.1365-2486.1998.00151.x

Smith P, Soussana JF, Angers D, Schipper L, Chenu C, Rasse DP, Batjes NH, van Egmond F, McNeill S, Kuhnert M, Arias-Navarro C, Olesen JE, Chirinda N, Fornara D, Wollenberg E, Álvaro-Fuentes J, Sanz-Cobena A, Klumpp K (2020) How to measure, report and verify soil carbon change to realize the potential of soil carbon sequestration for atmospheric greenhouse gas removal. Glob Change Biol 26(1):219–241. https://doi.org/10.1111/gcb.14815

Stenberg B, Rossel RAV, Mouazen AM, Wetterlind J (2010) Visible and near infrared spectroscopy in soil science 2113(December). https://doi.org/10.1016/S0065-2113(10)07005-7

Summers D, Lewis M, Ostendorf B, Chittleborough D (2011) Visible near-infrared reflectance spectroscopy as a predictive indicator of soil properties 11:123–131. https://doi.org/10.1016/j.ecolind.2009.05.001

Tao X, Liang S, He T, Jin H (2016) Estimation of fraction of absorbed photosynthetically active radiation from multiple satellite data: model development and validation. Remote Sens Environ 184:539–557. https://doi.org/10.1016/j.rse.2016.07.036

Tziolas N, Tsakiridis N, Chabrillat S, Demattê JAM, Ben-Dor E, Gholizadeh A, Zalidis G, van Wesemael B (2021) Earth observation data-driven cropland soil monitoring: a review. Remote Sens 13(21). https://doi.org/10.3390/rs13214439

Vaudour E, Gilliot JM, Bel L, Lefevre J, Chehdi K (2016) Regional prediction of soil organic carbon content over temperate croplands using visible near-infrared airborne hyperspectral imagery and synchronous field spectra. Int J Appl Earth Observ Geoinform 49:24–38. https://doi.org/10.1016/j.jag.2016.01.005

Vaudour E, Gomez C, Fouad Y, Lagacherie P (2019) Sentinel-2 image capacities to predict common topsoil properties of temperate and Mediterranean agroecosystems. Remote Sens Environ 223(January):21–33. https://doi.org/10.1016/j.rse.2019.01.006

Vaudour E, Gholizadeh A, Castaldi F, Saberioon M, Urbina-salazar D, Fouad Y, Arrouays D (2022) Satellite imagery to map topsoil organic carbon content over cultivated areas: an overview 1–22

Venter ZS, Hawkins H, Cramer MD, Mills AJ (2021) Science of the total environment mapping soil organic carbon stocks and trends with satellite-driven high resolution maps over South Africa. Sci Total Environ 771:145384. https://doi.org/10.1016/j.scitotenv.2021.145384

Viscarra Rossel RA, Behrens T, Ben-Dor E, Brown DJ, Demattê JAM, Shepherd KD, Shi Z, Stenberg B, Stevens A, Adamchuk V, Aïchi H, Barthès BG, Bartholomeus HM, Bayer AD, Bernoux M, Böttcher K, Brodský L, Du CW, Chappell A et al (2016) A global spectral library to characterize the world's soil. Earth Sci Rev 155(February):198–230. https://doi.org/10.1016/j.earscirev.2016.01.012

Wang W, Fang J (2009) Soil respiration and human effects on global grasslands. Glob Planet Change 67(1–2):20–28. https://doi.org/10.1016/j.gloplacha.2008.12.011

Wang B, Waters C, Orgill S, Gray J, Cowie A, Clark A, Liu DL (2018) High resolution mapping of soil organic carbon stocks using remote sensing variables in the semi-arid rangelands of eastern Australia. Sci Total Environ 630:367–378. https://doi.org/10.1016/j.scitotenv.2018.02.204

Wang X, Zhang Y, Atkinson PM, Yao H (2020) Predicting soil organic carbon content in Spain by combining Landsat TM and ALOS PALSAR images. Int J Appl Earth Obs Geoinf 92(March):102182. https://doi.org/10.1016/j.jag.2020.102182

Yang L, Cai Y, Zhang L, Guo M, Li A, Zhou C (2021) A deep learning method to predict soil organic carbon content at a regional scale using satellite-based phenology variables. Int J Appl Earth Obs Geoinf 102:102428. https://doi.org/10.1016/j.jag.2021.102428

Zhang Y, Hartemink AE, Huang J, Townsend PA (2021) Synergistic use of hyperspectral imagery, Sentinel-1 and LiDAR improves mapping of soil physical and geochemical properties at the farm-scale. Eur J Soil Sci 72(4):1690–1717. https://doi.org/10.1111/ejss.13086

Chapter 24
A Synthesis of Literature on the Grand Ethiopian Renaissance Dam (GERD) Using Text Mining Approaches

Wuletawu Abera, Melkamu Beyene, Aminu Mohammed, Teshome Alemu, Temtim Assefa, Miftah Hassen, Mekdelawit Messay, and Lulseged Tamane

Abstract The Grand Ethiopian Renaissance Dam (GERD) is a massive hydropower plant currently being constructed on the Upper Blue Nile River, known locally as Abbay, in Ethiopia. This project has attracted the attention of both political and scientific communities, resulting in numerous multidisciplinary studies over the past 11 years, covering a broad range of topics, including hydrology, geology, ecology, socioeconomics, politics, and legal issues associated with the dam and the basin. While most of the studies have been conducted using rigorous scientific approaches, there is concern that nationalism may influence water research. Therefore, the objective of this study is to synthesize and review GERD-related literature using a more objective, efficient, and automatic approach. By October 2022, over 340 peer-reviewed papers had been published on GERD, and traditional literature review approaches are not efficient for scanning, reading, and reviewing such a large volume of literature. The study aims to assess the knowledge and information generated by the scientific community since the start of GERD construction using text mining (TM) and natural language processing (NLP) techniques. The TM results explore and appraise key knowledge and findings and enable the identification of key topics related to GERD and its basin. Specifically, we focused on (1) conducting basic bibliometric statistics and discussing key topics related to GERD over time; (2) reviewing and synthesizing a large set of publications using TM and NLP approaches transforming the traditional review process; (3) employing the network of keywords to identify the most studied topic or thematic area related to GERD and vice versa;

W. Abera (✉)
International Center for Tropical Agriculture (CIAT), Accra, Ghana
e-mail: Wuletawu.Abera@cgiar.org

W. Abera · L. Tamane
International Center for Tropical Agriculture (CIAT), Addis Ababa, Ethiopia

M. Beyene · A. Mohammed · T. Alemu · T. Assefa · M. Hassen
Addis Ababa University, Addis Ababa, Ethiopia

M. Messay
Florida International University, Florida, USA

(4) identifying the sentiment orientation of the identified topics towards the GERD project; and (5) recommending future research directions regarding GERD and the basin. The study's findings indicate that the literature on GERD mainly focuses on seven central topics, namely, water conflict, law and negotiation for water sharing, temporal variability in access to water, land degradation, soil erosion, agricultural livelihood, sustainable ecology/ecosystem development, and the application of satellite and remote sensing technology in GERD analysis. The study also notes subtle variations in the focus of articles from riparian states and other countries. The sentiment orientation on these topics is found to be largely neutral, indicating that the scientific community takes a neutral position regarding GERD issues in academic writing. These findings can improve academic discourse on GERD and guide policymakers and government agencies in prioritizing their interventions regarding the GERD project.

Keywords GERD · Abbay · Topic modeling · Sentiment analysis · Water conflict · Ethiopia · Sudan · Egypt

Introduction

In 2011, Ethiopia started construction of the Grand Ethiopian Renaissance Dam (GERD) on the Blue Nile River, which is the main tributary of the Nile River (Basheer et al. 2021). The Nile is the longest river in the world and has a basin that extends over 11 African countries. The dam has a capacity to generate 5150 mW of hydroelectric power and will be the largest hydropower dam in Africa. Ethiopia started the initial filling of the dam in July 2020, with subsequent fillings in 2021 and 2022. It is widely believed that dams will improve the living standards of their citizens and maintain sustainable economic development (Kamara et al. 2022). The two downstream countries, Sudan and Egypt, have been the historic and current main beneficiaries of the Nile River for agriculture as well as hydroelectric power generation. Egypt takes approximately 55.5 BCM (billion cubic meters) of water from the Nile River annually according to the 1959 treaty (Aziz et al. 2019). While Ethiopia maintains that this project is beneficial to downstream countries, Egypt alleges that the dam will impinge on its "historic water use rights" over the Nile River.

Ethiopia sees the GERD as the centerpiece of its bid to become Africa's largest power exporter (Clapham 2018). Ethiopia also insists that the dam is constructed purely for electric power generation and does not affect the water supply of the two downstream countries. On the other hand, Egypt and Sudan expressed their concern about the construction of the dam, as it will affect their surface and groundwater supply (Aziz et al. 2019; Kamara et al. 2022). The concern mainly revolves around the size of the GERD reservoir and the time it will take to fill this reservoir (Kamara et al. 2022). There have been some contested studies exploring the potential impact of GERD on Egypt's agricultural land and the corresponding impact on macroeconomic variables such as food production, the cost of living, real GDP per capita, and general welfare (Aziz et al. 2019; Kamara et al. 2022). On the other hand, although they have concerns about the loss of traditional agriculture and clay-based artisan jobs, Sudan is expected to benefit from the GERD in terms of improved irrigation, water supply reliability, hydropower generation, and riverine flood control (Basheer et al. 2021).

The GERD is a massive project that touches upon the water, energy, food, and environmental sectors of the three countries. The water-energy-food and environment (WEFE) nexus approach is suggested to optimize resource use by enhancing cross-sectoral interactions, reducing trade-offs, and building synergies among different sectors and regions without compromising sustainability (Hoff 2011). Due to its socioeconomic and geopolitical importance, GERD has attracted significant attention from both scientific and political communities (Heggy et al. 2022; Abera et al. 2021). Despite the continuous negotiation among the three countries to resolve the conflict of interest on the construction of the GERD, they could not reach an agreement until now (Basheer et al. 2021). Many of the arguments posited by different parties about the project are largely based on media reports that lack scientific rigor (Kamara et al. 2022). While most studies on GERD have been conducted with utmost scientific ethics and attempts to provide valuable information on sustainable management of water resources, there are also large amounts of academic literature with poor rigor and scientific misconduct (Heggy et al. 2022; Abera et al. 2021). Recently, Wheeler and Hussein (2021) pinpointed the challenge of nationalism in water research and the new development of misinformation due to water nationalism and the impact that nationalism and partisanship have on research and reporting on the impact and benefits of GERD (Wheeler and Hussein 2021; Abera et al. 2021).

A synthesized summary of what has been published related to GERD has utility in understanding the research focuses and topics investigated and research gaps. Traditional literature review approaches suffer from "cherry picking", especially in dealing with a large body of literature with contrasting and highly opinionated inferences, such as the literature on GERD. This research explores key issues in the debate over GERD by analyzing the scientific literature using machine learning and natural language processing (NLP) techniques.

The objective of this study is to synthesize the scientific literature and provide a concise summary of GERD literature using reproducible, standardized machine learning approaches. Specifically, we are interested in (1) examining the overall trends in the literature regarding GERD in the last decade, (2) examining topics and issues investigated and discussed in the GERD literature landscape using topic modeling, (3) analyzing the sentiment orientation and biases that can be inferred in the literature using sentiment analysis, and (4) examining whether text sentiments vary by topic, country of the first author and other temporal variables. This approach provides up-to-date and objective evidence that can be used in resolving political conflicts on transboundary projects such as GERD. It can also be adapted to automatically analyze the sentiment of scientific outputs on topics of large-scale and/or contested projects beyond GERD.

Methods

Topic modeling is an unsupervised text mining method that is used to extract themes or patterns in a large text corpus. It is used to discover hidden patterns in documents, classify documents into discovered themes, and facilitate the search process for documents (Kulshrestha et al. 2019). Sentiment analysis is a text-classification task that tries to label the sentiment orientation (negative, neutral, or positive) of a text based on the opinion contained in it (Devika et al 2016). Sentiment analysis is also a process of extracting information about an entity and automatically classifying opinions of that entity as positive, negative, or neutral (Dang et al. 2020). Sentiment analysis has become the common method to evaluate public opinions on social media posts. It is also extensively used by businesses to evaluate the sentiment of their customers toward their brands. Although sentiment analysis is commonly applied in opinionated texts such as social media data, it is common to see very opposing viewpoints about a given aspect of very hot topics such as GERD. The research also used alternative rule-based and machine learning techniques for sentiment analysis to compare the performance and suitability of the approaches for sentiment analysis on topics of transboundary nature.

To extract topics and sentiments in the literary landscape from the collection of GERD literature, we conducted a chain of activities and methods, as outlined in Fig. 24.1. The first step of the synthetic review using the machine learning approach is to collate and extract all the scientific literature from reputable journals using different combinations of keywords and excluding scientific literature that is not in the English language. The second step is conducting text preprocessing over the articles using common natural language processing (NLP) techniques and removing noise. Relevant keywords that can represent the meaning of articles were selected, and term weighting of keywords was performed using the word2vec word embedding technique. The third step is building a topic modeling algorithm that can embed the high-dimensional term vector into a lower-dimensional latent topic space. Finally, an aspect/topic-level sentiment analysis model is built on top of a pretrained model by taking the most similar sentences to a topic as a context sentence. The similarity between a topic and the rest of the sentences in the corpus is computed using the cosine similarity metrics. The performance of the pretrained model is improved by adding domain-specific training samples from the GERD literature. These steps are detailed in the following subsequent sections.

Document Sourcing

A total of 341 scientific papers published in peer-reviewed journal articles between 2011 and 2022 were gathered from various publisher databases using a Boolean combination of keywords: ("GERD" OR "Grand Ethiopian Renaissance Dam" OR "Hidassie" OR "Millennium Dam") AND ("Ethiopia" OR "Egypt" OR "Sudan").

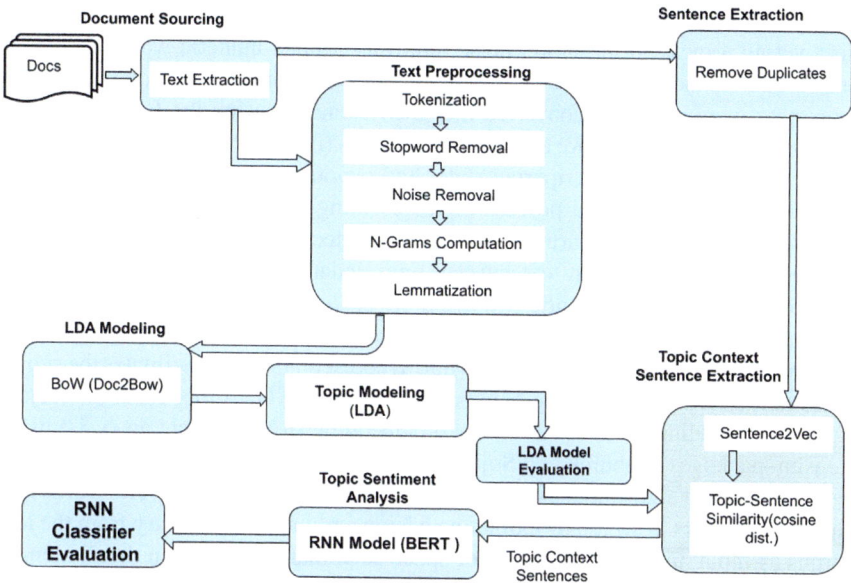

Fig. 24.1 Research workflow followed in the study

The papers were then further filtered based on specific inclusion criteria, including directly addressing an issue related to GERD, being written in English, being published during the active construction period of the GERD (2011-present), and being peer-reviewed. In total, 323 papers were utilized in this study for the purpose of text mining.

Text Preprocessing

Preprocessing has a sequence of activities to make a text ready for topic modeling and sentiment analysis. It includes tokenization, noise removal, stop word removal, and stemming. Tokenization is a process of breaking down or splitting textual data into smaller meaningful components called tokens (Sarkar 2019). In this study, word tokenization is performed based on an English word tokenization algorithm implemented in the NLTK library (Perkins 2010). After the text corpus is decomposed into words (i.e., the smallest individual pieces of tokens), stopword removal is conducted using the English stopword lists. Stop words have no value for the text mining task since they appear across articles. For the purpose of removing noises, texts with two characters that are not necessarily captured through the stop word removal process were removed. Words with outlier values for frequencies were also removed. In addition to well-known stopword lists, the researchers also filtered commonly appearing words in all literature (such as GERD, dam, river, Abay, water) in the domain as stopwords. These words frequently appear in all articles and do not have relevance to discovering topics. Furthermore, noise removal of alpha-numeric words, numbers, and invalid characters is made by writing regular expressions in Python scripts. The remaining words are also converted into their stems by using the Porter-Stemmer algorithm (Perkins 2010). We also created n-grams (for n = 2,3) to obtain frequently co-occurring words such as riparian-states and remote-sense. The last preprocessing step, word embedding, is a process of representing a text by numerical value for text processing. Machine learning algorithms and deep learning architectures are not capable of processing plain text but can only understand numbers (Ghosh 2021). It helps to capture the semantic and syntactic context of a word and calculates the similarity or dissimilarity of words in the document (Wu et al. 2020). Hence, the documents and sentences are converted into a representation that facilitates the subsequent tasks. Here, we used Bag of Words (BOW) using Gensim's doc2bow service (for topic modeling) and word2vec word embedding (for sentiment analysis) using Gensim–packages (Řehůřek and Sojka 2011). Word2vec is one of the most popular word embedding algorithms used for feature representation that converts words into numerical vectors. It works by extracting a large amount of vocabulary from the text corpus as input and generates a vector space such that each word in the dictionary maps itself to a unique vector (Wu et al. 2020). Word embeddings eventually help in establishing the association of a word with another word having a similar meaning through the created vectors (Sarkar 2019). It is based on the assumption that words located closer to each other or in similar contexts tend to have similar meanings (Li 2016).

Topic and Sentiment (Sentiment by Topic) Model Building

Topic modeling provides a method to automatically discover themes or ideas from a collection of scientific literature (Owa 2021). It is used to discover hidden patterns in documents and classify documents into themes or topics (Kulshrestha et al. 2019). While there are many topic extraction methods, such as latent semantic analysis (LSA), probabilistic latent semantic analysis (pLSA), and nonnegative matrix factorization (NMF), we used latent Dirichlet allocation (LDA) due to its high performance and its ability to include prior observations (Zvornicanin, 2022). LDA is one of the most popular models in topic modeling (Jelodar et al. 2019). Moreover, LDA or its variants have been used in aspect-based sentiment analysis for aspect/feature extraction (Rana et al. 2016), which is instrumental to our research. The original plate notation of LDA depicting the latent topic identification process is given in Fig. 24.2.

To extract human interpretable topics using LDA, setting the number of topics based on a coherence score is crucial (Yin et al. 2022). Some of the commonly used parameters are values for α and β. α controls the weights of topics in each document, while β is the parameter for the weights of words in each topic. A high value of α indicates a large mixture of topics in a document, while a low value of alpha shows

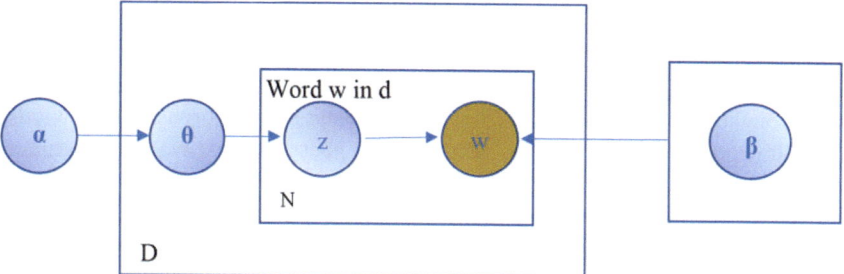

Fig. 24.2 LDA topic modeling process (Albalawi et al. 2020). Where α and β are LDA's Dirichlet parameters (hyperparameters), Θ is per-document topic proportions, z is the per-word topic assignment, w is an observed word in document D, and N is the total number of topics

that a document will be represented by a few topics (Yin et al. 2022). At the same time, a high value of Beta creates topics with a mixture of more words common to more topics, while a low-value results in topics with fewer terms that are relatively distinct, making the topics human interpretable. We used LDAVis tools to show the distribution of topics and their relationships (Sievert and Shirley 2014).

Once the topics are identified, the next task is building a sentiment analysis model for GERD-related topics based on the scientific text (Raza et al. 2019). However, during the latent topic extraction process, sentiment-indicating keywords were removed due to the preceding preprocessing phases and topic extraction process. Hence, the corpus for sentiment analysis should be represented in a way that preserves sentiment, indicating that words are maintained. To this end, given a topic K generated in the topic extraction phase, the top M sentences (based on a cosine distance) are taken as contexts for that topic K (Harris 1954). The assumption is that the sentiment orientation of a topic can be computed from the sentiments of sentences whose cosine distance to keywords that constitute a topic is short (i.e., are closer to each other in the vector space). Hence, the first task was topic-sentence mapping. The topic-to-sentence mapping was undertaken by computing the cosine distance between the topic vectors and sentence vectors. Hence, for a given topic K, the most similar sentences are considered context sentences to compute sentiment analysis of the topic.

There are two main approaches for sentiment analysis, namely, lexicon-based and machine-learning techniques (Zhang et al. 2018). The former is a dictionary and corpus-based approach (Dang et al. 2020). Dictionary-based sentiment classification is performed by using a dictionary of sentiment-indicating terms, such as those found in SentiWordNet and WordNet. On the other hand, corpus-based sentiment analysis is based on a statistical analysis of the contents of a collection of documents using techniques such as k-nearest neighbors (k-NN), conditional random field (CRF), and hidden Markov models (HMM) (Dang et al. 2020). Scientific evidence proves the superiority of deep learning language model techniques to improve the performance of sentiment analysis when compared with those of lexicon techniques in the presence of small or reduced datasets (Zhang et al. 2018). In this research, a corpus-based pretrained model known as BERT is used to detect sentiment analysis of context sentences. The BERT model is credited as one of the earliest pretrained algorithms to perform natural language processing, such as sentiment analysis. The BERT model is trained on the English Wikipedia (2500 M words) and Books Corpus (800 M words). It was trained by masking 15% of the tokens to guess the next word and achieved the best accuracy for some of the NLP tasks in 2018.

For the topic-level sentiment, we followed the following steps. The documents in the corpus were tokenized at the sentence level using nltk and sent_tokenize, resulting in 179,573 sentences. Preprocessing steps were applied to remove duplicate sentences and those containing nonvalues, leaving a final dataset of 169,785 context sentences. Context sentences were extracted for each topic using cosine similarity. A sentence-by-term matrix was constructed using TfIdfVectorizer, with rows representing sentence vectors and columns representing term vectors (Word2Vec embedding). The matrix had 178,4620 nonzero entries out of a possible 10,434,986,100, as not every term occurs in every sentence. Topic vectors were generated by projecting keywords as features, resulting in a K number of topics × 6154 matrix. The weight of each term in the topic vector was considered during similarity analysis to account for their varying contributions. The new weight for each keyword in each topic was calculated using the formula 1 nKwm/n, where k is the number of keywords in the topic, n is the weight in the topic vector space, m is the weight of the keyword in the LDA space, and n is the number of keywords in the topic. These new weights were then normalized and reassigned as the new weight of the topic keyword. Cosine similarity was computed between the weighted topic vectors and all other sentence vectors. A threshold technique was applied to determine the final context sentences for further sentiment analysis. For the topic-level sentiment analysis, we have experimented with the lexicon-based and deep learning-based approaches to sentiment analysis and selected the one that performs better based on expert feedback. Here, we used two pretrained models (Vader and BERT) for further evaluation.

Model Evaluation

The model developed is evaluated in two aspects. The first evaluation is the quality of the topics discovered from the unstructured text and whether they are understandable, coherent, and meet the purpose they are being used for using the state-of-the-art topic model evaluation methods (Kapadia 2019). For this, we used a combination of domain expert judgment and quantitative techniques. The expert evaluation is based on the top 'n' list of words generated by the system for each topic. Words listed under a topic with higher frequencies should be semantically similar. The word cloud generated by the algorithm is also evaluated by domain experts if the words assigned to the topic are semantically closer to each other. The expert is supported by a more comprehensive topic visualization tool, 'Termite' (Chuang et al. 2012). Termite is described as a visualization of the term-topic distributions produced by topic models.

Perplexity and topic coherence, which are the most widely used quantitative measures of model quality, were also used (Rüdiger et al. 2022). Perplexity is used to measure goodness of fit. In other words, it measures how surprising the LDA model will be on an unseen document. The lower the value of perplexity is, the better the model. Topic coherence measures the degree of semantic similarity between high-scoring words in the topic (Kapadia 2019). It distinguishes between topics that are semantically interpretable and topics that are artifacts of statistical inference. It uses conditional likelihood (rather than the log-likelihood) of the co-occurrence of words in a topic (Dagan et al. 1999).

We specifically used two different topic coherence metrics: C_V and C_umass. C_V is one of the best-performing topic coherence scores, which is a one-set segmentation of the top words and an indirect confirmation measure that uses normalized pointwise mutual information (NPMI) and cosine similarity (Röder et al. 2015). C_Umass, on the other hand, is based on document co-occurrence counts and is a one-preceding segmentation and a logarithmic conditional probability as a confirmation measure. C_Umass is also reported to be resilient to noise and helps to obtain clearer topics (Campagnolo et al. 2022). The second evaluation is used to measure the performance of the topic-level sentiment analysis model. This evaluation is conducted by calculating the model's accuracy in predicting topic context sentences into positive, negative, and neutral classes. Accuracy is estimated by the number of correct predictions divided by the total number of predictions. Topic context sentences are prelabeled by three experts independently. The value agreed upon by all three experts is taken as a test case to avoid expert bias.

Once the topics were defined, we also analyzed the variability of issues/aspects discussed on GERD and their sentiment orientation based on time and authors' nationality dimensions. We extracted the first author's details, including place of work and email domain name, to generate the first author's nationality. To achieve this, we performed a correlation analysis between distinctive topics extracted through the LDA model and the nationalities/affiliations of the first author of the paper. To simplify and improve the visualization, we dissected the entire dataset into four groups, Ethiopian, Egyptian, Sudanese, and the rest of the world, and conducted topic modeling on each group.

Results and Discussion

The scientific literature landscape of GERD is synthesized using machine learning and natural language processing techniques and presented in the following sections. The results are organized as bibliometric analysis, thematics/topics clustered using the automated text mining approach, and sentiment analysis of each topic extracted.

Bibliometrics Analysis

Based on the inclusion criteria stated in Sect. 2.1, a total of 323 peer-reviewed scientific papers with a total of 179,573 sentences were considered for this research. After conducting noise removal and preprocessing using NLP techniques and tools, 67,400 unique vocabularies were selected for further processing. The top ten most frequent keywords observed in the corpus are shown in Fig. 24.3. Based on the keyword distribution, "law" is the abundant keyword, followed by land surface process terminologies such as "sediment" and "erosion" (Fig. 24.3).

The number of scientific papers published on GERD has increased over time (i.e., the highest number of publications was recorded in 2021, while the lowest was recorded in 2011), as expected (Fig. 24.4). This indicates that GERD has attracted the attention of the scientific community and is becoming an important research agenda for many researchers.

Based on the leading authors of the papers, the vast majority of the publications come from four countries, Egypt, Ethiopia, the USA, and the UK, although there are many researchers around the world (Fig. 24.5). In addition, there are a few other countries with just one publication in our dataset (Sweden, Singapore, Senegal, Romania, Morocco, South Sudan, Qatar, New Zealand, Niger, Zimbabwe, Austria, Peru, Russia, South Africa, Hong Kong, and Iran—not shown in Fig. 24.5). Lead authors affiliated with Egypt have a significant number of papers published compared to all other countries, including the two riparian countries—Ethiopia and Sudan. Publications by lead authors from Sudan were small despite their active involvement in the political debates over the GERD.

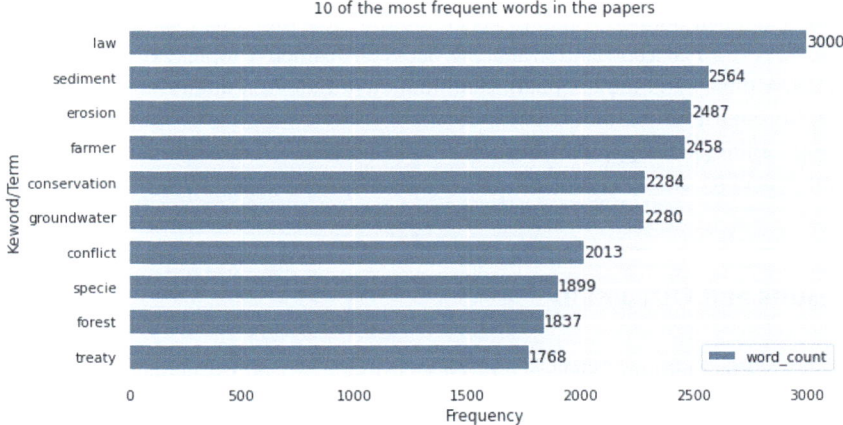

Fig. 24.3 Top 10 most frequent vocabularies

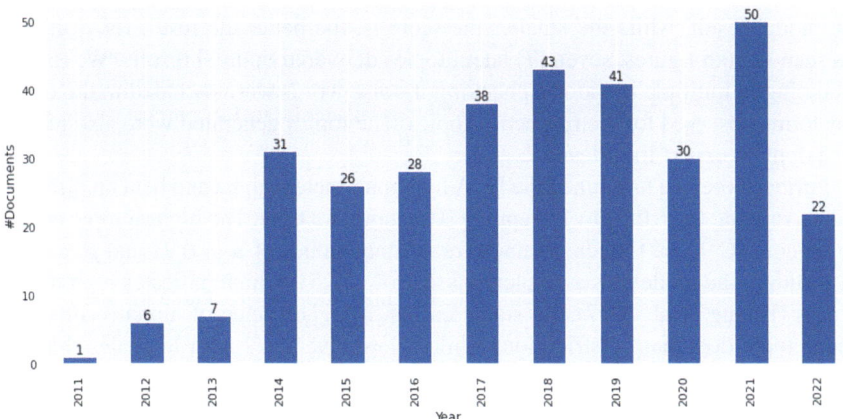

Fig. 24.4 Number of scientific papers published from 2011 to 2022. Please note that the number of papers in 2022 is not complete for the whole year, as the sourcing for this study was conducted in October 2022

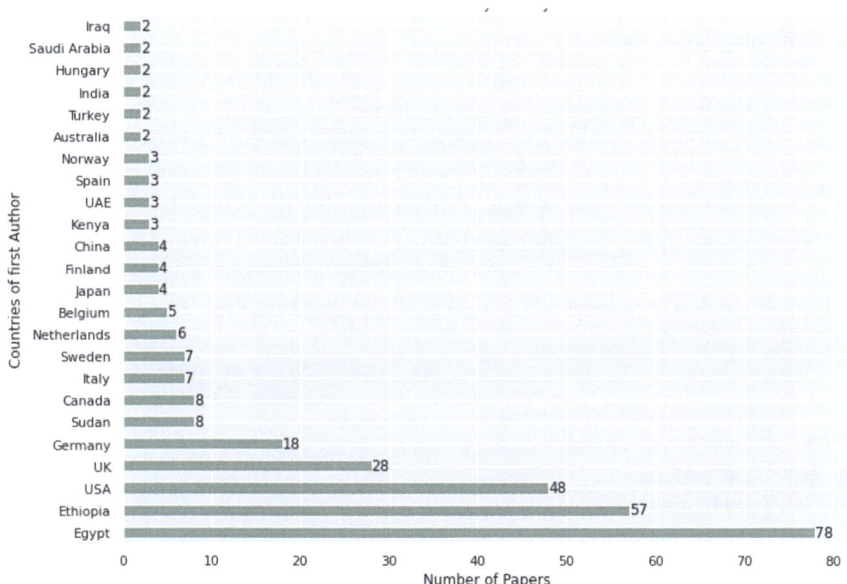

Fig. 24.5 Number of publications on GERD by countries of first author's affiliation

Topic Modeling

We presented the details of the LDA model used to synthesize key issues discussed in the scientific literature. The two coherence metrics (c_v and c_umass) scores are shown in Figs. 24.6 and 24.7, respectively. For C_V, the larger the score is, the better the model result, while the smaller the score is, the better the result for c_umass. As seen in both figures, seven (7) latent topics delivered optimal results. We ran the modeling in multiple iterations, checking if noise words were included in the list of key terms observed for the respective topics. The topics generated were also judged by a domain expert for relevance.

Furthermore, we fine-tuned the LDA hyperparameters alpha and beta on different values ranging from 0.01 to 1, stepping 0.03 units at a time. The highest topic coherence score (C_V = 0.55) is obtained for optimal values of $\alpha = 0.91$ and $\beta = 0.01$. In addition, the model has a perplexity score of -6.51, which indicates high model fitness (Huang et al. 2017). As stated earlier, a higher value of alpha results in a better topic-document distribution, while a lower value of beta indicates that the topics generated have relatively distinct terms, making them human-understandable. The optimal number of topics and terms ordered by their relevance are displayed in a two-dimensional plane where the center of the topics is determined based on their distance from each other (Fig. 24.8). The visualization shows that a good number of the topics are distinctly separated from each other. The frequency of the terms in the topic (in red) and in the entire corpus (in blue) is shown in the diagram. This helps

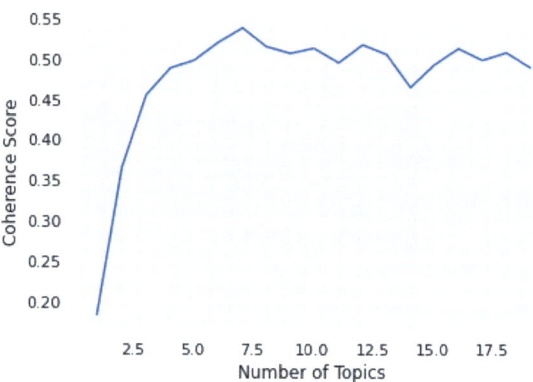

Fig. 24.6 Topic coherence score (C_V)

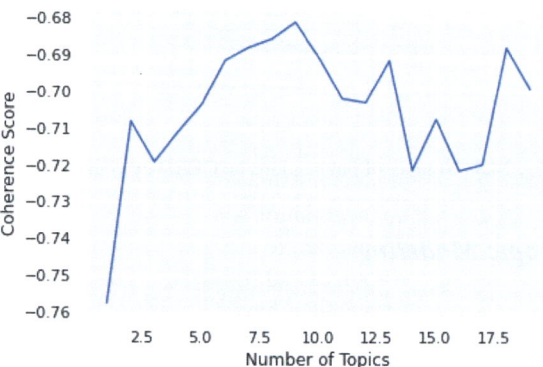

Fig. 24.7 Topic coherence score (C_V and U_mass)

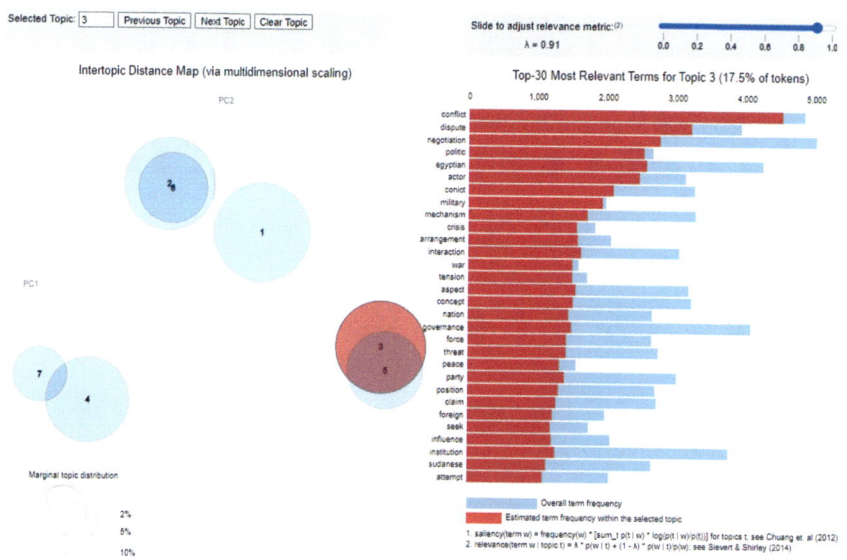

Fig. 24.8 LDA visualization of topics generated from GERD literature

to visualize the relevance of the terms to the topics and their prevalence in the entire corpus. The terms are listed in their order of relevance to describe or represent the topic. For example, conflict is ranked first in the diagram, indicating that it is the most relevant term to describe this topic.

One important point to consider in topic modeling is the degree of contribution of terms for topics generated. Some terms have more power in describing a topic than others. This information is important for the analysis of topics. Figure 24.9 shows the weight of keywords/terms for each topic. Hence, following this, a word cloud of the top 10 keywords that make up a topic is visualized so that an expert can label the topics based on the composition of terms. This makes the topics more convenient for further analysis.

The *water conflict* topic surrounding the Grand Ethiopian Renaissance Dam (GERD) is a highly significant and intriguing area of interest. It has garnered considerable attention as researchers have delved into the ramifications of water scarcity caused by climate change, inadequate water governance, and conflicting narratives regarding water rights and equity (Wheeler and Hussein 2021). The Nile water resource, in general, and the GERD project, in particular, have become focal points for discussions revolving around conflict and disputes, often overshadowing the possibilities of negotiation and resolution. The issue of water scarcity exacerbated by climate change has led to a surge in research exploring the complex relationship between water resources and conflict. Furthermore, deficiencies in water governance and management practices have further fueled tensions and disputes, particularly in regions where water scarcity is prevalent. Within the context of the GERD, which stands as a symbol of Ethiopia's aspirations for development and energy security,

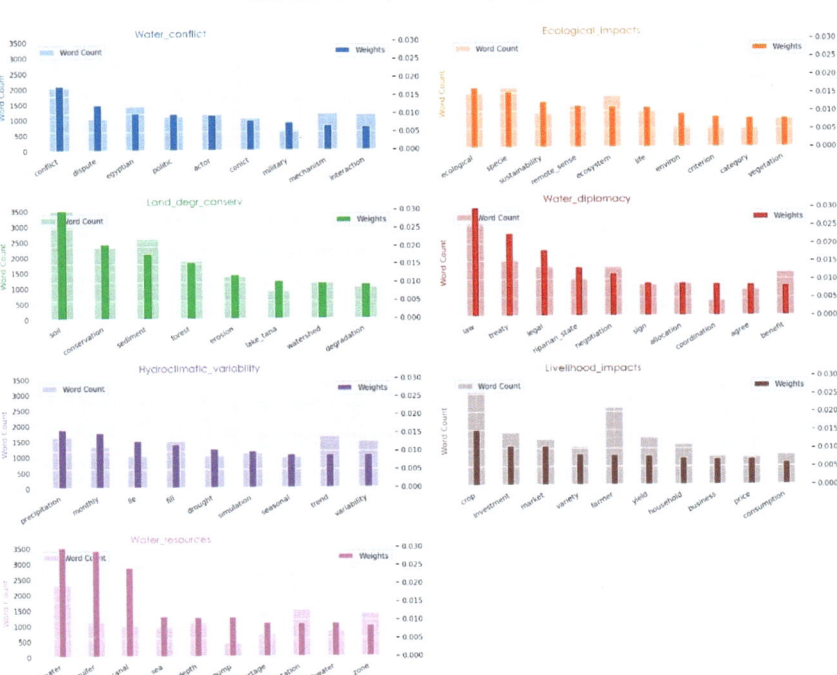

Fig. 24.9 Key term distribution in topics

conflicting narratives and perspectives on water rights and equity have dominated the discourse. The dam's potential impact on downstream countries, such as Sudan and Egypt, has raised concerns and sparked intense debates regarding the allocation and utilization of Nile waters. As a result, discussions around conflict and disputes have overshadowed the prospects of negotiation and collaborative solutions (Table 24.1).

The second cluster/topic of interest in GERD literature revolves around the dam's *ecological and environmental impacts*. Concerns regarding the potential consequences of hydropower plants on the environment and ecosystems are a common theme among stakeholders and researchers (Table 24.1, topic 2). Within the GERD literature, the focus lies on ecological aspects, particularly aquatic life, species, and overall ecosystem health. Remote sensing and satellite technology have emerged as crucial methods for data collection and analysis due to the lack of extensive ground-based observations. These tools provide valuable insights into changes in habitats, water quality, and the overall health of affected ecosystems.

The topic of *land degradation and conservation* in relation to GERD has received significant attention in research. Studies have focused on understanding the potential impacts of dams on erosion, sedimentation, and land degradation processes. Researchers have explored key aspects such as sediment flows, land productivity, and highland areas affected by dam operations. Conservation measures, including

Table 24.1 Topics identified by the LDA model and associated keywords

No.	Topic title	Keywords
1	Water conflict	"conflict", "dispute", "negotiation", "egyptian", "politic", "actor", "conict", "military", "mechanism", "interaction"
2	Ecological impacts	"Ecological", "species", "sustainability", "remote_sense", "life", "ecosystem", "environ", "criterion", "category", "vegetation"
3	Land degradation and conservation	"soil", "conservation", "sediment", "forest", "farmer", "crop", "erosion", "lake_tana", "watershed", "degradation"
4	Water diplomacy	"law", "treaty", "legal", "riparian_state", "negotiation", "sign", "allocation", "benefit", "coordination", "agree"
5	Hydroclimatic variability	"precipitation", "monthly", "fill", "drought", "simulation", "station", "seasonal", "trend", "variability"
6	Economic/livelihood impacts of GERD	"crop", "investment", "market", "variety", "yield", "farmer", "household", "business", "price", "consumption"
7	Status of water resources	"groundwater", "aquifer", "canal", "sea", "pump", "depth", "shortage", "freshwater", "station", "zone"

terracing, agroforestry, and sustainable land management practices, have been investigated to mitigate erosion and minimize the risks of siltation associated with GERD. The primary objective of this research cluster is to develop effective conservation strategies that address land degradation concerns and promote sustainable land management practices within the context of GERD.

Water diplomacy in the Nile basin and tripartite negotiations on dams have become distinct and prominent topics in the literature. With the launch of the Grand Ethiopian Renaissance Dam (GERD) in 2011 and subsequent concerns from riparian countries, researchers have focused on the legal dimensions, treaties, and negotiations associated with this issue. The cooperative framework agreement research encompasses keywords such as law, treaty, legal, and negotiation. Scholars aim to analyze the implications of international law and treaties, as well as the challenges and dynamics of tripartite negotiations, to develop sustainable and equitable water allocation approaches in the Nile basin.

The *hydroclimatic variability* surrounding GERD emerges as a significant topic identified through the topic modeling analysis. This topic revolves around key terms such as "precipitation", "monthly", "fill", "drought," "simulation," "station," "seasonal," "trend", and "variability". The studies within this domain primarily focus on leveraging diverse hydroclimatic information to inform the dam's filling strategy and evaluate the overall water availability. Furthermore, these investigations involve the simulation and assessment of hydrological conditions, providing valuable insights into the implications of droughts on water resources. Additionally, the topic encompasses the analysis of hydrodynamics and long-term trends, with a paramount goal of ensuring sustainable water resource management and implementing optimal dam-filling practices.

The *economic and livelihood impacts* of GERD form the fourth topic of interest in the literature. Key areas of research include the effects on agricultural prices, consumption, yield, households, rural markets, and export products. Scholars also investigate the potential benefits for agricultural productivity, employment, and other sectors, considering factors such as income distribution and socioeconomic disparities. By examining these impacts, researchers aim to understand the implications of GERD on the economy and livelihoods, encompassing aspects such as changes in commodity prices, crop yields, rural livelihoods, and the potential for economic growth and job creation.

The impact of the construction of GERD on *water resources*, particularly groundwater, is an area of significant interest for the last cluster of topics. These studies examine the effects of GERD's water-filling strategies on various water storage systems, including groundwater, the Nile Delta aquifer, and Lake Nasser's freshwater reserves. Additionally, this cluster of studies investigates the potential of groundwater and seawater desalination as alternative sources of freshwater for Egypt. Moreover, they focus on the development of groundwater management in the Nile Delta, considering challenges such as aquifer depletion, water quality degradation, pollution, and the intrusion of seawater.

Topic Correlation with Leading Author's Nationality

In this section, we analyzed the topic dynamics by the country of the lead authors and the temporal dynamics of the topics. The topic distribution across the countries of leading authors is presented in Fig. 24.10. Papers by authors affiliated with Egypt are more interested in water resource availability, hydroclimate variability, and water conflict issues and have limited interest in land degradation and conservation topics (Fig. 24.10). This could be because land degradation, ecological issues, and conservation efforts are targeted and facilitated by upstream riparian countries. It is possible that downstream Egypt might not feel they have a stake or a role in it. This could be indicative of the lack of basin-wide thinking in terms of water use as well as conservation. On the other hand, papers affiliated with Ethiopia have a wide coverage of topics with almost equal proportions across the topics, with relatively lower attention to the topics of "water resource availability" and "ecological impacts of the dam" (Fig. 24.10). Overall, authors affiliated with Ethiopia cover issues such as hydropower, land degradation and erosion, political negotiations, and water resources. Land degradation and erosion have tangible implications for GERD in terms of sedimentation, which might be one reason why it is dominant in Ethiopian

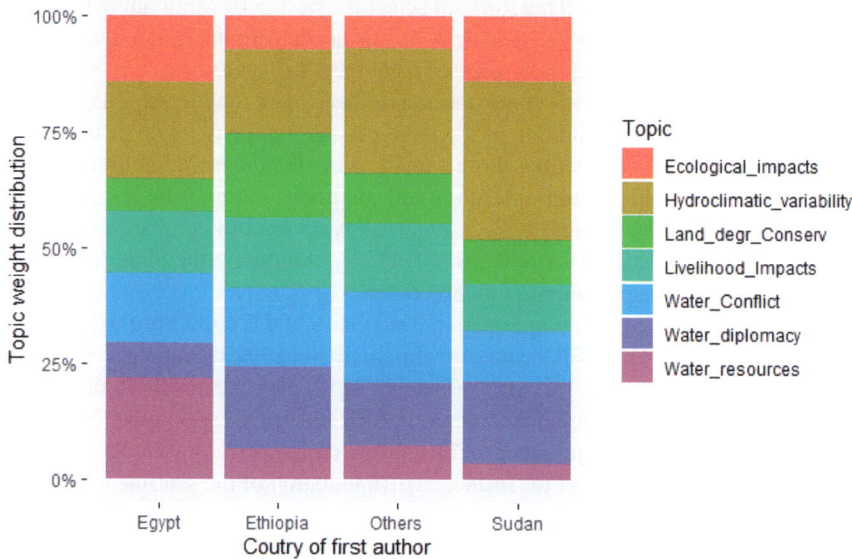

Fig. 24.10 Distributions of topics based on the leading authors' nationalities

writing. Moreover, the overall emphasis on environmental conservation and reclamation activities in the development trajectory of the country might put these issues at the forefront for Ethiopian authors. The hydropower and water resource focus is quite obvious with the main purpose of GERD, and the Blue Nile is a major resource for Ethiopia. Political negotiation is a shared interest in GERD writings. Papers led by authors affiliated with Sudan demonstrate an interest in hydroclimatic variability, water diplomacy, and ecological impacts (Fig. 24.10). It is evident that Sudanese authors tend to focus on the immediate geographical consequences of GERD, such as the Rosieres dam, phytoplankton, electric power generation, and water inflow.

Topic-Level Sentiment Analysis

The total number of selected context sentences for all topics was 3570. The highest number of sentences was for the hydroclimatic variability topic (935 sentences), followed by land degradation and conservation (742 sentences), ecological impacts (539 sentences), water resource availability (481 sentences), water diplomacy (445 sentences), and water conflict (343 sentences), and the lowest number was for the livelihood impacts topic (85 sentences). To provide topic-level sentiment analysis, we conducted various experiments using the Vader and BERT models. Both the Vader and BERT models classify 1360 as neutral, 43 as positive, and 241 as negative out of the total 3570 context sentences. However, the two models have different outcomes for 1952 context sentences. Therefore, to select the best-performing model, further evaluation of the models by expert judgment was needed. To this end, a total of 350 context sentences were randomly selected from the 1952 sentences.

The 350 randomly selected context sentences were sent to three experts for their review and classification into one of the three classes—positive, neutral, and negative. The results showed that the three experts provided the same sentiment value for 290 context sentences and differed in the sentiment value of the remaining 60 context sentences. Therefore, the 290 sentences were used to compare the performance of the two models (VADER and BERT) by computing the closeness of the model classification against expert judgment.

The results of the model comparison showed that VADER correctly classified 170 context sentences, while BERT correctly classified 243 context sentences out of the total 290 context sentences. This implies that VADER's classification is 58.62% (170/290) closer to the experts' judgment, whereas BERT's classification is 83.79% (243/290) closer to the experts' judgment. Thus, based on the performance obtained, the BERT model was considered for further experimentation of the sentiment analysis. Based on the above hyperparameters, the model is built and used for classification. Out of the 3570 context sentences selected in the previous process, 66% were classified as likely neutral about GERD in the sentiment classification analysis using the BERT pretrained model (Fig. 24.11). Only 9% of the total sentences had positive sentiments about GERD.

Further analysis is also conducted to examine the distribution of the three classes of sentiment across the seven topics. Figure 24.12 provides insights into how sentiments are distributed across different topics related to GERD, indicating the prevailing sentiment tendencies within scientific societies. A relatively higher proportion of neutral sentiment sentences is observed in all seven topics. The highest is 77.6% in the *water diplomacy* topic, and the smallest is 58.7% in the hydroclimatic variability topic. Among the seven topics, the highest proportion of negative sentiment context sentences, 33.9%, fall under hydroclimatic variability, while the smallest proportion, 10.6%, of the negative sentiment sentences are from livelihood impacts.

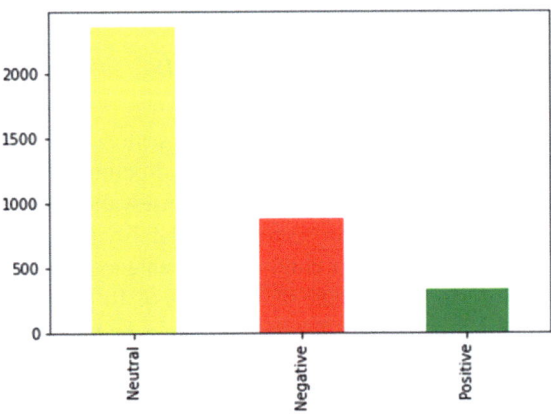

Fig. 24.11 BERT model sentiment classification for all topics combined

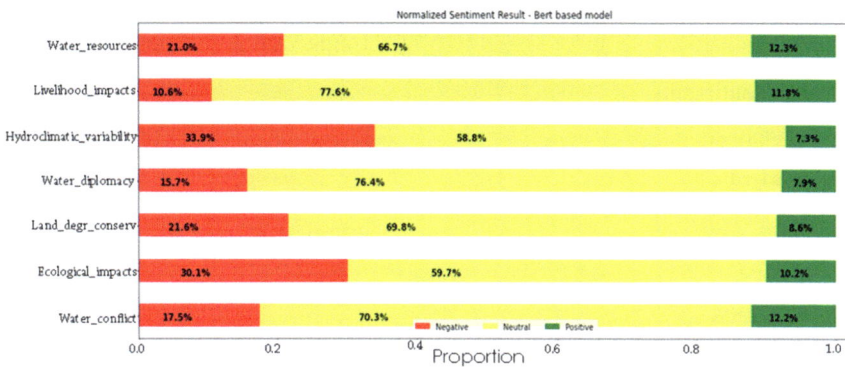

Fig. 24.12 Proportion of sentiment of context sentences within each topic

Conclusions

This research employed a pioneering approach to explore contentious topics in large-scale projects through text-mining techniques. A total of 323 scientific research papers were collected from reputable journals, and NLP methods were utilized for text preprocessing, involving the removal of stop words and lemmatization. Over time, the number of publications peaked in 2021, underscoring GERD's growing importance as a research agenda. Egypt emerged as the leading contributor in published works compared to other countries, including Ethiopia and Sudan, the riparian nations. The study applied topic modeling and sentiment analysis techniques to identify the key discussion themes in the GERD literature. The fine-tuning of LDA hyperparameters resulted in optimal topic coherence scores and offered valuable insights into the relevance and significance of each topic. Seven prominent topics were generated, encompassing water conflict, ecological impacts, land degradation, water diplomacy, hydroclimatic variability, economic/livelihood impacts, and the status of water resources. A detailed analysis of the LDA model was presented, unveiling significant issues discussed in the scientific literature concerning GERD. These topics are the main agendas among the scientific community as well as policy makers in riparian countries. The authors further conducted a comprehensive examination of topic dynamics based on lead authors' countries and temporal factors. Egyptian authors primarily focused on water resource availability, hydroclimate variability, and water conflict, with limited interest in land degradation and conservation topics. In contrast, Ethiopian authors covered a wide range of topics, with specific emphasis on hydropower, land degradation, political negotiations, and water resources. Sudanese authors showed interest in hydroclimatic variability, water diplomacy, and ecological impacts, with a focus on immediate geographical consequences of GERD. Sentiment analysis using the BERT model revealed that the majority of context sentences were classified as neutral, with only a small percentage exhibiting positive or negative sentiments across the various topics. The sentiment of the scientific community generally leaned toward neutrality, avoiding extreme positive or negative sentiments.

Acknowledgements The financial support of the CGIAR NEXUS Gains Initiative is gratefully acknowledged.

References

Abera W, Haregeweyn N, Dile Y, Fenta AA, Berihun ML, Demissie B, Mulatu CA, Nigussie TA, Billi P, Meaza H, Woldearegay K (2021) Scientific misconduct and partisan research on the stability of the grand Ethiopian Renaissance Dam: a critical review of a contribution to environmental remote sensing in Egypt (Springer, 2020). In: Melesse AM, Abtew W, Moges SA (eds) Nile and Grand Ethiopian Renaissance Dam. Springer Geography, Springer

Albalawi R, Yeap TH, Benyoucef M (2020) Using topic modeling methods for short-text data: a comparative analysis. Front Artif Intell 3:42

Aziz SA, Zeleňáková M, Mésároš P, Purcz P, Abd-Elhamid H (2019) Assessing the potential impacts of the Grand Ethiopian Renaissance Dam on water resources and soil salinity in the Nile Delta, Egypt. Sustainability 11(24):7050

Basheer M, Nechifor V, Calzadilla A, Siddig K, Etichia M, Whittington D et al (2021) Collaborative management of the Grand Ethiopian Renaissance Dam increases economic benefits and resilience. Nat Commun 12(1):5622

Campagnolo JM, Duarte D, Dal Bianco G (2022) Topic coherence metrics: how sensitive are they? J Inf Data Manag 13(4)

Chuang J, Manning CD, Heer J (2012) Termite: visualization techniques for assessing textual topic models. In: Proceedings of the international working conference on advanced visual interfaces, pp 74–77

Clapham C (2018) The Ethiopian developmental state. Third World Q 39(6):1151–1165

Dagan I, Lee L, Pereira FC (1999) Similarity-based models of word cooccurrence probabilities. Mach Learn 34:43–69

Dang NC, Moreno-García MN, De la Prieta F (2020) Sentiment analysis based on deep learning: a comparative study. Electronics 9(3):483

Devika MD, Sunitha C, Ganesh A (2016) Sentiment analysis: a comparative study on different approaches. Proced Comput Sci 87:44–49

Ghosh S (2021) Identifying click baits using various machine learning and deep learning techniques. Int J Inf Technol 13(3):1235–1242

Harris ZS (1954) Distributional structure. Word 10(2–3):146–162

Heggy E, Sharkawy Z, Abotalib AZ (2022) Reply to Comment on 'Egypt's water budget deficit and suggested mitigation policies for the Grand Ethiopian Renaissance Dam filling scenarios' by Kevin Wheeler et al'. Environ Res Lett 17(12):128001

Hoff H (2011) Understanding the nexus. Background paper for the Bonn2011 conference: the water, energy and food security Nexus. Stockholm Environment Institute, Stockholm

Huang L, Ma J, Chen C (2017) Topic detection from microblogs using T-LDA and perplexity. In: Proceedings of the 2017 24th Asia-Pacific software engineering conference workshops (APSECW), IEEE, pp 71–77

Jelodar H, Wang Y, Yuan C, Feng X, Jiang X, Li Y, Zhao L (2019) Latent Dirichlet allocation (LDA) and topic modelling: models, applications, a survey. Multim Tools Appl 78:15169–15211

Kamara A, Ahmed M, Benavides A (2022) Environmental and economic impacts of the Grand Ethiopian Renaissance Dam in Africa. Water 14(3):312

Kapadia S (2019) Evaluate topic models: Latent Dirichlet allocation (LDA). Towards Data Science

Kulshrestha J, Eslami M, Messias J, Zafar MB, Ghosh S, Gummadi KP, Karahalios K (2019) Search bias quantification: investigating political bias in social media and web search. Inf Retrieval J 22:188–227

Li I (2016) NLP 05: from Word2vec to Doc2vec: a simple example with Gensim. Available at: https://ireneli.eu/2016/07/27/nlp-05-from-word2vec-to-doc2vec-a-simple-example-with-gensim/. Accessed: 28 Oct 2022)

Owa DLM (2021) Identification of topics from scientific papers through topic modelling. Open J Appl Sci 10(04):541

Perkins J (2010) Python text processing with NLTK 2.0 cookbook. PACKT publishing

Rana TA, Cheah YN, Letchmunan S (2016) Topic modelling in sentiment analysis: a systematic review. J ICT Res Appl 10(1)

Raza H, Faizan M, Hamza A, Ahmed M, Akhtar N (2019) Scientific text sentiment analysis using machine learning techniques. Int J Adv Comput Sci Appl 10(12)

Řehůřek R, Sojka P (2011) Gensim—statistical semantics in python. Retrieved from genism. org

Röder M, Both A, Hinneburg A (2015) Exploring the space of topic coherence measures. In: Proceedings of the eighth ACM international conference on web search and data mining, pp 399–408

Rüdiger M, Antons D, Joshi AM, Salge TO (2022) Topic modelling revisited: new evidence on algorithm performance and quality metrics. PLoS ONE 17(4):e0266325

Sarkar D (2019) Text analytics with Python

Sievert C, Shirley K (2014) LDAvis: a method for visualizing and interpreting topics. In: Proceedings of the workshop on interactive language learning, visualization, and interfaces, pp 63–70

Wheeler KG, Hussein H (2021) Water research and nationalism in the posttruth era. Water Int 46(7–8):1216–1223

Wu F, Shi Z, Dong Z, Pang C, Zhang B (2020) Sentiment analysis of online product reviews based on SenBERT-CNN. In: Proceedings of the 2020 international conference on machine learning and cybernetics (ICMLC), IEEE, pp 229–234

Yin H, Song X, Yang S, Li J (2022) Sentiment analysis and topic modelling for COVID-19 vaccine discussions. Worldwide Web 25(3):1067–1083

Zhang L, Wang S, Liu B (2018) Deep learning for sentiment analysis: a survey. Wiley Interdiscipl Rev Data Min Knowl Discov 8(4):e1253

Zvornicanin E (2022) When coherence score is good or bad in topic modelling? Baeldung Comput Sci

Chapter 25
Future Research, Planning and Management Directions

Worku Zewdie and Berhan Gessesse

Abstract The Abbay basin is known for its diverse agroecology, from drylands to moist highlands, which is prone to climate shocks, exposing the environment and society. Several studies also indicated the variation of available resources due to expansion of agriculture, increase in population size and climate change impacts. This change resulted in transport of sediment to downstream countries. The loss of fertile soil significantly impacts the agricultural productivity of the basin. Hence, proper planning is vital to sustainably utilize the available natural resources. In this regard, the collection and organization of relevant data of the basin is crucial to mitigate the adverse social and natural impacts that may occur in the basin. Geospatial data and technology hugely contribute to the monitoring of the basin resources through indicating resource base sizes and places.

Keywords Abbay basin · Agroecology · Climate shocks · Geospatial data · Basin resources

Introduction

Several studies have been conducted in the Abbay basin focusing on assessing the natural resource conditions and climate change impacts, e.g., water resources (Mengistu et al. 2021), soil moisture conditions (Damtie et al. 2022), vegetation conditions (Teferi et al. 2015) and agriculture (Yalew et al. 2016a, b) of the basin. In this regard, monitoring changes and their consequences on the resource base of the basin indicated that there is a significant variation in the available resources. This is mainly linked to the expansion of agricultural activities, the increasing size of the population residing in the basin and climate change impacts (Belay and Mengistu 2021; Erkossa et al. 2014; Mengistu et al. 2021). Moreover, due to the loss of vegetation cover, a significant amount of sediment is transported to downstream countries.

W. Zewdie (✉) · B. Gessesse
Remote Sensing Department, Ethiopian Space Science and Geospatial Institute, Addis Ababa, Ethiopia
e-mail: wzewdie24@gmail.com

© The Author(s), under exclusive license to Springer Nature Switzerland AG 2025
A. Melesse et al. (eds.), *Abbay River Basin*, Springer Geography,
https://doi.org/10.1007/978-3-031-65241-7_25

This in turn exposed the cropland to loss of soil fertility with dwindling productivity in the basin. In addition, the relatively varying climatic conditions of the basin exposed the community to a significant change in their agricultural productivity (Endalamaw et al. 2021). There are times when crop failure is a challenge to farmers, leading them to abandon their farms.

Hydrological studies involving various modeling approaches have been conducted in the Abbay basin including application of hydrology of the basin (Melesse et al. 2009, 2010, 2011a, b, 2014; Abtew et al. 2009a, b; Abtew and Melesse 2014a, b, c; Yitayew and Melesse 2011; Chebud and Melesse 2009a, b, 2013; Setegn et al. 2009a, b, 2010; Melesse 2011), application of Soil and Water Assessment Tool (SWAT) (Setegn et al. 2009a, b, 2010, 2011; Getachew and Melesse 2012; Mohamed et al. 2015), land use land cover mapping (Getachew and Melesse 2012; Mohammed et al. 2013), soil erosion and sediment transport (Defersha and Melesse 2012a, b; Defersha et al. 2010, 2012; Setegn et al. 2010; Melesse et al. 2011a, b; Mekonnen and Melesse 2011; Setegn et al. 2009a, b; Mohammed et al. 2015), climate change impact on river flows (Setegn et al. 2011; Melesse et al. 2009, 2011a, b).

The basin is also known as a source of water for the downriparian countries that are dependent on the flow of the Abbay River. Proper management of the vegetation cover leads to the conservation of the available rainfall, avoiding excessive soil erosion. The main rain season (June to September) is known for providing water for agricultural activity in the basin. The loss of vegetation cover increases the prevalence of the region's soil erosion. This will also have a significant impact on soil moisture and the available water in the basin's reservoirs, leading to the accumulation of silt and minimizing the water-holding capacity of the dams. Hence, soil conservation practices are vital to reduce the transportation of soil organic matter to sustain agricultural activity and productivity in the basin.

Soil is a major factor that affects crop growth, agricultural production and productivity. The loss of soil could be a bottleneck to sustainable agriculture and food security (Endalamaw et al. 2021). Proper soil carbon management improves crop yield, maintains soil fertility and reduces carbon emissions from the soil. Improved soil management is ideal for carbon management and contributes to the digital agricultural transformation of the country. This also requires continuous monitoring of resources to understand the soil condition, particularly in croplands.

The United Nations announced the 17 Sustainable Development Goals (SDGs) in 2015 to end poverty, protect the environment and promise a better life by 2030. Most of the SDGs focus on the protection of the environment and sustainable utilization of available resources, including land and water, to act on climate change. The Abbay basin, with diverse agroecology that includes drylands to moist highlands, is prone to climate change that exposes the environment and society to different sizes of climate shocks. The basin is exposed to climate shocks that include droughts, floods and extended droughts (Gadisso 2007). This situation damages the sustainable utilization of the resources of the basin. Hence, strengthening the capacity of the vulnerable community, enhancing the climate information system and improving the institution's capacity to respond to climate shocks is necessary to sustain the natural resources of the basin.

Natural Resource Planning and Management Direction

The diverse agro-ecological conditions of the Abbay basin resulted in the endowments of several natural resources. It has water resources that flow to the downstream countries dependent on this resource for their agricultural and other water needs from this basin. It also covers a variety of landscape features that support diverse vegetation types and soil that supports the agricultural activities of the community in the basin. However, resource base depletion is visible in various forms, mainly land degradation, transportation of sediments, loss of vegetation cover, decline in agricultural productivity and varying climatic conditions (Amanuel et al. 2018). The variation in climate change may significantly affect the water resource availability and the agricultural activity of the basin (Cherinet et al. 2019).

Proper planning is crucial for sustainable utilization of the available natural resources. However, unsustainable use of resources may significantly impact the natural system to continue its replenishment (Malede et al. 2023). Population growth and decline in the productivity of land forces society to encroach into the natural vegetation to sustain their livelihood (Yalew et al. 2016a, b). On the other hand, agricultural activity in the basin demands intensive management with agricultural inputs to enhance and sustain crop production activity.

One of the major challenges of the country as a whole and the Abbay basin in particular is the unsustainable use of natural resources. The rapid population growth and inappropriate land use practices that encroach on natural resources cause degradation of landscapes, loss of biodiversity and decline in productivity. Several studies indicated a loss in vegetation cover and a decline in water resources with increasing temperature (Cherinet et al. 2019) in the basin and an increase in soil erosion (Belay and Mengistu 2021). However, the challenges regarding sustainable utilization of resources remain with minor attempts to curb the situation. There is competition among the various land use types, and natural vegetation commonly undergoes severe damage through conversion to other land use types. Hence, proper land use planning necessitates sustainably utilizing the available resources of the basin.

The adoption of land use planning is required among various land use types to ensure protection, restoration and sustainable utilization of the remaining natural resources. Available land is limited and needs to be balanced with the demand of the population and available land by reconciling among mutually exclusive land use types. There is always competition among land use types. The increase in population size urges the community to acquire new agricultural plots that mainly compete with the remaining natural vegetation areas. The use of land use planning compares conflicting land use types to create a balance for sustainable utilization. This planning activity will end in the allocation and zonation of land based on their specific uses, imposing legal and administrative regulations to implement the plan (World Bank 2010). Properly implemented land use plans maintain the conservation and sustainable use of resources, fulfilling the current and future demands of society.

Managing natural resources requires continuously measured data to investigate and understand the size and whereabouts of resources. Traditional methods of field data collection and monitoring of resources are costly, time-consuming and exposed to human errors. This approach also faces challenges in covering large areas and collecting continuous data over the years. Hence, the utilization of earth observation datasets allows continuous monitoring and observation of changes in the resource base. Geospatial technology denotes the tools and systems that capture, analyze and visualize the spatial distribution of resources over time. This technology allows data-driven decision-making through mapping resources and assessing their status to make options for allocating resources for managing resources. Spatial data and technology also allow us to monitor progress in conservation and identify critical areas that demand special attention. This technology also helps in collecting useful data and making comparisons for optimal land use planning to minimize competition among the land use types. However, planning and managing resources in the basin is not well equipped with a state-of-the-art geospatial monitoring system to continuously monitor and update the state of the resources. Hence, strengthening the monitoring system provides near real-time data and information for any remedial actions required in the basin.

Data Issues in Basin Management

The collection and organization of data relevant to basin management is crucial to mitigate any adverse social and natural impacts that may occur within the basin. Among the dominant data that can be used for basin management are demographic data that include population size, age and socioeconomic status. This helps us understand the social situation of the basin related to resource utilization. The hydrometric data are also vital in indicating the amount of river flow, water level, flood peak discharge and base flow. These data also explain the situation of the water quantity within the basin for better planning and sustainable utilization of the basin water. Meteorological data of the basin is another vital dataset for knowing air temperature, wind speed, humidity, evaporation and rainfall amount. These datasets significantly impact the water content of the basin for proper planning and management of the basin water content. The agricultural activity in the basin is also a vital data type that needs to be collected and properly managed in basin data management. This also helps to determine the type of crops growing in the basin and the areas covered by these crops. Industrial activities are one of the activities that may occur in the basin that consume available water in the basin. Hence, data on the industrial growth rate, available industries in the basin and the amount of water used by these industries were collected and organized. In addition, the hydropower capacity data within the basin are also critical for knowing the hydropower capacity, the amount of discharge and the timing. Surface water potential and groundwater potential data are also crucial in current and future basin water management. Land use and land cover data are other vital data types required to assess the temporal and spatial variation in land uses within the basin. The detection of variations in land use type significantly affects the basin management and conservation of resources. In particular, the loss in vegetation cover affects the infiltration capacity and removes topsoil due to the removal of vegetation cover. Hence, mapping and monitoring variations in land use changes contribute to better management of resources found within the basin.

Nevertheless, obtaining an organized and well-managed data handling system and storage that focuses on the basin is minimal. Many data were collected regarding resources of the basin but mostly sheltered by data collectors or not available for further usage and analysis. Certain datasets may be handled by respective government organizations. However, today's opportunity to obtain open datasets of various forms and organize them is not well utilized in the basin.

Remote Sensing and Future Needs in Basin Management

Geospatial datasets contribute immensely to the monitoring and management of resources. The Abbay Basin, with vast landscape variations and diverse agroecology, demands continuous monitoring to effectively manage the available resources. However, the continuous monitoring of these resources requires either field observations or earth observation data for monitoring changes in the basin. The open-access data policy and improvement in computational platforms are also other opportunities for monitoring the resource condition of the basin. Various types of satellite imagery can be freely accessed to be used for all types of applications required in the basin.

Optical remote sensing satellites collect data with different spatial, temporal and spectral resolutions depending on the platforms and sensors used for imaging. Accordingly, there is an enormous variation in products obtained from these satellites. Resource monitoring data can be acquired from ground platforms, airborne platforms and spaceborne platforms. The data obtained from ground and airborne platforms have high resolution to better characterize properties based on their high-resolution data products. The thermal bands of the optical sensors are useful for surface temperature estimation. On the other hand, spaceborne data products cover larger areas with repetitive coverage of an area of interest.

In addition to optical sensors, hyperspectral and microwave remote sensing are other earth observation data that can be used for mapping spatial resource information. Hyperspectral images have a higher number of spectral bands at a fine spatial resolution to distinguish diverse properties, particularly vegetation and soil texture descriptions. Hence, these sensors help to assess the distribution of soil properties. Radar sensors also measure the backscatter from the surfaces to provide the moisture content of soils and ecosystem structure. These sensors also have the potential to characterize various soil parameters, including soil roughness and texture. Another active sensor, lidar, is also useful in height measurement. This dataset is particularly important for assessing the diverse vegetation types found in the basin. It is helpful to measure the tree height and canopy cover of forests.

Satellite imagery provides data covering large areas with multiple temporal resolutions that fit the monitoring demands of basin management. It can be used for data collection, organization and analysis for evidence-based decisions. The basin areas and cover types in the basin can be monitored to determine the status of resources within the basin. The watershed and catchment area can be spatially mapped using open-access satellite images with various temporal and spatial resolutions. Images are also crucial for the monitoring and assessment of water resources.

The satellite data enable the gathering of data on the water and energy fluxes of the basin. The availability of data at various temporal and spatial resolutions for modeling the changing environment is linked to the uncertainties existing due to climate change. This analysis estimates how future climate conditions impact water resources and basin management systems. Climate change has resulted in prolonged drought and excess rainfall, which creates a burden for efficient decision-making.

The availability and open data access policy and the development of big data processing facilities and computing platforms create a large opportunity to use satellite images for basin-wide management. The free access to Landsat images from 2008 onwards created the chance to obtain data free of charge, particularly for developing countries such as Ethiopia. Sentinel series data are another type of remote sensing data that can be used in basin management. Sentinel data exist as optical and radar datasets with improved spatial, temporal and radiometric resolutions for assessing basin water changes and assessing the condition of vegetation cover.

The monitoring of surface water variability depends on in situ measurements that estimate the extent and discharge amount of water from the basin. Nonetheless, in situ measurements are not distributed in sufficient amounts specifically in inaccessible areas to measure and provide accurate data of the basin. Moreover, installing and maintaining the installed equipment is costly to use over longer periods of time. It is challenging to cover wider areas and to monitor other events, such as floods. Hence, satellite data are used to monitor basin-wide water cycles with continuous spatial and temporal coverage and acceptable accuracy (Chawla et al. 2020).

Studies have commenced using remote sensing data in the basin employing various types of applications. Land use/land cover change dominates with various levels of classification details and spatial and temporal coverage. These studies follow diverse types of classification approaches and land use classes to assess basin-wide situations of changes in land use. Sediment accumulation is another type of study performed in the basin. This also varies among the studied documents with the variation in data type and adopted algorithm. These challenges estimate basin-wide sediment accumulation and transportation. There are also few studies concerned about soil moisture in the basin. These studies used satellite data and in situ measurements to estimate soil moisture at different levels. A few studies also indicated the contribution of remote sensing in monitoring crop conditions and estimating crop yield. However, the studies focused only on certain crops and areas, and it was difficult to estimate the overall crop acreage and yield of the basin.

Vegetation studies are critical for the management of water resources and reservoirs in the basin. Satellite images are one of the key components in vegetation studies. There are investigations regarding the vegetation condition of the basin using satellite images. However, the studies are mostly performed by students to fulfill their degree requirements based on their time and budget availability. The temporal variation and area coverage are restricted to the needs of the researcher. It is difficult to estimate the overall situation and degree of disturbance or improvement of vegetation in the basin. Hence, it is vital to extensively utilize freely available satellite data and cloud computing platforms for a complete picture of the status of basin resources.

Goal six of the sustainable development goal (SDG 6) emphasizes ensuring access to water and sanitation for all. Among the targets mentioned in this goal, target six explains the importance of protecting and restoring water-related ecosystems, focusing on mountains, forests, wetlands, rivers, aquifers and lakes (SDG 2015). These sites cover vast areas and are also distributed all over the basin. The monitoring of these sites and understanding of the status of each land use type supports looking for mechanisms that improve the current condition. Hence, satellite images are one of the data sources that indicate the status of these natural resources in space and time for sustainable management of the resources. These can also provide information for decision-makers to better understand the conditions for devising appropriate policy options toward the SDGs. Satellite images are also one source for monitoring water quality. Satellites and their derived products measure water quality indicators such as temperature, turbidity, suspended matter and dissolved organic matter.

However, only minimal and mostly student researchers dominated the activities performed in the basin. Hence, using earth observation data in resource monitoring in the basin should be coordinated, particularly to monitor sediment yield, vegetation changes and hydrology of the basin. The availability of satellite data should be linked to capacity building and knowledge sharing to effectively utilize geospatial technology for resource monitoring, enhance collaboration and effectively plan land use options without harming the existing resources of the basin.

Concluding Remarks

Population growth and the scarcity of agricultural plots together with climate change impacts in the basin are challenges for future basin-wide resource management. This leads to competition among land use types that leads to the encroachment of forests and other vegetation to secure farming areas to sustain their livelihood. The loss of vegetation cover exposes the land to erosion and is also a source of greenhouse gas, mainly releasing the stored carbon in the woody vegetation and soil carbon. The cropland also suffers a loss of topsoil with a subsequent decline in soil fertility that lowers agricultural productivity.

Population growth also demands a higher volume of water for agricultural activities, daily consumption and livestock. Erosion also transports sediments toward the catchment, decreasing the water-holding capacity of rivers and reservoirs. The severe modification of natural water management triggers apparent ecological degradation. The amount of water that can be obtained from surface water and aquifers varies in space and time. Sustainable utilization of this finite resource is crucial to sustaining the demand for water for agriculture, drinking and industrial usage.

Geospatial data and technology enormously contribute to the monitoring of natural resources by showing resource base sizes and places. It also provides data indicating areas that have been facing depletion and need special treatments. Hence, a proper land use plan that adopts geospatial technology offers decision-makers proper data to make informed decisions. This technology also facilitates the revision of management plans by providing near real-time data that are mainly obtained from satellite imageries. The ever-growing demands for resources and emerging threats require the collection of all basin-wide data for integrated management. This management plan facilitates the organization of all resources of the basin supported with detailed data on the available resources. Therefore, it is vital to continuously collect technologically supported data that show the existing resources and their status to make timely decisions. This provides datasets for developing proper basin-wide sustainable plans for the efficient utilization of resources.

References

Abtew W, Melesse AM (2014a) Nile River basin hydrology. In: Melesse AM, Abtew W, Setegn S (eds) Nile River basin: ecohydrological challenges, climate change and hydropolitics, pp 7–22

Abtew W, Melesse AM (2014b) Climate teleconnections and water management. In: Nile River basin. Springer International Publishing, pp 685–705

Abtew W, Melesse AM (2014c) Transboundary rivers and the Nile. In: Nile River basin. Springer International Publishing, pp 565–579

Abtew W, Melesse A, Desalegn T (2009a) Spatial, inter and intra-annual variability of the Blue Nile River basin rainfall. Hydrol Process 23(21):3075–3082

Abtew W, Melesse A, Desalegn T (2009b) El Niño southern oscillation link to the Blue Nile River basin hydrology. Hydrol Process Spec Issue Nile Hydrol 23(26):3653–3660

Amanuel W, Yimer F, Karltun E (2018) Soil organic carbon variation in relation to land use changes: the case of Birr watershed, upper Blue Nile River basin, Ethiopia. J Ecol Environ 42(1):1–11. https://doi.org/10.1186/s41610-018-0076-1

Belay T, Mengistu DA (2021) Impacts of land use/land cover and climate changes on soil erosion in Muga watershed, Upper Blue Nile basin (Abay), Ethiopia. Ecol Process 10(1). https://doi.org/10.1186/s13717-021-00339-9

Chawla I, Karthikeyan L, Mishra AKA (2020) Review of remote sensing applications for water security: quantity, quality, and extremes. J Hydrol 585:124826. https://doi.org/10.1016/j.jhydrol.2020.124826

Chebud YA, Melesse AM (2009a) Numerical modeling of the groundwater flow system of the Gumera sub-basin in Lake Tana basin, Ethiopia. Hydrol Process Spec Issue Nile Hydrol 23(26):3694–3704

Chebud YA, Melesse AM (2009b) Modeling Lake stage and water balance of Lake Tana, Ethiopia. Hydrol Process 23(25):3534–3544

Chebud Y, Melesse AM (2013) Stage level, volume, and time-frequency change information content of Lake Tana using stochastic approaches. Hydrol Process 27(10):1475–1483. https://doi.org/10.1002/hyp.9291

Cherinet AA, Yan D, Wang H, Song X, Qin T, Kassa MT, Girma A, Dorjsuren B, Gedefaw M, Wang H, Yadamjav O (2019) Impacts of recent climate trends and human activity on the land cover change of the Abbay River basin in Ethiopia. Adv Meteorol 2019. https://doi.org/10.1155/2019/5250870

Damtie BB, Mengistu DA, Waktola DK, Meshesha DT (2022) Impacts of soil and water conservation practice on soil moisture in Debre Mewi and Sholit Watersheds, Abbay basin, Ethiopia. Agriculture (Switzerland) 12(3). https://doi.org/10.3390/agriculture12030417

Defersha MB, Melesse AM (2012a) Effect of rainfall intensity, slope and antecedent moisture content on sediment concentration and sediment enrichment ratio. CATENA 90:47–52

Defersha MB, Melesse AM (2012b) Field-scale investigation of the effect of land use on sediment yield and surface runoff using runoff plot data and models in the Mara River basin, Kenya. CATENA 89:54–64

Defersha MB, Quraishi S, Melesse AM (2010) Interrill erosion, runoff and sediment size distribution as affected by slope steepness and antecedent moisture content. Hydrol Earth Syst Sci Discuss 7(6447–6489):2010

Defersha MB, Melesse AM, McClain M (2012) Watershed scale application of WEPP and EROSION 3D models for assessment of potential sediment source areas and runoff flux in the Mara River basin, Kenya. CATENA 95:63–72

Endalamaw NT, Moges MA, Kebede YS, Alehegn BM, Sinshaw BG (2021) Potential soil loss estimation for conservation planning, upper Blue Nile basin, Ethiopia. Environ Challenges 5(July):100224. https://doi.org/10.1016/j.envc.2021.100224

Erkossa T, Haileslassie A, MacAlister C (2014) Enhancing farming system water productivity through alternative land use and water management in vertisol areas of Ethiopian Blue Nile basin (Abay). Agric Water Manage 132:120–128. https://doi.org/10.1016/j.agwat.2013.10.007

Gadisso BE (2007) Drought assessment for the Nile basin using meteosat second generation data with special emphasis on the upper Blue Nile region drought assessment for the Nile basin using meteosat second generation data with special emphasis on the upper Blue Nile region, p 91

Getachew HE, Melesse AM (2012) Impact of land use/land cover change on the hydrology of Angereb Watershed, Ethiopia. Int J Water Sci 1(4):1–7. https://doi.org/10.5772/56266

Malede DA, Alamirew T, Kosgie JR, Andualem TG (2023) Analysis of land use/land cover change trends over Birr River Watershed, Abbay basin, Ethiopia. Environ Sustain Indic 17(September 2022):100222. https://doi.org/10.1016/j.indic.2022.100222

Melesse AM (2011) Nile River basin: hydrology, climate and water use. Springer Science & Business Media

Mekonnen M, Melesse A (2011) Soil erosion mapping and hotspot area identification using GIS and remote sensing in northwest Ethiopian highlands, near Lake Tana. In: Melesse A (ed) Nile River basin: hydrology, climate and water use, chapter 10. Springer Science Publisher, pp 207–224. https://doi.org/10.1007/978-94-007-0689-7_10

Melesse AM, Loukas AG, Senay G, Yitayew M (2009) Climate change, land-cover dynamics and ecohydrology of the Nile River basin. Hydrol Process Spec Issue Nile Hydrol 23(26):3651–3652

Melesse AM, Abtew W, Desalegne T, Wang X (2010) Low and high flow analysis and wavelet application for characterization of the Blue Nile River system. Hydrol Process 24(3):241–252

Melesse AM, Ahmad S, McClain M, Wang X, Lim H (2011a) Sediment load prediction in large rivers: ANN approach. Agric Water Manage 98:855–866

Melesse AM, Abtew W, Setegn S, Dessalegne T (2011b) Hydrological variability and climate of the Upper Blue Nile River basin. In: Melesse A (ed) Nile River basin: hydrology, climate and water use, chapter 1. Springer Science Publisher, pp 3–37. https://doi.org/10.1007/978-94-007-0689-7_1

Melesse A, Abtew W, Setegn SG (2014) Nile River basin: ecohydrological challenges, climate change and hydropolitics. Springer Science & Business Media

Mengistu D, Bewket W, Dosio A, Panitz HJ (2021) Climate change impacts on water resources in the Upper Blue Nile (Abay) River basin, Ethiopia. J Hydrol 592:125614. https://doi.org/10.1016/j.jhydrol.2020.125614

Mohammed H, Alamirew A, Assen M, Melesse A (2013) Spatiotemporal mapping of land cover in Lake Hardibo drainage basin, Northeast Ethiopia: 1957–2007. In: Water conservation: practices, challenges and future implications. Nova Publishers, pp 147–164

Mohammed H, Alamirew T, Assen M, Melesse AM (2015) Modeling of sediment yield in Maybar gauged watershed using SWAT, northeast Ethiopia. CATENA 127:191–205

Setegn SG, Srinivasan R, Dargahi B, Melesse AM (2009a) Spatial delineation of soil erosion prone areas: application of SWAT and MCE approaches in the Lake Tana

Setegn SG, Srinivasan R, Melesse AM, Dargahi B (2009b) SWAT model application and prediction uncertainty analysis in the Lake Tana basin, Ethiopia. Hydrol Process 24(3):357–367

Setegn SG, Bijan Dargahi B, Srinivasan R, Melesse AM (2010) Modelling of sediment yield from Anjeni Gauged Watershed, Ethiopia using SWAT. JAWRA 46(3):514–526

Setegn S, Rayner D, Melesse AM, Dargahi B, Srinivasan R (2011) Impact of climate change on the hydro-climatology of Lake Tana basin, Ethiopia. Water Resour Res 47(W04511):13. https://doi.org/10.1029/2010WR009248s

Teferi E, Uhlenbrook S, Bewket W (2015) Interannual and seasonal trends of vegetation condition in the Upper Blue Nile (Abay) basin: dual-scale time series analysis. Earth Syst Dyn 6(2):617–636. https://doi.org/10.5194/esd-6-617-2015

Yalew SG, Mul ML, Van Griensven A, Teferi E, Priess J (2016a) Land-use change modelling in the Upper Blue Nile basin. Environments 1–16. https://doi.org/10.3390/environments3030021

Yalew SG, van Griensven A, Mul ML, van der Zaag P (2016b) Land suitability analysis for agriculture in the Abbay basin using remote sensing, GIS and AHP techniques. Model Earth Syst Environ 2(2):1–14. https://doi.org/10.1007/s40808-016-0167-x

Yitayew M, Melesse AM (2011) Critical water resources management issues in Nile River basin. In: Melesse A (ed) Nile River basin: hydrology, climate and water use, chapter 20. Springer Science Publisher, pp 401–416. https://doi.org/10.1007/978-94-007-0689-7_20

Index

A
Abbay Basin, 4–14, 21–26, 29, 33–39, 41, 43–49, 55–59, 67–71, 74–82, 109–113, 115, 117–122, 124–129, 133, 134, 137–139, 143, 145–154, 156, 163–169, 172–175, 177–186, 194, 201–203, 209–213, 215, 219, 222, 223, 225, 226, 228–230, 235, 237–239, 241–245, 248, 255, 261–275, 279–283, 344, 351, 354–356, 369, 370, 375, 377–379, 397–404, 406, 409, 410, 412, 418–421, 428–432, 434, 436, 438, 443–454, 457, 458, 460–463, 497, 498, 500–513, 515–517, 519, 520, 577–579, 581, 586, 588–591, 593, 621–623, 625
Abbay River, 4, 5, 7, 11, 12, 17, 18, 21, 24, 26, 117, 165, 168, 170, 172–174, 180, 181, 193–195, 201, 202, 204, 238, 246, 253, 262, 275, 323, 345, 352, 353, 369, 370, 375, 376, 381, 386, 389, 390, 392, 622
Abbay River Basin, 10, 11, 13, 17–29, 34, 69, 70, 72–74, 78–81, 83, 193–205, 223, 238, 321, 323, 324, 332–337, 343, 345, 347, 348, 350–352, 354–356, 375, 417, 530, 532, 533, 536, 546, 547, 549, 553, 556, 571
Adigrat sandstone, 9, 17, 20, 22–24, 26, 43
Agricultural drought, 11, 291–293, 299–304, 308–313, 316, 317, 321, 322, 327, 332, 334, 336, 337
Agriculture, 5, 6, 8, 9, 13, 67–69, 72, 74, 75, 78, 81, 83, 84, 125, 153, 167, 168, 196, 247, 262, 265, 276, 277, 298, 299, 316, 322, 328, 344, 346–348, 352–355, 365, 377, 397–399, 402–404, 409, 412, 420, 421, 424, 426, 428, 435, 437, 438, 444, 451, 471, 473, 474, 477, 482, 486, 511, 529, 530, 533, 547, 554, 556, 578, 584, 585, 593, 600, 601, 621, 622, 628
Agriculture expansion, 83, 402, 621
Agro-climatic classification, 165
Agroclimatic zones, 112, 116, 155
Agroecological zones, 68, 71, 111, 115, 150, 167, 399
Agroecology zone, 68
Automatic Weather Stations (AWS), 13, 533–537, 539–544, 547–549, 553, 556, 557, 562, 564, 565, 571
Available water per capita, 345

B
Bathymetry, 40
Biomass energy, 10, 163, 181
Biophysical setting, 5–7
Blue-Nile Basin, 36, 37, 40, 43

C
Catchment characterization, 446, 450, 452
Causes of soil erosion, 11, 272
Cenozoic era, 20
Cereal-based high production, 67, 72
Climate change, 5, 6, 9, 11, 40, 41, 56, 67, 82, 84, 110, 153, 179, 180, 182, 195, 201–204, 236, 242, 248, 249, 262, 263, 265, 276, 277, 280, 301,

321–323, 327, 328, 332, 336, 337, 447, 452, 458, 463, 504–506, 511, 553, 554, 584, 585, 611, 622, 623, 626
Climate change impact, 5, 11, 193, 203, 321–324, 328, 332, 621, 622, 627
Climate Change Initiative (CCI), 13, 506, 507, 510, 553–571
Climatology, 34, 35, 37, 292, 448, 449, 505, 508, 509, 512, 517
Commercially oriented production, 67, 82
Conservation tillage, 417, 419, 422, 430, 431, 435–438
Crop Growth Models (CGMs), 13, 471, 472, 484, 487
Crop monitoring, 13, 471, 472, 476, 477, 479, 486, 487
Cumulative Density Function (CDF), 13, 325, 326, 530–532, 537, 538, 540, 541, 544, 547–549

D

Deforestation, 70, 163, 167, 176, 181, 182, 186, 196, 204, 238, 239, 241, 242, 262, 274, 276, 277, 282, 298, 399–401, 403, 404, 408, 452, 508
Discharge areas, 210, 227, 228
Dominant soil types, 120, 298
Downscale, 478, 553, 555, 557, 559–570
Drivers of deforestation, 182
Drivers of land use land cover changes, 235, 238, 248
Drought monitoring, 292, 293, 299, 300, 302, 303, 309, 312, 316, 317, 448, 473, 509, 554
Droughts, 5, 11, 49, 56, 81, 180, 182, 183, 242, 247, 276, 291–293, 295, 296, 298–304, 306, 307, 309–314, 316, 317, 321–325, 327–329, 332, 333, 335–337, 344, 346, 351–353, 383, 400, 404, 405, 422, 423, 425, 427, 428, 448, 450, 459, 504, 554, 613, 614, 622, 626

E

Earth observation, 6, 12, 13, 84, 280, 443, 446, 471, 472, 579, 581, 582, 624, 625, 627
Earth observation application, 12
East African Orogen, 9, 19, 20, 33, 39, 41, 59
Eastern Abbay Basin, 11

Ecological benefits, 10, 163, 417, 419, 435
Ecotourism, 92, 99, 101
Effects of soil erostion, 11, 243, 261, 264, 273
Enhanced Vegetation Index (EVI), 270, 293, 473, 484
Ethiopia, 4, 5, 7, 11–13, 17, 18, 20, 21, 23, 26–28, 33–37, 39–42, 46, 47, 49, 51, 52, 54–59, 67–69, 71, 78, 79, 89, 91, 93, 94, 98, 99, 101, 110, 111, 118, 133, 137, 138, 144, 147, 150–155, 163–165, 168, 169, 175, 177, 179, 181, 183–186, 193–196, 201, 210–212, 214, 216, 219, 221, 236–238, 242, 243, 245, 248, 253, 262, 263, 266, 267, 270–277, 279–283, 291, 293, 295, 310, 316, 323, 343–349, 351–353, 364, 367, 369–371, 375–377, 379–391, 398–405, 409, 412, 417–424, 427, 434–436, 443, 445–447, 449–451, 455, 457, 458, 460, 462, 463, 471, 474, 475, 479, 481, 487, 497, 498, 500, 504, 530, 532, 534, 546, 549, 553, 556, 571, 584, 585, 592, 599, 600, 602, 608, 611, 614, 615, 617, 626
Ethiopian tourism, 91
Ethio-Sudanese Border, 212

F

Food-insecure areas, 67, 425, 427
Forest degradation, 182, 277
Forest resources, 10, 163, 167–169, 173, 179–181, 183, 184, 186, 277, 400, 404
Forest resource types, 177

G

Geographical overview, 5
Geolandforms, 111, 113, 116, 120, 121, 130, 155
Geology, 14, 17, 18, 29, 34, 35, 37, 40, 41, 43, 47, 51, 117, 177, 211, 214, 215, 220, 226, 434, 454, 587, 599
Geomorphic processes, 9, 33–35, 38, 39, 58, 412
Geomorphic regions, 33, 47, 58
Geomorphology, 9, 33–35, 37, 39, 40, 43, 51, 59, 211, 512, 515
Geoserver, 588
Geostatistical methods, 205, 580, 584

Index

Global Climate Models (GCMs), 82, 203, 204, 321, 322, 324–328, 336, 337
Gohatsion formation, 9, 17, 24, 25, 28
Google earth engine, 13, 280, 458, 497, 511, 512, 520
Grain yield, 360, 417, 430–433, 436–438, 482, 483
Grand Ethiopian Reminiscence Dam (GERD), 6, 9, 10, 14, 89, 92–94, 97–102, 118, 170, 179, 181, 185, 201, 243, 262, 274, 282, 283, 343, 351, 375, 376, 379, 382, 383, 387–391, 418, 455–457, 599–603, 605, 607–609, 611–617
Groundwater characterization, 446, 460
Groundwater flow system, 10, 210, 211, 214, 215, 217, 222–224, 229
Groundwater modelling, 219
Groundwater potential, 213, 445, 460, 461, 624

H
Hydroclimate data, 13, 497, 502, 503, 517
Hydrogeology, 212, 214, 216, 226
Hydro hegemony, 376
Hydrological balance, 40
Hydrological modelling, 193, 194, 201, 203, 205, 445, 446, 461–463, 502, 515, 517
Hydrological modelling parameters, 12
Hydrological models, 12, 193–195, 201–204, 266, 267, 434, 446, 448, 449, 452, 461–463, 502, 531
Hydrological parameters, 12, 443, 444, 446, 447, 463, 497, 499, 517
Hydrologic cycle, 236, 344, 345
Hydrology, 5, 6, 8, 10, 13, 14, 34, 35, 37, 40, 41, 68, 177, 193–195, 197, 203, 204, 216, 277, 328, 344, 376, 438, 450, 452, 462, 506, 508, 512–514, 517–519, 529, 554, 599, 622, 627
Hydrology research, 196, 201
Hydro-meteorological observation, 205
Hydropolitics, 34, 369, 377, 378

I
Implications of land use land cover changes, 235, 242
Interactive web map, 14, 577, 593
International law, 378, 381, 386, 389, 613
Irrigation, 5, 6, 8, 10, 11, 33, 34, 59, 67, 68, 74, 75, 78–80, 93, 136, 143, 173, 174, 193–195, 201, 204, 238, 243, 253, 262, 274, 282, 283, 323, 343–369, 371, 381, 382, 390, 404, 419, 422, 425, 427, 428, 444, 454, 473, 477, 516, 519, 554, 568, 601
Irrigation potential, 11, 34, 78, 80, 343, 345, 349, 352, 353

K
Khartoum City, 4

L
Lake Tana, 4, 5, 17, 18, 35, 38, 45–50, 52, 53, 56, 59, 69, 75, 170, 171, 182, 185, 201, 202, 212, 332, 343, 355, 356, 421, 425, 447, 454, 459, 461, 500
Lake Tana Basin, 9, 33, 45, 49, 51, 53, 59, 170–172, 198, 202–204, 268, 343, 355, 370, 449, 452
Land degradation, 5, 7, 12, 41, 67, 68, 71, 73, 81, 147, 163, 167, 173, 183, 186, 236, 237, 239, 241, 243, 245, 255, 263, 272–274, 276, 277, 298, 353, 397–406, 408, 412, 418, 419, 422, 427, 435, 578, 592, 600, 612–615, 617, 623
Land management, 7, 10–12, 40, 68, 72, 109, 118, 122, 155, 170, 181, 185, 237, 242–244, 253, 261–264, 272–276, 279, 281–283, 365, 397, 400–402, 404–406, 409–412, 427, 428, 458, 500, 579, 581, 584, 586, 587
Land productivity, 12, 75, 236, 262, 274, 279, 343, 357, 359, 360, 408–410, 612
Land resource, 7, 11, 12, 74, 182, 185, 235, 236, 238, 242, 255, 272, 276, 283, 397, 401, 408, 409, 412, 418, 422
Landsat 8, 175, 176
Land suitability, 75, 78, 511
Land surface evaporation, 446, 452, 453
Land Surface Temperature (LST), 170, 291–293, 300, 301, 304, 305, 309, 311–314, 316, 317, 504
Land use, 5, 10, 11, 40, 68, 69, 110, 117, 133, 136, 166–168, 170, 173, 174, 176, 177, 182, 196, 197, 235–244, 246–253, 255, 261–263, 267–271, 273, 275, 277–281, 283, 297, 298, 343, 347, 351, 353, 362, 364, 400,

402, 409, 419, 422, 437, 450, 458, 461–463, 511, 547, 577–581, 585–588, 590, 591, 622–628
Land use and land cover changes, 173, 177, 399
Leaf Area Index (LAI), 293, 476, 478, 480, 481, 485, 486
Lithology, 216, 225
Livelihood, 5, 12, 68, 69, 71, 72, 74–76, 83, 90, 94, 95, 100–102, 176, 181, 183, 184, 186, 236, 245, 264, 274, 279, 283, 299, 316, 367, 370, 371, 376, 397–401, 403, 405, 408, 417–419, 422, 424, 426, 427, 435, 438, 497, 556, 600, 613–617, 623, 627
Low soil fertility, 81, 282

M

Machine learning, 261, 280–283, 484, 498, 517–519, 580, 582–584, 601, 602, 604, 607
Major crops, 74, 77–79, 84, 166, 301, 317, 411
Management direction, 623
Mediterranean sea, 4, 39, 346, 387
Mesozoic era, 43, 44, 117
Microwave and optical sensors, 475
Modelling, 5, 10, 109, 110, 156, 193–197, 202–205, 210, 214, 217, 219, 225, 226, 282, 362, 370, 451, 454, 461, 462, 502, 507, 512, 514–516, 518, 547, 554, 587, 610, 622, 626
Monitoring system, 579, 624
Mugher Mudstone, 9, 17, 22, 26, 27, 43

N

Natural Language Processing (NLP), 599, 601, 602, 606–608, 617
New tourism destination development, 90
Nile Basin, 4, 5, 18, 20, 21, 29, 36, 37, 40, 43, 56, 68, 79, 81, 82, 93, 170, 185, 186, 261, 264, 292, 312, 348–350, 370, 371, 383, 386, 387, 389, 391, 392, 448, 449, 452, 458, 460, 498–500, 585, 613
Normalzed Difference Vegetion Index (NDVI), 169, 176, 269, 270, 291–293, 299–301, 303, 305, 308, 309, 311–317, 473, 478, 482, 484, 485, 487
Northwestern groundwater Basin, 209, 215, 216, 218, 223–225, 229, 230

P

Palaeozoic era, 43
Palmer Drought Severity Index (PDSI), 292, 505
Parent materials, 117, 122, 123, 133, 145, 147, 298
Phanerozoic, 9, 17, 20–23
Planning, 83, 84, 94, 99, 110, 167, 193, 196, 235, 238, 255, 261, 263–265, 268, 273, 358, 407, 423, 424, 427, 437, 443, 445, 453, 458, 502, 530, 623, 624
Podocarpus falcatus, 171, 179
Population pressure, 176, 236, 238, 239, 241, 242, 248, 250, 255, 262, 273, 277, 282, 399–402, 404, 407
Potentiometric surface access depth, 210
Poverty reduction, 95, 100, 398, 418
Precambrian basement, 9, 17, 19, 20, 23, 45, 49, 53
Precipitation, 5, 82, 93, 133, 215, 226, 227, 267, 291, 292, 297, 299, 301–303, 307, 310, 316, 317, 322–325, 327–332, 336, 337, 344, 443–455, 461–463, 498, 500, 504, 505, 517, 519, 520, 532, 536, 568, 613

Q

Quantile Mapping (QM), 204, 325
Quaternary sediments, 18, 213, 228

R

Rainfall Anomaly Index (RAI), 292, 299, 302, 307, 308, 313, 317
Rainfed systems, 13, 68, 201, 369
Recharge areas, 212, 230
Reference soil group, 118–121, 133, 154, 155
Remote sensing, 7–9, 11, 13, 43, 110, 267, 269, 270, 280, 291–293, 301, 316, 443, 446, 451, 453, 455, 459, 460, 471, 472, 476, 487, 501, 502, 506, 507, 509, 515, 529–532, 547, 549, 554, 555, 557, 570, 577, 580, 582, 600, 612, 625, 626
Residual moisture, 75, 547
Root-zone, 13, 531, 532, 534, 537, 543
R packages, 512, 513
Rural land certification, 75

Index 635

S

Securitization, 12, 375, 378–380, 387, 389, 390
Sedimentation, 5, 20–22, 27, 34, 38, 40, 41, 44, 55–59, 180, 181, 195, 251, 263, 270, 273, 274, 277, 280, 282, 283, 404, 405, 452, 612, 614
Sediment yield, 11, 170, 171, 235, 237–240, 243, 244, 248, 253, 255, 261–264, 266–270, 274, 275, 278–283, 457, 627
Sediment yield estimation, 261, 265, 267–270, 280, 283
Sedtiment analysis, 5, 272, 275, 279, 463
Sensor Allow Researchers (SAR), 13, 453, 474–476, 478–481, 483, 486, 487, 506–508, 510, 553–555, 558–564, 566, 569–571
Sentinel-1 & 2, 13, 453, 457, 471, 474, 475, 480, 481, 487
Sentinel Application Platforms (SNAP), 558, 559
Soil, 8, 10, 13, 14, 29, 33, 34, 47, 56, 57, 59, 67, 69, 70, 73, 75, 76, 78, 81, 83, 109–111, 117–129, 133–140, 143–156, 163, 177, 180–182, 196, 198, 202, 216, 218, 243–245, 253, 261, 264–274, 276–279, 281–283, 292, 298, 299, 327, 328, 344, 346, 348, 359, 361, 362, 366, 369, 399–401, 403–407, 409, 418, 422–425, 427, 429–433, 436–438, 452, 453, 458, 461–463, 499, 502, 505–508, 510, 511, 513, 515, 529–544, 546–549, 553–571, 577–590, 592, 593, 613, 622, 623, 625, 627
Soil and Water Analysis Tools (SWAT), 202–204, 261, 267, 268, 270, 274, 275, 362, 363, 452, 502, 622
Soil and water conservation, 33, 57, 181, 184, 255, 261, 263, 264, 268, 274–278, 405, 406, 419, 420, 423–425, 427–429, 432–434, 436–438, 578
Soil biological properties, 151
Soil carbon, 110, 577, 578, 583–585, 593, 622, 627
Soil chemical property, 128, 130, 153, 155
Soil degradation, 9, 153, 271, 277, 398, 399, 401, 404, 405
Soil erosion, 5, 11, 34, 41, 55–59, 70, 81, 133, 155, 171, 175, 181, 195, 196, 204, 235–240, 242–245, 248, 251–255, 261–268, 270–283, 298, 368, 399, 401, 403–405, 412, 419, 422, 428, 432, 437, 446, 457, 458, 511, 585, 600, 622, 623
Soil erosion estimation, 265, 280, 282
Soil forming factors, 111, 155
Soil loss, 58, 243, 252–254, 261–267, 271–274, 277–279, 281, 283, 398, 399, 403, 408, 417, 430–433, 435–438, 457, 458
Soil moisture, 13, 110, 244, 269, 279, 292, 296, 305, 310, 316, 322, 344, 345, 355, 365, 369, 405, 417, 429, 430, 437, 450, 453, 461–463, 484, 506, 510, 529–544, 546–549, 553–571, 580, 582, 583, 621, 622, 626
Soil moisture monitoring, 110, 446, 453, 530, 557
Soil physical property, 124
Soil properties, 110, 153, 156, 268, 278, 279, 471, 547, 577, 579, 580, 582, 583, 587, 625
Soil spatial information, 10, 109–111, 117, 118, 155, 156
Soil texture, 110, 125, 568, 580, 625
Soil types, 72, 78, 111, 117, 118, 196, 267, 275, 281, 298, 303, 399, 406, 578, 580, 582, 583
Standardized Precipitation Evapotranspiration Index (SPEI), 292
Standardized precipitation index, 292, 302, 321, 329–331
Sudan, 4, 5, 18, 34–38, 41, 45–47, 54–59, 68, 79, 165, 177, 179, 185, 194–196, 201, 262, 272, 275, 323, 343, 351, 370, 376, 379–383, 386, 388–391, 447, 479, 497, 498, 532, 600–602, 608, 612, 615, 617
Surface runoff, 11, 39, 55, 58, 170, 196, 244, 248, 253, 261, 264, 272, 274, 277, 279, 283, 292, 365, 417, 421, 430–433, 435–438, 445, 450, 452, 453, 462, 463, 499
Surface water extent, 446, 454–457
Surface water extent mapping, 455
Surface water extent monitoring and mapping, 453
Surface water potential, 624
Sustainable development, 100, 163, 180, 270, 351, 377, 399, 438, 586

Sustainable development goals, 10, 91, 100, 179, 581, 592, 622, 627
Sustainable land management, 11, 75, 81, 180, 261, 263, 264, 273, 276, 279, 282, 283, 398, 404, 406, 408, 409, 419, 420, 426, 427, 436, 613

T
Tepid moist mid-highlands, 69
Tepid submoist mid-highlands, 69
Terracing, 12, 269, 364, 397, 405, 417, 422, 423, 428, 429, 436, 437, 585, 613
Text mining, 14, 599, 602, 603, 607
Tolerable soil loss rate, 262
Topic modelling, 601–605, 607, 609, 611, 613, 617
Topography of the Abbay Basin, 196, 199, 268
Tourism infrastructure, 100
Tourist cottages, 100
Tourist houses, 100
Trap volcanics, 9, 17, 45

U
Universal soil loss equation, 57, 251, 261, 265–267, 458

V
Vegetation Health Index (VHI), 11, 291–293, 300, 304, 311, 312, 316, 317

Vegetation loss, 403

W
Water-based tourism, 95
Water conflict, 600, 611, 613–615, 617
Water productivity, 343, 356, 357, 359–361, 367, 368, 370, 435–437
Water quality analysis, 445, 446, 459, 516
Water resource development, 40, 55, 194, 195, 201, 238, 243
Water resources, 4–7, 10–12, 40, 56, 67, 93, 163–165, 169, 174, 180, 181, 193–195, 201, 203, 210, 212, 239, 244, 248, 261, 262, 282, 283, 292, 345, 346, 353, 370, 371, 377, 384, 387, 399, 403, 404, 418, 419, 421, 428, 435, 437, 443–446, 449, 450, 452, 453, 455, 461–463, 499, 500, 504, 515, 516, 554, 585, 601, 611, 613–615, 617, 621, 623, 626
Watershed management, 9, 10, 12, 185, 193, 195, 204, 267, 270, 272, 283, 355, 362, 363, 366–370, 417, 420, 422–429, 435–437
White Nile Basin, 4, 18, 447

Y
Yield estimation, 12, 13, 293, 471, 472, 477, 481–487
Yield gaps, 67, 68, 72, 73, 77–79, 83

GPSR Compliance

The European Union's (EU) General Product Safety Regulation (GPSR) is a set of rules that requires consumer products to be safe and our obligations to ensure this.

If you have any concerns about our products, you can contact us on ProductSafety@springernature.com

In case Publisher is established outside the EU, the EU authorized representative is:

Springer Nature Customer Service Center GmbH
Europaplatz 3
69115 Heidelberg, Germany

Batch number: 08758096

Printed by Printforce, the Netherlands